S0-ADC-676

Principles of Information Systems

A Managerial Approach

Ninth Edition

Ralph M. Stair
Professor Emeritus, Florida State University

George W. Reynolds

Australia · Canada · Mexico · Singapore · Spain · United Kingdom · United States

Principles of Information Systems, A Managerial Approach, Ninth Edition by Ralph M. Stair and George W. Reynolds

VP/Editorial Director: Jack Calhoun

Senior Acquisitions Editor: Charles McCormick, Jr.

Product Manager: Kate Hennessy

Development Editor: Lisa Ruffolo, The Software Resource

Editorial Assistant: Bryn Lathrop

Content Product Managers: Erin Dowler, Jennifer Goguen McGrail

Manufacturing Coordinator: Denise Powers

Marketing Manager: Bryant Chrzan

Marketing Coordinator: Suellen Ruttkay

Art Director: Stacy Shirley

Cover Image: Getty Images/Digital Vision

Compositor: Value Chain International

Copyeditor: Gary Spahl

Proofreader: Green Pen QA

Indexer: Rich Carlson

ISBN-13:978–0-324–66528–4
ISBN-10:0–324–66528–8

Instructor Edition:
ISBN-13:978–0-324–78141–0
ISBN-10:0–324–78141–5

Course Technology
25 Thomson Place
Boston, MA 02210
USA

Cengage Learning is a leading provider of customized learning solutions with office locations around the globe, including Singapore, the United Kingdom, Australia, Mexico, Brazil, and Japan. Locate your local office at:
international.cengage.com/region

Cengage Learning products are represented in Canada by Nelson Education, Ltd.

To learn more about Course Technology, visit www.cengage.com/coursetechnology
To learn more about Cengage Learning, visit www.cengage.com

Printed in the United States of America
1 2 3 4 5 6 7 13 12 11 10 09

For Lila and Leslie
—RMS

To my grandchildren: Michael, Jacob, Jared, Fievel, Aubrey, Elijah, Abrielle, Sophia, Elliot
—GWR

BRIEF CONTENTS

CONTENTS

PREFACE

As organizations continue to operate in an increasingly competitive and global marketplace, workers in all business areas including accounting, finance, human resources, marketing, operations management, and production must be well prepared to make significant contributions required for success. Regardless of your future role, you need to understand what information systems can and cannot do and be able to use them to help you accomplish your work. You will be expected to discover opportunities to use information systems and to participate in the design of solutions to business problems employing information systems. You will be challenged to identify and evaluate information systems options. To be successful, you must be able to view information systems from the perspective of business and organizational needs. For your solutions to be accepted, you must recognize and address their impact on fellow workers, customers, suppliers, and other key business partners. For these reasons, a course in information systems is essential for students in today's high-tech world.

Principles of Information Systems: A Managerial Approach, Ninth Edition, continues the tradition and approach of the previous editions. Our primary objective is to provide the best information systems text and accompanying materials for the first information technology course required of all business students. We want you to learn to use information technology to ensure your personal success in your current or future job and to improve the success of your organization. Through surveys, questionnaires, focus groups, and feedback that we have received from current and past adopters, as well as others who teach in the field, we have been able to develop the highest-quality set of teaching materials available to help you achieve these goals.

Principles of Information Systems: A Managerial Approach, Ninth Edition, stands proudly at the beginning of the IS curriculum and remains unchallenged in its position as the only IS principles text offering the basic IS concepts that every business student must learn to be successful. In the past, instructors of the introductory course faced a dilemma. On one hand, experience in business organizations allows students to grasp the complexities underlying important IS concepts. For this reason, many schools delayed presenting these concepts until students completed a large portion of the core business requirements. On the other hand, delaying the presentation of IS concepts until students have matured within the business curriculum often forces the one or two required introductory IS courses to focus only on personal computing software tools and, at best, merely to introduce computer concepts.

This text has been written specifically for the introductory course in the IS curriculum. *Principles of Information Systems: A Managerial Approach, Ninth Edition,* treats the appropriate computer and IS concepts together with a strong managerial emphasis on meeting business and organizational needs.

APPROACH OF THE TEXT

Principles of Information Systems: A Managerial Approach, Ninth Edition, offers the traditional coverage of computer concepts, but it places the material within the context of meeting business and organizational needs. Placing information system (IS) concepts in this context and taking a general management perspective has always set the text apart from general computer books thus making it appealing not only to MIS majors but also to students from other fields of study. The text isn't overly technical, but rather deals with the role that information systems play in an organization and the key principles a manager needs to grasp to be successful. These principles of IS are brought together and presented in a way that is both understandable and relevant. In addition, this book offers an overview of the entire IS discipline, while giving students a solid foundation for further study in advanced IS courses as programming, systems analysis and design, project management, database management, data communications, Web site and systems development, electronic commerce and mobile commerce applications, and decision support. As such, it serves the needs of both general business students and those who will become IS professionals.

The overall vision, framework, and pedagogy that made the previous editions so popular have been retained in the ninth edition, offering a number of benefits to students. We continue to present IS concepts with a managerial emphasis. While the fundamental vision of this market-leading text remains unchanged, the ninth edition more clearly highlights established principles and draws out new ones that have emerged as a result of business, organizational, and technological change.

IS Principles First, Where They Belong

Exposing students to fundamental IS principles is an advantage for students who do not later return to the discipline for advanced courses. Since most functional areas in business rely on information systems, an understanding of IS principles helps students in other course work. In addition, introducing students to the principles of information systems helps future business function managers employ information systems successfully and avoid mishaps that often result in unfortunate consequences. Furthermore, presenting IS concepts at the introductory level creates interest among general business students who may later choose information systems as a field of concentration.

Author Team

Ralph Stair and George Reynolds have teamed up again for the ninth edition. Together, they have more than sixty years of academic and industrial experience. Ralph Stair brings years of writing, teaching, and academic experience to this text. He wrote numerous books and a large number of articles while at Florida State University. George Reynolds brings a wealth of computer and industrial experience to the project, with more than thirty years of experience working in government, institutional, and commercial IS organizations. He has written numerous texts and has taught the introductory IS course at the University of Cincinnati and College of Mount St. Joseph. The Stair and Reynolds team brings a solid conceptual foundation and practical IS experience to students.

GOALS OF THIS TEXT

Because *Principles of Information Systems: A Managerial Approach, Ninth Edition,* is written for all business majors, we believe it is important not only to present a realistic perspective on IS in business but also to provide students with the skills they can use to be effective business leaders in their organization. To that end, *Principles of Information Systems: A Managerial Approach, Ninth Edition,* has four main goals:

1. To provide a core of IS principles with which every business student should be familiar
2. To offer a survey of the IS discipline that will enable all business students to understand the relationship of IS courses to their curriculum as a whole
3. To present the changing role of the IS professional
4. To show the value of the discipline as an attractive field of specialization

By achieving these goals, *Principles of Information Systems, Ninth Edition,* will enable students, regardless of their major, to understand and use fundamental information systems principles so that they can function more efficiently and effectively as workers, managers, decision makers, and organizational leaders.

IS Principles

Principles of Information Systems: A Managerial Approach, Ninth Edition, although comprehensive, cannot cover every aspect of the rapidly changing IS discipline. The authors, having recognized this, provide students an essential core of guiding IS principles to use as they face the career challenges ahead. Think of principles as basic truths or rules that remain constant regardless of the situation. As such, they provide strong guidance in the face of tough decisions. A set of IS principles is highlighted at the beginning of each chapter. The application of these principles to solve real-world problems is driven home from the opening vignettes to the end-of-chapter material. The ultimate goal of *Principles of Information Systems* is to develop effective, thinking, action-oriented employees by instilling them with principles to help guide their decision making and actions.

Survey of the IS Discipline

This text not only offers the traditional coverage of computer concepts but also provides a broad framework to impart students with a solid grounding in the business uses of technology. In addition to serving general business students, this book offers an overview of the entire IS discipline and solidly prepares future IS professionals for advanced IS courses and their careers in the rapidly changing IS discipline.

Changing Role of the IS Professional

As business and the IS discipline have changed, so too has the role of the IS professional. Once considered a technical specialist, today the IS professional operates as an internal consultant to all functional areas of the organization, being knowledgeable about their needs and competent in bringing the power of information systems to bear throughout the organization. The IS professional views issues through a global perspective that encompasses the entire organization and the broader industry and business environment in which it operates.

The scope of responsibilities of an IS professional today is not confined to just his/her employer but encompasses the entire interconnected network of employees, suppliers, customers, competitors, regulatory agencies, and other entities, no matter where they are located. This broad scope of responsibilities creates a new challenge: how to help an organization survive in a highly interconnected, highly competitive global environment. In accepting that challenge, the IS professional plays a pivotal role in shaping the business itself and ensuring its success. To survive, businesses must now strive for the highest level of customer satisfaction and loyalty through competitive prices and ever- improving product and service quality. The IS professional assumes the critical responsibility of determining the organization's approach to both overall cost and quality performance and therefore plays an important role in the ongoing survival of the organization. This new duality in the role of the IS employee—a professional who exercises a specialist's skills with a generalist's perspective—is reflected throughout the book.

IS as a Field for Further Study

Despite the downturn in the economy at the start of the 21st century, especially in technology-related sectors, the outlook for computer and information systems managers is optimistic. Indeed, employment of computer and information systems managers is expected to grow much faster than the average for all occupations through the year 2012. According to the Bureau of Labor Statistics, the number of information technology workers exceeded 4.1 million, an all-time high in the second quarter of 2008. Technological advancements will boost the employment of computer-related workers; in turn, this will boost the demand for managers to direct these workers. In addition, job openings will result from the need to replace managers who retire or move into other occupations.

A career in IS can be exciting, challenging, and rewarding! It is important to show the value of the discipline as an appealing field of study and that the IS graduate is no longer a technical recluse. Today, perhaps more than ever before, the IS professional must be able to align IS and organizational goals and to ensure that IS investments are justified from a business perspective. The need to draw bright and interested students into the IS discipline is part of our ongoing responsibility. Upon graduation, IS graduates at many schools are among the highest paid of all business graduates. Throughout this text, the many challenges and opportunities available to IS professionals are highlighted and emphasized.

CHANGES IN THE NINTH EDITION

We have implemented a number of exciting changes to the text based on user feedback on how the text can be aligned even more closely with how the IS principles and concepts course is now being taught. The following list summarizes these changes:

- **All new opening vignettes.** All of the chapter-opening vignettes are new, and continue to raise actual issues from foreign-based or multinational companies.
- **All new Information Systems @ Work special interest boxes.** Highlighting current topics and trends in today's headlines, these boxes show how information systems are used in a variety of business career areas.
- **All new Ethical and Societal Issues special interest boxes.** Focusing on ethical issues today's professionals face, these boxes illustrate how information systems professionals confront and react to ethical dilemmas.
- **New case studies.** Two new end-of-chapter cases provide a wealth of practical information for students and instructors. Each case explores a chapter concept or problem that a real-world company or organization has faced. The cases can be assigned as individual homework exercises or serve as a basis for class discussion.

Each chapter has been completely updated with the latest topics and examples. The following is a summary of the changes.

Chapter 1, An Introduction to Information Systems

The topics and sections in Chapter 1 create a framework for the entire book. As with all chapters, the opening material at the beginning of the chapter including the vignette, the Information Systems @ Work and Ethical and Societal Issues special interest boxes, the end-of-chapter cases, and all end-of-chapter material have been updated to reflect the changes in Chapter 1.

This chapter continues to emphasize the benefits of an information system, including speed, accuracy, reduced costs, and increased functionality. We have modified the last principle at the beginning of the chapter to emphasize the importance of global and international information systems. The Why Learn About Information Systems section has been updated to include a new example about financial advisor's use of information systems.

In the section on data, information, and knowledge, we have included a new definition of knowledge management system (KMS). New examples have been introduced in the section on computer-based information systems to give students a better understanding of these important components. The material on the Internet, for example, has been completely updated with the latest information, including Web 2.0 technologies. References of corporate IS usage have been stressed in the major section of business information systems.

The latest material on hardware, software, databases, telecommunications, and the Internet have been included. This material contains fresh, new examples of how organizations use computer-based information systems to their benefit. The best corporate users of IS, as reported by popular computer and business journals, has been explored. We continue to stress that ERP systems can replace many applications with one unified set of programs. The material on virtual reality includes new information on the use of this technology to design and manufacture Boeing's new Dreamliner 787 aircraft. The material on systems development has been updated with new examples of success and failures.

The section on ethical and social issues includes the latest threats and what is being done to prevent them. This material contains new examples on the dangers of identity theft, computer mistakes, and power consumption and computer waste. Some experts believe that computers waste up to half of the energy they consume and account for about 2% of worldwide energy usage. We also discuss legal actions in this section. For example, lawsuits have been filed against YouTube and other Internet sites to protect important copyrighted material from being posted and distributed.

We have trimmed material throughout the chapter to keep its length reasonable and consistent with previous editions. Table 1.3 from the 8th edition on uses of the Internet, for example, has been deleted. These types of applications are covered in detail in Chapter 7 on the Internet.

Chapter 2, Information Systems in Organizations

As with previous editions, Chapter 2 gives an overview of business organizations and presents a foundation for the effective and efficient use of IS in a business environment. Since its inception, the primary goal of this text has been to present a core of IS principles and concepts that every business student should know. Chapter 2 stresses the importance and usage of IS within the business organization.

As with all chapters, the opening material at the beginning of the chapter including the vignette, the Information Systems @ Work and Ethical and Societal Issues special interest boxes, the end-of-chapter cases, and all end-of-chapter material have been updated to reflect the changes in Chapter 2.

Chapter 2 gives an overview of business organizations and presents a foundation for the effective and efficient use of IS in a business environment. We have stressed that the traditional mission of the IS organization "to deliver the right information to the right person at the right time" has broadened to include how this information is used to improve overall performance and help people and organizations achieve their goals.

The section on why learn about information systems in organizations has been updated to include entrepreneurs. There are new photographs throughout the chapter that show how organizations use information systems to their benefit. New examples of how companies use information systems have been also introduced throughout the chapter. New rules for assembling and using virtual teams have been included. We have stressed that a competitive advantage can result in higher quality products, better customer service, and lower costs.

New examples of how companies use information systems have been included. There are new examples on the value chain, supply chain management, customer relationship management, employee empowerment, organizational change, reengineering, outsourcing, utility computing, return on investment, and more. For example, we provide new information and quotes about supply chain management and customer relationship management. New examples and references for just-in-time inventory have also been included. The section on performance-based information systems has been updated with fresh and recent examples of how organizations have reduced costs and improved performance.

The section on careers has been updated with a new list of top U.S. employers, what makes a satisfied IS worker on the job, and new information on various visas and their impact on the workforce. We have updated the information on the U.S. H-1B and L-1 visa programs. In the first few days that applications were available for the H-1B program in 2007, over 130,000 applications were filed for 65,000 positions. As in the past, some fear that the H-1B program is being abused to replace high-paid U.S. workers with less expensive foreign workers. In 2007, two U.S. senators on the Senate Judiciary Subcommittee on Immigration sent letters of concern to a number of Indian firms that were using the H-1B program to staff their U.S. operations with IS personnel from foreign countries.

We have also included information and a table about the best places to work in the information systems field. The changing role of the CIO has been highlighted with many new examples and quotes from CIOs at companies of all sizes. The changing roles include the involvement with strategic decisions and more involvement with customers. At the end of the section on IS careers, we have added a new section on finding a job in IS. This new section describes the many ways that students find jobs.

We have included information on IS certification and how jobs can be located using the Internet through sites such as Monster.com and Hotjobs.com. As with previous editions, this chapter continues to stress performance-based management with new examples of how companies can use information systems to improve productivity and increase return on investment.

Chapter 3, Hardware: Input, Processing, and Output Devices

Considerable changes were made to Chapter 3 starting with the chapter principles and objectives. The goal of these changes was to communicate the important principles that readers must understand about computer hardware and its evolution in an easily understood and interesting manner. We made every attempt to provide the most current information and realistic examples possible. As with all chapters, the opening material at the beginning of the chapter including the vignette, the Information Systems @ Work and Ethical and Societal Issues special interest boxes, the end-of-chapter cases, and all end- of-chapter material have been updated to reflect the changes in Chapter 3. Also included are more than three dozen new examples of applying business knowledge to reach critical hardware decisions.

In the "Computer Systems: Integrating the Power of Technology" section, Bosch Security Products and the Iowa Health System are provided as new examples of organizations applying business knowledge to reach critical hardware decisions.

In the "Processing and Memory Devices: Power, Speed, and Capacity" section, the new quad core processors from Intel and AMD are covered and Viiv (rhymes with five) is identified as an exciting new quad core application. The problems arising from excess heat generated by very fast CPUs are discussed. For example, in February 2007, a battery-related fire broke out in the overhead bin of a JetBlue Airways flight. In March 2007, a battery overheated or ignited on an American Airlines aircraft. In both cases, fast-acting flight attendants extinguished the fire and avoided disaster. Several approaches are offered that chip and computer manufacturers have taken to avoid heat problems in their new designs, including demand-based switching and direct jet impingement. Also discussed is the use of fuel cells by manufacturers of portable electronic devices such as computers and cell phones to provide more effective sources of energy as portable devices grow increasingly power hungry. The Cell Broadband Engine Architecture collaboration effort among IBM, Toshiba, and Sony to offer even greater computing capacity is examined. Also covered is Intel's new "tick-tock" manufacturing to speed up the introduction of new chips.

In the "Memory Characteristics and Functions" section, the various forms of RAM memory including SRAM, DRAM, and DDR RAM are covered, as are various forms of PROM memory, including EPROM, EEPROM, Flash, NOR Flash, NAND Flash, FeRAM, PCM, and MRAM.

The coverage of the important topic of multiprocessing is greatly expanded. Both the Intel and AMD quad core processors are examined. We point out that the processor manufacturers must work with software developers to create new multithreaded applications and next-generation games that will use the capabilities of the quad-core processor. For example, Viiv combines Intel products including the Core 2 Quad processor with additional hardware and software to build an extremely powerful multimedia computer capable of running the processing-intensive applications associated with high-definition entertainment. AMD's new quad-core Opteron processor and Fusion project to combine a graphics processing unit and a CPU on the same chip are also discussed.

In the section on parallel computing, new examples are given of grid computing. European and Asian researchers are using a grid consisting of some 40,000 computers spread across 45 countries to combat the deadly bird flu. Folding@home (a project to research protein folding and misfolding and to gain an understanding of how this protein behavior is related to diseases such as Alzheimer's, Parkinson's, and many forms of cancer) is provided as a new example of grid computing. Also covered is Chrysler's use of high-performance computers consisting of some 1,650 cores to simulate racecar performance and identify opportunities for improvement in the car's design and operation. The concept of cloud computing is introduced, highlighting the efforts of IBM, Google, and Amazon.com to offer exciting new services based on cloud computing. The potential is raised that some organizations may consider replacing part of their IS infrastructure with cloud computing.

The section on secondary storage is updated to identify the newest devices and the latest speeds and capacities. All data and comparisons of hardware devices have been updated. For example, Table 3.2 provides current cost comparisons of major forms of data storage. This section also discusses Wal-Mart's 1 petabyte database used to analyze in-store sales and determine the ideal mix of items and the optimal placement of products within each store to maximize sales. Medkinetics, a small business that automates collecting and submitting for approval of information about a doctor's qualifications, illustrates the use of RAID storage technology. The Girls Scouts of America is offered as an example of the use of virtual tape storage. In the section on enterprise storage options, the University of North Carolina Hospital is offered as a new example of network-attached storage. The Navy's Surface Combat Systems Center also provides a new example of the use of a storage area network.

In the section on input and output devices, speech recognition systems are explored, including the U.S. Department of Defense's recent award of $49 million to Johns Hopkins University to set up and run a Human Language Technology Center of Excellence to develop advanced technology and analyze a wide range of speech, text, and document image data in multiple languages. Dial Directions is offered as an example of an organization using speech recognition technology to provide customer service. The Large Synoptic Survey Telescope used for space research is offered as a new example of the use of digital cameras, as is a new example of the use of OCR technology to improve the payroll function of Con-way Inc. Also included is a new example of the use of RFID technology to track inventory for Boekhandels, a major book retailer in the Netherlands. Updated information is provided on the use of holographic disks. New photos include HP's new high-speed, high-volume CM8060 inkjet printer, the MacBook Air ultra thin laptop computer, and the Apple iPod Touch device.

In the section on computer system types, Table 3.3 depicting types of computer systems has been updated. Various types of new computer systems, including a pocket computer, ultra laptop computers, and ultra small desktop computers are illustrated and discussed. A new example is offered of the use of the Pocket PC (a handheld computer that runs the Microsoft Windows Mobile operating system) by the Coca-Cola field sales force to automate the collection of information about sales calls, customers, and prospects. Another new example is CSX Transportation, one of the nation's largest railroads, which uses DT Research's WebDT 360 to enable train conductors to monitor systems while onboard and communicate with stations for real-time updates. Also included is a brief discussion of how mainframe computers are used at the top 25 banks and retailers. Current information is provided on the IBM Blue Gene supercomputer. A table depicting the processing speed of supercomputers has been added.

Chapter 4, Software: Systems and Application Software

As with all chapters, the opening material including the vignette, the Information Systems @ Work and Ethical and Societal Issues special interest boxes, the end-of-chapter cases, and all end-of-chapter material have been updated to reflect the changes in Chapter 4.

The chapter includes descriptions, illustrations, and examples of the latest and greatest software shaping the way people live and work. A new and interesting vignette has been created to engage the student's attention from the first page of the chapter. New, thought-provoking boxes and cases have been created to help bring the chapter material to life for the reader. The operating system section has been updated to include detailed coverage of all editions of Windows Vista. Equally thorough coverage of Apple OS X Leopard and Linux (the various distributions) is also provided. These three operating systems are compared and contrasted. The section on workgroup and enterprise operating systems has been updated to include the latest systems including Windows Server 2008, Mac OS X Leopard Server, z/OS, and HP-UX. Smartphone operating systems are presented. Embedded operating systems are also discussed including Microsoft Sync, a popular feature on GM vehicles. Several new examples of utility software products are presented including security software, compression software, spam filters, and pop-up blockers. Virtualization software and VMware are introduced and described. Service-oriented architecture (SOA) and Software as a Service (SaaS) are introduced and discussed. Microsoft Office 2007 is covered thoroughly. Alternatives to Office 2007 are also provided. Online productivity software is introduced and discussed. This includes Google Docs, Zoho applications, Thinkfree, and Microsoft Office Live. A variety of software licensing options are presented. Open source is provided as an alternative to popular software.

Many new examples of systems and application software are cited, including the National Aquarium's use of Windows Vista, a lawyer's use of Mac OS X, radio station KRUU's use of Ubuntu Linux, Ebay's use of Solaris, the benefits of the software provided on RIM Blackberries to workers at the U.S. Department of Agriculture, Blue Cross Blue Shield's use of proprietary software for claims management, use of SaaS by The Improv for managing marketing and ticket sales, and the use of personal information management software by Greenfield online, a Web survey company.

Chapter 5, Database Systems and Business Intelligence

The database chapter has been updated to include descriptions, illustrations, and examples of the database technologies that shape the way people live and work. A new vignette, the Information Systems @ Work and Ethical and Societal Issues special interest boxes, the end-of-chapter cases, and all end-of-chapter material have been updated to reflect the changes in Chapter 5. The growth of digital information is discussed in the context of the importance of managing overwhelming amounts of data. The market leaders in database systems are introduced along with their market share information. After discussing traditional database technologies, Database as a Service (DaaS) is introduced as a new form of database management. New examples are provided for databases accessible on the Internet. The semantic Web and its relationship to databases are discussed.

Many new examples include a Hollywood talent agency's use of databases to store client information; the city of Albuquerque's use of a database to provide citizens with information on water bills, water usage, and other local information; a database security breach at an Ivy League college; the FBI's huge database of biometric data; Wal-Mart's medical database for use at its health clinics; Microsoft's use of OneNote for presenting management training classes; a New Delhi's lighting manufacturer's use of a DBMS; Morphbank, a special-purpose database for scientific data and photographs; the phenomenal growth of data centers around the world; databases used for medical records; 1–800-Flowers use of a data warehouse for customer data; the Defense Acquisition University's use of a data warehouse for student records; the use of data mining to forecast terrorist behavior; the use of data mining by MySpace for targeted marketing; the use of predictive analysis by police to forecast crimes; the use of business intelligence in the health industry; the use of object-oriented databases by King County Metro Transit system for routing buses; and the use of a virtual database system by Bank of America Prime Brokerage to reduce storage demands.

Chapter 6, Telecommunications and Networks

The entire chapter has been reorganized to incorporate the latest telecommunications and network developments and applications. The flow of the chapter has been changed from the previous edition to introduce the topics in a more logical progression.

The opening vignette, Ethical and Societal Issues, Information Systems @ Work, and chapter-ending cases are all new. New, realistic examples are sprinkled throughout the chapter to maintain the reader's interest and to demonstrate the actual application of the topics being discussed.

The presentation of wireless communications is divided into a discussion of the options for short range, medium range and long range communications. Two new options are covered: near field communications and Zigbee. A discussion is given of the potential impact of the reallocation of the 700 MHz frequency band made possible by the conversion to all-digital TV broadcasting. Google's new Android software development platform for mobile phones is also mentioned.

The discussion of both Wi-Fi and WiMAX are updated and expanded. The use of Wi-Fi to build public networks in various cities and also the plans of airlines to provide on-board Wi-Fi service is discussed. The collaborative effort between Sprint Nextel and Clearwire to build a nationwide WiMAX network as well as Intel's investment in this technology is discussed.

A section has been added that discusses likely future developments in wireless communications including the reuse of the 700 MHz frequency band freed up by the conversion to all digital TV. The use of a PBX and the Centrex PBX option to provide communications capabilities for an organization is discussed. A section has been added on Securing Data Transmission that addresses the needs and identifies two solutions: encryption and VPN.

A section has been added outlining the additional steps needed to secure wireless networks. An in-depth discussion has been added of Voice over Internet including advantages and disadvantages. The use of VoIP at Merrill Lynch is covered. Reverse 911, voice-to-text, shared workspaces and unified communications are also covered. The discussion of GPS applications has been updated and expanded.

Chapter 7, The Internet, Intranets, and Extranets

As in previous editions, updates to this chapter are significant as Internet and Web technologies are evolving at a rapid pace. This edition thoroughly covers Web 2.0 technologies including social networking, social bookmarking, media sharing, and other technologies from which Web 2.0 springs, including technologies such as blogging, podcasting, and wiki. A strong emphasis on cloud computing illustrates how many traditional computer applications and information systems are now served from remote servers and depend on the Web for delivery.

As with all chapters, the opening material including the vignette, the Information Systems @ Work and Ethical and Societal Issues special interest boxes, the end-of-chapter cases, and all end-of-chapter material have been updated to reflect the changes in Chapter 7. Dozens of new examples have been integrated into the chapter content. All statistics in the chapter have been updated.

The National LambdaRail is mentioned as a next-generation form of the Internet. The chapter describes how domain names are registered with ICANN, and how accredited domain registrars such as GoDaddy.com are used. Connecting to the Internet wirelessly has been added as a method of connection, both through cell phone networks and using Wi-Fi and WiMAX. Connect cards are introduced as a method of connecting a notebook computer to the Internet through a cell phone account.

The interplay among HTML, XHTML, CSS, and XML is explored as the current method for designing Web pages and sites. Cascading style sheets are explored more deeply as a powerful method of Web page design. The new section on Web2.0 includes coverage of social networking and rich Internet applications (RIAs). AJAX is introduced as a valuable tool for developing RIAs. Microsoft Silverlight is mentioned as a new competitor to Adobe Flash.

The techniques used by search engines are explored in depth. Search engine optimization (SEO) is introduced as a method for businesses to improve the rank of their Web sites in search results. Various forms of e-mail are discussed including Web mail, POP, IMAP, and Outlook. Push e-mail is also explained, as is the use of e-mail on Blackberries and other smartphones. Secure forms of telnet and ftp (SSH, SFTP) are encouraged. Wi-Fi phones and Internet phone services such as Skype are explored. The material on Internet TV and video has been expanded. A section on e-books and audio books has been added. Microsoft SharePoint is introduced.

Chapter 8, Electronic and Mobile Commerce

As with all other chapters, the opening vignette, Ethical and Societal Issues, Information Systems @ Work, and chapter-ending cases are all new. New, real-world examples are provided throughout the chapter to maintain the reader's interest and to demonstrate the actual application of the topics being discussed. New end-of-chapter questions and exercises are included.

Customer relationship management and supply chain management are included in the section on the multistage model for e-commerce. The section on e-commerce challenges has been rewritten to emphasize the issues of trust and privacy as well as identity theft. Many suggestions for online marketers to create specific trust-building strategies for their Web sites are offered, as are several tips for online shoppers to avoid problems.

The section on electronic and mobile commerce applications is fully updated with new material and examples of companies using electronic and mobile commerce to increase revenue and reduce costs. Many new and innovative applications of m-commerce are discussed.

The "Global Challenges for E-Commerce and M-Commerce" section has been rewritten to highlight challenges associated with all global systems: cultural, language, time and distance, infrastructure, currency, product and service, and state, regional and national laws. New examples have been added to illustrate the threats of e-commerce and m-commerce. The section on strategies for e-commerce has been expanded to include more material on m-commerce. Several examples of companies that can help online marketers to build a successful Web site are offered.

The technology infrastructure section has been expanded to include a discussion of the latest technologies used for both e-commerce and m-commerce.

The section on electronic payment systems has been updated with new solutions as well as solution providers. A table has been added to compare the various payment systems. Making payments using cell phones such as a credit card is also discussed.

Chapter 9, Enterprise Systems

The chapter has been greatly revised to emphasize ERP systems more than traditional transaction processing systems. Much of the discussion of traditional transaction processing systems has been reduced or placed in simple lists. In addition, the adoption and use of enterprise systems by SMBs is much more thoroughly covered.

A section has been added that discusses the use of transaction processing systems by SMBs and identifies a number of software packages available that provide integrated transaction processing system solutions for SMBs. The use of software packages to provide an integrated set of transaction processing systems by the city of Lexington, KY is mentioned.

The material on ERP and CRM has been updated with new material and examples. For example, the implementation of an ERP system by Gujarat Reclaim and Rubber Products, Amgen, BNSF Railway Company and the resulting benefits is discussed.

A new section has been added that covers the use of ERP by SMBs and highlights the special needs of those firms. Several open source ERP solutions, frequently preferred by SMBs, are identified. The adoption of ERP systems at SMBs Cedarlane, Vertex Distribution, Prevention Partners, Inc., and Galenicum is discussed. The 15 top-rated ERP and CRM packages for both large organizations and SMBs are presented.

As with all other chapters, the opening vignette, Ethical and Societal Issues, Information Systems @ Work, and chapter-ending cases are all new. New, real world examples are included throughout the chapter to maintain the reader's interest and to demonstrate the actual application of the topics being discussed. New end of chapter questions and exercises are included.

Chapter 10, Information and Decision Support Systems

All boxes, cases, and the opening vignette are new to this edition. We have added a new learning objective about the importance of special-purpose systems. This chapter includes many new examples of how managers and decision makers can use information and decision support systems to achieve personal and corporate goals. The section on problem solving and decision making has new examples. We have emphasized that many of the information and decision support topics discussed in the chapter can be built into some ERP systems, discussed in Chapter 9.

We have added a new section on the benefits of information and decision support systems at the end of the material on decision making and problem solving. This new section investigates the performance and costs of these systems and how they can benefit individuals, groups, and organizations by helping them make better decisions, solve and implement problems, and achieve their goals. We have also included new examples, including how British Airways used the problem solving approach to address flight delays.

The section on MIS has updated examples and references throughout. We explore how companies and local governments use software tools to help them in developing effective

MIS reporting systems. The section on demand reports provides new examples and references about how people can get medical records from the Internet. The section on exception reports has a new example of the use of mashups to get reports from different data sources. This section also includes a new example of the use of texting as a way to deliver exception reports. The "Financial MIS" section has new examples and material. We have shown how the Internet has been used to make microloans using social-networking sites such as Facebook. We have also investigated how financial companies use corporate news to help them make trading decisions. The section on manufacturing MISs has also been updated. We have included examples that show how companies, such as Toyota, continue to use JIT inventory control techniques. The material on marketing MISs has been updated with new material on the use of video advertising and social-networking sites to promote new products and services. This section also includes new examples of how the Internet is being used to auction radio ads. The material on the human resource (HR) MIS has also been updated with new information about just-in-time talent and the use of supply chain techniques in the HR area.

The use of digital dashboards and business activity monitoring continues to be stressed in the section on decision support systems along with examples and references. The section on optimization includes new examples, showing the huge cost savings of the technique. The use of data warehousing, data marts, and data mining, first introduced in Chapter 5, is emphasized in this chapter. New corporate examples of the use of data warehousing, data marts, and data mining that can be used to provide information and decision support have been placed throughout the section. The use of mashups to integrate data from different sources into a data-driven DSS has been included.

This chapter continues to highlight group support systems. We discuss additional features of groupware, including group monitoring, idea collection, and idea organizing and voting features. New examples and approaches have been explored, including the use of the Web to deliver group support. We have modified the section on parallel communication to include the use of unified communication in group decision making.

To keep the chapter length reasonable, we have trimmed a number of the sections. We have deleted the section on developing effective reports and moved some of the material into other sections. The material on the characteristics of a management information system has been summarized in a table. The characteristics of a decision support information system and those of an executive support system have also been put in new tables.

Chapter 11, Knowledge Management and Specialized Information Systems

We have updated all material in this chapter and added many new and exciting examples and references.

The overall purpose of this chapter is to cover knowledge management and specialized business information systems, including artificial intelligence, expert systems, and many other specialized systems. These systems are substantially different from more traditional information and decision support systems, and while they are not used to the same extent as more traditional information systems, they still have an important place in business.

As with the last edition, the section on knowledge management continues to be a natural extension of the material in Chapter 10 on information and decision support systems and leads to a discussion of some of the special-purpose systems discussed in the chapter, including expert systems and knowledge bases.

This chapter contains many new examples and references. We have new material on the importance of knowledge management for Tata, a large Indian company that uses it to retain and use knowledge from retiring employees. The Aerospace and Defense (A&D) organization also uses knowledge management to keep knowledge in its organization. This chapter also shows how Pratt & Whitney uses knowledge management systems to help it deliver information and knowledge about its jet engine parts to the company and airlines, including Delta and United. We highlight the importance of the knowledge manager and the chief knowledge officer with new references and a new quote. The material on knowledge management includes new examples of communities of practice (COP). A group of people from the International Conference on Knowledge Management in Nuclear Facilities has a COP to investigate the use of knowledge management systems in the development and control of nuclear facilities. We also have numerous new examples on how organizations can obtain, store, share, and use knowledge. The University of South Carolina, for example, has joined with Collexis to develop and deliver new knowledge management software, based on Collexis's Knowledge Discovery Platform. We have referenced a survey that estimates that American companies will spend about $70 billion on knowledge management technology in 2007. New material and examples have been included in the section on the technology to support knowledge management. This section contains information on research into the use and effectiveness of knowledge management.

The other topics discussed in this chapter also contain numerous new examples and references in robotics, vision systems, expert systems, virtual reality, and a variety of other special-purpose systems. New robotics examples include research done at the Robot Learning Laboratory at Carnegie Mellon University, robots by iRobot, medical uses of robots, and robots in the military. New examples of voice recognition have been included. New commercial and military examples of expert systems have also been included. We have also updated the information on expert systems tools and products. New material on computer vision has been included. One expert believes that in 10 years, computer vision systems may be able to recognize certain levels of emotions, expressions, gestures, and behaviors, all through vision. The section on virtual reality has many new examples and references. We have added a new section on business applications to reflect the increased use of virtual reality by businesses of all types and sizes. Kimberly-Clark Corporation has developed a virtual reality system to view store aisles carrying its products. Boeing uses virtual reality to help it design and manufacture airplane parts and new planes, including the 787 Dreamliner. Clothing and fashion companies, such as Neiman Marcus and Saks Fifth Avenue, are using virtual reality on the Internet to display and promote new products and fashions.

The section on other specialized systems also has new material, examples, and references. The U.S. Defense Advanced Research Projects Agency (DARPA) and other organizations are exploring mechanical computers that are energy efficient and can stand up to harsh environments. We have included information on Segway and 3VR Security that performs a video-face recognition test to identify people from pictures or images. Ford Motor Company and Microsoft have developed a voice-activated system called Sync that can plays music, make phones calls, and more from voice commands. The Advanced Warning System by Mobileye warns drivers to keep a safe distance from other vehicles and drivers. The Surface from Microsoft is a touch-screen computer that uses a glass-top display. Microsoft's Smart Personal Objects Technology (SPOT) allows small devices to transmit data and messages over the air. We also discuss that manufacturing is being done with inkjet printers to allow computers to "print" 3-D parts. More information has been included on RFID technology. Wearable computers used to monitor inventory levels and perform other functions have been introduced. We have also included new references in the material on game theory and informatics. The material on expert systems development has been deleted to trim chapter length. Much of this material is now covered in Chapters 12 and 13 on systems development.

Chapter 12, Systems Development: Investigation and Analysis

This chapter has new material, examples, and references. In addition, the end-of-chapter material has been completely updated with new questions and exercises. As with other editions, all cases are new to this edition.

In the introductory material, which provides an overview of systems development, we include new material on the importance of removing old systems. For example, Walt Disney developed the Virtual Magic Kingdom (VMK) game to celebrate the 50th anniversary of Disneyland. The VMK game used Disney avatars and offered virtual rewards to game players. When Disney decided to remove or terminate the game, some players were outraged and protested outside Disney offices in California.

This chapter has a new emphasis on entrepreneurs and small business owners. The systems development skills and techniques discussed in this chapter and the next can help people launch their own businesses. When Marc Mallow couldn't find off-the-shelf software to schedule workers, he took a few years to develop his own software. The software he created became the core of a company he founded, located in New York. We also have new material and examples on the use of systems development for nonprofit organizations. To stay competitive in today's global economy, some cities, including Chattanooga, Tennessee, are investing in high-speed fiber optic cables that have the potential to deliver greater speed compared to existing cable and phone company offerings.

We have increased the emphasis on long-range planning in systems development projects. It can result in getting the most from a systems development effort. It can also help make sure that IS goals are aligned with corporate goals. Hess Corporation, a large energy company with over 1,000 retail gasoline stations, used long-range planning to determine what computer equipment it needed and the IS personnel needed to run it.

The section on participants in systems development has new material and examples. Today, companies are using innovative ways to build new systems or modify existing ones without using in-house programmers. Constellation Energy, a $19 billion utility company, is using an approach that asks programmers from around the world to get involved. The approach, called crowd sourcing, asks programmers to contribute code to the project. Winning programmers that submit excellent code can be given from $500 to more than $2,000. Constellation is hoping to save time and money by using crowd sourcing, but neither result is guaranteed.

We've included new examples on initiating systems development projects. For example, systems development can be initiated when a vendor no longer supports an older system or older software. When this support is no longer available, companies are often forced to upgrade to new software and systems, which can be expensive and require additional training. Major system and application software companies, for example, often stop supporting their older software a few years after new software has been introduced. This lack of support is a dilemma for many companies trying to keep older systems operational and running. We also discuss that a company's customers or suppliers can cause the initiation of systems development. Daisy Brand, a dairy products company, was asked by one of its major customers, Wal-Mart, to start using special RFID tags. By putting RFID tags on every pallet of dairy products that it ships to customers, the company cut in half the time it used to take to load pallets onto delivery trucks.

The section on quality and standards includes a new example. Although many companies try to standardize their operations on one operating system or standard, others have multiple systems and platforms to take advantage of the strengths of each. In these cases, many IS managers seek one tool to manage everything. Today, many companies, including Microsoft, are developing software and systems that can be used to manage different operating systems and software products. We also describe how creative analysis can help organizations achieve their performance goals and include a new quote from Michael Hugos, principle at the Center for Systems Innovation and one of Computerworld's 2006 Premier 100 IT Leaders, about the importance of creativity.

The impact of laws and regulations, such as Sarbanes-Oxley, has been emphasized. Some of the disadvantages of new laws and regulations are also discussed, including the increased use of outsourcing by U.S. companies to reduce costs in complying with laws and regulations. The importance of user involvement and top management support has been included in the section on factors affecting systems development success.

New examples are provided in the section on IS planning and aligning corporate and IS goals. Procter & Gamble (P&G), for example, uses ROI to measure the success of its projects and systems development efforts. Providing outstanding service is another important corporate goal. Coca-Cola Enterprises, which is Coca Cola's largest bottler and distributor, decided to use online services from Microsoft and SharePoint to speed its systems development process.

We have updated the material on establishing objectives for systems development with new examples. Southern States, which sells farm equipment in over 20 states and is owned by about 300,000 farmers, decided to use Skyway Software, Inc.'s Visual Workplace to develop a new pricing application. The use of this service-oriented architecture (SOA) tool allowed Southern States to generate $1.4 million more in revenue the year after it was placed into operation.

The section on cost objectives also includes a new example that shows how Tridel Corporation used systems development to build a new invoicing application, called Invoice Zero, to save over $20 thousand in operating costs. The new invoicing application, which consolidated invoices and sent them out once a month, cut the number of monthly invoices from 2,400 to just 17. Reducing costs was also an important factor for Cincinnati Bell. By switching from dedicated PCs to thin client computers and virtualization software, Cincinnati Bell expects to see a large reduction in help desk costs. Some experts predict that help-desk costs could be reduced by 70 percent or more.

We continue to emphasize the importance of rapid and agile application development. Microsoft, for example, has adopted a more agile development process in its server development division. BT Group, a large British telecommunications company, uses agile systems development to substantially reduce development time and increase customer satisfaction. We also have a new example of the importance of information systems speed for Six Flags, one of the largest amusement parks in the world with about $1 billion in annual sales.

New examples of systems development failure have also been included. The United Airlines automated baggage systems development project, for example, failed to deliver baggage to airline passengers in good shape or on time. The $250 million systems development project costs United Airlines about $70 million to operate each year. Computers were overwhelmed with data from the cars carrying the baggage. United Airlines eventually abandoned the computerized system. In another case, a large $4 billion systems development effort ran into trouble. The objective was to convert older paper-based medical records to electronic records for a large healthcare company.

New examples of outsourcing are discussed. IBM, for example, has consultants located in offices around the world. In India, IBM has increased its employees from less than 10,000 people to more than 30,000. We have also included information on the scope of outsourcing, including hardware maintenance and management, software development, database systems, networks and telecommunications, Internet and Intranet operations, hiring and staffing, and the development of procedures and rules regarding the information system. Increasingly, companies are using several outsourcing services. GM, the large automotive company, uses six outsourcing companies after its outsourcing agreement with EDS expired. Using more than one outsourcing company can increase competition and reduce outsourcing costs. Small and medium-sized firms are also using outsourcing to cut costs and get needed technical expertise that would be difficult to afford with in-house personnel. Millennium Partners Sports Club Management, for example, used Center Beam to outsource many of its IS functions, including its helpdesk operations. The Boston-based company plans to spend about $30,000 a month on outsourcing services, which it estimates to be less than it would have to pay in salaries for additional employees. The market for outsourcing services for small and medium-sized firms is expected to increase by 15 percent annually through 2010 and beyond. Some of the disadvantages of outsourcing are also included. Some companies, such as J. Crew, are starting to reduce their outsourcing and bring systems development back in-house.

The section discussing on-demand computing has new material and examples. Amazon, the large online retailer of books and other products, will offer on-demand computing to individuals and other companies of all sizes, allowing them to use Amazon's computer expertise and database capacity. Individuals and companies will only pay for the computer services they use.

There are new project management references and examples. We also discuss new project management tools and software. As an academic exercise, Purdue University undertook a project to build a supercomputer using off-the-shelf PCs. The project was completed in a day and required more than 800 PCs. In what some people believe is the largest private construction project in the U.S., MGM Mirage and others used project management software to help them embark on an ambitious $8 billion construction project on 76 acres with over 4,000 hotel rooms, retail space, and other developments. To complete the project, managers selected Skire's Unifier, a powerful and flexible project management software package. The project management software should save the developer a substantial amount of money. We also provide new material on the problems facing project managers. Project escalation, where the size and scope of a new systems development effort greatly expands over time, is a major problem for project managers. Projects that are over budget and behind schedule are a typical result of project escalation.

We have stressed the need for cooperation and collaboration with the systems investigation team. The systems investigation team can be diverse, with members located around the world. When Nokia decided to develop a new cell phone, its investigation team members were from England, Finland, and the United States. Cooperation and collaboration are keys to successful investigation teams.

We have trimmed many sections to reduce the length of the chapter. The material in the section on outsourcing, for example, has been trimmed to some extent. The section on performance objectives has also been shortened. We have shortened and deemphasized the material on operational and non-operational prototypes. We have also trimmed the section on quality and standards. We have reduced some of the less important factors in Table 12.5.

Chapter 13, Systems Development: Design, Implementation, Maintenance, and Review

The section on design includes new material and examples. In the introductory material, we stress that system development should take advantage of the latest developments in technology. Many companies, for example, are looking into cloud computing, where applications are run on the Internet instead of being developed and run within the company or organization. Cloud computing is allowing individuals, like racecar driving instructor Tom Dyer, to do work while traveling or in a car. We also discuss that developing Web applications will become more important in the future. Microsoft's Live Mesh, for example, allows systems developers to seamlessly coordinate data among different devices and provide data backup on the Internet.

Designing new systems to reduce total costs is also discussed. The New York Stock Exchange, for example, decided to use the Linux operating system to lower total IS costs. With the high cost of many commodities today, some systems development efforts are saving money by avoiding copper wires and installing wireless telecommunications systems.

The security section describes how some small and medium sized corporations are buying unified threat management (UTM) products to protect their networks from security threats and breaches. In the section on disaster recovery, we continue to stress that the most cost-effective time to deal with potential errors is early in the design phase. Hanford Bros. Company, for example, had installed backup electrical generators in case of a power failure, but when a fuel truck crashed near its facility spilling its flammable cargo, the city shut down all power to the area and didn't let Hanford Bros use its electrical generators, fearing it could cause an explosion or severe fire. This minor incident shut down the IS center for the company until the spill could be cleaned up. We also stress that disaster planning can be expensive and many companies don't do an adequate job, even though many IS managers realize the importance of disaster planning. According to a Forrester Research study, only 34 percent of IS data center managers believed they were prepared for a disaster or data center failure. In another study, 71 percent of IS managers considered disaster planning recovery as important or critical.

This section also discusses the importance of personnel backup. Without IS employees, the IS department can't function. New material on commercially available disaster planning and recovery services has been included. EMC, for example, offers data backup in its RecoverPoint product.

This section also emphasizes how new federal regulations can result in designing new systems or modifying existing ones. Federal regulations that require companies to make e-mails, text messages, and other electronic communications available in some court hearings has resulted new systems development projects that can search and find electronic communications to meet federal requirements. In the section on evaluation techniques, we discuss how some companies are using the Internet to get important customer satisfaction information. Cabelas and Staples, for example, are using Web-style testimonials to get customer satisfaction information on its products and information systems. Other companies, such as Backcountry, use live online chat to get customer satisfaction information. We have also updated the section on systems controls to include more recent and important threats and controls to avoid them. California law enforcement officials busted a criminal ring that billed almost a million dollars in tests that were never performed. A major health care provider developed new software to combat medical identity theft by investigating unjustified spikes in medical charges. In the section on interface design and controls, we include new information about the importance of changing ID numbers and passwords. An IS worker for a large U.S. company operating in India was caught stealing about 4,000 sensitive corporate documents using the identification number and password of another employee.

We have a new section on environmental design, also called green design, that involves systems development efforts that slash power consumption, take less physical space, and result in systems that can be disposed in a way that doesn't negatively impact the environment. A *Computerworld* survey revealed that over 80 percent of IS managers considered energy efficiency when selecting new computer equipment. The Environmental Protection Agency (EPA) estimates that a 10 percent cut in data center electricity usage would be enough to power about a million U.S. homes every year. This new section discusses the companies that are developing products and services to help save energy. UPS developed its own software to route trucks more efficiently, helping UPS cut 30 million miles per year, slash fuel costs, and reduce carbon emissions by over 30,000 metric tons.

This section also shows how companies are developing systems to dispose of old equipment. Hewlett-Packard and Dell Computer have developed procedures and machines to dispose of old computers and computer equipment in environmentally friendly ways. Old computers and computer equipment are fed into machines that disintegrate them into small pieces and sort them into materials that can be reused. The process is often called green death. One study estimates that more than 130,000 PCs in the U.S. are thrown out every day. The U.S. government is also involved in environmental design. It has a plan to require federal agencies to purchase energy-efficient computer systems and equipment. The plan would require federal agencies to use the Electronic Product Environmental Assessment Tool (EPEAT) to analyze the energy usage of new systems. The U.S. Department of Energy rates products with the Energy Star designation to help people select products that save energy.

The material on systems implementation has been updated to include new references, material, and examples. Virtualization, first introduced in Chapter 3, has had a profound impact on many aspects of systems implementation. Virtualization software can make computers act like or simulate other computers. The result is often called a virtual machine. Using virtualization software, servers and mainframe computers can run software applications written for different operating systems. Virtualization is being used to implement hardware, software, databases, and other components of an information system. Virtualization can also be environmental friendly, reducing power consumption and requiring less space for equipment. Virtualization, however, introduces important implementation considerations, including security and backup procedures. The section on hardware acquisition has new material and examples, including material on hardware virtualization. Tellabs, for example, acquired virtualized Dell PowerEdge Servers for its operations to consolidate its hardware, increase utilization, and save space. We also include chipmakers in the list of hardware vendors. The title of software acquisition section has been changed to include software as a service (SaaS) to reflect new material and a new emphasis in the chapter. The Humane Society of the United States used SaaS to obtain a secure system to receive credit-card contributions from donors. We also stress the importance of software acquisition. Companies such as Google are delivering word processing, spreadsheet programs, and other office suite packages over the Internet. The SmartBeam IMRT software program from Varian Medical Systems focuses radiation beams to kill more cancer cells, spare good cells, and save lives. Allstate Insurance decided to develop or make a new software program, called Next Gen, to speed claims processing and reinforce its "You're in good hands" slogan. The company is expected to spend over $100 million on the new software. We also have new material and examples of software virtualization. Windows Server 2008, for example, provides virtualization tools that allow multiple and various operating systems to run on a single server. Virtualization software such as VMware is being used by businesses to safeguard private data. Kindred Healthcare, for example, uses VMware on its server to run hundreds of virtual Windows PC desktops that are accessed by mobile computers throughout the organization. Since the software tools used to access that data are running on the server, security measures are easy to implement.

We also stress the difficulties in developing in-house software, which is often constantly changing. The chief scientist at IBMs' Rational Software Corporation, for example, believes that software development can be hard and the software continuously evolving. The material on database acquisition includes the use of open-source databases. In addition, we have discussed that virtual databases and database as a service (DaaS) are also popular ways to acquire database capabilities. XM Radio, Bank of America, and Southwest Airlines, for example, use the DaaS approach to manage many of their database operations from the Internet. In another case, a brokerage company was able to reduce storage capacity by 50 percent by using database virtualization.

We've also updated the material on user preparation. When a new operating system or application software package is implemented, user training is essential. In some cases, companies decide not to install the latest software because the amount of time and money needed to train employees is too much. In one survey, over 70 percent of the respondents indicated that they were in no hurry to install a new operating system. Additional user training was a factor in delaying the installation of the new operating system. In the section on site preparation, we stress that developing IS sites today requires energy efficiency. We begin the section on testing by describing what can go wrong without adequate testing. A $13 million systems development effort to build a vehicle title and registration system, for example, had to be shut down because inaccurate data led to vehicles being pulled over or stopped by mistake. In some cases, one problem can cascade into other ones or cause multiple systems to fail. Problems with a project to consolidate data center servers, for example resulted in more than 160,000 Internet sites being shut down that the company hosted. Some were down for more than six days. Better testing may prevent these types of problems.

The section on maintenance and review contains updated material on a variety of topics. In the systems review section, we discuss that a review can result in halting a new system while it is being built because of problems. The section on systems review also stresses how problems with an existing system can trigger new systems development. A large insurance company operating in Louisiana was ordered by a Louisiana court to pay a client over $500,000 in wind damages and over $2 million in fines for not paying the claim in a timely fashion. This helped trigger an event-driven review that resulted in new software claims programs. The systems review section contains information about IS auditing. The section on systems maintenance also contains new examples.

We have trimmed many of the sections in this chapter to make room for new material and developments. Table 13.1, for example, has been deleted along with its discussion in the text. Some of the material in the section on generating systems design alternatives has been reduced to some extent. The material on commercial off-the-shelf (COTS) development in the section on software acquisition has been trimmed slightly. The material on the financial implications of maintenance and a related figure has been deleted. Some of the material in this section has been moved to the section on the relationship between maintenance and design. We have eliminated some of the information in the section on systems review.

Chapter 14, The Personal and Social Impact of Information Systems

As with all other chapters, the opening vignette, Ethical and Societal Issues, Information Systems @ Work, and chapter-ending cases are all new. New, real-world examples are sprinkled throughout the chapter to maintain the reader's interest and to demonstrate the actual application of the topics being discussed. New end-of-chapter questions and exercises are included.

The new opening vignette discusses how managing the largest online banking service and marketplace makes eBay a huge target for online hackers and fraudsters. The vignette goes on to outline many of the security measures and tools eBay uses to combat these threats.

Additional material is added on spam filters and a list of the most highly rated filters is given. In addition, it is mentioned that there is a potential problem using spam filters in that some require that first time e-mailers be verified else their e-mails will be rejected. Image-based spam is also covered.

In the "Computer-Related Mistakes" section, several new examples of waste are provided including Moody's Corporation, NASA, Nippon Airlines, United Airlines, and Wells Fargo. The mistake by Moody's was especially damaging and resulted in the firm's stock price dropping over 20%.

In the section on preventing computer waste and mistakes, the need for proper user training is illustrated with an example of the Maryland Department of Transportation preparing its users to use new business intelligence software. The Société Générale scandal in France is provided as a classic example of an individual employee circumventing internal policies and procedures. A low-level trader on the arbitrage desk at the French bank created a series of fraudulent and unauthorized investment transactions that built a $72 billion position in European stock index futures. Tokyo Electron, a global supplier of semiconductor production equipment, is offered as an excellent example of a firm thoroughly reviewing its policies and procedures.

In the section on computer crime, the point is made that even good IS policies might not be able to predict or prevent computer crime. Five new examples of hackers causing problems at Citibank, the Chilean government, MS Health, the Pennsylvania school district, and on-line brokers E-Trade and Schwab. Results of the 2007 FBI Internet Crime Report are summarized showing that 206,844 complaints of crime were perpetrated over the Internet during 2007 with a dollar value of $240 million in losses. Results of the 2007 Computer Crime and Survey are also highlighted.

In "The Computer as a Tool to Commit Crime" section, the growing problem of cyberterrorism is discussed. The International Multilateral Partnership Against Cyber Terrorism (IMPACT) is identified as a global public and privately supported initiative to counteract his threat. Examples of cyberterrorism against the small Baltic nation of Estonia and against the CNN news network are given. A new section on Internet gambling has been added. The point is made that although Internet gambling is legal in more than 70 countries, the legality of these online activities is far from clear in the U.S. The Internet Wire Act of 1961 and the Unlawful Internet Gambling Enforcement Act of 2006 along with various laws passed by the individual states make this a murky area indeed. CBSSports.com and Facebook are given as examples of organizations that were investigated briefly by the FBI for collaborating to make it easier for Facebook users to fill out brackets for the NCAA 2008 Basketball Tournament.

In "The Computer as the Object of Crime" section, it is discussed that some criminals have started phony VoIP phone companies and sold subscriptions for services to unsuspecting customers. Instead of establishing their own network, the criminals hack into the computers that route calls over the networks of legitimate VoIP providers and use this network to carry its customers' calls. The latest data is provided about the growth and spread of malware and several of the most current viruses are identified. New examples are provided of malware causing harm and destruction. Information has been added about the use of a rootkit, a set of programs that enable its user to gain administrator level access to a computer or network. Once installed, the attacker can gain full control of the system and even obscure the presence of the rootkit from legitimate system administrators. The highest rated antivirus software for 2007 is identified. It is mentioned that tests have shown that antivirus scans run significantly faster on computers with regularly defragmented files and free space reducing the time to do a complete scan by 18 to 58 minutes. A new section has been added that covers spyware—what it is, how you get it on your computer, and precautions to take to avoid spyware. Many new examples are provided of organizations that have lost valuable data by careless handling of laptops computers. Suggestions are offered on how to reduce the number of these incidences. The safe disposal of computers, which can be a risk because of toxic chemicals released at landfills or may result in the possibility that sensitive information that may be obtained by trash divers, is discussed. The use of disk-wiping software utilities that overwrite all sectors of your disk drive making all data unrecoverable is recommended. A definition has been added of copyright and patent to introduce the material on software piracy. New examples are offered of companies getting into trouble over software piracy. Also discussed is Operation

Copycat, an ongoing undercover investigation into warez groups, which are online organizations engaged in the illegal uploading, copying, and distribution of copyrighted works such as music, movies, games, and software, often even before they are released to the public. Examples of patent infringement involving Acer, Apple, Dell, Hewlett-Packard, Fujitsu, Firestar Software and DataTern are identified.

A new Ethical and Societal Issues special feature discusses international cyber espionage. A number of victims are identified as well as the tools used by the spies.

The "Preventing Computer Related Crime" section includes a discussion of the Terrorist Finance Tracking Program that relies on data in international money transfers from the Society for Worldwide Interbank Financial Telecommunications. The goal of the program was to track and combat terrorist financing. The program was credited with helping to capture at least two terrorists; however, revelation of the secret program's existence stirred up controversy and rendered the program ineffective. Mechanisms to reduce computer crime are discussed such as fingerprint authentication devices that provide security in the PC environment by using fingerprint recognition instead of passwords. Also the JetFlash 210 Fingerprint USB Flash Drive requires users to swipe their fingerprints and match them to one of up to 10 trusted users to access the data. The data on the flash drive can also be encrypted for further protection. New material is presented on the use of security dashboard software to provide a comprehensive display on a single computer screen of all the vital data related to an organization's security defenses including threats, exposures, policy compliance and incident alerts. Associated Newspapers is provided as an example of an organization that has implemented a successful security dashboard. Several attempts by the U.S. Congress to limit children's exposure to online pornography including the Communications Decency Act, the Child Online Protection Act (enacted 1998), and the Children's Internet Protection Act are discussed.

The privacy issues section has been updated and includes the latest information on computer monitoring of employees and others. Issues associated with instant messaging and privacy, personal sensing devices such as RFID chips, and social networking are discussed. A new Information Systems @ Work special feature discusses the issues associated with controlling the privacy of Finland's largest information system.

WHAT WE HAVE RETAINED FROM THE EIGHTH EDITION

The ninth edition builds on what has worked well in the past; it retains the focus on IS principles and strives to be the most current text on the market.

- **Overall principle.** This book continues to stress a single-all-encompassing theme: The right information, if it is delivered to the right person, in the right fashion, and at the right time, can improve and ensure organizational effectiveness and efficiency.
- **Information systems principles.** Information system principles summarize key concepts that every student should know. This important feature is a convenient summary of key ideas presented at the start of each chapter.
- **Global perspective.** We stress the global aspects of information systems as a major theme.
- **Learning objectives linked to principles.** Carefully crafted learning objectives are included with every chapter. The learning objectives are linked to the Information Systems Principles and reflect what a student should be able to accomplish after completing a chapter.
- **Opening vignettes emphasize international aspects.** All of the chapter-opening vignettes raise actual issues from foreign-based or multinational companies.

- **Why Learn About features.** Each chapter has a "Why Learn About" section at the beginning of the chapter to pique student interest. The section sets the stage for students by briefly describing the importance of the chapter's material to the students-whatever their chosen field.
- **Information Systems @ Work special interest boxes.** Each chapter has an entirely new "Information Systems @ Work" box that shows how information systems are used in a variety of business career areas.
- **Ethical and Societal Issues special interest boxes.** Each chapter includes an "Ethical and Societal Issues" box that presents a timely look at the ethical challenges and the societal impact of information systems
- **Current examples, boxes, cases, and references.** As we have in each edition, we take great pride in presenting the most recent examples, boxes, cases, and references throughout the text. Some of these were developed at the last possible moment, literally weeks before the book went into publication. Information on new hardware and software, the latest operating systems, mobile commerce, the Internet, electronic commerce, ethical and societal issues, and many other current developments can be found throughout the text. Our adopters have come to expect the best and most recent material. We have done everything we can to meet or exceed these expectations.
- **Summary linked to principles.** Each chapter includes a detailed summary with each section of the summary tied to an associated information system principle.
- **Self-assessment tests.** This popular feature helps students review and test their understanding of key chapter concepts.
- **Career exercises.** End-of-chapter career exercises ask students to research how a topic discussed in the chapter relates to a business area of their choice. Students are encouraged to use the Internet, the college library, or interviews to collect information about business careers.
- **End-of-chapter cases.** Two end-of-chapter cases provide a wealth of practical information for students and instructors. Each case explores a chapter concept of problem that a real-world company or organization has faced. The cases can be assigned as individual homework exercises or serve as a basis for class discussion.
- **Integrated, comprehensive, Web case.** The Whitmann Price Consulting cases at the end of each chapter provide an integrated and comprehensive case that runs throughout the text. The cases follow the activities of two individuals employed at the fictitious Whitmann Price Consulting firm as they are challenged to complete various IS-related projects. The cases provide a realistic fictional work environment in which students may imagine themselves in the role of systems analyst. Information systems problems are addressed utilizing the state of the art techniques discussed in the chapters.

STUDENT RESOURCES

Student Online Companion Web Site

We have created an exciting online companion, password protected for students to utilize as they work through the Ninth Edition. In the front of this text you will find a key code that provides full access to a robust Web site, located at *www.cengage.com/mis/stairreynolds*. This Web resource includes the following features:

- **PowerPoint slides**
  Direct access is offered to the book's PowerPoint presentations that cover the key points from each chapter. These presentations are a useful study tool.

- **Classic cases**
  A frequent request from adopters is that they'd like a broader selection of cases to choose from. To meet this need, a set of over 100 cases from the fifth, sixth, seventh, and eighth editions of the text are included here. These are the authors' choices of the "best cases" from these editions and span a broad range of companies and industries.

- **Links to useful Web sites**
  Chapters in *Principles of Information Systems, Ninth Edition* reference many interesting Web sites. This resource takes you to links you can follow directly to the home pages of those sites so that you can explore them. There are additional links to Web sites that the authors, Ralph Stair and George Reynolds, think you would be interested in checking out.

- **Hands-on activities**
  Use these hands-on activities to test your comprehension of IS topics and enhance your skills using Microsoft® Office applications and the Internet. Using these links, you can access three critical-thinking exercises per chapter; each activity asks you to work with an Office tool or do some research on the Internet.

- **Test yourself on IS**
  This tool allows you to access 20 multiple-choice questions for each chapter; test yourself and then submit your answers. You will immediately find out what questions you got right and what you got wrong. For each question that you answer incorrectly, you are given the correct answer and the page in your text where that information is covered. Special testing software randomly compiles 20 questions from a database of 50 questions, so you can quiz yourself multiple times on a given chapter and get some new questions each time.

- **Glossary of key terms**
  The glossary of key terms from the text is available to search.

- **Online readings**
  This feature provides you access to a computer database which contains articles relating to hot topics in Information Systems.

INSTRUCTOR RESOURCES

The teaching tools that accompany this text offer many options for enhancing a course. And, as always, we are committed to providing one of the best teaching resource packages available in this market.

Instructor's Manual

An all-new *Instructor's Manual* provides valuable chapter overviews; highlights key principles and critical concepts; offers sample syllabi, learning objectives, and discussion topics; and features possible essay topics, further readings and cases, and solutions to all of the end-of-chapter questions and problems, as well as suggestions for conducting the team activities. Additional end-of-chapter questions are also included. As always, we are committed to providing the best teaching resource packages available in this market.

Sample Syllabus

A sample syllabus with sample course outlines is provided to make planning your course that much easier.

Solutions

Solutions to all end-of-chapter material are provided in a separate document for your convenience.

Test Bank and Test Generator

ExamView® is a powerful objective-based test generator that enables instructors to create paper-, LAN- or Web-based tests from test banks designed specifically for their Course Technology text. Instructors can utilize the ultra-efficient QuickTest Wizard to create tests in less than five minutes by taking advantage of Course Technology's question banks or customizing their own exams from scratch. Page references for all questions are provided so you can cross-reference test results with the book.

PowerPoint Presentations

A set of impressive Microsoft PowerPoint slides is available for each chapter. These slides are included to serve as a teaching aid for classroom presentation, to make available to students on the network for chapter review, or to be printed for classroom distribution. Our presentations help students focus on the main topics of each chapter, take better notes, and prepare for examinations. Instructors can also add their own slides for additional topics they introduce to the class.

Figure Files

Figure files allow instructors to create their own presentations using figures taken directly from the text.

DISTANCE LEARNING

Course Technology, the premiere innovator in management information systems publishing, is proud to present online courses in WebCT and Blackboard.

- **Blackboard and WebCT Level 1 Online Content.** If you use Blackboard or WebCT, the test bank for this textbook is available at no cost in a simple, ready-to-use format. Go to *www.cengage.com/mis/stairreynolds* and search for this textbook to download the test bank.
- **Blackboard and WebCT Level 2 Online Content.** Blackboard 5.0 and 6.0 as well as Level 2 and WebCT Level 2 courses are also available for *Principles of Information Systems,* Ninth Edition. Level 2 offers course management and access to a Web site that is fully populated with content for this book.

For more information on how to bring distance learning to your course, instructors should contact their Course Technology sales representative.

ACKNOWLEDGMENTS

A book of this size and undertaking requires a strong team effort. We would like to thank all of our fellow teammates at Course Technology and the Software Resource for their dedication and hard work. We would like to thank Dave Boelio for helping us transition from Thomson Learning to Cengage Learning. Special thanks to Kate Hennessy, our Product Manager. Our appreciation goes out to all the many people who worked behind the scenes to bring this effort to fruition including Abigail Reip, our photo researcher, and Charles McCormick, our Senior Acquisitions Editor. We would like to acknowledge and thank Lisa Ruffolo, our development editor, who deserves special recognition for her tireless effort and help in all stages of this project. Erin Dowler and Jennifer Goguen McGrail, our Content Product Managers, shepherded the book through the production process.

We are grateful to the sales force at Course Technology whose efforts make this all possible. You helped to get valuable feedback from current and future adopters. As Course Technology product users, we know how important you are.

We would especially like to thank Ken Baldauf for his excellent help in writing the boxes and cases and revising several chapters for this edition. Ken also provided invaluable feedback on other aspects of this project.

Ralph Stair would like to thank the Department of Management Information Systems and its faculty members in the College of Business Administration at Florida State University for their support and encouragement. He would also like to thank his family, Lila and Leslie, for their support.

George Reynolds would like to thank his family, Ginnie, Tammy, Kim, Kelly, and Kristy, for their patience and support in this major project.

To Our Previous Adopters and Potential New Users

We sincerely appreciate our loyal adopters of the previous editions and welcome new users of *Principles of Information Systems: A Managerial Approach, Ninth Edition.* As in the past, we truly value your needs and feedback. We can only hope the Ninth Edition continues to meet your high expectations.

We are indebted to the following reviewers, both past and present, for their perceptive feedback on early drafts of this text:

Jill Adams, *Navarro College*
Robert Aden, *Middle Tennessee State University*
A.K. Aggarwal, *University of Baltimore*
Sarah Alexander, *Western Illinois University*
Beverly Amer, *University of Florida*
Noushin Asharfi, *University of Massachusetts*
Kirk Atkinson, Western Kentucky University
Yair Babad, *University of Illinois—Chicago*
Cynthia C. Barnes, *Lamar University*
Charles Bilbrey, *James Madison University*
Thomas Blaskovics, *West Virginia University*
John Bloom, *Miami University of Ohio*
Warren Boe, *University of Iowa*
Glen Boyer, *Brigham Young University*
Mary Brabston, *University of Tennessee*
Jerry Braun, *Xavier University*
Thomas A. Browdy, *Washington University*
Lisa Campbell, *Gulf Coast Community College*
Andy Chen, *Northeastern Illinois University*
David Cheslow, *University of Michigan—Flint*
Robert Chi, *California State University—Long Beach*
Carol Chrisman, *Illinois State University*
Phillip D. Coleman, *Western Kentucky University*
Miro Costa, *California State University—Chico*
Caroline Curtis, *Lorain County Community College*
Roy Dejoie, *USWeb Corporation*
Sasa Dekleva, *DePaul University*
Pi-Sheng Deng, *California State University—Stanislaus*
Roger Deveau, *University of Massachusetts—Dartmouth*
John Eatman, *University of North Carolina*
Gordon Everest, *University of Minnesota*
Juan Esteva, *Eastern Michigan University*
Badie Farah, *Eastern Michigan University*
Karen Forcht, *James Madison University*
Carroll Frenzel, *University of Colorado—Boulder*
John Gessford, *California State University—Long Beach*

Terry Beth Gordon, *University of Toledo*
Kevin Gorman, *University of North Carolina—Charlotte*
Costanza Hagmann, *Kansas State University*
Bill C. Hardgrave, *University of Arkansas*
Al Harris, *Appalachian State University*
William L. Harrison, *Oregon State University*
Dwight Haworth, *University of Nebraska—Omaha*
Jeff Hedrington, *University of Wisconsin—Eau Claire*
Donna Hilgenbrink, *Illinois State University*
Jack Hogue, *University of North Carolina*
Joan Hoopes, *Marist College*
Donald Huffman, *Lorain County Community College*
Patrick Jaska, *University of Texas at Arlington*
G. Vaughn Johnson, *University of Nebraska—Omaha*
Tom Johnston, *University of Delaware*
Grover S. Kearns, *Morehead State University*
Robert Keim, *Arizona State University*
Karen Ketler, *Eastern Illinois University*
Mo Khan, *California State University—Long Beach*
Chang E. Koh, *University of North Texas*
Michael Lahey, *Kent State University*
Jan de Lassen, *Brigham Young University*
Robert E. Lee, *New Mexico State University—Carlstadt*
Joyce Little, *Towson State University*
Herbert Ludwig, *North Dakota State University*
Jane Mackay, *Texas Christian University*
Al Maimon, *University of Washington*
Efrem Mallach, University of Massachusetts Dartmouth
James R. Marsden, *University of Connecticut*
Roger W. McHaney, *Kansas State University*
Lynn J. McKell, *Brigham Young University*
John Melrose, *University of Wisconsin—Eau Claire*
Michael Michaelson, *Palomar College*
Ellen Monk, *University of Delaware*
Bertrad P. Mouqin, *University of Mary Hardin-Baylor*
Bijayananda Naik, *University of South Dakota*
Zibusiso Ncube, *Concordia College*
Pamela Neely, *Marist College*
Leah R. Pietron, *University of Nebraska—Omaha*
John Powell, *University of South Dakota*
Maryann Pringle, *University of Houston*
John Quigley, *East Tennessee State University*
Mahesh S. Raisinghani, *University of Dallas*
Mary Rasley, *Lehigh-Carbon Community College*
Earl Robinson, *St. Joseph's University*
Scott Rupple, *Marquette University*
Dave Scanlon, *California State University—Sacramento*
Werner Schenk, *University of Rochester*
Larry Scheuermann, *University of Southwest Louisiana*
James Scott, *Central Michigan University*
Vikram Sethi, *Southwest Missouri State University*
Laurette Simmons, *Loyola College*
Janice Sipior, *Villanova University*
Anne Marie Smith, *LaSalle University*
Harold Smith, *Brigham Young University*
Patricia A. Smith, *Temple College*
Herb Snyder, *Fort Lewis College*
Alan Spira, *University of Arizona*
Tony Stylianou, *University of North Carolina*
Bruce Sun, *California State University—Long Beach*
Howard Sundwall, *West Chester University*
Hung-Lian Tang, *Bowling Green State University*
William Tastle, *Ithaca College*
Gerald Tillman, *Appalachian State University*
Duane Truex, *Georgia State University*
Jean Upson, *Lorain County Community College*
Misty Vermaat, *Purdue University—Calumet*
David Wallace, *Illinois State University*
Michael E. Whitman, *University of Nevada—Las Vegas*
David C. Whitney, *San Francisco State University*
Goodwin Wong, *University of California—Berkeley*
Amy Woszczynski, *Kennesaw State University*
Judy Wynekoop, *Florida Gulf Coast University*
Myung Yoon, *Northeastern Illinois University*

Focus Group Contributors for the Third Edition

Mary Brabston, *University of Tennessee*
Russell Ching, *California State University—Sacramento*
Virginia Gibson, *University of Maine*
Bill C. Hardgrave, *University of Arkansas*

Al Harris, *Appalachian State University*
Stephen Lunce, *Texas A & M International*
Merle Martin, *California State University—Sacramento*
Mark Serva, *Baylor University*
Paul van Vliet, *University of Nebraska—Omaha*

OUR COMMITMENT

We are committed to listening to our adopters and readers and to developing creative solutions to meet their needs. The field of IS continually evolves, and we strongly encourage your participation in helping us provide the freshest, most relevant information possible.

We welcome your input and feedback. If you have any questions or comments regarding *Principles of Information Systems: A Managerial Approach, Ninth Edition,* please contact us through Course Technology or your local representative, or via the Internet at *www.cengage.com/mis/stairreynolds.com.*

Ralph Stair
George Reynolds

PART 1

An Overview

CHAPTER • 1 •

An Introduction to Information Systems

PRINCIPLES	LEARNING OBJECTIVES
▪ The value of information is directly linked to how it helps decision makers achieve the organization's goals.	▪ Discuss why it is important to study and understand information systems. ▪ Distinguish data from information and describe the characteristics used to evaluate the value of data.
▪ Computers and information systems are constantly making it possible for organizations to improve the way they conduct business.	▪ Name the components of an information system and describe several system characteristics.
▪ Knowing the potential impact of information systems and having the ability to put this knowledge to work can result in a successful personal career and organizations that reach their goals.	▪ List the components of a computer-based information system. ▪ Identify the basic types of business information systems and discuss who uses them, how they are used, and what kinds of benefits they deliver.
▪ System users, business managers, and information systems professionals must work together to build a successful information system.	▪ Identify the major steps of the systems development process and state the goal of each.
▪ Information systems must be applied thoughtfully and carefully so that society, business, and industry around the globe can reap their enormous benefits.	▪ Describe some of the threats to security and privacy that information systems and the Internet can pose. ▪ Discuss the expanding role and benefits of information systems in business and industry.

Information Systems in the Global Economy Fossil, United States

Computer-Based Information Systems Support Best Business Practices

High-quality, up-to-date, well-maintained computer-based information systems are at the heart of today's most successful global corporations. For a business to succeed globally, it must be able to provide the right information to the right people in the organization at the right time, even if those people are located around the world. Increasingly, this means that decision makers can view the state of every aspect of the business in real time. For example, an executive in Paris can use an information system to see that a company product was purchased from a retailer in San Francisco three minutes ago. If a company's information system is not efficient and effective, the company will lose market share to a competitor with a better information system. For a deeper understanding of how information systems are used in business, consider Fossil.

You are probably familiar with the Fossil brand. Fossil is well known for its watches, handbags, jewelry, and fashion accessories that are sold in numerous retail and department stores around the world. Fossil was founded in 1984 when it set up wholesale distribution of its products to department stores in North America, Asia, and Europe. The company quickly grew and began manufacturing products for other brands such as Burberry, Diesel, DKNY, and Emporio Armani. As Fossil grew, the information it managed expanded until it threatened to be unmanageable, so Fossil invested in a corporate-wide information system developed by SAP Corporation and designed for wholesale companies. The SAP information system efficiently stored and organized all of Fossil's business information, which assisted Fossil management with important business decisions they needed to make.

An information system's ability to organize information so that it provides fuel for smart business decisions is the real value of computer-based information systems. SAP, IBM, Oracle, and other computer-based information systems developers do much more than provide hardware systems and databases. The systems they install are governed by software that implements best business practices. These systems assist managers in designing the best business solutions, which is why selecting the right computer-based information system is crucial to a company's success.

Using the SAP information system to manage its business, Fossil continued to prosper. The company linked its information system to those of its customers, such as Wal-Mart and Macy's, to automate the task of fulfilling orders. Fossil was one of the first companies to launch an online store on the Web, and managed its evolution from a wholesale business to a retail business. Another information system was developed for Web sales that worked with the core SAP corporate-wide information system.

More recently, Fossil began experimenting by opening its own retail stores, which have now blossomed into hundreds of Fossil stores across the United States and in 15 other countries. However, because managing a retail store is different from managing a wholesale company, Fossil again turned to SAP and IBM to design additional information systems that would service its retail needs. Because Fossil's retail and wholesale operations share production warehousing and shipping, the retail information system is designed to be integrated with its wholesale information system.

Fossil's information systems are all integrated, connecting to one central database. Using these information systems, the company can quickly react to market demands. For example, if Fossil sees that a particular style of watch is selling well at its retail store in

London, it can quickly ship more of that style to department stores operating in the same area. Fossil credits the information systems for simplifying its business infrastructure and supporting consistent best practices across its expanding global business.

As you read this chapter, consider the following:

- How might the information systems such as those used at Fossil make use of the various components of a computer-based information system: hardware, software, databases, telecommunications, people, and procedures?
- How do computer-based information systems like Fossil's help businesses implement best practices?

Why Learn About Information Systems?

Information systems are used in almost every imaginable profession. Entrepreneurs and small business owners use information systems to reach customers around the world. Sales representatives use information systems to advertise products, communicate with customers, and analyze sales trends. Managers use them to make multimillion-dollar decisions, such as whether to build a manufacturing plant or research a cancer drug. Financial planners use information systems to advise their clients to help them save for retirement or their children's education. From a small music store to huge multinational companies, businesses of all sizes could not survive without information systems to perform accounting and finance operations. Regardless of your college major or chosen career, information systems are indispensable tools to help you achieve your career goals. Learning about information systems can help you land your first job, earn promotions, and advance your career.

This chapter presents an overview of information systems. The overview sections on hardware, software, databases, telecommunications, e-commerce and m-commerce, transaction processing and enterprise resource planning, information and decision support, special purpose systems, systems development, and ethical and societal issues are expanded to full chapters in the book. Let's get started by exploring the basics of information systems.

People and organizations use information every day. Many retail chains, for example, collect data from their stores to help them stock what customers want and to reduce costs. The components that are used are often called an information system. An **information system (IS)** is a set of interrelated components that collect, manipulate, store, and disseminate data and information and provide a feedback mechanism to meet an objective. It is the feedback mechanism that helps organizations achieve their goals, such as increasing profits or improving customer service. Businesses can use information systems to increase revenues and reduce costs. This book emphasizes the benefits of an information system, including speed, accuracy, and reduced costs.

information system (IS)
A set of interrelated components that collect, manipulate, store, and disseminate data and information and provide a feedback mechanism to meet an objective.

We interact with information systems every day, both personally and professionally. We use automated teller machines at banks, access information over the Internet, select information from kiosks with touch screens, and scan the bar codes on our purchases at self-checkout lanes. Major *Fortune* 500 companies can spend more than $1 billion per year on information systems. Knowing the potential of information systems and putting this knowledge to work can help individuals enjoy a successful career and organizations reach their goals.

Today we live in an information economy. Information itself has value, and commerce often involves the exchange of information rather than tangible goods. Systems based on computers are increasingly being used to create, store, and transfer information. Using information systems, investors make multimillion-dollar decisions, financial institutions transfer billions of dollars around the world electronically, and manufacturers order supplies and distribute goods faster than ever before. Computers and information systems will continue to change businesses and the way we live. To prepare for these innovations, you need to be familiar with fundamental information concepts.

Information systems are everywhere. An air traveler checks in for a flight using a kiosk, which sends the check-in information to a network to verify the traveler's reservation and flight information. The terminal processes the information and prints a boarding pass, speeding airport check-in times.

(Source: Courtesy of Joshua Lott/ Bloomberg News/Landov.)

INFORMATION CONCEPTS

Information is a central concept of this book. The term is used in the title of the book, in this section, and in almost every chapter. To be an effective manager in any area of business, you need to understand that information is one of an organization's most valuable resources. This term, however, is often confused with *data*.

Data, Information, and Knowledge

Data consists of raw facts, such as an employee number, total hours worked in a week, inventory part numbers, or sales orders.[1] As shown in Table 1.1, several types of data can represent these facts. When facts are arranged in a meaningful manner, they become information. **Information** is a collection of facts organized so that they have additional value beyond the value of the individual facts.[2] For example, sales managers might find that knowing the total monthly sales suits their purpose more (i.e., is more valuable) than knowing the number of sales for each sales representative. Providing information to customers can also help companies increase revenues and profits. According to Frederick Smith, chairman and president of FedEx, "Information about the package is as important as the package itself... We care a lot about what's inside the box, but the ability to track and trace shipments, and therefore manage inventory in motion, revolutionized logistics."[3] FedEx is a worldwide leader in shipping packages and products around the world. Increasingly, information generated by FedEx and other organizations is being placed on the Internet. In addition, many universities are now placing course information and content on the Internet.[4] Using the Open Course Ware program, the Massachusetts Institute of Technology (MIT) places class notes and contents on the Internet for more than 1,500 of its courses.

data
Raw facts, such as an employee number, total hours worked in a week, inventory part numbers, or sales orders.

information
A collection of facts organized in such a way that they have additional value beyond the value of the individual facts.

Table 1.1

Types of Data

Data	Represented by
Alphanumeric data	Numbers, letters, and other characters
Image data	Graphic images and pictures
Audio data	Sound, noise, or tones
Video data	Moving images or pictures

Data represents real-world things. Hospitals and healthcare organizations, for example, maintain patient medical data, which represents actual patients with specific health situations. In many cases, hospitals and healthcare organizations are converting data to electronic form. Some have developed *electronic records management (ERM)* systems to store, organize, and control important data. However, data—raw facts—has little value beyond its existence. For example, consider data as pieces of railroad track in a model railroad kit. Each piece of track has limited inherent value as a single object. However, if you define a relationship among the pieces of the track, they will gain value. By arranging the pieces in a certain way, a railroad layout begins to emerge (see Figure 1.1a, top). Data and information work the same way. Rules and relationships can be set up to organize data into useful, valuable information.

The type of information created depends on the relationships defined among existing data. For example, you could rearrange the pieces of track to form different layouts. Adding new or different data means you can redefine relationships and create new information. For instance, adding new pieces to the track can greatly increase the value—in this case, variety and fun—of the final product. You can now create a more elaborate railroad layout (see Figure 1.1b, bottom). Likewise, a sales manager could add specific product data to his sales data to create monthly sales information organized by product line. The manager could use this information to determine which product lines are the most popular and profitable.

Figure 1.1

Defining and Organizing Relationships Among Data Creates Information

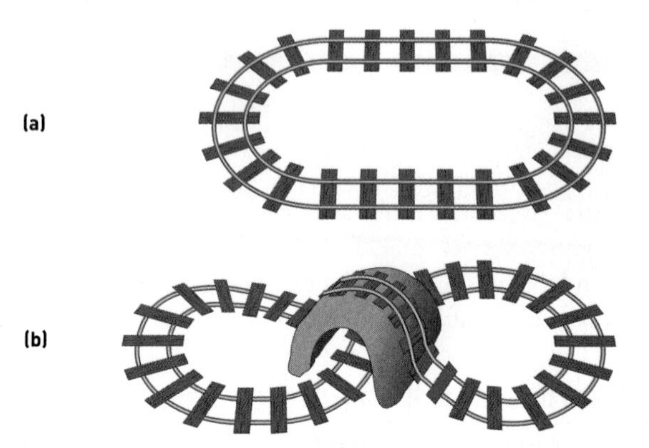

process
A set of logically related tasks performed to achieve a defined outcome.

knowledge
The awareness and understanding of a set of information and ways that information can be made useful to support a specific task or reach a decision.

Turning data into information is a **process**, or a set of logically related tasks performed to achieve a defined outcome. The process of defining relationships among data to create useful information requires knowledge. **Knowledge** is the awareness and understanding of a set of information and the ways that information can be made useful to support a specific task or reach a decision. Having knowledge means understanding relationships in information. Part of the knowledge you need to build a railroad layout, for instance, is the understanding of how much space you have for the layout, how many trains will run on the track, and how fast they will travel. Selecting or rejecting facts according to their relevance to particular tasks is based on the knowledge used in the process of converting data into information. Therefore, you can also think of information as data made more useful through the application of knowledge. *Knowledge workers (KWs)* are people who create, use, and disseminate knowledge, and are usually professionals in science, engineering, business, and other areas. A *knowledge management system (KMS)* is an organized collection of people, procedures, software, databases, and devices used to create, store, and use the organization's knowledge and experience.

In some cases, people organize or process data mentally or manually. In other cases, they use a computer. In the earlier example, the manager could have manually calculated the sum of the sales of each representative, or a computer could have calculated this sum. Where the data comes from or how it is processed is less important than whether the data is transformed into results that are useful and valuable. This transformation process is shown in Figure 1.2.

Figure 1.2

The Process of Transforming Data into Information

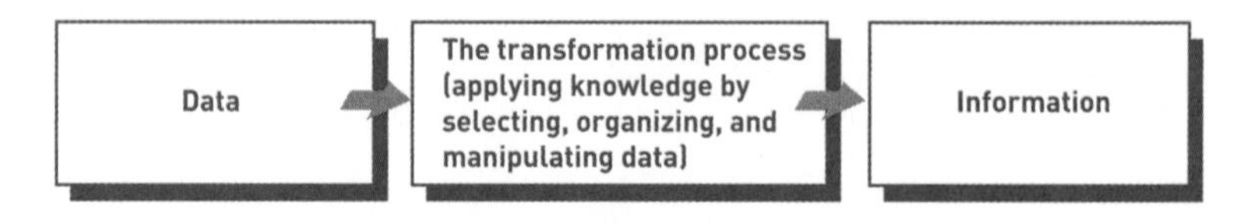

The Characteristics of Valuable Information

To be valuable to managers and decision makers, information should have the characteristics described in Table 1.2. These characteristics make the information more valuable to an organization. Many shipping companies, for example, can determine the exact location of inventory items and packages in their systems, and this information makes them responsive to their customers. In contrast, if an organization's information is not accurate or complete, people can make poor decisions, costing thousands, or even millions, of dollars. If an inaccurate forecast of future demand indicates that sales will be very high when the opposite is true, an organization can invest millions of dollars in a new plant that is not needed. Furthermore, if information is not relevant, not delivered to decision makers in a timely fashion, or too complex to understand, it can be of little value to the organization.

Characteristics	Definitions
Accessible	Information should be easily accessible by authorized users so they can obtain it in the right format and at the right time to meet their needs.
Accurate	Accurate information is error free. In some cases, inaccurate information is generated because inaccurate data is fed into the transformation process. (This is commonly called garbage in, garbage out [GIGO].)
Complete	Complete information contains all the important facts. For example, an investment report that does not include all important costs is not complete.
Economical	Information should also be relatively economical to produce. Decision makers must always balance the value of information with the cost of producing it.
Flexible	Flexible information can be used for a variety of purposes. For example, information on how much inventory is on hand for a particular part can be used by a sales representative in closing a sale, by a production manager to determine whether more inventory is needed, and by a financial executive to determine the total value the company has invested in inventory.
Relevant	Relevant information is important to the decision maker. Information showing that lumber prices might drop might not be relevant to a computer chip manufacturer.
Reliable	Reliable information can be trusted by users. In many cases, the reliability of the information depends on the reliability of the data-collection method. In other instances, reliability depends on the source of the information. A rumor from an unknown source that oil prices might go up might not be reliable.
Secure	Information should be secure from access by unauthorized users.
Simple	Information should be simple, not overly complex. Sophisticated and detailed information might not be needed. In fact, too much information can cause information overload, whereby a decision maker has too much information and is unable to determine what is really important.
Timely	Timely information is delivered when it is needed. Knowing last week's weather conditions will not help when trying to decide what coat to wear today.
Verifiable	Information should be verifiable. This means that you can check it to make sure it is correct, perhaps by checking many sources for the same information.

Table 1.2

Characteristics of Valuable Information

Depending on the type of data you need, some characteristics become more valuable than others. For example, with market-intelligence data, some inaccuracy and incompleteness is acceptable, but timeliness is essential. Sutter Health, for example, developed a real-time system for its intensive care units (ICUs) that can detect and prevent deadly infections, saving over 400 lives a year and millions of dollars in additional healthcare costs.[5] Market intelligence might alert you that competitors are about to make a major price cut. The exact details and timing of the price cut might not be as important as being warned far enough in advance to plan how to react. On the other hand, accuracy, verifiability, and completeness are critical for data used in accounting to manage company assets such as cash, inventory, and equipment.

The Value of Information

The value of information is directly linked to how it helps decision makers achieve their organization's goals.[6] Valuable information can help people and their organizations perform tasks more efficiently and effectively.[7] Consider a market forecast that predicts a high demand for a new product. If you use this information to develop the new product and your company makes an additional profit of $10,000, the value of this information to the company is $10,000 minus the cost of the information. Valuable information can also help managers decide whether to invest in additional information systems and technology. A new computerized ordering system might cost $30,000, but generate an additional $50,000 in sales. The *value added* by the new system is the additional revenue from the increased sales of $20,000. Most corporations have cost reduction as a primary goal. Using information systems, some manufacturing companies have slashed inventory costs by millions of dollars. Other companies have increased inventory levels to increase profits. Wal-Mart, for example, uses information about certain regions of the country and specific situations to increase needed inventory levels of certain products and improve overall profitability.[8] The giant retail store used valuable information about the needs of people in the path of Hurricane Ivan when it hit Florida. The store stocked strawberry Pop-Tarts and other food items that didn't need refrigeration or food preparation to serve people in the area and to increase its profits.

SYSTEM CONCEPTS

system
A set of elements or components that interact to accomplish goals.

Like information, another central concept of this book is that of a system. A **system** is a set of elements or components that interact to accomplish goals. The elements themselves and the relationships among them determine how the system works. Systems have inputs, processing mechanisms, outputs, and feedback (see Figure 1.3). For example, consider an automatic car wash. Tangible *inputs* for the process are a dirty car, water, and various cleaning ingredients. Time, energy, skill, and knowledge also serve as inputs to the system because they are needed to operate it. Skill is the ability to successfully operate the liquid sprayer, foaming brush, and air dryer devices. Knowledge is used to define the steps in the car wash operation and the order in which the steps are executed.

Figure 1.3

Components of a System

A system's four components consist of input, processing, output, and feedback.

The *processing mechanisms* consist of first selecting which cleaning option you want (wash only, wash with wax, wash with wax and hand dry, etc.) and communicating that to the operator of the car wash. A *feedback mechanism* is your assessment of how clean the car is. Liquid sprayers shoot clear water, liquid soap, or car wax depending on where your car is in the process and which options you selected. The *output* is a clean car. As in all systems, independent elements or components (the liquid sprayer, foaming brush, and air dryer) interact to create a clean car.

System Performance and Standards

efficiency
A measure of what is produced divided by what is consumed.

System performance can be measured in various ways. **Efficiency** is a measure of what is produced divided by what is consumed. It can range from 0 to 100 percent. For example, the efficiency of a motor is the energy produced (in terms of work done) divided by the energy consumed (in terms of electricity or fuel). Some motors have an efficiency of 50 percent or less because of the energy lost to friction and heat generation.

Efficiency is a relative term used to compare systems. For example, a hybrid gasoline engine for an automobile or truck can be more efficient than a traditional gasoline engine because, for the equivalent amount of energy consumed, the hybrid engine produces more energy and gets better gas mileage.

Effectiveness is a measure of the extent to which a system achieves its goals. It can be computed by dividing the goals actually achieved by the total of the stated goals. For example, a company might want to achieve a net profit of $100 million for the year using a new information system. Actual profits, however, might only be $85 million for the year. In this case, the effectiveness is 85 percent (85/100 = 85 percent).

effectiveness
A measure of the extent to which a system achieves its goals; it can be computed by dividing the goals actually achieved by the total of the stated goals.

Evaluating system performance also calls for using performance standards. A **system performance standard** is a specific objective of the system. For example, a system performance standard for a marketing campaign might be to have each sales representative sell $100,000 of a certain type of product each year (see Figure 1.4a). A system performance standard for a manufacturing process might be to provide no more than 1 percent defective parts (see Figure 1.4b). After standards are established, system performance is measured and compared with the standard. Variances from the standard are determinants of system performance.

system performance standard
A specific objective of the system.

Figure 1.4

System Performance Standards

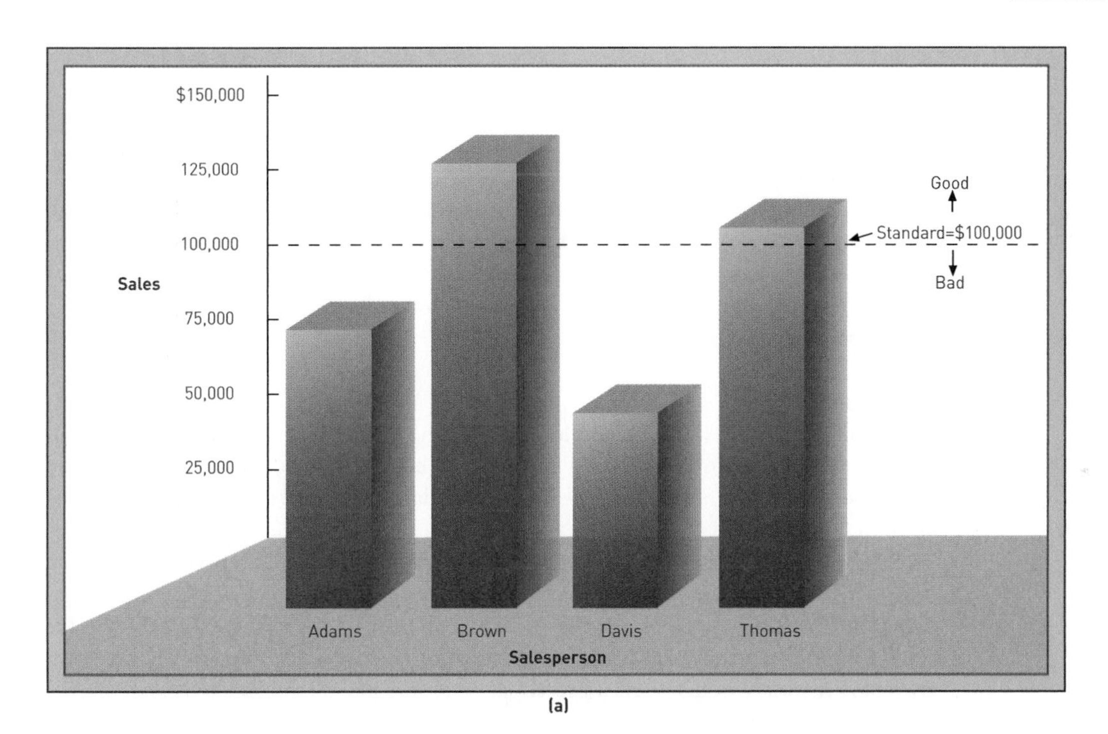

(a)

(b)

WHAT IS AN INFORMATION SYSTEM?

As mentioned previously, an information system (IS) is a set of interrelated elements or components that collect (input), manipulate (process), store, and disseminate (output) data and information, and provide a corrective reaction (feedback mechanism) to meet an objective (see Figure 1.5). The feedback mechanism is the component that helps organizations achieve their goals, such as increasing profits or improving customer service.

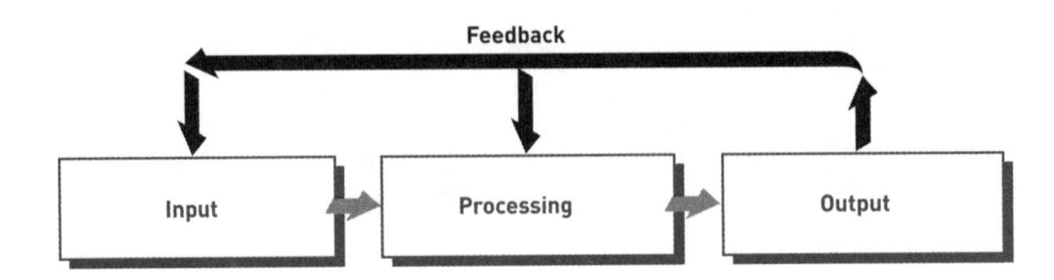

Figure 1.5

The Components of an Information System

Feedback is critical to the successful operation of a system.

Input, Processing, Output, Feedback

Input

In information systems, **input** is the activity of gathering and capturing raw data. In producing paychecks, for example, the number of hours every employee works must be collected before paychecks can be calculated or printed. In a university grading system, instructors must submit student grades before a summary of grades for the semester or quarter can be compiled and sent to the students.

input
The activity of gathering and capturing raw data.

Processing

In information systems, **processing** means converting or transforming data into useful outputs. Processing can involve making calculations, comparing data and taking alternative actions, and storing data for future use. Processing data into useful information is critical in business settings.

processing
Converting or transforming data into useful outputs.

Processing can be done manually or with computer assistance. In a payroll application, the number of hours each employee worked must be converted into net, or take-home, pay. Other inputs often include employee ID number and department. The processing can first involve multiplying the number of hours worked by the employee's hourly pay rate to get gross pay. If weekly hours worked exceed 40, overtime pay might also be included. Then deductions—for example, federal and state taxes, contributions to insurance or savings plans—are subtracted from gross pay to get net pay.

After these calculations and comparisons are performed, the results are typically stored. *Storage* involves keeping data and information available for future use, including output, discussed next.

Output

In information systems, **output** involves producing useful information, usually in the form of documents and reports. Outputs can include paychecks for employees, reports for managers, and information supplied to stockholders, banks, government agencies, and other groups. In some cases, output from one system can become input for another. For example, output from a system that processes sales orders can be used as input to a customer billing system.

output
Production of useful information, usually in the form of documents and reports.

Feedback

In information systems, **feedback** is information from the system that is used to make changes to input or processing activities. For example, errors or problems might make it necessary to correct input data or change a process. Consider a payroll example. Perhaps the number of hours an employee worked was entered as 400 instead of 40. Fortunately, most information

feedback
Output that is used to make changes to input or processing activities.

systems check to make sure that data falls within certain ranges. For number of hours worked, the range might be from 0 to 100 because it is unlikely that an employee would work more than 100 hours in a week. The information system would determine that 400 hours is out of range and provide feedback. The feedback is used to check and correct the input on the number of hours worked to 40. If undetected, this error would result in a very high net pay on the printed paycheck!

Feedback is also important for managers and decision makers. For example, a furniture maker could use a computerized feedback system to link its suppliers and plants. The output from an information system might indicate that inventory levels for mahogany and oak are getting low—a potential problem. A manager could use this feedback to decide to order more wood from a supplier. These new inventory orders then become input to the system. In addition to this reactive approach, a computer system can also be proactive—predicting future events to avoid problems. This concept, often called **forecasting**, can be used to estimate future sales and order more inventory before a shortage occurs. Forecasting is also used to predict the strength and landfall sites of hurricanes, future stock-market values, and who will win a political election.

forecasting
Predicting future events to avoid problems.

Manual and Computerized Information Systems

As discussed earlier, an information system can be manual or computerized. For example, some investment analysts manually draw charts and trend lines to assist them in making investment decisions. Tracking data on stock prices (input) over the last few months or years, these analysts develop patterns on graph paper (processing) that help them determine what stock prices are likely to do in the next few days or weeks (output). Some investors have made millions of dollars using manual stock analysis information systems. Of course, today many excellent computerized information systems follow stock indexes and markets and suggest when large blocks of stocks should be purchased or sold (called *program trading*) to take advantage of market discrepancies.

Program trading systems help traders monitor swift changes in stock prices and make better decisions for their investors.

(Source: Courtesy of REUTERS/ Allen Fredrickson/Landov.)

Computer-Based Information Systems

A **computer-based information system (CBIS)** is a single set of hardware, software, databases, telecommunications, people, and procedures that are configured to collect, manipulate, store, and process data into information. A company's payroll, order entry, or inventory-control system is an example of a CBIS. Lloyd's Insurance in London is starting to use a CBIS to reduce paper transactions and convert to an electronic insurance system.[9] The new CBIS allows Lloyd's to insure people and property more efficiently and effectively. Lloyd's often insures the unusual, including actress Betty Grable's legs, Rolling Stone Keith Richards' hands, and a possible appearance of the Lock Ness Monster (Nessie) in Scotland, which would result in a large payment for the person first seeing the monster. CBISs can also be embedded into products. Some new cars and home appliances include computer hardware, software, databases, and even telecommunications to control their operations and make them more useful. This is often called embedded, pervasive, or ubiquitous computing.

computer-based information system (CBIS)
A single set of hardware, software, databases, telecommunications, people, and procedures that are configured to collect, manipulate, store, and process data into information.

The components of a CBIS are illustrated in Figure 1.6. *Information technology (IT)* refers to hardware, software, databases, and telecommunications. A business's **technology infrastructure** includes all the hardware, software, databases, telecommunications, people, and procedures that are configured to collect, manipulate, store, and process data into information. The technology infrastructure is a set of shared IS resources that form the foundation of each computer-based information system.

technology infrastructure
All the hardware, software, databases, telecommunications, people, and procedures that are configured to collect, manipulate, store, and process data into information.

Figure 1.6

The Components of a Computer-Based Information System.

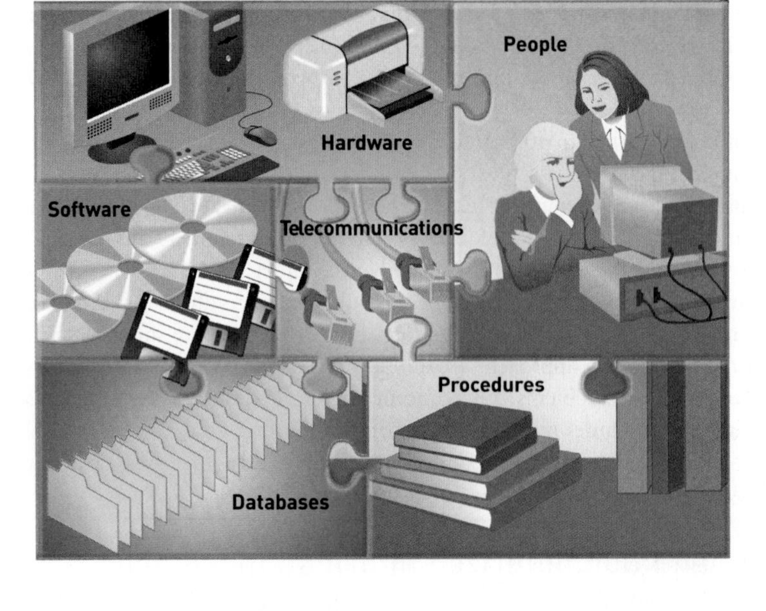

Hardware

hardware
Computer equipment used to perform input, processing, and output activities.

Hardware consists of computer equipment used to perform input, processing, and output activities.[10] Input devices include keyboards, mice and other pointing devices, automatic scanning devices, and equipment that can read magnetic ink characters.[11] Investment firms often use voice-response technology to allow customers to access their balances and other information with spoken commands. Processing devices include computer chips that contain the central processing unit and main memory.[12] Advances in chip design allow faster speeds, less power consumption, and larger storage capacity.[13] Processor speed is also important. Today's more advanced processor chips have the power of 1990s-era supercomputers that occupied a room measuring 10 feet by 40 feet.[14] A large IBM computer used by U.S. Livermore National Laboratories to analyze nuclear explosions might be the fastest in the world (up to 596 teraflops—596 trillion operations per second). The super fast computer, called Blue Gene, costs about $40 million. Computer imagery companies such as Mental Images of Germany and Pixar of the United States also need high processor speeds to produce award-winning images. Image technology is used to help design cars, such as the sleek shapes of Mercedes-Benz vehicles. Small, inexpensive computers are also becoming popular. The One Laptop Per Child (OLPC) computer, for example, costs under $200. The Classmate PC by Intel will cost about $300 and includes some educational software.[15] Both computers are intended for regions of the world that can't afford traditional personal computers.

The many types of output devices include printers and computer screens. Another type of output device is printer kiosks, which are located in some shopping malls. After inserting a disc or memory card from a computer or camera, you can print photos and some documents. Many special-purpose hardware devices have also been developed. Computerized event data recorders (EDRs) are now being placed into vehicles. Like an airplane's black box, EDRs record vehicle speed, possible engine problems, driver performance, and more. The technology is being used to document and monitor vehicle operation, determine the cause of accidents, and investigate whether truck drivers are taking required breaks. In one case, an EDR was used to help convict a driver of vehicular homicide.

Hardware consists of computer equipment used to perform input, processing, and output activities. The trend in the computer industry is to produce smaller, faster, and more mobile hardware.

(Source: © Alberto Pomares/iStockphoto.com.)

Software

Software consists of the computer programs that govern the operation of the computer. These programs allow a computer to process payroll, send bills to customers, and provide managers with information to increase profits, reduce costs, and provide better customer service. With software, people can work anytime at any place. Software that controls manufacturing tools, for example, can be used to fabricate parts almost anywhere in the world. For example, Fab Lab software controls tools such as cutters, milling machines, and other devices.[16] One Fab Lab system, which costs about $20,000, has been used to make radio frequency tags to track animals in Norway, engine parts to allow tractors to run on processed castor beans in India, and many other fabrication applications.

software
The computer programs that govern the operation of the computer.

The two types of software are *system software*, such as Microsoft Windows Vista, which controls basic computer operations including start-up and printing, and *applications software*, such as Microsoft Office 2007, which allows you to accomplish specific tasks including word processing or creating spreadsheets.[17] Software is needed for computers of all sizes, from small handheld computers to large supercomputers.[18] Although most software can be installed from CDs, many of today's software packages can be downloaded through the Internet.[19]

Sophisticated application software, such as Adobe Creative Suite 3, can be used to design, develop, print, and place professional-quality advertising, brochures, posters, prints, and videos on the Internet.[20]

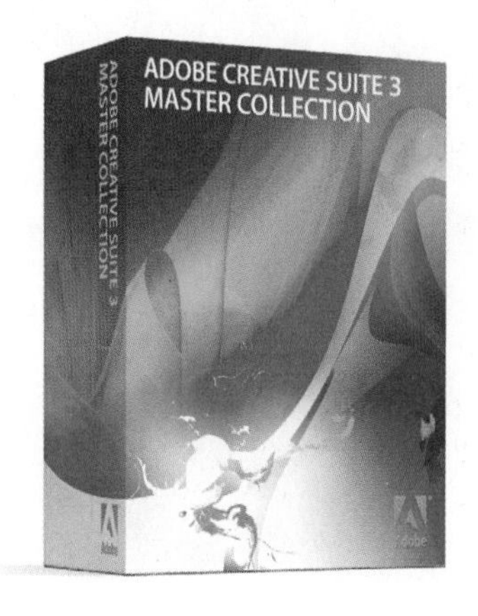

Adobe Creative Suite is an example of sophisticated application software, and is designed for producing professional-quality graphics for the Web, print, and video.

(Source: Courtesy of Adobe Systems Incorporated.)

Databases

A **database** is an organized collection of facts and information, typically consisting of two or more related data files. An organization's database can contain facts and information on customers, employees, inventory, competitors' sales, online purchases, and much more. Most managers and executives consider a database to be one of the most valuable parts of a computer-based information system.[21] A number of health insurance companies are now making their databases available to their customers through the Internet.[22] Aetna, for example, provides important health data to millions of its customers. Aetna customers can also place their own health information, such as blood pressure measurements taken at home, on the comprehensive database. However, making databases accessible can pose risks. The Department of Education decided to limit access to its database of college student loan information to banks and financial institutions.[23] The database contains over 50 million records on student loans that could be inappropriately used to market financial products to students and their families.

database
An organized collection of facts and information.

Telecommunications, Networks, and the Internet

Telecommunications is the electronic transmission of signals for communications, which enables organizations to carry out their processes and tasks through effective computer networks. The Associated Press was one of the first users of telecommunications in the 1920s, sending news over 103,000 miles of wire in the United States and almost 10,000 miles of cable across the ocean.[24] Today, telecommunications is used by organizations of all sizes and

telecommunications
The electronic transmission of signals for communications, which enables organizations to carry out their processes and tasks through effective computer networks.

individuals around the world. The U.S. government is expected to spend almost $50 billion on upgraded telecommunications systems and equipment in the next several years.[25] With telecommunications, people can work at home or while traveling.[26] This approach to work, often called *telecommuting*, allows a telecommuter living in England to send his or her work to the United States, China, or any location with telecommunications capabilities. Today, China is the largest provider of mobile phone and telecommunications services, with over 300 million subscribers.[27]

networks
Computers and equipment that are connected in a building, around the country, or around the world to enable electronic communications.

Networks connect computers and equipment in a building, around the country, or around the world to enable electronic communication. Investment firms can use wireless networks to connect thousands of investors with brokers or traders. Many hotels use wireless telecommunications to allow guests to connect to the Internet, retrieve voice messages, and exchange e-mail without plugging their computers or mobile devices into an Internet connector. Wireless transmission also allows aircraft drones, such as Boeing's Scan Eagle, to fly using a remote control system and monitor buildings and other commercial establishments. The drones are smaller and less-expensive versions of the Predator and Global Hawk drones that the U.S. military used in the Afghanistan and Iraq conflicts.

Internet
The world's largest computer network, consisting of thousands of interconnected networks, all freely exchanging information.

The **Internet** is the world's largest computer network, consisting of thousands of interconnected networks, all freely exchanging information. Research firms, colleges, universities, high schools, and businesses are just a few examples of organizations using the Internet.

People use the Internet wherever they are to research information, buy and sell products and services, make travel arrangements, conduct banking, download music and videos, and listen to radio programs.

(Source: © Bob Daemmrich / Photo Edit.)

People use the Internet to research information, buy and sell products and services, make travel arrangements, conduct banking, download music and videos, and listen to radio programs, among other activities.[28] Increasingly, the Internet is used for communications, collaboration, and information sharing.[29] Internet sites like MySpace (*www.myspace.com*) and FaceBook (*www.facebook.com*) have become popular places to connect with friends and colleagues.[30] Some people, however, fear that this increased usage can lead to problems, including criminals hacking into the Internet and gaining access to sensitive personal information.[31]

Large computers, personal computers, and today's cell phones, such as Apple's iPhone, can access the Internet.[32] This not only speeds communications, but also allows people to conduct business electronically. Some airline companies are providing Internet service on their flights so that travelers can send and receive e-mail, check investments, and browse the Internet.[33] Internet users can create *Web logs (blogs)* to store and share their thoughts and ideas with others around the world. Using *podcasting*, you can download audio programs or music from the Internet to play on computers or music players. One of the authors of this book uses podcasts to obtain information on information systems and technology. You can also record and store TV programs on computers or special viewing devices and watch them later.[34] Often called *place shifting*, this technology allows you to record TV programs at home and watch them at a different place when it's convenient.

The *World Wide Web (WWW)*, or the *Web*, is a network of links on the Internet to documents containing text, graphics, video, and sound. Information about the documents and access to them are controlled and provided by tens of thousands of special computers called Web servers. The Web is one of many services available over the Internet and provides access to millions of documents. New Internet technologies and increased Internet communications are collectively called *Web 2.0*.[35]

The technology used to create the Internet is also being applied within companies and organizations to create **intranets**, which allow people in an organization to exchange information and work on projects. Companies often use intranets to connect their employees around the globe. An **extranet** is a network based on Web technologies that allows selected outsiders, such as business partners and customers, to access authorized resources of a company's intranet. Companies can move all or most of their business activities to an extranet site for corporate customers. Many people use extranets every day without realizing it—to track shipped goods, order products from their suppliers, or access customer assistance from other companies. If you log on to the FedEx site (*www.fedex.com*) to check the status of a package, for example, you are using an extranet.

intranet
An internal network based on Web technologies that allows people within an organization to exchange information and work on projects.

extranet
A network based on Web technologies that allows selected outsiders, such as business partners and customers, to access authorized resources of a company's intranet.

People

People can be the most important element in most computer-based information systems. They make the difference between success and failure for most organizations. Information systems personnel include all the people who manage, run, program, and maintain the system. Large banks can hire IS personnel to speed the development of computer-related projects. Users are people who work with information systems to get results. Users include financial executives, marketing representatives, manufacturing operators, and many others. Certain computer users are also IS personnel.

Procedures

Procedures include the strategies, policies, methods, and rules for using the CBIS, including the operation, maintenance, and security of the computer. For example, some procedures describe when each program should be run. Others describe who can access facts in the database or what to do if a disaster, such as a fire, earthquake, or hurricane, renders the CBIS unusable. Good procedures can help companies take advantage of new opportunities and avoid potential disasters. Poorly developed and inadequately implemented procedures, however, can cause people to waste their time on useless rules or result in inadequate responses to disasters, such as hurricanes or tornadoes.

procedures
The strategies, policies, methods, and rules for using a CBIS.

Now that we have looked at computer-based information systems in general, we will briefly examine the most common types used in business today. These IS types are covered in more detail in Part 3.

BUSINESS INFORMATION SYSTEMS

The most common types of information systems used in business organizations are those designed for electronic and mobile commerce, transaction processing, management information, and decision support. In addition, some organizations employ special-purpose systems, such as virtual reality, that not every organization uses. Together, these systems help employees in organizations accomplish routine and special tasks—from recording sales, processing payrolls, and supporting decisions in various departments, to providing alternatives for large-scale projects and opportunities. Although these systems are discussed in separate sections in this chapter and explained in more detail later, they are often integrated in one product and delivered by the same software package (see Figure 1.7). For example, some enterprise resource planning packages process transactions, deliver information, and support decisions.

Figure 1.7

Business Information Systems

Business information systems are often integrated in one product and can be delivered by the same software package.

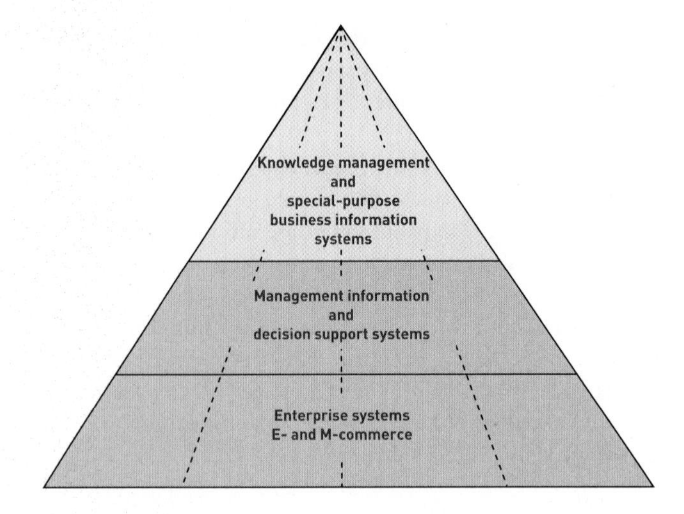

Figure 1.8 shows a simple overview of the development of important business information systems discussed in this section.

Figure 1.8

The Development of Important Business Information Systems

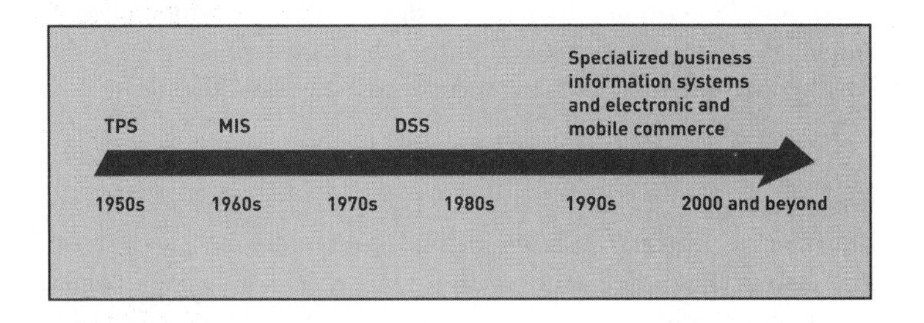

Electronic and Mobile Commerce

e-commerce
Any business transaction executed electronically between companies (business-to-business), companies and consumers (business-to-consumer), consumers and other consumers (consumer-to-consumer), business and the public sector, and consumers and the public sector.

E-commerce involves any business transaction executed electronically between companies (business-to-business, or B2B), companies and consumers (business-to-consumer, or B2C), consumers and other consumers (consumer-to-consumer, or C2C), business and the public sector, and consumers and the public sector. You might assume that e-commerce is reserved mainly for consumers visiting Web sites for online shopping. But Web shopping is only a small part of the e-commerce picture; the major volume of e-commerce—and its fastest-growing segment—is business-to-business (B2B) transactions that make purchasing easier for corporations.[36] This growth is being stimulated by increased Internet access, growing user confidence, rapidly improving Internet and Web security, and better payment systems. PayPal, an e-commerce payment system, for example, processes about $1.5 billion in e-commerce transactions annually.[37] E-commerce also offers opportunities for small businesses to market and sell at a low cost worldwide, allowing them to enter the global market.

mobile commerce (m-commerce)
Transactions conducted anywhere, anytime.

Mobile commerce (m-commerce) refers to transactions conducted anywhere, anytime. M-commerce relies on wireless communications that managers and corporations use to place orders and conduct business with handheld computers, portable phones, laptop computers connected to a network, and other mobile devices. Today, mobile commerce can use cell phones to pay for goods and services.[38] After an account is set up, text messages can be sent and received using a cell phone to authorize purchases. In South Korea, cell phones are used 70 percent of the time to pay for digital content, such as digital music.

INFORMATION SYSTEMS @ WORK

Welcome to Mobile Banking

Access to computer-based information systems is becoming increasingly pervasive; that is, the systems are available anywhere at anytime. Consider the banking industry. The earliest banks kept customers' money and valuables in vaults, and each bank dealt only with its customers' financial needs. Financial data networks were then created to support an interconnected banking system that allowed the transfer of funds electronically. Still, customers needed to visit the bank and speak with a cashier to deposit and retrieve funds. Next, automatic teller machines (ATMs) extended the electronic banking system to the customers and provided the convenience of banking in numerous locations, including out of town. More recently, banking services have extended to the Internet and Web, where a substantial number of bank transactions occur today. Because of online banking, ATMs, and direct deposits, bank customers rarely have to visit the bank.

The latest trend in computer-based information systems designed for banking is called mobile banking. Mobile banking provides banking services such as transferring funds, paying bills, and checking balances from cell phones. While mobile banking is well established in Japan, much of Europe, and elsewhere, it has been slow to catch on in the United States. Some analysts believe that this is due to banks' and wireless carriers' inability to agree on who should design and control the software. Others think that U.S. cell phone users simply aren't interested in the service. A study by Forrester Research found that only 10 percent of Americans were interested in mobile banking, while 35 percent already bank online.

Ready or not, mobile banking is coming to U.S. cell phone users. AT&T, a large telecommunications company, is now offering online banking applications in partnership with Wachovia and other banks. Citibank has designed its own mobile banking software that can be downloaded and installed on more than 100 handsets over any carrier's network. A new system called goDough has been designed by Jack Henry & Associates that delivers the same services offered at a bank's Web site from the small display of a cell phone. Most banks and cell phone service providers believe that the time for mobile banking in the United States has arrived and are making moves to set the standard. Chances are that by the time you read this, your bank will be offering cell phone banking services.

When considering mobile banking, many customers are concerned about security. Sending private financial data over wireless networks poses more risk than sending voice and text communications. Mobile banking systems address these risks with security measures. Typically, a six-digit PIN is required for accessing account information. Secondly, mobile banking software does not store account numbers or PINs on the handset. Lastly, mobile banking communications is secured with 128-bit encryption so that it cannot be intercepted and decoded easily.

Mobile banking provides an interesting case study for mobile computer-based information systems. It illustrates the difficulties of getting customers to adopt new systems and disproves the notion that "if you build it, they will come." Companies must invest time and resources to make consumers aware of the advantages and safety of mobile banking. If it catches on, mobile banking will pave the way for more electronic-wallet cell phone services. Countries that have a head start in mobile banking have moved on to use cell phones to pay cashiers in restaurants and stores, purchase items in vending machines, and buy a ride on the bus. Over the next decade, it is expected that the cell phone will become a user interface to thousands of different computer-based information systems.

Discussion Questions

1. Would you be comfortable using mobile banking for transferring funds, paying bills, and checking balances? Why or why not?
2. How might mobile banking attract the attention of hackers? Are the precautions discussed in this article enough to keep hackers at bay?

Critical Thinking Questions

1. One of the few services not available through online and mobile banking is depositing and withdrawing cash. What would have to change in society to do away with cash all together?
2. What additional dangers are there for making payments with a cell phone, that don't exist when making payments with credit cards? How might they be minimized?

SOURCES: Hamilton, Anita, "Banking Goes Mobile," *Time*, April 2, 2007, *www.time.com/time/business/article/0,8599,1605781,00.html*. Fagan, Mark, "Next Generation of Mobile Banking Draws Interest," *Ecommerce Times*, November 23, 2007, *www.ecommercetimes.com/story/60435.html*. Noyes, Katherine, "Qualcomm Beefs Up Mobile Banking with $210M Firethorn Buy," *Ecommerce Times*, November 14, 2007, *www.ecommercetimes.com/story/60318.html*.

With mobile commerce (m-commerce), people can use cell phones to pay for goods and services anywhere, anytime.

(Source: Courtesy of AP Photo/Itsuo Inouye.)

E-commerce offers many advantages for streamlining work activities. Figure 1.9 provides a brief example of how e-commerce can simplify the process of purchasing new office furniture from an office-supply company. In the manual system, a corporate office worker must get approval for a purchase that exceeds a certain amount. That request goes to the purchasing department, which generates a formal purchase order to procure the goods from the approved vendor. Business-to-business e-commerce automates the entire process. Employees go directly to the supplier's Web site, find the item in a catalog, and order what they need at a price set by their company. If management approval is required, the manager is notified automatically. As the use of e-commerce systems grows, companies are phasing out their traditional systems. The resulting growth of e-commerce is creating many new business opportunities.

Figure 1.9

E-Commerce Greatly Simplifies Purchasing

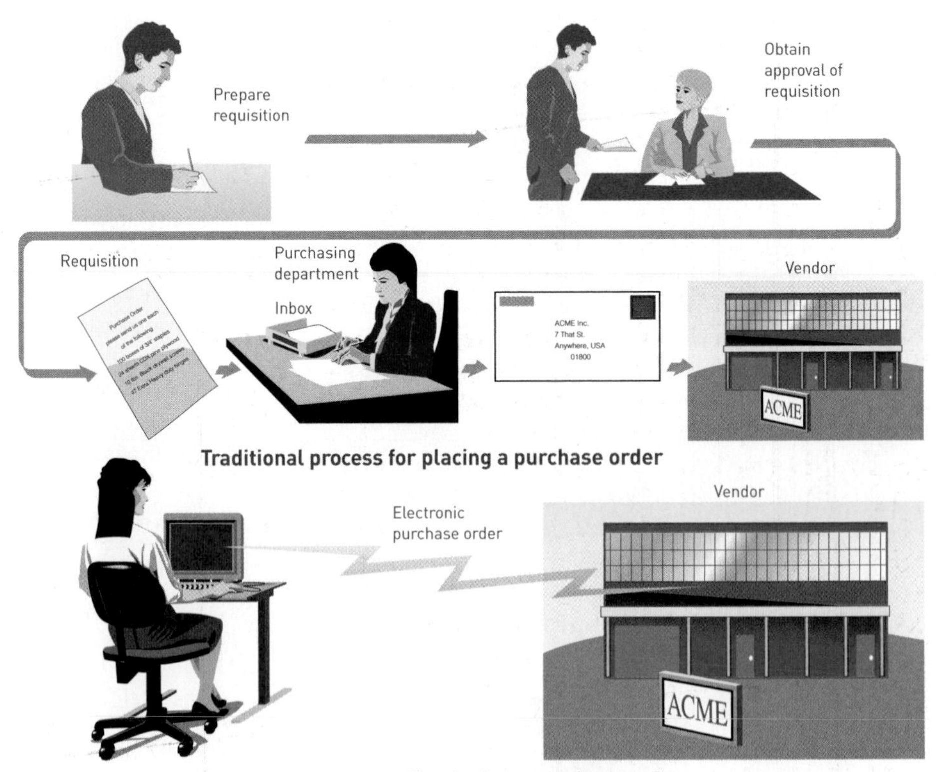

E-commerce can enhance a company's stock prices and market value. Today, several e-commerce firms have teamed up with more traditional brick-and-mortar businesses to draw from each other's strengths. For example, e-commerce customers can order products on a Web site and pick them up at a nearby store.

In addition to e-commerce, business information systems use telecommunications and the Internet to perform many related tasks. *Electronic procurement (e-procurement)*, for example, involves using information systems and the Internet to acquire parts and supplies. **Electronic business (e-business)** goes beyond e-commerce and e-procurement by using information systems and the Internet to perform all business-related tasks and functions, such as accounting, finance, marketing, manufacturing, and human resource activities. E-business also includes working with customers, suppliers, strategic partners, and stakeholders. Compared to traditional business strategy, e-business strategy is flexible and adaptable (see Figure 1.10).

electronic business (e-business)
Using information systems and the Internet to perform all business-related tasks and functions.

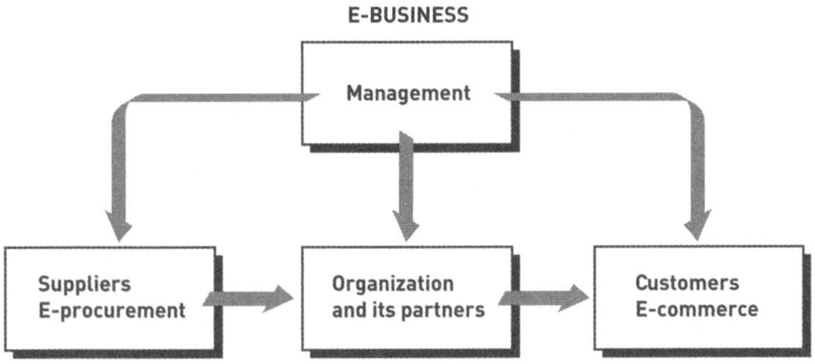

E-BUSINESS

Management

Suppliers
E-procurement

Organization
and its partners

Customers
E-commerce

Figure 1.10

Electronic Business

E-business goes beyond e-commerce to include using information systems and the Internet to perform all business-related tasks and functions, such as accounting, finance, marketing, manufacturing, and human resources activities.

Enterprise Systems: Transaction Processing Systems and Enterprise Resource Planning

Transaction Processing Systems

Since the 1950s, computers have been used to perform common business applications. Many of these early systems were designed to reduce costs by automating routine, labor-intensive business transactions. A **transaction** is any business-related exchange such as payments to employees, sales to customers, or payments to suppliers. Thus, processing business transactions was the first computer application developed for most organizations. A **transaction processing system** (TPS) is an organized collection of people, procedures, software, databases, and devices used to record completed business transactions. If you understand a transaction processing system, you understand basic business operations and functions.

transaction
Any business-related exchange, such as payments to employees, sales to customers, and payments to suppliers.

transaction processing system (TPS)
An organized collection of people, procedures, software, databases, and devices used to record completed business transactions.

One of the first business systems to be computerized was the payroll system (see Figure 1.11). The primary inputs for a payroll TPS are the number of employee hours worked during the week and the pay rate. The primary output consists of paychecks. Early payroll system produces employee paychecks and related reports required by state and federal agencies, such as the Internal Revenue Service. Other routine applications include sales ordering, customer billing and customer relationship management, and inventory control. Some automobile companies, for example, use their TPSs to buy billions of dollars of needed parts each year through Internet sites. Because these systems handle and process daily business exchanges, or transactions, they are all classified as TPSs.

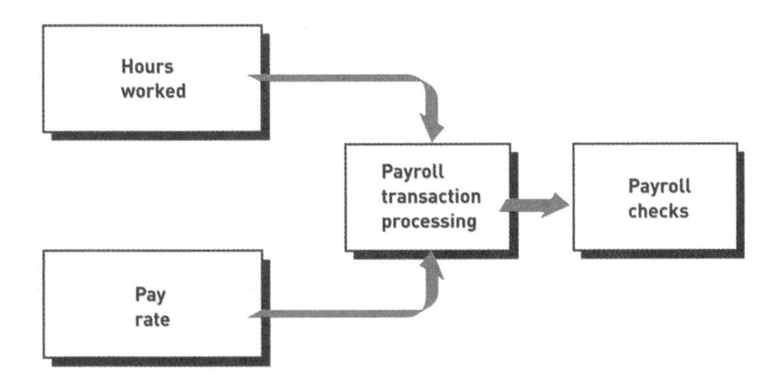

Figure 1.11

A Payroll Transaction Processing System

In a payroll TPS, the inputs (numbers of employee hours worked and pay rates) go through a transformation process to produce outputs (paychecks).

Enterprise systems help organizations perform and integrate important tasks, such as paying employees and suppliers, controlling inventory, sending invoices, and ordering supplies. In the past, companies accomplished these tasks using traditional transaction processing systems. Today, they are increasingly being performed by enterprise resource planning systems.

Enterprise Resource Planning

enterprise resource planning (ERP) system
A set of integrated programs capable of managing a company's vital business operations for an entire multisite, global organization.

An **enterprise resource planning (ERP) system** is a set of integrated programs that manages the vital business operations for an entire multisite, global organization. An ERP system can replace many applications with one unified set of programs, making the system easier to use and more effective.

Although the scope of an ERP system might vary from company to company, most ERP systems provide integrated software to support manufacturing and finance. In such an environment, a forecast is prepared that estimates customer demand for several weeks. The ERP system checks what is already available in finished product inventory to meet the projected demand. Manufacturing must then produce inventory to eliminate any shortfalls. In developing the production schedule, the ERP system checks the raw materials and packing materials inventories and determines what needs to be ordered to meet the schedule. Most ERP systems also have a purchasing subsystem that orders the needed items. In addition to these core business processes, some ERP systems can support functions such as customer service, human resources, sales, and distribution. The primary benefits of implementing an ERP system include easing adoption of improved work processes and increasing access to timely data for decision making.

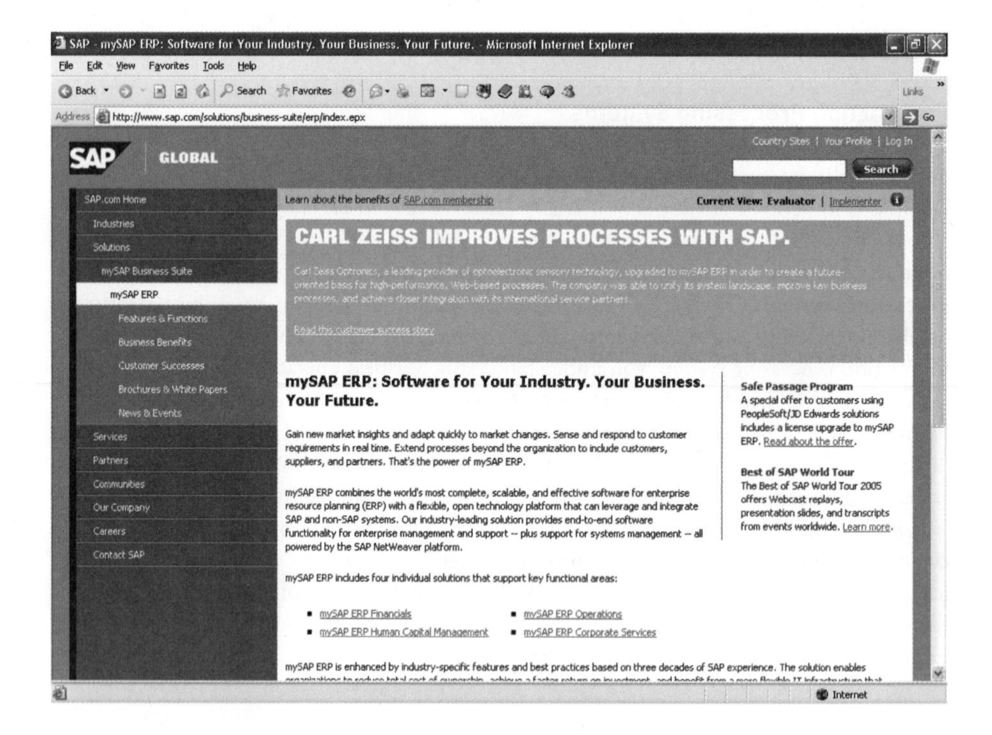

SAP AG, a German software company, is one of the leading suppliers of ERP software. The company employs more than 34,000 people in more than 50 countries.

Information and Decision Support Systems

The benefits provided by an effective TPS are tangible and justify their associated costs in computing equipment, computer programs, and specialized personnel and supplies. A TPS can speed business activities and reduce clerical costs. Although early accounting and financial TPSs were already valuable, companies soon realized that they could use the data stored in these systems to help managers make better decisions, whether in human resource management, marketing, or administration. Satisfying the needs of managers and decision makers continues to be a major factor in developing information systems.

ETHICAL AND SOCIETAL ISSUES

Green Data Centers

Midsized to large businesses maintain powerful computers called servers that store data and run software to provide information system services to users on the corporate network and Internet. Large corporations might maintain hundreds or even thousands of servers in large facilities called data centers. For example, Microsoft is building a 400,000 square foot data center in San Antonio at a cost of $550 million. Google is investing $750 million in a data center facility in Goose Creek, South Carolina and $600 million in a facility in Lenoir, North Carolina.

Because businesses rely on information and its management, the demand for powerful data centers is rapidly growing. Unfortunately, data centers require a lot of power to run and to cool. It is estimated that the money required to cool a data center is equal to the cost of the servers themselves. With the increased awareness of global warming and the contributions of coal-burning power plants to this problem, data centers are drawing the attention of environmentalists and others who want to save energy. The electricity required to run data centers worldwide doubled between 2000 and 2005. This trend is expected to continue; one report estimates that by 2010, the world will require at least 10 new 1,000 megawatt power plants to support the increased demands of data centers.

Governments and corporations, wanting to do what they can to minimize the impact of data centers on the environment and gain some good publicity in the process, are taking action. At the end of 2006, President Bush signed a law authorizing the U.S. Environmental Protection Agency (EPA) to analyze the effect of data centers on the environment. The U.S. federal government has plans to consolidate its own many data centers into smaller, more efficient facilities. The United Kingdom is evaluating its data centers and moving to greener technologies and techniques to comply with new environmental policies and laws passed in the United Kingdom and the European Union.

The report from the EPA projects that data center power consumption could be cut by as much as 20 percent if data center managers take simple steps such as using power management systems, turning off unused servers, and consolidating resources. Manufacturers are working on new technologies to minimize power consumption in servers. One company is experimenting with building a data center in an abandoned coal mine underground, where cooling requirements will be minimal. It is estimated that the subterranean data center will save $9 million per year.

Clearly the current power requirements of information systems and the concern over global warming are at odds. Technology companies are well aware of these concerns and are directing the power of technology at finding solutions to the problem.

Discussion Questions

1. In what ways do information systems negatively affect the environment? Are there positive effects as well? If so, what are they?
2. What can be done to minimize the effect of data centers on the environment?

Critical Thinking Questions

1. Consider our rapidly growing dependence on data centers. What is the risk of this dependence on our society?
2. How might the geography of our planet change if the growth of data centers continues increasing? Will there come a time when the growth levels out?

SOURCES: Bushell, Sue, "British Government Turns Green," *CIO*, December 20, 2007, *www.cio.com.au/index.php/id;1300344377*. Levine, Barry, "Data Center Study Looks at Global Trends," *Top Tech News*, December 14, 2007, *www.toptechnews.com/story.xhtml?story_id=57242*. Brodkin, Jon, "Server electricity use doubled from 2000 to 2005," *itWorld Canada*, December 10, 2007, *www.itworldcanada.com/a/Green-IT/149cf7ef-2d04-41b5-a23d-0732d23c5e40.html*. Mullins, Robert, "Bush signs law to study data center energy usage," *Computerworld*, December 22, 2006, *www.computerworld.com/action/article.do?command=viewArticleBasic&articleId=9006698*. Gittlen, Sandra, "Data center land grab: How to get ready for the rush," *Computerworld*, March 12, 2007, *www.computerworld.com/action/article.do?command=viewArticleBasic&articleId=9012963*. Mellor, Chris, "Sun to set up underground data center to save on power," Computerworld, November 16, 2007, *www.computerworld.com/action/article.do?command=viewArticleBasic&articleId=9047478*.

Management Information Systems

management information system (MIS)
An organized collection of people, procedures, software, databases, and devices that provides routine information to managers and decision makers.

A **management information system (MIS)** is an organized collection of people, procedures, software, databases, and devices that provides routine information to managers and decision makers. An MIS focuses on operational efficiency. Marketing, production, finance, and other functional areas are supported by MISs and linked through a common database. MISs typically provide standard reports generated with data and information from the TPS (see Figure 1.12). Producing a report outlining inventory that should be ordered is an example.

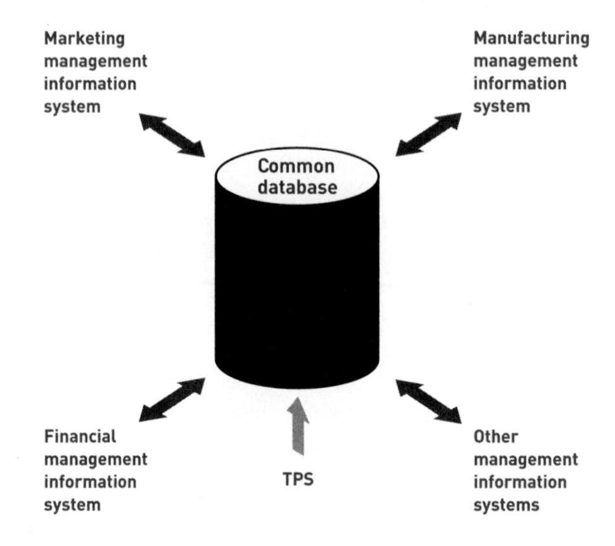

Figure 1.12

Management Information System

Functional management information systems draw data from the organization's transaction processing system.

MISs were first developed in the 1960s and typically use information systems to produce managerial reports. In many cases, these early reports were produced periodically—daily, weekly, monthly, or yearly. Because of their value to managers, MISs have proliferated throughout the management ranks. For instance, the total payroll summary report produced initially for an accounting manager might also be useful to a production manager to help monitor and control labor and job costs.

Decision Support Systems

By the 1980s, dramatic improvements in technology resulted in information systems that were less expensive but more powerful than earlier systems. People at all levels of organizations began using personal computers to do a variety of tasks; they were no longer solely dependent on the IS department for all their information needs. People quickly recognized that computer systems could support additional decision-making activities. A **decision support system (DSS)** is an organized collection of people, procedures, software, databases, and devices that support problem-specific decision making. The focus of a DSS is on making effective decisions. Whereas an MIS helps an organization "do things right," a DSS helps a manager "do the right thing."

decision support system (DSS)
An organized collection of people, procedures, software, databases, and devices used to support problem-specific decision making.

In addition to assisting in all aspects of problem-specific decision making, a DSS can support customers by rapidly responding to their phone and e-mail inquiries. A DSS goes beyond a traditional MIS by providing immediate assistance in solving problems. Many of these problems are unique and complex, and key information is often difficult to obtain. For instance, an auto manufacturer might try to determine the best location to build a new manufacturing facility. Traditional MISs are seldom used to solve these types of problems; a DSS can help by suggesting alternatives and assisting in final decision making.

Decision support systems are used when the problem is complex and the information needed to determine appropriate action is difficult to obtain and use. Consequently, a DSS also involves managerial judgment and perspective. Managers often play an active role in developing and implementing the DSS. A DSS recognizes that different managerial styles and decision types require different systems. For example, two production managers in the same position trying to solve the same problem might require different information and support. The overall emphasis is to support, rather than replace, managerial decision making.

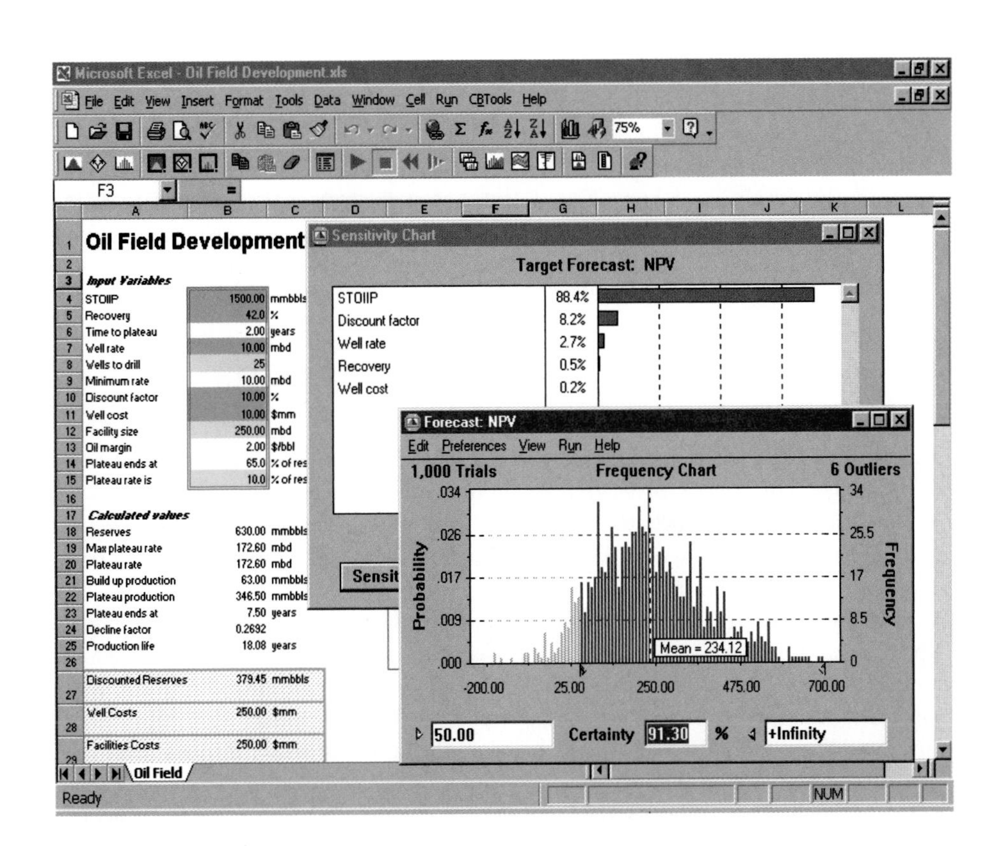

Decisioneering provides decision support software called Crystal Ball, which helps businesspeople of all types assess risks and make forecasts. Shown here is the Standard Edition being used for oil field development.

(Source: Crystal Ball screenshot courtesy of Decisioneering, Inc.)

A DSS can include a collection of models used to support a decision maker or user (model base), a collection of facts and information to assist in decision making (database), and systems and procedures (user interface or dialogue manager) that help decision makers and other users interact with the DSS (see Figure 1.13). Software is often used to manage the database—the database management system (DBMS)—and the model base—the model management system (MMS). Not all DSSs have all of these components.

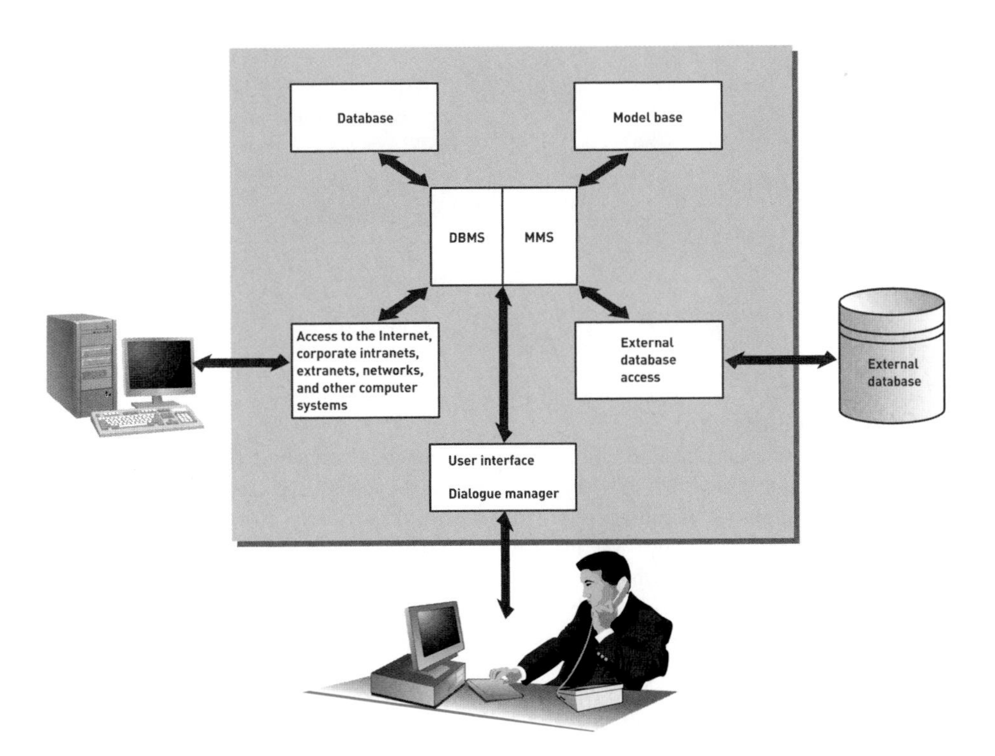

Figure 1.13

Essential DSS Elements

In addition to DSSs for managers, other systems use the same approach to support groups and executives. A *group support system* includes the DSS elements just described as well as software, called *groupware*, to help groups make effective decisions. An executive support system, also called an *executive information system*, helps top-level managers, including a firm's president, vice presidents, and members of the board of directors, make better decisions. An executive support system can assist with strategic planning, top-level organizing and staffing, strategic control, and crisis management.

Specialized Business Information Systems: Knowledge Management, Artificial Intelligence, Expert Systems, and Virtual Reality

In addition to TPSs, MISs, and DSSs, organizations often rely on specialized systems. Many use *knowledge management systems (KMSs)*, an organized collection of people, procedures, software, databases, and devices to create, store, share, and use the organization's knowledge and experience.[39] A shipping company, for example, can use a KMS to streamline its transportation and logistics business.

In addition to knowledge management, companies use other types of specialized systems. Experimental specialized systems in cars can help prevent accidents.[40] These new systems allow cars to communicate with each other using radio chips installed in their trunks. When two or more cars move too close together, the specialized systems sound alarms and brake in some cases. Some specialized systems are based on the notion of **artificial intelligence (AI)**, in which the computer system takes on the characteristics of human intelligence. The field of artificial intelligence includes several subfields (see Figure 1.14). Some people predict that in the future we will have nanobots, small molecular-sized robots, traveling throughout our bodies and in our bloodstream, keeping us healthy. Other nanobots will be embedded in products and services, making our lives easier and creating new business opportunities.

artificial intelligence (AI)
A field in which the computer system takes on the characteristics of human intelligence.

A Nissan Motor Company car swerves back into its lane on its own shortly after it ran off the track during a test of the Lane Departure Prevention feature, which also sounds a warning when the car veers out of its lane.

(Source: Courtesy of AP Photo/ Katsumi Kasahara.)

Artificial Intelligence

Robotics is an area of artificial intelligence in which machines take over complex, dangerous, routine, or boring tasks, such as welding car frames or assembling computer systems and components. Vision systems allow robots and other devices to "see," store, and process visual images. Natural language processing involves computers understanding and acting on verbal or written commands in English, Spanish, or other human languages. Learning systems allow computers to learn from past mistakes or experiences, such as playing games or making business decisions, and neural networks is a branch of artificial intelligence that allows computers to recognize and act on patterns or trends. Some successful stock, options, and futures traders use neural networks to spot trends and improve the profitability of their investments.

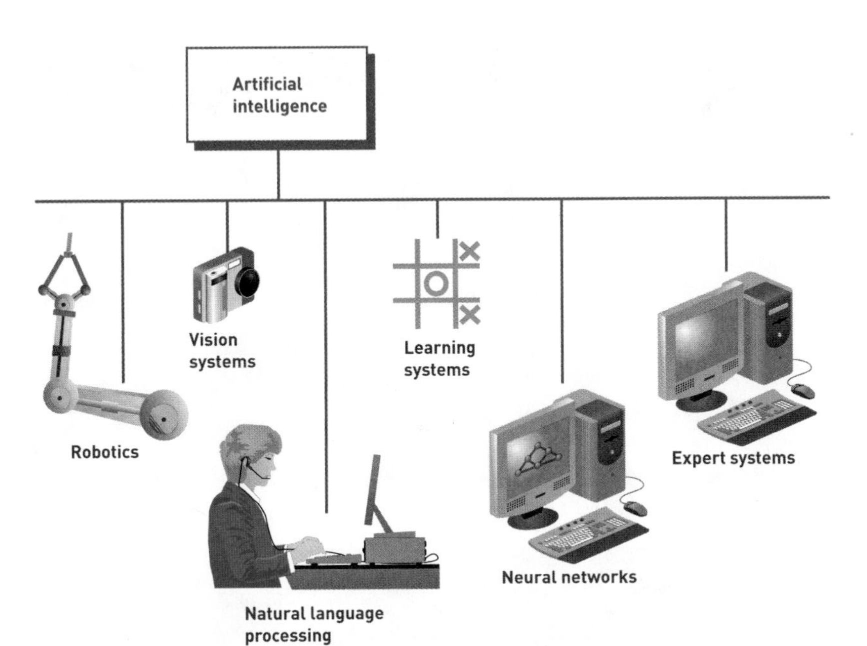

Figure 1.14

The Major Elements of Artificial Intelligence

Expert Systems

Expert systems give the computer the ability to make suggestions and function like an expert in a particular field, helping enhance the performance of the novice user. The unique value of expert systems is that they allow organizations to capture and use the wisdom of experts and specialists. Therefore, years of experience and specific skills are not completely lost when a human expert dies, retires, or leaves for another job. Expert systems can be applied to almost any field or discipline. They have been used to monitor nuclear reactors, perform medical diagnoses, locate possible repair problems, design and configure IS components, perform credit evaluations, and develop marketing plans for a new product or investment strategy. The collection of data, rules, procedures, and relationships that must be followed to achieve value or the proper outcome is contained in the expert system's **knowledge base.**

expert system
A system that gives a computer the ability to make suggestions and function like an expert in a particular field.

knowledge base
The collection of data, rules, procedures, and relationships that must be followed to achieve value or the proper outcome.

Virtual Reality

Virtual reality is the simulation of a real or imagined environment that can be experienced visually in three dimensions. Cigna Healthcare, for example, is experimenting with a virtual reality game designed to help treat cancer in young adults and children.[41] Developed by HopeLab (*www.hopelab.org*), the virtual reality game called Re-Mission shows young adults and children how to combat cancer.

virtual reality
The simulation of a real or imagined environment that can be experienced visually in three dimensions.

Originally, virtual reality referred to immersive virtual reality, which means the user becomes fully immersed in an artificial, computer-generated 3-D world. The virtual world is presented in full scale and relates properly to the human size. Virtual reality can also refer to applications that are not fully immersive, such as mouse-controlled navigation through a 3-D environment on a graphics monitor, stereo viewing from the monitor via stereo glasses, stereo projection systems, and others. Boeing, for example, used virtual reality and computer simulation to help design and build its Dreamliner 787.[42] According to Kevin Fowler, Boeing's vice president of process integration, "A breakthrough program like the 787 Dreamliner needed to lead the way in performance, quality, cost, and schedule supported by efficient and effective production planning." Boeing used 3-D models from Dassault Systems to design and manufacture the new aircraft. Retail stores like Saks Fifth Avenue and Neiman-Marcus are using virtual reality to help advertise high-end products on the Internet.[43] In one virtual ad campaign, about $500,000 of orders from over 20 countries were received in less than a week.

A variety of input devices, such as head-mounted displays (see Figure 1.15), data gloves, joysticks, and handheld wands, allow the user to navigate through a virtual environment and to interact with virtual objects. Directional sound, tactile and force feedback devices, voice recognition, and other technologies enrich the immersive experience. Because several people can share and interact in the same environment, virtual reality can be a powerful medium for communication, entertainment, and learning.

Boeing used virtual reality and computer simulation when designing and building its Dreamliner 787 aircraft.

(Source: Frank Brandmaier/dpa/Landov.)

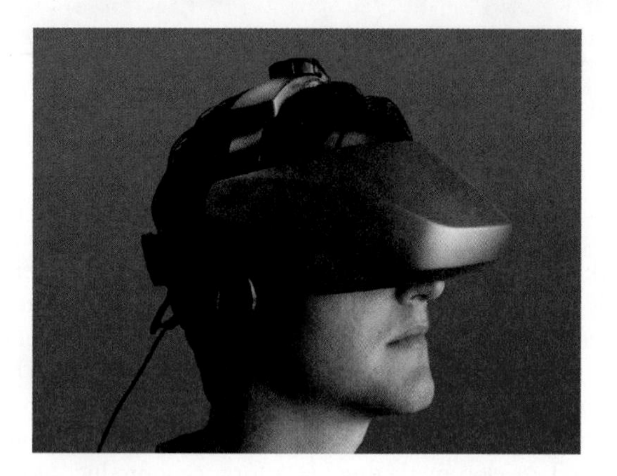

Figure 1.15

A Head-Mounted Display

The head-mounted display (HMD) was the first device to provide the wearer with an immersive experience. A typical HMD houses two miniature display screens and an optical system that channels the images from the screens to the eyes, thereby presenting a stereo view of a virtual world. A motion tracker continuously measures the position and orientation of the user's head and allows the image-generating computer to adjust the scene representation to the current view. As a result, the viewer can look around and walk through the surrounding virtual environment.

(Source: Courtesy of 5DT, Inc. *www.5dt.com*.)

It is difficult to predict where information systems and technology will be in 10 to 20 years. It seems, however, that we are just beginning to discover the full range of their usefulness. Technology has been improving and expanding at an increasing rate; dramatic growth and change are expected for years to come. Without question, having knowledge of the effective use of information systems will be critical for managers both now and in the long term. Now, let's examine how information systems are created.

SYSTEMS DEVELOPMENT

systems development
The activity of creating or modifying business systems.

Systems development is the activity of creating or modifying business systems. Systems development projects can range from small to very large and are conducted in fields as diverse as stock analysis and video game development. Some systems development efforts are a huge success. Wachovia Corporation and Investment Bank, for example, used systems development to create a new computer-trading platform that increased processing capacity by a factor of three, while dramatically reducing costs.[44] According to Tony Bishop, senior vice president of the firm, "We looked at the current system and said, 'Where can we build standardized frameworks, components, and services…?' We now do pricing in milliseconds, not seconds, for either revenue protection or revenue gain." Other systems development efforts fail to meet their cost or schedule goals. A large federal database designed to track hundreds of millions of dollars of money transfers in an effort to curb terrorism was delayed by several years.[45] Scheduled to be implemented by 2007, the system may not be ready until 2010. Some also question whether the $32 million budget for the new system will be met. In another case, Colorado Governor Bill Ritter ordered that all centralized systems development efforts be placed under the state's chief information officer to curb expensive computer systems that don't work correctly.[46] In previous years, some have claimed that over $300 million was spent on systems that couldn't pay welfare benefits on time, issue overtime payments to road crews, issue license plates correctly, or accurately track unemployment benefits. A new Colorado vehicle registration system may also have to be abandoned because of faulty systems development procedures, costing Colorado taxpayers over $10 million.[47] Systems development failures can be a result of poor planning and scheduling, insufficient management of risk, poor requirements determination, and lack of user involvement.[48] Training people to use a new or modified system can be critical to the successful implementation of these systems and can help avoid systems development failures.[49]

People inside a company can develop systems, or companies can use *outsourcing*, hiring an outside company to perform some or all of a systems development project.[50] Outsourcing allows a company to focus on what it does best and delegate other functions to companies with expertise in systems development. Outsourcing, however, is not the best alternative for all companies.

Developing information systems to meet business needs is highly complex and difficult—so much so that it is common for IS projects to overrun budgets and exceed scheduled completion dates. One strategy for improving the results of a systems development project is to divide it into several steps, each with a well-defined goal and set of tasks to accomplish (see Figure 1.16). These steps are summarized next.

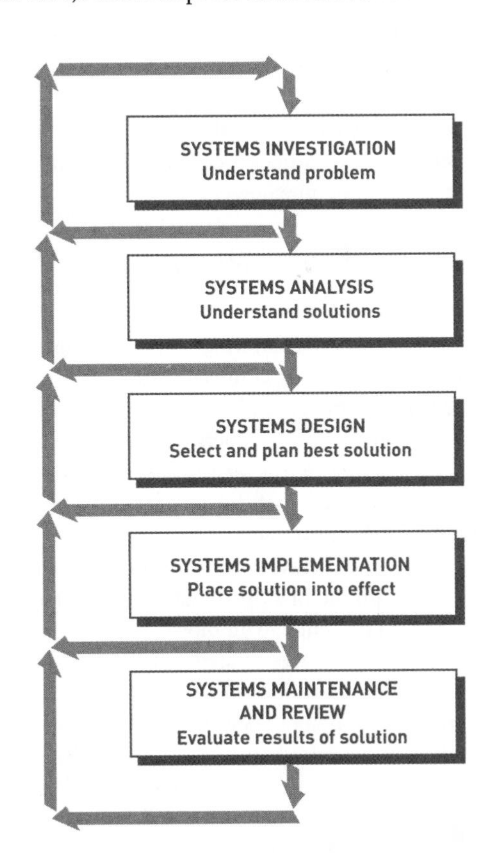

Figure 1.16

An Overview of Systems Development

Systems Investigation and Analysis

The first two steps of systems development are systems investigation and analysis. The goal of the *systems investigation* is to gain a clear understanding of the problem to be solved or opportunity to be addressed. After an organization understands the problem, the next question is, "Is the problem worth solving?" Given that organizations have limited resources—people and money—this question deserves careful consideration. If the decision is to continue with the solution, the next step, *systems analysis*, defines the problems and opportunities of the existing system. During systems investigation and analysis, as well as design maintenance and review, discussed next, the project must have the complete support of top-level managers and focus on developing systems that achieve business goals.

Systems Design, Implementation, and Maintenance and Review

Systems design determines how the new system will work to meet the business needs defined during systems analysis. *Systems implementation* involves creating or acquiring the various system components (hardware, software, databases, etc.) defined in the design step, assembling them, and putting the new system into operation. The purpose of *systems maintenance and review* is to check and modify the system so that it continues to meet changing business needs. Increasingly, companies are hiring outside companies to do their design, implementation, maintenance, and review functions.

INFORMATION SYSTEMS IN SOCIETY, BUSINESS, AND INDUSTRY

Information systems have been developed to meet the needs of all types of organizations and people. The speed and widespread use of information systems, however, opens users to a variety of threats from unethical people. Computer criminals and terrorists, for example, have used the Internet to steal millions of dollars or promote terrorism and violence.[51] Some organizations are using information systems to prevent or thwart such attacks. Because of the tragic shooting deaths at Virginia Tech in the spring of 2007, for example, some universities are developing security alert systems that allow the school to send text messages to students' cell phones in case of an emergency.[52] Students can also send text messages to the university if they see or encounter a problem or an emergency.

Security, Privacy, and Ethical Issues in Information Systems and the Internet

Although information systems can provide enormous benefits, they do have drawbacks. Some drawbacks are minor, such as always being connected at work and to your boss through the Internet, minimizing free time.[53] Others can be more severe, where people's personal data, including Social Security and credit card numbers, can be lost or stolen, resulting in credit card fraud and ruined credit.[54] The U.S. Secret Service arrested a man in Turkey who may have been responsible for tens of millions of dollars of fraud from identity theft by stealing tens of thousands of credit and debit card numbers.[55] The Federal Trade Commission (FTC) reported that 25 percent of identity theft victims from this individual resulted in credit card fraud, 16 percent encountered phone or utility theft, another 16 percent had bank fraud committed against them, and 14 percent had employee-related fraud committed using identity theft. In the United States, it has been estimated that about 150 million computer records have been stolen or exposed to fraud.[56] According or a Forrester Research report, this type of data loss has cost companies from about $90 to $300 per lost record. Some companies have spent millions of dollars to investigate and counteract stolen computer records.

Computer-related mistakes and waste are also a concern. In Japan, a financial services firm had trading losses of $335 million due to a typing mistake in entering a trade. Another computer mistake stranded hundreds of United Airline flights.[57] Similar computer-related errors have caused airline delays or cancellations with other airlines around the world.[58] Unwanted e-mail, called "spam," can be a huge waste of people's time.[59] Many individuals and organizations are trying to find better ways to block spam.

Most computer monitors are now manufactured and sold with an Energy Star rating, including many Dell monitors, such as the UltraSharp 19-inch flat panel LCD monitor.

(Source: © 2008 Dell. Inc. All rights reserved.)

Some computers that burn a large amount of energy are wasteful. One survey reported that almost half of European computer managers were concerned about power consumption and computer-related waste.[60] Thirty-three percent of American computer managers were concerned about power consumption and waste. Some experts believe that computers waste up to half of the energy they consume and account for about two percent of worldwide energy usage.[61]

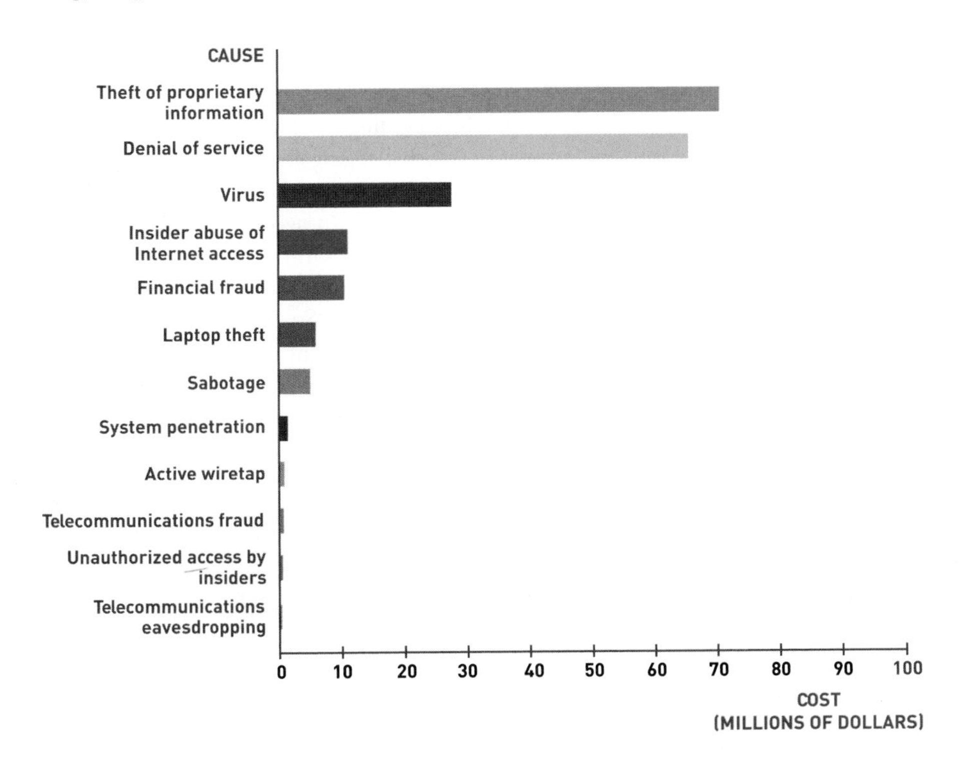

Figure 1.17

The Cost and Cause of Computer Attacks

(Source: Data from Riva Richmond, "How to Find Your Weak Spots," *The Wall Street Journal*, September 29, 2003, p. R3.)

Increasingly, the legal and ethical use of systems has been highlighted in the news. Music publishers and other companies, for example, have sued YouTube, the popular video-sharing Web site, for copyright violations.[62] The suit claims that YouTube has illegally posted copyrighted material and video content. Ethical issues concern what is generally considered right or wrong. Some IS professionals believe that computers may create new opportunities for unethical behavior. For example, a faculty member of a medical school falsified computerized research results to get a promotion—and a higher salary. In another case, a company was charged with using a human resource information system to time employee layoffs and firings to avoid paying pensions. More and more, the Internet is associated with unethical behavior. Unethical investors have placed false rumors or incorrect information about a company on the Internet and tried to influence its stock price to make money.

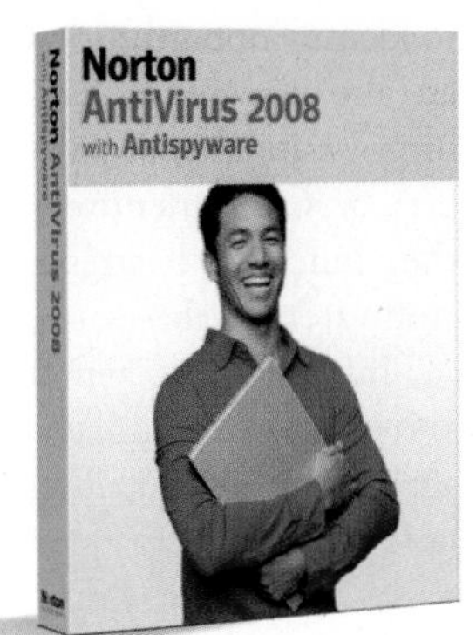

Norton AntiVirus is a popular virus-detection program.

(Source: Courtesy of Norton by Symantec.)

To protect against threats to your privacy and data, you can install security and control measures. For example, many software products can detect and remove viruses and spam from computer systems. Barclays, an international bank, is using handheld identity authorization devices to prevent bank fraud.[63] The new devices will help eliminate the problem of stolen identification numbers and passwords.

You can install *firewalls* (software and hardware that protect a computer system or network from outside attacks) to avoid viruses and prevent unauthorized people from gaining access to your computer system. You can also use identification numbers and passwords. Some security experts propose installing Web cameras and hiring "citizen spotters" to monitor the Webcams. In response to possible abuses, a number of laws have been passed to protect people from invasion of their privacy, including The Privacy Act, enacted in the 1970s.

Use of information systems also raises work concerns, including job loss through increased efficiency and some potential health problems from making repetitive motions. *Ergonomics*, the study of designing and positioning workplace equipment, can help you avoid health-related problems of using computer systems.

Computer and Information Systems Literacy

Whatever your college major or career path, understanding computers and information systems will help you cope, adapt, and prosper in this challenging environment. Some colleges are requiring a certain level of computer and information systems literacy before students are admitted or accepted into the college.[64] The University of Chicago's School of Business, for example, requires entering students be able to use Microsoft Office PowerPoint, a presentation and graphics program. While at school, you might connect with friends and other students using a social networking Internet site, such as MySpace (*www.myspace.com*) or FaceBook (*www.facebook.com*).[65] When you graduate, you might find yourself interviewing for a job using the TMP Island in Second Life (*www.second.life.com*). The TMP Island in Second Life is a site where job candidates and corporate recruiters can meet and conduct a virtual job interview.

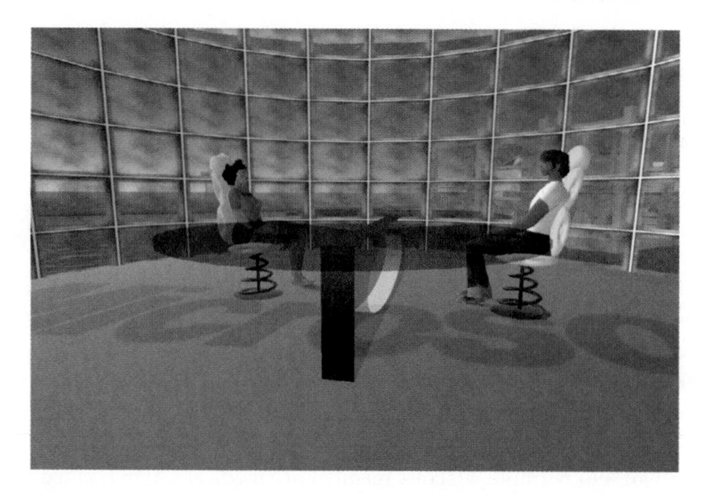

The TMP Island allows you to conduct a job interview in the virtual world.

(Source: Courtesy of TMP Worldwide Advertising & Communications, LLC.)

A knowledge of information systems will help you make a significant contribution on the job. It will also help you advance in your chosen career or field. Managers are expected to identify opportunities to implement information systems to improve their business. They are also expected to lead IS projects in their areas of expertise. To meet these personal and organizational goals, you must acquire both computer literacy and information systems literacy. **Computer literacy** is a knowledge of computer systems and equipment and the ways they function. It stresses equipment and devices (hardware), programs and instructions (software), databases, and telecommunications.

computer literacy
Knowledge of computer systems and equipment and the ways they function; it stresses equipment and devices (hardware), programs and instructions (software), databases, and telecommunications.

Information systems literacy goes beyond knowing the fundamentals of computer systems and equipment. **Information systems literacy** is the knowledge of how data and information are used by individuals, groups, and organizations. It includes knowledge of computer technology and the broader range of information systems. Most important, however, it encompasses *how* and *why* this technology is applied in business. Knowing about various types of hardware and software is an example of computer literacy. Knowing how to use hardware and software to increase profits, cut costs, improve productivity, and increase customer satisfaction is an example of information systems literacy. Information systems literacy can involve recognizing how and why people (managers, employees, stockholders, and others) use information systems; being familiar with organizations, decision-making approaches, management levels, and information needs; and understanding how organizations can use

information systems literacy
Knowledge of how data and information are used by individuals, groups, and organizations.

computers and information systems to achieve their goals. Knowing how to deploy transaction processing, management information, decision support, and special-purpose systems to help an organization achieve its goals is a key aspect of information systems literacy.

Information Systems in the Functional Areas of Business

Information systems are used in all functional areas and operating divisions of business. In *finance* and *accounting,* information systems forecast revenues and business activity, determine the best sources and uses of funds, manage cash and other financial resources, analyze investments, and perform audits to make sure that the organization is financially sound and that all financial reports and documents are accurate. *Sales* and *marketing* use information systems to develop new goods and services (product analysis), select the best location for production and distribution facilities (place or site analysis), determine the best advertising and sales approaches (promotion analysis), and set product prices to get the highest total revenues (price analysis). In *manufacturing,* information systems process customer orders, develop production schedules, control inventory levels, and monitor product quality. In addition, information systems help to design products (*computer-assisted design,* or *CAD*), manufacture items (*computer-assisted manufacturing,* or *CAM*), and integrate machines or pieces of equipment (*computer-integrated manufacturing,* or *CIM*). *Human resource management* uses information systems to screen applicants, administer performance tests to employees, monitor employee productivity, and more. *Legal information systems* analyze product liability and warranties and help to develop important legal documents and reports.

Festo, a global manufacturer of components and controls for industrial automation, uses a CAD system when developing its products.

(Source: Courtesy of Festo AG & Co. KG.)

Information Systems in Industry

In addition to being used in every department in a company, information systems are used in almost every industry or field in business. The *airline industry* develops Internet auction sites to offer discount fares and increase revenue. *Investment firms* use information systems to analyze stocks, bonds, options, the futures market, and other financial instruments, and provide improved services to their customers. *Banks* use information systems to help make sound loans and good investments as well as to provide online check payment for account holders. The *transportation industry* uses information systems to schedule trucks and trains to deliver goods and services at the lowest cost. *Publishing companies* use information systems to analyze markets and to develop and publish newspapers, magazines, and books. *Healthcare organizations* use information systems to diagnose illnesses, plan medical treatment, track patient records, and bill patients. Health maintenance organizations (HMOs) use Web technology to access patients' insurance eligibility and other information stored in databases to cut patient costs. *Retail companies* are using the Web to take orders and provide customer service support. Retail companies also use information systems to help market products and services, manage inventory levels, control the supply chain, and forecast demand. *Power management* and *utility companies* use information systems to monitor and control power generation and usage. *Professional services* firms employ information systems to improve the speed and quality of services they provide to customers. Management consulting firms use intranets and extranets to offer information on products, services, skill levels, and past engagements to their consultants. These industries are discussed in more detail as we continue through the book.

GLOBAL CHALLENGES IN INFORMATION SYSTEMS

Changes in society as a result of increased international trade and cultural exchange, often called globalization, have always had a significant impact on organizations and their information systems. In his book, *The World Is Flat,* Thomas Friedman describes three eras of globalization[66] (see Table 1.4). According to Friedman, we have progressed from the globalization of countries to the globalization of multinational corporations and individuals. Today, people in remote areas can use the Internet to compete with and contribute to other people, the largest corporations, and entire countries. These workers are empowered by high-speed Internet access, making the world flatter. In the Globalization 3.0 era, designing a new airplane or computer can be separated into smaller subtasks and then completed by a person or small group that can do the best job. These workers can be located in India, China, Russia, Europe, and other areas of the world. The subtasks can then be combined or reassembled into the complete design. This approach can be used to prepare tax returns, diagnose a patient's medical condition, fix a broken computer, and many other tasks.

Table 1.4

Eras of Globalization

Era	Dates	Characterized by
Globalization 1.0	Late 1400–1800	Countries with the power to explore and influence the world
Globalization 2.0	1800–2000	Multinational corporations that have plants, warehouses, and offices around the world
Globalization 3.0	2000–today	Individuals from around the world who can compete and influence other people, corporations, and countries by using the Internet and powerful technology tools

Today's information systems have led to greater globalization. High-speed Internet access and networks that can connect individuals and organizations around the world create more international opportunities. Global markets have expanded. People and companies can get products and services from around the world, instead of around the corner or across town. These opportunities, however, introduce numerous obstacles and issues, including challenges involving culture, language, and many others.

- **Cultural challenges.** Countries and regional areas have their own cultures and customs that can significantly affect individuals and organizations involved in global trade.
- **Language challenges.** Language differences can make it difficult to translate exact meanings from one language to another.
- **Time and distance challenges.** Time and distance issues can be difficult to overcome for individuals and organizations involved with global trade in remote locations. Large time differences make it difficult to talk to people on the other side of the world. With long distance, it can take days to get a product, a critical part, or a piece of equipment from one location to another location.

- **Infrastructure challenges.** High-quality electricity and water might not be available in certain parts of the world. Telephone services, Internet connections, and skilled employees might be expensive or not readily available.
- **Currency challenges.** The value of different currencies can vary significantly over time, making international trade more difficult and complex.
- **Product and service challenges.** Traditional products that are physical or tangible, such as an automobile or bicycle, can be difficult to deliver to the global market. However, *electronic products (e-products)* and *electronic services (e-services)* can be delivered to customers electronically, over the phone, through networks, through the Internet, or by other electronic means. Software, music, books, manuals, and advice can all be delivered globally and over the Internet.
- **Technology transfer issues.** Most governments don't allow certain military-related equipment and systems to be sold to some countries. Even so, some believe that foreign companies are stealing intellectual property, trade secrets, and copyrighted materials, and counterfeiting products and services.
- **State, regional, and national laws.** Each state, region, and country has a set of laws that must be obeyed by citizens and organizations operating in the country. These laws can deal with a variety of issues, including trade secrets, patents, copyrights, protection of personal or financial data, privacy, and much more. Laws restricting how data enters or exits a country are often called *transborder data-flow* laws. Keeping track of these laws and incorporating them into the procedures and computer systems of multinational and transnational organizations can be very difficult and time consuming, requiring expert legal advice.
- **Trade agreements.** Countries often enter into trade agreements with each other. The North American Free Trade Agreement (NAFTA) and the Central American Free Trade Agreement (CAFTA) are examples. The European Union (EU) is another example of a group of countries with an international trade agreement.[67] The EU is a collection of mostly European countries that have joined together for peace and prosperity. Additional trade agreements include the Australia-United States Free Trade Agreement (AUSFTA), signed into law in 2005, and the Korean-United States Free Trade Agreement (KORUS-FTA), signed into law in 2007.[68] Free trade agreements have been established between Bolivia and Mexico, Canada and Costa Rica, Canada and Israel, Chile and Korea, Mexico and Japan, the United States and Jordan, and many others.[69]

SUMMARY

Principle

The value of information is directly linked to how it helps decision makers achieve the organization's goals.

Information systems are used in almost every imaginable career area. Regardless of your college major or chosen career, you will find that information systems are indispensable tools to help you achieve your career goals. Learning about information systems can help you get your first job, earn promotions, and advance your career.

Data consists of raw facts; information is data transformed into a meaningful form. The process of defining relationships among data requires knowledge. Knowledge is an awareness and understanding of a set of information and the way that information can support a specific task. To be valuable, information must have several characteristics: It should be accurate, complete, economical to produce, flexible, reliable, relevant, simple to understand, timely, verifiable, accessible, and secure. The value of information is directly linked to how it helps people achieve their organization's goals.

Principle

Computers and information systems are constantly making it possible for organizations to improve the way they conduct business.

A system is a set of elements that interact to accomplish a goal or set of objectives. The components of a system include inputs, processing mechanisms, and outputs. A system uses feedback to monitor and control its operation to make sure that it continues to meet its goals and objectives.

System performance is measured by its efficiency and effectiveness. Efficiency is a measure of what is produced divided by what is consumed; effectiveness measures the extent to which a system achieves its goals. A systems performance standard is a specific objective.

Principle

Knowing the potential impact of information systems and having the ability to put this knowledge to work can result in a successful personal career and organizations that reach their goals.

Information systems are sets of interrelated elements that collect (input), manipulate and store (process), and disseminate (output) data and information. Input is the activity of capturing and gathering new data, processing involves converting or transforming data into useful outputs, and output involves producing useful information. Feedback is the output that is used to make adjustments or changes to input or processing activities.

The components of a computer-based information system (CBIS) include hardware, software, databases, telecommunications and the Internet, people, and procedures. The types of CBISs that organizations use can be classified into four basic groups: (1) e-commerce and m-commerce, (2) TPS and ERP systems, (3) MIS and DSS, and (4) specialized business information systems. The key to understanding these types of systems begins with learning their fundamentals.

E-commerce involves any business transaction executed electronically between parties such as companies (business-to-business), companies and consumers (business-to-consumer), business and the public sector, and consumers and the public sector. The major volume of e-commerce and its fastest-growing segment is business-to-business transactions that make purchasing easier for big corporations. E-commerce also offers opportunities for small businesses to market and sell at a low cost worldwide, thus allowing them to enter the global market right from start-up. M-commerce involves anytime, anywhere computing that relies on wireless networks and systems.

The most fundamental system is the transaction processing system (TPS). A transaction is any business-related exchange. The TPS handles the large volume of business transactions that occur daily within an organization. An enterprise resource planning (ERP) system is a set of integrated programs that can manage the vital business operations for an entire multisite, global organization. A management information system (MIS) uses the information from a TPS to generate information useful for management decision making.

A decision support system (DSS) is an organized collection of people, procedures, databases, and devices that help make problem-specific decisions. A DSS differs from an MIS in the support given to users, the emphasis on decisions, the development and approach, and the system components, speed, and output.

Specialized business information systems include knowledge management, artificial intelligence, expert, and virtual reality systems. Knowledge management systems are organized collections of people, procedures, software, databases, and devices used to create, store, share, and use the organization's knowledge and experience. Artificial intelligence (AI) includes a wide range of systems in which the computer takes on the characteristics of human intelligence. Robotics is an area of artificial intelligence in which machines perform complex, dangerous, routine, or boring tasks, such as welding car frames or assembling computer systems and components. Vision systems allow robots and other devices to have "sight" and to store and process visual images. Natural language processing involves computers interpreting and acting on verbal or written commands in English, Spanish, or other human languages. Learning systems let computers learn from past mistakes or experiences, such as playing games or making business decisions, while neural networks is a branch of artificial intelligence that allows computers to recognize and act on patterns or trends. An expert system (ES) is designed to act as an expert consultant to a user who is seeking advice about a specific situation. Originally, the term *virtual reality* referred to immersive virtual reality, in which the user becomes fully immersed in an artificial, computer-generated 3-D world. Virtual reality can also refer to applications that are not fully immersive, such as mouse-controlled navigation through a 3-D environment on a graphics monitor, stereo viewing from the monitor via stereo glasses, and stereo projection systems.

Principle

System users, business managers, and information systems professionals must work together to build a successful information system.

Systems development involves creating or modifying existing business systems. The major steps of this process and their goals include systems investigation (gain a clear understanding of what the problem is), systems analysis (define what the system must do to solve the problem), systems design (determine exactly how the system will work to meet the business needs), systems implementation (create or acquire the various system components defined in the design step), and systems maintenance and review (maintain and then modify the system so that it continues to meet changing business needs).

Principle

Information systems must be applied thoughtfully and carefully so that society, business, and industry around the globe can reap their enormous benefits.

Information systems play a fundamental and ever-expanding role in society, business, and industry. But their use can also raise serious security, privacy, and ethical issues. Effective information systems can have a major impact on corporate strategy and organizational success. Businesses around the globe are enjoying better safety and service, greater efficiency and effectiveness, reduced expenses, and improved decision making and control because of information systems. Individuals who can help their businesses realize these benefits will be in demand well into the future.

Computer and information systems literacy are prerequisites for numerous job opportunities, and not only in the IS field. Computer literacy is knowledge of computer systems and equipment; information systems literacy is knowledge of how data and information are used by individuals, groups, and organizations. Today, information systems are used in all the functional areas of business, including accounting, finance, sales, marketing, manufacturing, human resource management, and legal information systems. Information systems are also used in every industry, such as airlines, investment firms, banks, transportation companies, publishing companies, healthcare, retail, power management, professional services, and more.

Changes in society as a result of increased international trade and cultural exchange, often called globalization, have always had a significant impact on organizations and their information systems. In his book, *The World Is Flat,* Thomas Friedman describes three eras of globalization, spanning the globalization of countries to the globalization of multinational corporations and individuals. Today, people in remote areas can use the Internet to compete with and contribute to other people, the largest corporations, and entire countries. People and companies can get products and services from around the world, instead of around the corner or across town. These opportunities, however, introduce numerous obstacles and issues, including challenges involving culture, language, and many others.

CHAPTER 1: SELF-ASSESSMENT TEST

The value of information is directly linked to how it helps decision makers achieve the organization's goals.

1. A(n) ________________ is a set of interrelated components that collect, manipulate, and disseminate data and information and provide a feedback mechanism to meet an objective.
2. Numbers, letters, and other characters are represented by ____________.
   a. image data
   b. numeric data
   c. alphanumeric data
   d. symmetric data
3. Knowledge workers are usually professionals in science, engineering, business, and other areas. True or False?

Computers and information systems are constantly making it possible for organizations to improve the way they conduct business.

4. A(n) ______________ is a set of elements or components that interact to accomplish a goal.

5. A measure of what is produced divided by what is consumed is known as __________.
   a. efficiency
   b. effectiveness
   c. performance
   d. productivity
6. A specific objective of a system is called effectiveness. True or False?

Knowing the potential impact of information systems and having the ability to put this knowledge to work can result in a successful personal career and organizations that reach their goals.

7. A(n) ____________ consists of hardware, software, databases, telecommunications, people, and procedures.
8. Computer programs that govern the operation of a computer system are called ____________.
   a. feedback
   b. feedforward
   c. software
   d. transaction processing systems
9. Payroll and order processing are examples of a computerized management information system. True or False?
10. What is an organized collection of people, procedures, software, databases, and devices used to create, store, share, and use the organization's experience and knowledge?
    a. TPS (transaction processing system)
    b. MIS (management information system)
    c. DSS (decision support system)
    d. KM (knowledge management)
11. ______________ involves anytime, anywhere commerce that uses wireless communications.

System users, business managers, and information systems professionals must work together to build a successful information system.

12. What defines the problems and opportunities of the existing system?
    a. systems analysis
    b. systems review
    c. systems development
    d. systems design

Information systems must be applied thoughtfully and carefully so that society, business, and industry around the globe can reap their enormous benefits.

13. ____________ literacy is a knowledge of how data and information are used by individuals, groups, and organizations.

CHAPTER 1: SELF-ASSESSMENT TEST ANSWERS

(1) information system (2) c (3) True (4) system (5) a (6) False (7) computer-based information system (CBIS) (8) c (9) False (10) d (11) Mobile commerce (m-commerce) (12) a (13) Information systems

REVIEW QUESTIONS

1. What is an information system? What are some of the ways information systems are changing our lives?
2. How would you distinguish data and information? Information and knowledge?
3. Identify at least six characteristics of valuable information.
4. What is the difference between efficiency and effectiveness?
5. What are the components of any information system?
6. What is feedback? What are possible consequences of inadequate feedback?
7. How is system performance measured?
8. What is a knowledge management system? Give an example.
9. What is a computer-based information system? What are its components?
10. Identify the functions of a transaction processing system.
11. What is the difference between an intranet and an extranet?
12. What is m-commerce? Describe how it can be used.
13. What are the most common types of computer-based information systems used in business organizations today? Give an example of each.
14. Describe three applications of virtual reality.
15. What are computer literacy and information systems literacy? Why are they important?
16. What are some of the benefits organizations seek to achieve through using information systems?
17. Identify the steps in the systems development process and state the goal of each.

DISCUSSION QUESTIONS

1. Why is the study of information systems important to you? What do you hope to learn from this course to make it worthwhile?
2. Describe how information systems are used at school or work.
3. What is the value of software? Give several examples of software you use at school or home.
4. Why is a database an important part of a computer-based information system?
5. What is the difference between e-commerce and m-commerce?
6. What is the difference between DSS and knowledge management?
7. Suppose that you are a teacher assigned the task of describing the learning processes of preschool children. Why would you want to build a model of their learning processes? What kinds of models would you create? Why might you create more than one type of model?
8. Describe the "ideal" automated vehicle license plate renewal system for the drivers in your state. Describe the input, processing, output, and feedback associated with this system.
9. What computer application needs the most improvement at your college or university? Describe how systems development could be used to develop it.
10. Discuss how information systems are linked to the business objectives of an organization.
11. What are your career goals and how can a computer-based information system be used to achieve them?

PROBLEM-SOLVING EXERCISES

1. Prepare a data disk and a backup disk for the problem-solving exercises and other computer-based assignments you will complete in this class. Create one directory for each chapter in the textbook (you should have 14 directories). As you work through the problem-solving exercises and complete other work using the computer, save your assignments for each chapter in the appropriate directory. On the label of each disk or USB flash drive, be sure to include your name, course, and section. On one disk write "Working Copy"; on the other write "Backup."
2. Search through several business magazines (*Business Week, Computerworld, PC Week,* etc.) for a recent article that discusses the use of information systems to deliver significant business benefits to an organization. Now use other resources to find additional information about the same organization (*Reader's Guide to Periodical Literature*, online search capabilities available at your school's library, the company's public relations department, Web pages on the Internet, etc.). Use word processing software to prepare a one-page summary of the different resources you tried and their ease of use and effectiveness.
3. Create a table that lists ten or more possible career areas, annual salaries, and brief job descriptions, and rate how much you would like the career area on a scale from 1 (don't like) to 10 (like the most). Print the results. Sort the table according to annual salaries from high to low and print the resulting table. Sort the table from the most liked to least liked and print the results.
4. Do some research to obtain estimates of the rate of growth of social networking sites like MySpace and FaceBook. Use the plotting capabilities of your spreadsheet or graphics software to produce a bar chart of that growth over a number of years. Share your findings with the class.

TEAM ACTIVITIES

1. Before you can do a team activity, you need a team! As a class member, you might create your own team, or your instructor might assign members to groups. After your group has been formed, meet and introduce yourselves to each other. Find out the first name, hometown, major, e-mail address, and phone number of each member. Find out one interesting fact about each member of your team as well. Brainstorm a name for your team. Put the information on each team member into a database and print enough copies for each team member and your instructor.

2. With the other members of your group, use word processing software to write a one-page summary of what your team hopes to gain from this course and what you are willing to do to accomplish these goals. Send the report to your instructor via e-mail.

WEB EXERCISES

1. Throughout this book, you will see how the Internet provides a vast amount of information to individuals and organizations. We will stress the World Wide Web, or simply the Web, which is an important part of the Internet. Most large universities and organizations have an address on the Internet, called a Web site or home page. The address of the Web site for this publisher is *www.course.com*. You can gain access to the Internet through a browser, such as Microsoft Internet Explorer or Netscape. Using an Internet browser, go to the Web site for this publisher. What did you find? Try to obtain information on this book. You might be asked to develop a report or send an e-mail message to your instructor about what you found.
2. Go to an Internet search engine, such as *www.google.com*, and search for information about virtual reality. Write a brief report that summarizes what you found.
3. Using the Internet, search for information on the use of information systems in a company or organization that interests you. How does the organization use technology to help it accomplish its goals?

CAREER EXERCISES

1. In the Career Exercises found at the end of every chapter, you will explore how material in the chapter can help you excel in your college major or chosen career. Write a brief report on the career that appeals to you the most. Do the same for two other careers that interest you.
2. Research careers in accounting, marketing, information systems, and two other career areas that interest you. Describe the job opportunities, job duties, and the possible starting salaries for each career area in a report.

CASE STUDIES

Case One

New York City Cabbies Strike Over New Information System

New York City's Taxi & Limousine Commission has mandated that all licensed city cab owners install new information systems in their cabs. The new state-of-the-art system connects the cabs to a wireless data network. The new system will not be used for dispatching cabs (most New York City cabs are hailed from the curb) but will provide text messages informing cabbies of nearby opportunities. It includes global positioning system technology that provides an interactive map that passengers can use to see their current location, destination, and routes. The new system will allow customers to pay via credit card, and will provide music and other forms of entertainment. The system also automates the process of keeping business records of fares and trips and spits out receipts for customers; cabbies will no longer need to maintain records with pencil and paper.

Sounds like a win-win situation, right? Many of the cabbies don't agree. They are concerned that the GPS system will track their movements after hours. About 85 percent of New York City cabbies are independent contractors who own their cab and use it for their personal transportation while off duty. They are also upset about the $1,300 that they need to pay for the system and a five percent required fee for every credit card transaction made by their customers. Mostly, they are upset that they were not a part of the decision and design process that led to the citywide mandate. In a headline-making move, the New York Taxi Workers Alliance, which represents 10,000 of the city's 13,000 cabbies, called a two-day strike to pressure the city to rethink its new system. Unfortunately, the strike had little effect as many cabbies stayed on duty, unable to afford the time off of work.

The city is moving forward with the deployment of the new cab information system. It has justified the expense to the cab drivers by pointing out the increase of cab fares over the past two years that doubled driver wages to $28 an hour on average; an increase that was imposed with promises to customers for better service and in-cab technologies.

New York City's Taxi & Limousine Commission versus the New York Taxi Workers Alliance provides several lessons about business interactions and information systems. Today's businesses have an incredible amount of pressure on them to implement the latest and smartest technologies and information systems. Often it is a business's information system that gives a business an important advantage over its competition. The City of New York no doubt feels pressure to provide visitors and locals with high-quality service and smooth running transportation systems. If it fails in this regard, visitors and locals may begin to find other cities more attractive to visit and reside in. Information systems also allow businesses to operate more efficiently and effectively. The new system for New York City cabs ultimately saves drivers and passengers time and energy.

Finally, this story illustrates a resistance to change that is often experienced when implementing sweeping new information systems in large businesses and organizations. Many businesses today are revamping entire corporate systems and investing in retraining employees and winning them over to the new system. This is an integral and challenging part of new system implementation. So should you visit New York City anytime soon, and experience the cool technology built into the cabs, remember the struggle that went into bringing this new system to life, and give your driver a generous tip.

Discussion Questions

1. Were New York City cabbies justified in their concern over GPS tracking in their cabs? Why or why not?
2. What might New York City's Taxi & Limousine Commission have done to create a smoother transition to the new system?

Critical Thinking Questions

1. Many cabbies expressed the opinion that the technology being installed in their cabs was inevitable, no matter what their opinion. Do you agree with this assessment? Why or why not?
2. As a customer of a cab service, would you appreciate the benefits provided by the new system in New York City? Would it help provide you with a favorable view of the city in general? List the benefits and provide your view of each.

SOURCES: Hamblen, Matt, "N.Y. taxi agency says cabs will get GPS technology, despite strike threat," *Computerworld*, August 7, 2007, *www.computerworld.com/action/article.do?command=viewArticleBasic&articleId=9032482&source=rss_news10.* Hamblen, Matt, "N.Y. taxi drivers set strike date to protest GPS systems," *Computerworld*, August 23, 2007, *www.computerworld.com/action/article.do?command=viewArticleBasic&articleId=9032482&source=rss_news10.* Lopez, Elias, "City Cabdrivers Strike Again, but Protest Gets Little Notice," *New York Times*, October 23, 2007, *www.nytimes.com/2007/10/23/nyregion/23taxi.html?_r=1&ref=nyregion&oref=slogin.*

Case Two

Yansha Leans on IS to Stay Competitive

More than ever before, Chinese retailers are facing local competition from foreign companies. China's highly regulated economy has insulated businesses from competition. Now that China is loosening its regulations in an effort to benefit from international trade, its own businesses must work harder to become more efficient and effective and keep customers.

Yansha is one of China's biggest retailers. It sells upscale designer clothes from around the world along with other fine merchandise. One of its largest retail stores occupies 215,000 square feet in Beijing's famous Youyi Shopping City.

Yansha has long experienced market leadership in China, but in recent years has felt increasing competition from international companies. Yansha's management team was aware that its methods of communication with suppliers—the placing and receiving of orders—was less than efficient. It was also aware of other inefficiencies in communication throughout the organization. For Yansha to maintain its leadership role in the market, it would need to cut the waste and become lean and mean in its application of information system technology.

Yansha turned to IBM China Research Lab to evaluate its information systems and recommend the latest technologies to bring it up to date. IBM implemented a massive system upgrade across the entire enterprise: an enterprise resource planning (ERP) system. The ERP allows Yansha executives and managers to view real-time performance data, such as sales across all locations, in certain regions, or in one particular store. Using this system, managers could, for example, determine the success of a particular marketing approach. The new ERP interfaces with a new supply chain management (SCM) system that provides close communication between Yansha and its suppliers. These two systems working together, the ERP system and the SCM system, allow for Yansha and its suppliers to work as one tightly knit organization.

The new systems required a substantial investment of time and money for Yansha, but the benefits have vastly overshadowed the costs. The new systems reduced the time it takes for suppliers to ship merchandise to Yansha (order lead time) from 2.5 days to 4.5 hours. The order acknowledgment rate has increased from 80 percent to 99 percent. Order errors have been reduced from 9 percent to 1 percent.

The money saved by Yansha receiving the right merchandise at the right time has saved the company enough money to pay for its expensive new information systems within nine months of rolling them out. Achieving a return on investment (ROI) in such a short time is something any chief information officer (CIO) would be proud of.

Discussion Questions

1. China is experiencing a rapidly evolving economy. Why do you think most international businesses are looking to China as both an opportunity and a threat?
2. Yansha has raced to catch up with the latest information systems and technologies, but remains a local Chinese business. What might Yansha's next move be in order to increase its revenue? How might that move be most successfully taken?

Critical Thinking Questions

1. What benefits do local businesses such as Yansha have over foreign businesses that come to China to compete. How can those advantages be used to maintain leadership in a market?
2. What benefits do international businesses have over smaller local businesses? How can they be used to infiltrate a new market and take over leadership?

SOURCES: Staff, "Yansha department store embraces supplier collaboration to streamline processes," IBM Success Stories, November 29, 2007, *www-01.ibm.com/software/success/cssdb.nsf/CS/JSTS-79BMT3?OpenDocument&Site=corp&cty=en_us*. Yansha's Web site, accessed December 21, 2007, *www.yansha.com.cn*.

Questions for Web Case

See the Web site for this book to read about the Whitmann Price Consulting case for this chapter. The following questions cover this Web case.

Whitmann Price Consulting: A New Systems Initiative

Discussion Questions

1. What advantages would the proposed Advanced Mobile Communications and Information System provide for Whitmann Price Consulting? What problems might it assist in eliminating?
2. Why do you think Josh and Sandra have been asked to interview the managers of the six business units within WPC as a first step? As IT professionals, Josh, Sandra, and their boss Matt know much more about technology and information systems than the heads of the business units. Shouldn't they be able to design the system without suggestions from amateurs? Including more people in the planning stage is sure to complicate the process.

Critical Thinking Questions

1. If you were Josh or Sandra, what questions would you ask the heads of the six business units?
2. If you were Josh or Sandra, what additional research might you request of your IT staff at this point?

NOTES

Sources for the opening vignette: Staff, "Fossil integrates retail and wholesale operations with IBM and SAP," IBM Success Stories, *www-01.ibm.com/software/success/cssdb.nsf/CS/STRD-795MNK?OpenDocument&Site=corp&cty=en_us*. Fossil Web site, accessed December 26, 2007, *www.fossil.com*. SAP Web site, accessed December 26, 2007, *www.sap.com/about/press/press.epx?pressid=4514*.

1 Mills, Steven, "News," *Computerworld*, June 4, 2007, p. 20.
2 Dhar, V. and Sundararajan, A., "Information Technologies in Business: A Blueprint for Education and Research," *Information Systems Research*, June 2007, p. 125.
3 Smith, Frederick, "A Budding Network," *Forbes*, May 7, 2007, p. 64.
4 Chaker, Anne Marie, "Yale on $0 Per Day," *The Wall Street Journal*, February 15, 2007, p. D1.
5 Hoffman, Thomas, "Saving Lives Via Video at Sutter Health's eICU," *Computerworld*, June 25, 2007, p. 28.
6 Oh, Wonseok and Pinsonneault, Alian, "On the Assessment of the Strategic Value of Information Technologies," *MIS Quarterly*, June 2007, p. 239.
7 Peppard, J. and Daniel, E. "Managing the Realization of Business Benefits from IT Investments," *MIS Quarterly Executive*, June 2007, p. 1.
8 Ayres, I. and Nalebuff, B., "Experiment," *Forbes*, September 3, 2007, p. 130.
9 McDonald, Ian, "Making Paperless Trails at Lloyd's," *The Wall Street Journal*, August 13, 2007, p. B1.
10 Wingfield, Nick, "Hide the Button," *The Wall Street Journal*, July 25, 2007, p. D1.
11 Lawton, Christopher, "Dumb Terminals Can Be a Smart Move," *The Wall Street Journal*, January 30, 2007, p. B3.
12 Clark, Don, "Intel Scores Speed Breakthrough," *The Wall Street Journal*, July 25, 2007, p. B4.
13 Wildstrom, Stephen, "How Flash Will Change PCs," *Business Week*, June 11, 2007.
14 Gonsalves, Antone, "A Supercomputer on a Chip," *Information Week*, June 25, 2007, p. 26.
15 Einhorn, Bruce, "Intel Inside the Third World," *Business Week*, July 9, 2007, p. 38.
16 Fab Lab, *http://fab.cba.mit.edu/*, accessed August 25, 2007.
17 Spanbauer, Scott, "The Right Operating System For You," *PC World*, April 2007, p. 102.
18 Dunn, Scott, "Sync Your PC's Tunes with Windows Mobile Devices," *PC World*, March 2007, p. 124.
19 Arar, Yardena, "Say So Long to Shrink-Wrapped Software," *PC World*, February 2007, p. 37.

20 "Adobe Creative Suite 3," *www.adobe.com/products/creativesuite/*, accessed August 20, 2007.

21 Nickum, Chris, "The BI Prescription," *Optimize Magazine*, April 2007, p. 45.

22 Havenstein, Heather, "Aetna Clients to Get Access to Online Health Data," *Computerworld*, January 15, 2007, p. 7.

23 Greenemeier, Larry, "Data on Loan," *Information Week*, April 23, 2007, p. 33.

24 Schupak, Amanda, "90 Years of Networks," *Forbes*, May 7, 2007, p. 106.

25 Malykhina, E. and Gardner, D., "A $48 Billion Upgrade," *Information Week*, April 2, 2007, p. 22.

26 Gomes, Lee, "Paradoxes Abound in Telecommuting," *The Wall Street Journal*, January 23, 2007, p. B3.

27 Malykhina, Elena, "China Picks Google," *Information Week*, January 8, 2007, p. 17.

28 McBride, Sarah, "Online-Radio Fight Reaches New Pitch," *The Wall Street Journal*, May 31, 2007, p. A13.

29 Ali, Sarmad, "When I Was Your Age, We Didn't Have Sites for Writing Our Bios," *The Wall Street Journal*, May 31, 2007, p. B1.

30 Hardy, Quentin, "Better Than YouTube," *Forbes*, May 21, 2007, p. 72.

31 McMillan, Robert, "Is Web 2.0 Safe?" *PC World*, July 2007, p. 18.

32 Borrows, P. and Crockett, R., "Turning Cell Phones on Their Ear," *Business Week*, January 22, 2007, p. 40.

33 McCartney, Scott, "WiFi in the Sky," *The Wall Street Journal*, April 3, 2007, p. B9.

34 Green, Heather, "Don't Quit Your Day Job, Podcasters," *Business Week*, April 9, 2007, p. 72.

35 Neville, Jeffery, "Web 2.0s Wild Blue Yonder," *Information Week*, January 8, 2007, p. 45.

36 Buckman, Rebecca, "Backers Thrive on B2B Firms," *The Wall Street Journal*, July 16, 2007, p. C1.

37 Holahan, Catherine, "Going, Going, Everywhere," *Business Week*, June 18, 2007, p. 62.

38 Buckman, Rebecca, "Just Charge It to Your Cell Phone," *The Wall Street Journal*, June 21, 2007, p. B3.

39 Ma, M. and Agarwal, R., "Through a Glass Darkly," *Information Systems Research*, March 2007, p. 42.

40 Fahey, Jonathan, "Car Talk," *Forbes*, January 29, 2007, p. 52.

41 Greenemeier, Larry, "A Cancer-Zapping Heroine," *Information Week*, June 4, 2007, p. 22.

42 Boeing Simulates and Manufactures 787 Dreamliner, *www.3ds.com/news-events/press-room/release/1357/1/*, accessed August 16, 2007.

43 Peng, Tina, "Why Trunk Shows Are Going Virtual," *The Wall Street Journal*, August 11, 2007, p. P1.

44 Anthes, Gary, "Split-Second Trading," *Computerworld*, May 21, 2007, p. 40.

45 Vijayan, J., "Feds Delay Database for Money Transfer Records," *Computerworld*, January 22, 2007, p. 1.

46 Imse, Ann and Gathright, Alan, "Governor Seeks to Bring Order to Computer Chaos," *Rocky Mountain News*, July 23, 2007, p. 4.

47 Bartels, Lynn, "State Might Scrap DMV Computers," *Rocky Mountain News*, August 14, 2007, p. 5.

48 Nelson, Ryan, "IT Project Management: Infamous Failures, Classic Mistakes, and Best Practices," *MIS Quarterly Executive*, June 2007, p. 67.

49 Rajeev, Sharma and Yetton, Philip, "The Contingent Effects of Training, Technical Complexity, and Task Interdependence on Successful Information Systems Implementation," *MIS Quarterly*, June 2007, p. 219.

50 Flandez, Raymond, "In Search of Help," *The Wall Street Journal*, March 19, 2007, p. R8.

51 Katz, R. and Devon, J., "Web of Terror," *Forbes*, May 7, 2007, p. 184.

52 Gonzalez, Erika, "CU Sets Up Alert System," *Rocky Mountain News*, August 8, 2007, p. 16.

53 Turkle, Sherry, "Can You Hear Me Now?" *Forbes*, May 7, 2007, p. 176.

54 Vijayan, J., "Pfizer Breech Illustrates Risks of Sharing Files," *Computerworld*, June 18, 2007, p. 20.

55 Bryan-Low, Cassell, "Turkish Police Hold Data-Theft Suspect," *The Wall Street Journal*, August 10, 2007, p. A6.

56 Banhan, Russ, "Personal Data for Sale," *The Wall Street Journal*, June 5, 2007, p. A18.

57 Walsh, Chris, "Computer Glitch Delays Hundreds of United Flights," *Rocky Mountain News*, June 21, 2007, p. 5.

58 Staff, "Computer Glitch Cancels Japan Flights," *Computerworld*, June 4, 2007, p. 14.

59 Murphy-Barret, Victoria, "Spam Hunter," *Forbes*, July 23, 2007, p. 54.

60 Mitchell, Robert, "Green Data Centers," *Computerworld*, June 25, 2007, p. 26.

61 Clark, Don, "Computer Power Waste Targeted," *The Wall Street Journal*, June 13, 2007, p. B6.

62 Smith, Ethan and Delaney, Kevin, "Music Publishers Join YouTube Copyright Suit," *The Wall Street Journal*, August 7, 2007, p. A2.

63 Babcock, Charles, "Barclays Attacks Phishers," *Information Week*, April 23, 2007, p. 27.

64 Pope, Justin, "PowerPoint Skills Needed to Apply to Business School," *Rocky Mountain News*, August 8, 2007, p. 5.

65 Postrel, Virginia, "A Small Circle of Friends," *Forbes*, May 7, 2007, p. 204.

66 Friedman, Thomas, "The World Is Flat," Farrar, Straus and Giroux, 2005, p. 488.

67 *www.europa.eu.int*, accessed August 15, 2007.

68 *www.ustr.gov/Document_Library/Press_Releases/2007/June/United_States_the_Republic_of_Korea_Sign_Lmark_Free_Trade_Agreement.html*, accessed November 8, 2007.

69 *www.sice.oas.org/tradee.asp*, accessed August 15, 2007.

CHAPTER • 2 •

Information Systems in Organizations

PRINCIPLES

- The use of information systems to add value to the organization is strongly influenced by organizational structure, culture, and change.

LEARNING OBJECTIVES

- Identify the value-added processes in the supply chain and describe the role of information systems within them.
- Provide a clear definition of the terms *organizational structure, culture,* and *change* and discuss how they affect the implementation of information systems.

- Because information systems are so important, businesses need to be sure that improvements or completely new systems help lower costs, increase profits, improve service, or achieve a competitive advantage.

- Identify some of the strategies employed to lower costs or improve service.
- Define the term c*ompetitive advantage* and discuss how organizations are using information systems to gain such an advantage.
- Discuss how organizations justify the need for information systems.

- Cooperation between business managers and IS personnel is the key to unlocking the potential of any new or modified system.

- Define the types of roles, functions, and careers available in the field of information systems.

Information Systems in the Global Economy FedEx, United States

Information Systems Connect People with Packages

International shipping companies like FedEx might appear to be fairly simple operations. After all, how much technology does it take to transport a package from one location to another? In truth, however, FedEx is counted among the most technologically-advanced corporations in the world.

FedEx manages two intimately joined networks: its shipping network composed of distribution centers, trucks, planes, and delivery personnel, and its information system composed of scanners, computers of all kinds, telecommunications equipment, databases, software, and the people that manage and use it. Both systems are equally important to the service that FedEx provides and to the company's leadership role in the industry. The shipping network is the primary service, and the information systems allow the shipping to function efficiently and effectively.

Consider, for example, the new information system recently deployed across offices of FedEx Canada. FedEx realizes the importance of being able to provide information to its customers regarding package deliveries without hesitation or delay. With this in mind, FedEx Canada implemented a state-of-the-art customer relationship management (CRM) system designed for use by customer service representatives and sales staff. Representatives use the new system to find the location of any package in the global distribution system at any moment in time. The system also prompts FedEx staff on how to manage the dialogue with the customer based on the scenario. For example, if the customer is interested in having a shipment picked up, the information system will provide the sequence of questions for the representative to ask the customer. In this way, FedEx personnel can handle any foreseeable request with a minimum of training.

To make this type of service possible, information about packages is collected at every stop along the route. A package label may be scanned a dozen times prior to its delivery at its final destination. Each time it is scanned, the record in the central FedEx database is updated. The database acts as a source of information that feeds many information systems designed for varying services across the company. Personnel and customers can then access that information from the FedEx network and the Internet.

Managing a global business is a complex job with many considerations. For example, the FedEx Web site is available in 25 languages. The site isn't simply translated into other languages, but rather redesigned from the ground up for each language to satisfy and appeal to the culture for which it is created. FedEx must also understand and work with numerous currencies and international laws. For example, packages traveling across international borders are often subject to customs inspections.

FedEx's unique experience as a global shipping company has provided it with valuable insight into the interplay between businesses around the world. By understanding the flow of packages, products, and parts between global organizations, FedEx has positioned itself as one of the most important components of the burgeoning global economy. FedEx is much more than a delivery service; it is a logistics manager that connects the many components of the supply chain—that is, the process that takes a product from the raw materials from which it is built to the retail store at which it is sold, and increasingly to the customer's doorstep. The supply chain is part the value chain, which includes methods for adding value to products and services. FedEx and other global shipping companies have become key components in business value chains.

As you read this chapter, consider the following:

- How do FedEx information systems contribute to the value of the services the company provides?
- How can FedEx itself act as a component of the information systems used by its business customers?
- Why is FedEx more valuable today than it was ten years ago?

Why Learn About Information Systems in Organizations?

Organizations of all types use information systems to cut costs and increase profits. After graduating, a management major might be hired by a shipping company to help design a computerized system to improve employee productivity. A marketing major might work for a national retailer using a network to analyze customer needs in different areas of the country. An accounting major might work for an accounting or consulting firm using a computer to audit other companies' financial records. A real estate major might use the Internet and work in a loose organizational structure with clients, builders, and a legal team located around the world. A biochemist might conduct research for a drug company and use a computer to evaluate the potential of a new cancer treatment. An entrepreneur might use information systems to advertise and sell products and bill customers.

Although your career might be different from your classmates', you will almost certainly work with computers and information systems to help your company or organization become more efficient, effective, productive, and competitive in its industry. In this chapter, you will see how information systems can help organizations produce higher-quality products and services to increase their return on investment. We begin by investigating organizations and information systems.

Information systems have changed the way organizations work in recent years. While it once was used primarily to automate manual processes, information technology has transformed the nature of work and the shape of organizations themselves. In this chapter and throughout the book, you will explore the benefits and drawbacks of information systems in today's organizations.

ORGANIZATIONS AND INFORMATION SYSTEMS

organization
A formal collection of people and other resources established to accomplish a set of goals.

An **organization** is a formal collection of people and other resources established to accomplish a set of goals. The primary goal of a for-profit organization is to maximize shareholder value, often measured by the price of the company stock. Nonprofit organizations include social groups, religious groups, universities, and other organizations that do not have profit as their goal.

An organization is a system, which means that it has inputs, processing mechanisms, outputs, and feedback. An organization constantly uses money, people, materials, machines and other equipment, data, information, and decisions. As shown in Figure 2.1, resources such as materials, people, and money serve as inputs to the organizational system from the environment, go through a transformation mechanism, and then are produced as outputs to the environment. The outputs from the transformation mechanism are usually goods or services, which are of higher relative value than the inputs alone. Through adding value or worth, organizations attempt to achieve their goals.

How does the organizational system increase the value of resources? In the transformation mechanism, subsystems contain processes that help turn inputs into goods or services of increasing value. These processes increase the relative worth of the combined inputs on their way to becoming final outputs. Let's reconsider the simple car wash example from Chapter 1 (see Figure 1.3). The first process is washing the car. The output of this

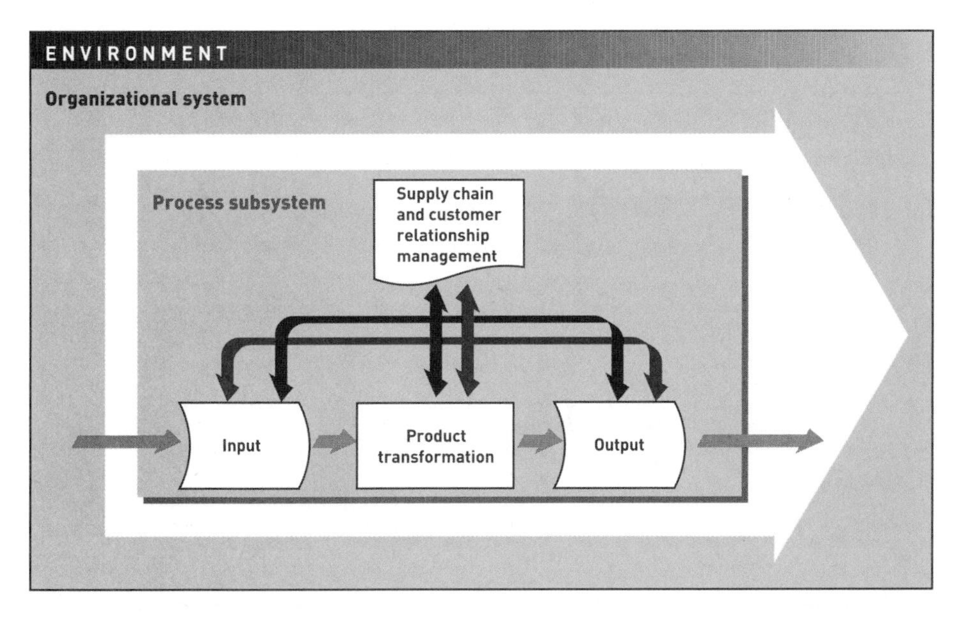

Figure 2.1

A General Model of an Organization

Information systems support and work within all parts of an organizational process. Although not shown in this simple model, input to the process subsystem can come from internal and external sources. Just prior to entering the subsystem, data is external. After it enters the subsystem, it becomes internal. Likewise, goods and services can be output to either internal or external systems.

system—a clean but wet car—is worth more than the mere collection of ingredients (soap and water), as evidenced by the popularity of automatic car washes. Consumers are willing to pay for the skill, knowledge, time, and energy required to wash their car. The second process is drying—transforming the wet car into a dry one with no water spotting. Again, consumers are willing to pay for the additional skill, knowledge, time, and energy required to accomplish this transformation.

Providing value to a stakeholder—customer, supplier, manager, shareholder, or employee—is the primary goal of any organization. The value chain, first described by Michael Porter in a 1985 *Harvard Business Review* article, reveals how organizations can add value to their products and services. The **value chain** is a series (chain) of activities that includes inbound logistics, warehouse and storage, production, finished product storage, outbound logistics, marketing and sales, and customer service (see Figure 2.2). You investigate each activity in the chain to determine how to increase the value perceived by a customer. Depending on the customer, value might mean lower price, better service, higher quality, or uniqueness of product. The value comes from the skill, knowledge, time, and energy that the company invests in the product or activity. The value chain is just as important to companies that don't manufacture products, such as tax preparers, legal firms, and other service providers. By adding a significant amount of value to their products and services, companies ensure success. Combining a value chain with just-in-time (JIT) inventory means companies can deliver materials or parts when they are needed. Ball Aerospace, for example, uses JIT to help reduce inventory costs and enhance customer satisfaction.[1]

value chain
A series (chain) of activities that includes inbound logistics, warehouse and storage, production, finished product storage, outbound logistics, marketing and sales, and customer service.

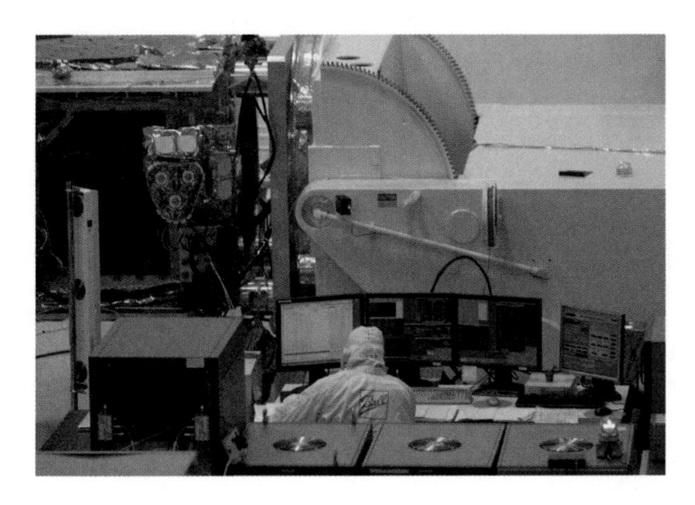

Combining a value chain with just-in-time (JIT) inventory means companies can deliver materials or parts when they are needed. Ball Aerospace uses JIT to help reduce inventory costs and enhance customer satisfaction.

(Source: AP Photo/Denver Post, R. J. Sangosti.)

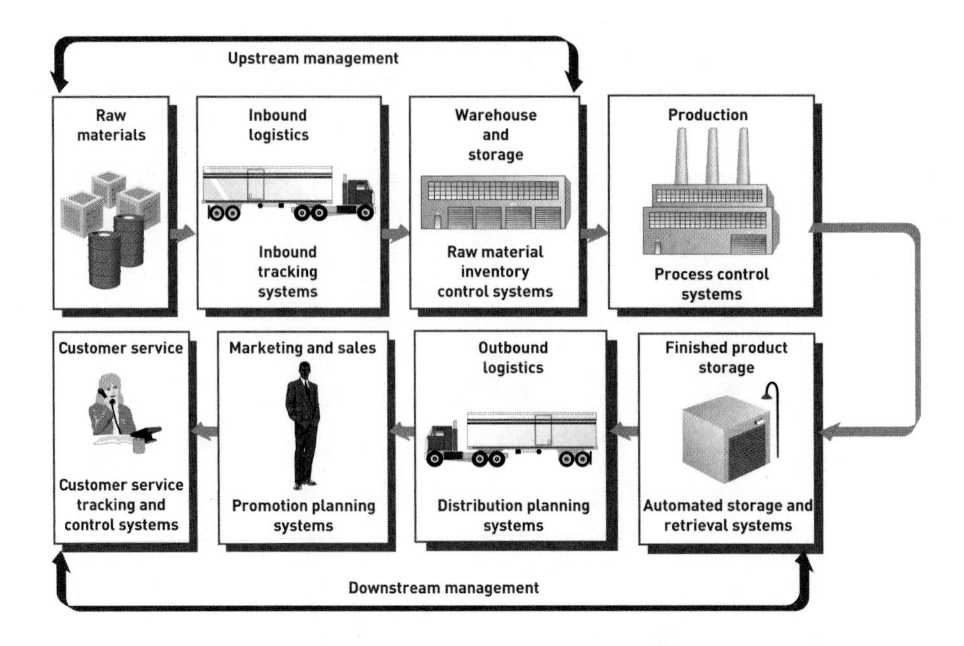

Figure 2.2

The Value Chain of a Manufacturing Company

Managing raw materials, inbound logistics, and warehouse and storage facilities is called *upstream management*. Managing finished product storage, outbound logistics, marketing and sales, and customer service is called *downstream management*.

Wal-Mart's use of information systems is an integral part of its operation. The company gives suppliers access to its inventory system, so the suppliers can monitor the database and automatically send another shipment when stocks are low, eliminating the need for purchase orders. This speeds delivery time, lowers Wal-Mart's inventory carrying costs, and reduces stockout costs.

(Source: *www.walmart.com*.)

Managing the supply chain and customer relationships are two key elements of managing the value chain. *Supply chain management (SCM)* helps determine what supplies are required for the value chain, what quantities are needed to meet customer demand, how the supplies should be processed (manufactured) into finished goods and services, and how the shipment of supplies and products to customers should be scheduled, monitored, and controlled.[2] For example, in an automotive company, SCM can identify key supplies and parts, negotiate with vendors for the best prices and support, make sure that all supplies and parts are available to manufacture cars and trucks, and send finished products to dealerships around the country when they are needed. Increasingly, SCM is accomplished using the Internet and electronic marketplaces (e-marketplaces). When an organization has many suppliers, it can use Internet exchanges to negotiate favorable prices and service. SCM is becoming a global practice as companies have parts and products manufactured around the world.[3] According to Jean Philippe Thenoz, vice president of CMA-CGM, a worldwide shipper, "The client wants to know where the blue socks in size medium are that he ordered two weeks ago from China."

Customer relationship management (CRM) programs help a company manage all aspects of customer encounters, including marketing and advertising, sales, customer service after the sale, and programs to retain loyal customers. CRM can help a company collect customer data, contact customers, educate them about new products, and actively sell products to existing and new customers. Often, CRM software uses a variety of information sources, including sales from retail stores, surveys, e-mail, and Internet browsing habits, to compile comprehensive customer profiles. CRM systems can also get customer feedback to help design new products and services. Tesco, Britain's largest retail operation, uses a CRM Clubcard program to provide outstanding customer service and deliver loyalty rewards and perks to valued customers. See Figure 2.3. Customers can earn services such as meals out, travel, dry cleaning, and car maintenance. The Clubcard loyalty program also extends to Tesco's business partners, introducing Tesco customers to other businesses. To be of most benefit, CRM programs must be tailored for each company or organization.[4] According to Amanda Zuniga, senior research analyst at pharmaceutical intelligence firm Cutting Edge Information, "customer relationship management programs must be individually tailored to meet the programs' specific objectives."

Figure 2.3

Tesco Web Site

Tesco uses its Web site to help with customer relationship management.

What role does an information system play in these processes? A traditional view of information systems holds that organizations use them to control and monitor processes and ensure effectiveness and efficiency. An information system can turn feedback from the subsystems into more meaningful information for employees. This information might summarize the performance of the subsystems and be used to change how the system operates. Such changes could involve using different raw materials (inputs), designing new assembly-line procedures (product transformation), or developing new products and services (outputs). In this view, the information system is external to the process and serves to monitor or control it.

A more contemporary view, however, holds that information systems are often so intimately involved that they are *part of* the process itself. From this perspective, the information system plays an integral role in the process, whether providing input, aiding product transformation, or producing output. Consider a phone directory business that creates phone books for international corporations. A corporate customer requests a phone directory listing all steel suppliers in Western Europe. Using its information system, the directory business can sort files to find the suppliers' names and phone numbers and organize them into an alphabetical list. The information system itself is an integral part of this process. It does not just monitor the process externally but works as part of the process to transform raw data into a product. In this example, the information system turns input (names and phone numbers) into a salable output (a phone directory). The same system might also provide the input (data files) and output (printed pages for the directory).

The latter view provides a new perspective on how and why businesses can use information systems. Rather than attempting to understand information systems independent of the organization, we consider the potential role of information systems within the process itself, often leading to the discovery of new and better ways to accomplish the process.

Organizational Structures

organizational structure
Organizational subunits and the way they relate to the overall organization.

Organizational structure refers to organizational subunits and the way they relate to the overall organization. An organization's structure depends on its goals and approach to management, and can affect how it views and uses information systems. The types of organizational structures typically include traditional, project, team, and virtual. Organizational structure can have a direct impact on the organization's information system.

ETHICAL AND SOCIETAL ISSUES

Fighting Global Poverty with Information Systems

The World Bank is not a typical bank, but two financial institutions owned by 185 member countries. The International Bank for Reconstitution and Development (IBRD) is the part of the bank that focuses on middle-income and creditworthy poor countries, and the International Development Association (IDA) focuses on the poorest countries. Together, these institutions provide "low-interest loans and interest-free credit and grants to developing countries for education, health, infrastructure, communications, and many other purposes," according to the World Bank Web site. The bank has about 10,000 employees worldwide, with loans of about $20 billion annually.

In recent years, the World Bank has suffered from front page scandals regarding suspected improprieties with its senior-level officials. In 2007, World Bank president Paul Wolfowitz was pressured to resign, and in 2005, vice president and CIO Mohammed Muhsen retired under a cloud of suspicion. However, while the press and the world were focused on corruption in the World Bank, some very positive developments were taking place with World Bank infrastructure and information systems that went relatively unnoticed.

The World Bank has traditionally been run as a top-down hierarchy, which is a traditional organizational structure. In recent years, through the use of global information systems, the World Bank has transformed into a "decentralized, front-line, matrix organization," observes a recent article in *Baseline* magazine. Rather than controlling information systems from the top, the World Bank has been investing to empower its clients with the information systems they need locally to participate in the global economy.

The effort to distribute economic knowledge to World Bank customers began in the mid-1990s with then-president James Wolfensohn. In a 1996 speech to the Bank Board of Governors, Wolfensohn said, "The revolution in information technology increases the potential value of [the bank's development] efforts by vastly extending their reach. We need to invest in systems that will enhance our ability to gather information and experience and share it with our clients." Ex-World Bank CIO, Mohammed Muhsen, embraced that mandate to revamp the World Bank's information infrastructure and communications networks to create a global knowledge-sharing network, which has been highly praised in the industry.

Wolfensohn and Muhsen were among the first to formalize what is now referred to as a knowledge management information system. Muhsen defined his mission as follows: "We position ourselves at a major intersection of the network economy where we help to connect global learning opportunities with investment assistance to governments. Put another way, it's about having two currencies: the currency of money and the currency of knowledge. We believe our work in bringing knowledge and information to developing countries is as important as the capital and investments that we provide as an engine for development."

Muhsen's project cost the World Bank hundreds of millions of dollars over several years. It included many information system packages including an SAP ERP system; an Oracle Record Integrated Information System; a multilingual, natural-language system from Teregram for document management; a custom-designed, Web-based dashboard interface; Lotus Notes for e-mail, online collaboration, and content storage; and IBM WebSphere. While these systems cost the World Bank nearly $100 million, the bulk of its investment went to building its own global, high-speed network infrastructure complete with regional satellites and miles of fiber optics to provide network connectivity to remote and poor regions of the world.

Discussion Questions

1. Why do you think Muhsen believes that the knowledge provided to its customers by the World Bank's information system is equal in value to the money loaned? Do you believe this to be true? Why?
2. What unique challenges does the World Bank face when designing an organization-wide information system?

Critical Thinking Questions

1. Once the network infrastructure is created that connects the World Bank to its customers, what types of services might it support?
2. Mohammed Muhsen invested 17 years of his life in developing the World Bank's international knowledge network, but retired under a "cloud of suspicion" over investments he allegedly made in an information system company that was contracted by the World Bank. What lessons can be learned from the unfortunate circumstances of Muhsen's retirement?

SOURCES: McCartney, Laton and Watson, Brian, "World Bank: Behind the IT Transformation," Baseline, August 5, 2007, *www.baselinemag.com/print_article2/0,1217,a=212463,00.asp*. World Bank Web site, *www.worldbank.org*, accessed December 27, 2007.

traditional organizational structure
An organizational structure in which major department heads report to a president or top-level manager.

Traditional Organizational Structure

A **traditional organizational structure**, also called a *hierarchical structure,* is like a managerial pyramid where the hierarchy of decision making and authority flows from the strategic management at the top down to operational management and nonmanagement employees. Compared to lower levels, the strategic level, including the president of the company and vice presidents, has a higher degree of decision authority, more impact on corporate goals, and more unique problems to solve (see Figure 2.4). In most cases, major department heads report to a president or top-level manager. The major departments are usually divided according to function and can include marketing, production, information systems, finance and accounting, research and development, and so on (see Figure 2.5). The positions or departments that are directly associated with making, packing, or shipping goods are called *line positions.* A production supervisor who reports to a vice president of production is an example of a line position. Other positions might not be directly involved with the formal chain of command but instead assist a department or area. These are *staff positions,* such as a legal counsel reporting to the president.

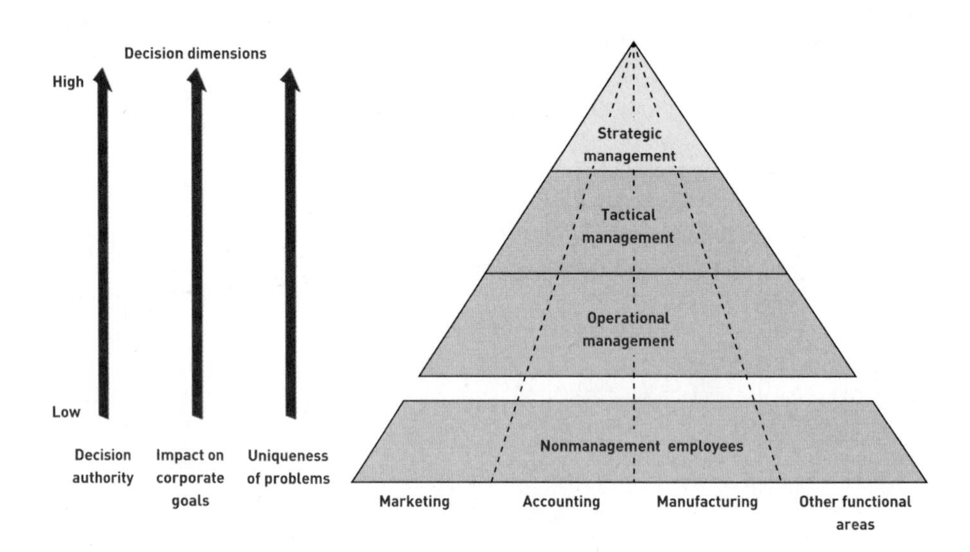

Figure 2.4

A simplified model of the organization, showing the managerial pyramid from top-level managers to nonmanagement employees.

Figure 2.5

A Traditional Organizational Structure

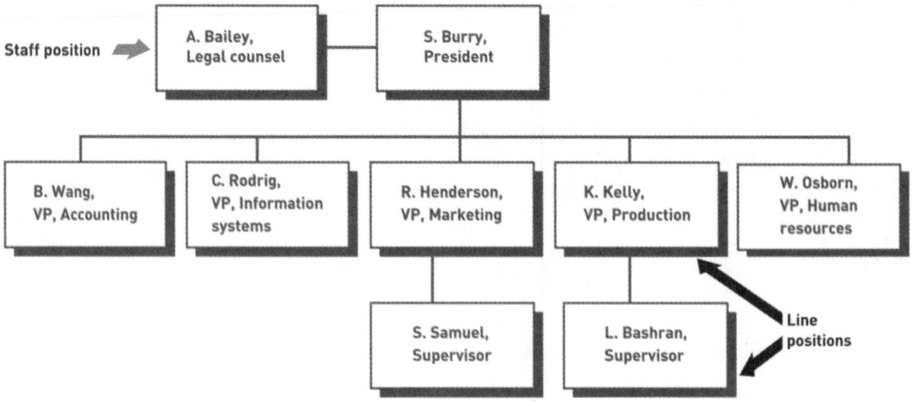

Staff position
A. Bailey, Legal counsel
S. Burry, President
B. Wang, VP, Accounting
C. Rodrig, VP, Information systems
R. Henderson, VP, Marketing
K. Kelly, VP, Production
W. Osborn, VP, Human resources
S. Samuel, Supervisor
L. Bashran, Supervisor
Line positions

flat organizational structure
An organizational structure with a reduced number of management layers.

empowerment
Giving employees and their managers more responsibility and authority to make decisions, take certain actions, and have more control over their jobs.

Today, the trend is to reduce the number of management levels, or layers, in the traditional organizational structure. This type of structure, often called a **flat organizational structure**, empowers employees at lower levels to make decisions and solve problems without needing permission from midlevel managers. **Empowerment** gives employees and their managers more responsibility and authority to make decisions, take action, and have more control over their jobs. For example, an empowered sales clerk could respond to certain customer requests or problems without needing permission from a supervisor. Policies and programs that let employees share ownership in a company flatten the organizational structure. The Clark County School District in Nevada, for example, is using empowerment to give school principals and teachers more control over budgets, programs, and classes.[5] According to school Superintendent Walt Rulffes, "There are powerful indicators from the first-year empowerment school results that, given the right conditions and some additional funding, empowerment schools can dramatically improve student performance."

Information systems can be a key element in empowering employees because they provide the information employees need to make decisions. The employees might also be empowered to develop or use their own personal information systems, such as a simple forecasting model or spreadsheet.

Project and Team Organizational Structures

A **project organizational structure** is centered on major products or services. For example, in a manufacturing firm that produces baby food and other baby products, each line is produced by a separate unit. Traditional functions such as marketing, finance, and production are positioned within these major units (see Figure 2.6). Many project teams are temporary—when the project is complete, the members go on to new teams formed for another project.

project organizational structure
A structure centered on major products or services.

Figure 2.6
A Project Organizational Structure

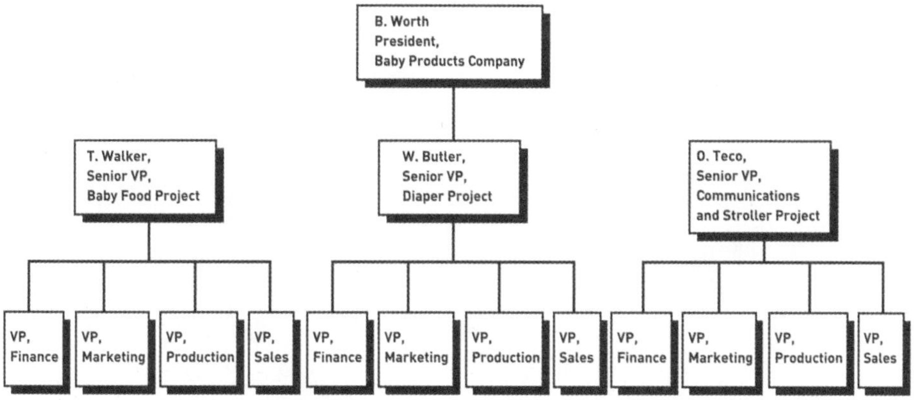

The **team organizational structure** is centered on work teams or groups. In some cases, these teams are small; in others, they are very large. Typically, each team has a leader who reports to an upper-level manager. Depending on its tasks, the team can be temporary or permanent. A healthcare company, for example, can form small teams to organize its administrators, physicians, and others to work with individual patients.

team organizational structure
A structure centered on work teams or groups.

Virtual Organizational Structure and Collaborative Work

A **virtual organizational structure** employs individuals, groups, or complete business units in geographically dispersed areas that can last for a few weeks or years, often requiring telecommunications or the Internet.[6] Virtual teams are employed to ensure the participation of the best available people to solve important organizational problems.

virtual organizational structure
A structure that employs individuals, groups, or complete business units in geographically dispersed areas that can last for a few weeks or years, often requiring telecommunications or the Internet.

These people might be in different countries, operating in different time zones. In other words, virtual organizational structures allow work to be separated from location and time. Work can be done anywhere, anytime. People might never meet physically, which explains the use of the word *virtual,* and highlights the difference between virtual organizations and traditional ones that have operations in more than one location—a virtual organization is geographically distributed, and uses information technology to communicate and coordinate the work. In some cases, a virtual organization is temporary, lasting only a few weeks or months. In others, it can last for years or decades.

Successful virtual organizational structures share key characteristics. One strategy is to have in-house employees concentrate on the firm's core businesses and use virtual employees, groups, or businesses to do everything else. Using information systems to manage the activities of a virtual structure is essential, often requiring specialized software to coordinate joint work. Even with sophisticated IS tools, teams still need face-to-face meetings, especially at the beginning of new projects. Some virtual workers that travel around the country or the globe to different client sites, however, experience increased stress and difficulty in handling

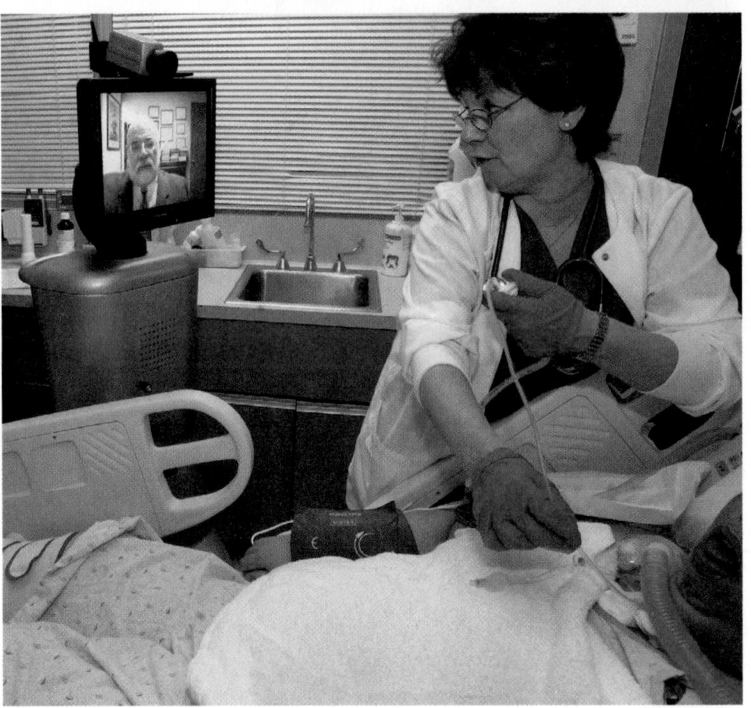

Virtual teams let people consult with experts no matter their physical location, which is especially useful in the healthcare industry.

(Source: Courtesy of AP Photo/Paul Sancya.)

both job and family obligations.[7] This can result in these employees seeking different, more traditional positions to reduce the stress. Some experts have proposed the following for virtual teams.[8]

- When possible, use virtual team members that know each other or use technology that helps virtual team members quickly get to know each other.
- Use virtual team members who are already connected with other people and resources throughout the organization.
- Virtual team projects should be as independent as possible from other projects, so that a delay or problem with one team doesn't affect the progress or success of another virtual team.
- Develop Internet resources that help virtual teams communicate and collaborate on important projects.
- Make sure that virtual team projects are challenging, meaningful, and interesting.

A virtual organizational structure allows *collaborative* work, in which managers and employees can effectively work in groups, even those composed of members from around the world. A management team, for example, can include executives from Australia and England. A programming team can consist of people in the United States and India. Collaborative work can also include all aspects of the supply chain and customer relationship management. An automotive design team, for example, can include critical parts suppliers, engineers from the company, and important customers.

A virtual organizational structure allows collaborative work, in which managers and employees can effectively work in groups, even those composed of members from around the world.

(Source: © Jon Feingersh/Getty Images.)

Organizational Culture and Change

Culture is a set of major understandings and assumptions shared by a group, such as within an ethnic group or a country. **Organizational culture** consists of the major understandings and assumptions for a business, corporation, or other organization. The understandings, which can include common beliefs, values, and approaches to decision making, are often not stated or documented as goals or formal policies. For example, Procter & Gamble has an organizational culture that places an extremely high value on understanding its customers and their needs. For marketing recommendations to be accepted, they must be based on facts known about customers. As another example, employees might be expected to be clean-cut, wear conservative outfits, and be courteous in dealing with all customers. Sometimes organizational culture is formed over years. In other cases, top-level managers can form it rapidly by starting a "casual Friday" dress policy. Organizational culture can also have a positive affect on the successful development of new information systems that support the organization's culture.[9]

culture
A set of major understandings and assumptions shared by a group.

organizational culture
The major understandings and assumptions for a business, corporation, or other organization.

Organizational change deals with how for-profit and nonprofit organizations plan for, implement, and handle change. Change can be caused by internal factors, such as those initiated by employees at all levels, or external factors, such as activities wrought by competitors, stockholders, federal and state laws, community regulations, natural occurrences (such as hurricanes), and general economic conditions. Many European countries, for example, adopted the euro, a single European currency, which changed how financial companies do business and use their information systems. Organizational change also occurs when two or more organizations merge. When organizations merge, however, integrating their information systems can be critical to future success.[10] Unfortunately, many organizations only consider the integration of their different information systems late in the merger process.

organizational change
How for-profit and nonprofit organizations plan for, implement, and handle change.

Change can be sustaining or disruptive.[11] *Sustaining change* can help an organization improve the supply of raw materials, the production process, and the products and services it offers. Developing new manufacturing equipment to make disk drives is an example of a sustaining change for a computer manufacturer. The new equipment might reduce the costs of producing the disk drives and improve overall performance. *Disruptive change,* on the other hand, often harms an organization's performance or even puts it out of business. In general, disruptive technologies might not originally have good performance, low cost, or even strong

If approved, a merger of Sirius Satellite Radio and XM Satellite Radio could significantly increase the number of subscribers for both companies.

(Source: © Dennis Van Tine/Landov.)

demand. Over time, however, they often replace existing technologies. They can cause profitable, stable companies to fail when they don't change or adopt the new technology.

change model
A representation of change theories that identifies the phases of change and the best way to implement them.

The dynamics of change can be viewed in terms of a change model. A **change model** represents change theories by identifying the phases of change and the best way to implement them. Kurt Lewin and Edgar Schein propose a three-stage approach for change (see Figure 2.7). *Unfreezing* is ceasing old habits and creating a climate that is receptive to change. *Moving* is learning new work methods, behaviors, and systems. *Refreezing* involves reinforcing changes to make the new process second nature, accepted, and part of the job.[12] When a company introduces a new information system, a few members of the organization must become agents of change to confront and overcome possible resistance to change. They are champions of the new system and its benefits. Understanding the dynamics of change can help them confront and overcome resistance from employees and others so that the new system can be used to maximum efficiency and effectiveness.

Figure 2.7

A Change Model

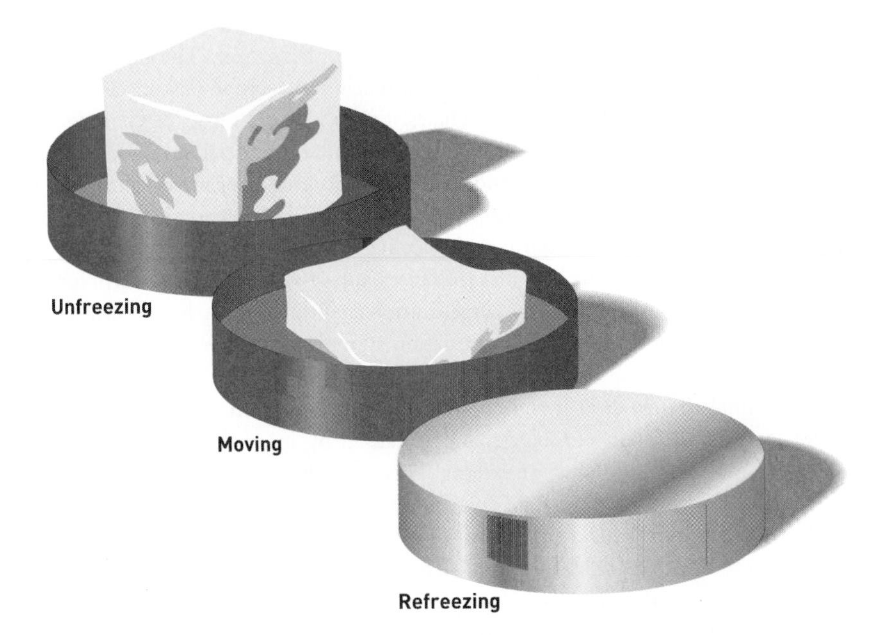

organizational learning
The adaptations to new conditions or alterations of organizational practices over time.

Organizational learning is closely related to organizational change. According to the concept of **organizational learning**, organizations adapt to new conditions or alter their practices over time. Assembly-line workers, secretaries, clerks, managers, and executives learn better ways of doing business and incorporate them into their day-to-day activities. Collectively, these adjustments based on experience and ideas are called organizational learning. In some cases, the adjustments can be a radical redesign of business processes, often called *reengineering*. In other cases, these adjustments can be more incremental, a concept called *continuous improvement*. Both adjustments reflect an organization's *strategy*, the long-term plan of action for achieving their goals.

Reengineering and Continuous Improvement

To stay competitive, organizations must occasionally make fundamental changes in the way they do business. In other words, they must change the activities, tasks, or processes they use to achieve their goals. **Reengineering**, also called **process redesign** and *business process reengineering (BPR),* involves the radical redesign of business processes, organizational structures, information systems, and values of the organization to achieve a breakthrough in business results. Union Bank of California, for example, decided to reengineer many of its tasks and functions.[13] According to CIO Jim Yee, "That the bank asked me to head up the reengineering process and put operations and IT in the same organization is a real statement about the bank's foresight about how IT can be an enabler and is such a critical function for the bank to continue to be successful and competitive in the marketplace." Reengineering can cause changes of an organization's values and information systems (see Figure 2.8). Reengineering can reduce delivery time, increase product and service quality, enhance customer satisfaction, and increase revenues and profitability.

reengineering (process redesign)
The radical redesign of business processes, organizational structures, information systems, and values of the organization to achieve a breakthrough in business results.

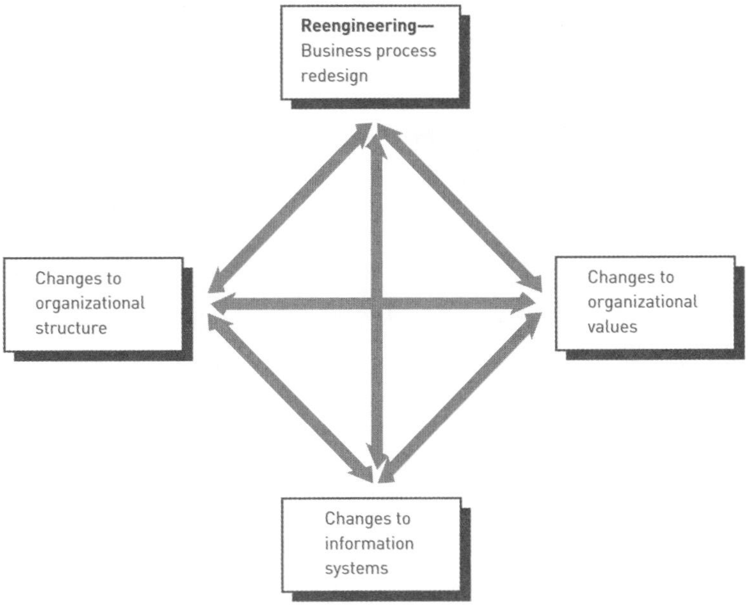

Reengineering—
Business process
redesign

Changes to
organizational
structure

Changes to
organizational
values

Changes to
information
systems

Figure 2.8

Reengineering

Reengineering involves the radical redesign of business processes, organizational structure, information systems, and values of the organization to achieve a breakthrough in business results.

In contrast to simply automating the existing work process, reengineering challenges the fundamental assumptions governing their design. It requires finding and vigorously challenging old rules blocking major business process changes. These rules are like anchors weighing down a firm and keeping it from competing effectively. Table 2.1 provides some examples of such rules.

Table 2.1

Selected Business Rules That Affect Business Processes

Rule	Original Rationale	Potential Problem
Hold small orders until full-truckload shipments can be assembled.	Reduce delivery costs.	Customer delivery is slow—lost sales.
Do not accept an order until customer credit is approved.	Reduce potential for bad debt.	Customer service is poor—lost sales.
Let headquarters make all merchandising decisions.	Reduce number of items carried in inventory.	Customers perceive organization has limited product selection—lost sales.

Union Bank of California, a full-service commercial bank headquartered in San Francisco, recently decided to reengineer many of its IS tasks and functions.

(Source: Courtesy of AP Photo/ Anjum Naveed.)

continuous improvement
Constantly seeking ways to improve business processes to add value to products and services.

In contrast to reengineering, the idea of **continuous improvement** is to constantly seek ways to improve business processes and add value to products and services. This continual change will increase customer satisfaction and loyalty and ensure long-term profitability. Manufacturing companies make continual product changes and improvements. Service organizations regularly find ways to provide faster and more effective assistance to customers. By doing so, these companies increase customer loyalty, minimize the chance of customer dissatisfaction, and diminish the opportunity for competitive inroads. The $200 million annual market for continuous improvement software and systems is expected to grow by almost 13 percent annually through 2010, according to some experts.[14] Table 2.2 compares these two strategies.

Table 2.2

Comparing Business Process Reengineering and Continuous Improvement

Business Process Reengineering	Continuous Improvement
Strong action taken to solve serious problem	Routine action taken to make minor improvements
Top-down change driven by senior executives	Bottom-up change driven by workers
Broad in scope; cuts across departments	Narrow in scope; focus is on tasks in a given area
Goal is to achieve a major breakthrough	Goal is continuous, gradual improvements
Often led by outsiders	Usually led by workers close to the business
Information system integral to the solution	Information systems provide data to guide the improvement team

User Satisfaction and Technology Acceptance

To be effective, reengineering and continuous improvement efforts must result in satisfied users and be accepted and used throughout the organization. Over the years, IS researchers have studied user satisfaction and technology acceptance as they relate to IS attitudes and usage. Although user satisfaction and technology acceptance started as two separate theories, some believe that they can be integrated into one.[15]

User satisfaction with a computer system and the information it generates often depend on the quality of the system and the information.[16] A quality information system is usually flexible, efficient, accessible, and timely. Recall that quality information is accurate, reliable, current, complete, and delivered in the proper format.[17]

The **technology acceptance model** (TAM) specifies the factors that can lead to better attitudes about the information system, along with higher acceptance and usage of the system in an organization.[18] These factors include the perceived usefulness of the technology, the ease of its use, the quality of the information system, and the degree to which the organization supports its use.[19]

technology acceptance model (TAM)
A model that describes the factors leading to higher levels of acceptance and usage of technology.

You can determine the actual usage of an information system by the amount of technology diffusion and infusion.[20] **Technology diffusion** is a measure of how widely technology is spread throughout an organization. An organization in which computers and information systems are located in most departments and areas has a high level of technology diffusion.[21] Some online merchants, such as Amazon.com, have a high diffusion and use computer systems to perform most of their business functions, including marketing, purchasing, and billing. **Technology infusion**, on the other hand, is the extent to which technology permeates an area or department. In other words, it is a measure of how deeply embedded technology is in an area of the organization. Some architectural firms, for example, use computers in all aspects of designing a building from drafting to final blueprints. The design area, thus, has a high level of infusion. Of course, a firm can have a high level of infusion in one part of its operations and a low level of diffusion overall. The architectural firm might use computers in all aspects of design (high infusion in the design area), but not to perform other business functions, including billing, purchasing, and marketing (low diffusion). Diffusion and infusion often depend on the technology available now and in the future, the size and type of the organization, and the environmental factors that include the competition, government regulations, suppliers, and so on. This is often called the technology, organization, and environment (TOE) framework.[22]

technology diffusion
A measure of how widely technology is spread throughout the organization.

technology infusion
The extent to which technology is deeply integrated into an area or department.

Although an organization might have a high level of diffusion and infusion, with computers throughout the organization, this does not necessarily mean that information systems are being used to their full potential. In fact, the assimilation and use of expensive computer technology throughout organizations varies greatly.[23] Companies hope that a high level of diffusion, infusion, satisfaction, and acceptance will lead to greater performance and profitability.[24]

Quality

The definition of the term *quality* has evolved over the years. In the early years of quality control, firms were concerned with meeting design specifications—that is, conforming to standards. If a product performed as designed, it was considered a high-quality product. A product can perform its intended function, however, and still not satisfy customer needs. Today, **quality** means the ability of a product (including services) to meet or exceed customer expectations. For example, a computer that not only performs well but is easy to maintain and repair would be considered a high-quality product. Increasingly, customers expect helpful support after the sale. This view of quality is completely customer oriented. A high-quality product satisfies customers by functioning correctly and reliably, meeting needs and expectations, and being delivered on time with courtesy and respect.

quality
The ability of a product (including services) to meet or exceed customer expectations.

Quality often becomes critical for foreign suppliers and manufacturers.[25] In 2007, a major U.S. toy company had to recall about 1.5 million toys because lead was found in the products likely manufactured in China. According to a representative for the Hong Kong Toys Council, "Everyone is trying to find out which paint supplier is getting into trouble, because once they know, they can stay away from that supplier." In another case, a Russian car magazine asked for a $9,000 Chinese car to be recalled from the Russian market because of

Table 2.3

Total Quality Management and Six Sigma

poor performance in a crash test.[26] Organizations now use techniques to ensure quality, including total quality management and Six Sigma. Six Sigma, for example, was used by Ford-Otosan, a company that makes commercial vehicles, to help save millions of dollars.[27] See Table 2.3.

Technique	Description	Examples
Total Quality Management (TQM)	Involves developing a keen awareness of customer needs, adopting a strategic vision for quality, empowering employees, and rewarding employees and managers for producing high-quality products.[28]	The U.S. Postal Service uses the Mail Preparation Total Quality Management (MPTQM) program to certify leading mail sorting and servicing companies. MAA Bozell, a communications company in India, used TQM to improve the quality for all of its business processes.
Six Sigma	A statistical term that means products and services will meet quality standards 99.9997% of the time. In a normal distribution curve used in statistics, six standard deviations (Six Sigma) is 99.9997% of the area under the curve. Six Sigma was developed at Motorola, Inc. in the mid 1980s.[29]	Transplace, a $57 million trucking and logistics company, uses Six Sigma to improve quality by eliminating waste and unneeded steps. There are a number of training and certification programs for Six Sigma.[30] Six Sigma, however, has been criticized by some.[31]

Outsourcing, On-Demand Computing, and Downsizing

A significant portion of an organization's expenses are used to hire, train, and compensate talented staff. So organizations try to control costs by determining the number of employees they need to maintain high-quality goods and services. Strategies to contain costs are outsourcing, on-demand computing, and downsizing.

outsourcing
Contracting with outside professional services to meet specific business needs.

Outsourcing involves contracting with outside professional services to meet specific business needs. Often, companies outsource a specific business process, such as recruiting and hiring employees, developing advertising materials, promoting product sales, or setting up a global telecommunications network. Organizations often outsource a process to focus more closely on their core business—and target limited resources to meet strategic goals. A KPMG survey on global outsourcing revealed that over 40 percent of the survey respondents believe that outsourcing definitely improved their financial performance, and almost 50 percent reported that outsourcing brought business experience to their companies that they didn't have previously.[32]

Allergan, a developer of pharmaceuticals and medical devices, outsources its IS services, including data center operations and network monitoring and management.

(Source: Courtesy of AP Photo/Chris Carlson.)

Companies that are considering outsourcing to cut the cost of their IS operations need to review this decision carefully, however. A growing number of organizations are finding that outsourcing does not necessarily lead to reduced costs. One of the primary reasons for cost increases is poorly written contracts that tack on charges from the outsourcing vendor for each additional task. Other potential drawbacks of outsourcing include loss of control and flexibility, overlooked opportunities to strengthen core competency, and low employee morale.

on-demand computing
Contracting for computer resources to rapidly respond to an organization's varying workflow. Also called on-demand business and utility computing.

On-demand computing is an extension of the outsourcing approach, and many companies offer it to business clients and customers. **On-demand computing,** also called *on-demand business* and *utility computing,* involves rapidly responding to the organization's flow of work as the need for computer resources varies. It is often called utility computing because the organization pays for computing resources from a computer or consulting company, just as it pays for electricity from a utility company. This approach treats the information system—including hardware, software, databases, telecommunications, personnel, and other components—more as a service than as separate products. In other words, instead of purchasing hardware, software, and database systems, the organization only pays a fee for the systems it needs at peak times. This approach can save money because the organization does not pay for systems that it doesn't routinely need. It also allows the organization's IS staff to concentrate on more-strategic issues.

Downsizing involves reducing the number of employees to cut costs. The term *rightsizing* is also used. Rather than pick a specific business process to downsize, companies usually look to downsize across the entire company. Downsizing clearly reduces total payroll costs, though employee morale can suffer.[33]

downsizing
Reducing the number of employees to cut costs.

Employers need to be open to alternatives for reducing the number of employees but use layoffs as the last resort. It's simpler to encourage people to leave voluntarily through early retirement or other incentives. Voluntary downsizing programs often include a buyout package offered to certain classes of employees (for example, those over 50 years old). The buyout package offers employees certain benefits and cash incentives if they voluntarily retire from the company. Other options are job sharing and transfers.

IBM Global Services provides consulting, technology, and outsourcing services.

(Source: *www.ibm.com*.)

COMPETITIVE ADVANTAGE

A **competitive advantage** is a significant and (ideally) long-term benefit to a company over its competition, and can result in higher-quality products, better customer service, and lower costs. Establishing and maintaining a competitive advantage is complex, but a company's survival and prosperity depend on its success in doing so. An organization often uses its information system to help achieve a competitive advantage. According to Meg McCarthy, "At Aetna, the IT organization is critical to enabling the implementation of our business strategy. I report to the chairman of our company and I am a member of the executive

competitive advantage
A significant and (ideally) long-term benefit to a company over its competition.

committee. In that capacity, I participate in all of the key business conversations/decisions that impact the company strategy and the technology strategy."[34] In his book *Good to Great,* Jim Collins outlines how technology can be used to accelerate companies to greatness.[35] Table 2.4 shows how a few companies accomplished this move. Ultimately, it is not how much a company spends on information systems but how it makes and manages investments in technology. Companies can spend less and get more value.

Table 2.4

How Some Companies Used Technology to Move from Good to Great

(Source: Data from Jim Collins, *Good to Great*, Harper Collins Books, 2001, p. 300.)

Company	Business	Competitive Use of Information Systems
Circuit City	Consumer electronics	Developed sophisticated sales and inventory-control systems to deliver a consistent experience to customers
Gillette	Shaving products	Developed advanced computerized manufacturing systems to produce high-quality products at low cost
Walgreens	Drug and convenience stores	Developed satellite communications systems to link local stores to centralized computer systems
Wells Fargo	Financial services	Developed 24-hour banking, ATMs, investments, and increased customer service using information systems

Factors That Lead Firms to Seek Competitive Advantage

A number of factors can lead to attaining a competitive advantage. Michael Porter, a prominent management theorist, suggested a now widely accepted competitive forces model, also called the **five-forces model**. The five forces include (1) the rivalry among existing competitors, (2) the threat of new entrants, (3) the threat of substitute products and services, (4) the bargaining power of buyers, and (5) the bargaining power of suppliers. The more these forces combine in any instance, the more likely firms will seek competitive advantage and the more dramatic the results of such an advantage will be.

five-forces model
A widely accepted model that identifies five key factors that can lead to attainment of competitive advantage, including (1) the rivalry among existing competitors, (2) the threat of new entrants, (3) the threat of substitute products and services, (4) the bargaining power of buyers, and (5) the bargaining power of suppliers.

Rivalry Among Existing Competitors

Typically, highly competitive industries are characterized by high fixed costs of entering or leaving the industry, low degrees of product differentiation, and many competitors. Although all firms are rivals with their competitors, industries with stronger rivalries tend to have more firms seeking competitive advantage. To gain an advantage over competitors, companies constantly analyze how they use their resources and assets. This *resource-based view* is an approach to acquiring and controlling assets or resources that can help the company achieve a competitive advantage. For example, a transportation company might decide to invest in radio-frequency technology to tag and trace products as they move from one location to another.

Threat of New Entrants

A threat appears when entry and exit costs to an industry are low and the technology needed to start and maintain a business is commonly available. For example, a small restaurant is threatened by new competitors. Owners of small restaurants do not require millions of dollars to start the business, food costs do not decline substantially for large volumes, and food processing and preparation equipment is easily available. When the threat of new market entrants is high, the desire to seek and maintain competitive advantage to dissuade new entrants is also usually high.

In the restaurant industry, competition is fierce because entry costs are low. Therefore, a small restaurant that enters the market can be a threat to existing restaurants.

(Source: © Sergio Pitamitz/Getty Images.)

Threat of Substitute Products and Services

Companies that offer one type of goods or services are threatened by other companies that offer similar goods or services. The more consumers can obtain similar products and services that satisfy their needs, the more likely firms are to try to establish competitive advantage. For example, consider the photographic industry. When digital cameras became popular, traditional film companies had to respond to stay competitive and profitable. Traditional film companies, such as Kodak and others, started to offer additional products and enhanced services, including digital cameras, the ability to produce digital images from traditional film cameras, and Web sites that could be used to store and view pictures.

Bargaining Power of Customers and Suppliers

Large customers tend to influence a firm, and this influence can increase significantly if the customers can threaten to switch to rival companies. When customers have a lot of bargaining power, companies increase their competitive advantage to retain their customers. Similarly, when the bargaining power of suppliers is strong, companies need to improve their competitive advantage to maintain their bargaining position. Suppliers can also help an organization gain a competitive advantage. Some suppliers enter into strategic alliances with firms and eventually act as a part of the company. Suppliers and companies can use telecommunications to link their computers and personnel to react quickly and provide parts or supplies as necessary to satisfy customers. Government agencies are also using strategic alliances. The investigative units of the U.S. Customs and Immigration and Naturalization Service entered into a strategic alliance to streamline investigations.

Strategic Planning for Competitive Advantage

To be competitive, a company must be fast, nimble, flexible, innovative, productive, economical, and customer oriented. It must also align its IS strategy with general business strategies and objectives.[36] Given the five market forces previously mentioned, Porter and others have proposed a number of strategies to attain competitive advantage, including cost leadership, differentiation, niche strategy, altering the industry structure, creating new products and services, and improving existing product lines and services.[37] In some cases, one of these strategies becomes dominant. For example, with a cost leadership strategy, cost can be the key consideration, at the expense of other factors if need be.

- **Cost leadership.** Deliver the lowest possible cost for products and services. Wal-Mart and other discount retailers have used this strategy for years. Cost leadership is often achieved by reducing the costs of raw materials through aggressive negotiations with suppliers, becoming more efficient with production and manufacturing processes, and reducing warehousing and shipping costs. Some companies use outsourcing to cut costs when making products or completing services.

Wal-Mart and other discount retailers have used a cost leadership strategy to deliver the lowest possible price for products and services.

(Source: © Jeff Zelevansky/Getty Images.)

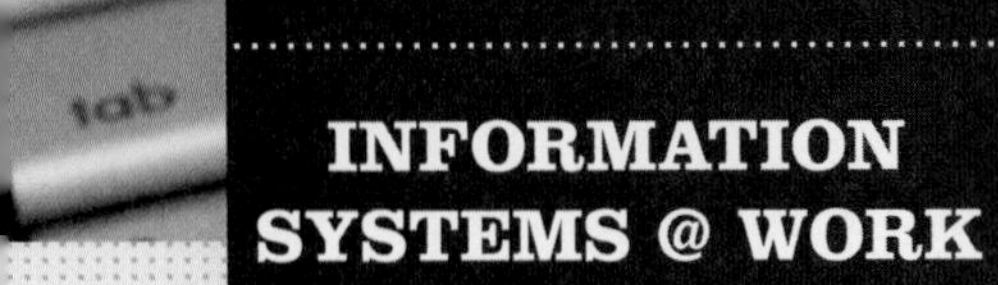

Grand & Toy Seeks Competitive Advantage by Tracking Key Performance Indicators

The landscape of today's global business environment is one of mergers, acquisitions, and relationships. Companies combine and form partnerships to share benefits and extend their reach into new regions and markets. OfficeMax, a dominant global force in retail and commercial office supplies, with close to 1,000 superstores worldwide, extended its reach into Canada by acquiring Grand & Toy, Canada's largest commercial office products company. Rather than rebranding and reorganizing Grand & Toy (named after James Grand and Samuel Toy), OfficeMax allowed the company to operate independently, a wise move since Grand & Toy has proved itself to be a smart company, implementing cutting-edge information systems with impressive results.

Recently, Grand & Toy has been experimenting with new information management tools in hopes of better understanding the ebb and flow of market demand and weaknesses within its organization. Grand & Toy uses two tools from Clarity Systems—a Key Performance Indicator (KPI) reporting application that tracks important data in the system over time, and a tool called the Defector Detector, which indicates when a customer reduces order quantity. Using these tools, Grand & Toy can closely monitor its customers and business to determine when and where it has successes and problems.

Grand & Toy's new information system from Clarity is an example of a trend in business management called corporate performance management (CPM). CPM is an extension of ERP that allows a business to use the information it collects to analyze and improve business practices and processes. Analyzing business processes often depends on analyzing key indicators—specific information within the system that indicates broader trends. For example, many senior managers look at growth, profitability, productivity, and satisfaction as methods of determining the health of a business. Key indicator information can be combined to compile scores for each of these areas. Tracking these areas over time helps managers and employees evaluate the health of the company. Information systems are used to analyze these areas at a variety of levels.

Using corporate performance management tools, decision makers at Grand & Toy can evaluate the company's performance at many levels with little effort. Using the Clarity system, a manager can study sales for a specific geographic area, on a given day, or for the entire corporation for the previous year. Tracking key indicators and adjusting corporate performance accordingly is one way businesses are working to win a competitive advantage.

Discussion Questions

1. What types of data do you think go in to scoring the growth, profitability, productivity, and satisfaction of a company like Grand & Toy?
2. How can Grand & Toy maintain a competitive advantage over other office supply companies that use the same information system from Clarity?

Critical Thinking Questions

1. What key indicators would you watch if you managed an office supply business? Which are internal and which are external?
2. Three hundred employees of Grand & Toy use the Clarity system on a daily basis. Why do you think it is important to provide system access to so many people in the company?

SOURCES: Ruffolo, Rafael, "Grand & Toy Gets Business Performance Clarity," *itWorldCanada*, July 6, 2007, *www.itworldcanada.com/Pages/Docbase/ViewArticle.aspx?ID=idgml-36107f64-e342-4b25-8968-5ef53dd8433e&Portal=2e6e7040-2373-432d-b393-91e487ee7d70&ParaStart=0&ParaEnd=10&direction=next&Next=Next.* Staff, "Retail Case Study: Grand & Toy," Clarity Systems Case Studies, *www.claritysystems.com/Resources/CaseStudies.aspx*, accessed December 26, 2007. Grand & Toy Web site, *www.grandandtoy.com*, accessed December 26, 2007.

- **Differentiation.** Deliver different products and services. This strategy can involve producing a variety of products, giving customers more choices, or delivering higher-quality products and services. Many car companies make different models that use the same basic parts and components, giving customers more options. Other car companies attempt to increase perceived quality and safety to differentiate their products and appeal to consumers who are willing to pay higher prices for these features. Companies that try to differentiate their products often strive to uncover and eliminate counterfeit products produced and delivered by others.[38] Some believe counterfeit products cost companies about $600 billion annually. To distinguish their products from fakes, microscopic particles or other markers are inserted to allow companies, government regulators, and law enforcement agencies to distinguish genuine products from bogus ones.
- **Niche strategy.** Deliver to only a small, niche market. Porsche, for example, doesn't produce inexpensive station wagons or large sedans. It makes high-performance sports cars and SUVs. Rolex only makes high-quality, expensive watches. It doesn't make inexpensive, plastic watches that can be purchased for $20 or less.

Porsche is an example of a company with a niche strategy, producing only high-performance SUVs and sports cars, such as the Carrera.

(Source: © Sajjad Hussain/AFP/ Getty Images.)

- **Altering the industry structure.** Change the industry to become more favorable to the company or organization. The introduction of low-fare airline carriers, such as Southwest Airlines, has forever changed the airline industry, making it difficult for traditional airlines to make high profit margins. To fight back, airlines such as United launched their own low-fare flights. Creating strategic alliances can also alter the industry structure. A **strategic alliance**, also called a **strategic partnership**, is an agreement between two or more companies that involves the joint production and distribution of goods and services.

strategic alliance (strategic partnership)
An agreement between two or more companies that involves the joint production and distribution of goods and services.

- **Creating new products and services.** Introduce new products and services periodically or frequently. This strategy always helps a firm gain a competitive advantage, especially for the computer industry and other high-tech businesses. If an organization does not introduce new products and services every few months, the company can quickly stagnate, lose market share, and decline. Companies that stay on top are constantly developing new products and services. A large U.S. credit-reporting agency, for example, can use its information system to help it explore new products and services in different markets.
- **Improving existing product lines and services.** Make real or perceived improvements to existing product lines and services. Manufacturers of household products are always advertising new and improved products. In some cases, the improvements are more perceived than actual refinements; usually, only minor changes are made to the existing product, such as to reduce the amount of sugar in breakfast cereal. Some direct-order companies are improving their service by using Radio Frequency Identification (RFID) tags to identify and track the location of their products as they are shipped from one location to another. Customers and managers can instantly locate products as they are shipped from suppliers to the company, to warehouses, and finally to customers.
- **Other strategies.** Some companies seek strong *growth* in sales, hoping that it can increase profits in the long run due to increased sales. Being the *first to market* is another competitive strategy. Apple Computer was one of the first companies to offer complete

and ready-to-use personal computers. Some companies offer *customized* products and services to achieve a competitive advantage. Dell, for example, builds custom PCs for consumers. *Hire the best people* is another example of a competitive strategy. The assumption is that the best people will determine the best products and services to deliver to the market and the best approach to deliver these products and services. Companies can also combine one or more of these strategies. In addition to customization, Dell attempts to offer low-cost computers (cost leadership) and top-notch service (differentiation).

PERFORMANCE-BASED INFORMATION SYSTEMS

Businesses have passed through at least three major stages in their use of information systems. In the first stage, organizations focused on using information systems to reduce costs and improve productivity. The National ePrescribing Patient Safety Initiative, for example, offers powerful software to doctors that is used to reduce medication errors and costs.[39] United Airlines uses flight-planning software to help it schedule flights and compute fuel needs based on distance traveled and wind currents.[40] The company estimates that it saves almost $1,500 on fuel for many long-distance flights. In this stage, companies generally ignored the revenue potential, not looking for opportunities to use information systems to increase sales. The second stage was defined by Porter and others. It was oriented toward gaining a competitive advantage. In many cases, companies spent large amounts on information systems and downplayed the costs. Today, companies are shifting from strategic management to performance-based management of their information systems. In this third stage, companies carefully consider both strategic advantage and costs. They use productivity, return on investment (ROI), net present value, and other measures of performance to evaluate the contributions their information systems make to their businesses. Figure 2.9 illustrates these stages. This balanced approach attempts to reduce costs and increase revenues.

Figure 2.9

Three Stages in the Business Use of Information Systems

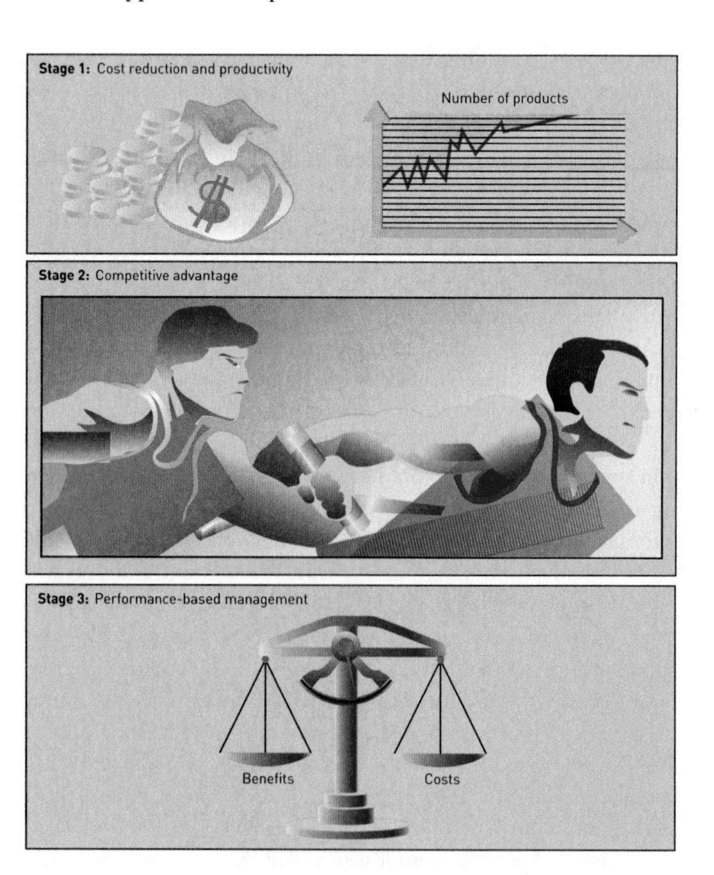

Productivity

Developing information systems that measure and control productivity is a key element for most organizations. **Productivity** is a measure of the output achieved divided by the input required. A higher level of output for a given level of input means greater productivity; a lower level of output for a given level of input means lower productivity. The numbers assigned to productivity levels are not always based on labor hours—productivity can be based on factors such as the amount of raw materials used, resulting quality, or time to produce the goods or service. The value of the productivity number is not as significant as how it compares with other time periods, settings, and organizations. A number of politicians and healthcare professionals hope that keeping electronic medical records (EMRs) on computerized databases will increase the productivity of doctors and healthcare professionals as well as reduce healthcare costs.[41]

productivity
A measure of the output achieved divided by the input required.

Productivity = (Output / Input) × 100%

After a basic level of productivity is measured, an information system can monitor and compare it over time to see whether productivity is increasing. Then, a company can take corrective action if productivity drops below certain levels. An automotive company, for example, might use robots in assembling new cars to increase its labor productivity and reduce costs. In addition to measuring productivity, an information system can be used within a process to significantly increase productivity. Thus, improved productivity can result in faster customer response, lower costs, and increased customer satisfaction. A study of Canadian productivity increases, for example, showed that more than half of the country's productivity gains were caused by improvements in equipment and machinery.[42] Twenty percent was caused by worker improvements.

Return on Investment and the Value of Information Systems

One measure of IS value is **return on investment (ROI)**. This measure investigates the additional profits or benefits that are generated as a percentage of the investment in IS technology. A small business that generates an additional profit of $20,000 for the year as a result of an investment of $100,000 for additional computer equipment and software would have a return on investment of 20 percent ($20,000/$100,000). Because of the importance of ROI, many computer companies provide ROI calculators to potential customers. ROI calculators are typically provided on a vendor's Web site and can be used to estimate returns. According to Megan Burns, an analyst for Forrester Research, "What ROI models allow you to do is run through the what-if scenarios..."[43]

Paris-based PPR, France's biggest clothing retailer, recently acquired Puma AG, Europe's second-biggest sporting goods maker. The merger can help the company create a global brand that straddles sports and fashion.

(Source: Courtesy of AP Photo/ Christof Stache.)

return on investment (ROI)
One measure of IS value that investigates the additional profits or benefits that are generated as a percentage of the investment in IS technology.

Earnings Growth

Another measure of IS value is the increase in profit, or earnings growth, the system brings. For instance, a mail-order company might install an order-processing system that generates a seven percent earnings growth compared with the previous year.

Market Share and Speed to Market

Market share is the percentage of sales that a product or service has in relation to the total market. If installing a new online catalog increases sales, it might help a company increase its market share by 20 percent. Information systems can also help organizations bring new products and services to customers in less time. This is often called speed to market. A music producer, for example, can bring a new song or record to the market by placing it on an online music site faster than it can produce CDs and ship them to retail music stores.

Customer Awareness and Satisfaction

Although customer satisfaction can be difficult to quantify, about half of today's best global companies measure the performance of their information systems based on feedback from internal and external users. Some companies and nonprofit organizations use surveys and questionnaires to determine whether the IS investment has increased customer awareness and satisfaction. Researchers at the University of Auckland, for example, developed surveys and other measurements of an electronic learning system, called CECIL, to determine the satisfaction and experience of students using the electronic learning approach.[44]

Total Cost of Ownership

total cost of ownership (TCO)
The measurement of the total cost of owning computer equipment, including desktop computers, networks, and large computers.

Another way to measure the value of information systems was developed by the Gartner Group and is called the **total cost of ownership (TCO)**. TCO is the sum of all costs over the life of the information system and includes the cost to acquire the technology, technical support, administrative costs, end-user operations, etc. Market research groups often use TCO to compare products and services. For example, a survey of large global enterprises ranked messaging and collaboration software products using the TCO model.[45] TransUnion Interactive, a credit reporting company, uses TCO to rate and select hardware.[46] According to TransUnion's Chief Technology Officer, "Looking at the total cost of ownership and short implementation cycle, Azul's hardware was the best alternative for us, providing minimal downside risk." TCO is also used by many other companies to rate and select hardware, software, databases, and other computer-related components.

Return on investment, earnings growth, market share, customer satisfaction, and TCO are only a few measures that companies use to plan for and maximize the value of their IS investments. Regardless of the difficulties, organizations must attempt to evaluate the contributions that information systems make to assess their progress and plan for the future. Information systems and personnel are too important to leave to chance.

Risk

In addition to the return-on-investment measures of a new or modified system discussed in Chapter 1 and this chapter, managers must also consider the risks of designing, developing, and implementing these systems. Information systems can sometimes be costly failures. Some companies, for example, have attempted to implement ERP systems and failed, costing them millions of dollars. In other cases, e-commerce applications have been implemented with little success. The costs of development and implementation can be greater than the returns from the new system. The risks of designing, developing, and implementing new or modified systems are covered in more detail in Chapters 12 and 13, which discuss systems development.

CAREERS IN INFORMATION SYSTEMS

Realizing the benefits of any information system requires competent and motivated IS personnel, and many companies offer excellent job opportunities. As mentioned in Chapter 1, *knowledge workers (KWs)* are people who create, use, and disseminate knowledge. They are usually professionals in science, engineering, business, and other areas that specialize in information systems. Numerous schools have degree programs with such titles as information systems, computer information systems, and management information systems. These programs are typically offered by information schools, business schools, and within computer science departments. Graduating students with degrees in information systems have attracted high starting salaries. In addition, students are increasingly completing business degrees with a global or international orientation. Skills that some experts believe are important for IS workers to have include the following:[47]

1. Machine learning
2. Mobilizing applications
3. Wireless networking
4. Human-computer interface
5. Project management
6. General networking skills
7. Network convergence technology
8. Open-source programming
9. Business intelligence systems
10. Embedded security
11. Digital home technology integration
12. Languages, including C#, C++, and Java

The IS job market in the early 2000s was tight. Many jobs were lost in U.S. companies as firms merged, outsourced certain jobs overseas, or went bankrupt. Today, demand for IS personnel is on the rise, along with salaries.[48] The U.S. Department of Labor's Bureau of Labor Statistics predicts that many technology jobs will increase through 2012 or beyond. Today, the median salary for IS personnel is almost $80,000, while the average IS manager makes slightly more than $100,000.[49] Table 2.5 summarizes some of the best places to work as an IS professional.[50]

Table 2.5

Best Places to Work as an IS Professional

Source: Brandel, Mary, "Best Places to Work in IT 2007," *Computerworld,* June 18, 2007, p. 34.

Company	Description
Quick Loans	The Internet loan company offers many training opportunities and has excellent promotion practices, benefits, and high retention.
University of Miami	The university is highly rated for diversity. It has good training and career development.
Sharp Health Care	The company has a training budget for IS professionals of about $3,500 annually per person. About half of its managers are women.
The Capital Groups Companies	This investment management firm has a low turnover rate for IS professionals. The company has a reputation for business managers working closely with IS professionals.
The Mitre Corporation	This is a nonprofit organization started at the Massachusetts Institute of Technology (MIT). The organization has good work/life balance programs with flextime, part-time work, and telecommuting.
BAE Systems, Inc.	This large defense company offers many opportunities for IS professionals to work on the latest technologies and computer systems. The company has a training budget for IS professionals of about $30 million.
General Mills, Inc.	As a maker of such brands as Green Giant and Betty Crocker, the company has a successful "Women in IS" program that helps women advance in their IS careers. The company also has an IS Manager's forum that helps IS managers with global issues, project management, and recruitment issues.
University of Pennsylvania	The university offers an excellent child-support program for its IS workers, including its Baby Prep 101 seminar. The university also offers other programs and seminars, including retirement planning.
Anheuser-Busch Companies	This popular brewer offers good vacation programs for its IS professionals. Eligible employees also get free beer or nonalcoholic beverages every month.
Fairfax County Public Schools	This large public school system helps young IS professionals advance in their careers by offering the IT Leadership Development Cohort Program. The school also gives out a number of awards, like the Going The Extra Mile award.

Opportunities in information systems are also available to people from foreign countries, including Russia and India. The U.S. H-1B and L-1 visa programs seek to allow skilled employees from foreign lands into the United States.[51] These programs, however, are limited and in high demand. The L-1 visa program is often used for intracompany transfers for multinational companies. The H-1B program can be used for new employees. In the first few days that applications were available for the H-1B program in 2007, over 130,000 applications were filed for 65,000 positions.[52] The number of H-1B visas offered annually can be political and controversial.[53] Some fear that the H-1B program is being abused to replace high-paid U.S. workers with less expensive foreign workers. Some believe that companies pretend to seek U.S. workers, while actually seeking less expensive foreign workers.[54] In 2007, two U.S. senators on the Senate Judiciary Subcommittee on Immigration sent letters of concern to a number of Indian firms that were using the H-1B program to staff their U.S. operations with IS personnel from foreign countries.[55] One IS professional, concerned about abuses in the H1-B program, got a law degree and is now suing companies that he feels are violating H-1B rules.[56] Others, however, believe the H-1B program and similar programs are invaluable to the U.S. economy and competitiveness.

Roles, Functions, and Careers in IS

IS offers many exciting and rewarding careers, as reported by the Bureau of Labor Statistics and research done by a variety of organizations.[57] Professionals with careers in information systems can work in an IS department or outside a traditional IS department as Web developers, computer programmers, systems analysts, computer operators, and many other positions. In addition to technical skills, they need skills in written and verbal communication, an understanding of organizations and the way they operate, and the ability to work with people and in groups. Today, many good information, business, and computer science schools require these business and communications skills of their graduates.

In general, IS professionals are charged with maintaining the broadest perspective on organizational goals. Most medium to large organizations manage information resources through an IS department. In smaller businesses, one or more people might manage information resources, with support from outsourced services. (Recall that outsourcing is also popular with larger organizations.) As shown in Figure 2.10, the IS organization has three primary responsibilities: operations, systems development, and support.

Operations

People in the operations component of a typical IS department work with information systems in corporate or business unit computer facilities. They tend to focus more on the *efficiency* of IS functions rather than their effectiveness.

System operators primarily run and maintain IS equipment, and are typically trained at technical schools or through on-the-job experience. They are responsible for starting, stopping, and correctly operating mainframe systems, networks, tape drives, disk devices, printers, and so on. Other operations include scheduling, hardware maintenance, and preparing input and output. Data-entry operators convert data into a form the computer system can use. They can use terminals or other devices to enter business transactions, such as sales orders and payroll data. Increasingly, data entry is being automated—captured at the source of the transaction rather than entered later. In addition, companies might have local area network and Web operators who run the local network and any Web sites the company has.

Systems Development

The systems development component of a typical IS department focuses on specific development projects and ongoing maintenance and review. Systems analysts and programmers, for example, address these concerns to achieve and maintain IS effectiveness. The role of a systems analyst is multifaceted. Systems analysts help users determine what outputs they need from the system and construct plans for developing the necessary programs that produce these outputs. Systems analysts then work with one or more programmers to make sure that the appropriate programs are purchased, modified from existing programs, or developed. A computer programmer uses the plans created by the systems analyst to develop or adapt one

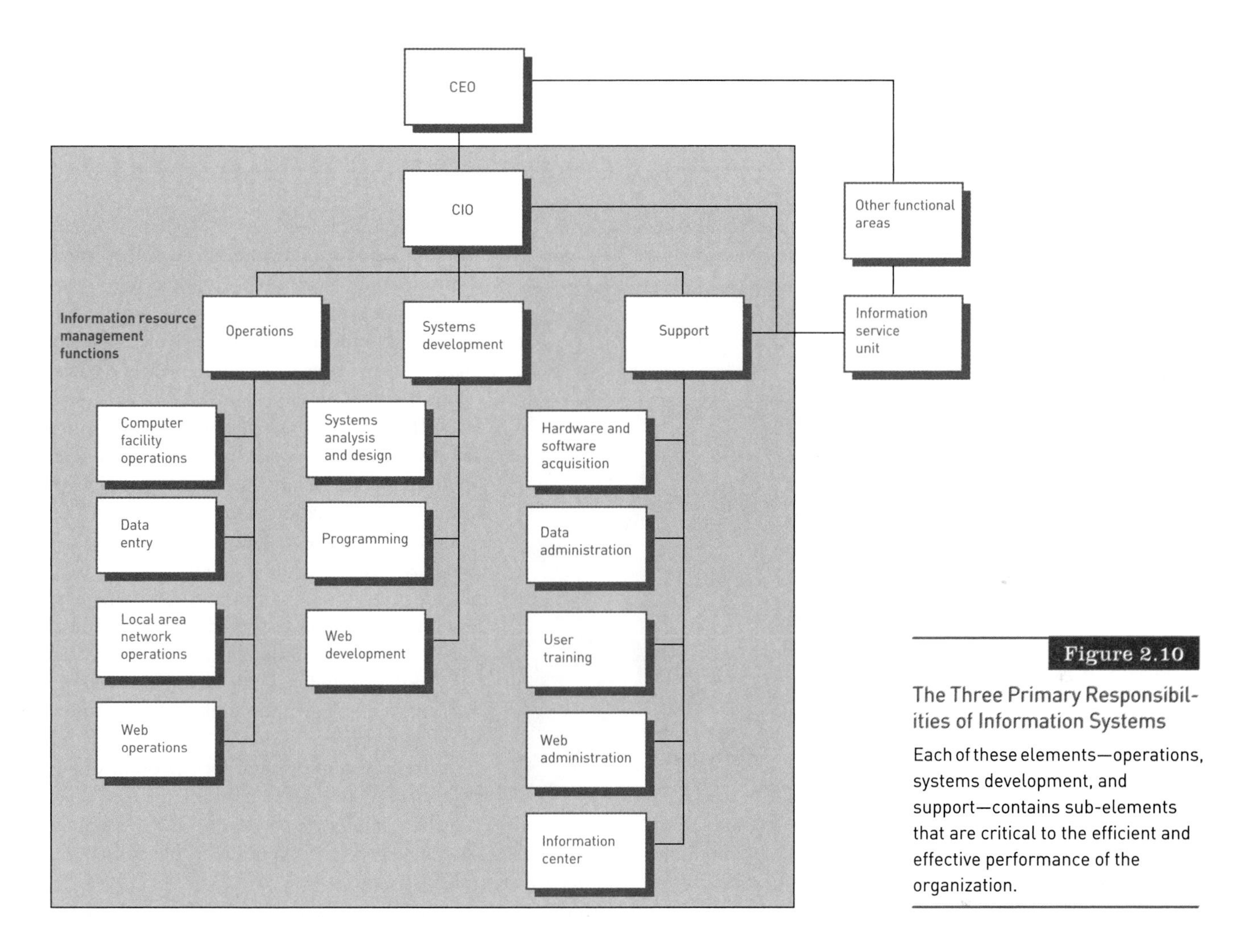

Figure 2.10

The Three Primary Responsibilities of Information Systems

Each of these elements—operations, systems development, and support—contains sub-elements that are critical to the efficient and effective performance of the organization.

or more computer programs that produce the desired outputs. In some cases, foreign companies are actively recruiting IS professionals in the United States. Katrina Anderson, for example, was hired by Infosys, an Indian company, to train in India and return to the U.S. as a programmer.[58] According to Anderson, "The opportunity to train in India was eye-opening, as I came to realize how respected and prominent Infosys is within the country." To help businesses select the best analysts and programmers, companies such as TopCoder offer tests to evaluate the proficiency and competence of current IS employees or job candidates. TopCoder Collegiate Challenge allows programming students to compete with other programmers around the world.[59] Some companies, however, are skeptical of the usefulness of these types of tests.[60]

System operators focus on the efficiency of IS functions, rather than their effectiveness. Their primary responsibility is to run and maintain IS equipment.

(Source: Courtesy of iStockphoto.com.)

With the dramatic increase in the use of the Internet, intranets, and extranets, many companies have Web or Internet developers who create effective and attractive Web sites for customers, internal personnel, suppliers, stockholders, and others who have a business relationship with the company. The Internet is also being used to help with systems development projects.[61] Professor Luis von Ahn of Carnegie Mellon University, for example, has developed an Internet game that shows images to two or more players and asks them to type keywords that describe the image. If the keywords match, points are given in the game. Professor Ahn uses the matches to develop keywords to describe and categorize the images on the Internet. In other words, the game players are doing the work of developing names and categories for images on the Web, an important function that is difficult to do without human judgment and work.

Support

The support component of a typical IS department provides user assistance in hardware and software acquisition and use, data administration, user training and assistance, and Web administration. Increasingly, training is done using the Internet.[62] Microsoft, for example, offers free training in areas including time management, marketing, sales, and other areas (*office.microsoft.com/en-us/officelive/FX102119031033.aspx*). Other companies, such as Hewlett Packard (*www.hp.com/sbso*), also offer online training. In many cases, support is delivered through an information center.

IS personnel provide assistance in hardware and software acquisition, data administration, user training and assistance, and Web administration.

(Source: Courtesy of Christ Schmidt/iStockPhoto.com.)

Because IS hardware and software are costly, a specialized support group often manages computer hardware and software acquisitions. This group sets guidelines and standards for the rest of the organization to follow in making purchases. They must gain and maintain an understanding of available technology and develop good relationships with vendors.

A database administrator focuses on planning, policies, and procedures regarding the use of corporate data and information. For example, database administrators develop and disseminate information about the corporate databases for developers of IS applications. In addition, the database administrator monitors and controls database use.

User training is a key to get the most from any information system, and the support area ensures that appropriate training is available. Training can be provided by internal staff or from external sources. For example, internal support staff can train managers and employees in the best way to enter sales orders, to receive computerized inventory reports, and to submit expense reports electronically. Companies also hire outside firms to help train users in other areas, including the use of word processing, spreadsheets, and database programs.

Web administration is another key area for support staff. With the increased use of the Internet and corporate Web sites, Web administrators are sometimes asked to regulate and monitor Internet use by employees and managers to make sure that it is authorized and appropriate. Web administrators also maintain the corporate Web site to keep it accurate and current, which can require substantial resources.

information center
A support function that provides users with assistance, training, application development, documentation, equipment selection and setup, standards, technical assistance, and troubleshooting.

The support component typically operates the information center. An **information center** provides users with assistance, training, application development, documentation, equipment selection and setup, standards, technical assistance, and troubleshooting. Although many firms have attempted to phase out information centers, others have changed their focus from technical training to helping users find ways to maximize the benefits of the information resource.

Information Service Units

information service unit
A miniature IS department.

An **information service unit** is basically a miniature IS department attached and directly reporting to a functional area in a large organization. Notice the information service unit shown in Figure 2.10. Even though this unit is usually staffed by IS professionals, the project assignments and the resources necessary to accomplish these projects are provided by the functional area to which it reports. Depending on the policies of the organization, the salaries of IS professionals staffing the information service unit might be budgeted to either the IS department or the functional area.

Typical IS Titles and Functions

The organizational chart shown in Figure 2.10 is a simplified model of an IS department in a typical medium-sized or large organization. Many organizations have even larger departments, with increasingly specialized positions such as librarian or quality assurance manager. Smaller firms often combine the roles shown in Figure 2.10 into fewer formal positions.

The Chief Information Officer

The role of the chief information officer (CIO) is to employ an IS department's equipment and personnel to help the organization attain its goals.[63] The CIO is usually a vice president concerned with the overall needs of the organization, and sets corporate-wide policies and plans, manages, and acquires information systems. In one survey, more than 60 percent of CIOs reported directly to the president of the company or the chief executive officer (CEO).[64] Some of the CIO's top concerns include integrating IS operations with corporate strategies, keeping up with the rapid pace of technology, and defining and assessing the value of systems development projects. According to a survey, almost 80 percent of CIOs are actively involved in or consulted on most major decisions.[65] Tom Shelman, CIO for Northrop Grumman Corporation, for example, changed his job description to be more strategic, meet with customers, and to help win new business.[66]

The high level of the CIO position reflects that information is one of the organization's most important resources. A CIO works with other high-level officers of an organization, including the chief financial officer (CFO) and the chief executive officer (CEO), in managing and controlling total corporate resources. CIOs must also work closely with advisory committees, stressing effectiveness and teamwork and viewing information systems as an integral part of the organization's business processes—not an adjunct to the organization. Thus, CIOs need both technical and business skills. The CIO is also becoming more involved with the customers of their company.[67] According to Tom Gosnell, CIO and senior vice president at CUNA Mutual Group, "Helping customers to become successful is very much the job of today's IT department." For federal agencies, the Clinger-Cohen Act of 1996 requires that a CIO coordinate the purchase and management of information systems.

A company's CIO is usually a vice president who sets corporate-wide policies, and plans, manages, and acquires information systems.

(Source: © Click Productions/Getty Images.)

Depending on the size of the IS department, several people might work in senior IS managerial levels. Some job titles associated with IS management are the CIO, vice president of information systems, manager of information systems, and chief technology officer (CTO). A central role of all these people is to communicate with other areas of the organization to determine changing needs. Often these employees are part of an advisory or steering committee that helps the CIO and other IS managers make decisions about the use of information systems. Together they can best decide what information systems will support corporate goals. The CTO, for example, typically works under a CIO and specializes in networks and related equipment and technology.

LAN Administrators

Local area network (LAN) administrators set up and manage the network hardware, software, and security processes. They manage the addition of new users, software, and devices to the network. They also isolate and fix operations problems. LAN administrators are in high demand and often solve both technical and nontechnical problems.

Internet Careers

The bankruptcy of some Internet start-up companies in the early 2000s, called the *dot-gone era* by some, has resulted in layoffs for some firms. Executives of these bankrupt start-up Internet companies lost hundreds of millions of dollars in a few months. Yet, the use of the Internet to conduct business continues to grow and has stimulated a steady need for skilled personnel to develop and coordinate Internet usage. As shown in Figure 2.10, these careers are in the areas of Web operations, Web development, and Web administration. As with other areas in IS, many top-level administrative jobs are related to the Internet. These career opportunities are found in both traditional companies and those that specialize in the Internet.

Internet jobs within a traditional company include Internet strategists and administrators, Internet systems developers, Internet programmers, and Internet or Web site operators. Some companies suggest a new position, chief Internet officer, with responsibilities and a salary similar to the CIO's.

In addition to traditional companies, Internet companies offer exciting career opportunities. These companies include Amazon.com, Yahoo!, eBay, and many others. Systest, for example, specializes in finding and eliminating digital bugs that could halt the operation of a computer system.[68]

certification
A process for testing skills and knowledge, which results in a statement by the certifying authority that confirms an individual is capable of performing a particular kind of job.

Often, the people filling IS roles have completed some form of certification. **Certification** is a process for testing skills and knowledge resulting in an endorsement by the certifying authority that an individual is capable of performing a particular job. Certification frequently involves specific, vendor-provided or vendor-endorsed coursework. Popular certification programs include Microsoft Certified Systems Engineer, Certified Information Systems Security Professional (CISSP), Oracle Certified Professional, Cisco Certified Security Professional (CCSP), and many others.[69]

Other IS Careers

To respond to the increase in attacks on computers, new and exciting careers have developed in security and fraud detection and prevention. Today, many companies have IS security positions such as a chief information security officer or a chief privacy officer. Some universities offer degree programs in security or privacy. The National Insurance Crime Bureau, a nonprofit organization supported by roughly 1,000 property and casualty insurance companies, uses computers to join forces with special investigation units and law enforcement agencies as well as to conduct online fraud-fighting training to investigate and prevent these types of crimes.[70] The University of Denver has offered a program in video-game development.[71] It is even possible to work from home in an IS field. Programmers, systems developers, and others are also working from home in developing new information systems.

In addition to working for an IS department in an organization, IS personnel can work for large consulting firms, such as Accenture (*www.accenture.com*), IBM (*www.ibm.com/services*), EDS (*www.eds.com*), and others.[72] Some consulting jobs can entail frequent travel because consultants are assigned to work on various projects wherever the client is. Such roles require excellent project management and people skills in addition to IS technical skills.

Other IS career opportunities include being employed by technology companies, such as Microsoft (*www.microsoft.com*), Google (*www.google.com*), Dell (*www.dell.com*), and many others. Such a role enables an individual to work on the cutting edge of technology, which can be extremely challenging and exciting. As some computer companies cut their services to customers, new companies are being formed to fill the need. With names such as Speak with a Geek and Geek Squad located in many Best Buy stores, these companies are helping people and organizations with their computer-related problems that computer vendors are no longer solving.

Finding a Job in IS

There are many traditional approaches to finding a job in the information systems area, including on-campus visits from recruiters and referrals from professors, friends, and family members. Many colleges and universities have excellent programs to help students develop résumés and conduct job interviews. Developing an online résumé can be critical to finding a good job. Many companies accept résumés online and use software to search for keywords and skills used to screen job candidates.[73] Thus, having the right keywords and skills can mean the difference between getting a job interview and not being considered.

Increasingly, students are using the Internet and other sources to find IS jobs. Many Web sites, such as Monster.com, post job opportunities for Internet careers and more traditional careers. Most large companies list job opportunities on their Web sites. These sites allow prospective job hunters to browse job opportunities, locations, salaries, benefits, and other factors. In addition, some sites allow job hunters to post their résumés. Many of the social networking sites, including MySpace and Facebook, can be used to help get job leads. Facebook and Jobster are investigating a joint venture to launch a job site available on Facebook's Internet site.[74] Corporate recruiters also use the Internet or Web logs (blogs) to gather information on existing job candidates or to locate new job candidates.[75] According to Ryan Loken, an executive at Wal-Mart Stores, Inc., "Blogs are a tool in the tool kit." Loken has used the Internet and blogs to fill over 100 corporate positions. To fill more than 5,000 intern positions and entry-level jobs, Ernst & Young LLP uses the Internet, blogs, and social networking sites such as Facebook and MySpace. The company has a page on the Facebook Internet site to attract job candidates.[76] "It's a very good thing for communicating with potential job seekers. You're reaching the student in their lair," says Mark Mehler of CareerXroads.

Internet job sites such as Monster.com allow job hunters to browse job opportunities and post their résumés.

(Source: *www.monster.com*.)

SUMMARY

Principle

The use of information systems to add value to the organization is strongly influenced by organizational structure, culture, and change.

Organizations use information systems to support their goals. Because information systems typically are designed to improve productivity, organizations should devise methods for measuring the system's impact on productivity.

An organization is a formal collection of people and other resources established to accomplish a set of goals. The primary goal of a for-profit organization is to maximize shareholder value. Nonprofit organizations include social groups, religious groups, universities, and other organizations that do not have profit as the primary goal.

Organizations are systems with inputs, transformation mechanisms, and outputs. Value-added processes increase the relative worth of the combined inputs on their way to becoming final outputs of the organization. The value chain is a series (chain) of activities that includes (1) inbound logistics, (2) warehouse and storage, (3) production, (4) finished product storage, (5) outbound logistics, (6) marketing and sales, and (7) customer service.

Organizational structure refers to how organizational subunits relate to the overall organization. Several basic organizational structures include traditional, project, team, and virtual. A virtual organizational structure employs individuals, groups, or complete business units in geographically dispersed areas. These can involve people in different countries operating in different time zones and different cultures.

Organizational culture consists of the major understandings and assumptions for a business, corporation, or organization. Organizational change deals with how profit and nonprofit organizations plan for, implement, and handle change. Change can be caused by internal or external factors. The stages of the change model are unfreezing, moving, and refreezing. According to the concept of organizational learning, organizations adapt to new conditions or alter practices over time.

Principle

Because information systems are so important, businesses need to be sure that improvements or completely new systems help lower costs, increase profits, improve service, or achieve a competitive advantage.

Business process reengineering involves the radical redesign of business processes, organizational structures, information systems, and values of the organization to achieve a breakthrough in results. Continuous improvement to business processes can add value to products and services.

The extent to which technology is used throughout an organization can be a function of technology diffusion, infusion, and acceptance. Technology diffusion is a measure of how widely technology is in place throughout an organization. Technology infusion is the extent to which technology permeates an area or department. User satisfaction with a computer system and the information it generates depends on the quality of the system and the resulting information. The technology acceptance model (TAM) investigates factors—such as the perceived usefulness of the technology, the ease of use of the technology, the quality of the information system, and the degree to which the organization supports the use of the information system—to predict IS usage and performance.

Total quality management consists of a collection of approaches, tools, and techniques that fosters a commitment to quality throughout the organization. Six Sigma is often used in quality control. It is based on a statistical term that means products and services will meet quality standards 99.9997 percent of the time.

Outsourcing involves contracting with outside professional services to meet specific business needs. This approach allows the company to focus more closely on its core business and to target its limited resources to meet strategic goals. Downsizing involves reducing the number of employees to reduce payroll costs; however, it can lead to unwanted side effects.

Competitive advantage is usually embodied in either a product or service that has the most added value to consumers and that is unavailable from the competition or in an internal system that delivers benefits to a firm not enjoyed by its competition. A five-forces model covers factors that lead firms to seek competitive advantage: the rivalry among existing competitors, the threat of new market entrants, the threat of substitute products and services, the bargaining power of buyers, and the bargaining power of suppliers. Strategies to address these factors and to attain competitive advantage include cost leadership, differentiation, niche strategy, altering the industry structure, creating new products and services, improving existing product lines and services, and other strategies.

Developing information systems that measure and control productivity is a key element for most organizations. A useful measure of the value of an IS project is return on investment (ROI). This measure investigates the additional profits or benefits that are generated as a percentage of the investment in IS technology. Total cost of ownership (TCO) can also be a useful measure.

Principle

Cooperation between business managers and IS personnel is the key to unlocking the potential of any new or modified system.

Information systems personnel typically work in an IS department that employs a chief information officer, chief technology officer, systems analysts, computer programmers, computer operators, and other personnel. The chief information officer (CIO) employs an IS department's equipment and personnel to help the organization attain its goals. The chief technology officer (CTO) typically works under a CIO and specializes in hardware and related equipment and technology. Systems analysts help users determine what outputs they need from the system and construct the plans needed to develop the necessary programs that produce these outputs. Systems analysts then work with one or more programmers to make sure that the appropriate programs are purchased, modified from existing programs, or developed. The major responsibility of a computer programmer is to use the plans developed by the systems analyst to build or adapt one or more computer programs that produce the desired outputs.

Computer operators are responsible for starting, stopping, and correctly operating mainframe systems, networks, tape drives, disk devices, printers, and so on. LAN administrators set up and manage the network hardware, software, and security processes. Trained personnel are also needed to set up and manage a company's Internet site, including Internet strategists, Internet systems developers, Internet programmers, and Web site operators. Information systems personnel can also support other functional departments or areas.

In addition to technical skills, IS personnel need skills in written and verbal communication, an understanding of organizations and the way they operate, and the ability to work with people (users). In general, IS personnel are charged with maintaining the broadest enterprise-wide perspective.

Besides working for an IS department in an organization, IS personnel can work for a large consulting firm, such as Accenture, IBM, EDS, and others. Developing or selling products for a hardware or software vendor is another IS career opportunity.

CHAPTER 2: SELF-ASSESSMENT TEST

The use of information systems to add value to the organization is strongly influenced by organizational structure, culture, and change.

1. Customer relationship management can help a company determine what supplies and equipment are required for the value chain. True or False?
2. A(n) ________________ is a formal collection of people and other resources established to accomplish a set of goals.
3. User satisfaction with a computer system and the information it generates often depends on the quality of the system and the resulting information. True or False?
4. The concept in which organizations adapt to new conditions or alter their practices over time is called ________________.
   a. organizational learning
   b. organizational change
   c. continuous improvement
   d. reengineering

Because information systems are so important, businesses need to be sure that improvements or completely new systems help lower costs, increase profits, improve service, or achieve a competitive advantage.

5. ________________ involves contracting with outside professional services to meet specific business needs.
6. Today, quality means ________________.
   a. achieving production standards
   b. meeting or exceeding customer expectations
   c. maximizing total profits
   d. meeting or achieving design specifications
7. Technology infusion is a measure of how widely technology is spread throughout an organization. True or False?
8. Reengineering is also called ________________.
9. What is a measure of the output achieved divided by the input required?
   a. efficiency
   b. effectiveness
   c. productivity
   d. return on investment
10. ________________ is a measure of the additional profits or benefits generated as a percentage of the investment in IS technology.

Cooperation between business managers and IS personnel is the key to unlocking the potential of any new or modified system.

11. Who is involved in helping users determine what outputs they need and constructing the plans required to produce these outputs?
    a. CIO
    b. applications programmer
    c. systems programmer
    d. systems analyst
12. An information center provides users with assistance, training, and application development. True or False?
13. The ________________ is typically in charge of the IS department or area in a company.

CHAPTER 2: SELF-ASSESSMENT TEST ANSWERS

(1) False (2) organization (3) True (4) a (5) Outsourcing (6) b (7) False (8) process redesign (9) c (10) Return on investment (11) d (12) True (13) chief information officer (CIO)

REVIEW QUESTIONS

1. What is the value chain?
2. What is supply chain management?
3. What role does an information system play in today's organizations?
4. What is reengineering? What are the potential benefits of performing a process redesign?
5. What is user satisfaction?
6. What is the difference between reengineering and continuous improvement?
7. What is the difference between technology infusion and technology diffusion?
8. What is quality? What is total quality management (TQM)?
9. What are organizational change and organizational learning?
10. List and define the basic organizational structures.
11. Sketch and briefly describe the three-stage organizational change model.
12. What is downsizing? How is it different from outsourcing?
13. What are some general strategies employed by organizations to achieve competitive advantage?
14. What are several common justifications for implementing an information system?
15. Define the term *productivity*. How can a company best use productivity measurements?
16. What is on-demand computing? What two advantages does it offer to a company?
17. What is the total cost of ownership?
18. Describe the role of the CIO.

DISCUSSION QUESTIONS

1. You have been hired to work in the IS area of a manufacturing company that is starting to use the Internet to order parts from its suppliers and offer sales and support to its customers. What types of Internet positions would you expect to see at the company?
2. You have decided to open an Internet site to buy and sell used music CDs to other students. Describe the supply chain for your new business.
3. What sort of IS career would be most appealing to you—working as a member of an IS organization, consulting, or working for an IT hardware or software vendor? Why?
4. What are the advantages of using a virtual organizational structure? What are the disadvantages?
5. How would you measure user satisfaction with a registration program at a college or university? What are the important features that would make students and faculty satisfied with the system?
6. You have been asked to participate in preparing your company's strategic plan. Specifically, your task is to analyze the competitive marketplace using Porter's five-forces model. Prepare your analysis, using your knowledge of a business you have worked for or have an interest in working for.
7. Based on the analysis you performed in Discussion Question 6, what possible strategies could your organization adopt to address these challenges? What role could information systems play in these strategies? Use Porter's strategies as a guide.
8. There are many ways to evaluate the effectiveness of an information system. Discuss each method and describe when one method would be preferred over another method.
9. Assume you are the manager of a retail store and need to hire a CIO to run your new computer system. What characteristics would you want in a new CIO?

PROBLEM-SOLVING EXERCISES

1. Identify three companies that make the highest quality products or services for an industry of your choice. Find the number of employees, total sales, total profits, and earnings growth rate for these three firms. Using a database program, enter this information for the last year. Use the database to generate a report of the three companies with the highest earnings growth rate. Use your word processor to create a document that describes these firms. What other measures would you use to determine which is the best company in terms of future profit potential? Does high quality always mean high profits?
2. A new IS project has been proposed that will produce not only cost savings but also an increase in revenue. The initial costs to establish the system are estimated to be $500,000. The remaining cash flow data is presented in the following table.

	Year 1	Year 2	Year 3	Year 4	Year 5
Increased Revenue	$0	$100	$150	$200	$250
Cost Savings	$0	$ 50	$ 50	$ 50	$ 50
Depreciation	$0	$ 75	$ 75	$ 75	$ 75
Initial Expense	$500				

Note: All amounts are in 000s.

   a. Using a spreadsheet program, calculate the return on investment (ROI) for this project. Assume that the cost of capital is 7 percent.
   b. How would the rate of return change if the project delivered $50,000 in additional revenue and generated cost savings of $25,000 in the first year?
3. Using a word processing program, write a detailed job description of a CIO for a medium-sized manufacturing company. Use a graphics program to make a presentation on the requirements for the new CIO.

TEAM ACTIVITIES

1. With your team, interview one or more managers at a local manufacturing company. Describe the company's supply chain. How does the company handle customer relationship management?
2. With your team, research a firm that has achieved a competitive advantage. Write a brief report that describes how the company was able to achieve its competitive advantage.

WEB EXERCISES

1. This book emphasizes the importance of information. You can get information from the Internet by going to a specific address, such as *www.ibm.com*, *www.whitehouse.gov*, or *www.fsu.edu*. This will give you access to the home page of the IBM corporation, the White House, or Florida State University, respectively. Note that "com" is used for businesses or commercial operations, "gov" is used for governmental offices, and "edu" is used for educational institutions. Another approach is to use a search engine, which is a Web site that allows you to enter keywords or phrases to find information. Yahoo!, developed by two Tulane University students, was one of the first search engines on the Internet. You can also locate information through lists or menus. The search engine will return other Web sites (hits) that correspond to a search request. Using Yahoo! at *www.yahoo.com*, search for information about a company or topic discussed in Chapter 1 or 2. You might be asked to develop a report or send an e-mail message to your instructor about what you found.
2. Use the Internet to search for information about a systems development failure. You can use a search engine, such as Google, or a database at your college or university. Write a brief report describing what you found. What were the mistakes that caused the failure?

CAREER EXERCISES

1. Organizations can use traditional, project, virtual, and other organizational structures. For the career of your choice, describe which organizational structure or structures are likely to be used and how computers and information systems can help you communicate and work with others in this structure.
2. Pick the five best companies for your career. Describe the quality of the products and services offered by each company. Are the best companies always the ones with the highest quality products and services?

CASE STUDIES

Case One

Customer Service Drives Information Systems at Volvo Cars Belgium

You can often gauge the role of information systems within a business by evaluating how the chief information officer (CIO), or head of information systems, interacts professionally with the president, chief executive officer (CEO), and other high-ranking decision makers. At Volvo Cars Belgium, the company's Brussels-based subsidiary, the head of information systems, Michelangelo Adamo, who goes by the title "IT supervisor," reports directly to the customer service manager. This relationship is unique since most individuals in Michelangelo's position would report to the CEO, president, or even chief operating officer (COO) or chief financial officer (CFO).

The relationship of Volvo's IT supervisor to the customer service manager indicates Volvo's strong emphasis on customer service as a primary goal of the business. It also suggests a belief that information systems should always be designed in a manner that provides additional service to the customer. While Michelangelo is the IT supervisor for the 65-dealer network, his ultimate goal is to sell more cars, and to do so, he needs to make and keep customers happy. Consider the following example of customer service-driven information systems.

Recently, Volvo Cars Belgium completed a massive information system implementation that connects the 65 Belgium Volvo dealers over a common network, and to Volvo headquarters in Goteborg, Sweden. The overarching goal of the project was to improve customer service. Through the new system, Volvo Cars Belgium is able to track a single car through its entire lifecycle; from the moment the order is placed, to delivery, to after-sale maintenance and customer service. This allows Volvo to ensure that customers get the right service at the right time, regardless of what dealership they visit. The system also speeds up repairs by allowing a dealership to order parts for a vehicle from Volvo suppliers at the same time the service appointment is made.

The system also allows Volvo Cars Belgium to easily collect sales and service data and apply standardized performance metrics across the entire dealer network. This data is passed on to Volvo headquarters where it is used to target dealerships that need improvement and reward dealerships that excel.

Volvo Cars Belgium's new integrated information system illustrates the emphasis that the company places on customer service. The time and investment in the new system has paid off, with customer-approval ratings increasing for the first time in years.

Discussion Questions

1. Compare how the goals of information systems may differ if driven by the CEO, CIO, customer service, or equally by all business areas.
2. How did joining the dealerships with a common data network help Volvo Cars Belgium improve customer satisfaction?

Critical Thinking Questions

1. What other information systems could assist an auto dealer in improving customer service?
2. How can an auto dealer judge customer satisfaction, and how can information systems be used to compile an overall customer-approval rating for a dealership?

SOURCES: Briody, Dan, "How SOA Boosts Customer Service," *CIO Insight*, March 7, 2007, *www.cioinsight.com/article2/0,1540,2104339,00.asp*. Volvo Cars Belgium Web site, *www.volvocars.be*, accessed December 27, 2007.

Case Two

CIO Plays Important Role at J&J Philippines

Johnson & Johnson is the world's most "comprehensive and broadly based manufacturer of health care products, as well as a provider of related services, for the consumer, pharmaceutical, and medical devices and diagnostic markets" according to its Web site. Consider what it must be like to be the chief information officer (CIO) of a division of Johnson & Johnson located in a developing country. On one hand, you are affiliated with one of the most technologically-advanced corporations in the world. On the other, you are working to support operations in a poor and technologically-young environment. Such is the task of Sadiq Rowther, regional IT director for Johnson & Johnson ASEAN, Phillipines.

The Phillipines division of Johnson & Johnson is well aware of the importance of information systems to a business's success. Sadiq Rowther is involved in making all of the company's key business decisions. While some business executives still perceive an IS department as a back-office operation that provides support functions, smart businesses, like Johnson & Johnson are including the CIO in top-level decisions.

CIO Sadiq Rowther participates in customer interfaces in order to better understand the issues the company faces. He believes that a CIO must think like a business owner. "The value I bring to the leadership team is really how IT can seize a business opportunity and bring about a solution that uses a combination of both business and IT skills," Rowther explained in a *Computerworld* interview.

Johnson & Johnson Philippines utilizes information systems to automate its core functions including supply chain, order processing, and finance. An enterprise resource planning (ERP) system from SAP ties together all the systems into one cohesive system. Unfortunately, much of the corporation's IT budget is used in systems maintenance, and it is difficult to find funding for innovative advances. Sadiq Rowther implements new systems by tying them to organizational goals. "Whatever we are doing in IT has to help grow the business," Rowther is quoted as saying. In fact, rather than calling them information system or IT projects, Johnson & Johnson calls them IT-enabled business projects.

An example of some IT-enabled business projects includes the recently launched Neutrogena Philippines Web site (*www.neutrogena.com.ph*). Rowther looks at the project as a "direct-to-consumer approach for targeted marketing," and collaborated closely with the brand team to make sure the site was effective. In another project, Rowther worked closely with sales and marketing to streamline the ordering process for Johnson & Johnson distributors. The resulting system provided faster and more reliable order processing. Rowther is currently working on a project to get smart phones in the hands of the sales force and provide mobile software tools to help Johnson & Johnson sales and service representatives be more effective.

Working on a shoestring budget, Rowther has to constantly prioritize the company's information system projects and decide which are financially feasible. He teaches his team to continuously check for the "value-add to the business" when considering information system expenditures. Each project has to show a return on investment (ROI) in the near term. "At the end of the day, it's all about aligning with the priorities of the business and ensuring that it is not your choice of projects to keep or throw but a decision jointly made with the rest of the business," concludes Rowther.

Discussion Questions

1. Why have CIOs become important contributors to corporate strategies?
2. How might a CIO with a larger budget have an advantage over Sadiq Rowther at Johnson & Johnson Philippines?

Critical Thinking Questions

1. If you were CIO at Johnson & Johnson Philippines, how might you convince the corporation to invest in a project you designed?
2. Johnson & Johnson Philippines, like the country, is growing rapidly. What considerations should Rowther take into account when planning the information system budget for the next five years? What about considerations for current investments in information systems?

SOURCES: Rubio, Jenalyn, "Reinforcing IT," *Computerworld Philippines*, July 16, 2007, *www.computerworld.com.ph/Default.aspx?_s=4&_ss=P&P=3&PN=4775&L=S&II=33 3&ID=S,333,BYB,BYB-13*. Johnson & Johnson Web site, *www.jnj.com/our_company/index.htm*, accessed December 26, 2007. Neutrogena Philippines Web site, *www.neutrogena.com.ph*, accessed December 26, 2007.

Questions for Web Case

See the Web site for this book to read about the Whitmann Price Consulting case for this chapter. The following questions concern this Web case.

Whitmann Price Consulting: Addressing the Needs of the Organization

Discussion Questions

1. Compare and contrast the benefits and shortcomings of notebook PCs and handheld PCs in accessing information on the road.
2. If Josh and Sandra are able to provide a solution that meets all of the requirements, how might it impact the effectiveness and efficiency of Whitmann Price?
3. Why is it important for Josh and Sandra to build a list of requirements through interviews with stakeholders?

Critical Thinking Questions

1. The six business units of Whitmann Price have some needs for this new system in common and others that are unique. Do you think it will be possible to satisfy all of them with one solution? What are the possibilities for customizing a solution that meets all the requirements?
2. Why might Whitmann Price prefer one common device for all business units rather than unique devices for each unit?

NOTES

Sources for the opening vignette: Lau, Kathleen, "FedEx Canada's got the package," *itWorld Canada*, July 5, 2007, *www.itworldcanada.com/a/Enterprise-Business-Applications/162f033b-5dd7-4362-abdb-1155a90b0a37.html*. Brandel, Mary, "Global Web Sites Go Native: How to Make Yours Work for the Locals," *Computerworld*, November 19, 2007, *www.computerworld.com/action/article.do?command=viewArticleBasic&taxonomyName=internet_applications&articleId=306870&taxonomyl d=168&intsrc=kc_feat*. Staff, "Inside FedEx," FedEx Web site, *news.van.fedex.com/inside*.

1 McKenzie, P. and Jayanthi, S., "Ball Aerospace Explores Operational and Financial Trade-Offs in Batch Sizing in Implementing JIT," *Interfaces*, March 2007, p. 108.
2 Craighead, C., et al., "The Severity of Supply Chain Disruptions," *Decision Sciences*, February 2007, p. 131.
3 Abboud, Leila, "Global Suppliers Play Catch-Up in Information Age," *The Wall Street Journal*, January 4, 2007, p. B3.
4 Staff, "Average Customer Relationship Management Program Takes 16.2 Months to Develop," *Science Letter*, May 1, 2007, p. 600.
5 Planas, Antonio, "Results Cause for Optimism," *Las Vegas Review-Journal*, June 14, 2007, p. 1A.
6 Konana, Bin, et al., "Competition Among Virtual Communities and User Valuation," *Information Systems Research*, March 2–7, p. 68.
7 Ahuja, M. and Chudoba, K., "IT Road Warriors: Balancing Work-Family Conflict," *MIS Quarterly*, March 2007, p. 1.
8 Gratton, Lynda, "Working Together ... When Apart," *The Wall Street Journal*, June 16, 2007, p. R4.
9 Livari, J. and Huisman, M., "The Relationship Between Organizational Culture and the Deployment of Systems Development Methodologies," *MIS Quarterly*, March 2007, p. 35.
10 Mehta, M. and Hirschheim, R., "Strategic Alignment in Mergers and Acquisitions," *Journal of the Association of Information Systems*, March 2007, p. 143.
11 Christensen, Clayton, *The Innovator's Dilemma*, Harvard Business School Press, 1997, p. 225 and *The Inventor's Solution*, Harvard Business School Press, 2003.
12 Schein, E. H., *Process Consultation: Its Role in Organizational Development*, (Reading, MA: Addison-Wesley, 1969). *See also* Keen, Peter G. W., "Information Systems and Organizational Change," *Communications of the ACM*, Vol. 24, No. 1, January 1981, pp. 24–33.
13 Valentine, Lisa, "Reengineering With a Customer Focus," *Bank Systems and Technology*, June 1, 2007, p. 22.
14 Katz, Jonathan, "Continuous Improvement Technology Market to Double," *Industry Week*, January 2007, p. 35.
15 Wixom, Barbara and Todd, Peter, "A Theoretical Integration of User Satisfaction and Technology Acceptance," *Information Systems Research*, March 2005, p. 85.
16 Bailey, J. and Pearson, W., "Development of a Tool for Measuring and Analyzing Computer User Satisfaction," *Management Science*, 29(5), 1983, p. 530.
17 Chaparro, Barbara, et al., "Using the End-User Computing Satisfaction Instrument to Measure Satisfaction with a Web Site," *Decision Sciences*, May 2005, p. 341.
18 Schwarz, A. and Chin, W., "Toward an Understanding of the Nature and Definition of IT Acceptance," *Journal of the Association for Information Systems*, April 2007, p. 230.
19 Davis, F., "Perceived Usefulness, Perceived Ease of Use, and User Acceptance of Information Technology," *MIS Quarterly*, 13(3) 1989, p. 319. Kwon, et al., "A Test of the Technology Acceptance Model," *Proceedings of the Hawaii International Conference on System Sciences*, January 4–7, 2000.
20 Barki, H., et al., "Information System Use-Related Activity," *Information Systems Research*, June 2007, p. 173.
21 Loch, Christoph and Huberman, Bernardo, "A Punctuated-Equilibrium Model of Technology Diffusion," *Management Science*, February 1999, p. 160.
22 Tornatzky, L. and Fleischer, M., "The Process of Technological Innovation," *Lexington Books*, Lexington, MA, 1990; Zhu, K. and Kraemer, K., "Post-Adoption Variations in Usage and Value of E-Business by Organizations," *Information Systems Research*, March 2005, p. 61.
23 Armstrong, Curtis and Sambamurthy, V., "Information Technology Assimilation in Firms," *Information Systems Research*, April 1999, p. 304.
24 Agarwal, Ritu and Prasad, Jayesh, "Are Individual Differences Germane to the Acceptance of New Information Technology?" *Decision Sciences*, Spring 1999, p. 361.
25 Casey, N. and Zamiska, N., "Chinese Factors Identified in Tainted-Toy Recall," *The Wall Street Journal*, August 8, 2007, p. A4.
26 Osborn, Andrew, "Crash Course in Quality for Chinese Car," *The Wall Street Journal*, August 8, 2007, p. B1.
27 Denizel, M., et al., "Ford-Otosan Optimizes its Stocks Using a Six Sigma Framework," *Interfaces*, March 2007, p. 97.
28 Bobadilla, Cherry, "Promoting Patient-Centric Culture in Hospitals," *Business World*, March 16, 2007, p. S5/1.
29 Richardson, Karen, "The Six Sigma Factor for Home Depot," *The Wall Street Journal*, January 4, 2007, p. C3.
30 *www.sixsigma.com* and *www.6sigma.us*, accessed November 30, 2007.
31 Hindo, Brian, "Six Sigma: So Yesterday?" *IN*, June 2007, p. 11.
32 Preston, Rob, "Customers Send Some Mixed Signals," *Information Week*, February 19, 2007, p. 56.
33 Surowiecki, James, "It's the Workforce, Stupid," *The New Yorker*, April 30, 2007, p. 32.
34 Squeri, Steve, "What's Next for IT," *The Wall Street Journal*, July 30, 2007, p. R6.
35 Collins, Jim, *Good to Great*, Harper Collins Books, 2001, p. 300.
36 Porter, M. E., *Competitive Advantage: Creating and Sustaining Superior Performance*, New York: Free Press, 1985; *Competitive Strategy: Techniques for Analyzing Industries and Competitors*, The Free Press, 1980; and *Competitive Advantage of Nations*, The Free Press, 1990.
37 Porter, M. E. and Millar, V., "How Information Systems Give You Competitive Advantage," *Journal of Business Strategy*, Winter 1985. *See also* Porter, M. E., *Competitive Advantage* (New York: Free Press, 1985).
38 Staff, "Technologies to Spot Fakes Are Short Lived," *The Wall Street Journal*, May 26, 2007, p. A6.
39 Havenstein, Heather, "U.S. Doctors Offered Free Prescribing Software," *Computerworld*, January 22, 2007, p. 6.
40 Carey, Susan, "Calculating Costs in the Clouds," *The Wall Street Journal*, March 6, 2007, p. B1.
41 McGee, M., "Urgent Care," *Information Week*, May 26, 2007, p. 40.
42 Beauchesne, Eric, "Machines, Not People, Boosted Productivity," *The Gazette (Montreal)*, June 26, 2007, p. B5.
43 Pratt, Mary, "Web Site ROI," *Computerworld*, May 28, 2007, p. 30.
44 Davis, Robert and Wong, Don, "Conceptualizing and Measuring the Optimal Experience of the eLearning Environment," *Decision Sciences Journal of Innovation Education*, January 2007, p. 97.
45 Staff, "Detailed Comparison of a Range of Factors Affecting Total Cost of Ownership (TCO) for Messaging and Collaboration," *M2 Presswire*, September 5, 2003.
46 Metzer, Scott, "The Allure of New IT," *Information Week*, August 20, 2007, p. CIO5.

47 Brandel, Mary, "12 IT Skills That Employers Can't Say No To," *Computerworld,* July 11, 2007.
48 Levina, N. and Xin, M., "Comparing IT Workers' Compensation Across Country Contexts," *Information Systems Research,* June 2007, p. 193.
49 Murphy, Chris, "Six-Figure Club," *Business Week,* May 7, 2007, p. 86.
50 Brandel, Mary, "Best Places to Work in IT 2007," *Computerworld,* June 18, 2007, p. 34.
51 McDougall, P. and Murphy, C., "Another Round of H-1B," *Information Week,* March 26, 2007, p. 17.
52 McGee, Marianne, "The H-1B Limit," *Information Week,* April 9, 2007, p. 29.
53 Thibodeau, P., "Another H-1B Fight Looms in Congress," *Computerworld,* January 8, 2007, p. 9.
54 Herbst, Moira, "Americans Need Not Apply," *Business Week,* July 9, 2007, p. 10.
55 McGee, M. and Murphy, C., "Feds Cast Wary Eye on Indian Outsourcers' Use of H-1B," *Information Week,* May 21, 2007, p. 32.
56 Miano, John, "Founder of Programmers Guild," *Information Week,* February 5, 2005, p. 15.
57 The Bureau of Labor Statistics, *www.bls.gov*, accessed January 11, 2008.
58 Brandel, Mary, "Three IT Pros Talk About the Flip Side of Outsourcing," *Computerworld,* May 28, 2007, p. 33.
59 Top Coder Collegiate Challenge, *www.topcoder.com*, accessed September 2, 2007.
60 TopCoder, *www.topcoder.com*, accessed August 29, 2007.
61 Gomes, Lee, "Computer Scientists Pull a Tom Sawyer to Finish Grunt Work," *The Wall Street Journal,* June 27, 2007, p. B1.
62 Flandez, Raymund, "Firms Go Online to Train Employees," *The Wall Street Journal,* August 14, 2007, p. B4.
63 Martin, Richard, "The CIO Dilemma," *Information Week,* March 26, 2007, p. 38.
64 Gillooly, Brian, "CIO Role Revs Up," *Optimize Magazine,"* June, 2007, p. 22.
65 Gillooly, Brian, "CIO: Time To Step Up," *Information Week,* June 4, 2007, p. 45.
66 Tam, P., "CIO Jobs Morph from Tech Support into Strategy," *The Wall Street Journal,* February 20, 2007, p. B1.
67 Gosnel, Tom, "The Outward Facing CIO," *Optimize Magazine,* June 2007, p. 45.
68 *www.systest.com*, accessed November 9, 2007.
69 Kelly, C.J., "Getting Certified and Just a Bit Certifiable," *Computerworld,* January 29, 2007, p. 29.
70 *www.nicb.org*, accessed November 9, 2007.
71 *www.du.edu/newsroom/releases/media/2007-08-01-gamecamp.html*, accessed November 9, 2007.
72 *www.ibm.com/services*, *www.eds.com*, and *www.accenture.com*, accessed November 9, 2007
73 MacMillan, Douglas, "The Art of the Online Resume," *Business Week,* May 7, 2007, p. 86.
74 Athavaley, A., "Facebook, Jobster to Partner on Site for Job-Seeker Profiles," *The Wall Street Journal,* February 8, 2007, p. D2.
75 Needleman, Sarah, "How Blogging Can Help You Get a New Job," *The Wall Street Journal,* April 10, 2007, p. B1.
76 White, Erin, "Employers Are Putting New Face on Web Recruiting," *The Wall Street Journal,* January 8, 2007, p. B3.

PART

Information Technology Concepts

CHAPTER • 3 •

Hardware: Input, Processing, and Output Devices

PRINCIPLES	LEARNING OBJECTIVES
▪ Computer hardware must be carefully selected to meet the evolving needs of the organization and its supporting information systems.	▪ Describe the role of the central processing unit and main memory. ▪ State the advantages of multiprocessing and parallel computing systems, and provide examples of the types of problems they address. ▪ Describe the access methods, capacity, and portability of various secondary storage devices. ▪ Identify and discuss the speed, functionality, and importance of various input and output devices. ▪ Identify the characteristics of and discuss the usage of various classes of single-user and multiuser computer systems.
▪ The computer hardware industry is rapidly changing and highly competitive, creating an environment ripe for technological breakthroughs.	▪ Describe Moore's Law and discuss its implications for future computer hardware developments. ▪ Give an example of recent innovations in computer CPU chips, memory devices, and input/output devices.

Information Systems in the Global Economy UB Spirits, India

UB Spirits Serves Up Success

The UB Group is a successful and growing international conglomerate based in Bangalore, India. The bright future of the conglomerate is illustrated by the rising steel and glass structures in the heart of Bangalore and known as UB City—a 7-acre high-rise business, shopping, and living community. The UB Group is invested in several industries including aviation, fertilizers, engineering, information technology, pharmaceuticals, and alcoholic beverages, or spirits. While this may sound like an unusual and even dangerous combination, the diverse business portfolio is making the UB Group billions of dollars.

The alcoholic beverage division of UB, called UB Spirits, is itself a successful international corporation made up of several successful distilleries: McDowell, Herbertsons, and Triumph Distillers and Vintners. Combined, these companies produce 140 brands, 15 of which are top-shelf classics. The brands are produced in 75 locations across India. UB Spirits dominates the Indian marketplace and ranks as the world's second largest distilled spirits corporation, with sales that exceed 60 million cases a year.

To comply with laws that control the distribution of alcoholic beverages in India and around the world, UB Spirits often must negotiate a selling price for their products with a national government. Because of this constraint, UB Spirits has to improve its profits by minimizing overhead and streamlining operations. One obvious waste of resources in the business was its outdated and bloated computer-based information systems. Each UB Spirits facility across India used its own information systems on its own servers——111 Microsoft Windows-based servers in all, which required constant attention. At the close of each month, the data from the disparate systems would be merged in a costly and lengthy process to produce corporate reports. It was clear to the decision makers at UB Spirits that if the company was going to grow, it would have to invest in a new system with state-of-the-art hardware.

As with many system overhauls, UB Spirits began by choosing an ERP software package before it decided on hardware. This is smart because different ERP solutions have varying hardware requirements. Working with IBM, the company selected SAP R/3, which includes information systems for every area of the business.

After selecting an ERP system, the company built the infrastructure on which the system would run. The main goal of the overhaul was to reduce the work required to maintain and administer the system. The company selected an IBM System i550 because it is powerful and simple to manage. The i550 server has multiple processors, up to 64 GB of memory, and up to 77 TB of disk storage. The i550 easily supports 400 UB Spirits users and 1.5 TB of corporate data. It can run all of the corporate systems and be accessed by 64 UB Spirits offices over a network.

By switching to a centralized server system, UB Spirits was able to replace 111 servers spread across India with one server located in its home office. The new system assisted UB Spirits in lowering manufacturing costs. It standardized business processes across the enterprise and accelerated monthly financial reporting. UB Spirits also reassigned 20 of its IT personnel from server administration to more productive projects.

UB Spirits found that the path to higher profits lay in centralizing its operations and investing in large, powerful servers. Other companies have found that they can accomplish more by distributing systems over many workstations in what is called grid computing. The hardware in which a business invests is intimately linked to the type of information system it implements. Creating an underlying infrastructure from hardware should support the evolving needs of the business and its information systems.

As you read this chapter, consider the following:

- How does the type of hardware a company purchases—the size and amount of computers—affect the way the company operates?
- Businesses are in constant fluctuation—growing, diversifying, acquiring, and always working to reduce costs. How do these conditions and requirements affect the purchase of the hardware on which information systems run?

Why Learn About Hardware?

Organizations invest in computer hardware to improve worker productivity, increase revenue, reduce costs, provide better customer service, speed up time-to-market, and enable collaboration among employees. Organizations that don't make wise hardware investments will be stuck with outdated equipment that is unreliable and cannot take advantage of the latest software advances. Such obsolete hardware can place an organization at a competitive disadvantage. Managers, no matter what their career field and educational background, are expected to help define the business needs that the hardware must support. In addition, managers must be able to ask good questions and evaluate options when considering hardware investments for their area of the business. Managers in marketing, sales, and human resources often help IS specialists assess opportunities to apply computer hardware and evaluate the options and features specified for the hardware. Managers in finance and accounting especially must keep an eye on the bottom line, guarding against overspending, yet be willing to invest in computer hardware when and where business conditions warrant it.

Today's use of technology is practical—it's intended to yield real business benefits, as demonstrated by UB Spirits. Employing information technology and providing additional processing capabilities can increase employee productivity, expand business opportunities, and allow for more flexibility. This chapter concentrates on the hardware component of a computer-based information system (CBIS). **Hardware** consists of any machinery (most of which uses digital circuits) that assists in the input, processing, storage, and output activities of an information system. When making hardware decisions, the overriding consideration of a business should be how hardware can support the objectives of the information system and the goals of the organization.

hardware
Any machinery (most of which uses digital circuits) that assists in the input, processing, storage, and output activities of an information system.

COMPUTER SYSTEMS: INTEGRATING THE POWER OF TECHNOLOGY

To assemble an effective and efficient system, you should select and organize components while understanding the trade-offs between overall system performance and cost, control, and complexity. For instance, in building a car, manufacturers try to match the intended use of the vehicle to its components. Racecars, for example, require special types of engines, transmissions, and tires. Selecting a transmission for a racecar requires balancing how much engine power can be delivered to the wheels (efficiency and effectiveness) with how expensive the transmission is (cost), how reliable it is (control), and how many gears it has (complexity). Similarly, organizations assemble computer systems so that they are effective, efficient, and well suited to the tasks that need to be performed.

People involved in selecting their organization's computer hardware must clearly understand current and future business requirements so they can make informed acquisition decisions. Consider the following examples of applying business knowledge to reach critical hardware decisions.

As auto manufacturers must match the intended use of a vehicle to its components, so too must business managers select the hardware components of an effective information system.

(Source: © Mark Jenkinson/ CORBIS.)

- Bosch Security Products provides and maintains physical security systems to control access and detect intrusions for the secure locations of large organizations such as the Dutch Army. For a single customer, Bosch might perform preventive maintenance on thousands of components to ensure that its system is working properly. Service technicians used to rely on a paper-intensive process that required them to fill out forms on the status of each device. The process was error prone and could not provide customers with the comprehensive reports they needed to verify that all necessary repairs and replacements had been completed. Maintenance technicians and their managers defined the requirements for a new and improved solution to meet the business needs. They then consulted with IS experts and chose a PDA-based device for data entry instead of a laptop device because it was lighter, easier to handle, and more resistant to rough handling.[1]
- The Iowa Health System is a network of physicians, hospitals, civic leaders, and local volunteers who serve more than 100 communities. To support patient care, it operates a multifacility Picture Archiving and Communications System in which they store and manage image data such as magnetic resonance images (MRIs). The storage capacity requirements of this system are rapidly growing to 1 million exams per year. Staff and administrators identified additional data storage needs for secure and redundant storage of patient data and to meet the United States Health Insurance Portability and Accountability Act of 1996 (HIPAA) requirements for data integrity. Taking these requirements, the IS staff selected and implemented an appropriate data storage solution that met all needs in a cost-effective manner.[2]

As these examples demonstrate, choosing the right computer hardware requires understanding its relationship to the information system and the needs of the organization. Furthermore, hardware objectives are subordinate to, but supportive of, the information system and the current and future needs of the organization.

Hardware Components

Computer system hardware components include devices that perform input, processing, data storage, and output (see Figure 3.1). To understand how these hardware devices work together, consider an analogy from a paper-based office. Imagine a one-room office occupied by a single person named George. George (the processing device) can organize and manipulate data. George's mind (register storage) and his desk (primary storage) are places to temporarily store data. Filing cabinets fill the need for more permanent storage (secondary storage). In this analogy, the incoming and outgoing mail trays are sources of new data (input) or places to put the processed paperwork (output).

Figure 3.1

Hardware Components

These components include the input devices, output devices, primary and secondary storage devices, and the central processing unit (CPU). The control unit, the arithmetic/logic unit (ALU), and the register storage areas constitute the CPU.

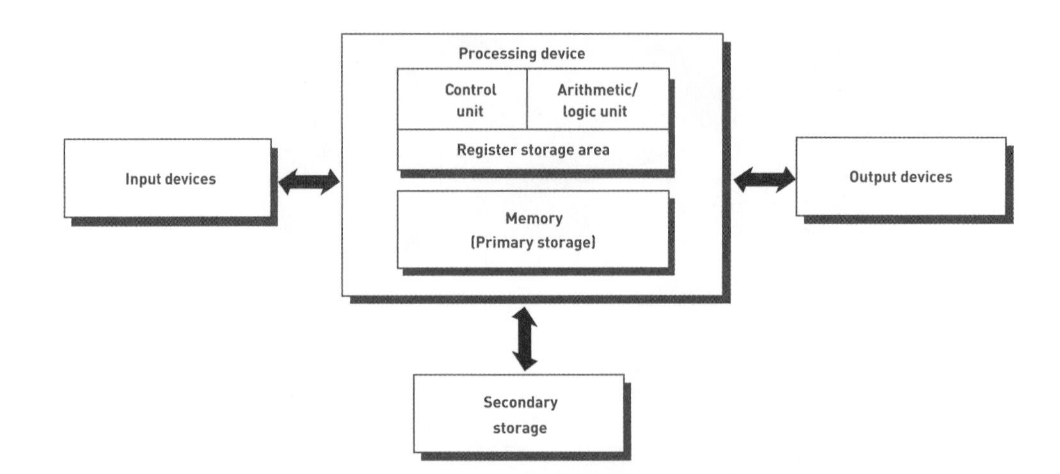

central processing unit (CPU)
The part of the computer that consists of three associated elements: the arithmetic/logic unit, the control unit, and the register areas.

arithmetic/logic unit (ALU)
The part of the CPU that performs mathematical calculations and makes logical comparisons.

control unit
The part of the CPU that sequentially accesses program instructions, decodes them, and coordinates the flow of data in and out of the ALU, registers, primary storage, and even secondary storage and various output devices.

register
A high-speed storage area in the CPU used to temporarily hold small units of program instructions and data immediately before, during, and after execution by the CPU.

primary storage (main memory; memory)
The part of the computer that holds program instructions and data.

instruction time (I-time)
The time it takes to perform the fetch-instruction and decode-instruction steps of the instruction phase.

execution time (E-time)
The time it takes to execute an instruction and store the results.

Recall that any system must be able to process (organize and manipulate) data, and a computer system does so through an interplay between one or more central processing units and primary storage. Each **central processing unit** (**CPU**) consists of three associated elements: the arithmetic/logic unit, the control unit, and the register areas. The **arithmetic/logic unit** (**ALU**) performs mathematical calculations and makes logical comparisons. The **control unit** sequentially accesses program instructions, decodes them, and coordinates the flow of data in and out of the ALU, registers, primary storage, and even secondary storage and various output devices. **Registers** are high-speed storage areas used to temporarily hold small units of program instructions and data immediately before, during, and after execution by the CPU.

Primary storage, also called **main memory** or **memory**, is closely associated with the CPU. Memory holds program instructions and data immediately before or after the registers. To understand the function of processing and the interplay between the CPU and memory, let's examine the way a typical computer executes a program instruction.

Hardware Components in Action

Executing any machine-level instruction involves two phases: instruction and execution. During the instruction phase, a computer performs the following steps:

- **Step 1: Fetch instruction.** The computer reads the next program instruction to be executed and any necessary data into the processor.
- **Step 2: Decode instruction.** The instruction is decoded and passed to the appropriate processor execution unit. Each execution unit plays a different role: The arithmetic/logic unit performs all arithmetic operations, the floating-point unit deals with noninteger operations, the load/store unit manages the instructions that read or write to memory, the branch processing unit predicts the outcome of a branch instruction in an attempt to reduce disruptions in the flow of instructions and data into the processor, the memory-management unit translates an application's addresses into physical memory addresses, and the vector-processing unit handles vector-based instructions that accelerate graphics operations.

The time it takes to perform the instruction phase (Steps 1 and 2) is called the **instruction time (I-time)**.

The second phase is execution. During the execution phase, a computer performs the following steps:

- **Step 3: Execute instruction.** The hardware element, now freshly fed with an instruction and data, carries out the instruction. This could involve making an arithmetic computation, logical comparison, bit shift, or vector operation.
- **Step 4: Store results.** The results are stored in registers or memory.

The time it takes to complete the execution phase (Steps 3 and 4) is called the **execution time (E-time)**.

After both phases have been completed for one instruction, they are performed again for the second instruction, and so on. Completing the instruction phase followed by the execution phase is called a **machine cycle** (see Figure 3.2). Some processing units can speed processing by using **pipelining**, whereby the processing unit gets one instruction, decodes another, and executes a third at the same time. The Pentium 4 processor, for example, uses two execution unit pipelines. This means the processing unit can execute two instructions in a single machine cycle.

machine cycle
The instruction phase followed by the execution phase.

pipelining
A form of CPU operation in which multiple execution phases are performed in a single machine cycle.

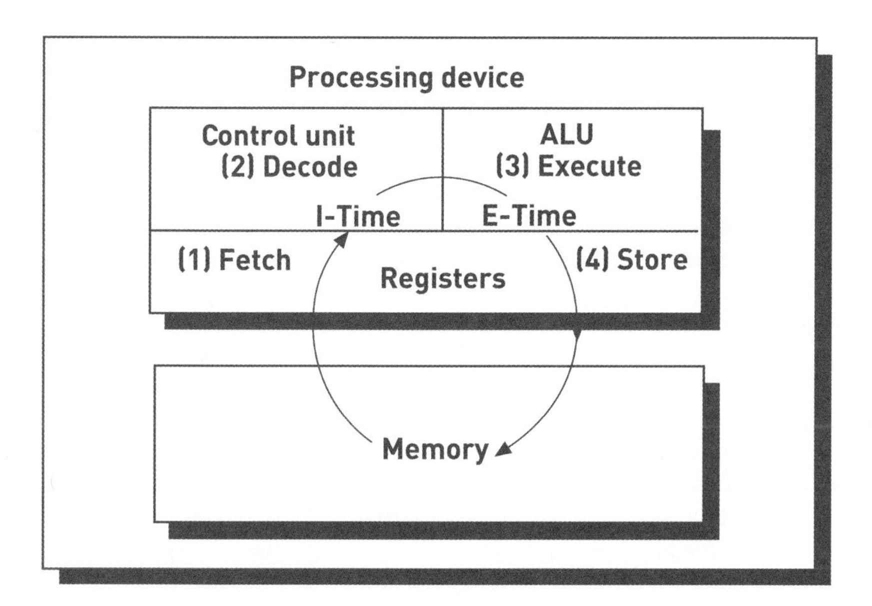

Figure 3.2

Execution of an Instruction

In the instruction phase, a program's instructions and any necessary data are read into the processor (1). Then the instruction is decoded so the central processor can understand what to do (2). In the execution phase, the ALU does what it is instructed to do, making either an arithmetic computation or a logical comparison (3). Then the results are stored in the registers or in memory (4). The instruction and execution phases together make up one machine cycle.

PROCESSING AND MEMORY DEVICES: POWER, SPEED, AND CAPACITY

The components responsible for processing—the CPU and memory—are housed together in the same box or cabinet, called the *system unit.* All other computer system devices, such as the monitor, secondary storage, and keyboard, are linked directly or indirectly into the system unit housing. In this section, we investigate the characteristics of these important devices.

Processing Characteristics and Functions

Because organizations want efficient processing and timely output, they use a variety of measures to gauge processing speed. These measures include the time it takes to complete a machine cycle and clock speed.

Machine Cycle Time

As you have seen, a computer executes an instruction during a machine cycle. The time in which a machine cycle occurs is measured in *nanoseconds* (one-billionth of 1 second) and *picoseconds* (one-trillionth of 1 second). Machine cycle time also can be measured by how many instructions are executed in 1 second. This measure, called **MIPS**, stands for millions of instructions per second. MIPS is another measure of speed for computer systems of all sizes.

MIPS
Millions of instructions per second, a measure of machine cycle time.

Clock Speed

Each CPU produces a series of electronic pulses at a predetermined rate, called the **clock speed**, which affects machine cycle time. The control unit in the CPU manages the stages of

clock speed
A series of electronic pulses produced at a predetermined rate that affects machine cycle time.

microcode
Predefined, elementary circuits and logical operations that the processor performs when it executes an instruction.

the machine cycle by following predetermined internal instructions, known as **microcode**. You can think of microcode as predefined, elementary circuits and logical operations that the processor performs when it executes an instruction. The control unit executes the microcode in accordance with the electronic cycle, or pulses of the CPU "clock." Each microcode instruction takes at least the same amount of time as the interval between pulses. The shorter the interval between pulses, the faster each microcode instruction can be executed.

Because the number of microcode instructions needed to execute a single program instruction—such as performing a calculation or printing results—can vary, the clock speed is not directly related to the true processing speed of the computer.

megahertz (MHz)
Millions of cycles per second.

gigahertz (GHz)
Billions of cycles per second.

Clock speed is often measured in **megahertz** (MHz, millions of cycles per second) or **gigahertz** (GHz, billions of cycles per second). One of the earliest microprocessors was the Intel 8080 with a clock speed of only 2 MHz. This microprocessor was used in the first IBM PC circa 1982. Twenty years later, the Pentium 4 processor had a clock speed of 3.2 GHz—1,600 times faster. Unfortunately, the faster the clock speed of the CPU, the more heat that is generated. This heat must be dissipated to avoid corruption of the data and instructions the computer is trying to process. Also, chips that run at higher temperatures need bigger heat sinks, fans, and other components to eliminate the excess heat. This increases the size of the computer, a problem for manufacturers of portable devices.

The excess heat created by a fast CPU can also be a safety issue. In the summer of 2006, Dell and Apple Computer, in conjunction with the U.S. Consumer Product Safety Commission, announced large recalls of laptop computer batteries. Additional recalls of batteries in Toshiba and Lenovo laptop computers followed. Under certain circumstances, these batteries could overheat and cause a fire or even an explosion.[3] For example, in February 2007, a battery-related fire broke out in the overhead bin of a JetBlue Airways flight. In March 2007, a battery overheated or ignited on an American Airlines aircraft. In both cases, fast-acting flight attendants quickly extinguished the fire and avoided disaster. In response to these accidents, some airlines now require that laptop users remove the computer's battery when plugged into the aircraft's power supply. Other airlines are asking passengers to be sure all loose batteries are stored in insulated bags or otherwise capped to prevent being shorted.

Chip and computer manufacturers are exploring various means to avoid heat problems in their new designs. Demand-based switching is a power management technology developed by Intel that varies the clock speed of the CPU so that it runs at the minimum speed necessary to allow optimum performance of the required operations. IBM and Hewlett-Packard (HP) are also experimenting with direct jet impingement, a technique that deploys an array of tiny nozzles and a distributed return architecture (spray water on, funnel it off) to spray cooling water on the back of the processor.

Manufacturers of portable electronic devices such as computers and cell phones are also seeking more effective sources of energy as portable devices grow increasingly power hungry. A number of companies are exploring the substitution of fuel cells for lithium ion batteries to provide additional, longer-lasting power. Fuel cells generate electricity by consuming fuel (often methanol) while traditional batteries store electricity and release it through a chemical reaction. A spent fuel cell is replenished in moments by simply refilling its reservoir or by replacing the spent fuel cartridge with a fresh one. The use of micro fuel cells based on volatile alcohol will be limited until regulatory restrictions against transporting them on aircraft are lifted.

Cell Broadband Engine Architecture (or simply Cell) is a microprocessor architecture that provides power-efficient, cost-effective, and high-performance processing for a wide range of applications. The Cell is an example of innovation across multiple organizations. A team from IBM Research joined forces with teams from IBM Systems Technology Group, Sony, and Toshiba to provide a breakthrough in performance for consumer applications. The Cell can be used as a component in high-definition displays, recording equipment, and computer entertainment systems. The first commercial application of the Cell was in the Sony PlayStation 3 game console, where it performs at the rate of 2 trillion calculations per second. Toshiba plans to incorporate Cell in high-definition (HD) TV sets. IBM plans to use Cell processors as add-on cards to enhance the performance of the IBM System z9 mainframe computers.[4]

Physical Characteristics of the CPU

Most CPUs are collections of digital circuits imprinted on silicon wafers, or chips, each no bigger than the tip of a pencil eraser. To turn a digital circuit on or off within the CPU, electrical current must flow through a medium (usually silicon) from point A to point B. The speed the current travels between points can be increased by reducing the distance between the points or reducing the resistance of the medium to the electrical current.

Reducing the distance between points has resulted in ever smaller chips, with the circuits packed closer together. In the 1960s, shortly after patenting the integrated circuit, Gordon Moore, former chairman of the board of Intel (the largest microprocessor chip maker), hypothesized that progress in chip manufacturing ought to make it possible to double the number of transistors (the microscopic on/off switches) on a chip roughly every two years. When actual results bore out his idea, the doubling of transistor densities on a single chip every two years became known as **Moore's Law**, and this "rule of thumb" has become a goal that chip manufacturers have met for over four decades. As shown in Figure 3.3, the number of transistors on a chip continues to climb.

Moore's Law
A hypothesis stating that transistor densities on a single chip double every two years.

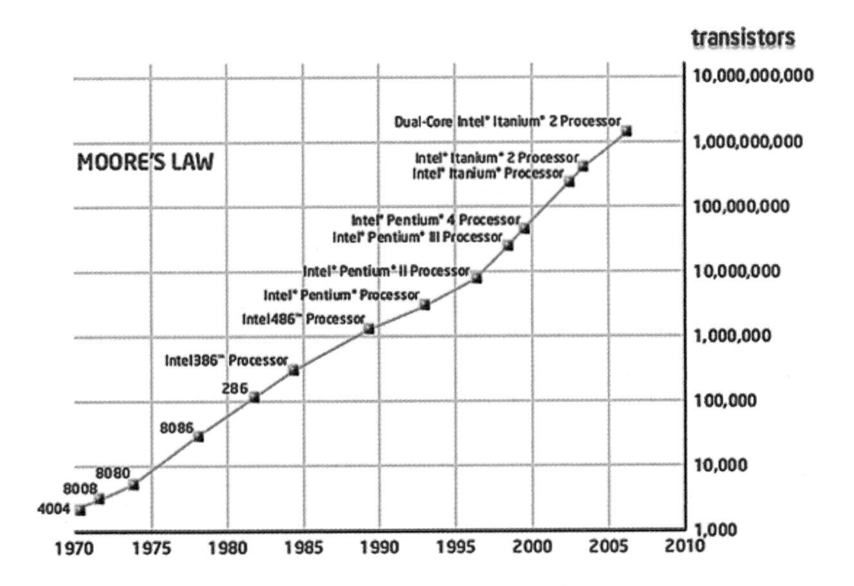

Figure 3.3

Moore's Law

Transistor densities on a single chip double about every two years.

(Source: Intel Web site Moore's Law: Made Real by Intel Innovation, *www.intel.com/technology/mooreslaw/?iid=search*, accessed January 9, 2008.)

In what Gordon Moore touted as "the biggest change in transistor technology in 40 years,"[5] Intel created its next generation 45-nanometer (one-billionth of a meter) Penryn chip—so small that more than 2 million such transistors can fit in the period at the end of this sentence. The new Core 2 processor chips enable extremely high processing speeds at very low power usage. The design is based on a new material called "high-k." The material replaces the thin layer of silicon dioxide insulation that electrically isolates the transistor's gate from the channel through which current flows when the transistor is on.[6] Intel is on track to deliver a 32-nanometer chip in 2009.

Moore's Law enables chip makers to improve performance by putting more transistors on the same size chip and at the same time reduce the amount of power required to get work done. Furthermore, since the chips are smaller, chip manufacturers can cut more chips from a single silicon wafer and thus reduce the cost per chip. As silicon-based components and computers gain in performance, they become cheaper to produce, and therefore more plentiful, more powerful, and more a part of our everyday lives.

Intel has defined a new manufacturing strategy to introduce chips and manufacturing technologies every two years. Intel refers to this as their "tick-tock" strategy, which will drive Intel to make smaller changes to its chip designs more frequently. Intel needs this kind of continuous improvement effort to avoid getting caught off guard as it did early in the twenty-first century when AMD introduced a new chip architecture that yielded significant improvement and power efficiency over Intel chips.[7]

The Aeroflex MIP7000 family of MIPS processors provides high-performance and low-power solutions for microprocessor products.

(Source: Courtesy of Aeroflex Incorporated.)

Researchers are taking many approaches to continue to improve the performance of computers including using sophisticated tri-gate transistors, forming tiny computer circuits from carbon nanotubes only a nanometer in diameter, and extreme miniaturization using radio waves to manipulate atoms into executing a simple computer program. IBM developed technology to send information between processors using light pulses instead of electrical signals. If the technology proves practical, it could lead to microprocessors that are 100 times faster and that require one-tenth the power of today's processors.[8]

Memory Characteristics and Functions

Main memory is located physically close to the CPU, but not on the CPU chip itself. It provides the CPU with a working storage area for program instructions and data. The chief feature of memory is that it rapidly provides the data and instructions to the CPU.

Storage Capacity

byte (B)
Eight bits that together represent a single character of data.

Like the CPU, memory devices contain thousands of circuits imprinted on a silicon chip. Each circuit is either conducting electrical current (on) or not conducting current (off). Data is stored in memory as a combination of on or off circuit states. Usually, 8 bits are used to represent a character, such as the letter *A*. Eight bits together form a **byte (B)**. In most cases, storage capacity is measured in bytes, with 1 byte equivalent to one character of data. The contents of the Library of Congress, with over 126 million items and 530 miles of bookshelves, would require about 20 petabytes of digital storage. Table 3.1 lists units for measuring computer storage.

Table 3.1

Computer Storage Units

Name	Abbreviation	Number of Bytes
Byte	B	1
Kilobyte	KB	2^{10} or approximately 1,024 bytes
Megabyte	MB	2^{20} or 1,024 kilobytes (about 1 million)
Gigabyte	GB	2^{30} or 1,024 megabytes (about 1 billion)
Terabyte	TB	2^{40} or 1,024 gigabytes (about 1 trillion)
Petabyte	PB	2^{50} or 1,024 terabytes (about 1 quadrillion)
Exabyte	EB	2^{60} or 1,024 petabytes (about 1 quintillion)

Types of Memory

random access memory (RAM)
A form of memory in which instructions or data can be temporarily stored.

Computer memory can take several forms, as shown in Table 3.2. Instructions or data can be temporarily stored in and read from **random access memory (RAM)**. With the current design of RAM chips, they are volatile storage devices, meaning they lose their contents if the current is turned off or disrupted (as in a power surge, brownout, or electrical noise generated by lightning or nearby machines). RAM chips are mounted directly on the computer's main circuit board or in other chips mounted on peripheral cards that plug into the main circuit board. These RAM chips consist of millions of switches that are sensitive to changes in electric current.

Memory Type	Abbreviation	Name	Description
Volatile	**RAM**	**Random access memory**	**Volatile storage devices that lose their contents if the current is turned off or disrupted.**
	SRAM	Static Random Access Memory	Byte-addressable storage used for high-speed registers and caches.
	DRAM	Dynamic Random Access Memory	Byte-addressable storage used for the main memory in a computer.
	DDR SDRAM	Double Data Rate Synchronous Dynamic Random Access Memory	An improved form of DRAM.
Nonvolatile	**ROM**	**Read-only memory**	**Nonvolatile storage devices that do not lose their contents if the current is turned off or disrupted.**
	PROM	Programmable read-only memory	Memory used to hold data and instructions that can never be changed. PROMs are programmed in an external device like EPROMs.
	EPROM	Erasable programmable read-only memory	Programmable ROM that can be erased and reused. Erasure is caused by shining an intense ultraviolet light through a window that is designed into the memory chip. EPROM chips are initially written in an external programmer device and must be removed from the circuit board and placed back in the device for reprogramming.
	EEPROM	Electrically erasable programmable read-only memory	User-modifiable read-only memory that can be erased and reprogrammed repeatedly through the application of higher than normal electrical voltage.
	Flash		Used for storage modules for USB drives and digital camera memory cards. Able to erase a block of data in a flash.
	NOR Flash		Flash memory that supports 1-byte random access so that machine instructions can be fetched and executed directly from the flash chip just like computers fetch instructions from main memory.
	NAND Flash		Flash Translation Layer software enables NAND flash memory cards and USB drives to look like a regular disk drive to the operating system.
	FeRAM		Can hold data in memory even when the power is disconnected and offers the higher speed of SDRAM.
	PCM	Phase Change Memory	One of a number of new memory technologies that may eventually replace flash memory.
	MRAM	Magnetoresistive random access memory	A nonvolatile random access memory chip based on magnetic polarization that reads and writes data faster than flash memory.

Table 3.2

Types of Memory Chips

RAM comes in many varieties. Static Random Access Memory (SRAM) is byte-addressable storage used for high-speed registers and caches. Dynamic Random Access Memory (DRAM) is byte-addressable storage used for the main memory in a computer. Double Data Rate Synchronous Dynamic Random Access Memory (DDR SDRAM) is an improved form of DRAM that effectively doubles the rate at which data can be moved in and out of main memory. Other forms of memory include DDR2 SDRAM and DDR3 SDRAM.

Read-only memory (**ROM**), another type of memory, is nonvolatile, meaning that its contents are not lost if the power is turned off or interrupted. ROM provides permanent storage for data and instructions that do not change, such as programs and data from the computer manufacturer, including the instructions that tell the computer how to start up when power is turned on.

read-only memory (ROM)
A nonvolatile form of memory.

Chip manufacturers are competing to develop a nonvolatile memory chip that requires minimal power, offers extremely fast write speed, and can store data accurately even after a large number of write-erase cycles. Such a chip could eliminate the need for RAM and simplify and speed up memory processing.[9] PCM, FeRAM, and MRAM are three potential approaches to provide such a memory device.

Although microprocessor speed has doubled every 24 months over the past decade, memory performance has not kept pace. In effect, memory has become the principal bottleneck to system performance. The use of **cache memory**, a type of high-speed memory that a processor can access more rapidly than main memory (see Figure 3.4), helps to ease this bottleneck. Frequently used data is stored in easily accessible cache memory instead of slower memory such as RAM. Because cache memory holds less data, the CPU can access the desired data and instructions more quickly than selecting from the larger set in main memory. Thus, the CPU can execute instructions faster, improving the overall performance of the computer system. Cache memory is available in three forms. The Level 1 (L1) cache is on the CPU chip. The Level 2 (L2) cache memory can be accessed by the CPU over a high-speed dedicated interface. The latest processors go a step further and place the L2 cache directly on the CPU chip itself and provide high-speed support for a tertiary Level 3 (L3) external cache.

cache memory
A type of high-speed memory that a processor can access more rapidly than main memory.

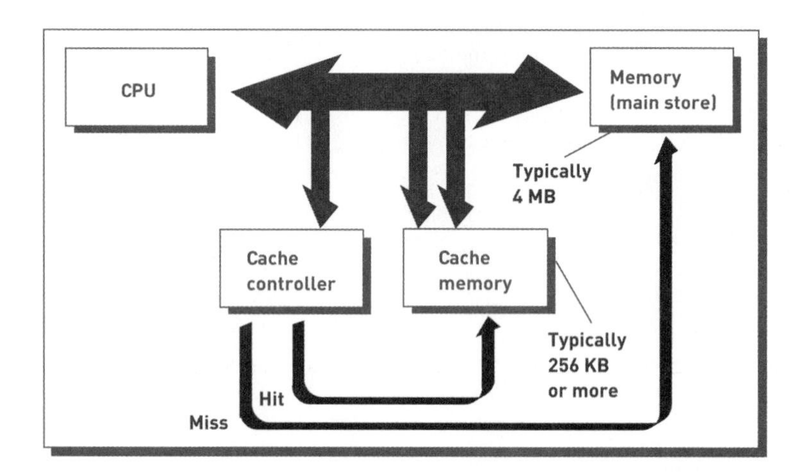

Figure 3.4

Cache Memory

Processors can access this type of high-speed memory faster than main memory. Located on or near the CPU chip, cache memory works with main memory. A cache controller determines how often the data is used, transfers frequently used data to cache memory, and then deletes the data when it goes out of use.

When the processor needs to execute an instruction, the instruction's operation code indicates whether the data will be in a register or in memory. If the operation code specifies a register as the source, it is taken from there. Otherwise, the processor looks for the data in the L1 cache, then the L2 cache, and then the L3 cache. If the data is not in any cache, the CPU requests the data from main memory. If the data is not even stored in main memory, the system has to retrieve the data from secondary storage. It can take from one to three clock cycles to fetch information from the L1 cache, while the CPU waits and does nothing. It takes 6 to 12 cycles to get data from an L2 cache on the processor chip. It can take dozens of cycles to fetch data from an L3 cache and hundreds of cycles to fetch data from secondary storage. Because this hierarchical arrangement of memory helps the CPU find data faster, it bridges a widening gap between processor speeds, which are increasing at roughly 50 percent per year, and DRAM access rates, which are climbing at only 5 percent per year.

Memory capacity contributes to the effectiveness of a CBIS. The specific applications of a CBIS determine the amount of memory required for a computer system. For example, complex processing problems, such as computer-assisted product design, require more memory than simpler tasks such as word processing. Also, because computer systems have different types of memory, they might need other programs to control how memory is accessed and used. In other cases, the computer system can be configured to maximize memory usage. Before purchasing additional memory, an organization should address all these considerations.

Multiprocessing

Generally, **multiprocessing** involves the simultaneous execution of two or more instructions at the same time. One form of multiprocessing uses coprocessors. A **coprocessor** speeds processing by executing specific types of instructions while the CPU works on another processing activity. Coprocessors can be internal or external to the CPU and can have different clock speeds than the CPU. Each type of coprocessor performs a specific function. For example, a math coprocessor chip speeds mathematical calculations, and a graphics coprocessor chip decreases the time it takes to manipulate graphics.

multiprocessing
The simultaneous execution of two or more instructions at the same time.

coprocessor
The part of the computer that speeds processing by executing specific types of instructions while the CPU works on another processing activity.

A **multicore microprocessor** combines two or more independent processors into a single computer so that they can share the workload and boost processing capacity. In addition, a dual-core processor enables people to perform multiple tasks simultaneously, such as playing a game and burning a CD. The use of low clock speed, multicore processors with a shared cache (rather than separate dedicated caches for each processor core) is another approach to reduce the heat generated by the computer without reducing its processing power. For example, the Intel Dual Core processor runs at 1.66 MHz compared to the single core Pentium 4 processor, which runs at 3.2 GHz. AMD and Intel are battling for leadership in the multicore processor marketplace.

multicore microprocessor
A microprocessor that combines two or more independent processors into a single computer so they can share the workload and improve processing capacity.

Both Intel and AMD have improved on dual processors by introducing new quad-core chips. However, a major need for basic research in computer science is to develop software that can actually take advantage of four processors. The processor manufacturers are working with software developers to create new multithreaded applications and next-generation games that will use the capabilities of the quad-core processor. "Industry has basically thrown a Hail Mary," warns David Patterson, a pioneering computer scientist at the University of California, Berkeley, referring to the hardware shift. "The whole industry is betting on parallel computing. They've thrown it, but the big problem is catching it."[10]

Multicore systems are most effective when they run programs that can split their workload among multiple CPUs. Such applications include working with large databases and multimedia. Intel has introduced Viiv (rhymes with five), a marketing initiative that combines Intel products including the Core 2 Quad processor with additional hardware and software to build an extremely powerful multimedia computer capable of running the highly processing-intensive applications associated with high-definition entertainment, including the following:[11]

- Full 1080P video playback of movie clips, media streams, and HD video cameras
- High-quality audio for surround-sound capabilities from movies and music
- Fast, extremely high-quality photo editing, retouching, and publishing
- Capability to watch, record, and pause live TV

AMD has countered with its new quad-core Opteron processor. AMD also plans to catch up to Intel by building its first 45-nanometer chips in 2008.[12] AMD is counting on a project it calls Fusion, which will combine a graphics processing unit and a CPU on the same chip, and is expected in late 2008.[13]

When selecting a CPU, organizations must balance the benefits of processing speed with energy requirements and cost. CPUs with faster clock speeds and shorter machine cycle times require more energy to dissipate the heat generated by the CPU and are bulkier and more expensive than slower ones.

parallel computing
The simultaneous execution of the same task on multiple processors to obtain results faster.

Parallel Computing

Parallel computing is the simultaneous execution of the same task on multiple processors to obtain results faster. Systems with thousands of such processors are known as **massively parallel processing systems**. The processors might communicate with one another to coordinate when executing a computer program, or they might run independently of one another but under the direction of another processor that distributes the work to the other processors and collects their processing results. The dual-core processors mentioned earlier are a simple form of parallel computing.

massively parallel processing systems
A form of multiprocessing that speeds processing by linking hundreds or thousands of processors to operate at the same time, or in parallel, with each processor having its own bus, memory, disks, copy of the operating system, and applications.

In response to higher fuel prices and the desire to reduce carbon dioxide emissions, auto manufacturers are introducing smaller models. However, consumers are concerned about the safety of these smaller cars. To address these concerns, automobile engineers use finite element modeling and massively parallel processing computer systems to simulate crashes. Such simulations are much less expensive than using actual cars and crash dummies. Also, the speed and accuracy of the computer simulations allows for many more crash tests at an earlier stage in the design than using physical crashes with crash dummies. As a result, engineers gain confidence earlier in the design process that their car will pass federal safety standards, enabling them to bring the car to market sooner.[14]

grid computing
The use of a collection of computers, often owned by multiple individuals or organizations, to work in a coordinated manner to solve a common problem.

Grid computing is the use of a collection of computers, often owned by multiple individuals or organizations, to work in a coordinated manner to solve a common problem. Grid computing is a low-cost approach to parallel computing. The grid can include dozens, hundreds, or even thousands of computers that run collectively to solve extremely large processing problems. Key to the success of grid computing is a central server that acts as the grid leader and traffic monitor. This controlling server divides the computing task into subtasks and assigns the work to computers on the grid that have (at least temporarily) surplus processing power. The central server also monitors the processing, and if a member of the grid fails to complete a subtask, it restarts or reassigns the task. When all the subtasks are completed, the controlling server combines the results and advances to the next task until the whole job is completed.

European and Asian researchers are using a grid consisting of some 40,000 computers spread across 45 countries to combat the deadly bird flu. Ulf Dahlsten, a member of the Information Society and Media Directorate-General of the European Commission, used the success of grid computing in battling this potential pandemic to point out the breakthroughs that are being made in drug discovery. "Computer grids have achieved a productivity increase of more than 6,000 percent in the identification of potential new drugs. Three hundred thousand molecules have already been screened using the grid. Of these, 123 potential inhibitors were identified, of which seven have now been shown to act as inhibitors in in-vitro laboratory tests. This is a 6 percent success rate compared to typical values of around 0.1 percent using classical drug discovery methods."[15]

Folding@home is a grid computing project with more than 1 million people around the world downloading and running software to form one of the largest supercomputers in the world. To carry out their various functions, proteins self-assemble into a particular shape in a process called "folding." The goal of the Folding@home project is to research protein folding and misfolding and gain an understanding of how this protein behavior is related to diseases such as Alzheimer's, Parkinson's, and many forms of cancer. It takes a single computer about one day to simulate a nanosecond (1/1,000,000th of a second) in the life of a protein. The folding process takes about 10,000 nanoseconds. Thus, 10,000 days (30 years) are required to simulate a single folding! The Folding@home group has developed ways to speed up the simulation of protein folding by dividing the work among over 100,000 processors.[16] In September 2007, with more than half a million PlayStation 3 consoles participating on the grid, the combined computing power exceeded 1×10^{15} floating-point operations per second—more than twice the speed of the world's fastest stand-alone supercomputer.[17]

The most frequent uses for parallel computing include modeling, simulation, and analyzing large amounts of data. Chrysler uses high-performance computers consisting of some 1,650 cores to simulate racecar performance and identify opportunities for improvement in the car's design and operation. The ability to develop more complete fluid dynamic models of the extreme conditions associated with vehicles traveling at 190 mph has led to improvements not only in racecars but in passenger cars as well. For example, simulations show how a racecar traveling behind another car receives restricted airflow, which can affect engine performance. This finding can be reapplied to the design of passenger cars to deal with the restricted airflow they receive when traveling behind a large truck.[18]

cloud computing
Using a giant cluster of computers to serve as a host to run applications that require high-performance computing.

Cloud computing involves using a giant cluster of computers that serves as a host to run applications that require high-performance computing. Cloud computing supports a wider variety of applications than grid computing and pools computing resources so they can be managed primarily by software rather than people. IBM and Google have provided hardware, software, and services to many universities so that students and faculty can explore cloud

By installing the Folding@home screen saver program on their personal computers, more than 1 million people around the world contribute their idle CPU time to research diseases such as Alzheimer's, Parkinson's, and many forms of cancer.

(Source: *http://folding.stanford.edu/*)

computing and massively parallel computing. For example, University of Washington students use the Google cluster to scan millions of changes made to the online encyclopedia Wikipedia to identify spam.[19] Amazon.com offers the Elastic Compute Cloud as a service that provides users with a highly expandable computing capacity for Web site development and operations.[20] Some industry observers think that Google is planning a bold move into the delivery of applications for businesses and consumers. Indeed, CEO Eric Schmidt has stated that "Most people who run small businesses would like to throw out their infrastructure and use ours for $50 per year." Several *Fortune* 1000 companies are running Google Apps pilots to test the concept of running their applications on the world's most efficient supercomputer—the Google cloud computer. Companies are asking whether Google, IBM, Amazon.com, or others can provide a cheaper, more reliable, more secure, more flexible, and more powerful alternative through cloud computing. Would effective cloud computing encourage companies to abandon large components of their IT infrastructure?[21]

SECONDARY STORAGE

Driven by factors such as needing to retain more data longer to meet government regulatory concerns, store new forms of digital data such as audio and video, and keep systems running under the onslaught of increasing volumes of e-mail, the amount of data that companies store digitally is increasing at a rate of close to 100 percent per year. As an extreme example, the Large Hadron Collider (LHC) located near Geneva, Switzerland, will be the world's largest and highest-energy particle accelerator when it becomes operational mid-2008. The LHC is an essential tool of physicists in their search for a Grand Unified Theory that would unify three of the four fundamental forces of the universe: electromagnetism, the strong force, and the weak force. Proving this theory would also explain why the remaining force, gravitation, is so weak compared to the other three forces. The LHC project is expected to produce 15 petabytes of data each year.[22]

Storing data safely and effectively is critical to an organization's success. Recognizing this, Wal-Mart operates one of the largest collections of customer data in the world to hold data from 800 million transactions generated by its 30 million customers each day. The amount of data stored is about 1 petabyte and is used to analyze in-store sales to determine the ideal mix of items and the optimal placement of products within each store to maximize sales.[23]

Jim Scantlin, Director of Enterprise Information Management, states: "At Wal-Mart, we never underestimate the importance of investing in innovative solutions that will improve our ability to understand and anticipate our customers' needs."[24]

When determining the best method of data storage, the best overall solution is likely a combination of different storage options. **Secondary storage**, also called *permanent storage*, serves this purpose.

secondary storage (permanent storage)
Devices that store larger amounts of data, instructions, and information more permanently than allowed with main memory.

Compared with memory, secondary storage offers the advantages of nonvolatility, greater capacity, and greater economy. On a cost-per-megabyte basis, most forms of secondary storage are considerably less expensive than primary memory (see Table 3.3). The selection of secondary storage media and devices requires understanding their primary characteristics—access method, capacity, and portability.

Table 3.3

Cost Comparison for Various Forms of Storage

All forms of secondary storage cost considerably less per megabyte of capacity than SDRAM, although they have slower access times. A data cartridge costs about $.21 per gigabyte, while SDRAM can cost around $49 per gigabyte—over 200 times more expensive.

(Source: Office Depot Web site, *www.officedepot.com*, January 18, 2008.)

Description	Cost	Storage Capacity (GB)	Cost Per GB
72 GB DAT 72 data cartridge	$14.95	72	$0.21
10 - 4.7 GB DVD+R disks	$9.95	47	$0.21
20 GB 4 MM backup data tape	$16.99	20	$0.85
120 GB portable hard drive	$139.99	120	$1.16
25 GB Rewritable Blu-ray disk	$29.99	25	$1.20
9.1 GB Write Once Read Many optical disk	$69.95	9.1	$7.69
1 GB flash drive	$7.99	1	$7.99
512 MB DDR2 SDRAM memory upgrade	$24.99	0.512	$48.81

As with other computer system components, the access methods, storage capacities, and portability required of secondary storage media are determined by the information system's objectives. An objective of a credit card company's information system, for example MasterCard or Visa, might be to rapidly retrieve stored customer data to approve customer purchases. In this case, a fast access method is critical. In other cases, such as equipping the Coca-Cola field salesforce with pocket-sized personal computers, portability and storage capacity might be major considerations in selecting and using secondary storage media and devices.

Storage media that allow faster access are generally more expensive than slower media. The cost of additional storage capacity and portability vary widely, but they are also factors to consider. In addition to cost and portability, organizations must address security issues to allow only authorized people to access sensitive data and critical programs. Because the data and programs kept in secondary storage devices are so critical to most organizations, all of these issues merit careful consideration.

Access Methods

sequential access
A retrieval method in which data must be accessed in the order in which it is stored.

direct access
A retrieval method in which data can be retrieved without the need to read and discard other data.

sequential access storage device (SASD)
A device used to sequentially access secondary storage data.

direct access storage device (DASD)
A device used for direct access of secondary storage data.

Data and information access can be either sequential or direct. **Sequential access** means that data must be accessed in the order in which it is stored. For example, inventory data might be stored sequentially by part number, such as 100, 101, 102, and so on. If you want to retrieve information on part number 125, you must read and discard all the data relating to parts 001 through 124.

Direct access means that data can be retrieved directly, without the need to pass by other data in sequence. With direct access, it is possible to go directly to and access the needed data—for example, part number 125—without having to read through parts 001 through 124. For this reason, direct access is usually faster than sequential access. The devices used only to access secondary storage data sequentially are simply called **sequential access storage devices** (SASDs); those used for direct access are called **direct access storage devices** (DASDs).

Devices

The most common forms of secondary storage include magnetic tapes, magnetic disks, virtual tapes, and optical discs. In general, magnetic tapes are the oldest storage medium, while optical discs are the most recent. Some of these media (magnetic tape) allow only sequential access, while others (magnetic and optical discs) provide direct and sequential access. Figure 3.5 shows one type of secondary storage media.

Figure 3.5

Secondary Storage

Secondary storage devices such as magnetic tapes and disks, optical discs, CD-ROMs, and DVDs are used to store data for easy retrieval at a later date.

(Source: Courtesy of Imation Corp.)

Magnetic Tape

Magnetic tape is a type of sequential secondary storage medium, now used primarily for storing backups. Similar to the tape found in audio- and videocassettes, magnetic tape is a Mylar film coated with iron oxide. Portions of the tape are magnetized to represent bits. If the computer needs to read data from the middle of a reel of tape, it must first pass all the tape before the desired piece of data—one disadvantage of magnetic tape. When information is needed, it can take time to retrieve the proper tape and mount it on the tape reader to get the relevant data into the computer. Despite the falling prices of hard drives, tape storage is still a popular choice for low-cost data backup for off-site storage in the event of a disaster. Not surprisingly, the U.S. federal government is the largest user of magnetic tape in the world, buying over 1 million reels of tape each year for use by such organizations as the Internal Revenue Service, National Oceanic and Atmospheric Administration, the Federal Reserve Bank, and the various branches of the military.[25]

magnetic tape
A secondary storage medium; Mylar film coated with iron oxide with portions of the tape magnetized to represent bits.

Technology is improving to provide tape storage devices with greater capacities and faster transfer speeds. In addition, the bulky tape drives used to read and write on large reels of tapes in the early days of computing have been replaced with tape cartridge devices measuring a few millimeters in diameter, requiring much less floor space and allowing hundreds of tapes to be stored in a small area.

Magnetic Disks

Magnetic disks are also coated with iron oxide; they can be thin metallic platters (hard disks, see Figure 3.6) or Mylar film (diskettes). As with magnetic tape, magnetic disks represent bits using small magnetized areas. When reading from or writing to a disk, the disk's read/write head can go directly to the desired piece of data. Thus, the disk is a direct-access storage medium. Because direct access allows fast data retrieval, this type of storage is ideal for companies that need to respond quickly to customer requests, such as airlines and credit card firms. For example, if a manager needs information on the credit history of a customer or the seat availability on a particular flight, the information can be obtained in seconds if the data is stored on a direct access storage device.

magnetic disk
A common secondary storage medium, with bits represented by magnetized areas.

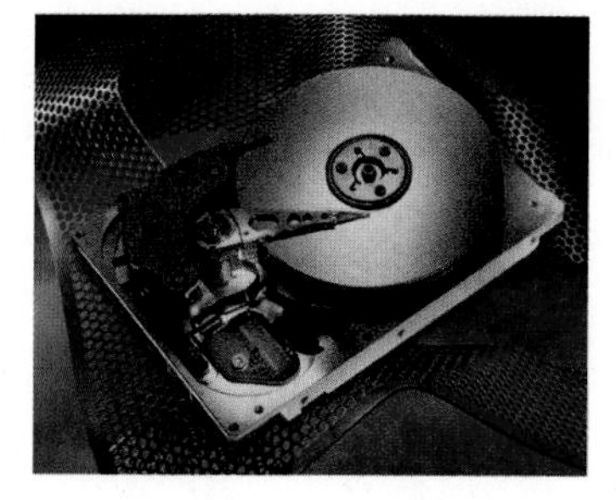

Figure 3.6

Hard Disk

Hard disks provide direct access to stored data. The read/write head can move directly to the location of a desired piece of data, dramatically reducing access times as compared with magnetic tape.

(Source: Courtesy of Seagate Technology.)

Magnetic disk storage varies widely in capacity and portability. Removable magnetic disks, such as diskettes or Zip disks, are nearly obsolete. Hard disks, though more costly and less portable, are more popular because of their greater storage capacity and quicker access time. The Iomega REV Loader 560, shown in Figure 3.7, is a storage device that holds up to eight removable 70 GB discs, each housed in a shock-resistant plastic case about the size of a 3.5-inch floppy disk. The device works well for anyone rotating off-site backups of critical data. Tape has been giving way to the external hard drive as the preferred backup medium for small businesses. However, if you need off-site copies of your data, hauling home a heavy, fragile hard drive seems less than ideal. You could use an online backup service, but for more than a few gigabytes, speed and expense become problems. Burning to DVDs might offer a solution, except a single disc holds only 4.7 GB (or 9 GB for the few people who have dual-layer drives), making capacity an issue.[26] Hitachi has announced a 500 GB 2.5-inch hard drive for portable computers.[27]

Figure 3.7

Iomega REV Loader 560

(Source: Courtesy of Iomega Corporation.)

RAID

Putting an organization's data online involves a serious business risk—the loss of critical data can put a corporation out of business. The concern is that the most critical mechanical components inside a disk storage device—the disk drives, the fans, and other input/output devices—can break (like most things that move).

Organizations now require that their data-storage devices be fault tolerant—they can continue with little or no loss of performance if one or more key components fails. A **redundant array of independent/inexpensive disks (RAID)** is a method of storing data that generates extra bits of data from existing data, allowing the system to create a "reconstruction map" so that if a hard drive fails, it can rebuild lost data. With this approach, data is split and stored on different physical disk drives using a technique called *striping* to evenly distribute the data. RAID technology has been applied to storage systems to improve system performance and reliability.

redundant array of independent/inexpensive disks (RAID)
A method of storing data that generates extra bits of data from existing data, allowing the system to create a "reconstruction map" so that if a hard drive fails, the system can rebuild lost data.

RAID can be implemented in several ways. In the simplest form, RAID subsystems duplicate data on drives. This process, called **disk mirroring**, provides an exact copy that protects users fully in the event of data loss. However, to keep complete duplicates of current backups, organizations need to double the amount of their storage capacity. Thus, disk mirroring is expensive. Other RAID methods are less expensive because they only partly duplicate the data, allowing storage managers to minimize the amount of extra disk space (or overhead) they must purchase to protect data. Optional second drives for personal computer users who need to mirror critical data are available for less than $100.

disk mirroring
A process of storing data that provides an exact copy that protects users fully in the event of data loss.

Medkinetics is a small (12 employees) business that automates collecting and submitting for approval of information about a doctor's qualifications. Jim Cox, founder and president of Medkinetics, says: "The high availability of data is also really important to us." The firm employs terabytes of inexpensive but secure RAID storage.[28]

Virtual Tape

Virtual tape is a storage technology that manages less frequently needed data so that it appears to be stored entirely on tape cartridges, although some parts might actually be located on faster hard disks. The software associated with a virtual tape system is sometimes called a *virtual tape server*. Virtual tape can be used with a sophisticated storage-management system that moves data to slower but less costly forms of storage media as people use the data less often. Virtual tape technology can decrease data access time, lower the total cost of ownership, and reduce the amount of floor space consumed by tape operations. IBM and Storage Technology are well-established vendors of virtual tape systems. One organization that uses a virtual tape system is the Girl Scouts of the USA, which operates a major data center that holds data on 4 million active members of the organization. The amount of data is roughly half a terabyte but is expected to grow at a rate of 25 percent per year as the length of time that data is kept is expanded from ten years to indefinite. The organization uses an REO 9500D virtual tape library from Overland Storage Inc. at a cost of $65,400 for 3.75 terabytes of storage capacity.[29]

virtual tape
A storage device that manages less frequently needed data so that it appears to be stored entirely on tape cartridges, although some parts of it might actually be located on faster hard disks.

Optical Discs

Another type of secondary storage medium is the **optical disc**. An optical disc is simply a rigid disk of plastic onto which data is recorded by special lasers that physically burn pits in the disk. Data is directly accessed from the disc by an optical disc device, which operates much like a stereo's compact disc player. This optical disc device uses a low-power laser that measures the difference in reflected light caused by a pit (or lack thereof) on the disc.

optical disc
A rigid disc of plastic onto which data is recorded by special lasers that physically burn pits in the disc.

A common optical disc is the **compact disc read-only memory (CD-ROM)** with a storage capacity of 740 MB of data. After data is recorded on a CD-ROM, it cannot be modified—the disc is "read-only." A CD burner, the informal name for a CD recorder, is a device that can record data to a compact disc. *CD-recordable (CD-R)* and *CD-rewritable (CD-RW)* are the two most common types of drives that can write CDs, either once (in the case of CD-R) or repeatedly (in the case of CD-RW). CD-rewritable (CD-RW) technology allows PC users to back up data on CDs.

compact disc read-only memory (CD-ROM)
A common form of optical disc on which data, once it has been recorded, cannot be modified.

Digital Video Disc

A **digital video disc (DVD)** looks like a CD but can store about 135 minutes of digital video or several gigabytes of data (see Figure 3.8). Software, video games, and movies are often stored or distributed on DVDs. At a data transfer rate of 1.352 MB/second, the access speed of a DVD drive is faster than that of the typical CD-ROM drive.

digital video disc (DVD)
A storage medium used to store digital video or computer data.

DVDs have replaced recordable and rewritable CD discs (CD-R and CD-RW) as the preferred format for sharing movies and photos. Whereas a CD can hold about 740 MB of data, a single-sided DVD can hold 4.7 GB, with double-sided DVDs having a capacity of 9.4 GB. Unfortunately, DVD manufacturers haven't agreed on a recording standard, so several types of recorders and discs are currently in use. Recordings can be made on record-once discs (DVD-R and DVD+R) or on rewritable discs (DVD-RW, DVD+RW, and DVD-RAM). Not all types of rewritable DVDs are compatible with other types.

Figure 3.8

Digital Video Disc and Player

DVDs look like CDs but have a greater storage capacity and can transfer data at a faster rate.

(Source: Courtesy of Toshiba America Information Systems.)

The Blu-ray high-definition video-disc format based on blue-laser technology stores at least three times as much data as a DVD now holds. The primary use for this new format is in home entertainment equipment to store high-definition video, though this format can also store computer data.

Holographic Disc

Holographic Versatile Disc (HVD) is an advanced optical disc technology still in the research stage that would store more data than even the Blu-ray optical disc system. One approach to HVD records data through the depth of the storage media in three dimensions by splitting a laser beam in two—the signal beam carries the data, and the reference beam positions where the data is written and reads it. HVD can transfer data at the rate of 1 Gigabit per second and store 1 terabyte (TB) of data on a single optical disk.[30]

Enterprise Storage Options

Businesses increasingly need to store large amounts of data created throughout the organization. Such large secondary storage is called *enterprise storage* and comes in three forms: attached storage, network-attached storage (NAS), and storage area networks (SANs).

Attached Storage

Attached storage methods include the tape, hard disks, and optical devices discussed previously, which are connected directly to a single computer. Attached storage methods, though simple and cost-effective for single users and small groups, do not allow systems to share storage, and they make it difficult to back up data.

Because of the limitations of attached storage, firms are turning to network-attached storage (NAS) and storage area networks (SANs). These alternative forms of enterprise data storage enable an organization to share data-storage resources among a much larger number of computers and users, resulting in improved storage efficiency and greater cost-effectiveness. In addition, they simplify data backup and reduce the risk of downtime. Nearly one-third of system downtime is a direct result of data-storage failures, so eliminating storage problems as a cause of downtime is a major advantage.

Network-Attached Storage

network-attached storage (NAS)
Storage devices that attach to a network instead of to a single computer.

Network-attached storage (NAS) employs storage devices that attach to a network instead of to a single computer. NAS includes software to manage storage access and file management and relieve the users' computers of those tasks. The result is that both application software and files can be served faster because they are not competing for the same processor resources. Computer users can share and access the same information, even if they are using different types of computers. Common applications for NAS include consolidated storage, Internet and e-commerce applications, and digital media.

The University of North Carolina (UNC) Hospital employs 2,000 physicians, over 5,000 staff members, and operates on a $600 million budget. It uses a state-of-the-art Picture Archiving and Communications System (PACS) to manage x-rays, CAT scans, and MRIs in a digital form instead of more traditional x-ray film. The system improves patient care and facilitates teaching by streamlining access to critical information. However, the sheer volume of data was causing UNC to struggle with its inefficient local storage devices. Data from over 200,000 radiology procedures each year requiring 4–5 TB of data storage was overwhelming the existing system. UNC recently converted to a centralized NAS data solution that now allows it to consolidate data onto fewer servers and storage devices and reduce the effort required to manage data. More importantly, the NAS solution enables rapid retrieval of patient information, saving doctors time, which can mean the difference between life and death in the operating room.[31]

Storage Area Network

storage area network (SAN)
The technology that provides high-speed connections between data-storage devices and computers over a network.

A **storage area network** (SAN) is a special-purpose, high-speed network that provides direct connections between data-storage devices and computers across the enterprise (see Figure 3.9). A SAN also integrates different types of storage subsystems, such as multiple RAID storage devices and magnetic tape backup systems, into a single storage system. Use of a SAN loads the network traffic associated with storage onto a separate network. The data can then be copied to a remote location, making it easier for companies to create backups and implement disaster recovery policies.

Using a SAN, an organization can centralize the people, policies, procedures, and practices for managing storage, and a data-storage manager can apply the data consistently across an enterprise. This centralization eliminates inconsistent treatment of data by different system administrators and users, providing efficient and cost-effective data-storage practices.

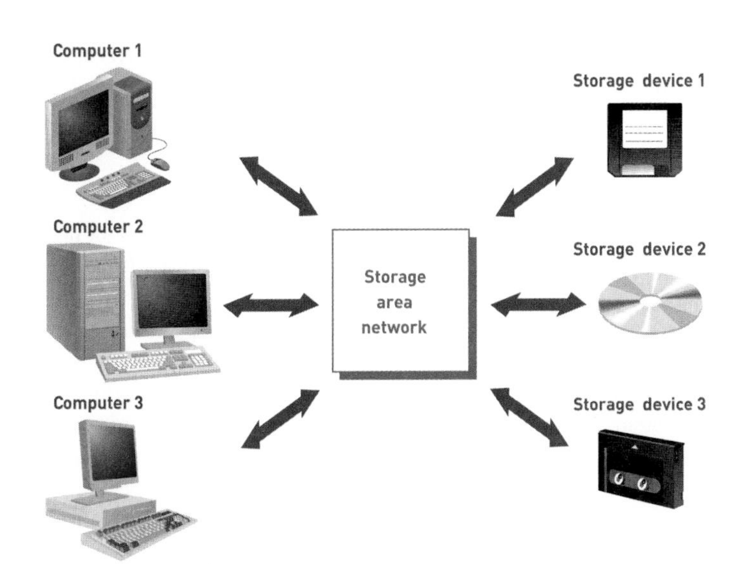

Figure 3.9

Storage Area Network

A SAN provides high-speed connections between data-storage devices and computers over a network.

The Navy's Surface Combat Systems Center in Wallops Island, Virginia, uses a SAN to hold 168 TB of data with data spread across more than 200 hard drives.[32] The Bombay Company designs and markets home furnishings and decorative accessories via 422 retail outlets, catalogs, and the Internet. The firm implemented 1.5 TB of SAN data storage to effectively hold its inventory data.[33]

A fundamental difference between NAS and SAN is that NAS uses file input/output, which defines data as complete containers of information, while SAN deals with block input/output, which is based on subsets of data smaller than a file. SAN manufacturers include EMC, Hitachi Data Systems Corporation, Xiotech, and IBM.

As organizations set up large-scale SANs, they use more computers and network connections, which become difficult to manage. In response, software tools designed to automate storage using previously defined policies are finding a place in the enterprise. Known as **policy-based storage management**, the software products from industry leaders such as Veritas Software Corporation, Legato Systems, Inc., EMC, and IBM automatically allocate storage space to users, balance the loads on servers and disks, and reroute networks when systems go down—all based on policies set up by system administrators.

policy-based storage management
Automation of storage using previously defined policies.

The trend in secondary storage is toward higher capacity, increased portability, and automated storage management. Organizations should select a type of storage based on their needs and resources. In general, storing large amounts of data and information and providing users with quick access makes an organization more efficient. Businesses can also choose pay-per-use services, where they rent space on massive storage devices housed either at a service provider (e.g., Hewlett-Packard or IBM) or on the customer's premises, paying only for the amount of storage they use. This approach is sensible for organizations with wildly fluctuating storage needs, such as those involved in the testing of new drugs or developing software.

ᑭUT AND OUTPUT DEVICES: THE GATEWAY TO COMPUTER SYSTEMS

Your first experience with computers is usually through input and output devices. These devices are the gateways to the computer system—you use them to provide data and instructions to the computer and receive results from it. Input and output devices are part of a computer's user interface, which includes other hardware devices and software that allow you to interact with a computer system.

As with other computer system components, an organization should keep their business goals in mind when selecting input and output devices. For example, many restaurant chains use handheld input devices or computerized terminals that let food servers enter orders efficiently and accurately. These systems have also cut costs by helping to track inventory and market to customers.

Characteristics and Functionality

In general, businesses want input devices that let them rapidly enter data into a computer system, and they want output devices that let them produce timely results. When selecting input and output devices, businesses also need to consider the form of the output they want, the nature of the data required to generate this output, and the speed and accuracy they need for both. Some organizations have very specific needs for output and input, requiring devices that perform specific functions. The more specialized the application, the more specialized the associated system input and output devices.

The speed and functions of input and output devices should be balanced with their cost, control, and complexity. More specialized devices might make it easier to enter data or output information, but they are generally more costly, less flexible, and more susceptible to malfunction.

The Nature of Data

Getting data into the computer—input—often requires transferring human-readable data, such as a sales order, into the computer system. "Human-readable" means data that people can read and understand. A sheet of paper containing inventory adjustments is an example of human-readable data. In contrast, machine-readable data can be understood and read by computer devices (e.g., the universal bar code that grocery scanners read) and is typically stored as bits or bytes. Inventory changes stored on a disk is an example of machine-readable data.

Some data can be read by people and machines, such as magnetic ink on bank checks. Usually, people begin the input process by organizing human-readable data and transforming it into machine-readable data. Every keystroke on a keyboard, for example, turns a letter symbol of a human language into a digital code that the machine can understand.

Data Entry and Input

data entry
Converting human-readable data into a machine-readable form.

data input
Transferring machine-readable data into the system.

Getting data into the computer system is a two-stage process. First, the human-readable data is converted into a machine-readable form through **data entry**. The second stage involves transferring the machine-readable data into the system. This is **data input**.

Today, many companies are using online data entry and input—they communicate and transfer data to computer devices directly connected to the computer system. Online data entry and input places data into the computer system in a matter of seconds. Organizations in many industries require the instantaneous updating offered by this approach. For example, when ticket agents need to enter a request for concert tickets, they can use online data entry and input to record the request as soon as it is made. Ticket agents at other terminals can then access this data to make a seating check before they process another request.

Source Data Automation

source data automation
Capturing and editing data where it is initially created and in a form that can be directly input to a computer, thus ensuring accuracy and timeliness.

Regardless of how data gets into the computer, it should be captured and edited at its source. **Source data automation** involves capturing and editing data where it is originally created and in a form that can be directly input to a computer, thus ensuring accuracy and timeliness. For example, using source data automation, salespeople enter sales orders into the computer at the time and place they take the order. Any errors can be detected and corrected immediately. If an item is temporarily out of stock, the salesperson can discuss options with the customer. Prior to source data automation, orders were written on paper and entered into the computer later (usually by a clerk, not the person who took the order). Often the handwritten information wasn't legible or, worse yet, got lost. If problems occurred during data entry, the clerk had to contact the salesperson or the customer to "recapture" the data needed for order entry, leading to further delays and customer dissatisfaction.

Input Devices

You can use hundreds of devices for data entry and input. They range from special-purpose devices that capture specific types of data to more general-purpose input devices. Some of the special-purpose data entry and input devices are discussed later in this chapter. First, we focus on devices used to enter and input general types of data, including text, audio, images, and video for personal computers.

Personal Computer Input Devices

A keyboard and a computer mouse are the most common devices used for entry and input of data such as characters, text, and basic commands. Some companies are developing keyboards that are more comfortable, more easily adjusted, and faster to use than standard keyboards. These ergonomic keyboards, such as the split keyboard by Microsoft and others, are designed to avoid wrist and hand injuries caused by hours of typing. Other keyboards include touchpads that let you enter sketches on the touchpad and text using the keys. Another innovation is wireless mice and keyboards, which keep a physical desktop free from clutter.

You use a computer mouse to point to and click symbols, icons, menus, and commands on the screen. The computer takes a number of actions in response, such as placing data into the computer system.

A keyboard and mouse are two of the most common devices for computer input. Wireless mice and keyboards are now readily available.

(Source: Courtesy of Hewlett-Packard Company.)

Speech-Recognition Technology

speech-recognition technology Input devices that recognize human speech.

Using **speech-recognition technology**, a computer equipped with a source of speech input such as a microphone can interpret human speech as an alternative means of providing data or instructions to the computer. The most basic systems require you to train the system to recognize your speech patterns or are limited to a small vocabulary of words. More advanced systems can recognize continuous speech without requiring you to break your speech into discrete words. The U.S. Department of Defense awarded $49 million to Johns Hopkins University to set up and run a Human Language Technology Center of Excellence to develop advanced technology and analyze a wide range of speech, text, and document image data in multiple languages. According to Gary Strong, the executive director of the center, "We need a better way to sort, filter, interpret, and call attention to important material that's buried within the enormous amount of multilingual data being produced every day in other nations. The government does not have nearly enough people with the multiple language skills needed to review this material. We need to develop technology to help."[34]

Companies that must constantly interact with customers are eager to reduce their customer support costs while improving the quality of their service. One company, Dial Directions, offers a free cell phone direction service. Users dial 347-328-4667 and tell the voice-activated service their originating location and desired destination and receive instant text messages with MapQuest driving directions on their cell phone.[35]

Digital Cameras

digital camera
An input device used with a PC to record and store images and video in digital form.

Digital cameras record and store images or video in digital form (see Figure 3.10). When you take pictures, the images are electronically stored in the camera. You can download the images to a computer either directly or by using a flash memory card. After you store the images on the computer's hard disk, you can edit and print them, send them to another location, or paste them into another application. For example, you can download a photo of your project team captured by a digital camera and then post it on a Web site or paste it into a project status report. Digital cameras have eclipsed film cameras used by professional photographers for photo quality and features such as zoom, flash, exposure controls, special effects, and even video-capture capabilities. With the right software, you can add sound and handwriting to the photo.

Figure 3.10

A Digital Camera

Digital cameras save time and money by eliminating the need to buy and process film.

(Source: Courtesy of Casio, Inc.)

More than two dozen camera manufacturers offer at least one digital camera model for under $225 with sufficient resolution to produce high-quality 5 × 7-inch photos. Many manufacturers offer a video camera that records full-motion video.

The primary advantage of digital cameras is saving time and money by eliminating the need to process film. In fact, digital cameras that can easily transfer images to CDs have made the consumer film business of Kodak and Fujitsu nearly obsolete. Until film-camera users switch to digital cameras, Kodak is allowing photographers to have it both ways. When you want to develop print film, Kodak offers the option of placing pictures on a CD in addition to the traditional prints. After the photos are stored on the CD, they can be edited, placed on a Web site, or sent electronically to business associates or friends around the world.

Organizations use digital cameras for research as well as for business purposes. Microsoft chairman Bill Gates and philanthropist Charles Simonyi donated $30 million to build the Large Synoptic Survey Telescope on a mountain in Chile. When operational in 2014, the 8.4 meter telescope will include a 3,200 megapixel digital camera that captures up to 30 TB of image data per night. The images from deep space will be loaded onto the Web and made available to the public.[36]

Terminals

Inexpensive and easy to use, terminals are input and display devices that perform data entry and input at the same time. A terminal is connected to a complete computer system, including a processor, memory, and secondary storage. After you enter general commands, text, and other data via a keyboard or mouse, it is converted into machine-readable form and transferred to the processing portion of the computer system. Terminals, normally connected directly to the computer system by telephone lines or cables, can be placed in offices, in warehouses, and on the factory floor.

Scanning Devices

You can input image and character data using a scanning device. A page scanner is like a copy machine. You typically insert a page you want to input into the scanner or place it face down on the glass plate of the scanner, cover it, and then scan it. With a handheld scanner, you manually move or roll the scanning device over the image you want to scan. Both page and handheld scanners can convert monochrome or color pictures, forms, text, and other images into machine-readable digits. Considering that U.S. enterprises generate an estimated 1 billion pieces of paper daily, many companies are looking to scanning devices to help them manage their documents and reduce the high cost of using and processing paper.

Optical Data Readers

You can also use a special scanning device called an *optical data reader* to scan documents. The two categories of optical data readers are for optical mark recognition (OMR) and optical character recognition (OCR). You use OMR readers for test scoring and other purposes when test takers use pencils to fill in boxes on OMR paper, which is also called a "mark sense form." OMR systems are used in standardized tests, including the SAT and GMAT tests, and are being considered as a means to capture voters' choices on Election Day. In comparison, most OCR readers use reflected light to recognize and scan various characters. With special software, OCR readers can convert handwritten or typed documents into digital data. After being entered, this data can be shared, modified, and distributed over computer networks to hundreds or thousands of people.

Con-way Inc. is a $4.7 billion company that offers freight transportation and logistics services. Not long ago, the company had an antiquated and expensive payroll system that required its 15,000 drivers to fill out timesheets, which were then manually processed at several document management centers. The centers forwarded the timesheets to Portland, where data-entry clerks keyed the information into the payroll system. The process was awkward, error prone, and expensive—Con-way paid $300,000 per year just to have the forms shipped to Portland. The firm converted to an OCR-based system so that the timesheets can be processed and scanned at any of 38 locations in the United States and Canada, and then forwarded electronically to the payroll office. Initially, the OCR scans were 85 percent accurate, but over time and with a few improvements, the scans are now 99.9 percent accurate. Con-way eliminated the cost of shipping the forms to Portland along with cost of three full-time positions in the payroll department. In addition, the OCR system increased the speed of the entire process and made it more reliable.[37]

Magnetic Ink Character Recognition (MICR) Devices

In the 1950s, the banking industry became swamped with paper checks, loan applications, bank statements, and so on. The result was the development of magnetic ink character recognition (MICR), a system for reading banking data quickly. With MICR, data is placed on the bottom of a check or other form using a special magnetic ink. Using a special character set, data printed with this ink is readable by people and computers (see Figure 3.11).

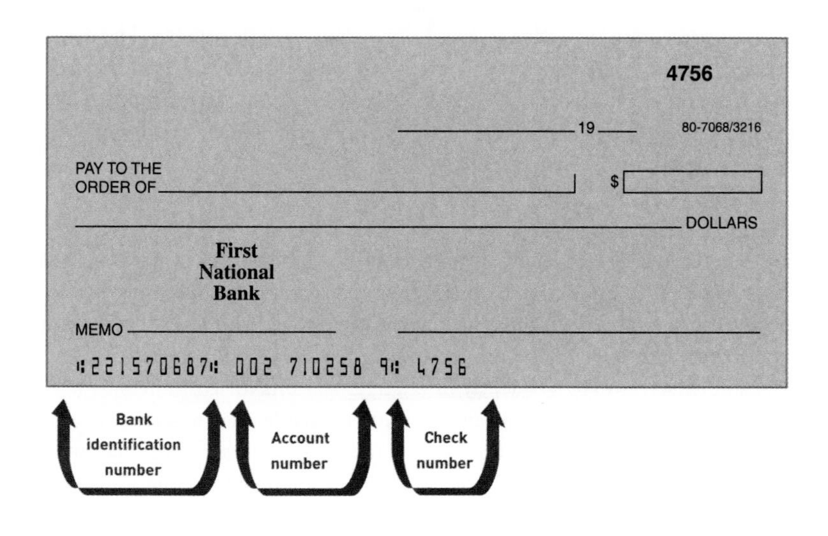

Figure 3.11

MICR Device

Magnetic ink character recognition technology codes data on the bottom of a check or other form using special magnetic ink, which is readable by people and computers. For an example, look at the bottom of a bank check.

(Source: Courtesy of NCR Corporation.)

Magnetic Stripe Card

magnetic stripe card
A type of card that stores limited amounts of data by modifying the magnetism of tiny iron-based particles contained in a band on the card.

A **magnetic stripe card** stores limited amounts of data by modifying the magnetism of tiny iron-based particles contained in a band on the card. The magnetic stripe is read by physically swiping the card past a reading head. Magnetic stripe cards are commonly used in credit cards, transportation tickets, and driver's licenses. The Revolution Card credit card is being touted as more secure than traditional credit cards such as those from Discover, Visa, and MasterCard. The cardholder's name does not appear on the card nor does the card contain any information about the cardholder in the magnetic stripe. Instead, the user must enter a personal ID number to use the card.[38]

Point-of-Sale Devices

point-of-sale (POS) device
A terminal used in retail operations to enter sales information into the computer system.

Point-of-sale (POS) **devices** are terminals used in retail operations to enter sales information into the computer system. The POS device then computes the total charges, including tax. Many POS devices also use other types of input and output devices, such as keyboards, bar-code readers, printers, and screens. Much of the money that businesses spend on computer technology involves POS devices. First Data, Hewlett-Packard, and Microsoft have collaborated to create a combined hardware and software point-of-sale solution for small retailers called First Data POS Value Exchange. The system can handle all forms of payment including cash, check, credit, debit, and gift cards. The software comes installed on Hewlett-Packard's rp5000 computer, complete with a touch screen interface.[39]

Automated Teller Machine (ATM) Devices

Another type of special-purpose input/output device, the automated teller machine (ATM) is a terminal that bank customers use to perform withdrawals and other transactions with their bank accounts. The ATM, however, is no longer used only for cash and bank receipts. Companies use various ATM devices, sometimes called *kiosks,* to support their business processes. Some can dispense tickets, such as for airlines, concerts, and soccer games. Some colleges use them to produce transcripts. AT&T and Wireless Advocates (a provider of mobile phones and services) sell mobile phones from manufacturers such as Samsung, Nokia, and Motorola plus services at kiosks inside Costco stores.[40]

Pen Input Devices

By touching the screen with a pen input device, you can activate a command or cause the computer to perform a task, enter handwritten notes, and draw objects and figures. Pen input requires special software and hardware. Handwriting recognition software can convert handwriting on the screen into text. The Tablet PC from Microsoft and its hardware partners can transform handwriting into typed text and store the "digital ink" just the way a person writes it. Users can use a pen to write and send e-mail, add comments to Word documents, mark up PowerPoint presentations, and even hand-draw charts in a document. The data can then be moved, highlighted, searched, and converted into text. If perfected, this interface is likely to become widely used. Pen input is especially attractive if you are uncomfortable using a keyboard. The success of pen input depends on how accurately handwriting can be read and translated into digital form and at what cost.

Touch-Sensitive Screens

Advances in screen technology allow display screens to function as input as well as output devices. By touching certain parts of a touch-sensitive screen, you can start a program or trigger other types of action. Touch-sensitive screens are popular input devices for some small computers because they do not require a keyboard, which conserves space and increases portability. Touch screens are frequently used at gas stations for customers to select grades of gas and request a receipt, on photocopy machines to enable users to select various options, at fast-food restaurants for order clerks to enter customer choices, at information centers in hotels to allow guests to request facts about local eating and drinking establishments, and at amusement parks to provide directions to patrons. They also are used in kiosks at airports and department stores. Touch-sensitive screens are also being considered as a technology to use in capturing voter choices.

Bar-Code Scanners

A bar-code scanner employs a laser scanner to read a bar-coded label. This form of input is used widely in grocery store checkouts and warehouse inventory control. Often, bar-code technology is combined with other forms of technology to create innovative ways for capturing data.

Radio Frequency Identification

The purpose of a **Radio Frequency Identification** (**RFID**) system is to transmit data by a mobile device, called a tag (see Figure 3.12), which is read by an RFID reader and processed according to the needs of an IS program. One popular application of RFID is to place a microchip on retail items and install in-store readers that track the inventory on the shelves to determine when shelves should be restocked. Recall that the RFID tag chip includes a special form of EPROM memory that holds data about the item to which the tag is attached. A radio frequency signal can update this memory as the status of the item changes. The data transmitted by the tag might provide identification, location information, or details about the product tagged, such as date manufactured, retail price, color, or date of purchase.

Radio Frequency Identification (RFID)
A technology that employs a microchip with an antenna that broadcasts its unique identifier and location to receivers.

Figure 3.12

RFID Tag

An RFID tag is small compared to current bar-code labels used to identify items.

(Source: Courtesy of Intermec Technologies Corporation.)

Boekhandels Groep Nederland (BGN) is a major book retailer with 40 stores in the Netherlands that sell to roughly 30,000 customers per day. BGN implemented item-level RFID tagging to track the movement of books along with new software to create a tightly integrated warehouse-to-consumer supply chain. With this solution, BGN simplified the inventory process, reduced errors in inventory, and improved the entire supply chain process.[41]

Read the Ethical and Societal Issues special feature to learn about the various approaches being taken to capture votes in an accurate and verifiable manner.

ETHICAL AND SOCIETAL ISSUES

Collecting Accurate and Verifiable Data Where It Counts

Imagine having to design or choose an input device that will satisfy every person's needs: the young, elderly, intelligent, illiterate, sighted, or blind. Then imagine that this device has to provide a 100 percent guarantee that it is easy to use for all and collects accurate data—exactly what the user wants to enter. Sound challenging? That's the struggle that countries around the world are facing as they continue to create the perfect voting machine.

As the technology revolution races ahead, those responsible for voting systems are trying to harness technology to streamline the voting process. Submitting paper ballots now seems prehistoric in this day of movie downloads and cell phone text messaging. It was only natural that the touch screen would make its way into the voting process—with disastrous results. Touch screen machines, also called Direct Recording Electronic (DRE) units, allow the voter to press the name of the person for whom they want to vote. Each vote is either stored in the machine's storage device to be collected later and batch processed, or sent directly to a central database over a private network.

The use of touch screen machines has led to numerous questionable elections and accusations of scandal. The most prominent are the 2000 and 2004 United States presidential elections, where the close results were questioned due to voting irregularities caused by electronic voting machines.

To overcome the problems with touch screen voting machines, many experts feel that a paper backup of a citizen's vote should be generated along with the electronic vote. By providing a "paper trail" of votes, questionable elections can be easily checked. At the time of this writing, 12 U.S. states still have no paper record requirements.

Many voting administrators have given up on touch screen voting systems altogether. In the 2008 primary presidential elections, the state of New Hampshire relied on optical mark recognition (OMR) technology for their voting. Voters fill in the circle next to a candidate's name on a card. The voter's card is scanned to record the votes, and is filed away as a backup in case a recount is needed. Some precincts in New Hampshire provide voters with simple paper ballots that are counted by hand at the polling place. Visually impaired and disabled can use a touch-tone phone to place their votes. In this way, New Hampshire uses several methods to collect votes.

Other states are experimenting with other systems. Oregon holds its votes by mail. Citizens do not have to travel to a precinct center to cast their votes; instead, they simply mark their ballots, stamp them, and put them in the mailbox. The state claims record voting turnouts and little strife.

Some states seem committed to touch screen systems. Despite a report describing several methods of compromising the vote records of its voting machines, the Crawford County, Ohio, county commissioner tells the citizens that there is nothing to worry about. Since only officials from the county are provided with access codes to the inner workings of the machines, the system should be secure.

The voting machine debate extends beyond the United States to every other voting country. In Germany, a group of computer experts collected signatures to request that a court grant an injunction stopping the use of electronic voting machines. They wanted the system switched to a paper ballot system. They argue that the system had security flaws that allowed a hacker to manipulate voting outcomes. The group contended that the government didn't have the technical understanding to ensure an accurate vote count using the electronic system.

As the search for the perfect voting machine continues, one thing is clear: Collecting data into a system that is verifiably accurate, using easy and fast methods, can be a challenge. Security experts put forth three requirements for touch screen machines: They should produce a voter-verifiable paper trail, use software that is open to examination by the public, and provide verifiable ballots to safeguard against machine failure.

Discussion Questions

1. Do you think that one method of collecting data into a voting system can satisfy all the different types of voters? Or are multiple methods required?
2. What would be your concerns about elections by mail, such as the system used in Oregon?

Critical Thinking Questions

1. Of the systems described in this feature, which would you most like to use? In other words, describe your ideal voting method.
2. What are the security risks of your ideal voting method?

SOURCES: Weiss, Todd, "As primary season ramps up, an e-voting snapshot," *Computerworld*, January 8, 2008, *www.computerworld.com/action/article.do?command=viewArticleBasic&articleId=9056098*. Smith, Jane, "Officials confident voting machines pose no problems," The Meadville Tribune, January 12, 2008, *www.meadvilletribune.com/local/local_story_009222956.html*. Kirk, Jeremy, "German activists move to block e-voting," *NetworkWorld*, January 8, 2008, *www.networkworld.com/news/2008/010808-german-activists-move-to-block.html?fsrc=rss-security*. Kim, Myung, "Most clerks pushing for mail ballots," Rocky Mountain News, January 12, 2008, *www.rockymountainnews.com/news/2008/jan/12/most-clerks-pushing-for-mail-ballots*.

Output Devices

Computer systems provide output to decision makers at all levels of an organization so they can solve a business problem or capitalize on a competitive opportunity. In addition, output from one computer system can provide input into another computer system. The desired form of this output might be visual, audio, or even digital. Whatever the output's content or form, output devices are designed to provide the right information to the right person in the right format at the right time.

Display Monitors

The display monitor is a device similar to a TV screen that displays output from the computer. Because early monitors used a cathode-ray tube to display images, they were sometimes called *CRTs.* Such a monitor works much the same way a traditional TV screen does—the cathode-ray tubes generate one or more electron beams. As the beams strike a phosphorescent compound (phosphor) coated on the inside of the screen, a dot on the screen called a **pixel** lights up. A pixel is a dot of color on a photo image or a point of light on a display screen. It appears in one of two modes: on or off. The electron beam sweeps across the screen so that as the phosphor starts to fade, it is struck and lights up again.

pixel
A dot of color on a photo image or a point of light on a display screen.

With today's wide selection of monitors, price and overall quality can vary tremendously. The quality of a screen image is often measured by the number of horizontal and vertical pixels used to create it. The more pixels per square inch, the higher the resolution, or clarity and sharpness, of the image. For example, a screen with a 1,024 × 768 resolution (786,432 pixels) has a higher sharpness than one with a resolution of 800 × 600 (480,000 pixels). Another way to measure image quality is the distance between one pixel on the screen and the next nearest pixel, which is known as *dot pitch.* The common range of dot pitch is from .25 mm to .31 mm. The smaller the dot pitch, the better the picture. A dot pitch of .28 mm or smaller is considered good. Greater pixel densities and smaller dot pitches yield sharper images of higher resolution.

The characteristics of screen color depend on the quality of the monitor, the amount of RAM in the computer system, and the monitor's graphics adapter card. Digital Video Interface (DVI) is a video interface standard designed to maximize the visual quality of digital display devices such as flat-panel LCD computer displays.

Companies are competing on the innovation frontier to create thinner display devices for computers, cell phones, and other mobile devices. In its effort to gain an edge, LG Phillips has developed an extremely thin display that is only .15 mm thick, or roughly as thick as a human hair. The display is also flexible so that it can be bent or rolled up without being damaged. This flexible display opens up some exciting possibilities for manufacturers to make cell phones, PDAs, and laptops with significantly larger displays but without increasing the size of the device itself, as the screen could be rolled up or folded and tucked away into a pocket.[42]

Plasma Displays

A **plasma display** uses thousands of smart cells (pixels) consisting of electrodes and neon and xeon gases that are electrically turned into plasma (electrically charged atoms and negatively charged particles) to emit light. The plasma display lights up the pixels to form an image based on the information in the video signal. Each pixel is made up of three types of light—red, green, and blue. The plasma display varies the intensities of the lights to produce a full range of colors. Plasma displays can produce high resolution and accurate representation of colors to create a high-quality image.

plasma display
A plasma display uses thousands of smart cells (pixels) consisting of electrodes and neon and xeon gases which are electrically turned into plasma (electrically charged atoms and negatively charged particles) to emit light.

Liquid Crystal Displays (LCDs)

LCD displays are flat displays that use liquid crystals—organic, oil-like material placed between two polarizers—to form characters and graphic images on a backlit screen. These displays are easier on your eyes than CRTs because they are flicker-free, brighter, and don't emit the type of radiation that makes some CRT users worry. In addition, LCD monitors take up less space and use less than half of the electricity required to operate a comparably sized CRT monitor. *Thin-film transistor (TFT) LCDs* are a type of liquid crystal display that

LCD display
Flat display that uses liquid crystals—organic, oil-like material placed between two polarizers—to form characters and graphic images on a backlit screen.

assigns a transistor to control each pixel, resulting in higher resolution and quicker response to changes on the screen. TFT LCD monitors have displaced the older CRT technology and are commonly available in sizes from 12 to 30 inches. A number of companies are capable of providing multimonitor solutions that enable users to see a wealth of related information at a single glance, as shown in Figure 3.13.

Figure 3.13

A Four-Screen Wide Display

(Source: © Justin Pumfrey/Getty Images.)

Organic Light-Emitting Diodes

Organic light-emitting diode (OLED) technology is based on research by Eastman Kodak Company and is appearing on the market in small electronic devices. OLEDs use the same base technology as LCDs, with one key difference: Whereas LCD screens contain a fluorescent backlight and the LCD acts as a shutter to selectively block that light, OLEDs directly emit light. OLEDs can provide sharper and brighter colors than LCDs and CRTs, and because they don't require a backlight, the displays can be half as thick as LCDs and used in flexible displays. Another big advantage is that OLEDs don't break when dropped. OLED technology can also create three-dimensional (3-D) video displays by taking a traditional LCD monitor and then adding layers of transparent OLED films to create the perception of depth without the need for 3-D glasses or laser optics.[43]

Printers and Plotters

One of the most useful and popular forms of output is called *hard copy*, which is simply paper output from a printer. The two main types of printers are laser printers and inkjet printers, and they are available with different speeds, features, and capabilities. Some can be set up to accommodate paper forms, such as blank check forms and invoice forms. Newer printers allow businesses to create customized printed output for each customer from standard paper and data input using full color. Ticket-receipt printers such as those used in restaurants, ATMs, and point-of-sale systems are in wide-scale use.

The speed of the printer is typically measured by the number of pages printed per minute (ppm). Like a display screen, the quality, or resolution, of a printer's output depends on the number of dots printed per inch (dpi). A 600-dpi printer prints more clearly than a 300-dpi printer. A recurring cost of using a printer is the inkjet or laser cartridge that must be replaced periodically—every few thousand pages for laser printers and every 500 to 900 pages for inkjet printers. Figure 3.14 shows a laser printer.

Laser printers are generally faster than inkjet printers and can handle more volume than inkjet printers. Laser printers print 15 to 50 pages per minute (ppm) for black and white and 4 to 20 ppm for color. Inkjet printers print 10 to 30 ppm for black and white and 2 to 10 ppm for color.

For color printing, inkjet printers print vivid hues and with an initial cost much less than color laser printers. Inkjet printers can produce high-quality banners, graphics, greeting cards, letters, text, and prints of photos. Hewlett-Packard introduced the CM8060 inkjet printers with a stationary print head that uses 60,000 nozzles to spray ink as the paper moves.

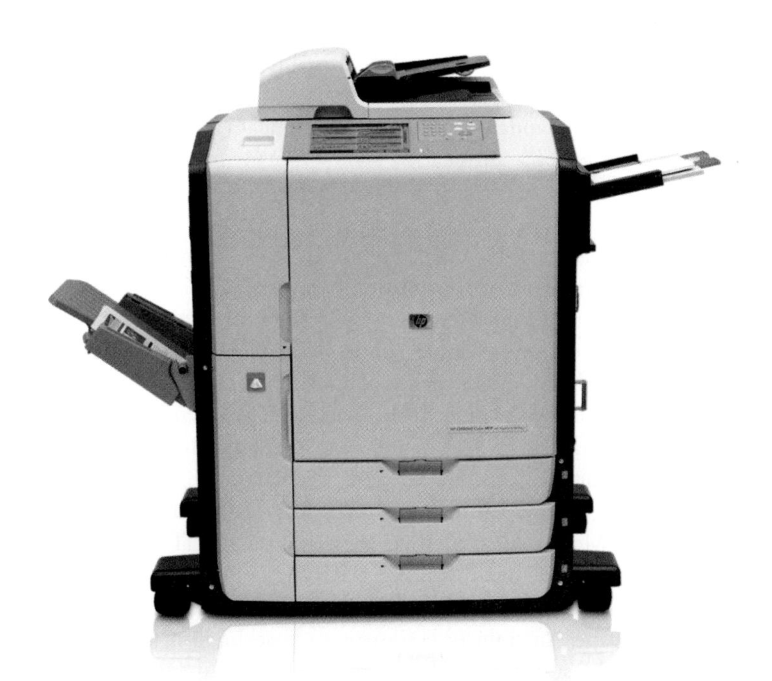

Figure 3.14

The Hewlett-Packard CM8060 Inkjet Printer

(Source: Courtesy of Hewlett-Packard Company.)

Traditional inkjet printers rely on a moving print head and many fewer nozzles. The advantage of the new technology is the ability to print pages much faster (50 pages per minute for color and 60 pages per minute for black and white). In addition, the printer uses less ink.[44]

A number of manufacturers offer multiple-function printers that can copy, print (in color or black and white), fax, and scan. Such multifunctional devices are often used when people need to do a relatively low volume of copying, printing, faxing, and scanning. The typical price of multifunction printers ranges from $150 to $500, depending on features and capabilities. Because these devices take the place of more than one piece of equipment, they are less expensive to acquire and maintain than a stand-alone fax, plus a stand-alone printer, plus a stand-alone copier, and so on. Also, eliminating equipment that was once located on a countertop or desktop clears a workspace for other work-related activities. As a result, such devices are popular in homes and small office settings.

3-D printers can be used to turn three-dimensional computer models into three-dimensional objects. See Figure 3.15. One form of 3-D printer uses an inkjet printing system to print an adhesive in the shape of a cross-section of the model. Next, a fine powder is sprayed onto the adhesive to form one layer of the object. This process is repeated thousands of times until the object is completed. 3-D printing is commonly used in aerospace companies, auto manufacturers, and other design-intensive companies. It is especially valuable during the conceptual stage of engineering design, when the exact dimensions and material strength of the prototype are not critical.

Figure 3.15

The Spectrum Z510 3D Printer

(Source: Courtesy of Z Corporation.)

Plotters are a type of hard-copy output device used for general design work. Businesses typically use plotters to generate paper or acetate blueprints, schematics, and drawings of buildings or new products. Standard plot widths are 24 inches and 36 inches, and the length can be whatever meets the need—from a few inches to many feet.

Digital Audio Player

digital audio player
A device that can store, organize, and play digital music files.

MP3
A standard format for compressing a sound sequence into a small file.

A **digital audio player** is a device that can store, organize, and play digital music files. **MP3** (MPEG-1 Audio Layer-3) is a popular format for compressing a sound sequence into a very small file while preserving the original level of sound quality when it is played. By compressing the sound file, it requires less time to download the file and less storage space on a hard drive.

You can use many different music devices about the size of a deck of cards to download music from the Internet and other sources. These devices have no moving parts and can store hours of music. Apple expanded into the digital music market with an MP3 player (the iPod) and the iTunes Music Store, which allows you to find music online, preview it, and download it in a way that is safe, legal, and affordable. Other MP3 manufacturers include Dell, Sony, Samsung, Iomega, and Motorola, whose Rokr product is the first iTunes-compatible phone.

The Apple iPod Touch 3.5-inch widescreen lets the user watch movies and TV shows and view photos and album art. The display automatically adjusts the view when it's rotated from portrait to landscape. An ambient light sensor adjusts brightness to match the current lighting conditions. It also supports wireless networking so that the user can access the Internet, view YouTube videos, and purchase music from the iTunes Wi-Fi Music Store.

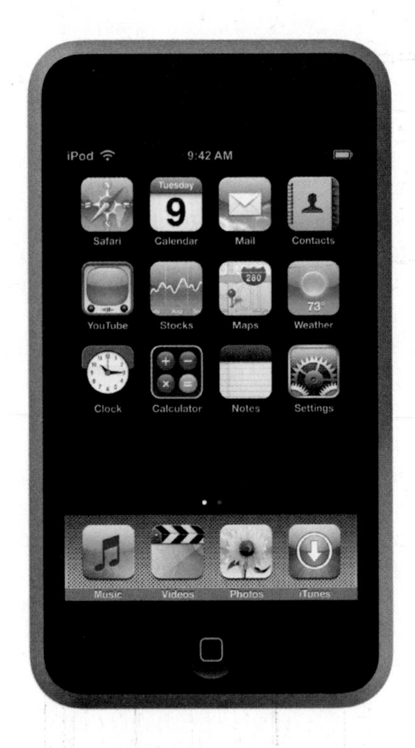

The Apple's iPod Touch
(Source: Courtesy of Apple.)

Special-Purpose Input and Output Devices

Many additional input and output devices are used for specialized or unique applications. Two examples of such devices are discussed in the following sections.

E-Books

The digital media equivalent of a conventional printed book is called an e-book (short for electronic book). The Project Gutenberg Online Book Catalog lists over 20,000 free e-books and a total of more than 100,000 e-books available. E-books can be downloaded from this source or many others onto personal computers or dedicated hardware devices known as e-book readers. A number of e-book hardware devices are available including the Kindle from Amazon.com and the Pepper Pad, Cybook, Franklin eBookMan, Easyread, Personal Digital Reader, and Hanlin Reader, all priced from around $150 to $400. The Sony Reader Digital Book sells for just under $300 and has a 6-inch display that uses e-Ink technology and is almost paper-like, making it easy to read even in bright sunshine. The text can also be magnified for readers with impaired vision. The Reader Digital Book weighs 9 ounces, is only half an inch thin, and can store up to 160 e-Books. It is more compact than most paperbacks so it can be easily held in one hand.[45]

Eyebud Screens

Eyebud screens are portable media devices that display video in front of one eye. They employ optical technology that provides very high resolution and "enlarges" the video or images. With the proximity of the screen to the eye and the magnifying effect of the optical technology, using an eyebud screen is like watching a 105-inch display from 12 feet away. Such devices enable users of portable media devices to capture the big-screen, movie-screen, or home-theater experience, wherever they are.

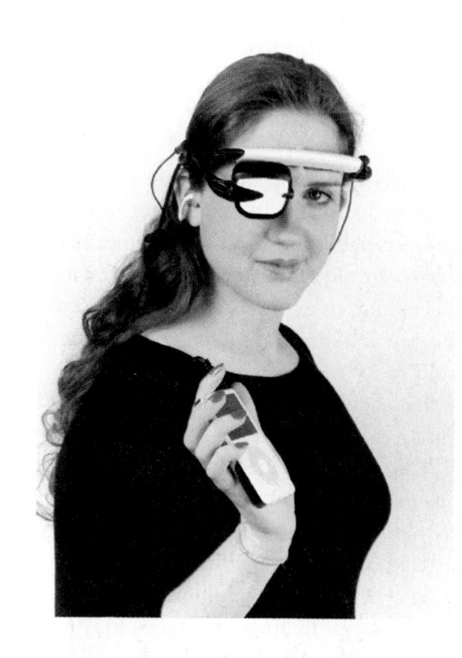

The eyebud screen displays "enlarged" and high-resolution video.

(Source: Courtesy of eMagin Corp.)

COMPUTER SYSTEM TYPES

In general, computers can be classified as either special purpose or general purpose. *Special-purpose computers* are used for limited applications by military and scientific research groups such as the CIA and NASA. Other applications include specialized processors found in appliances, cars, and other products. For example, automobile repair shops connect special-purpose computers to your car's engine to identify specific performance problems.

General-purpose computers are used for a variety of applications and to execute the business applications discussed in this text. General-purpose computer systems can be divided into two major groups: systems used by one user at a time and systems used by multiple concurrent users. Table 3.4 shows the general ranges of capabilities for various types of computer systems.

	Single-User Systems					Multiuser System			
Factor	**Handheld**	**Ultra Laptop**	**Portable**	**Thin Client**	**Desktop**	**Workstation**	**Server**	**Mainframe**	**Supercomputer**
Cost Range	$90 to $900	$700 to $2250	$500 to $3,000	$300 to $900	$400 to $2,500	$3,000 to $40,000	$500 to $50,000	>$100,000	>$250,000
Weight	<24 oz.	<3 lbs.	<7 lbs.	<15 lbs.	<25 lbs.	<25 lbs.	>25 lbs.	>200 lbs.	>200 lbs.
Typical Size	Palm size	Size of a notebook	Size of a notebook	Fits on desktop	Fits on desktop	Fits on desktop	Three-drawer filing cabinet	Refrigerator	Refrigerator and larger
Typical Use	Organize personal data	Improve productivity of highly mobile worker	Improve worker productivity	Enter data and access the Internet	Improve worker productivity	Perform engineering, CAD, and software development	Perform network and Internet applications	Perform computing tasks for large organizations and provide massive data storage	Run scientific applications; perform intensive number crunching
Example	HP iPAQ Pocket PC	Fujitsu Lifebook Q2010	Dell Inspiron T5450	Wyse V90LE Thin Client	Mac Pro	Sun Ultra 40 M2 workstation	Hewlett-Packard HP ProLiant BL	Unisys Clear Path	IBM RS/6000 SP

Table 3.4

Types of Computer Systems

Computer System Types

Computer systems can range from small handheld computers to massive supercomputers that fill an entire room. We start first with the smallest computers.

Handheld Computers

handheld computer
A single-user computer that provides ease of portability because of its small size.

Handheld computers are single-user computers that provide ease of portability because of their small size—some are as small as a credit card. These systems often include a variety of software and communications capabilities. Most can communicate with desktop computers over wireless networks. Some even add a built-in GPS receiver with software that can integrate location data into the application. For example, if you click an entry in an electronic address book, the device displays a map and directions from your current location. Such a computer can also be mounted in your car and serve as a navigation system. One of the shortcomings of handheld computers is that they require a lot of power relative to their size.

PalmOne is the company that invented the Palm Pilot organizer in 1996. The Palm personal digital assistant (PDA) lets you track appointments, addresses, and tasks. PalmOne has now signed licensing agreements with Handspring, IBM, Sony, and many other manufacturers, permitting them to make what amounts to Palm clones. As a result of the popularity of the Palm PDA, all handheld computers are often referred to as PDAs.

The U.S. Census Bureau awarded a $600 million contract to the Harris Corporation to integrate multiple automated systems to obtain field data for the 2010 census in an efficient and secure manner. It is anticipated that 500,000 PDAs will be used by census takers for address canvassing and nonresponse followup.[46]

smartphone
A phone that combines the functionality of a mobile phone, personal digital assistant, camera, Web browser, e-mail tool, and other devices into a single handheld device.

A **smartphone** combines the functionality of a mobile phone, personal digital assistant, camera, Web browser, e-mail tool, MP3 player, and other devices into a single handheld device. Smartphones will continue to evolve as new applications are defined and installed on the device. The applications might be developed by the manufacturer of the handheld device, by the operator of the communications network on which it operates, or by any other third-party software developer.

Portable Computers

portable computer
A computer small enough to be carried easily.

Many computer manufacturers offer a variety of **portable computers**, those that can be carried easily—from pocket or handheld computers to laptops, to notebooks, to subnotebooks, to tablet computers.

The pocket computer is a device smaller than the smallest laptop that can perform most of the common functions of a PC. The Nokia 770 Internet Tablet is a $360 pocket computer that you can use to surf the Web, send and receive e-mail and instant messages, view images and videos, and play music and simple games. It weighs 8.1 ounces and is 5.5 inches long and 0.7 inches thick. It has a vivid, bright display with a resolution of 800 × 480, which is very good for a handheld digital device.[47]

The Coca-Cola field salesforce uses a Pocket PC (a handheld computer that runs the Microsoft Windows Mobile operating system) to automate the collection of information about sales calls, customers, and prospects. They chose a Pocket PC over a laptop because of the cost savings and because it is easier to point and click using radio buttons and drop-down menus on the Pocket PC than to fumble with a keyboard and mouse on a much heavier laptop.[48]

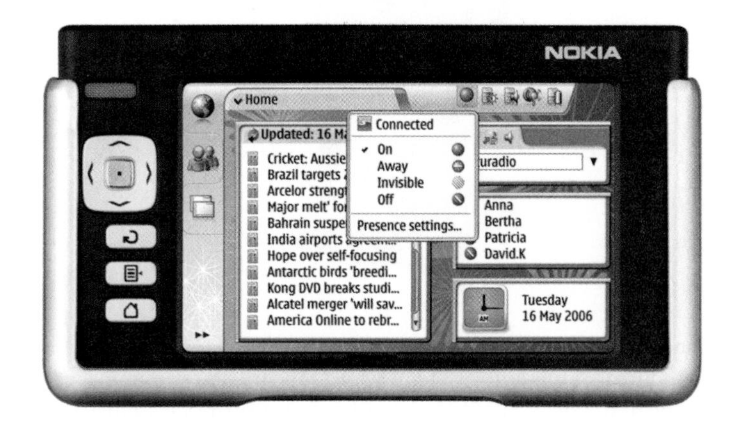

The Nokia 770 Internet Tablet is designed for wireless Internet browsing and electronic mail, and includes software such as Internet radio, an RSS news reader, and audio and video players.

(Source: Courtesy of Nokia.)

An ultra laptop is a laptop computer weighing less than 3 pounds (1.4 kg) and is usually targeted for use by business travelers. Such laptops typically have a screen that measures 12 inches (30 cm) or less diagonally with a less than full-size keyboard. Many ultra laptops come with extended battery life and energy-efficient CPUs. Popular ultra laptop computers include the Fujitsu Lifebook Q2010, Lenovo ThinkPad X60, Sony VAIO VGN-TXN15P, Samsung Q1 Ultra, and Apple MacBook Air. See Figure 3.16. The Q1 uses tablet PC technology with a built-in keyboard and can accept handwritten notes on its computer screen. These devices cost between $700 and $2,250 as of January 2008.[49]

Figure 3.16

The MacBook Air

The MacBook Air is an ultraportable laptop that measures 0.76 inches deep at the back and tapers down to 0.16 inches at the front. It weighs 3 pounds and includes a 13-inch LED screen and full-size keyboard.

(Source: Courtesy of Apple.)

Tablet PCs (introduced earlier) are portable, lightweight computers that allow you to roam the office, home, or factory floor carrying the device like a clipboard. Recall that you can enter text with a writing stylus directly on the screen thanks to built-in handwriting recognition software. Other input methods might include an on-screen (virtual) keyboard, speech recognition, or a physical keyboard. Tablet PCs that only support input via a writing stylus are called *slates*. The *convertible tablet PC* comes with a swivel screen and can be used as a traditional notebook or as a pen-based tablet PC.

Tablet computers are especially popular and useful in the healthcare, retail, insurance, and manufacturing industries because of their versatility. DT Research provides portable tablet personal computers that come with optional input devices including an integrated bar-code scanner, a card reader, and a camera. The bar-code scanner can capture data from retail, patient, or shipping labels. The card reader can capture data from any card with a magnetic stripe, such as a credit card or driver's license. These devices capture data quickly and accurately and can input the data directly into applications running on the computer.[50] CSX

The Nokia N80 smartphone can function as a phone, camera, FM radio, or e-mail device, and allows you to send text messages with audio and video clips.

(Source: Courtesy of Nokia.)

Transportation, one of the nation's largest railroads, uses DT Research's WebDT 360 to enable train conductors to monitor systems while onboard and communicate with stations for real-time updates. The WebDT 360 has improved operations efficiency and worker productivity.[51]

Low-Cost Laptops The mission of the nonprofit One Laptop per Child (OLPC) association is to provide children around the world with new opportunities to explore, experiment, and express themselves with the help of a low-cost laptop priced at about $100. OLPC was founded by Nicholas Negroponte of the Massachusetts Institute of Technology and includes a wide variety of members from academia, the arts, business, and the information technology industry. Negroponte states: "It's an education project, not a laptop project."[52] The first version of the laptop, the OLPC XO, was made available to third-world countries in 2007. OLPC launched a "give one, get one" campaign in North America, asking consumers to pay $400 for an XO for themselves and a "free" XO to be given to a child in a developing country. The bright green computer is designed to be extremely rugged and durable with child-friendly features including an easy-to-use interface. See Figure 3.17.

Figure 3.17

The OLPC XO Laptop Computer

(Source: Courtesy of fuseproject.)

For-profit computer manufacturers have also recognized the tremendous market for low-cost computers. Taiwan's Asus produces a $300–$400 basic laptop called the Eee. Intel, in collaboration with local manufacturers in the developing world, is producing a $300 laptop for use in schools called the Classmate PC.[53] Mary Lou Jepsen, founder of low-cost laptop company Pixel Xi, says, "The computer industry has been able to keep the price flat by focusing on gazillion-gigahertz machines running really bloated software, and that's worked for years since the IBM PC revolution."[54] Will the development of low-cost, limited-capability laptops drive software manufacturers to develop simpler, less resource-intensive programs? What will happen to the current market for relatively high-priced laptops?

Thin Client

A **thin client** is a low-cost, centrally managed computer with no extra drives, such as a CD or DVD drive, or expansion slots. These computers have limited capabilities and perform only essential applications, so they remain "thin" in terms of the client applications they include. These stripped-down versions of desktop computers do not have the storage capacity or computing power of typical desktop computers, nor do they need it for the role they play. With no hard disk, they never pick up viruses or suffer a hard disk crash. Unlike personal computers, thin clients download software from a network when needed, making support, distribution, and updating of software applications much easier and less expensive. Thin-client manufacturers include Hewlett-Packard, Wyse, BOSaNOVA, and DTR Research.

thin client
A low-cost, centrally managed computer with essential but limited capabilities and no extra drives, such as a CD or DVD drive, or expansion slots.

Amerisure is a mutual insurance company specializing in worker's compensation policies, with 800 employees spread across eight U.S. locations. The firm had 700 personal computers of varying types and manufacturers. The diverse set of computers was so difficult to support that the resulting lack of operational reliability threatened the firm's quality of service. To improve the situation, Amerisure converted every PC into a thin client capable of running the same software, but with processing taking place on centralized servers rather than on users' desktops. This conversion resulted in a much more stable environment that led to savings of nearly $1 million per year from reduced hardware and support costs.[55]

Desktop Computers

Desktop computers are relatively small, inexpensive single-user computer systems that are highly versatile. Named for their size—the parts are small enough to fit on or beside an office desk—*desktop computers* can provide sufficient memory and storage for most business computing tasks.

desktop computer
A relatively small, inexpensive, single-user computer that is highly versatile.

Ultrasmall desktop personal computers are much smaller than traditional desktop computers yet often as powerful in terms of processor speed and hard drive capacity. Although they may have fewer expansion slots for RAM and other devices, they are highly energy efficient. These small computers are available from most personal computer manufacturers. For example, the Mac mini is the smallest desktop computer released from Apple. It measures only 6.5 × 6.5 × 2 inches and weighs 2.9 pounds. You can purchase a Mac mini with a 2-GHz Intel Core 2 Duo chip set, a 120 GB hard drive, and 1 GB of RAM for under $1,000. The Mac mini does not come with a mouse, keyboard, or monitor but it works with surplus equipment from Windows personal computers. The Mac mini comes with software that lets the user enhance, organize, and share photos—on the Mac mini itself or stored on other computers.[56] See Figure 3.18.

Figure 3.18
The Apple Mac mini
(Source: Courtesy of Apple.)

Workstations

Workstations are more powerful than personal computers but still small enough to fit on a desktop. They are used to support engineering and technical users who perform heavy mathematical computing, computer-aided design (CAD), and other applications requiring a high-end processor. Such users need very powerful CPUs, large amounts of main memory, and extremely high-resolution graphic displays.

workstation
A more powerful personal computer that is used for technical computing, such as engineering, but still fits on a desktop.

Servers

server
A computer designed for a specific task, such as network or Internet applications.

A **server** is a computer used by many users to perform a specific task, such as running network or Internet applications. Servers typically have large memory and storage capacities, along with fast and efficient communications abilities. A Web server handles Internet traffic and communications. An Internet caching server stores Web sites that a company uses frequently. An enterprise server stores and provides access to programs that meet the needs of an entire organization. A file server stores and coordinates program and data files. A transaction server processes business transactions. Server systems consist of multiuser computers, including supercomputers, mainframes, and other servers. Often an organization will house a large number of servers in the same room where access to the machines can be controlled and authorized support personnel can more easily manage and maintain them from this single location. Such a facility is called a *server farm.*

scalability
The ability to increase the capability of a computer system to process more transactions in a given period by adding more, or more powerful, processors.

Servers offer great **scalability**, the ability to increase the processing capability of a computer system so that it can handle more users, more data, or more transactions in a given period. Scalability is increased by adding more, or more powerful, processors. *Scaling up* adds more powerful processors, and *scaling out* adds many more equal (or even less powerful) processors to increase the total data-processing capacity.

A virtual server is a method of logically dividing the resources of a single physical server to create multiple logical servers, with each acting as if it is running on its own dedicated machine. Often a single physical Web server is divided into two virtual private servers. One of the virtual servers hosts the live Web site while the other hosts a copy of the Web site. The second private virtual server is used to test and verify updates to software before changes are made to the live Web site. The U.S. Marine Corps has adopted server virtualization to reduce the number of its data centers from 300 to 30 plus 100 mobile platforms. They will do this by setting up virtual servers capable of running half a dozen or so applications, eliminating the need to dedicate one server to one application.[57]

blade server
A server that houses many individual computer motherboards that include one or more processors, computer memory, computer storage, and computer network connections.

A **blade server** houses many computer motherboards that include one or more processors, computer memory, computer storage, and computer network connections. These all share a common power supply and air-cooling source within a single chassis. By placing many blades into a single chassis, and then mounting multiple chassis in a single rack, the blade server is more powerful but less expensive than traditional systems based on mainframes or server farms of individual computers. In addition, the blade server approach requires much less physical space than traditional server farms.

The city of Burbank, California upgraded to IBM blade servers to provide the speed and flexibility it needed in its computer hardware. The servers run an Oracle ERP system and a geographic information system that maps the city's infrastructure of streets, gas and power lines, and sewers. Converting to new blade servers from a collection of different stand-alone computers and regular servers reduced the total cost of ownership by 40 percent and made it easy for the city to add more blades when it needed additional processing power.[58] Read the Information Systems @ Work special feature to learn about another interesting use of blade server technology.

The Dell Power Edge 1855 Chassis can hold up to ten blade servers.

(Source: Courtesy of Dell Inc.)

INFORMATION SYSTEMS @ WORK

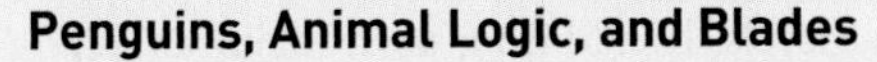

Penguins, Animal Logic, and Blades

And the Oscar goes to... *Happy Feet*! Perhaps you saw *Happy Feet*, the motion picture that won the 2007 Oscar for best animated feature. The movie included animated shots of groups of thousands of picture-perfect Emperor penguins, including Mumble, a young penguin with an uncanny ability to dance.

Happy Feet is the brainchild of George Miller with animation by Australia's Animal Logic. *Happy Feet* was Animal Logic's first full-length animated movie, although the company had its hand in many other popular films including the *Harry Potter* series, *Moulin Rouge*, the *Matrix* trilogy, and *The Lord of the Rings*.

If you did see *Happy Feet*, you were sure to be impressed by the detail of the photo-realistic animation. For example, the star of the film, Mumble, had 6 million picture-perfect feathers. Several shots included more than 400,000 realistic-looking penguins. This type of artisanship requires processing power, and lots of it.

Animators at Animal Logic realized that they would require more processing power than the company currently owned. Producing 3-D animated films requires a process called rendering, where defined 3-D objects in a scene, along with the lighting, shadings, shadows, and reflections, are created on a computer based on commands from the artist and the laws of physics. *Happy Feet* required the rendering of 140,000 frames, with each frame taking hours to render. Using a PC, one rendering of the film would take around 17 years. Xavier Desdoigts, director of technical operations at Animal Logic, calculated that nine months of production would require 17 million CPU hours. Animal Logic turned to IBM for help.

IBM installed a rendering server farm built from blade servers, each containing two processors—2,000 of them for a total of 4,000 processors. The installation of the system posed some challenges. The density of the blade centers that housed the servers produced a higher amount of power consumption and heat generation than standard servers. Animal Logic had to work with IBM to create a suitable environment for the system. IBM provided management tools that allowed one technician to handle the day-to-day maintenance of the system. Desdoigts says, "Sometimes we forget that we have 2,000 CPUs doing their job every day. There's one person who looks after all of them....But that's what we were aiming for. It was part of choosing a vendor that could provide that level of service and support so we could focus on creating movies. We didn't want to get bogged down in technology issues; it just had to work every day."

Animal Logic's new system gives it the power of industry leaders like Pixar and Sony Pictures. *Happy Feet* proved to be a great leap into the big league for Animal Logic, grossing more than $41 million on its opening weekend, and beating out the top-tier companies for an Oscar.

Discussion Questions

1. Why did Animal Logic want a system that was easy to maintain? How did this requirement contribute to the company's ability to meet its goals?
2. Purchasing a 4,000-processor blade server required a huge investment from Animal Logic and a leap of faith. How do you think they justified the expense?

Critical Thinking Questions

1. The IBM Case Study on which this article is partly based states that "in specialized areas such as weather forecasting, scientific and financial research, and digital media production, there can never be enough processing power." Why do you think this is?
2. Based on what you have read in this chapter, why do you think IBM recommended blade servers instead of other types of servers for Animal Logic's needs?

SOURCES: Rossi, Sandra, "And the Oscar goes to ... jovial penguins and 2,000 blade servers," *Computerworld*, March 6, 2007, *www.computerworld.com/action/article.do?command=viewArticleBasic&taxonomyName=Servers_and_Data_Center&articleId=9012400&taxonomyId=154&intsrc=kc_li_story*. Staff, "Animal Logic builds rendering farm with IBM eServer BladeCenter," IBM Success Story, October 11, 2005, *www-01.ibm.com/software/success/cssdb.nsf/CS/MCAG-6H2SR2?OpenDocument&Site=corp&cty=en_us*. Animal Logic Web Site, *www.animallogic.com*, accessed January 12, 2008.

Mainframe Computers

mainframe computer
A large, powerful computer often shared by hundreds of concurrent users connected to the machine via terminals.

A **mainframe computer** is a large, powerful computer shared by dozens or even hundreds of concurrent users connected to the machine over a network. The mainframe computer must reside in a data center with special heating, ventilating, and air-conditioning (HVAC) equipment to control temperature, humidity, and dust levels. In addition, most mainframes are kept in a secure data center with limited access to the room. The construction and maintenance of a controlled-access room with HVAC can add hundreds of thousands of dollars to the cost of owning and operating a mainframe computer.

The role of the mainframe is undergoing some remarkable changes as lower-cost, single-user computers become increasingly powerful. Many computer jobs that used to run on mainframe computers have migrated onto these smaller, less-expensive computers. This information-processing migration is called *computer downsizing*.

Mainframe computers have been the workhorses of corporate computing for more than 50 years. They can support hundreds of users simultaneously and handle all of the core functions of a corporation.

(Source: Courtesy of IBM Corporation.)

The new role of the mainframe is as a large information-processing and data-storage utility for a corporation—running jobs too large for other computers, storing files and databases too large to be stored elsewhere, and storing backups of files and databases created elsewhere. The mainframe can handle the millions of daily transactions associated with airline, automobile, and hotel/motel reservation systems. It can process the tens of thousands of daily queries necessary to provide data to decision support systems. Its massive storage and input/output capabilities enable it to play the role of a video computer, providing full-motion video to multiple, concurrent users. IBM mainframe computer customers include the top 25 banks and the top 25 retailers in the world who use the machines for processing large amounts of transactions. For example, the Bank of China houses 350 million accounts with 3 billion transaction histories and processes 30 million transactions in less than an hour using IBM's System z mainframe computer.[59]

Supercomputers

supercomputers
The most powerful computer systems with the fastest processing speeds.

Supercomputers are the most powerful computers with the fastest processing speed and highest performance. They are *special-purpose machines* designed for applications that require extensive and rapid computational capabilities. Originally, supercomputers were used primarily by government agencies to perform the high-speed number crunching needed in weather forecasting and military applications. With recent reductions in the cost of these machines, they are now used more broadly for commercial purposes.

The IBM Blue Gene/L supercomputer was designed and built in collaboration with the Department of Energy and the Lawrence Livermore National Laboratory. This basketball court-size, number-crunching monster supports a wide range of research projects, ranging from detailed simulations of nuclear weapons programs to human biological processes such as protein folding. When comparing the speed of supercomputers, the metric used is floating-point operating instructions per second, or FLOPS. The Blue Gene/L supercomputer has a peak computational speed of 596 TeraFLOPS (1 TeraFLOP = 10^{12} floating-point operations per second) and is currently the fastest single computer in the world.[60] Three additional

Blue Gene computers are in development including the BlueGene/C (sister project to BlueGene L and scheduled to be released in 2007 but delayed), BlueGene/P (designed to run at a speed of 1 to 3 PetaFLOPS), and BlueGene/Q (targeted to achieve 10 PetaFLOPS by 2012). Table 3.5 compares the processing speeds of supercomputers.

Table 3.5

Supercomputer Processing Speeds

Speed	Meaning
GigaFLOPS	1 x 10^9 FLOPS
TeraFLOPS	1 x 10^{12} FLOPS
PetaFLOPS	1 x 10^{15} FLOPS

Scientists say that these supercomputers will enable more realistic computer simulations that will provide new insights into new drug development, geology, climate change, dark matter, and other mysteries of the universe. "They are a tool that really helps stimulate the imagination of scientists and engineers in ways that weren't previously possible," according to David Turek, vice president of supercomputing at IBM. "Nature is the final arbiter of truth," says Mark Seager, a Lawrence Livermore computer scientist, but "rather than doing experiments, a lot of times now we're actually simulating those experiments and getting the data that way."[61]

IBM's Blue Gene/L System at the Lawrence Livermore National Laboratory is the fastest supercomputer in the world and can perform 596 trillion floating-point operations per second.

(Source: Courtesy of IBM Corporation.)

In an effort to keep the United States in the forefront of supercomputing, the Defense Advanced Research Projects Agency (DARPA) awarded $250 million to both Cray and IBM to develop so-called Petascale computers by 2010 capable of achieving 2 PetaFLOPS of sustained performance. According to Dr. William Harrod, DARPA program manager, "High-productivity computing is a key technology enabler for meeting our national security and economic competitiveness requirements. High-productivity computing contributes substantially to the design and development of advanced vehicles and weapons, planning and execution of operational military scenarios, the intelligence problems of cryptanalysis and image processing, the maintenance of our nuclear stockpile, and is a key enabler for science and discovery in security-related fields."[62]

SUMMARY

Principle

Computer hardware must be carefully selected to meet the evolving needs of the organization and its supporting information systems.

Computer hardware should be selected to meet specific user and business requirements. These requirements can evolve and change over time.

The central processing unit (CPU) and memory cooperate to execute data processing. The CPU has three main components: the arithmetic/logic unit (ALU), the control unit, and the register areas. Instructions are executed in a two-phase process called a machine cycle that includes the instruction phase and the execution phase.

Computer system processing speed is affected by clock speed, which is measured in gigahertz (GHz). As the clock speed of the CPU increases, heat is generated that can corrupt the data and instructions the computer is trying to process. Bigger heat sinks, fans, and other components are required to eliminate the excess heat. This excess heat can also raise safety issues.

Primary storage, or memory, provides working storage for program instructions and data to be processed and provides them to the CPU. Storage capacity is measured in bytes.

A common form of memory is random access memory (RAM). RAM is volatile; loss of power to the computer erases its contents. RAM comes in many different varieties including dynamic RAM (DRAM), synchronous DRAM (SDRAM), Double Data Rate SDRAM, and DDR2 SDRAM.

Read-only memory (ROM) is nonvolatile and contains permanent program instructions for execution by the CPU. Other nonvolatile memory types include programmable read-only memory (PROM), erasable programmable read-only memory (EPROM), electrically erasable PROM, and flash memory.

Cache memory is a type of high-speed memory that CPUs can access more rapidly than RAM.

A multicore microprocessor is one that combines two or more independent processors into a single computer so they can share the workload. Intel and AMD have introduced quad-core processors that are effective in working on problems involving large databases and multimedia.

Parallel computing is the simultaneous execution of the same task on multiple processors to obtain results faster. Massively parallel processing involves linking many processors to work together to solve complex problems.

Grid computing is the use of a collection of computers, often owned by multiple individuals or organizations, to work in a coordinated manner to solve a common problem.

Computer systems can store larger amounts of data and instructions in secondary storage, which is less volatile and has greater capacity than memory. The primary characteristics of secondary storage media and devices include access method, capacity, portability, and cost. Storage media can implement either sequential access or direct access. Common forms of secondary storage include magnetic tape, magnetic disk, virtual tape, optical disc, digital video disc (DVD), and holographic versatile disc (HVD).

Redundant array of independent/inexpensive disks (RAID) is a method of storing data that generates extra bits of data from existing data, allowing the system to more easily recover data in the event of a hardware failure.

Network-attached storage (NAS) and storage area networks (SAN) are alternative forms of data storage that enable an organization to share data resources among a much larger number of computers and users for improved storage efficiency and greater cost-effectiveness.

The overall trend in secondary storage is toward direct-access methods, higher capacity, increased portability, and automated storage management. Interest in renting space on massive storage devices is increasing.

Input and output devices allow users to provide data and instructions to the computer for processing and allow subsequent storage and output. These devices are part of a user interface through which human beings interact with computer systems.

Data is placed in a computer system in a two-stage process: Data entry converts human-readable data into machine-readable form; data input then transfers it to the computer. Common input devices include a keyboard, a mouse, speech recognition, digital cameras, terminals, scanning devices, optical data readers, magnetic ink character recognition devices, magnetic stripe cards, point-of-sale devices, automated teller machines, pen input devices, touch-sensitive screens, bar-code scanners, and Radio Frequency Identification tags.

Display monitor quality is determined by size, color, and resolution. Liquid crystal display and organic light-emitting diode technology is enabling improvements in the resolution and size of computer monitors. Other output devices include printers, plotters, and digital audio players. E-books and multiple-function printers are common forms of special-purpose input/output devices.

Computer systems are generally divided into two categories: single user and multiple users. Single-user systems include handheld, ultra laptop, portable, thin client, desktop, and workstation computers.

Multiuser systems include servers, blade servers, mainframes, and supercomputers.

Principle

The computer hardware industry is rapidly changing and highly competitive, creating an environment ripe for technological breakthroughs.

CPU processing speed is limited by physical constraints such as the distance between circuitry points and circuitry materials. Moore's Law is a hypothesis stating that the number of transistors on a single chip will double every two years. This hypothesis has been accurate since it was introduced in 1970 and has led to smaller, faster, less expensive computer hardware.

Advances in tri-gate transistors, carbon nanotubes, and extreme miniaturization will result in faster CPUs.

Cell Broadband Engine Architecture is a microprocessor architecture developed by IBM, Sony, and Toshiba to provide more power-efficient, cost-effective, and higher-performance processing. This technology has numerous applications.

Manufacturers are competing to develop a nonvolatile memory chip that requires minimal power, offers extremely fast write speed, and can store data accurately even after it has been stored and written over many times. Such a chip could eliminate the need for RAM forms of memory. PCM, FeRAM, and MRAM are three potential solutions.

Cloud computing involves the use of a giant cluster of computers that serves as a host to run applications that require high-performance computing. Some organizations are exploring the use of cloud computing to replace major components of their infrastructure.

CHAPTER 3: SELF-ASSESSMENT TEST

Computer hardware must be carefully selected to meet the evolving needs of the organization and its supporting information systems.

1. Organizations typically make a one-time investment in the computer hardware necessary to meet their needs with little need for future changes and upgrades. True or False?
2. The computer hardware that most nonprofit organizations choose is virtually identical. True or False?
3. The overriding consideration for a business making hardware decisions should be how the hardware meets specific ________ and ________ requirements.
4. Which represents a larger amount of data—a terabyte or a gigabyte?
5. Which of the following components performs mathematical calculations and makes logical comparisons?
   a. control unit
   b. register
   c. ALU
   d. main memory
6. Executing an instruction by the CPU involves two phases: the instruction phase and the ___________ phase.
7. ___________ involves capturing and editing data when it is originally created and in a form that can be directly input to a computer, thus ensuring accuracy and timeliness.

The computer hardware industry is rapidly changing and highly competitive, creating an environment ripe for technological breakthroughs.

8. Some organizations are exploring the use of ___________ to replace major components of their infrastructure.
9. There are few examples of companies in the computer industry collaborating to create a new product or service. True or False?

CHAPTER 3: SELF-ASSESSMENT TEST ANSWERS

(1) False (2) False (3) user and business requirements (4) terabyte (5) c (6) execution (7) Source data automation (8) cloud computing (9) False

REVIEW QUESTIONS

1. When determining the appropriate hardware components of a new information system, what role must the end user of the system play?
2. What is a blade server? What advantages does it offer over an ordinary server?
3. Identify three basic characteristics of RAM and ROM.
4. What is RFID technology? Identify three practical uses for this technology.
5. What issues can arise when the CPU runs at a very fast rate?

6. What advantages do fuel cells offer over batteries for use in portable electronic devices? Do they have any disadvantages?
7. What is the difference between data entry and data input?
8. What is RAID storage technology?
9. Explain the two-phase process for executing instructions.
10. Why are the components of all information systems described as interdependent?
11. Identify the three components of the CPU and explain the role of each.
12. What is the difference between sequential and direct access of data?
13. Identify several types of secondary storage media in terms of access method, capacity, portability, and cost per GB of storage.
14. Identify and briefly describe the various classes of personal computers.
15. What is the difference between cloud computing and grid computing?
16. What is source data automation?
17. What is the overall trend in secondary storage devices?

DISCUSSION QUESTIONS

1. Briefly describe how RFID technology works. Why is RFID technology being used to track product inventory in retail stores?
2. What would be the advantages for a university computer lab to install thin clients rather than standard desktop personal computers? Can you identify any disadvantages?
3. What is a quad-core processor? What advantages does it offer users?
4. Describe a practical business problem that could be solved through the use of a multiple-monitor solution.
5. Briefly describe Moore's Law. What are the implications of this law? Are there any practical limitations to Moore's Law?
6. Identify and briefly describe the three fundamental approaches to data storage.
7. Discuss the potential impact of converting to cloud computing on an individual organization's IS infrastructure.
8. Briefly discuss several data-storage issues that face the modern organization.
9. If cost was not an issue, describe the characteristics of your ideal computer. What would you use it for? Would you choose a handheld, portable, desktop, or workstation computer? Why?
10. How should organizations allocate grid computing resources so they address only important research projects?

PROBLEM-SOLVING EXERCISES

1. Use word processing software to document what your needs are as a computer user and your justification for selecting either a desktop or laptop computer. Find a Web site that allows you to order and customize a computer, and select those options that best meet your needs in a cost-effective manner. Assume that you have a budget of $1,250. Enter the computer specifications and associated costs into an Excel spreadsheet that you cut and paste into the document defining your needs. E-mail the document to your instructor.
2. Develop a spreadsheet that compares the features, initial purchase price, and ongoing operating costs for three laser printers. Now do the same for three inkjet printers. Write a brief memo on which printer you would choose and why. Cut and paste the spreadsheet into a document.
3. Enter data from Figure 3.3 (Moore's Law) into a spreadsheet program. Use the forecasting capabilities of the program to estimate the number of transistors on a chip for the next six years, and draw a chart depicting this. Are there basic limitations that can keep this forecast from being met? If so, what are they?

TEAM ACTIVITIES

1. With one or two of your classmates, visit a retail store that employs Radio Frequency Identification chips to track inventory. Interview an employee involved in inventory control, and document the advantages and disadvantages they see in this technology.
2. With two or three of your classmates, visit a computer retail store and identify the most popular ultra laptop computers. Interview members of the sales staff to find out why they think this particular laptop is popular.

WEB EXERCISES

1. Do research on the Web to document the current state of computer hardware disposal. What are some of the issues? What solutions are there to this problem? Write a brief report summarizing your findings.
2. Do research on the Web to identify the current status of the use of cloud computing. What companies are offering this service? What companies are exploring use of the service? Write a brief report summarizing your findings.

CAREER EXERCISES

1. Imagine that you are going to buy a single handheld device to improve your communication and organizational abilities. What tasks do you need it to perform? What features would you look for in this device? Visit a computer store or a consumer electronics store and see whether you can purchase such a device for under $400.
2. Your company's finance department plans to acquire 25 new computers and monitors, plus several new printers. The finance vice president has asked you to lead a project team assigned to define users' computer hardware needs. Who else (role, department) and how many people would you select to be a member of the team? How would you go about defining users' needs? Do you think that one hardware configuration (computer, monitor, and printer) will meet everyone's needs? Should you define multiple configurations based on the needs of various classes of end user? What business justification can you define to substantiate this expenditure of roughly $50,000?

CASE STUDIES

Case One

Advance America Implements Grid Computing

Chances are you have seen places that offer payday loans in your town. Payday loans are short-term loans designed for people that run out of money before payday, but can repay the loan when their paycheck arrives. Advance America is the leading payday loan company in the United States. It includes 3,000 centers in 37 states, and employs nearly 7,000 people, according to its Web site. Advance America is big, and growing bigger every day. Its growth in recent years is straining the capabilities of its client-server information system infrastructure and holding the company back from further growth.

Advance America used a system in which each center was equipped with an independent hardware and software environment. Installation and maintenance costs were high, and compiling data for all centers was time consuming and difficult. Each night the thousands of centers would upload their data to the main server for consolidation. With the growing number of centers, there wasn't enough time in the night to process all of the incoming data. Advance America's system had run up against a wall. It was time for a change.

Advance America decided to invest in a new system based on a grid computing architecture. They installed thin client machines to run in each center, connecting via the Web to a fault-tolerant server cluster running Oracle database software. The server cluster consists of a four-node cluster of

IBM P5 series servers, which include four processors per node for a total of 16 processors. The servers in the cluster work as a grid by sharing the work load of the entire organization equally among them. A pair of Cisco load balancers make sure that processing is distributed evenly among the servers for maximum performance. The new system includes a 2 TB storage area network (SAN) that uses an IBM disk array controlled by the Oracle Automatic Storage Management (ASM) software.

System IT managers at headquarters use a central grid-management console to oversee the entire nationwide network. Problems are easily identified and fixed through the centrally managed system. So far, the system has provided 100 percent uptime at the cash-advance centers.

Advance America took a chance with its $3.8 million investment in this new technology, but it has paid off. Center managers can now tap into "a continuously updated central database and generate reports in near real time." The new system has decreased the time it takes to open a new Advance America center. Managers are getting information much more quickly, making it easier for them to analyze business performance and customer trends. The new system is also easy to expand as the business grows. It is estimated that the new system will provide total net benefits of almost $3 million over five years for an ROI of 131 percent.

Discussion Questions

1. How does grid computing provide Advance America managers with faster access to data?
2. How did grid computing assist Advance America in breaking through the wall that held it back from growth?

Critical Thinking Questions

1. Why is the new grid computing system at Advance America much easier to install, manage, and maintain than its old system?
2. How might Advance America expand its system as the company outgrows it?

SOURCES: Staff, "Advance America Grows with Oracle Enterprise Grid," *Computerworld Honors Program*, 2007, *www.cwhonors.org/viewCaseStudy.asp?NominationID=104*. Advance America Web Site, *www.advanceamerica.net/values.php*, accessed January 12, 2008. Staff, "Integrated Data Infrastructure Pays Off for Advance America, Cash Advance Centers, Inc," Oracle Customer Snapshot, *www.oracle.com/customers/snapshots/advance_america.pdf*, accessed January 12, 2008.

Case Two

Mayo Clinic Turns to Game Processor to Save Lives

The Mayo Clinic and IBM have partnered in a venture to improve medical imaging technology. The clinic's current technologies aren't keeping up with the intense processing demands required to analyze digital medical images such as x-rays, CT scans, and MRIs.

You've learned in this chapter that transistor densities on a chip double roughly every two years, a rule of thumb referred to as Moore's Law. Bradley Erickson, chairman of radiology at the Rochester-based Mayo Clinic, was quoted in Computerworld as saying, "We are facing significant problems in medical imaging because the number of images produced in CT scanners basically tracks Moore's Law. My eyes and brain can't keep up. I see more and more images I have to interpret. ... The innovation here is to take computer chips and extract the information in these increasing number of images and help present it usefully to the radiologist."

This is a case of technology outpacing the human ability to manage the information it produces. In such cases, we turn to technology for solutions. For doctors and radiologists at the Mayo Clinic, standard computer processors cannot keep up with their need to analyze digital images. So they are turning to the Cell processor from IBM in hopes that it will provide a solution. The Cell processor is the chip that makes Sony's PlayStation video-game console the most powerful console in the industry, according to many game enthusiasts. The Cell processor was created in a joint effort by IBM, Sony, and Toshiba, with an architecture that is specially designed to accelerate graphics processing. Researchers at IBM and Mayo believe that it could turn a 10-minute CT image analysis into a four-second job.

One of the tasks in which the Cell processor could be useful is in comparing scan images of a patient over time. For example, to track the progression or regression of cancer in a patient, physicians compare CT scans of the tumor over time to look for change. Changes are often too subtle for the human eye to notice, so software that implements a complex algorithm is used to analyze the photos. Using a standard PC processor, the algorithm may take several minutes to complete. While this may not sound like much, typically a physician needs to run several analyses in sequence, consuming significant amounts of time. The process of transforming 2-D images into 3-D—something the Cell processor was designed for—also requires significant time using traditional processors. With the Cell processor, these tasks might be completed in a matter of a seconds.

Mayo Clinic and its effort to speed up the analysis of image scanning illustrate the importance of time when it comes to processing. Whether it's working to save a life, to finish design specifications for a new product, or to analyze stock market trends, the difference between a minute and a second can mean success or failure. For professionals in most industries, having the best processor for the task at hand, and matching it with the best hardware and software, provides them with a winning solution.

Discussion Questions

1. Why is the Cell processor the best processor for the Mayo Clinic's tasks? How might it empower physicians to save more lives?
2. If the Cell processor is so much faster than typical PC processors, why isn't it being used in PCs?

Critical Thinking Questions

1. In what other industry and scenario might time play an important role when it comes to processing? Explain how reducing minutes to seconds has an impact in that scenario.
2. What other processing technologies presented in this chapter might assist the Mayo Clinic in speeding up its computations?

SOURCES: Gaudin, Sharon, "IBM, Mayo Clinic team up to improve medical imaging," *Computerworld*, January 9, 2007, *http://computerworld.com/action/article.do?command=viewArticleBasic&taxonomyName=hardware&articleId=9056618&taxonomyId=12&intsrc=kc_top*. Barrett, Larry, "IBM, Mayo Clinic Open Imaging Research Center," *internetnews.com*, January 9, 2008, *www.internetnews.com/ent-news/article.php/3720686*. Staff, "Mayo Clinic, IBM Establish Medical Imaging Research Center," *Medical News Today*, January 9, 2007, *www.medicalnewstoday.com/articles/93454.php*. IBM's Cell Processor Web site, *www.research.ibm.com/cell*, accessed January 10, 2007.

Questions for Web Case

See the Web site for this book to read about the Whitmann Price Consulting case for this chapter. Following are questions concerning this Web case.

Whitmann Price Consulting: Choosing Hardware

Discussion Questions

1. What considerations led Josh and Sandra to lean towards a Blackberry as the handheld device on which to run the AMCIS?
2. How did Josh and Sandra organize their hardware considerations for the new system?

Critical Thinking Questions

1. What role does system compatibility play in Josh and Sandra's decision?
2. What device(s) might have been chosen if the system requirements called for a 12-inch display and the ability to take handwritten notes and communicate through voice, text, and video?

NOTES

Sources for the opening vignette: Staff, "UB Spirits cuts 111 servers and boosts capacity with IBM and SAP," *IBM Case Study*, July 24, 2006, *www-01.ibm.com/software/success/cssdb.nsf/CS/STRD-6RZJS8?OpenDocument&Site=eservermain*. UB Spirits Web site, *ww.theubgroup.com/beveragealcohol.html*, accessed January 12, 2008. Club McDowell Web site, *www.clubmcdowell.com/brands/ub_main.htm*, accessed January 12, 2008. IBM i-Series Web site, *www.clubmcdowell.com/brands/ub_main.htm*, accessed January 12, 2008.

1 "Bosch Security Projects – The 2007 Computerworld Honors Program," *Computerworld*, *www.cwhonors.org.viewCaseStudy.asp?NominationID=65*, accessed January 27, 2008.
2 "Iowa Health System – The 2007 ComputerWorld Honors Program," *Computerworld*, *www.cwhonors.org/viewCaseStudy.asp?NominatonID=347*, accessed January 27, 2008.
3 Keizer, Gregg, "Feds Warn Travelers about Batteries after Two In-Flight Fires," *Computerworld*, March 23, 2007.
4 "The Cell Project at IBM Research," IBM Research Web site, *research.ibm/com/cell*, accessed January 4, 2008.
5 Staff, "Innovation That Breaks the Performance Barrier," *www.intel.com/technology/architecture*, accessed January 30, 2008.
6 Schwartz, Ephraim, "Breaking News – Intel Makes Chip Breakthrough," *InfoWorld*, January 27, 2007.
7 Krazit, Tom, "Intel Shows Off Penryn Chips," *C/Net News*, January 29, 2007.
8 Levine, Barry, "IBM Announces Optical Chip Breakthough," *NewsFactor Network*, December 10, 2007.
9 Kay, Russell, "QuickStudy: Phase-Change Memory," *Computerworld*, May 7, 2007.
10 John Markoff, "Faster Chips Are Leaving Programmers in Their Dust," *New York Times Technology*, December 17, 2007.
11 "Intel Core 2 Processor with Viiv Technology," Intel Web site, *www.intel.com/products/viiv/index.html*, accessed January 6, 2008.
12 Krazit, Tom, "Intel Shows Off Penryn Chips," *C/Net News*, January 29, 2007.
13 Gardiner, Bryan, "AMD Pins Hopes on Barcelona Quad-Core Chip," *Wired*, September 7, 2007.
14 Brown, Stuart F., "Crashing Cars When They're Still a Gleam in the Designer's Eye," *The New York Times*, June 17, 2007.
15 "Grid Computing Offers New Hope in Race Against Bird Flu," *Science Daily*, October 9, 2007.
16 "The Science Behind Folding@home," Folding@home Web site, *folding.stanford.edu/English/Science*, accessed January 4, 2008.
17 Blue Gene," IBM Research Web site, *domino.reseaarch.ibm.co/comm./research_projectsw.nsf/pages/bluegene.index.html*, accessed January 4, 2008.
18 Thibodeau, Patrick, "NASCAR Drivers Get HPC Help with Performance Extremes," *Computerworld*, November 14 2007.
19 "IBM and Google Float in Parallel Cloud," Grant Gross Blog, *Computerworld*, October 9, 2007.
20 "Amazon Elastic Compute Cloud (Amazon EC2) – Beta," *www.amazon.com/gp/browse.html?node=201590011*, accessed January 14, 2008.
21 Carr, Nicholas, Rough Type blog, "Google's Cloud," October 26, 2007, *www.roughtype.com/archives/2007/10/googles_cloud.php*, accessed January 15, 2008.
22 Large Hadron Collider, Wikipedia, *http://en.wikipedia.org/wiki/Large_Hadron_Collider*, accessed January 30, 2008.
23 Weier, Mary Haynes, "Hewlett-Packard Data Warehouse Lands in Wal-Mart's Shopping Cart, *InformationWeek*, August 4, 2007.
24 "Wal-Mart Picks HP," *Byte and Switch*, August 1, 2007.
25 "Government Sale of Used Magnetic Tape Storage Not a Big Security Risk, GAO Reports," *Network World*, September 21, 2007.

26 Lipschutz, Robert P., "Iomega REV Loader 560," *PC Magazine*, February 28, 2007.

27 Perenson, Melissa, J., "Mobile Hard Drive Hits 500 GB," *PC World*, January 2, 2008.

28 Quain, John, R., "2007 Small Business Awards: The Winners," *PC Magazine*, September 26, 2007.

29 Fonseca, Brian, "Girl Scouts Box Up New IT Flavor: Data De-Duplication," *Computerworld*, November 7, 2007.

30 "Holographic Versatile Disk," *Express Computer*, November 14–16, 2007.

31 "UNC Hospitals Team with Agfa to Improve Access to Patient Information with Network Appliance Storage," Customer Success Story at NetApp Web site, *www.netapp.com/library/cs/unc.pdf*, accessed January 30, 2008.

32 Schultz, Beth, "Sleeping SAN for an Energy-Conscious World," *Network World*, November 26, 2007.

33 "Remodeling the Data Center: A Centralized Data Storage Management Model," accessed at whitepapers.zdNet.com on January 30, 2008.

34 Flynn, Mary Kathleen, "DOD Invests $484 in Speech Recognition," *Tech Confidential*, June 26, 2007.

35 Jones, K.C., "Free Cell Phone Direction Service Gets Voice Recognition Boost," *InformationWeek*, November 13, 2007.

36 McDougall, Paul, "Gates, Simonyi Give $30 Million to Build Giant Telescope," *InformationWeek*, January 4, 2008.

37 Martin, Richard, "InformationWeek 500: Con-way Earns No. 1 Ranking in Information Week 500," *InformationWeek*, September 18, 2007.

38 Gonsalves, Antone, "Revolution Online Money Transfer Service Pits Itself Against PayPal," *InformationWeek*, November 8, 2007.

39 Gonsalves, Antone, "Microsoft, HP, and First Data Partner on Point-of-Sale System," *InformationWeek*, January 16, 2007.

40 Malykhina, Elena, "AT&T, Costco Ink Retail Kiosk Deal," *InformationWeek*, November 12, 2007.

41 "Boekhandels Groep Nederland (BGN)" The 2007 Computerworld Honors Program, *www.cwhonors.org/viewCaseStudy.asp?NominationID=119*, accessed January 16, 2008.

42 Cheng, Jacqui, "Samsung and LG. Phillips Announce Super-Thin OLEDs," *arts technical*, May 17, 2007.

43 Criswell, Chad, "OLED Monitors and 3D Display Tech," *Suite 101.com*, December 4, 2007.

44 Gonsalves, Antone, "New HP Printers Sport 60,000 Nozzles," *InformationWeek*, April 11, 2007.

45 Sony Reader Digital Book, Sony Web page, *www.sonystyle.com*, accessed January 17, 2008.

46 Mosquera, Mary, "Census Gets Demo Run of PDA for 2010 Census," *Government Computer News*, January 4, 2007.

47 Mossberg, Walter, "Not Yet the Holy Grail: Nokia's Tiny Computer Is Crisp But Slow," *Wall Street Journal*, July 20, 2006.

48 "Pocket PC Development Provides Sales Force Tool for Coca-Cola," Syware Web site, *www.syware.com*, accessed January 31, 2008.

49 Dejean, Davi, "Review: Three Ultralight Laptops Get the Job Done," *InformationWeek*, March 17, 2007.

50 Malykhina, Elena, "DT Research Adds Barcode Scanners, Card Readers, Cameras to Tablet PCs," *InformationWeek*, November 16, 2007.

51 Industry Solutions – CSX, DT Research Web site, *www.dtresearch.com/industry/case.htm*, accessed January 31, 2008.

52 OLPC Web site, *laptop.org/vision/index.shtml*, accessed January 31, 2008.

53 Nutall, Chris, "Non-profits Lead the Way to the Age of the Low-Cost Laptop," *Financial Times*, January 31, 2008.

54 Nutall, Chris, "Non-profits Lead the Way to the Age of the Low-Cost Laptop," *Financial Times*, January 31, 2008.

55 "Amerisure: How Thin Computing Saved $1 Million in Annual Costs by Boosting IT Performance, Improving Security, and Enhancing Customer Service," Case Study: Financial Services, *www.wyse.com/resources/casestudies/CS_Amerisure_register.asp*, accessed January 17, 2008.

56 Haskin, David, "Review Roundup: Invasion of the Ultrasmall Desktop PCs," *Computerworld*, December 5, 2007.

57 Thibodeau, Patrick, "Marines Look for a Few Less Servers, Via Virtualization," *Network World*, November 6, 2007.

58 Gonsalves, Antone, "Burbank Gets Higher Performance in Move to Blade Servers," *InformationWeek*, June 25, 2007.

59 Malykhina, Elena, "IBM: The Mainframe Is Alive and Going Strong, *InformationWeek*, June 21, 2007.

60 Blue Gene," IBM Research Web site, *domino.research.ibm.co/comm./research_projectsw.nsf/pages/bluegene.index.html*, accessed January 4, 2008.

61 Lee, Christopher, "Faster Computers Accelerate Pace of Discovery," *Washington Post*, pg. A07, December 3, 2007.

62 Vance, Ashlee, "IBM and Cray Win $500m DARPA Handout," *The Register*, November 21, 2006.

CHAPTER • 4 •

Software: Systems and Application Software

PRINCIPLES	LEARNING OBJECTIVES
▪ Systems and application software are critical in helping individuals and organizations achieve their goals.	▪ Identify and briefly describe the functions of the two basic kinds of software. ▪ Outline the role of the operating system and identify the features of several popular operating systems.
▪ Organizations should not develop proprietary application software unless doing so will meet a compelling business need that can provide a competitive advantage.	▪ Discuss how application software can support personal, workgroup, and enterprise business objectives. ▪ Identify three basic approaches to developing application software and discuss the pros and cons of each.
▪ Organizations should choose a programming language whose functional characteristics are appropriate for the task at hand, considering the skills and experience of the programming staff.	▪ Outline the overall evolution and importance of programming languages and clearly differentiate among the generations of programming languages.
▪ The software industry continues to undergo constant change; users need to be aware of recent trends and issues to be effective in their business and personal life.	▪ Identify several key software issues and trends that have an impact on organizations and individuals.

Information Systems in the Global Economy General Motors, United States

GM Changes Focus from Gears and Mechanics to Software and Electronic Systems

The automotive industry is experiencing perhaps the most significant evolutionary transition since Henry Ford designed the first production line. Faced with formidable pressures that include international competition, environmental concerns, increasing traffic, and driving-related fatalities, automotive companies are rethinking the way they design and build cars. For General Motors (GM), this means changing its focus from gears and pistons to electronic systems and software.

Addressing an audience of technology innovators, GM researcher Robert Baillargeon explained that GM is embracing what it calls "a new automotive DNA." Although the automotive industry historically focused on mechanical innovation, GM is now turning its attention to electronic propulsion, steering systems, and the software that controls them. Baillargeon suggested that an increasing number of GM researchers will have backgrounds in software engineering. Overseas competitors such as Toyota have relied on technology to streamline production processes and offer lower prices to consumers. Now GM is countering with its own technological innovations.

The new automotive DNA that Baillargeon described uses dozens of software systems to control some vehicle operations and work together by communicating over a network. Not only will various systems within a car communicate with each other, but each car on the road will communicate with other cars. For example, cars a mile ahead of your car could warn you of icy conditions, a pothole, or heavy traffic, allowing you to prepare by slowing down or choosing an alternate route. Software in the car will also provide information about the cost of travel routes based on fuel consumption and tolls. Software will empower cars with new levels of intelligence, creating smart cars that provide the driver with helpful travel information. Eventually, cars will be able to drive themselves using vehicle-to-vehicle communications, GPS, 360-degree sensing, and swarm intelligence (the ability to solve traffic problems as a group) to deliver passengers to their destination safely, quickly, and with minimum impact on the environment.

Companies such as GM are taking the first steps to realize this automotive vision. GM engineers are selecting software platforms on which to base these systems and determining how to distribute the software systems throughout the car's components. They are relying on state-of-the-art software development techniques such as object-oriented design and programming to define how software systems interact within the car, and using the Unified Modeling Language (UML) to map the entire automotive system. The new electronics paradigm of the automotive industry will dramatically change the way we think of cars and transportation.

As you read this chapter, consider the following:

- What types of activities can we entrust to software? In systems where life is at stake, how can we ensure safety when software fails?
- What should companies consider when designing software systems that need to interact with similar systems designed by competitors?

Why Learn About Software?

Software is indispensable for any computer system and the people using it. In this chapter, you will learn about systems and application software. Without systems software, computers would not be able to input data from a keyboard, make calculations, or print results. Application software is the key to helping you achieve your career goals. Sales representatives use software to enter sales orders and help their customers get what they want. Stock and bond traders use software to make split-second decisions involving millions of dollars. Scientists use software to analyze the threat of global warming. Regardless of your job, you most likely will use software to help you advance in your career and earn higher wages. Today, most organizations could not function without accounting software to print payroll checks, enter sales orders, and send out bills. You can also use software to help you prepare your personal income taxes, keep a budget, and play entertaining games. Software can truly advance your career and enrich your life. We begin with an overview of software.

Software has a profound impact on individuals and organizations. It can make the difference between profits and losses, and between financial health and bankruptcy. As Figure 4.1 shows, companies recognize this impact and spend more on software than on computer hardware.

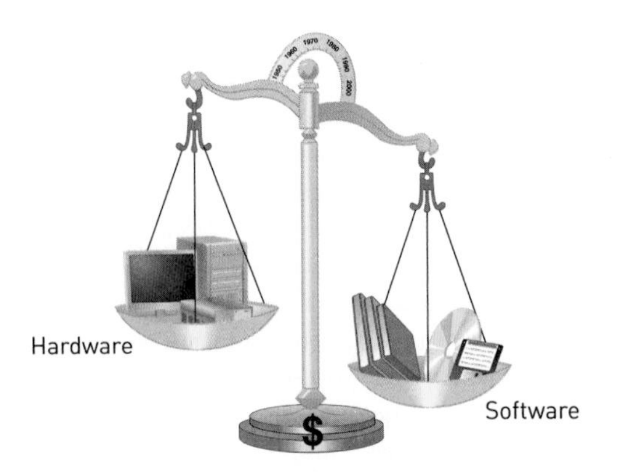

Figure 4.1

The Importance of Software in Business

Since the 1950s, businesses have greatly increased their expenditures on software compared with hardware.

AN OVERVIEW OF SOFTWARE

computer programs
Sequences of instructions for the computer.

documentation
The text that describes the program functions to help the user operate the computer system.

As you learned in Chapter 1, software consists of computer programs that control the workings of computer hardware. **Computer programs** are sequences of instructions for the computer. **Documentation** describes the program functions to help the user operate the computer system. The program displays some documentation on screen, while other forms appear in external resources, such as printed manuals. People using commercially available software are usually asked to read and agree to End-User License Agreements (EULAs). After reading the EULA, you normally have to click an "I agree" button before you can use the software, which can be one of two basic types: systems software and application software.

Systems Software

Systems software is the set of programs that coordinates the activities and functions of the hardware and other programs throughout the computer system. Each type of systems software is designed for a specific CPU and class of hardware. The combination of a hardware configuration and systems software is known as a computer system platform.

Application software has the greatest potential to affect processes that add value to a business because it is designed for specific organizational activities and functions.

(Source: © Jim West / Alamy.)

Application Software

Application software consists of programs that help users solve particular computing problems. In most cases, application software resides on the computer's hard disk before it is brought into the computer's memory and run. Application software can also be stored on CDs, DVDs, and even flash or keychain storage devices that plug into a USB port. Before a person, group, or enterprise decides on the best approach for acquiring application software, they should analyze their goals and needs carefully.

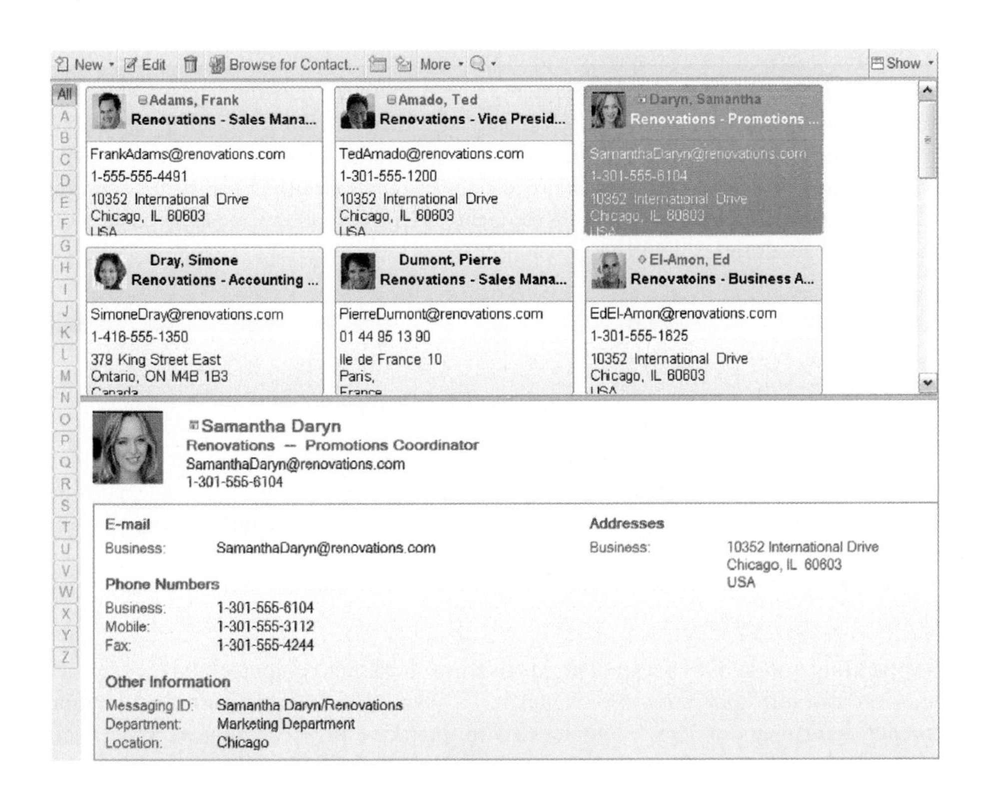

Lotus Notes is an application that enables a workgroup to schedule meetings and coordinate activities.

(Source: Courtesy of IBM Corporation.)

Supporting Individual, Group, and Organizational Goals

Every organization relies on the contributions of people, groups, and the entire enterprise to achieve its business objectives. Conversely, the organization also supports people, groups, and the enterprise with application software and information systems. One useful way of classifying the many potential uses of information systems is to identify the scope of the problems and opportunities that an organization addresses. This scope is called the sphere of influence. For most companies, the spheres of influence are personal, workgroup, and enterprise. Table 4.1 shows how software can support these three spheres.

Table 4.1

Software Supporting Individuals, Workgroups, and Enterprises

Software	Personal	Workgroup	Enterprise
Systems software	Personal computer and workstation operating systems	Network operating systems	Midrange computer and mainframe operating systems
Application software	Word processing, spreadsheet, database, graphics	Electronic mail, group scheduling, shared work, collaboration	General ledger, order entry, payroll, human resources

personal sphere of influence
The sphere of influence that serves the needs of an individual user.

personal productivity software
The software that enables users to improve their personal effectiveness, increasing the amount of work they can perform and enhancing its quality.

workgroup
Two or more people who work together to achieve a common goal.

workgroup sphere of influence
The sphere of influence that serves the needs of a workgroup.

Information systems that operate within the **personal sphere of influence** serve the needs of an individual user. These information systems help users improve their personal effectiveness, increasing the amount and quality of work they can do. Such software is often called **personal productivity software.** When two or more people work together to achieve a common goal, they form a **workgroup**. A workgroup might be a large, formal, permanent organizational entity, such as a section or department, or a temporary group formed to complete a specific project. An information system in the **workgroup sphere of influence** helps a workgroup attain its common goals. Users of such applications must be able to communicate, interact, and collaborate to be successful.

Information systems that operate within the **enterprise sphere of influence** support the firm in its interaction with its environment. The surrounding environment includes customers, suppliers, shareholders, competitors, special-interest groups, the financial community, and government agencies. This means the enterprise sphere of influence includes business partners such as suppliers that provide raw materials, retail companies that store and sell a company's products, and shipping companies that transport raw materials to the plant and finished goods to retail outlets.

SYSTEMS SOFTWARE

enterprise sphere of influence
The sphere of influence that serves the needs of the firm in its interaction with its environment.

Controlling the operations of computer hardware is one of the most critical functions of systems software. Systems software also supports the application programs' problem-solving capabilities. Types of systems software include operating systems, utility programs, and middleware.

Operating Systems

operating system (OS)
A set of computer programs that controls the computer hardware and acts as an interface with application programs.

An **operating system (OS)** is a set of programs that controls the computer hardware and acts as an interface with applications (see Figure 4.2). Operating systems can control one or more computers, or they can allow multiple users to interact with one computer. The various combinations of OSs, computers, and users include the following:

- **Single computer with a single user.** This system is commonly used in a personal computer or a handheld computer that allows one user at a time.
- **Single computer with multiple users.** This system is typical of larger, mainframe computers that can accommodate hundreds or thousands of people, all using the computer at the same time.
- **Multiple computers.** This system is typical of a network of computers, such as a home network with several computers attached or a large computer network with hundreds of computers attached around the world.

- **Special-purpose computers.** This system is typical of a number of computers with specialized functions, such as those that control sophisticated military aircraft, the space shuttle, and some home appliances.

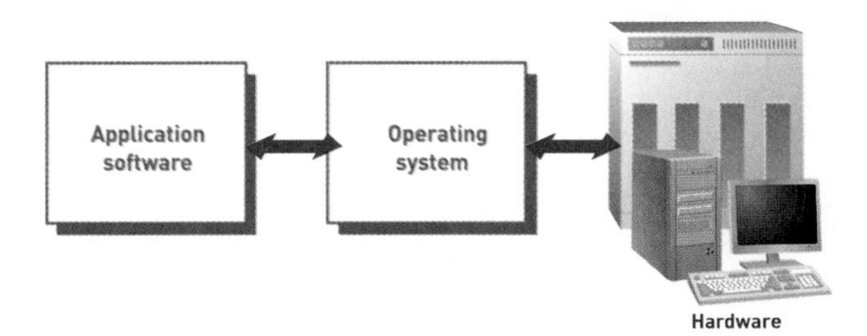

Figure 4.2

The Role of Operating Systems

The role of the operating system is to act as an interface or buffer between application software and hardware.

The OS, which plays a central role in the functioning of the complete computer system, is usually stored on disk. After you start, or "boot up," a computer system, portions of the OS are transferred to memory as they are needed. You can also boot a computer from a CD, DVD, or even a thumb drive that plugs into a USB port. A storage device that contains some or all of the OS is often called a "rescue disk" because you can use it to start the computer if you have problems with the primary hard disk.

Some OSs for handheld computers and notebooks that use solid-state hard drives have an "Instant On" feature that significantly reduces the time needed to boot a computer. The set of programs that make up the OS performs a variety of activities, including the following:

- Performing common computer hardware functions
- Providing a user interface and input/output management
- Providing a degree of hardware independence
- Managing system memory
- Managing processing tasks
- Providing networking capability
- Controlling access to system resources
- Managing files

The **kernel**, as its name suggests, is the heart of the OS and controls the most critical processes. The kernel ties all of the OS components together and regulates other programs.

kernel
The heart of the operating system, which controls the most critical processes.

Common Hardware Functions

All applications must perform certain hardware-related tasks, such as the following:

- Get input from the keyboard or another input device
- Retrieve data from disks
- Store data on disks
- Display information on a monitor or printer

Each of these tasks requires a detailed set of instructions. The OS converts a basic request into the instructions that the hardware requires. In effect, the OS acts as an intermediary between the application and the hardware. The typical OS performs hundreds of such tasks, translating each task into one or more instructions for the hardware. The OS notifies the user if input or output devices need attention, if an error has occurred, and if anything abnormal happens in the system.

User Interface and Input/Output Management

One of the most important functions of any OS is providing a **user interface.** A user interface allows people to access and command the computer system. The first user interfaces for mainframe and personal computer systems were command based. A **command-based user interface** requires you to give text commands to the computer to perform basic activities (see Figure 4.3). For example, the command ERASE 00TAXRTN would cause the computer to erase a file called 00TAXRTN. RENAME and COPY are other examples of commands used to rename files and copy files from one location to another. Many operating systems that use a graphical user interface, discussed next, also have powerful command-based features.

user interface
The element of the operating system that allows you to access and command the computer system.

command-based user interface
A user interface that requires you to give text commands to the computer to perform basic activities.

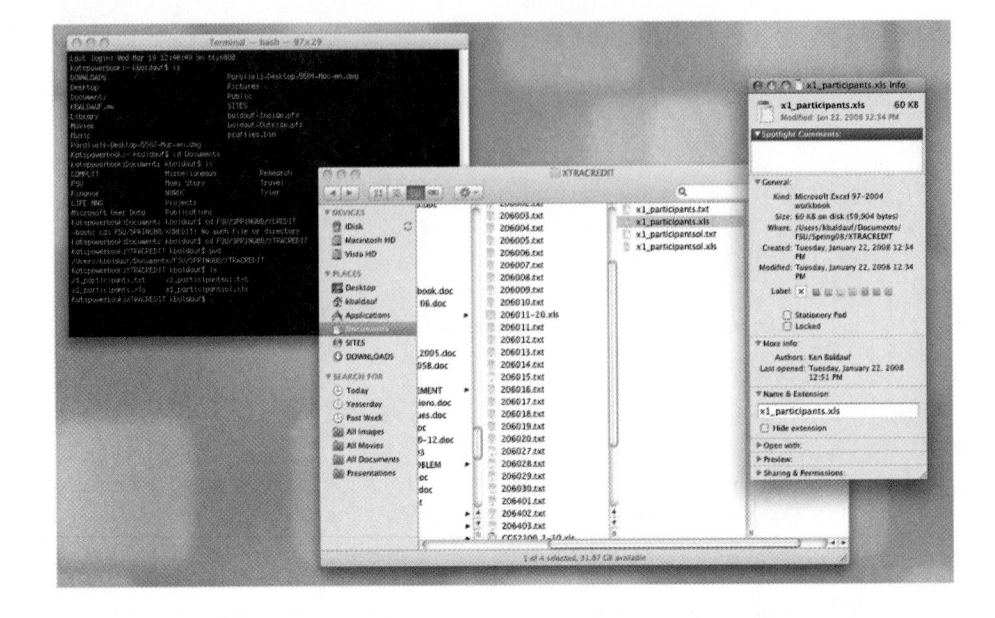

Figure 4.3

Command-Based and Graphical User Interfaces

While a command-based user interface provides only a prompt for text commands, a GUI provides icons, menus, and dialog boxes to support many forms of input.

graphical user interface (GUI)
An interface that uses icons and menus displayed on screen to send commands to the computer system.

Figure 4.3 also shows a **graphical user interface (GUI)**, which uses pictures (called icons) and menus displayed on screen to send commands to the computer system. Many people find that GUIs are easier to use because they intuitively grasp the functions. Today, the most widely used graphical user interface is Microsoft Windows. Alan Kay and others at Xerox PARC (Palo Alto Research Center, located in California) were pioneers in investigating the use of overlapping windows and icons as an interface. As the name suggests, Windows is based on the use of a window, or a portion of the display screen dedicated to a specific application. The screen can display several windows at once. GUIs have contributed greatly to the increased use of computers because users no longer need to know command-line syntax to accomplish tasks.

Hardware Independence

application program interface (API)
An interface that allows applications to make use of the operating system.

To run, applications request services from the OS through a defined **application program interface (API)**, as shown in Figure 4.4. Programmers can use APIs to create application software without having to understand the inner workings of the OS.

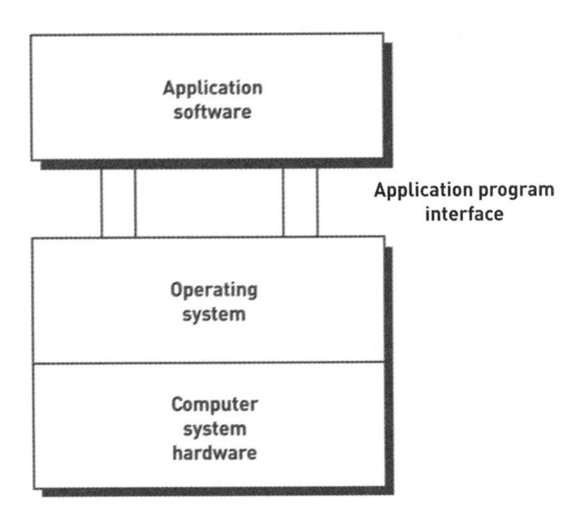

Figure 4.4

Application Program Interface Links Application Software to the Operating System

Suppose that a computer manufacturer designs new hardware that can operate much faster than before. If the same OS for which an application was developed can run on the new hardware, the application will require minimal (or no) changes to enable it to run on the new hardware. If APIs did not exist, the application developers might have to completely rewrite the application to take advantage of the new, faster hardware.

Memory Management

The OS also controls how memory is accessed and maximizes available memory and storage. Newer OSs typically manage memory better than older OSs. The memory-management feature of many OSs allows the computer to execute program instructions effectively and to speed processing. One way to increase the performance of an old computer is to upgrade to a newer OS and increase the amount of memory.

Most OSs support virtual memory, which allocates space on the hard disk to supplement the immediate, functional memory capacity of RAM. Virtual memory works by swapping programs or parts of programs between memory and one or more disk devices—a concept called paging. This reduces CPU idle time and increases the number of jobs that can run in a given time span.

Processing Tasks

The task-management features of today's OSs manage all processing activities. Task management allocates computer resources to make the best use of each system's assets. Task-management software can permit one user to run several programs or tasks at the same time (multitasking) and allow several users to use the same computer at the same time (time-sharing).

An OS with multitasking capabilities allows a user to run more than one application at the same time. Without having to exit a program, you can work in one application, easily pop into another, and then jump back to the first program, picking up where you left off. Better still, while you're working in the *foreground* in one program, one or more other applications can be churning away, unseen, in the *background*, sorting a database, printing a document, or performing other lengthy operations that otherwise would monopolize your computer and leave you staring at the screen unable to perform other work. Multitasking can save users a considerable amount of time and effort.

Time-sharing allows more than one person to use a computer system at the same time. For example, 15 customer service representatives might be entering sales data into a computer system for a mail-order company at the same time. In another case, thousands of people might be simultaneously using an online computer service to get stock quotes and valuable business news.

The ability of the computer to handle an increasing number of concurrent users smoothly is called *scalability*. This feature is critical for systems expected to handle a large number of users, such as a mainframe computer or a Web server. Because personal computer OSs usually are oriented toward single users, they do not need to manage multiple-user tasks often.

Networking Capability

Most operating systems include networking capabilities so that computers can join together in a network to send and receive data and share computing resources. PCs running Mac, Windows, or Linux operating systems allow users to easily set up home or business networks for sharing Internet connections, printers, storage, and data. Operating systems for larger server computers are designed specifically for computer networking environments.

Access to System Resources and Security

Because computers often handle sensitive data that can be accessed over networks, the OS needs to provide a high level of security against unauthorized access to the users' data and programs. Typically, the OS establishes a logon procedure that requires users to enter an identification code, such as a user name, and a matching password. If the identification code is invalid or if the password does not match the identification code, the user cannot gain access to the computer. Some OSs require that user passwords change frequently—such as every 20 to 40 days. If the user successfully logs on to the system, the OS restricts access to only portions of the system for which the user has been cleared. The OS records who is using the system and for how long, and reports any attempted breaches of security.

File Management

The OS manages files to ensure that files in secondary storage are available when needed and that they are protected from access by unauthorized users. Many computers support multiple users who store files on centrally located disks or tape drives. The OS keeps track of where each file is stored and who can access it. The OS must determine what to do if more than one user requests access to the same file at the same time. Even on stand-alone personal computers with only one user, file management is needed to track where files are located, what size they are, when they were created, and who created them.

Current Operating Systems

Early OSs were very basic. Recently, however, more advanced OSs have been developed, incorporating sophisticated features and impressive graphics effects. Table 4.2 classifies a few current OSs by sphere of influence.

Table 4.2

Popular Operating Systems Cross All Three Spheres of Influence

Personal	Workgroup	Enterprise
Microsoft Windows Vista, Windows XP, Windows Mobile, Windows Automotive, and Windows Embedded	Microsoft Windows Server 2003 and Server 2008	Microsoft Windows Server 2003 and Server 2008
Mac OS X	Mac OS X Server	
UNIX	UNIX	UNIX
Solaris	Solaris	Solaris
Linux	Linux	Linux
Red Hat Linux	Red Hat Linux	Red Hat Linux
Palm OS	Netware	
	IBM i5/OS and z/OS	IBM i5/OS and z/OS
	HP-UX 11i	HP-UX 11i

Microsoft PC Operating Systems

Since a small company called Microsoft developed PC-DOS and MS-DOS to support the IBM personal computer introduced in the 1980s, personal computer OSs have steadily evolved. *PC-DOS* and *MS-DOS* had command-driven interfaces that were difficult to learn and use. Each new version of OS has improved the ease of use, processing capability, reliability, and ability to support new computer hardware devices.

Windows XP (XP reportedly stands for the positive experience that you will have with your personal computer) was released in fall 2001. Previous consumer versions of Windows were notably unstable and crashed frequently, requiring frustrating and time-consuming reboots. With XP, Microsoft sought to bring reliability to the consumer.

In 2007, Microsoft released Windows Vista to the public, introducing it as the most secure version of Windows ever. Windows Vista includes design improvements that make it attractive and easy to use. The most advanced editions of Windows Vista include a 3-D graphics interface called Aero. However, the system requirements for Windows Vista with Aero require many users to purchase new, more powerful PCs. Windows Vista also suffered some negative press when early adopters found that some software and hardware designed for Windows XP did not run on Vista.

Windows Vista is available in five editions. Windows Vista Home Basic provides improved security, but otherwise has features similar to those included in Windows XP. Windows Vista Home Premium includes enhanced security, the Aero interface, and other improvements such as home media, but lacks business features. Windows Vista Business includes all of the above except the home media features plus business features such as a backup and restore tool, a scan and fax tool, and easy access to business networks from home.

Windows Vista Ultimate includes all of these features (see Figure 4.5). A fifth version of Windows Vista, Vista Enterprise, is designed for use on business networks. It includes encryption technology to keep stored data secure, and the ability to deliver a Windows desktop environment from an enterprise server. Today, Microsoft has over 90 percent of the PC OS market. Apple holds 7.3 percent of the market, and Linux publishers and other companies account for the rest of the PC OS market.[1]

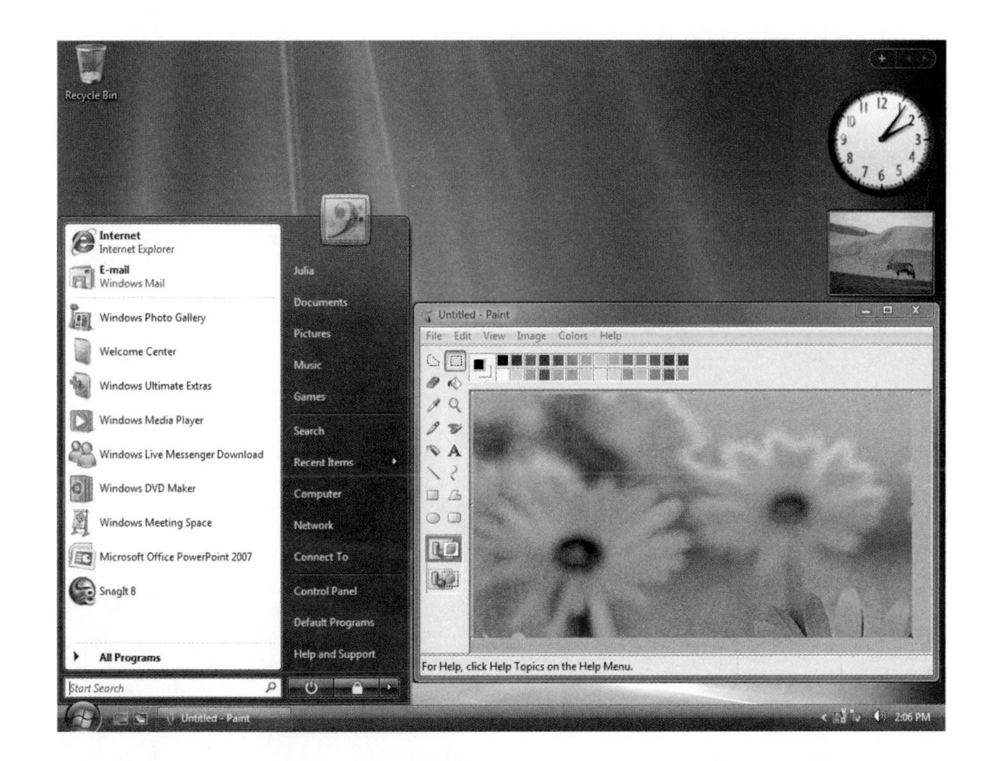

Figure 4.5

Microsoft Windows Vista

The National Aquarium in Baltimore decided to upgrade to Microsoft Vista Home Premium to improve data security and staff productivity.[2] The staff that manages the 16,000 aquatic specimens has little time for computer work. Staff members share PC workstations placed strategically around the 250,000 square foot facility. Before upgrading to Windows Vista, various versions of operating systems were installed around the aquarium, and staff members found the logon time a test of their patience. It was not uncommon for users to forget to log out and leave secure data open to others. The aquarium chose Windows Vista for two important features: Fast User Switching, which automatically logs users out after a period of inactivity, but allows them to return to their work in seconds, and Windows built-in desktop search, which saves time when looking for data. The staff estimates that it has doubled its computing productivity since switching to Windows Vista.

Apple Computer Operating Systems

Although IBM system platforms traditionally use one of the Windows OSs and Intel microprocessors (often called *Wintel* for this reason), Apple computers have used non-Intel microprocessors designed by Apple, IBM, and Motorola which run a proprietary Apple OS—the Mac OS. Newer Apple computers, however, use Intel chips. Although Wintel computers hold the largest share of the business PC market, Apple computers are also popular, especially in the fields of publishing, education, graphic arts, music, movies, and media. Software developed for the Macintosh often provides cutting-edge options for creative people. GarageBand, for example, is Macintosh software that allows you to create your own music the way a professional does, and it can sound like a small orchestra. Pro Tools is another software program used to edit digital music.

The Apple OSs have also evolved over a number of years and often provide features not available from Microsoft. Starting in July 2001, the Mac OS X was installed on all new Macs. It includes an entirely new user interface, which provides a new visual appearance for users—including luminous and semitransparent elements, such as buttons, scroll bars, windows, and fluid animation to enhance the user's experience.

Since its first release, Apple has upgraded OS X several times. Leopard is the most recent version of OS X, released in 2007 to compete with Windows Vista (see Figure 4.6). OS X Leopard includes an attractive 3-D graphical user interface that Apple claims is more intuitive than Windows. Leopard includes Time Machine, a powerful backup tool that allows users to view their system as it looked in the past and resurrect deleted files. Leopard also includes multiple desktops, a video chat program that allows users to pose in front of imaginary landscapes, a powerful system search utility, and other updated software. Because Mac OS X runs on Intel processors, Mac users can set up their PC to run both Windows Vista and Mac OS X and select which platform they want to work with when they boot their PC. Macs are also considered very secure, with no widespread virus or spyware infections to date.

Figure 4.6

Mac OS X Leopard

(Source: Courtesy of Apple Computer, Inc.)

When attorney Renee Mancino decided to leave her Las Vegas law firm and start her own home-based practice, she chose an Apple MacBook Pro with the Mac OS as her mobile office.[3] She appreciates the Mac's organizational features that help her to manage and sift through the thousands of documents associated with her cases.

Linux

Linux is an OS developed by Linus Torvalds in 1991 as a student in Finland. The OS is distributed under the GNU General Public License, and its source code is freely available to everyone. It is, therefore, called an open-source operating system. This doesn't mean, however, that Linux and its assorted distributions are necessarily free—companies and developers can charge money for a distribution as long as the source code remains available. Linux is actually only the kernel of an OS, the part that controls hardware, manages files, separates processes, and so forth. Several combinations of Linux are available, with various sets of capabilities and applications to form a complete OS. Each of these combinations is called a *distribution* of Linux. Many distributions are available as free downloads.

Linux is available on the Internet and from other sources, including Red Hat Linux and Caldera OpenLinux. Many people and organizations use Linux.

In addition, several large computer vendors, including IBM, Hewlett-Packard, and Intel, support the Linux operating system. For example, IBM has more than 500 programmers working with Linux, primarily because of its security features. Many CIOs are considering switching to Linux and open-source software because of security concerns with Microsoft software.

Linux is making inroads to the consumer PC market with their GUI distributions. Both Dell and Lenovo sell notebook computers running Ubuntu and SuSE Linux.[4] Ubuntu is a user-friendly Linux distribution that is free to download and includes dozens of free software packages (see Figure 4.7). Wal-Mart and Sears are selling Linux PCs for $200, making them a popular alternative to other types of computers.[5] (Wal-Mart is selling the PCs only online.) New ultra-compact notebooks such as the ASUS Eee PC are preinstalled with Linux to make the most of their limited system resources.

Figure 4.7

Ubuntu Linux Operating System

(Source: Courtesy of Ubuntu.)

Radio station KRUU, "the Voice of Fairfield" in Iowa, is a nonprofit, community-based radio station. It broadcasts locally every day, 24 hours a day, and online to 30 countries (*www.kruufm.com*). The station supports about 100 hosts and 75 programs, broadcasting programs ranging from bedtime stories to death-metal music. When shopping for an operating system to use in the studio, KRUU selected Linux Ubuntu.[6] "Our requirements were quite complex and our decision to go with Ubuntu was based on three factors and Ubuntu won hands down," stated Sundar Raman, a presenter at the station. The three factors were: (1) Ubuntu looks good and is simple for both Windows and Mac users to use, (2) Ubuntu is reliable and easy to manage both locally and remotely, and (3) Ubuntu software supports professional audio editing and mixing software and hardware. One benefit of using Linux Ubuntu is the user community support. KRUU found all the answers it needed regarding running a professional studio on Linux from the "Ubuntu-Studio" community. Communicating with other Linux professionals helped the station use Linux computers for all of their computing tasks, including recording and mixing consoles.

Workgroup Operating Systems

To keep pace with user demands, the technology of the future must support a world in which network usage, data-storage requirements, and data-processing speeds increase at a dramatic rate. This rapid increase in communications and data-processing capabilities pushes the boundaries of computer science and physics. Powerful and sophisticated OSs are needed to run the servers that meet these business needs for workgroups. Small businesses, for example, often use workgroup OSs to run networks and perform critical business tasks.

Windows Server

Microsoft designed *Windows Server* to perform a host of tasks that are vital for Web sites and corporate Web applications. For example, Microsoft Windows Server can be used to coordinate large data centers. The OS also works with other Microsoft products. It can be used to prevent unauthorized disclosure of information by blocking text and e-mails from being copied, printed, or forwarded to other people. Microsoft *Windows Server 2008* is the most recent version of Windows Server and delivers benefits such as a powerful Web server management system, virtualization tools that allow various operating systems to run on a single server, advanced security features, and robust administrative support.

UNIX

UNIX is a powerful OS originally developed by AT&T for minicomputers. UNIX can be used on many computer system types and platforms, from personal computers to mainframe systems. UNIX also makes it much easier to move programs and data among computers or to connect mainframes and personal computers to share resources. There are many variants of UNIX—including HP/UX from Hewlett-Packard, AIX from IBM, UNIX SystemV from UNIX Systems Lab, Solaris from Sun Microsystems, and SCO from Santa Cruz Operations. Sun Microsystems hopes that its open-source Solaris will attract developers to make the software even better.

The online marketplace eBay uses Sun Microsystems servers, software, storage, and services to run its operations.[7] Sun's Solaris operating system manages eBay's systems, including database servers, Web servers, tape libraries, and identity management systems. The online auction company found that when they switched to Sun and Solaris, system performance increased by 20 percent. The Idaho National Laboratory also uses Solaris to conduct research in their work to design more efficient and safe nuclear reactors.[8]

NetWare

NetWare is a network OS sold by Novell that can support users on Windows, Macintosh, and UNIX platforms. NetWare provides directory software to track computers, programs, and people on a network, helping large companies to manage complex networks. NetWare users can log on from any computer on the network and use their own familiar desktop with all their applications, data, and preferences.

Red Hat Linux

Red Hat Software offers a Linux network OS that taps into the talents of tens of thousands of volunteer programmers who generate a steady stream of improvements for the Linux OS. The *Red Hat Linux* network OS is very efficient at serving Web pages and can manage a cluster of up to eight servers. Linux environments typically have fewer virus and security problems than other OSs. Distributions such as SuSE and Red Hat have proven Linux to be a very stable and efficient OS.

Mac OS X Server

The *Mac OS X Server* is the first modern server OS from Apple Computer and is based on the UNIX OS. The most recent version is OS X Server 10.5 Leopard. It includes features that allow the easy management of network and Internet services such as e-mail, Web site hosting, calendar management and sharing, wikis, and podcasting.

Enterprise Operating Systems

New mainframe computers provide the computing and storage capacity to meet massive data-processing requirements and offer many users high performance and excellent system availability, strong security, and scalability. In addition, a wide range of application software has been developed to run in the mainframe environment, making it possible to purchase software to address almost any business problem. As a result, mainframe computers remain the computing platform of choice for mission-critical business applications for many companies. Examples of mainframe OSs include z/OS from IBM, HP-UX from Hewlett-Packard, and Linux.

z/OS

The *z/OS* is IBM's first 64-bit enterprise OS. It supports IBM's z900 and z800 lines of mainframes that can come with up to sixteen 64-bit processors. (The z stands for zero downtime.) The OS provides several new capabilities to make it easier and less expensive for users to run large mainframe computers. The OS has improved workload management and advanced e-commerce security. The IBM zSeries mainframe, like previous generations of IBM mainframes, lets users subdivide a single computer into multiple smaller servers, each of which can run a different application. In recognition of the widespread popularity of a competing OS, z/OS allows partitions to run a version of the Linux OS. This means that a company can upgrade to a mainframe that runs the Linux OS.

HP-UX and Linux

The *HP-UX* is a robust UNIX-based OS from Hewlett-Packard designed to handle a variety of business tasks, including online transaction processing and Web applications. It supports Internet, database, and business applications on server and mainframe enterprise systems. It can work with Java programs and Linux applications. The OS comes in five versions: foundation, enterprise, mission critical, minimal technical, and technical. HP-UX supports Hewlett-Packard's computers and those designed to run Intel's Itanium processors. *Red Hat Enterprise Linux* for IBM mainframe computers is another example of an enterprise operating system.

Operating Systems for Small Computers, Embedded Computers, and Special-Purpose Devices

New OSs and other software are changing the way we interact with personal digital assistants (PDAs), smartphones, cell phones, digital cameras, TVs, and other appliances. These OSs are also called *embedded operating systems* because they are typically embedded within a device, such as an automobile or TV recorder. Embedded software is a multibillion dollar industry. Some of these OSs allow you to synchronize handheld devices with PCs using cradles, cables, and wireless connections. Cell phones also use embedded OSs (see Figure 4.8). In addition, some OSs have been developed for special-purpose devices, such as TV set-top boxes, computers on the space shuttle, computers in military weapons, and computers in some home appliances. Some of the more popular OSs for devices are described in the following section.

Figure 4.8

Mobile Phones Have Embedded Operating Systems

Many cell phones and smartphones, such as this BlackBerry, have an embedded OS that can support access to communications, media, and information.

(Source: Courtesy of PRNewsFoto/Verizon Wireless.)

An IT group within the United States Department of Agriculture recently deployed BlackBerries to their IT staff.[9] The high-speed network connection between BlackBerry and the organization's private network allowed system support staff to troubleshoot problems on Linux, UNIX, and Microsoft servers located in the home office from any location.

Palm OS

ACCESS Systems makes the Palm operating system, which is used in over 30 million handheld computers and smartphones manufactured by Palm, Inc. and other companies. Palm also develops and supports applications, including business, multimedia, games, productivity, reference and education, hobbies and entertainment, travel, sports, utilities, and wireless applications. Today, the smartphone market is overtaking the PDA market, as mobile users prefer to combine phone and information services in one device. OSs for this market are also provided by Research in Motion, Microsoft, Symbian, Apple (for the iPhone), and others.

Windows Embedded

Windows Embedded is a family of Microsoft OSs included with or embedded into small computer devices. Windows Embedded includes several versions that provide computing power for TV set-top boxes, automated industrial machines, media players, medical devices, digital cameras, PDAs, GPS receivers, ATMs, gaming devices, and business devices such as cash registers. Microsoft Auto provides a computing platform for automotive software such as Ford Sync. The Ford Sync system uses an in-dashboard display and wireless networking technologies to link automotive systems with cell phones and portable media players (see Figure 4.9).

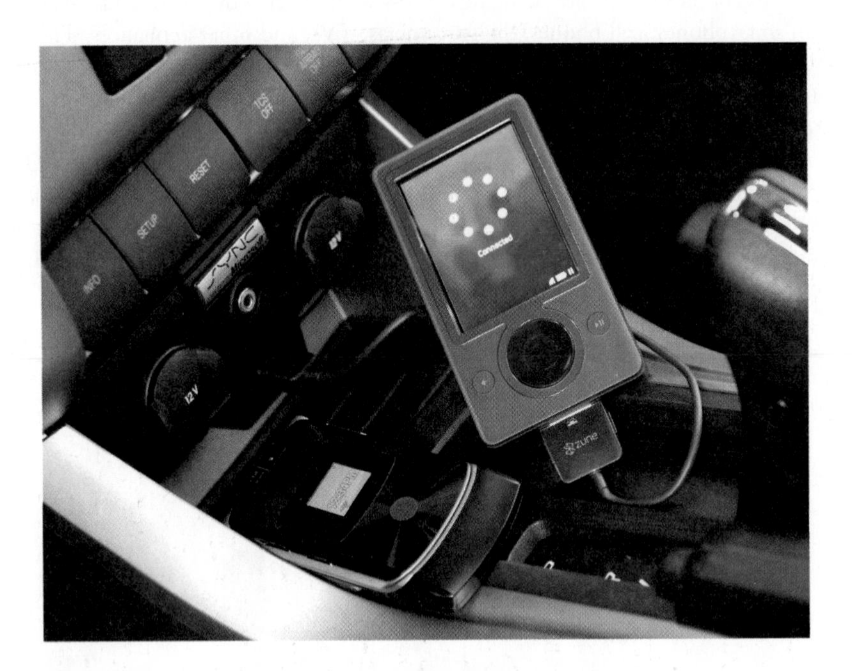

Figure 4.9

Microsoft Auto and Ford Sync

The Ford Sync system, developed on the Microsoft Auto operating system, allows drivers to wirelessly connect cell phones and media devices to automotive systems.

(Source: Courtesy of Microsoft Corporation and Ford Motor Company.)

Windows Mobile

Windows Mobile is an operating system designed for smartphones and PDAs. Different versions of Windows Mobile support either a touch screen interface or a menu-driven interface. In addition to supporting typical cellular services, Windows Mobile provides handwriting recognition, instant messaging technology, support for more secure Internet connections, and the ability to beam information to other devices. The OS also has advanced telecommunications capabilities, discussed in more detail in Chapter 6. Dozens of phones provided by all of the major carriers run Windows Mobile.

Utility Programs

utility programs
Programs that help to perform maintenance or correct problems with a computer system.

Utility programs help to perform maintenance or correct problems with a computer system. For example, some utility programs merge and sort sets of data, keep track of computer jobs being run, compress files of data before they are stored or transmitted over a network (thus saving space and time), and perform other important tasks. Some utility programs can help computer systems run better and longer without problems.

Another type of utility program allows people and organizations to take advantage of unused computer power over a network. Often called *grid computing*, the approach can be very efficient and less expensive than purchasing additional hardware or computer equipment. Financial services firm Wachovia Corporation uses grid computing to combine the power of 10,000 CPUs located on computers around the world for processing transactions. [10] In the future, grid computing could become a common feature of OSs and provide inexpensive, on-demand access to computer power and resources.

Utility programs can also help to secure and safeguard data. For example, the recording and motion picture industry uses digital rights management (DRM) technologies to prevent copyright-protected movies and music from being unlawfully copied. Music and media files are encoded so that software running on players recognizes and plays only legally obtained copies. DRM has been criticized for infringing on the freedom and rights of customers. Record companies are experimenting with DRM-free music to see if it increases sales.

Although many PC utility programs come installed on computers (see Figure 4.10), you can also purchase utility programs separately. The following sections examine some common types of utilities.

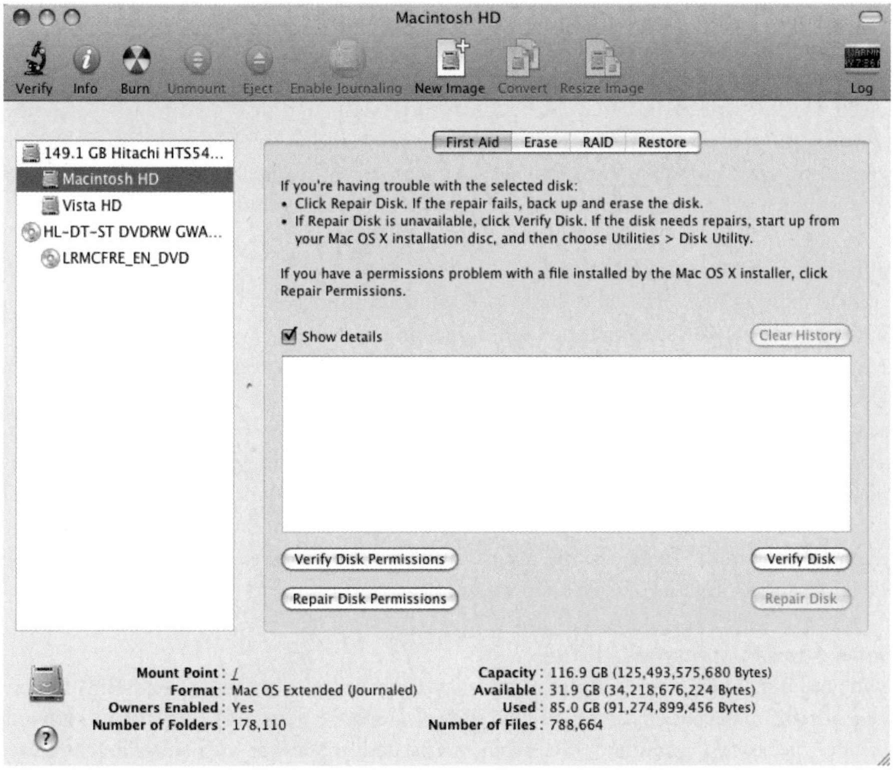

Figure 4.10

Mac Disk Utility

The Apple Mac Disk Utility is packaged with OS X and provides tools for repairing disks, backing up disks, creating disk images, and burning CDs and DVDs.

Hardware Utilities

Some hardware utilities are available from companies such as Symantec, which produces Norton Utilities. Hardware utilities can check the status of all parts of the PC, including hard disks, memory, modems, speakers, and printers. Disk utilities check the hard disk's boot sector, file allocation tables, and directories, and analyze them to ensure that the hard disk is not damaged. Disk utilities can also optimize the placement of files on a crowded disk.

Security Utilities

Computer viruses and spyware from the Internet and other sources can be a nuisance—and sometimes can completely disable a computer. Antivirus and antispyware software can be installed to constantly monitor and protect the computer. If a virus or spyware is found, often times it can be removed. This software runs continuously in the background to keep new viruses and spyware from entering the system. To keep current and make sure that the software checks for the latest threats, it can be easily updated over the Internet. It is also a good idea to protect computer systems with firewall software. Firewall software filters incoming and outgoing packets making sure that hackers or their tools are not attacking the system. Some software assists in keeping private data from being accessed from a computer system, in order to protect you from scams and fraud. Symantec, McAfee, and Microsoft are the most popular providers of security software.

File-Compression Utilities

File-compression programs can reduce the amount of disk space required to store a file or reduce the time it takes to transfer a file over the Internet. A popular program on Windows PCs is WinZip (*www.winzip.com*), which generates zip files, which are collections of one or more compressed files. A zip file has a .zip extension, and its contents can be easily unzipped to their original size. Windows Vista includes utilities for compressing and uncompressing files. *MP3 (Motion Pictures Experts Group-Layer 3)* is a popular file-compression format used to store, transfer, and play music and audio files, such as podcasts—audio programs that can be downloaded from the Internet. MP3 can compress files ten times smaller than the original file with near-CD-quality sound. Software, such as iTunes from Apple, can be used to store, organize, and play MP3 music files.

Spam and Pop-Up Blocker Utilities

Getting unwanted e-mail (spam) and having annoying and unwanted ads pop up on your screen while you are on the Web can be a frustrating waste of time. You can install a number of utility programs to help block unwanted e-mail spam and pop-up ads. Most Internet service providers and Web-based e-mail systems provide a spam-blocking service, and Web browsers such as Internet Explorer and Firefox include pop-up blocking utilities.

Network and Internet Utilities

A broad range of network- and systems-management utility software is available to monitor hardware and network performance and trigger an alert when a Web server is crashing or a network problem occurs. Although these general management features are helpful, what is needed is a way to pinpoint the cause of the problem. Topaz from Mercury Interactive is an example of software called an *advanced Web-performance monitoring utility*. It is designed to sound an alarm when it detects problems and let network administrators isolate the most likely causes of the problems. Its Auto RCA (root-cause analysis) module uses statistical analysis with built-in rules to measure system and Web performance. Actual performance data is compared with the rules, and the results can help pinpoint where trouble originated—in the application software, database, server, network, or the security features.

Server and Mainframe Utilities

Some utilities enhance the performance of servers and mainframe computers. IBM has created systems-management software that allows a support person to monitor the growing number of desktop computers in a business attached to a server or mainframe computer. With this software, the support people can sit at their personal computers and check or diagnose problems, such as a hard disk failure on a network computer. The support people can even repair individual systems anywhere on the organization's network, often without having to leave their desks. The direct benefit is to the system manager, but the business also gains from having a smoothly functioning information system. Utility programs can meet the needs of a single user, workgroup, or enterprise, as listed in Table 4.3. These programs perform useful tasks—from tracking jobs to monitoring system integrity.

Personal	Workgroup	Enterprise
Software to compress data so that it takes less hard disk space	Software to provide detailed reports of work-group computer activity and status of user accounts	Software to archive contents of a database by copying data from disk to tape
Screen saver	Software that manages an uninterruptible power supply to do a controlled shutdown of the workgroup computer in the event of a loss of power	Software that compares the content of one file with another and identifies any differences
Antivirus and antispyware software	Software that reports unsuccessful user logon attempts	Software that reports the status of a particular computer job

Table 4.3

Examples of Utility Programs

Virtualization software can make computers simulate other computers. The result is often called a *virtual machine*. Using virtualization software, servers and mainframe computers can run software applications written for different operating systems. For example, you can use a server or mainframe to test and run a number of PC applications simultaneously, such as spreadsheets, word processors, and databases. Virtualization software such as VMWare is being used by businesses to safeguard private data. For example, Kindred Healthcare uses VMWare on its server to run hundreds of virtual Windows PC desktops that are accessed by mobile computers throughout the organization.[11] Because the patient data and the software tools used to access that data are running on the server, security measures are easy to implement.

Other Utilities

Utility programs are available for almost every conceivable task or function. For example, you can use Microsoft Windows Rights Management Services with Microsoft Office programs to manage and protect important corporate documents. ValueIT is a utility that can help a company verify the value of investments in information systems and technology. Widgit Software has developed an important software utility that helps people with visual disabilities use the Internet. The software converts icons and symbols into plain text that can be easily seen. Another software utility allows a manager to see every keystroke a worker makes on a computer system. Monitoring software can catalog the Internet sites that employees visit and the time that employees are working at their computer.

In addition, you can use many search tools to find important files and documents. Most of these desktop search tools are free and available from a number of popular Internet sites. Yahoo! Desktop Search, Google Desktop, Mac Spotlight, and Windows Search are examples (see Figure 4.11).

Middleware

Middleware is software that allows different systems to communicate and exchange data. Middleware can also be used as an interface between the Internet and older legacy systems. (Legacy software is a previous, major version that continues to be used.) For example, middleware can be used to transfer a request for information from a corporate customer on the corporate Web site to a traditional database on a mainframe computer and return the results to the customer on the Internet.

middleware
Software that allows different systems to communicate and exchange data.

The use of middleware to connect disparate systems has evolved into an approach for developing software and systems called SOA. A **service-oriented architecture**, or SOA, uses modular application services to allow users to interact with systems, and systems to interact with each other. Systems developed with SOA are flexible and ideal for businesses that need a system to expand and evolve over time. SOA modules can be reused for a variety of purposes, which reduces development time. Because SOA modules are designed using programming standards so they can interact with other modules, rigid custom-designed middleware software is not needed to connect systems. However, Southside Electric Cooperative, Inc. in Virginia found SOA to be the perfect solution to eliminating time-consuming paperwork, reducing response time to customer needs, and doubling the rate that it is able to collect delinquent payments.[12] The system uses Qualcomm OmniTRACS wireless communications and IBM's SOA-based WebSphere software.

service-oriented architecture (SOA)
A modular method of developing software and systems that allows users to interact with systems, and systems to interact with each other.

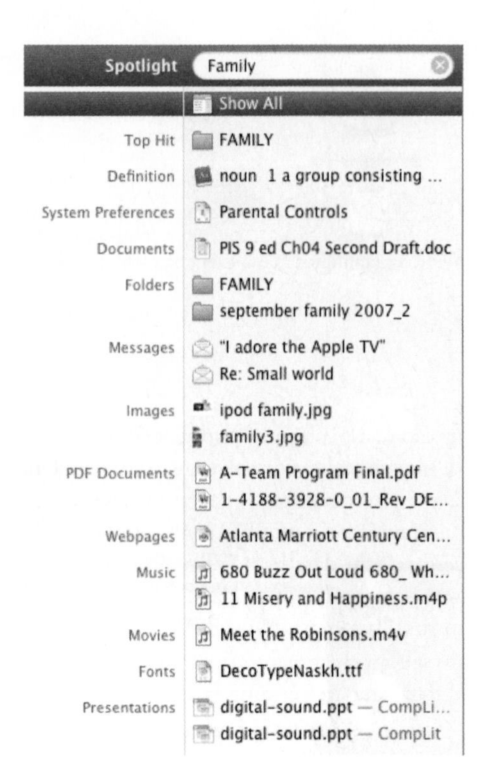

Figure 4.11

Desktop Search Tool

With a desktop search tool, you can search through files on your computer to instantly find files of all types that have something to do with the specified keyword.

APPLICATION SOFTWARE

As discussed earlier in this chapter, the primary function of application software is to apply the power of the computer to give people, workgroups, and the entire enterprise the ability to solve problems and perform specific tasks. When you need the computer to do something, you use one or more application programs. The application programs interact with systems software, and the systems software then directs the computer hardware to perform the necessary tasks. Applications help you perform common tasks, such as creating and formatting text documents, performing calculations, or managing information, though some applications are more specialized. A pharmaceutical company, for example, has developed application software to detect the early signs of Parkinson's disease. The new software detects slight trembling in speech patterns not detectable by the human ear that predicts the disease. Application software is used throughout the medical profession to save and prolong lives. For example, Oregon Health & Science University uses iRecruitment software from Oracle to match employees to job openings.[13]

The functions performed by application software are diverse, and range from personal productivity to business analysis. For example, application software can help sales managers track sales of a new item in a test market. Software from IntelliVid monitors video feeds from store security cameras and notifies security when a shopper is behaving suspiciously.[14] Most of the computerized business jobs and activities discussed in this book involve application software. We begin by investigating the types and functions of application software.

INFORMATION SYSTEMS @ WORK

Software Helps Target Radiation Treatment for Cancer

Doctors have been using radiation therapy as a treatment for cancer since the 1940s. The treatment has saved countless lives, yet has been somewhat imprecise until recently. The original method of treating a tumor with radiation used a linear accelerator that delivered radiation in rectangular beams. Doctors used lead blocks to prevent the beams from harming healthy tissue. The process was cumbersome and only partially effective. Surrounding tissue was often destroyed along with the tumor.

In the 1980s, a machine called an MLC, for multileaf collimator, was invented. The MLC had motorized leaves to disrupt the beam of radiation and focus it more closely on where it was needed. Still, the treatment was imprecise, lacking real-time control of the radiation intensity and direction.

Until the mid-1990s, most of the development of radiation treatment technologies focused on hardware. Varian Medical Systems decided that devising a more effective system would require a heavy investment in software development. Computing processors and hardware were advanced enough to precisely control beams of radiation, but the software to empower the hardware had yet to be developed. Varian transformed itself from a hardware company to a software company to get the job done.

Varian hired experts in programming embedded controls, user interfaces, treatment planning, and databases. It proceeded incrementally over many years to develop a trustworthy and powerful system called the SmartBeam IMRT (for Intensity Modulated Radiation Therapy), which is now in use at thousands of medical facilities around the world.

The SmartBeam IMRT combines an x-ray and radiation technology into one device that rotates around the patient delivering radiation at precise intensities from any angle. The machine is the first that allows physicians to examine and treat a tumor at the same time. The on-board imager produces "high-resolution images of tumors and tracks changes in a tumor's shape, size, and position... that when coupled with SmartBeam IMRT, allows clinicians to be even more precise when targeting tumors," according to *Computerworld*. The magazine awarded Varian the top prize for information systems in manufacturing in its 2007 Computerworld Honors Program.

Discussion Questions

1. What role does software play in the SmartBeam IMRT medical system?
2. Why couldn't Varian produce the SmartBeam IMRT before it did?

Critical Thinking Questions

1. What additional safeguards must be programmed into the software that runs the SmartBeam IMRT that aren't necessary in typical PC software?
2. How do you think the development of the SmartBeam IMRT launched Varian to the top of the market in cancer treatment systems?

SOURCES: Pratt, Mary K., "Software Helps Target Radiation Treatment for Cancer," *Computerworld*, December 3, 2007, *www.computerworld.com/action/article.do?command=viewArticleBasic&articleId=30486 5&pageNumber=1*. Varian Medical Systems Web site, *www.varian.com*, accessed February 3, 2007.

Overview of Application Software

Proprietary software and off-the-shelf software are important types of application software (see Figure 4.12). A company can develop a one-of-a-kind program for a specific application (called **proprietary software**). Proprietary software is not in the public domain. A company can also purchase or acquire an existing software program (sometimes called **off-the-shelf software** because it can literally be purchased or acquired "off the shelf" in a store). The relative advantages and disadvantages of proprietary software and off-the-shelf software are summarized in Table 4.4.

Figure 4.12

Types of Application Software

Some off-the-shelf software can be customized to suit user needs.

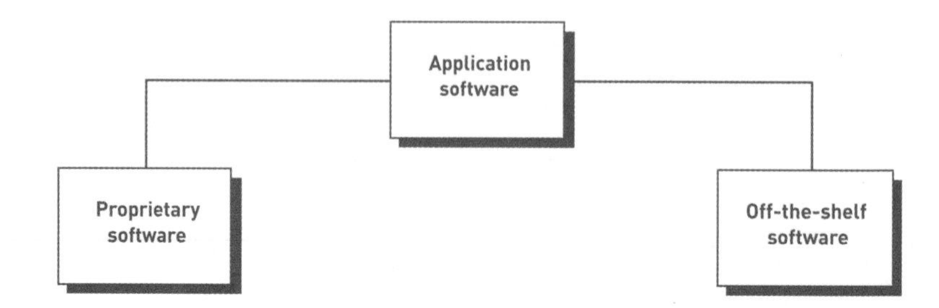

Proprietary Software		Off-the-Shelf Software	
Advantages	**Disadvantages**	**Advantages**	**Disadvantages**
You can get exactly what you need in terms of features, reports, and so on.	It can take a long time and significant resources to develop required features.	The initial cost is lower because the software firm can spread the development costs over many customers.	An organization might have to pay for features that are not required and never used.
Being involved in the development offers control over the results.	In-house system development staff may become hard pressed to provide the required level of ongoing support and maintenance because of pressure to move on to other new projects.	The software is likely to meet the basic business needs—you can analyze existing features and the performance of the package before purchasing.	The software might lack important features, thus requiring future modification or customization. This can be very expensive because users must adopt future releases of the software as well.
You can modify features that you might need to counteract an initiative by competitors or to meet new supplier or customer demands. A merger with or acquisition of another firm also requires software changes to meet new business needs.	The features and performance of software that has yet to be developed presents more potential risk.	The package is likely to be of high quality because many customer firms have tested the software and helped identify its bugs.	The software might not match current work processes and data standards.

Table 4.4

A Comparison of Proprietary and Off-the-Shelf Software

Many companies use off-the-shelf software to support business processes. For example, the lawyers at Ferwick & West LLP use a combination of several off-the-shelf software packages to help "cull data" from millions of legal documents.[15] One case required sorting through over 100 million legal files. The system they developed is called FIND, for File Identification Narrowed by Definition, and it combines the power of 75 software tools, most of them off-the-shelf applications. Key questions for selecting off-the-shelf software include the following: (1) Will the software run on the OS and hardware you have selected? (2) Does the software meet the essential business requirements that have been defined? (3) Is the software manufacturer financially solvent and reliable? and (4) Does the total cost of purchasing, installing, and maintaining the software compare favorably to the expected business benefits?

Some off-the-shelf programs can be modified, in effect blending the off-the-shelf and customized approaches. For example, police officers and dispatchers in Dover, N.H., use a customized off-the-shelf software package that provides a map view of the jurisdiction. Dispatchers can easily identify the location of patrol cars and crime scenes on the map, and quickly route the nearest car to the desired location.[16] In another example of the blended approach, Blue Cross and Blue Shield worked with Sun Microsystems to customize a claims management system for its customers to access over the Web.[17]

Another approach to obtaining a customized software package is to use an application service provider. An **application service provider (ASP)** is a company that can provide the software, support, and computer hardware on which to run the software from the user's facilities over a network. Some vendors refer to the service as *on-demand software*. An ASP can also simplify a complex corporate software package so that it is easier for the users to set up and manage. ASPs provide contract customization of off-the-shelf software, and they speed deployment of new applications while helping IS managers avoid implementation headaches, reducing the need for many skilled IS staff members and decreasing project start-up expenses. Such an approach allows companies to devote more time and resources to more important tasks. For example, Avanax, a Silicon Valley company that develops intelligent photonic solutions for optical networks, uses a Product Lifecycle Management system provided by SAP. The system runs on SAP servers, which has helped Avanax reduce costs and provide much higher levels of service.[18]

application service provider (ASP)
A company that provides software, support, and the computer hardware on which to run the software from the user's facilities over a network.

Using an ASP makes the most sense for relatively small, fast-growing companies with limited IS resources. It is also a good strategy for companies that want to deploy a single, functionally focused application quickly, such as setting up an e-commerce Web site or supporting expense reporting. Contracting with an ASP might make less sense, however, for larger companies that have major systems and their technical infrastructure already in place.

Using an ASP involves some risks—sensitive information could be compromised in a number of ways, including unauthorized access by employees or computer hackers; the ASP might not be able to keep its computers and network up and running as consistently as necessary; or a disaster could disable the ASP's data center, temporarily putting an organization out of business. These are legitimate concerns that an ASP must address.

The high overhead of an ASP designing, running, managing, and supporting many customized applications for many businesses has led to a new form of software distribution known as software as a service. **Software as a service (SaaS)** allows businesses to subscribe to Web-delivered business application software by paying a monthly service charge or a per-use fee. Like ASP, SaaS providers maintain software on their own servers and provide access to it over the Internet. SaaS usually uses a Web browser-based user interface. SaaS can reduce expenses by sharing its running applications among many businesses. For example, Sears, JCPenney, and Wal-Mart might use customer relationship management software provided by a common SaaS provider. Providing one high-quality SaaS application to thousands of businesses is much more cost-effective than custom designing software for each business.

software as a service (SaaS)
A service that allows businesses to subscribe to Web-delivered business application software by paying a monthly service charge or a per-use fee.

Customer relationship management (CRM) and other general business systems are good candidates for SaaS. For example, The Improv, "America's Original Comedy Showcase," turned to a SaaS CRM system from salesforce.com to manage marketing and sales of event space to businesses wanting to use its theaters.[19] SaaS is becoming popular for information security as well, as described in the Ethical and Societal Issues sidebar.

ETHICAL AND SOCIETAL ISSUES

Imperial Chemical Turns to SaaS Security Tools

Imperial Chemical Industries is a very large paint and chemicals manufacturer based in London. The company was recently purchased by Akzo Nobel for $16 billion. With a research budget of around $60 million annually, and research data spread geographically over many computer systems at a variety of locations, Imperial Chemical works hard to keep its valuable data protected and secure.

Securing data over large distributed systems can be a costly, time-consuming affair. It becomes more complex when one company's systems are merged with another company's systems over a network. In today's global information economy, it is not unusual for a corporation to join its network with several partners and suppliers. To secure such networks would require a large suite of security software continuously running on all computers and a team of security experts working around the clock.

Rather than incur these costs, Imperial Chemical decided to outsource much of its information security to online companies offering security SaaS. SaaS makes sense for many security applications because the scanning of systems can take place from any network-connected system.

Imperial uses three SaaS security providers:

- Qualys provides a vulnerability management service that includes network discovery and mapping, asset prioritization, vulnerability assessment reporting, and remediation tracking according to business risk.
- Veracode provides a service that scans all binary executable files on the system, looking for bugs and viruses.
- Message Labs protects Imperial's e-mail systems from spam and viruses. It can also be used to filter out unauthorized and inappropriate content.

As securing corporate and customer data becomes increasingly regulated, many companies are turning to security SaaS vendors to make sure that they are in compliance with the law. For example, the three companies above insure that their customers are in compliance with the PCI DSS, the Payment Card Industry Data Security Standard. This standard is required by certain companies and banks that wish to insure their customers' privacy.

SaaS security systems are ideal for large organizations that have thousands of computers to secure. However, it is also easy to imagine how such services could provide a security solution for individual personal computers as well. Currently hundreds or thousands of home PCs are infected by spyware and serving as bots being controlled by hackers to send spam and attack other systems. Internet service providers do what they can to keep their users safe, but they can't stop a user from running an infected file or wandering to an infected Web site. Incorporating SaaS security systems through Internet service providers to personal PCs would clear up most of the infections that plague the Internet. As with most security practices, there would probably be some tradeoff in convenience and privacy.

Discussion Questions

1. Why does it make sense for a large corporation to outsource information security to a SaaS provider?
2. What are the dangers of trusting corporate information systems to an outside security firm?

Critical Thinking Questions

1. Would you be willing to allow a security company to guard your PC remotely while you are connected to the Internet? Why or why not?
2. Currently, PC users must run about four different security applications to keep their computers safe: a firewall, virus protection, spyware protection, and Windows Update. The user is responsible for making sure these systems are operational and up to date. Whose responsibility should it be to secure a PC? How might this system be simplified for users?

SOURCES: Hines, Matt, "Security SaaS offerings growing up fast," *Computerworld*, August 23, 2007, *www.computerworld.com/action/article.do?command=viewArticleBasic&taxonomyName =saas&articleId=9032321&taxonomyId=170&intsrc=kc_feat*. Qualys Web site, *www.qualys.com*, accessed February 2, 2008. MessageLabs Web site, *www.messagelabs.com*, accessed February 2, 2008. PCI (Payment Card Industry) Security Standards Council Web site, *https://www.pcisecuritystandards.org*, accessed February 2, 2008.

Personal Application Software

Hundreds of computer applications can help people at school, home, and work. New computer software under development and existing GPS technology, for example, will allow people to see 3-D views of where they are, along with directions and 3-D maps to where they would like to go. The features of personal application software are summarized in Table 4.5. In addition to these general-purpose programs, thousands of other personal computer applications perform specialized tasks: to help you do your taxes, get in shape, lose weight, get medical advice, write wills and other legal documents, repair your computer, fix your car, write music, and edit your pictures and videos. This type of software, often called *user software* or *personal productivity software*, includes the general-purpose tools and programs that support individual needs.

Table 4.5

Examples of Personal Productivity Software

Type of Software	Explanation	Example	Vendor
Word processing	Create, edit, and print text documents	Word WordPerfect Google Docs Pages Writer	Microsoft Corel Google Apple Sun
Spreadsheet	Provide a wide range of built-in functions for statistical, financial, logical, database, graphics, and date and time calculations	Excel Lotus 1-2-3 Spreadsheet Numbers Calc	Microsoft Lotus/IBM Google Apple Sun
Database	Store, manipulate, and retrieve data	Access Approach dBASE Base	Microsoft Lotus/IBM Borland Sun
Graphics	Develop graphs, illustrations, and drawings	Illustrator FreeHand	Adobe Macromedia
Project management	Plan, schedule, allocate, and control people and resources (money, time, and technology) needed to complete a project according to schedule	Project for Windows On Target Project Schedule Time Line	Microsoft Symantec Scitor Symantec
Financial management	Provide income and expense tracking and reporting to monitor and plan budgets (some programs have investment portfolio management features)	Quicken Money	Intuit Microsoft
Desktop publishing (DTP)	Use with personal computers and high-resolution printers to create high-quality printed output, including text and graphics; various styles of pages can be laid out; art and text files from other programs can also be integrated into "published" pages	QuarkXPress Publisher PageMaker Ventura Publisher Pages	Quark Microsoft Adobe Corel Apple
Creativity	Generate innovative and creative ideas and problem solutions. The software does not propose solutions, but provides a framework conducive to creative thought. The software takes users through a routine, first naming a problem, then organizing ideas and "wishes," and offering new information to suggest different ideas or solutions	Organizer Notes	Macromedia Lotus

Word Processing

Word processing applications are installed on most PCs today. These applications come with a vast array of features, including those for checking spelling, creating tables, inserting formulas, creating graphics, and much more (see Figure 4.13). This book (and most like it) was entered into a word processing application using a personal computer.

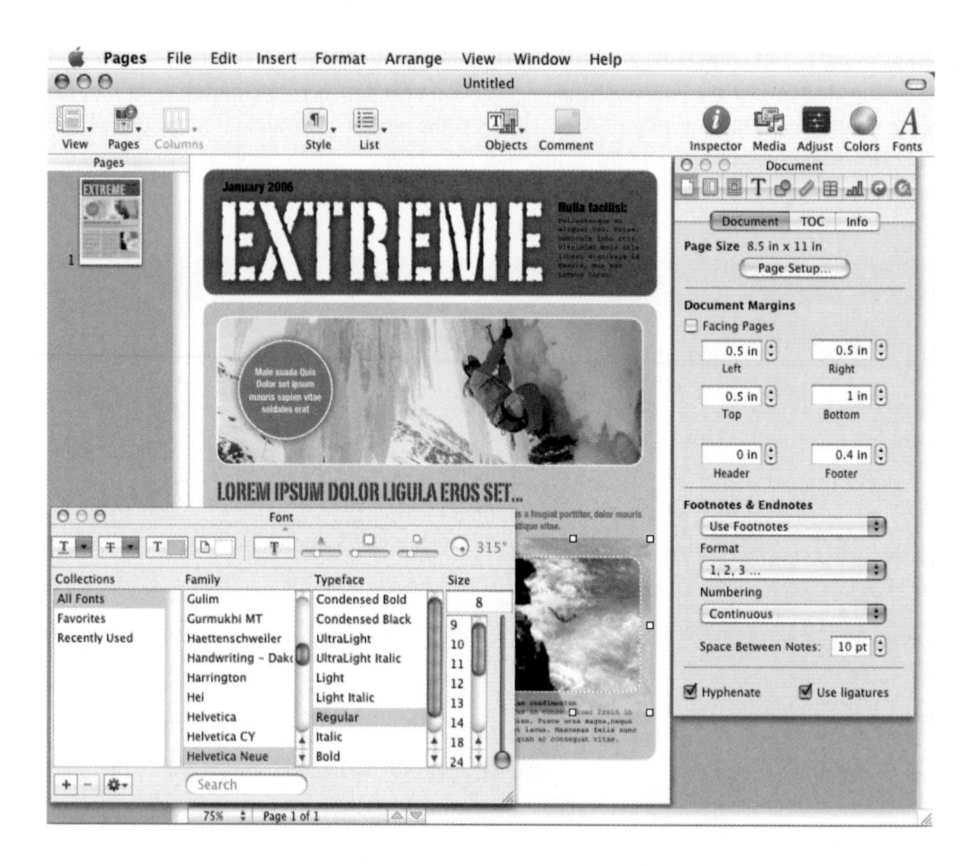

Figure 4.13

Word Processing Program

Word processing applications can be used to write letters, professional documents, work reports, and term papers.

A team of people can use a word processing program to collaborate on a project. The authors and editors who developed this book, for example, used the Track Changes and Reviewing features of Microsoft Word to track and make changes to chapter files. You can add comments or make revisions to a document that a coworker can review and either accept or reject.

Professional chef JoAnna Minneci runs her own catering and in-home cooking services in Los Angeles, California. She believes in treating all of her customers as though they were celebrities. Minneci uses Word for Mac to design colorful and artistic menus and gift certificates.[20] The cross-platform compatibility of Microsoft Office for Mac allows her to deliver materials such as menus, contracts, and budgets to clients working on Macintosh or Windows-based PCs.

Spreadsheet Analysis

Spreadsheets are powerful tools for individuals and organizations. Features of spreadsheets include graphics, limited database capabilities, statistical analysis, built-in business functions, and much more (see Figure 4.14). The business functions include calculation of depreciation, present value, internal rate of return, and the monthly payment on a loan, to name a few. Optimization is another powerful feature of many spreadsheet programs. *Optimization* allows the spreadsheet to maximize or minimize a quantity subject to certain constraints. For example, a small furniture manufacturer that produces chairs and tables might want to maximize its profits. The constraints could be a limited supply of lumber, a limited number of workers who can assemble the chairs and tables, or a limited amount of various hardware fasteners that might be required. Using an optimization feature, such as Solver in Microsoft

Excel, the spreadsheet can determine what number of chairs and tables to produce with labor and material constraints to maximize profits.

Figure 4.14

Spreadsheet Program

Spreadsheet programs should be considered when calculations are required.

Database Applications

Database applications are ideal for storing, manipulating, and retrieving data. These applications are particularly useful when you need to manipulate a large amount of data and produce reports and documents. Database manipulations include merging, editing, and sorting data. The uses of a database application are varied. You can keep track of a CD collection, the items in your apartment, tax records, and expenses. A student club can use a database to store names, addresses, phone numbers, and dues paid. In business, a database application can help process sales orders, control inventory, order new supplies, send letters to customers, and pay employees. Database management systems can be used to track orders, products, and customers; analyze weather data to make forecasts for the next several days; and summarize medical research results. A database can also be a front end to another application. For example, you can use a database application to enter and store income tax information, then export the stored results to other applications, such as a spreadsheet or tax-preparation application (see Figure 4.15).

Graphics Program

It is often said that a picture is worth a thousand words. With today's graphics programs, it is easy to develop attractive graphs, illustrations, and drawings (see Figure 4.16). Graphics programs can be used to develop advertising brochures, announcements, and full-color presentations, and to organize and edit photographic images. If you need to make a presentation at school or work, you can use a special type of graphics program called a presentation application to develop slides and then display them while you are speaking. Because of their popularity, many colleges and departments require students to become proficient at using presentation graphics programs.

Many graphics programs, such as Microsoft Office PowerPoint, consist of a series of slides. Each slide can be displayed on a computer screen, printed as a handout, or (more commonly) projected onto a large viewing screen for audiences. Powerful built-in features allow you to develop attractive slides and complete presentations. You can select a template for a type of presentation, such as recommending a strategy for managers, communicating news to a sales force, giving a training presentation, or facilitating a brainstorming session.

Figure 4.15

Database Program

After being entered into a database application, information can be manipulated and used to produce reports and documents.

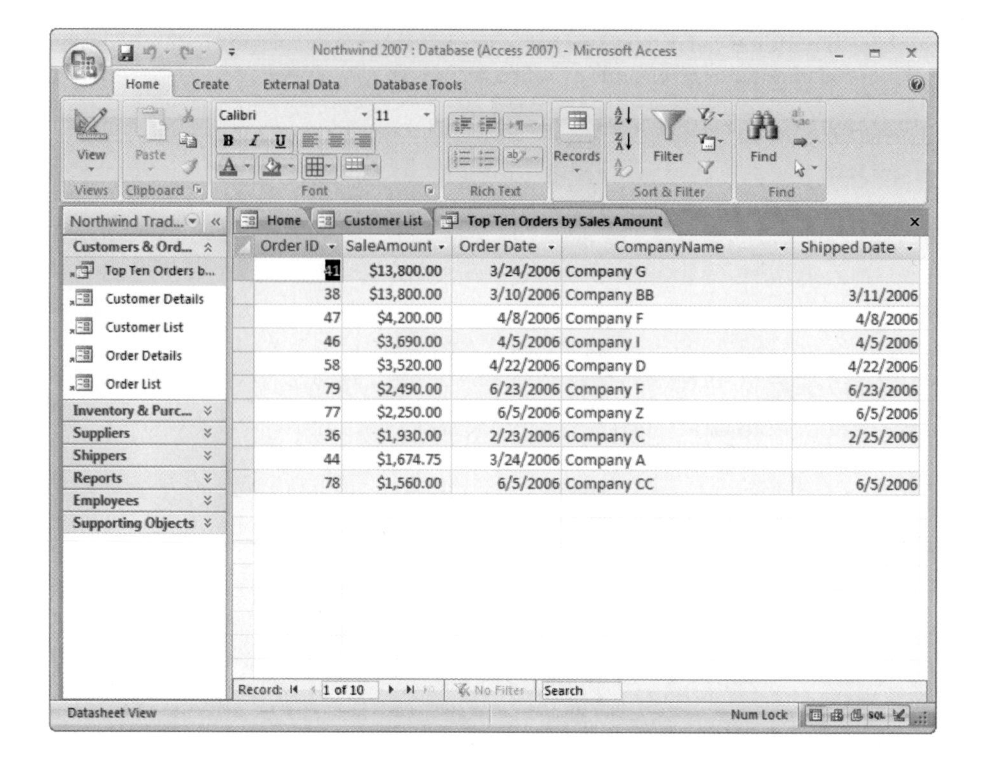

Figure 4.16

Presentation Graphics Program

Graphics programs can help you make a presentation at school or work.

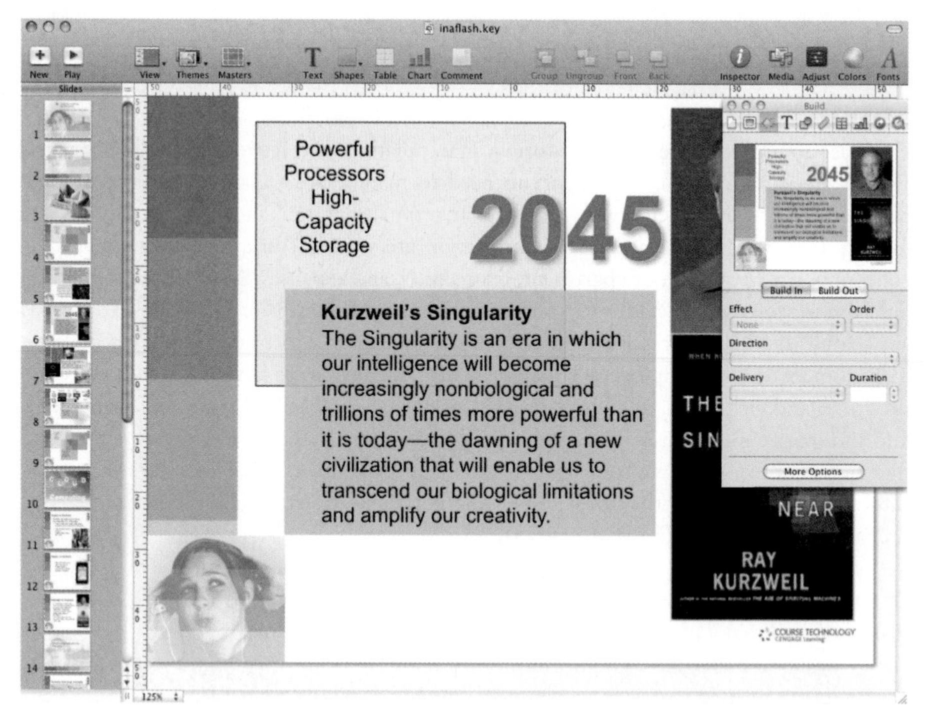

The presentation graphics program takes you through the presentation step by step, including applying color and attractive formatting. You can also design a custom presentation using the many types of charts, drawings, and formatting available. Most presentation graphics programs come with many pieces of *clip art*, such as drawings and photos of people meeting, medical equipment, telecommunications equipment, entertainment, and much more.

Personal Information Managers

Personal information managers (PIMs) help people, groups, and organizations store useful information, such as a list of tasks to complete or a set of names and addresses. PIMs usually provide an appointment calendar and a place to take notes. In addition, information in a

PIM can be linked. For example, you can link an appointment with a sales manager in the calendar to information on the sales manager in the address book. When you click the appointment in the calendar, a window opens displaying information on the sales manager from the address book. Google provides PIM software to integrate e-mail, appointment, and address book tasks.

Figure 4.17

Personal Information Manager

PIM software assists individuals, groups, and organizations with organizing appointments, schedules, contacts, and to-do lists.

Some PIMs allow you to schedule and coordinate group meetings. If a computer or handheld device is connected to a network, you can upload the PIM data and coordinate it with the calendar and schedule of others using the same PIM software on the network. You can also use some PIMs to coordinate e-mails sent and received over the Internet.

Consider Greenfield Online as an example of one collaborative PIM system. Greenfield Online is a Web survey solution provider that has about 500 employees in 12 countries and 30 cities. Employees were having a difficult time scheduling and preparing for meetings. Using Microsoft Live Meeting and Outlook, the company reduced meeting time by 60 percent.[21] Now employees schedule Web conferences directly using Microsoft Office Outlook. Documents are distributed to meeting participants who receive meeting requests through Outlook to attend the Web conferences. When it is time for a meeting, all participants click a URL in the meeting request and the link takes them to the Web-conferencing area.

Software Suites and Integrated Software Packages

A **software suite** is a collection of single application programs packaged in a bundle. Software suites can include word processors, spreadsheets, database management systems, graphics programs, communications tools, organizers, and more. Some suites support the development of Web pages, note taking, and speech recognition, where applications in the suite can accept voice commands and record dictation. Software suites offer many advantages. The software programs have been designed to work similarly, so after you learn the basics for one application, the other applications are easy to learn and use. Buying software in a bundled suite is cost-effective; the programs usually sell for a fraction of what they would cost individually.

software suite
A collection of single application programs packaged in a bundle.

Microsoft Office, Corel's WordPerfect Office, Lotus SmartSuite, and Sun Microsystems's StarOffice are examples of popular general-purpose software suites for personal computer users. Microsoft Office has the largest market share. The Free Software Foundation offers software similar to Sun Microsystems's StarOffice that includes word processing, spreadsheet, database, presentation graphics, and e-mail applications for the Linux OS. OpenOffice is another Office suite for Linux. Wine, software designed for Linux and Unix, can run any Windows application, including those in Microsoft Office, on Linux, although some features might not work as well as a Microsoft OS. Each of these software suites includes a spreadsheet program, word processor, database program, and graphics package with the ability to move documents, data, and diagrams among them (see Table 4.6). Thus, a user can create a spreadsheet and then cut and paste that spreadsheet into a document created using the word processing application.

Table 4.6

Major Components of Leading Software Suites

Personal Productivity Function	Microsoft Office	Lotus Symphony	Corel WordPerfect Office	Sun StarOffice	Apple iWork	Google
Word Processing	Word	Documents	WordPerfect	Writer	Pages	Docs
Spreadsheet	Excel	Spreadsheets	Quattro Pro	Calc	Numbers	Spreadsheet
Presentation Graphics	PowerPoint	Presentations	Presentations	Impress	Keynote	Presentation
Database	Access		Paradox	Base		

More than a hundred million people worldwide use the Microsoft Office software suite, with Office 2007 representing the latest version of the productivity software. Office 2007 uses new file formats that are more compatible with Web standards. It also provides a revolutionary new interface, moving from menus and toolbars to a Ribbon with tabs. Office 2007 is available in seven editions: Professional, Standard, Home and Student, Small Business, Ultimate, Professional Plus, and Enterprise. Each edition includes a subset of 15 applications

In addition to suites, some companies produce *integrated application packages* that contain several programs. For example, *Microsoft Works* is one program that contains basic word processing, spreadsheet, database, address book, calendar, and other applications. Although not as powerful as stand-alone software included in software suites, integrated software packages offer a range of capabilities for less money. Some integrated packages cost about $100.

Some companies are offering Web-based productivity software suites that require no installation, only a Web browser. Zoho, Google, and Thinkfree offer free online word processing, spreadsheet, presentation, and other software that require no installation on the PC. Documents created with the software can be stored on the Web server. Currently these online applications are not as powerful and robust as installed software such as Microsoft Office. However, it is likely that as the technology becomes more powerful, and network connection speeds increase, users will need to install less software on their own PCs and turn instead to using software online.

Microsoft has observed the trend towards Web-based software and is migrating its software towards the Web as well. Microsoft Windows Live provides several Web-based services such as a Live Search for searching the Web, Windows Live Messenger for instant messaging, Windows Live Hotmail for e-mail, and Windows Live OneCare for PC security. Windows Live Spaces provides Windows users with online storage for sharing files with others on the Web. Microsoft Office Live provides tools for sharing Office documents on the Web. The difference between Office Live and Google's applications is that Microsoft requires users to have its software installed on their PCs. Microsoft also provides Xbox Live for online multiplayer gaming.

Other Personal Application Software

In addition to the software already discussed, people can use many other interesting and powerful application software tools. In some cases, the features and capabilities of these applications can more than justify the cost of an entire computer system. TurboTax, for example, is a popular tax-preparation program. Other exciting software packages have been developed for training and distance learning. University professors often believe that colleges and universities must invest in distance learning for their students. Using this type of software, some universities offer complete degree programs over the Internet. Engineers, architects, and designers often use computer-aided design (CAD) software to design and develop buildings, electrical systems, plumbing systems, and more. Autosketch, CorelCAD, and AutoCad are examples of CAD software. Other programs perform a wide array of statistical tests. Colleges and universities often have a number of courses in statistics that use this type of application software. Two popular applications in the social sciences are SPSS and SAS.

Workgroup Application Software

Workgroup application software is designed to support teamwork, whether people are in the same location or dispersed around the world. This support can be accomplished with software known as *groupware* that helps groups of people work together effectively. Microsoft Exchange Server, for example, has groupware and e-mail features. Also called *collaborative software*, the approach allows a team of managers to work on the same production problem, letting them share their ideas and work via connected computer systems. The "Three Cs" rule for successful implementation of groupware is summarized in Table 4.7.

workgroup application software
Software that supports teamwork, whether in one location or around the world.

Table 4.7

Ernst & Young's "Three Cs" Rule for Groupware

Quality	Description
Convenient	If it's too hard to use, it's not used; it should be as easy to use as the telephone.
Content	It must provide a constant stream of rich, relevant, and personalized content.
Coverage	If it isn't easy to access, it might never be used.

Examples of workgroup software include group scheduling software, electronic mail, and other software that enables people to share ideas. Lotus Notes from IBM, for example, lets companies use one software package and one user interface to integrate many business processes. Lotus Notes can allow a global team to work together from a common set of documents, have electronic discussions using threads of discussion, and schedule team meetings. As the program matured, Lotus added services to it and renamed it Domino (Lotus Notes is now the name of the e-mail package), and now an entire third-party market has emerged to build collaborative software based on Domino.

The Web-based software described in the previous section is ideal for group use. Because documents are stored on an Internet server, anyone with an Internet connection can access them easily. Google provides options in its online applications that allow users to share documents, spreadsheets, presentations, calendars, and notes with other specified users or everyone on the Web (see Figure 4.18). This makes it convenient for several people to contribute to a document without concern for software compatibility or storage.

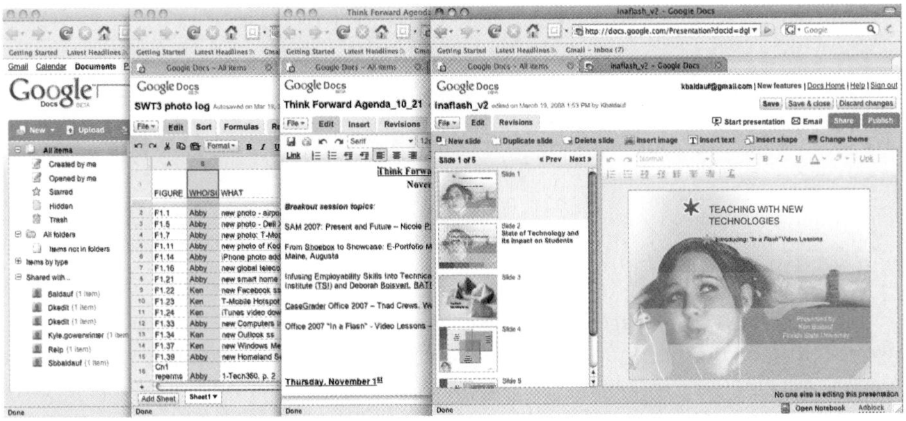

Figure 4.18

Google's Online Applications

Google applications are designed to share documents, presentations, spreadsheets, calendars, and notes with specific users or everyone on the Web.

An increasing number of software applications are moving online to support group document and information sharing. Google applications let users share notes, calendars, documents, spreadsheets, and presentations. At *tadalists.com*, users can share to-do lists with others in a group. Microsoft offers Office Live Workspace for sharing documents, spreadsheets, and other Office files with Office users online. If you have digital information you wish to share, it is likely that some online service has been set up to allow you to put it online and control who can access it.

Enterprise Application Software

Software that benefits an entire organization can also be developed or purchased. Some software vendors, such as SAP, specialize in developing software for enterprises. A fast-food chain, for example, might develop a materials ordering and distribution program to make sure that each of its franchises gets the necessary raw materials and supplies during the week. This program can be developed internally using staff and resources in the IS department or purchased from an external software company. Boeing and DaimlerChrysler use enterprise software to design new airplanes and automotive products. The software simulates the effectiveness and safety of designs, allowing the companies to save time and money compared to developing physical prototypes of airplanes and vehicles. Dunkin Brands, owner of Dunkin' Donuts, Baskin-Robbins, and Togo's, uses enterprise software to help it locate new stores. iSite from geoVue (*www.geovue.com*) is site-selection software that lets companies analyze factors to help determine the location of new stores.

One of the first enterprise applications was a payroll program for Lyons Bakeries in England, developed in 1954 on the Leo 1 computer. Table 4.8 lists some applications that can be addressed with enterprise software. Many organizations are moving to integrated enterprise software that supports supply chain management (movement of raw materials from suppliers through shipment of finished goods to customers), as shown in Figure 4.19.

Table 4.8

Examples of Enterprise Application Software

Type of Software	Description
Accounts receivable	Sales ordering
Accounts payable	Order entry
Airline industry operations	Payroll
Automatic teller systems	Human resource management
Cash-flow analysis	Check processing
Credit and charge card administration	Tax planning and preparation
Manufacturing control	Receiving
Distribution control	Restaurant management
General ledger	Retail operations
Stock and bond management	Invoicing
Savings and time deposits	Shipping
Inventory control	Fixed asset accounting

Enterprise Resource Planning (ERP) Software

A set of integrated programs that manage a company's vital business operations for an entire multisite, global organization.

Organizations can no longer respond to market changes using nonintegrated information systems based on overnight processing of yesterday's business transactions, conflicting data models, and obsolete technology. Wal-Mart and many other companies have sophisticated information systems to speed processing and coordinate communications between stores and their main offices. Many corporations are turning to **enterprise resource planning (ERP)** software, a set of integrated programs that manage a company's vital business operations for an entire multisite, global organization. Thus, an ERP system must be able to support many legal entities, languages, and currencies. Although the scope can vary from vendor to vendor, most ERP systems provide integrated software to support manufacturing and finance. In addition to these core business processes, some ERP systems might support business functions such as human resources, sales, and distribution. The primary benefits of implementing ERP include eliminating inefficient systems, easing adoption of improved work processes, improving access to data for operational decision making, standardizing technology vendors

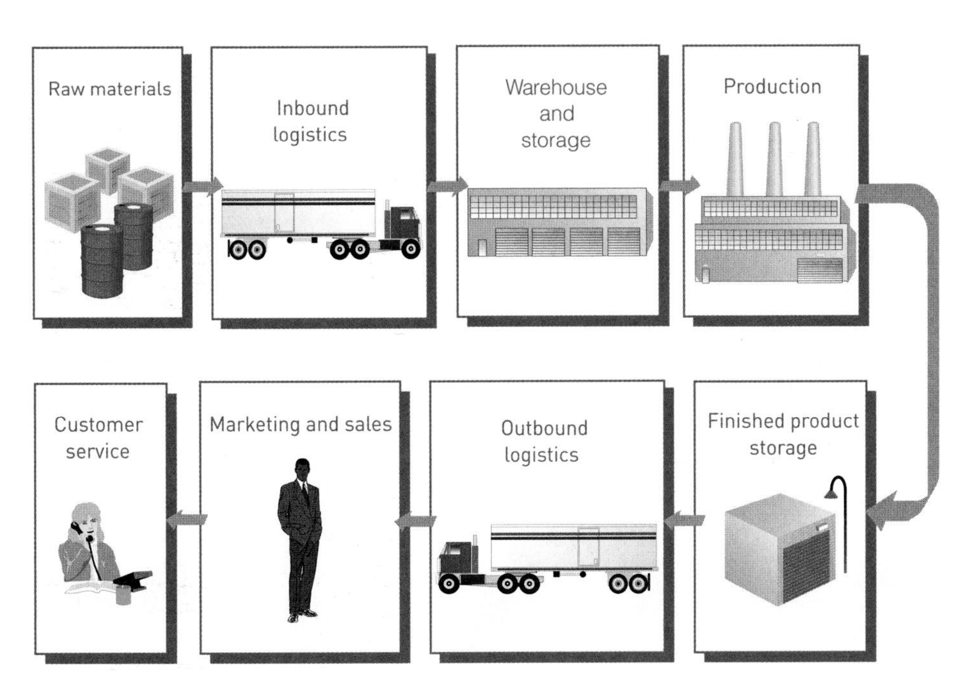

Figure 4.19

Use of Integrated Supply Chain Management Software

and equipment, and enabling supply chain management. Even small businesses can benefit from enterprise application software. Intuit's QuickBooks and Microsoft's Office Small Business Accounting are accounting and recording-keeping programs for small businesses and organizations.

Application Software for Information, Decision Support, and Specialized Purposes

Specialized application software for information, decision support, and other purposes is available in every industry. Genetic researchers, for example, are using software to visualize and analyze the human genome. Music executives use decision support software to help pick the next hit song. Sophisticated decision support software is also being used to increase the cure rate for cancer by analyzing about 100 scans of a cancerous tumor to create a 3-D view of the tumor. Software can then consider thousands of angles and doses of radiation to determine the best program of radiation therapy. The software analysis takes only minutes, but the results can save years or decades of life for the patient. As you will see in future chapters, information, decision support, and specialized systems are used in businesses of all sizes and types to increase profits or reduce costs. But how are all these systems actually developed or built? The answer is through the use of programming languages, discussed next.

PROGRAMMING LANGUAGES

Both OSs and application software are written in coding schemes called *programming languages*. The primary function of a programming language is to provide instructions to the computer system so that it can perform a processing activity. IS professionals work with **programming languages**, which are sets of keywords, symbols, and rules for constructing statements by which people can communicate instructions to be executed by a computer. Programming involves translating what a user wants to accomplish into a code that the computer can understand and execute. *Program code* is the set of instructions that signal the CPU to perform circuit-switching operations. In the simplest coding schemes, a line of code typically contains a single instruction such as, "Retrieve the data in memory address X." As discussed in Chapter 3, the instruction is then decoded during the instruction phase of the

programming languages
Sets of keywords, symbols, and a system of rules for constructing statements by which humans can communicate instructions to be executed by a computer.

machine cycle. Like writing a report or a paper in English, writing a computer program in a programming language requires the programmer to follow a set of rules. Each programming language uses symbols that have special meaning. Each language also has its own set of rules, called the **syntax** of the language. The language syntax dictates how the symbols should be combined into statements capable of conveying meaningful instructions to the CPU. A rule that "Variable names must start with a letter" is an example. A variable is a quantity that can take on different values. Program variable names such as SALES, PAYRATE, and TOTAL follow the rule because they start with a letter, whereas variables such as %INTEREST, $TOTAL, and #POUNDS do not.

syntax
A set of rules associated with a programming language.

The Evolution of Programming Languages

The desire to use the power of information processing efficiently in problem solving has pushed the development of newer programming languages. The evolution of programming languages is typically discussed in terms of generations of languages (see Table 4.9).

Table 4.9

The Evolution of Programming Languages

Generation	Language	Approximate Development Date	Sample Statement or Action
First	Machine language	1940s	00010101
Second	Assembly language	1950s	MVC
Third	High-level language	1960s	READ SALES
Fourth	Query and database languages	1970s	PRINT EMPLOYEE NUMBER IF GROSS PAY>1000
Beyond Fourth	Natural and intelligent languages	1980s	IF gross pay is greater than 40, THEN pay the employee overtime pay

Visual, Object-Oriented, and Artificial Intelligence Languages

Today, programmers often use visual and object-oriented languages. In the future, they will likely be using artificial intelligence languages to a greater extent. In general, these languages are easier for nonprogrammers to use compared with older generation languages.

Visual languages use a graphical or visual interface for program development. Unlike earlier languages that depended on writing detailed programming statements, visual languages allow programmers to "drag and drop" programming elements and icons onto the computer screen. Many of these languages are used to develop Web applications. *Visual Basic* was one of the first visual programming languages. Microsoft Visual Studio is a set of object-oriented programming languages and tools to develop Windows and Web-based applications. You can develop applications that can range from a simple Web-based program for displaying your résumé to complex business applications that process customer orders, control inventory, and send out bills—using languages such as Visual Basic .NET, Visual C++ .NET, Visual C#, and Visual J#. *C++* is a powerful and flexible programming language used mostly by computer systems professionals to develop applications. *Java* is an object-oriented programming language developed by Sun Microsystems that can run on any OS and on the Internet. Java can be used to develop complete applications or smaller applications, called *Java applets.* Many of these languages are also examples of object-oriented languages, which are discussed next.

The preceding programming languages separate data elements from the procedures or actions that will be performed on them, but another type of programming language ties them together into units called *objects.* An object consists of data and the actions that can be performed on the data. For example, an object could be data about an employee and all the operations (such as payroll calculations) that might be performed on the data. Programming languages that are based on objects are called *object-oriented programming languages.*

Building programs and applications using object-oriented programming languages is like constructing a building using prefabricated modules or parts. The object containing the data, instructions, and procedures is a programming building block. The same objects (modules or parts) can be used repeatedly. One of the primary advantages of an object is that it contains reusable code. In other words, the instruction code within that object can be reused in different programs for a variety of applications, just as the same basic prefabricated door can be used in two different houses. An object can relate to data on a product, an input routine, or an order-processing routine. An object can even direct a computer to execute other programs or to retrieve and manipulate data. So, a sorting routine developed for a payroll application could be used in both a billing program and an inventory control program. By reusing program code, programmers can write programs for specific application problems more quickly (see Figure 4.20). By combining existing program objects with new ones, programmers can easily and efficiently develop new object-oriented programs to accomplish organizational goals.

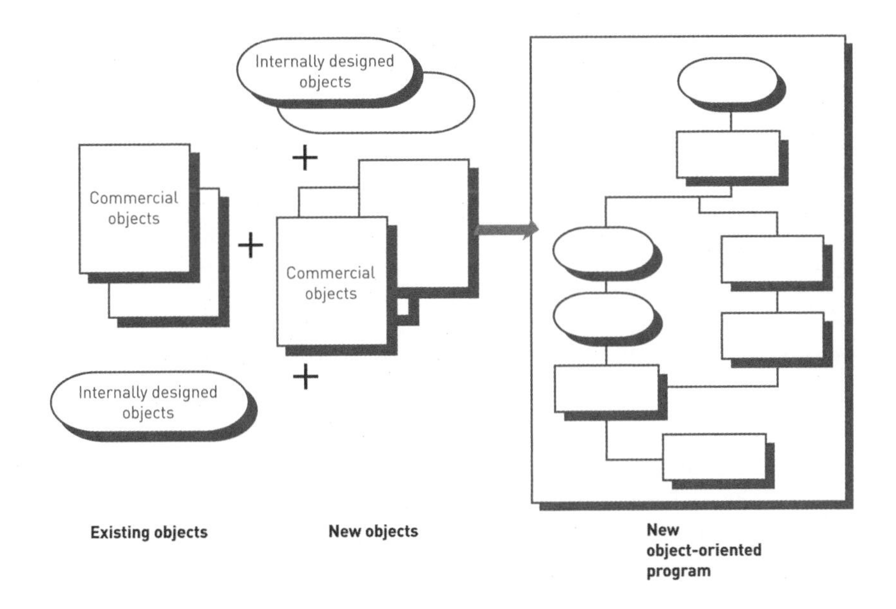

Figure 4.20

Reusable Code in Object-Oriented Programming

By combining existing program objects with new ones, programmers can easily and efficiently develop new object-oriented programs to accomplish organizational goals. Note that these objects can be either commercially available or designed internally.

Programmers often start writing object-oriented programs by developing one or more user interfaces, usually in a Windows or Web environment. You can create programs to run in a Windows or Web environment by using forms to design and develop the type of interface you want. You can select and drag text boxes to add descriptions, buttons that can be clicked and executed, a list box that contains several choices that can be selected, and other input/output features. After creating the Windows interface, you can write programming code to convert tasks a user selects in the interface into actions the computer performs.

Some of the most popular object-oriented programming languages include Smalltalk, Visual Basic .NET, C++, and Java (see Figure 4.21). Some old languages, such as COBOL, have been modified to support the object-oriented approach. As mentioned earlier, Java is an Internet programming language from Sun Microsystems that can run on a variety of computers and OSs, including UNIX, Windows, and Macintosh OSs.

Object-oriented programs often use *methods*, which are instructions to perform a specific task in the program. The following instructions in C++ use a method named ComputeArea to compute the area of a rectangle, given the width and length.

```
// Method to Compute the Area of a Rectangle Given the Width and
Length
double Rectangle::ComputeArea()
{
   return width * length;
}
// End of the ComputeArea Method
```

Figure 4.21

Microsoft Visual Basic

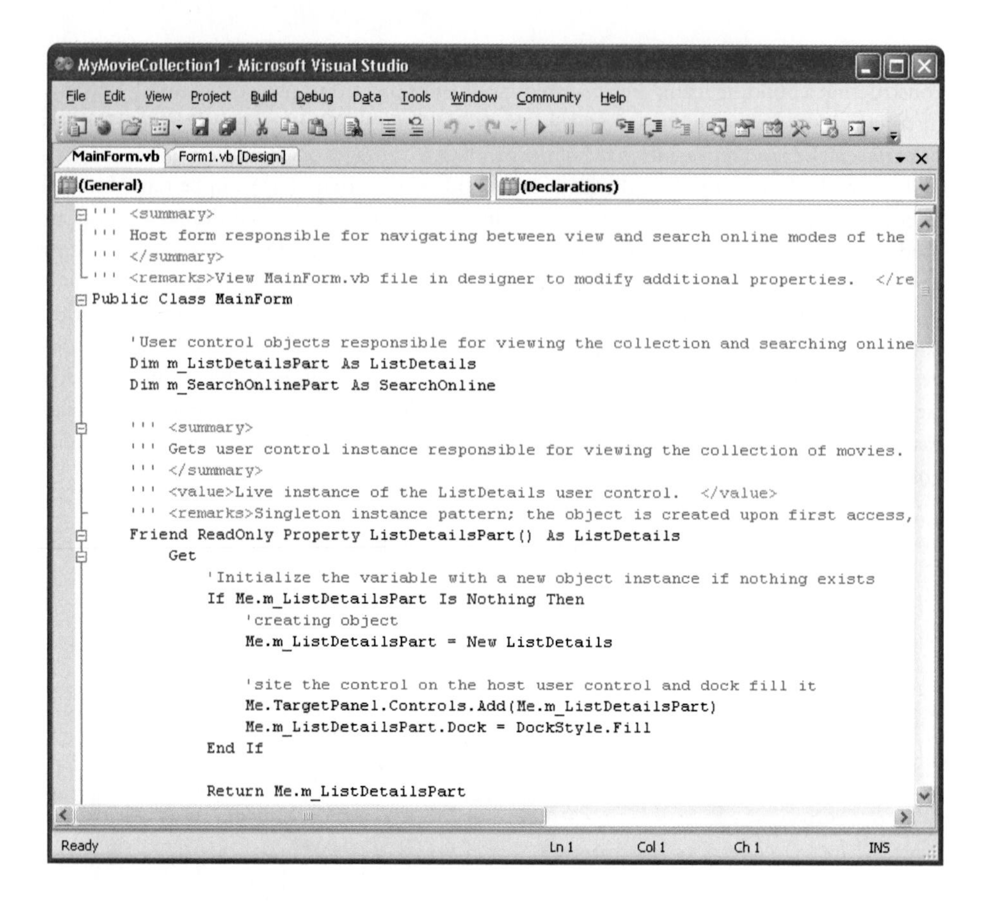

After they are developed as part of a C++ program, the instructions or method can be used in other programs to compute the area of a picture frame, a living room, a front lawn, or any other application that requires the area of a rectangle. Following are a few instructions in another C++ program that show how to use the ComputeArea method to compute the area of a picture frame.

```
//Assign Data and Compute Area
frameObject -> SetDimensions (frameWidth, frameLength);
frameArea = frameObject -> ComputeArea();
```

Programming languages used to create artificial intelligence or expert systems applications are often called *fifth-generation languages (5GLs).* FLEXPERT, for example, is an expert system used to perform plant layout and helps companies determine the best placement for equipment and manufacturing facilities. Fifth-generation languages are sometimes called *natural languages* because they use even more English-like syntax than 4GLs. They allow programmers to communicate with the computer by using normal sentences. For example, computers programmed in fifth-generation languages can understand queries such as, "How many athletic shoes did our company sell last month?"

compiler
A special software program that converts the programmer's source code into the machine-language instructions consisting of binary digits.

With third-generation and higher-level programming languages, each statement in the language translates into several instructions in machine language. A special software program called a **compiler** converts the programmer's source code into the machine-language instructions consisting of binary digits, as shown in Figure 4.22. A compiler creates a two-stage process for program execution. First, the compiler translates the program into a machine language; second, the CPU executes that program. Another approach is to use an *interpreter,* which is a language translator that carries out the operations called for by the source code. An interpreter does not produce a complete machine-language program. After the statement executes, the machine-language statement is discarded, the process continues for the next statement, and so on.

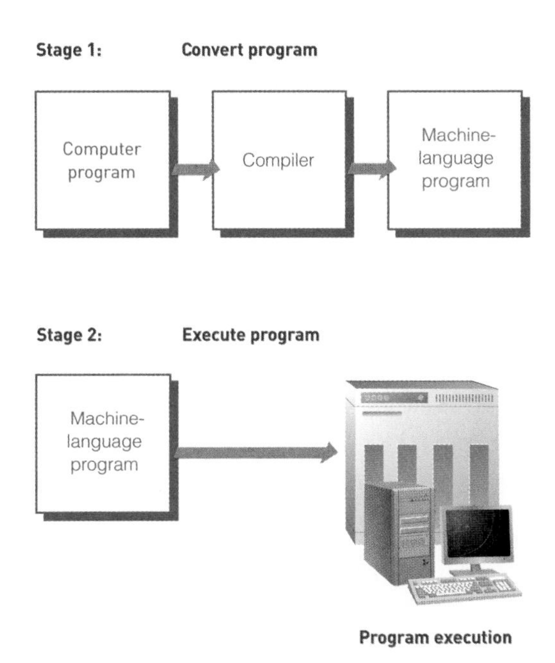

Figure 4.22

How a Compiler Works

A compiler translates a complete program into a complete set of binary instructions (Stage 1). After this is done, the CPU can execute the converted program in its entirety (Stage 2).

SOFTWARE ISSUES AND TRENDS

Because software is such an important part of today's computer systems, issues such as software bugs, licensing, upgrades, and global software support have received increased attention. We highlight several major software issues and trends in this section: software bugs, copyright, software licensing, open-source software, shareware and public domain software, multiorganizational software development, software upgrades, and global software support.

Software Bugs

A software bug is a defect in a computer program that keeps it from performing as it is designed to perform. Some software bugs are obvious and cause the program to terminate unexpectedly. Other bugs are subtler and allow errors to creep into your work. For example, a bug discovered in Microsoft Office Excel 2007 caused the equation 850 x 77.1 to display 100,000 rather than the correct result of 65,535.[22] Computer and software vendors say that as long as people design and program hardware and software, bugs are inevitable. In fact, according to the Pentagon and the Software Engineering Institute at Carnegie Mellon University, there are typically 5 to 15 bugs in every 1,000 lines of code. The following list summarizes tips for reducing the impact of software bugs.

- Register all software so that you receive bug alerts, fixes, and patches.
- Check the manual or read-me files for solutions to known problems.
- Access the support area of the manufacturer's Web site for patches.
- Install the latest software updates.
- Before reporting a bug, make sure that you can re-create the circumstances under which it occurs.
- After you can re-create the bug, call the manufacturer's tech support line.
- Avoid buying the latest release of software for several months or a year until the software bugs have been discovered and removed.

Copyrights and Licenses

Most software products are protected by law using copyright or licensing provisions. Those provisions can vary, however. In some cases, you are given unlimited use of software on one or two computers. This is typical with many applications developed for personal computers.

In other cases, you pay for your usage—if you use the software more, you pay more. This approach is becoming popular with software placed on networks or larger computers. Most of these protections prevent you from copying software and giving it to others without restrictions. Some software now requires that you *register* or *activate* it before it can be fully used. Registration and activation sometimes put software on your hard disk that monitors activities and changes your computer system.

When people purchase software, they don't actually own the software, but rather are licensed to use the software on a computer. This is called a single-user license. A **single-user license** permits you to install the software on one computer, or sometimes two computers, used by one person. A single-user license does not allow you to copy and share the software with others. Table 4.10 describes different types of software licenses. Licenses that accommodate multiple users are usually provided at a discounted price.

single-user license
A software license that permits only one person to use the software, typically on only one computer.

Table 4.10
Software Licenses

License	Description
Single-user license	Permits you to install the software on one computer, or sometimes two computers, used by one person.
Multiuser license	Specifies the number of users allowed to use the software, and can be installed on each user's computer. For example, a 20-user license can be installed on 20 computers for 20 users.
Concurrent-user license	Designed for network-distributed software, this license allows any number of users to use the software, but only a specific number of users to use it at the same time.
Site license	Permits the software to be used anywhere on a particular site, such as a college campus, by everyone on the site.

Open-Source Software

open-source software
Software that is freely available to anyone in a form that can be easily modified.

Open-source software is freely available to anyone in a form that can be easily modified. The Open Source Initiative (OSI) is a nonprofit corporation dedicated to the development and promotion of open-source software (see the OSI Web site at *www.opensource.org* for more information on the group's efforts). Users can download the source code and build the software themselves, or the software developers can make executable versions available along with the source. Open-source software development is a collaborative process—developers around the world use the Internet to keep in close contact via e-mail and to download and submit new software. Major software changes can occur in days rather than weeks or months. Many open-source software packages are widely used, including the Linux OS; Free BSD, another OS; Apache, a popular Web server; Sendmail, a program that delivers e-mail for most systems on the Internet; and Perl, a programming language used to develop Internet application software. See Table 4.11 for some examples of open-source software.

Why would an organization run its business using software that's free? Can something that's given away over the Internet be stable or reliable or sufficiently supported to place at the core of a company's day-to-day operations? The answer is surprising—many believe that open-source software is often *more* reliable and secure than commercial software. How can this be? First, by making a program's source code readily available, users can fix any problems they discover. A fix is often available within hours of the problem's discovery. Second, with the source code for a program accessible to thousands of people, the chances of a bug being discovered and fixed before it does any damage are much greater than with traditional software packages. Of course, open-source software is usually much less expensive than traditional software that is purchased from a software vendor. The auditor of one state estimated that the cost savings using open-source software could be as high as $10 million compared to developing software internally, when legal and project delays are included. Some companies are also starting to reveal their source code, including IBM, Microsoft, and others.

Software Type	Example
Operating system	Linux
Application software	Open Office
Database software	MySQL
Internet browser	Firefox
Photo editing	Gimp
Project management	OpenProj
Personal accounting	Grisbi
E-mail	Thunderbird

Table 4.11

Examples of Open-Source Software

However, using open-source software does have some disadvantages. Although open-source systems can be obtained for next to nothing, the up-front costs are only a small piece of the total cost of ownership that accrues over the years that the system is in place. Some claim that open-source systems contain many hidden costs, particularly for user support or solving problems with the software. Licensed software comes with guarantees and support services that open-source software does not. Still, many businesses appreciate the additional freedom that open-source software provides. The question of software support is the biggest stumbling block to the acceptance of open-source software at the corporate level. Getting support for traditional software packages is easy—you call a company's toll-free support number or access its Web site. But how do you get help if an open-source package doesn't work as expected? Because the open-source community lives on the Internet, you look there for help. Through use of Internet discussion areas, you can communicate with others who use the same software, and you might even reach someone who helped develop it. Users of popular open-source packages can get correct answers to their technical questions within a few hours of asking for help on the appropriate Internet forum. Another approach is to contact one of the many companies emerging to support and service such software—for example, Red Hat for Linux, C2Net for Apache, and Sendmail, Inc., for Sendmail. These companies offer high-quality, for-pay technical assistance.

Shareware, Freeware, and Public Domain Software

Many software users are doing what they can to minimize software costs. Some are turning to **shareware** and **freeware**—software that is very inexpensive or free, usually for use in personal computers, but whose source code cannot be modified. Freeware can be used to perform a variety of tasks. StarOffice is a freeware office suite that contains word processor, spreadsheet, database, drawing, and presentation programs. PhotoPlus 6 is a photo-editing program, and Picasa is a photo-editing and management program. The Web site *www.SourceForge.net* is a resource for programmers to freely exchange programs and program code. It allows programmers to create, collaborate on, and evaluate program code. Over 80,000 programs are at various stages of completion.

shareware and freeware
Software that is very inexpensive or free, but whose source code cannot be modified.

Shareware might not be as powerful as commercial software, but it provides what some people need at a good price. In some cases, you can try the software before sending a nominal fee to the software developer. Some shareware and freeware is in the public domain, often called *public domain software.* This software is not protected by copyright laws and can be freely copied and used. Although shareware and freeware can be free or inexpensive to acquire, it can be more expensive to use and maintain over time compared with software that is purchased. If the software is hard to use and doesn't perform all the required functions, the cost of wasted time and lost productivity can be far greater than the cost of purchasing better software. Shareware, freeware, and public domain software is often not open source—that is, the source code is not available and cannot be modified.

Software Upgrades

Software companies revise their programs and sell new versions periodically. In some cases, the revised software offers new and valuable enhancements. In other cases, the software uses complex program code that offers little in terms of additional capabilities. In addition, revised software can contain bugs or errors. When software companies stop supporting older software versions or releases, some customers feel forced to upgrade to the newer software. Deciding whether to purchase the newest software can be a problem for corporations and people with a large investment in software. Should the newest version be purchased when it is released? Some users do not always get the most current software upgrades or versions, unless it includes significant improvements or capabilities. Instead, they might upgrade to newer software only when it offers vital new features. Software upgrades usually cost much less than the original purchase price.

Global Software Support

Large global companies have little trouble persuading vendors to sell them software licenses for even the most far-flung outposts of their company. But can those same vendors provide adequate support for their software customers in all locations? Supporting local operations is one of the biggest challenges IS teams face when putting together standardized, company-wide systems. Slower technology growth markets, such as Eastern Europe and Latin America, might not have any official vendor presence. Instead, large vendors such as Sybase, IBM, and Hewlett-Packard typically contract with local providers to provide support for their software.

One approach that has been gaining acceptance in North America is to outsource global support to one or more third-party distributors. The user company can still negotiate its license with the software vendor directly, but it then hands the global support contract to a third-party supplier. The supplier acts as a middleman between software vendor and user, often providing distribution, support, and invoicing. American Home Products Corporation handles global support for both Novell NetWare and Microsoft Office applications this way throughout the 145 countries in which it operates. American Home Products, a pharmaceutical and agricultural products company, negotiated the agreements directly with the vendors for both purchasing and maintenance, but fulfillment of the agreement is handled exclusively by Philadelphia-based Softsmart, an international supplier of software and services.

In today's computer systems, software is an increasingly critical component. Whatever approach people and organizations take to acquire software, everyone must be aware of the current trends in the industry. Informed users are wiser consumers, and they can make better decisions.

SUMMARY

Principle

Systems and application software are critical in helping individuals and organizations achieve their goals.

Software consists of programs that control the workings of the computer hardware. The two main categories of software are systems software and application software. Systems software is a collection of programs that interacts between hardware and application software, and includes operating systems, utility programs, and middleware. Application software can be proprietary or off the shelf, and enables people to solve problems and perform specific tasks.

An operating system (OS) is a set of computer programs that controls the computer hardware to support users' computing needs. An OS converts an instruction from an application into a set of instructions needed by the hardware. This intermediary role allows hardware independence. An OS also manages memory, which involves controlling storage access and use by converting logical requests into physical locations and by placing data in the best storage space, including virtual memory.

An OS manages tasks to allocate computer resources through multitasking and time-sharing. With multitasking, users can run more than one application at a time. Time-sharing allows more than one person to use a computer system at the same time.

The ability of a computer to handle an increasing number of concurrent users smoothly is called *scalability*, a feature critical for systems expected to handle a large number of users.

An OS also provides a user interface, which allows users to access and command the computer. A command-based user interface requires text commands to send instructions; a graphical user interface (GUI), such as Windows, uses icons and menus.

Software applications use the OS by requesting services through a defined application program interface (API). Programmers can use APIs to create application software without having to understand the inner workings of the OS. APIs also provide a degree of hardware independence so that the underlying hardware can change without necessarily requiring a rewrite of the software applications.

Over the years, several popular OSs have been developed. These include several proprietary OSs used primarily on mainframes. MS-DOS is an early OS for IBM-compatibles. Older Windows OSs are GUIs used with DOS. Newer versions, such as Windows Vista and XP, are fully functional OSs that do not need DOS. Apple computers use proprietary OSs such as the Mac OS and Mac OS X. UNIX is a powerful OS that can be used on many computer system types and platforms, from personal computers to mainframe systems. UNIX makes it easy to move programs and data among computers or to connect mainframes and personal computers to share resources. Linux is the kernel of an OS whose source code is freely available to everyone. Several variations of Linux are available, with sets of capabilities and applications to form a complete OS, for example, Red Hat Linux. z/OS and HP-UX are OSs for mainframe computers. Some OSs, such as Palm OS, Windows Mobile, Windows Embedded, Pocket PC, and variations of Linux, have been developed to support mobile communications and consumer appliances.

Utility programs can perform many useful tasks and often come installed on computers along with the OS. This software is used to merge and sort sets of data, keep track of computer jobs being run, compress files of data, protect against harmful computer viruses, and monitor hardware and network performance. Middleware is software that allows different systems to communicate and transfer data back and forth. A service-oriented architecture (SOA) uses modular application services to allow users to interact with systems, and systems to interact with each other.

Principle

Organizations should not develop proprietary application software unless doing so will meet a compelling business need that can provide a competitive advantage.

Application software applies the power of the computer to solve problems and perform specific tasks. One useful way of classifying the many potential uses of information systems is to identify the scope of problems and opportunities addressed by a particular organization or its sphere of influence. For most companies, the spheres of influence are personal, workgroup, and enterprise.

User software, or personal productivity software, includes general-purpose programs that enable users to improve their personal effectiveness, increasing the quality and amount of work that can be done. Software that helps groups work together is often called workgroup application software, and includes group scheduling software, electronic mail, and other software that enables people to share ideas. Enterprise software that benefits the entire organization can also be developed or purchased. Many organizations are turning to enterprise resource planning software, a set of integrated programs that manage a company's vital business operations for an entire multisite, global organization.

Three approaches to developing application software are to build proprietary application software, buy existing programs off the shelf, or use a combination of customized and off-the-shelf application software. Building proprietary software (in-house or on contract) has the following advantages: The organization will get software that more closely matches its needs; by being involved with the development, the organization has further control over the results; and the organization has more flexibility in making changes. The disadvantages include the following: It is likely to take longer and cost more to develop, the in-house staff will be hard pressed to provide ongoing support and maintenance, and there is a greater risk that the software features will not work as expected or that other performance problems will occur.

Purchasing off-the-shelf software has many advantages. The initial cost is lower, there is a lower risk that the software will fail to work as expected, and the software is likely to be of higher quality than proprietary software. Some disadvantages are that the organization might pay for features it does not need, the software might lack important features requiring expensive customization, and the system might require process reengineering.

Some organizations have taken a third approach—customizing software packages. This approach usually involves a mixture of the preceding advantages and disadvantages and must be carefully managed.

An application service provider (ASP) is a company that can provide the software, support, and computer hardware on which to run the software from the user's facilities over a network. ASPs customize off-the-shelf software on contract and speed deployment of new applications while helping IS managers avoid implementation headaches. Use of ASPs reduces the need for many skilled IS staff members and also lowers a project's start-up expenses. Software as a service (SaaS) allows businesses to subscribe to Web-delivered business application software by paying a monthly service charge or a per-use fee.

Although hundreds of computer applications can help people at school, home, and work, the primary applications are word processing, spreadsheet analysis, database, graphics, and online services. A software suite, such as SmartSuite, WordPerfect, StarOffice, or Office, offers a collection of powerful programs.

Principle

Organizations should choose a programming language whose functional characteristics are appropriate for the task at hand, considering the skills and experience of the programming staff.

All software programs are written in coding schemes called *programming languages*, which provide instructions to a computer to perform some processing activity. The several classes of programming languages include machine, assembly, high-level, query and database, object-oriented, and visual programming languages.

Programming languages have changed since their initial development in the early 1950s. In the first generation, computers were programmed in machine language, and the second generation of languages used assembly languages. The third generation consists of many high-level programming languages that use English-like statements and commands. They also must be converted to machine language by special software called a compiler, and include BASIC, COBOL, FORTRAN, and others. Fourth-generation languages include database and query languages such as SQL.

Fifth-generation programming languages combine rules-based code generation, component management, visual programming techniques, reuse management, and other advances. Visual and object-oriented programming languages—such as Smalltalk, C++, and Java—use groups of related data, instructions, and procedures called *objects*, which serve as reusable modules in various programs. These languages can reduce program development and testing time. Java can be used to develop applications on the Internet.

Principle

The software industry continues to undergo constant change; users need to be aware of recent trends and issues to be effective in their business and personal life.

Software bugs, software licensing and copyrighting, open-source software, shareware and freeware, multiorganizational software development, software upgrades, and global software support are all important software issues and trends.

A software bug is a defect in a computer program that keeps it from performing in the manner intended. Software bugs are common, even in key pieces of business software.

Open-source software is software that is freely available to anyone in a form that can be easily modified. Open-source software development and maintenance is a collaborative process, with developers around the world using the Internet to keep in close contact via e-mail and to download and submit new software. Shareware and freeware can reduce the cost of software, but sometimes they might not be as powerful as commercial software. Also, their source code usually cannot be modified.

Multiorganizational software development is the process of extending software development beyond a single organization by finding others who share the same business problem and involving them in a common development effort.

Software upgrades are an important source of increased revenue for software manufacturers and can provide useful new functionality and improved quality for software users.

Global software support is an important consideration for large, global companies putting together standardized, company-wide systems. A common solution is outsourcing global support to one or more third-party software distributors.

CHAPTER 4: SELF-ASSESSMENT TEST

Systems and application software are critical in helping individuals and organizations achieve their goals.

1. Which of the following is an example of a command-driven operating system?
   a. Windows XP
   b. Leopard
   c. MS-DOS
   d. Windows Vista
2. Application software such as Microsoft Office Excel manipulates the computer hardware directly. True or False?
3. Today's operating systems support ______________, the ability to run multiple processes seemingly simultaneously.
4. The file manager component of the OS controls how memory is accessed and maximizes available memory and storage. True or False?

Organizations should not develop proprietary application software unless doing so will meet a compelling business need that can provide a competitive advantage.

5. The primary function of system software is to apply the power of the computer to give people, workgroups, and the entire enterprise the ability to solve problems and perform specific tasks. True or False?
6. Software that enables users to improve their personal effectiveness, increasing the amount of work they can do and its quality, is called ______________.
   a. personal productivity software
   b. operating system software
   c. utility software
   d. graphics software
7. Optimization can be found in which type of application software?
   a. spreadsheets
   b. word processing programs
   c. database programs
   d. presentation graphics programs
8. Software used to solve a unique or specific problem that is usually built in-house but can also be purchased from an outside company is called ______________.
9. A program to detect and eliminate viruses is an example of what type of software?
   a. personal productivity software
   b. operating system software
   c. utility software
   d. applications software

Organizations should choose a programming language whose functional characteristics are appropriate for the task at hand, considering the skills and experience of the programming staff.

10. Most software purchased to run on a personal computer uses a __________ license.
    a. site
    b. concurrent-user
    c. multiuser
    d. single-user
11. A class of application software that helps groups work together and collaborate is called __________.
12. Each programming language has its own set of rules, called the __________ of the language.
13. A special software program called an *interpreter* performs the conversion from the programmer's source code into the machine-language instructions consisting of binary digits, and results in a machine-language program. True or False?

CHAPTER 4: SELF-ASSESSMENT TEST ANSWERS

(1) c (2) False (3) multitasking (4) False (5) False (6) a (7) a (8) proprietary software (9) c (10) d (11) workgroup application software (12) syntax (13) False

REVIEW QUESTIONS

1. What is the difference between systems and application software? Give four examples of personal productivity software.
2. What steps can a user take to correct software bugs?
3. Identify and briefly discuss two types of user interfaces provided by an operating system.
4. What is a software suite? Give several examples.
5. Name four operating systems that support the personal, workgroup, and enterprise spheres of influence.
6. What is a service-oriented architecture (SOA)?
7. What is multitasking?
8. Define the term *utility* software and give two examples.

9. Identify the two primary sources for acquiring application software.
10. What is an application service provider? What issues arise in considering the use of an ASP?
11. What is open-source software? What is the biggest stumbling block with the use of open-source software?
12. What does the acronym API stand for? What is the role of an API?
13. Briefly discuss the advantages and disadvantages of frequent software upgrades.
14. Describe the term *enterprise resource planning (ERP) system*. What functions does such a system perform?
15. What is freeware? Give two examples.

DISCUSSION QUESTIONS

1. Assume that you must take a computer-programming course next semester. What language do you think would be best for you to study? Why? Do you think that a professional programmer needs to know more than one programming language? Why or why not?
2. You are going to buy a personal computer. What operating system features are important to you? What operating system would you select?
3. Identify the fundamental types of application software. Discuss the advantages and disadvantages of each type.
4. You are using a new release of an application software package. You think that you have discovered a bug. Outline the approach that you would take to confirm that it is indeed a bug. What actions would you take if it truly was a bug?
5. How can application software improve the effectiveness of a large enterprise? What are some of the benefits associated with implementation of an enterprise resource planning system? What are some of the issues that could keep the use of enterprise resource planning software from being successful?
6. Define the term *software as a service (SaaS)*. What are some of the advantages and disadvantages of employing a SaaS? What precautions might you take to minimize the risk of using one?
7. Describe three personal productivity software packages you are likely to use the most. What personal productivity software packages would you select for your use?
8. Contrast and compare three popular OSs for personal computers.
9. If you were the IT manager for a large manufacturing company, what issues might you have with the use of open-source software? What advantages might there be for use of such software?
10. Identify four types of software licenses frequently used. Which approach does the best job of ensuring a steady, predictable stream of revenue from customers? Which approach is most fair for the small company that makes infrequent use of the software?

PROBLEM-SOLVING EXERCISES

1. Choose an application software package that might be useful for a career that interests you and develop a six-slide presentation of its history, current level of usage, typical applications, ease of use, and so on.
2. Use a spreadsheet package to prepare a simple monthly budget and forecast your cash flow—both income and expenses for the next six months (make up numbers rather than using actual ones). Now use a graphics package to plot the total monthly income and monthly expenses for six months. Cut and paste both the spreadsheet and the graph into a word processing document that summarizes your financial condition.
3. Use a database program to enter five software products you are likely to use at work. List the name, vendor or manufacturer, cost, and features in the columns of a database table. Use a word processor to write a report on the software. Copy the database table into the word processing program.

TEAM ACTIVITIES

1. Form a group of three or four classmates. Find articles from business periodicals, search the Internet, or interview people on the topic of software bugs. How frequently do they occur, and how serious are they? What can software users do to encourage defect-free software? Compile your results for an in-class presentation or a written report.
2. Form a group of three or four classmates. Identify and contact an employee of a local firm. Interview the individual and describe the application software the company uses and the importance of the software to the organization. Write a brief report summarizing your findings.
3. Divide your team into two groups. The first group should prepare a report using a word processing program. Make sure to include a large number of spelling, grammatical, and similar errors in the document. The second group should use the word processing program's features to locate and eliminate the errors. The entire team should write a report on the advantages and limitations of the spelling and grammar checking features of the word processing program you used. What additional features would you like to see in future word processing programs?

WEB EXERCISES

1. Use the Web to research four productivity software suites from various vendors (see *http://en.wikipedia.org/wiki/Office_Suite*). Create a table in a word processing document to show what applications are provided by the competing suites. Write a few paragraphs on which suite you think best matches your needs and why.
2. Use the Internet to search for three popular freeware utilities that you would find useful. Write a report that describes the features of these three utility programs.
3. Do research on the Web and develop a two-page report summarizing the latest consumer appliance OSs. Which one seems to be gaining the most widespread usage? Why do you think this is the case?
4. Do research on the Web about application software that is used in an industry and is of interest to you. Write a brief report describing how the application software can be used to increase profits or reduce costs.

CAREER EXERCISES

1. What personal computer OS would help you the most in the first job you would like to have after you graduate? Why? What features are the most important to you?
2. Think of your ideal job. Describe five application software packages that could help you advance in your career. If the software package doesn't exist, describe the kinds of software packages that could help you in your career.

CASE STUDIES

Case One

Systems Management Software Helps Fight Crime

The York regional police protect 1,800 square kilometers north of Toronto, Canada. Until recently, the force has been challenged trying to keep the rugged Panasonic laptops in its 200 cruisers, boats, and helicopters up-to-date, secure, and synchronized. Although the notebooks were wirelessly networked, the data, software, and systems were not necessarily well synchronized.

To apply system updates and patches every few months, officers were required to check in at the main station where they had to wait for a few hours while the notebook was updated. With over 200 laptops on the force, this cost the department hundreds of working hours every few months.

Not only were human resources wasted, but the internal IT department was pushed past its limit. Staff members sometimes unknowingly worked to solve the same problems. They spent too much time coordinating applications, running backups, and trying to keep up with new law enforcement applications.

Recently the York police installed system management software from Microsoft called System Center Configuration Manager 2007. The software allows system administrators at the main station to access notebooks remotely over the network for system upgrades, patch management, software distribution, and hardware and software inventory. No longer do officers need to spend hours waiting on their PC updates. PCs are updated as needed over the wireless network. New software and system changes are pushed out to all notebooks simultaneously so all officers have the same information and services at all times.

The new system software allows the department to come close to its paperless ideal. An e-ticketing system allows officers to swipe a driver's license, run a background check, and issue a ticket in minutes. Officers receive daily briefings online and submit reports directly from their notebooks, which allows them to stay on the road rather than at a desk. The new system has freed up the IT staff to concentrate on delivering new and useful services rather than just maintaining the old services.

The York regional police are looking forward to the next edition of System Center Configuration Manager, which promises to support streaming media. They would like to use it to stream video from the helicopter to cruisers on the road.

Discussion Questions

1. What unique challenges did the York regional police IT staff have to overcome?
2. How did Microsoft System Center Configuration Manager resolve the issues for the York regional police?

Critical Thinking Questions

1. What other types of industries would benefit from products like Microsoft System Center Configuration Manager? Why?
2. What general lesson regarding information system administration can you take from this case?

SOURCES: Smith, Briony, "Cops roll out remote patch updates, e-ticketing," *ITWorld Canada*, November 29, 2007, *www.itworldcanada.com/a/Enterprise-Business-Applications/50132083-0699-401d-b7ce-d6f1c16193b5.html*. Microsoft System Center Configuration Manager Web site, *www.microsoft.com/smserver*, accessed February 3, 2008. York Regional Police Web site, *www.yrp.on.ca*, accessed February 3, 2008.

Case Two

Energy Giant Valero Turns to SOA Software

Valero Energy is North America's largest refiner. When they recently acquired some competing refineries, Valero tripled its annual revenue to $90 billion. While Valero's rapid growth has been good for its shareholders, it has been a nightmare for the company's information systems management professionals.

By acquiring several companies that were themselves products of multiple acquisitions, Valero found itself with dozens of incompatible software systems that somehow had to find a way to communicate and share data. Although one traditional solution would have been to design or purchase middleware to bridge the gap, Valero chose a cutting edge software development technique called a service-oriented architecture (SOA).

Valero software engineers began designing software services to provide users with an interface to its disparate systems. They developed the services based on industry standards and designed them to be flexibly reused and recombined. By using the SOA approach, Valero planned to pull together the various information systems, conduct business more efficiently, and reduce operating costs. Over time, the company has introduced over 100 services built on SAP's NetWeaver Application Server Development Environment. Many are composite services built by combining several smaller services. Roughly 22,000 employees and 5,000 customers now use Valero's SOA services.

When approaching a new service request, rather than programming from scratch, Valero's software engineers consider the services that have already been developed to find one that can be reused or refashioned. Organizing and cataloging services for reuse helps Valero save time, effort, and money. Developing services with the NetWeaver platform lets Valero engineers develop 300 services quickly and easily. Nayaki Nayyar, Valero's director of enterprise architecture and technology services, stepped in to reduce the number of services to 50, isolating the best core services before moving forward with new development. "Unless you can catalog, find, and use these services," she said in a *CIO Insight* interview, "you just end up with a virtual junk drawer of services."

The result of their SOA approach has saved Valero millions of dollars. One system designed to provide visibility into tanker transportation schedules saved the company a half-million dollars in penalties for ships that sit idle at the dock. Other savings are incurred from management being able to view corporate data from across the enterprise in real time.

Valero is working on tools that will allow managers to design their own SOA services. If managers can access the information they need without the usual system request process, the business becomes more streamlined, and the information system staff is freed to focus on bigger projects.

Discussion Questions

1. Why did Valero choose an SOA to integrate its systems rather than creating middleware?
2. What benefits does the SOA provide to businesses such as Valero?

Critical Thinking Questions

1. Why is it important to maintain a reasonably sized catalog of services rather than a large amount of services?
2. Is it wise for Valero to empower their managers to create their own services? How might this system benefit the company? Are there any dangers?

SOURCES: Zalno, Jennifer, "Valero Pumped on SOA," *CIO Insight*, July 11, 2007, *www.cioinsight.com/c/a/Case-Studies/Valero-Pumped-On-SOA*. SAP NetWeaver Web site, *www.sap.com/platform/netweaver*, accessed February 3, 2008. Valero Web site, *www.valero.com*, accessed February 3, 2008.

Questions for Web Case

See the Web site for this book to read about the Whitmann Price Consulting case for this chapter. The following questions concern this Web case.

Whitmann Price Consulting: Software Considerations

Discussion Questions

1. What three types of software made the BlackBerry ideal for meeting the needs of Whitmann Price?
2. How did the choice of hardware affect options for software solutions? If Sandra and Josh picked a newly developed handheld device unknown in industry, how would that change the solutions?

Critical Thinking Questions

1. For software other than the BlackBerry software, should Sandra and Josh look to BlackBerry Alliance Program members or should they plan to have their own software engineers develop the software? What are the benefits and drawbacks of either option?
2. What process would you use to evaluate software for the Advanced Mobile Communications and Information System?

NOTES

Sources for the opening vignette: Ruffolo, Rafael, "GM presents vision of software-based cars," *ITWorld Canada*, October 26, 2007, *www.itworldcanada.com/a/Enterprise-Business-Applications/6b437baf-31d2-4844-89cd-068c0b5b8221.html*. GM Technology Web site, *www.gm.com/explore/technology*, accessed February 3, 2008. IBM CASCON Web site, *https://www-927.ibm.com/ibm/cas/archives/2007/speakers*, accessed February 3, 2008.

1 Elmer-DeWitt, Philip, "Survey: Mac OS hit record 7.3% share in December; iPhone up 33%," *Fortune*, January 1, 2008, *http://apple20.blogs.fortune.cnn.com/2008/01/01/survey-mac-os-hit-record-73-share-in-december-iphone-up-33*.

2 Microsoft Case Study Web site, "National Aquarium in Baltimore," *www.microsoft.com/casestudies*, accessed June 14, 2007.

3 Eaton, Nancy, "Renee Mancino: What Happens in Vegas Stays on Her Mac," Apple Small Business Profile, *www.apple.com/business/profiles/mancino*, accessed June 14, 2007.

4 Gruener, Wolfgang, "Linux Returns to Thinkpads," *tgdaily*, January 15, 2008, *www.tgdaily.com/content/view/35615/145*.

5 McDougall, Paul, "$199 Linux PC Sells Out At Wal-Mart," *InformationWeek*, November 13, 2007, *www.informationweek.com/software/showArticle.jhtml?articleID=202805941*.

6 Ubuntu success stories, "Ubuntu proved the optimal choice for a non-profit radio station," *www.ubuntu.com/products/casestudies/KRUU*, accessed June 14, 2007.

7 Sun Customer Snapshot Web site, "eBay Inc," *www.sun.com/customers/servers/ebay.xml*, accessed June 7, 2007.

8 Sun Customer Snapshot Web site, "Idaho National Lab," *www.sun.com/customers/servers/inl.xml*, accessed June 7, 2007.

9 BlackBerry Customer Success, "IT Group Deploys Remote Systems Management on BlackBerry for Business-Critical Systems," *http://na.BlackBerry.com/eng/solutions/types/government/#tab_tab_customersuccess*, accessed June 14, 2007.

10 Thibodeau, Patrick, "Wachovia uses grid technology to speed up transaction apps," *Computerworld*, May 15, 2006, *www.computerworld.com/action/article.do?command=viewArticleBasic&articleId=9000476*.

11 Staff, "Kindred Healthcare leverages VI3 and VDI to provide physicians with secure bedside access to patient records using wirelessly connected thin clients," *Computerworld VMWare Success Stories* (video), *www.vmware.com/customers/stories/success_video.html?id=khealthcarevdi*, accessed February 2, 2008.

12 Brodkin, Jon, "Electricity provider cites SOA for efficiency gains," *Computerworld*, March 7, 2007, *www.computerworld.com/action/article.do?command=viewArticleBasic&taxonomyName=Enterprise_Applications &articleId=9012428&taxonomyId=87&intsrc=kc_li_story*.

13 Hildreth, Sue, "Electricity provider cites SOA for efficiency gains," *Computerworld*, February 5, 2007, *www.computerworld.com/action/article.do?command=viewArticleBasic&taxonomyName=Enterprise_Applications &articleId=279408&taxonomyId=87&intsrc=kc_li_story*.

14 Adams, Steve, "Boston-area firm offers video intelligence to merchants," *The Patriot Ledger*, November 17, 2007, *http://ledger.southofboston.com/articles/2007/11/17/business/biz02.txt*.

15 Pratt, Mary, "Law firm develops in-house system to deal with discovery," *Computerworld*, March 17, 2008, *www.computerworld.com/action/article.do?command=viewArticleBasic&articleId=312605&source=rss_news10*.

16 Hildreth, Suzanne, "GPS and GIS: On the Corporate Radar," *Computerworld*, April 2, 2007, *www.computerworld.com/action/article.do?command=viewArticleBasic&taxonomyName=Enterprise_Applications &articleId=284307&taxonomyId=87&intsrc=kc_li_story*.

17 Martens, China, "Insurer urges flexible SOA approach," *Computerworld*, January 12, 2007, *www.computerworld.com/action/article.do?command=viewArticleBasic&taxonomyName=Servers_and_Data_Cent er&articleId=9007942&taxonomyId=154&intsrc=kc_li_story*.
18 SAP Case Study, "Avanex Inc.," *www.sap.com/industries/hightech/midsize/customers/stories/avanex.epx*, accessed February 15, 2008.
19 Staff, "Improv,"salesforce.com customer snapshot, *www.salesforce.com/customers/communications-media/improv.jsp*, accessed February 2, 2008.
20 Microsoft Case Study, "Chef JoAnna," *www.microsoft.com/casestudies*, accessed June 15, 2007.
21 Microsoft Case Study, "Greenfield Online," *www.microsoft.com/casestudies*, accessed June 14, 2007.
22 Shankland, Stephen, "Microsoft: Excel 2007 bug is skin deep," CNET, September 26, 2007, *www.news.com/8301-13580_3-9785728-39.html*.

CHAPTER • 5 • Database Systems and Business Intelligence

PRINCIPLES	LEARNING OBJECTIVES
▪ Data management and modeling are key aspects of organizing data and information.	▪ Define general data management concepts and terms, highlighting the advantages of the database approach to data management. ▪ Describe the relational database model and outline its basic features.
▪ A well-designed and well-managed database is an extremely valuable tool in supporting decision making.	▪ Identify the common functions performed by all database management systems, and identify popular database management systems.
▪ The number and types of database applications will continue to evolve and yield real business benefits.	▪ Identify and briefly discuss current database applications.

Information Systems in the Global Economy Wal-Mart, United States

Warehousing and Mining Data on a Grand Scale

One company that really needs to know how to manage data is Wal-Mart. With a total of over 800 million transactions per day in over 7,000 stores around the world, Wal-Mart produces more data in a day than many businesses produce in a lifetime. No matter what size the business, databases and the systems that manage them provide the foundation on which business decisions are made.

Wal-Mart is successful due to its ability to learn from the data it collects. In a nutshell, Wal-Mart owes its success to its databases and business intelligence tools—software tools that manipulate data to provide useful information to Wal-Mart decision makers.

At Wal-Mart headquarters in Arkansas, massive amounts of data are collected every day from its stores around the world, and stored in a data warehouse over a petabyte in size—which is a quadrillion bytes or a million gigabytes. A data warehouse is a large database that collects data from many sources, which can then be analyzed to guide business decisions. Wal-Mart uses HP's Neoview technology for its data warehouse. The system integrates data warehousing hardware, software, and services to manage large amounts of data. It is ideal for a company looking for a powerful database tool that is easy to administer.

Neoview offers "next generation business intelligence" features that embed useful information mined from the data directly into the systems that executives, managers, and employees use every day. "A lot of people in the company are asking for quicker and easier access to data. We want to make sure it's readable and usable by internal customers," says Jim Scantlin, the director of enterprise information management at Wal-Mart.

Specifics about how Wal-Mart uses business intelligence are corporate secrets that the company works hard to keep from its competitors. Clearly one of the top goals of the system is to determine which products are selling well at various locations so that Wal-Mart can manage inventory and promotions. When asked about the role of business intelligence in Wal-Mart's business strategies, Wal-Mart CTO Nancy Stewart says, "Business intelligence is huge. It is huge." Without sophisticated data analyses, making decisions regarding business strategy would be like running through the woods wearing a blindfold. A data warehouse not only allows a company to navigate through current market conditions, but in many cases provides information that allows the business to predict and plan for the future.

Wal-Mart uses its databases, data warehouse, and business intelligence tools to collect, analyze, and disseminate massive amounts of data across its networks every day. Top-level executives, regional managers, store managers, and associates are provided with custom-designed reports, charts, and graphs presented in easy-to-read dashboard software that lets users understand the state of the business at any time so they can do their jobs more effectively. As a pilot watches and analyzes the gauges and meters on the control panel of a jumbo jet to provide a smooth flight, Wal-Mart executives and managers watch and analyze the dashboard of Wal-Mart's data warehouse to keep the business running smoothly.

As you read this chapter, consider the following:

- What role do databases play in the overall effectiveness of information systems?
- What techniques do businesses use to maximize the value of the information provided from databases?

Why Learn About Database Systems and Business Intelligence?

A huge amount of data is entered into computer systems every day. Where does all this data go and how is it used? How can it help you on the job? In this chapter, you will learn about database systems and business intelligence tools that can help you make the most effective use of information. If you become a marketing manager, you can access a vast store of data on existing and potential customers from surveys, their Web habits, and their past purchases. This information can help you sell products and services. If you become a corporate lawyer, you will have access to past cases and legal opinions from sophisticated legal databases. This information can help you win cases and protect your organization legally. If you become a human resource (HR) manager, you will be able to use databases and business intelligence tools to analyze the impact of raises, employee insurance benefits, and retirement contributions on long-term costs to your company. Regardless of your field of study in school, using database systems and business intelligence tools will likely be a critical part of your job. In this chapter, you will see how you can use data mining to extract valuable information to help you succeed. This chapter starts by introducing basic concepts of database management systems.

A database is an organized collection of data. Like other components of an information system, a database should help an organization achieve its goals. A database can contribute to organizational success by providing managers and decision makers with timely, accurate, and relevant information based on data. For example, at Creative Artists Agency (CAA), a successful Hollywood talent agency, a database helps agents organize information about clients.[1] With clients such as Tom Cruise, Julia Roberts, and Brad Pitt, a talent agency must prevent mistakes and misunderstandings. CAA's database can store various types of information about each client. For example, the database informs agents about movies in which Tom Cruise is acting, movies he is producing, products he is endorsing, and any other pertinent information about the actor's career. Using the database, an agent could find all clients that are associated with a particular product or film, or all the products and films associated with one client. Databases also help companies generate information to reduce costs, increase profits, track past business activities, and open new market opportunities. In some cases, organizations collaborate in creating and using international databases. Six organizations, including the Organization of Petroleum Exporting Countries (OPEC), International Energy Agency (IEA), and the United Nations, use a database to monitor the global oil supply.

A database provides an essential foundation for an organization's information and decision support system. Without a well-designed, accurate database, executives, managers, and others do not have access to the information they need to make good decisions. For example, the city of Albuquerque, New Mexico, provides its citizens with access to a database that provides information on "water bills and usage, crime statistics in specific neighborhoods, and election campaign contributions."[2] The database provides citizens with direct access to valuable information and frees city workers from having to supply the information.

A database is also the foundation of most systems development projects. If the database is not designed properly, the systems development effort can be like a house of cards, collapsing under the weight of inaccurate and inadequate data. Because data is so critical to an organization's success, many firms develop databases to help them access data more efficiently and use it more effectively. This typically requires a well-designed database management system and a knowledgeable database administrator.

database management system (DBMS)
A group of programs that manipulate the database and provide an interface between the database and the user of the database and other application programs.

database administrator (DBA)
A skilled IS professional who directs all activities related to an organization's database.

A **database management system (DBMS)** consists of a group of programs that manipulate the database and provide an interface between the database and its users and other application programs. Usually purchased from a database company, a DBMS provides a single point of management and control over data resources, which can be critical to maintaining the integrity and security of the data. A database, a DBMS, and the application programs that use the data make up a database environment. A **database administrator (DBA)** is a skilled and trained IS professional who directs all activities related to an organization's database, including providing security from intruders. A security breach at an Ivy

League college provided an intruder with access to a database that stored students' private information.[3] Such data breaches have become commonplace for businesses and organizations because many databases are now accessible from the Internet. Data quality and accuracy also continue to be important issues for DBAs. A database error in the United Kingdom left 400,000 people without paychecks in March, 2007.[4]

Databases and database management systems are becoming even more important to businesses as they deal with increasing amounts of digital information. A report from IDC, called "The Diverse and Exploding Digital Universe," estimates the size of the digital universe to be 281 exabytes, or 281 billion gigabytes. By 2011, there will be 1,800 exabytes of electronic data in existence, or 1.8 zettabytes.[5] If a tennis ball were one byte of information, a zettabyte-sized ball would be around the size of one earth. IDC recommends that businesses and organizations move now to create policies, tools, and standards to accommodate the approaching tidal wave of digital data and information.[6]

DATA MANAGEMENT

Without data and the ability to process it, an organization could not successfully complete most business activities. It could not pay employees, send out bills, order new inventory, or produce information to assist managers in decision making. Recall that data consists of raw facts, such as employee numbers and sales figures. For data to be transformed into useful information, it must first be organized in a meaningful way.

The Hierarchy of Data

Data is generally organized in a hierarchy that begins with the smallest piece of data used by computers (a bit) and progresses through the hierarchy to a database. A bit (a binary digit) represents a circuit that is either on or off. Bits can be organized into units called *bytes*. A byte is typically eight bits. Each byte represents a **character**, which is the basic building block of information. A character can be an uppercase letter (A, B, C... Z), lowercase letter (a, b, c... z), numeric digit (0, 1, 2... 9), or special symbol (., !, [+], [-], /, ...).

character
A basic building block of information, consisting of uppercase letters, lowercase letters, numeric digits, or special symbols.

Characters can be combined to form a field. A **field** is typically a name, number, or combination of characters that describes an aspect of a business object (such as an employee, a location, or a truck) or activity (such as a sale). In addition to being entered into a database, fields can be computed from other fields. *Computed fields* include the total, average, maximum, and minimum values. A collection of related data fields is a **record**. By combining descriptions of the characteristics of an object or activity, a record can provide a complete description of the object or activity. For instance, an employee record is a collection of fields about one employee. One field includes the employee's name, another field contains the address, and still others the phone number, pay rate, earnings made to date, and so forth. A collection of related records is a **file**—for example, an employee file is a collection of all company employee records. Likewise, an inventory file is a collection of all inventory records for a particular company or organization. Some database software refers to files as tables.

field
Typically a name, number, or combination of characters that describes an aspect of a business object or activity.

record
A collection of related data fields.

file
A collection of related records.

At the highest level of this hierarchy is a *database*, a collection of integrated and related files. Together, bits, characters, fields, records, files, and databases form the **hierarchy of data** (see Figure 5.1). Characters are combined to make a field, fields are combined to make a record, records are combined to make a file, and files are combined to make a database. A database houses not only all these levels of data but also the relationships among them.

hierarchy of data
Bits, characters, fields, records, files, and databases.

Data Entities, Attributes, and Keys

Entities, attributes, and keys are important database concepts. An **entity** is a generalized class of people, places, or things (objects) for which data is collected, stored, and maintained. Examples of entities include employees, inventory, and customers. Most organizations organize and store data as entities.

entity
A generalized class of people, places, or things for which data is collected, stored, and maintained.

Figure 5.1

The Hierarchy of Data

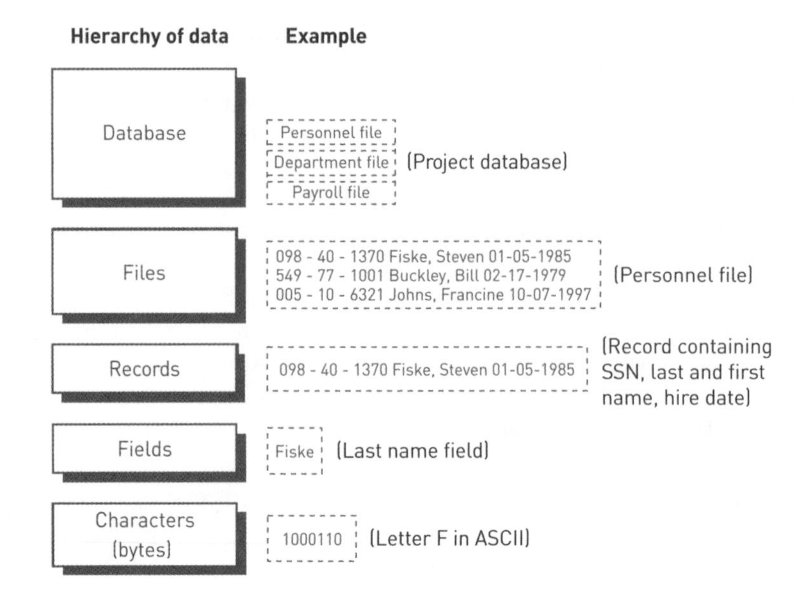

attribute
A characteristic of an entity.

data item
The specific value of an attribute.

An **attribute** is a characteristic of an entity. For example, employee number, last name, first name, hire date, and department number are attributes for an employee (see Figure 5.2). The inventory number, description, number of units on hand, and location of the inventory item in the warehouse are attributes for items in inventory. Customer number, name, address, phone number, credit rating, and contact person are attributes for customers. Attributes are usually selected to reflect the relevant characteristics of entities such as employees or customers. The specific value of an attribute, called a **data item**, can be found in the fields of the record describing an entity.

Figure 5.2

Keys and Attributes

The key field is the employee number. The attributes include last name, first name, hire date, and department number.

Employee #	Last name	First name	Hire date	Dept. number
005-10-6321	Johns	Francine	10-07-1997	257
549-77-1001	Buckley	Bill	02-17-1979	632
098-40-1370	Fiske	Steven	01-05-1985	598

ENTITIES (records)

KEY FIELD

ATTRIBUTES (fields)

Most organizations use attributes and data items. Many governments use attributes and data items to help in criminal investigations. The United States Federal Bureau of Investigation is building the "world's largest computer database of peoples' physical characteristics."[7] At a cost of $1 billion, the database management system named Next Generation Identification will catalog digital images of faces, fingerprints, and palm prints of U.S. citizens and visitors. Each person in the database is an entity, each biometric category is an attribute, and each image is a data item. The information will be used as a forensics tool and to increase homeland security.

key
A field or set of fields in a record that is used to identify the record.

primary key
A field or set of fields that uniquely identifies the record.

As discussed, a collection of fields about a specific object is a record. A **key** is a field or set of fields in a record that identifies the record. A **primary key** is a field or set of fields that uniquely identifies the record. No other record can have the same primary key. The primary key is used to distinguish records so that they can be accessed, organized, and manipulated. For an employee record, such as the one shown in Figure 5.2, the employee number is an example of a primary key.

Locating a particular record that meets a specific set of criteria might be easier and faster using a combination of secondary keys. For example, a customer might call a mail-order company to place an order for clothes. If the customer does not know the correct primary key (such as a customer number), a secondary key (such as last name) can be used. In this case, the order clerk enters the last name, such as Adams. If several customers have a last name of Adams, the clerk can check other fields, such as address, first name, and so on, to find the correct customer record. After locating the correct customer record, the order can be completed and the clothing items shipped to the customer.

The Database Approach

At one time, applications used specific files. For example, a payroll application would use a payroll file. In other words, each application used files dedicated to that application. This approach to data management, whereby separate data files are created and stored for each application program, is called the **traditional approach to data management**.

traditional approach to data management
An approach whereby separate data files are created and stored for each application program.

Today, most organizations use the **database approach to data management**, where multiple application programs share a pool of related data. A database offers the ability to share data and information resources. Federal databases, for example, often include the results of DNA tests as an attribute for convicted criminals. The information can be shared with law enforcement officials around the country.

database approach to data management
An approach whereby a pool of related data is shared by multiple application programs.

To use the database approach to data management, additional software—a database management system (DBMS)—is required. As previously discussed, a DBMS consists of a group of programs that can be used as an interface between a database and the user of the database and application programs. Typically, this software acts as a buffer between the application programs and the database itself. Figure 5.3 illustrates the database approach.

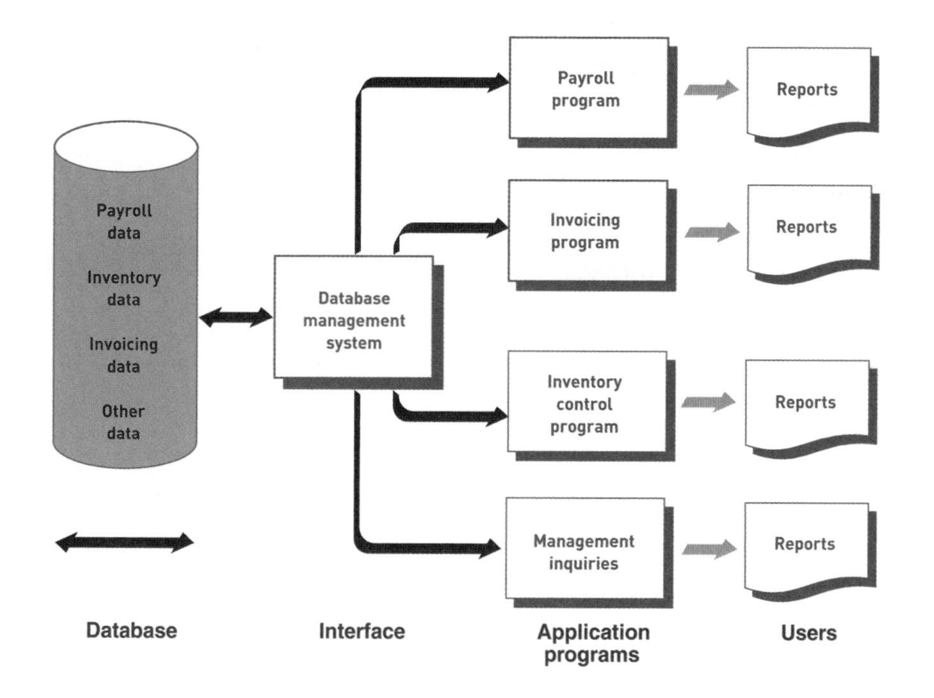

Figure 5.3
The Database Approach to Data Management

Table 5.1 lists some of the primary advantages of the database approach, and Table 5.2 lists some disadvantages.

Advantages	Explanation
Improved strategic use of corporate data	Accurate, complete, up-to-date data can be made available to decision makers where, when, and in the form they need it. The database approach can also give greater visibility to the organization's data resource.
Reduced data redundancy	Data is organized by the DBMS and stored in only one location. This results in more efficient use of system storage space.
Improved data integrity	With the traditional approach, some changes to data were not reflected in all copies of the data kept in separate files. The database approach prevents this problem because no separate files contain copies of the same piece of data.
Easier modification and updating	The DBMS coordinates data modifications and updates. Programmers and users do not have to know where the data is physically stored. Data is stored and modified once. Modification and updating is also easier because the data is commonly stored in only one location.
Data and program independence	The DBMS organizes the data independently of the application program, so the application program is not affected by the location or type of data. Introduction of new data types not relevant to a particular application does not require rewriting that application to maintain compatibility with the data file.
Better access to data and information	Most DBMSs have software that makes it easy to access and retrieve data from a database. In most cases, users give simple commands to get important information. Relationships between records can be more easily investigated and exploited, and applications can be more easily combined.
Standardization of data access	A standardized, uniform approach to database access means that all application programs use the same overall procedures to retrieve data and information.
A framework for program development	Standardized database access procedures can mean more standardization of program development. Because programs go through the DBMS to gain access to data in the database, standardized database access can provide a consistent framework for program development. In addition, each application program need address only the DBMS, not the actual data files, reducing application development time.
Better overall protection of the data	Accessing and using centrally located data is easier to monitor and control. Security codes and passwords can ensure that only authorized people have access to particular data and information in the database, thus ensuring privacy.
Shared data and information resources	The cost of hardware, software, and personnel can be spread over many applications and users. This is a primary feature of a DBMS.

Table 5.1

Advantages of the Database Approach

Disadvantages	Explanation
More complexity	DBMSs can be difficult to set up and operate. Many decisions must be made correctly for the DBMS to work effectively. In addition, users have to learn new procedures to take full advantage of a DBMS.
More difficult to recover from a failure	With the traditional approach to file management, a failure of a file affects only a single program. With a DBMS, a failure can shut down the entire database.
More expensive	DBMSs can be more expensive to purchase and operate. The expense includes the cost of the database and specialized personnel, such as a database administrator, who is needed to design and operate the database. Additional hardware might also be required.

Table 5.2

Disadvantages of the Database Approach

Many modern databases serve entire enterprises, encompassing much of the data of the organization. Often, distinct yet related databases are linked to provide enterprise-wide databases. For example, many Wal-Mart stores include in-store medical clinics for customers. Wal-Mart uses a centralized electronic health records database that stores the information of all patients across all stores.[8] The database is interconnected with the main Wal-Mart database to provide information about customers' interactions with the clinics and stores. The Ethical and Societal Issues box provides more information about databases used for electronic health record systems.

Web-Based Electronic Health Record Systems

The United States federal government is pushing for most Americans to have their medical records stored in electronic form by 2014. Electronic health record (EHR) systems store patient records in a central database that can be accessed by many physicians at more than one location. Such a system eliminates problems caused by duplicate records at different physician offices, avoids having to fill out a new patient history with each new physician visited by the patient, and reduces errors made by incorrectly deciphering handwritten notes and prescriptions. Electronic records can make for a better and healthier world. However, the cost of moving to electronic systems is prohibitive, especially for small medical practices. At this point, only ten percent of small medical offices and five percent of solo practitioners have moved to EHR systems.

Although the government is introducing financial incentives to encourage physicians to use EHR systems, some big companies that aren't typically associated with healthcare are becoming involved—particularly Microsoft and Google. Approximately 52 percent of adults look to the Web when seeking health advice. Google and Microsoft believe that they can better assist health consumers by providing them with a robust tool for managing their health records. Microsoft's tool is named HealthVault, while Google's is named Google Health. The companies see their EHR systems as a solution to the government's problem for finding a low-cost records system designed for both physicians and patients.

John D. Halamka, a doctor and CIO of the Harvard Medical School, thinks systems in which the patient manages the information, such as those proposed by Microsoft and Google, are the inevitable future. "Patients will ultimately be the stewards of their own information," Halamka stated. "In the future, healthcare will be a much more collaborative process between patients and doctors."

Google agrees that patients should be in charge. A statement at Google Health's welcome page reads, "At Google, we feel patients should be in charge of their health information, and they should be able to grant their healthcare providers, family members, or whomever they choose, access to this information. Google Health was developed to meet this need."

But just how private and secure will our medical records be when stored in Web-accessible databases, protected only by one password? Privacy and security concerns are raised both by corporate access to private records by Microsoft and Google and outsider access by hackers. It is likely that both companies will use automated systems to target advertising at individuals based on medical records, just as Google's Gmail places ads next to e-mail messages based on the message contents. Unauthorized users might also be able to access records stored on a network that billions of users around the world use.

Another problem that complicates Google and Microsoft's involvement is that third-party medical record services are not covered by the Health Insurance Portability and Accountability Act (HIPAA). HIPAA provides strict standards for keeping medical records private. If a patient chooses to use Microsoft or Google to store medical records, those records would no longer be protected by the standards imposed by HIPAA in its current form.

As in similar cases, patients should weigh the costs in terms of privacy and security against the benefits of convenience and data reliability. Meanwhile, the software vendors need to work to build higher levels of security, privacy assurances, and customer trust.

Discussion Questions

1. Why does the U.S. federal government want to move health records to electronic systems?
2. What benefits and risks are offered by Web-based health records management systems like Google Health?

Critical Thinking Questions

1. How might Google and Microsoft reassure users about the privacy and security issues posed in this sidebar?
2. Would you consider registering for Google Health? Why or why not?

Sources: Lohr, Steve, "Google and Microsoft Look to Change Health Care," *New York Times*, August 14, 2007, *www.nytimes.com/2007/08/14/technology/14healthnet.html*, AP Staff, "Google ventures into health records biz," *CNN.com*, February 21, 2008, *www.cnn.com/2008/TECH/02/21/google.records.ap*.

DATA MODELING AND DATABASE CHARACTERISTICS

Because today's businesses have so many elements, they must keep data organized so that it can be used effectively. A database should be designed to store all data relevant to the business and provide quick access and easy modification. Moreover, it must reflect the business processes of the organization. When building a database, an organization must carefully consider these questions:

- **Content.** What data should be collected and at what cost?
- **Access.** What data should be provided to which users and when?
- **Logical structure.** How should data be arranged so that it makes sense to a given user?
- **Physical organization.** Where should data be physically located?

Data Modeling

Key considerations in organizing data in a database include determining what data to collect in the database, who will have access to it, and how they might want to use the data. After determining these details, an organization can create a database. Building a database requires two different types of designs: a logical design and a physical design. The *logical design* of a database is an abstract model of how the data should be structured and arranged to meet an organization's information needs. The logical design involves identifying relationships among the data items and grouping them in an orderly fashion. Because databases provide both input and output for information systems throughout a business, users from all functional areas should assist in creating the logical design to ensure that their needs are identified and addressed. *Physical design* starts from the logical database design and fine-tunes it for performance and cost considerations (such as improved response time, reduced storage space, and lower operating cost). For example, the database administrator at Intermountain Healthcare in Salt Lake City, Utah, combined the databases of 21 hospitals and 100 clinics into one integrated system, saving the organization the cost of dozens of servers, and providing new and improved services.[9] The person who fine-tunes the physical design must have an in-depth knowledge of the DBMS. For example, the logical database design might need to be altered so that certain data entities are combined, summary totals are carried in the data records rather than calculated from elemental data, and some data attributes are repeated in more than one data entity. These are examples of **planned data redundancy**, which improves the system performance so that user reports or queries can be created more quickly.

planned data redundancy
A way of organizing data in which the logical database design is altered so that certain data entities are combined, summary totals are carried in the data records rather than calculated from elemental data, and some data attributes are repeated in more than one data entity to improve database performance.

One of the tools database designers use to show the logical relationships among data is a data model. A **data model** is a diagram of entities and their relationships. Data modeling usually involves understanding a specific business problem and analyzing the data and information needed to deliver a solution. When done at the level of the entire organization, this is called enterprise data modeling. **Enterprise data modeling** is an approach that starts by investigating the general data and information needs of the organization at the strategic level, and then examines more specific data and information needs for the various functional areas and departments within the organization. Various models have been developed to help managers and database designers analyze data and information needs. An entity-relationship diagram is an example of such a data model.

data model
A diagram of data entities and their relationships.

enterprise data modeling
Data modeling done at the level of the entire enterprise.

Entity-relationship (ER) diagrams use basic graphical symbols to show the organization of and relationships between data. In most cases, boxes in ER diagrams indicate data items or entities contained in data tables, and diamonds show relationships between data items and entities. In other words, ER diagrams show data items in tables (entities) and the ways they are related.

entity-relationship (ER) diagrams
Data models that use basic graphical symbols to show the organization of and relationships between data.

ER diagrams help ensure that the relationships among the data entities in a database are correctly structured so that any application programs developed are consistent with business operations and user needs. In addition, ER diagrams can serve as reference documents after a database is in use. If changes are made to the database, ER diagrams help design them. Figure 5.4 shows an ER diagram for an order database. In this database design, one salesperson serves many customers. This is an example of a one-to-many relationship, as indicated by the one-to-many symbol (the "crow's-foot") shown in Figure 5.4. The ER diagram also shows that each customer can place one-to-many orders, each order includes one-to-many line items, and many line items can specify the same product (a many-to-one relationship). This database can also have one-to-one relationships. For example, one order generates one invoice.

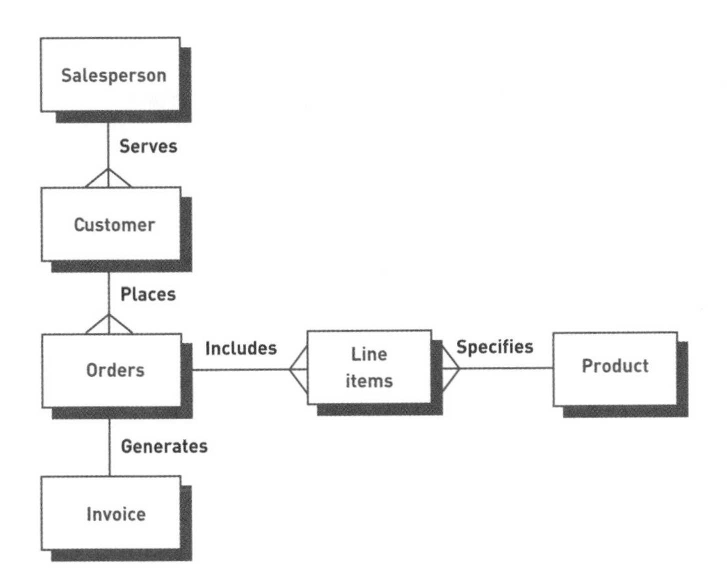

Figure 5.4

An Entity-Relationship (ER) Diagram for a Customer Order Database

Development of ER diagrams helps ensure that the logical structure of application programs is consistent with the data relationships in the database.

The Relational Database Model

Although there are a number of different database models, including flat files, hierarchical, and network models, the **relational model** has become the most popular, and use of this model will continue to increase. The relational model describes data using a standard tabular format. In a database structured according to the relational model, all data elements are placed in two-dimensional tables, called *relations*, which are the logical equivalent of files. The tables in relational databases organize data in rows and columns, simplifying data access and manipulation. It is normally easier for managers to understand the relational model (see Figure 5.5) than other database models.

relational model
A database model that describes data in which all data elements are placed in two-dimensional tables, called *relations*, which are the logical equivalent of files.

Databases based on the relational model include IBM DB2, Oracle, Sybase, Microsoft SQL Server, Microsoft Access, and MySQL. Oracle is currently the market leader in general-purpose databases, with over 40 percent of the $16.5 billion database market. IBM comes in second with about 21 percent, and Microsoft third with about 19 percent.[10]

In the relational model, each row (or record) of a table represents a data entity, with the columns (or fields) of the table representing attributes. Each attribute can accept only certain values. The allowable values for these attributes are called the **domain**. The domain for a particular attribute indicates what values can be placed in each column of the relational table. For instance, the domain for an attribute such as gender would be limited to male or female. A domain for pay rate would not include negative numbers. In this way, defining a domain can increase data accuracy.

domain
The allowable values for data attributes.

Manipulating Data

After entering data into a relational database, users can make inquiries and analyze the data. Basic data manipulations include selecting, projecting, and joining. **Selecting** involves eliminating rows according to certain criteria. Suppose a project table contains the project number, description, and department number for all projects a company is performing. The president of the company might want to find the department number for Project 226, a sales manual project. Using selection, the president can eliminate all rows but the one for Project 226 and see that the department number for the department completing the sales manual project is 598.

selecting
Manipulating data to eliminate rows according to certain criteria.

Figure 5.5

A Relational Database Model

In the relational model, all data elements are placed in two-dimensional tables, or relations. As long as they share at least one common element, these relations can be linked to output useful information. Note that some organizations might use employee number instead of Social Security number (SSN) in Data Tables 2 and 3.

Data Table 1: Project Table

Project	Description	Dept. number
155	Payroll	257
498	Widgets	632
226	Sales manual	598

Data Table 2: Department Table

Dept.	Dept. name	Manager SSN
257	Accounting	005-10-6321
632	Manufacturing	549-77-1001
598	Marketing	098-40-1370

Data Table 3: Manager Table

SSN	Last name	First name	Hire date	Dept. number
005-10-6321	Johns	Francine	10-07-1997	257
549-77-1001	Buckley	Bill	02-17-1979	632
098-40-1370	Fiske	Steven	01-05-1985	598

projecting
Manipulating data to eliminate columns in a table.

joining
Manipulating data to combine two or more tables.

linking
Data manipulation that combines two or more tables using common data attributes to form a new table with only the unique data attributes.

Projecting involves eliminating columns in a table. For example, a department table might contain the department number, department name, and Social Security number (SSN) of the manager in charge of the project. A sales manager might want to create a new table with only the department number and the Social Security number of the manager in charge of the sales manual project. The sales manager can use projection to eliminate the department name column and create a new table containing only department number and SSN.

Joining involves combining two or more tables. For example, you can combine the project table and the department table to create a new table with the project number, project description, department number, department name, and Social Security number for the manager in charge of the project.

As long as the tables share at least one common data attribute, the tables in a relational database can be **linked** to provide useful information and reports. Being able to link tables to each other through common data attributes is one of the keys to the flexibility and power of relational databases. Suppose the president of a company wants to find out the name of the manager of the sales manual project and the length of time the manager has been with the company. Assume that the company has the manager, department, and project tables shown in Figure 5.5. A simplified ER diagram showing the relationship between these tables is shown in Figure 5.6. Note the crow's-foot by the project table. This indicates that a department can have many projects. The president would make the inquiry to the database, perhaps via a personal computer. The DBMS would start with the project description and search the project table to find out the project's department number. It would then use the department number to search the department table for the manager's Social Security number. The department number is also in the department table and is the common element that links the project table to the department table. The DBMS uses the manager's Social Security number to search the manager table for the manager's hire date. The manager's Social Security number is the common element between the department table and the manager table. The final result is that the manager's name and hire date are presented to the president as a response to the inquiry (see Figure 5.7).

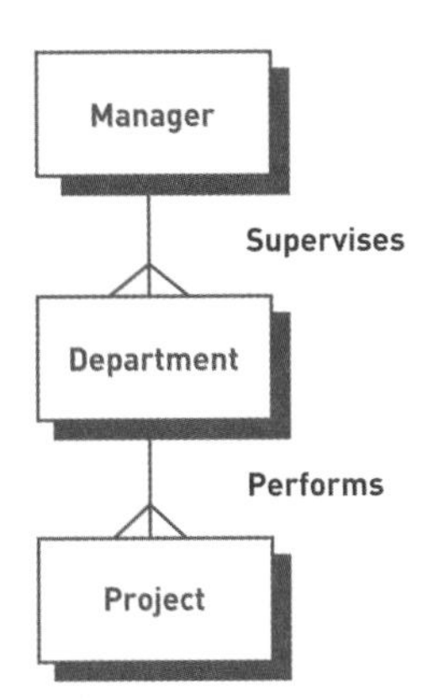

Figure 5.6

A Simplified ER Diagram Showing the Relationship Between the Manager, Department, and Project Tables

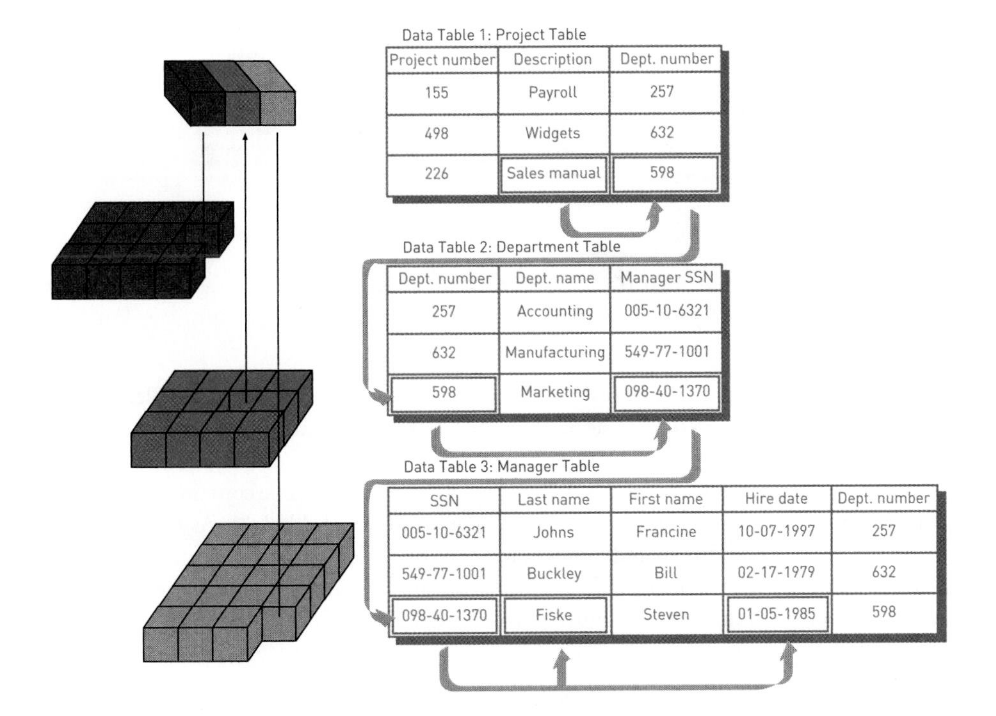

Data Table 1: Project Table

Project number	Description	Dept. number
155	Payroll	257
498	Widgets	632
226	Sales manual	598

Data Table 2: Department Table

Dept. number	Dept. name	Manager SSN
257	Accounting	005-10-6321
632	Manufacturing	549-77-1001
598	Marketing	098-40-1370

Data Table 3: Manager Table

SSN	Last name	First name	Hire date	Dept. number
005-10-6321	Johns	Francine	10-07-1997	257
549-77-1001	Buckley	Bill	02-17-1979	632
098-40-1370	Fiske	Steven	01-05-1985	598

Figure 5.7

Linking Data Tables to Answer an Inquiry

In finding the name and hire date of the manager working on the sales manual project, the president needs three tables: project, department, and manager. The project description (Sales manual) leads to the department number (598) in the project table, which leads to the manager's SSN (098-40-1370) in the department table, which leads to the manager's name (Fiske) and hire date (01-05-1985) in the manager table. Again, note that some organizations might use employee number instead of Social Security number (SSN).

One of the primary advantages of a relational database is that it allows tables to be linked, as shown in Figure 5.7. This linkage is especially useful when information is needed from multiple tables. For example, the manager's Social Security number is maintained in the manager table. If the Social Security number is needed, it can be obtained by linking to the manager table.

The relational database model is by far the most widely used. It is easier to control, more flexible, and more intuitive than other approaches because it organizes data in tables. As shown in Figure 5.8, a relational database management system, such as Access, provides tips and tools for building and using database tables. In this figure, the database displays information about data types and indicates that additional help is available. The ability to link relational tables also allows users to relate data in new ways without having to redefine complex relationships. Because of the advantages of the relational model, many companies use it for large corporate databases, such as those for marketing and accounting. The relational model can also be used with personal computers and mainframe systems. A travel reservation company, for example, can develop a fare-pricing system by using relational database technology that can handle millions of daily queries from online travel companies, such as Expedia, Travelocity, and Orbitz.

Figure 5.8

Building and Modifying a Relational Database

Relational databases provide many tools, tips, and shortcuts to simplify the process of creating and modifying a database.

(Source: Courtesy of Microsoft Corporation.)

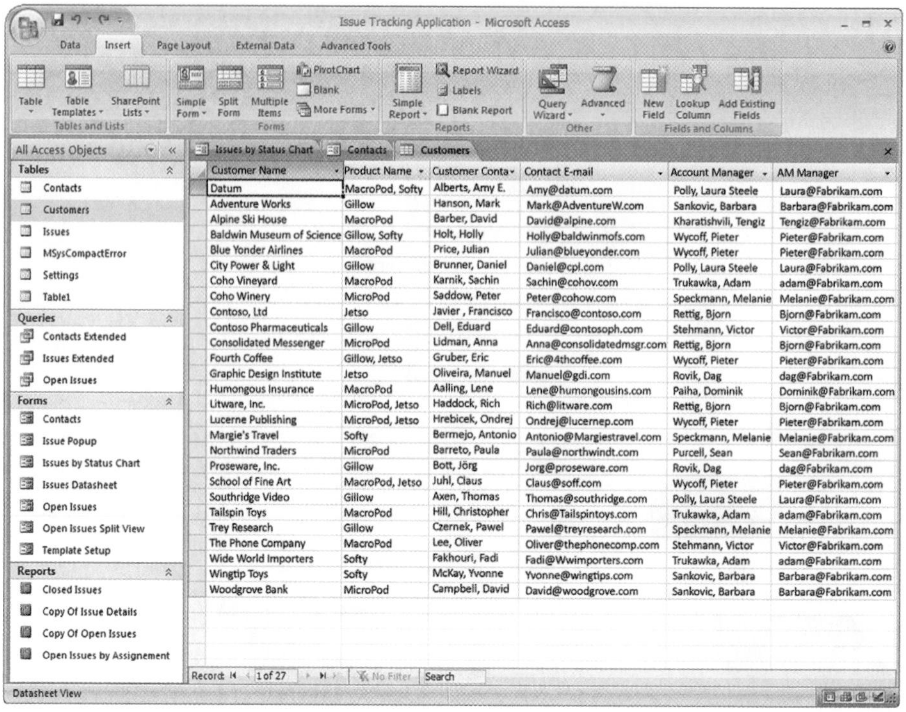

Data Cleanup

data cleanup
The process of looking for and fixing inconsistencies to ensure that data is accurate and complete.

As discussed in Chapter 1, valuable data is accurate, complete, economical, flexible, reliable, relevant, simple, timely, verifiable, accessible, and secure. The database must also be properly designed. The purpose of **data cleanup** is to develop data with these characteristics. Consider a database for a fitness center designed to track member dues. The table contains the attribute name, phone number, gender, dues paid, and date paid (see Table 5.3). As the records in Table 5.3 show, Anita Brown and Sim Thomas have paid their dues in September. Sim has paid his dues in two installments. Note that no primary key uniquely identifies each record. As you will see next, this problem must be corrected.

Table 5.3

Fitness Center Dues

Name	Phone	Gender	Dues Paid	Date Paid
Brown, A.	468-3342	Female	$30	September 15
Thomas, S.	468-8788	Male	$15	September 15
Thomas, S.	468-5238	Male	$15	September 25

Because Sim Thomas has paid dues twice in September, the data in the database is now redundant. The name, phone number, and gender for Thomas are repeated in two records. Notice that the data in the database is also inconsistent: Thomas has changed his phone number, but only one of the records reflects this change. Further reducing this database's reliability is the lack of a primary key to uniquely identify Sim Thomas's record. The first Thomas could be Sim Thomas, but the second might be Steve Thomas. These problems and irregularities in data are called *anomalies.* Data anomalies often result in incorrect information, causing database users to be misinformed about actual conditions. Anomalies must be corrected.

To solve these problems in the fitness center's database, we can add a primary key, such as member number, and put the data into two tables: a Fitness Center Members table with gender, phone number, and related information, and a Dues Paid table with dues paid and date paid (see Tables 5.4 and 5.5). Both tables include the member number attribute so that they can be linked.

Table 5.4
Fitness Center Members

Member No.	Name	Phone	Gender
SN123	Brown, A.	468-3342	Female
SN656	Thomas, S.	468-5238	Male

Table 5.5
Dues Paid

Member No.	Dues Paid	Date Paid
SN123	$30	September 15
SN656	$15	September 15
SN656	$15	September 25

The relations in Table 5.4 and Table 5.5 reduce the redundancy and eliminate the potential problem of having two different phone numbers for the same member. Also note that the member number gives each record in the Fitness Center Members table a primary key. Because the Dues Paid table lists two payment entries ($15 each) with the same member number (SN656), one person clearly made the payments, not two different people. Formalized approaches, such as *database normalization*, are often used to clean up problems with data.

DATABASE MANAGEMENT SYSTEMS

Creating and implementing the right database system ensures that the database will support both business activities and goals. But how do we actually create, implement, use, and update a database? The answer is found in the database management system. As discussed earlier, a DBMS is a group of programs used as an interface between a database and application programs or a database and the user. The capabilities and types of database systems, however, vary considerably. For example, visitors to the Baseball Hall of Fame in Cooperstown, New York, use a DBMS to search baseball highlight films from famous games and plays.[11] DBMSs are used to manage all kinds of data for all kinds of purposes.

Overview of Database Types

Database management systems can range from small, inexpensive software packages to sophisticated systems costing hundreds of thousands of dollars. The following sections discuss a few popular alternatives. See Figure 5.9 for one example.

Flat File

A flat file is a simple database program whose records have no relationship to one another. Flat file databases are often used to store and manipulate a single table or file, and do not use any of the database models discussed previously, such as the relational model. Many spreadsheet and word processing programs have flat file capabilities. These software packages can sort tables and make simple calculations and comparisons. Microsoft OneNote is designed to let people put ideas, thoughts, and notes into a computer file. In OneNote, each note can be placed anywhere on a page or in a box on a page, called a *container*. Pages are organized into sections and subsections that appear as colored tabs. After you enter a note, you can retrieve, copy, and paste it into other applications, such as word processing and spreadsheet programs. Microsoft uses OneNote as the primary technology for its management training classes. OneNote allows managers-in-training to collect photos, handwritten notes, online content, and audio recordings in one flat file.[12] OneNote enables Microsoft to offer training to a larger number of managers, while saving $360,000 per year in printed training materials.

Figure 5.9

Microsoft OneNote

Microsoft OneNote lets you gather any type of information and then retrieve, copy, and paste the information into other applications, such as word processing and spreadsheet programs.

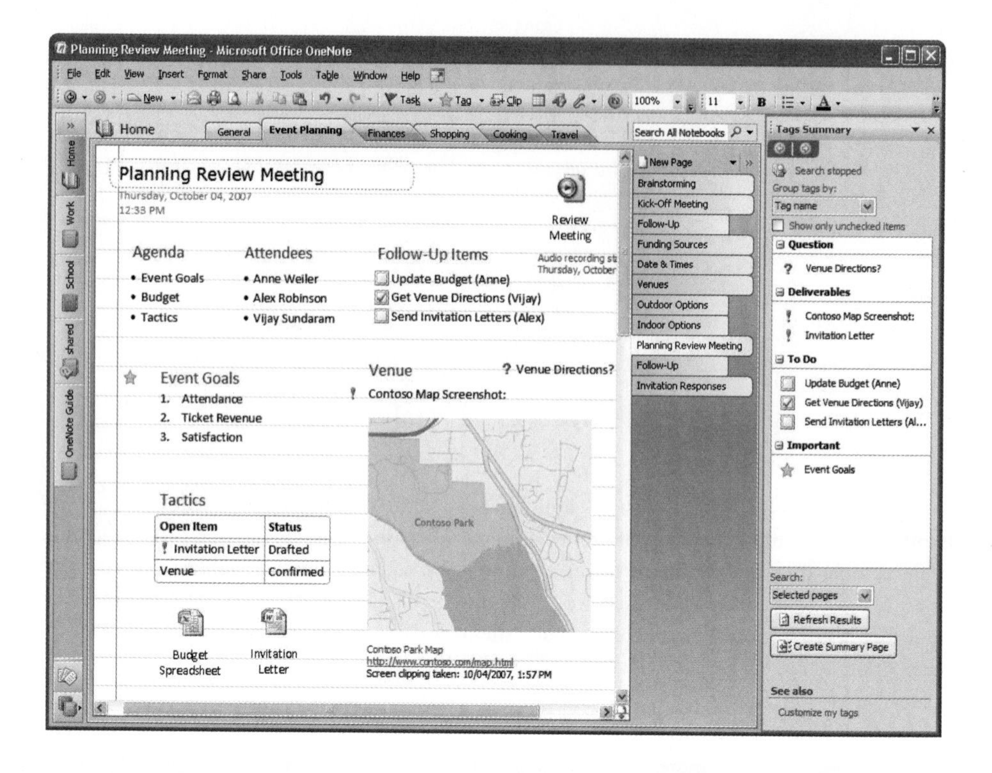

Similar to OneNote, Evernote is a free database that can store notes and other pieces of information. Considering the amount of information today's high-capacity hard disks can store, the popularity of databases that can handle unstructured data will continue to grow.

Single User

A database installed on a personal computer is typically meant for a single user. Microsoft Office Access and FileMaker Pro are designed to support single-user implementations. Microsoft InfoPath is another example of a database program that supports a single user. This software is part of the Microsoft Office suite, and it helps people collect and organize information from a variety of sources. InfoPath has built-in forms that can be used to enter expense information, timesheet data, and a variety of other information.

Multiple Users

Small, midsize, and large businesses need multiuser DBMSs to share information throughout the organization over a network. These more powerful, expensive systems allow dozens or hundreds of people to access the same database system at the same time. Popular vendors for multiuser database systems include Oracle, Microsoft, Sybase, and IBM. Many single-user databases, such as Microsoft Access, can be implemented for multiuser support over a network, though they often are limited in the amount of users they can support.

All DBMSs share some common functions, such as providing a user view, physically storing and retrieving data in a database, allowing for database modification, manipulating data, and generating reports. These DBMSs can handle the most complex data-processing tasks, and because they are accessed over a network, one database can serve many locations around the world. For example, Surya Roshni Ltd is a major manufacturer of lighting products based in New Delhi, India, with a global reach. One Oracle database stored on servers in New Delhi provides corporate information to associates around the world.[13]

Providing a User View

schema
A description of the entire database.

Because the DBMS is responsible for access to a database, one of the first steps in installing and using a large database involves telling the DBMS the logical and physical structure of the data and relationships among the data in the database for each user. This description is called a **schema** (as in schematic diagram). Large database systems, such as Oracle, typically use schemas to define the tables and other database features associated with a person or user. A schema can be part of the database or a separate schema file. The DBMS can reference a schema to find where to access the requested data in relation to another piece of data.

Creating and Modifying the Database

Schemas are entered into the DBMS (usually by database personnel) via a data definition language. A **data definition language (DDL)** is a collection of instructions and commands used to define and describe data and relationships in a specific database. A DDL allows the database's creator to describe the data and relationships that are to be contained in the schema. In general, a DDL describes logical access paths and logical records in the database. Figure 5.10 shows a simplified example of a DDL used to develop a general schema. The *Xs* in Figure 5.10 reveal where specific information concerning the database should be entered. File description, area description, record description, and set description are terms the DDL defines and uses in this example. Other terms and commands can be used, depending on the particular DBMS employed.

data definition language (DDL)
A collection of instructions and commands used to define and describe data and relationships in a specific database.

```
SCHEMA DESCRIPTION
SCHEMA NAME IS XXXX
AUTHOR       XXXX
DATE         XXXX
FILE DESCRIPTION
    FILE NAME IS XXXX
      ASSIGN XXXX
    FILE NAME IS XXXX
      ASSIGN XXXX
AREA DESCRIPTION
    AREA NAME IS XXXX
RECORD DESCRIPTION
    RECORD NAME IS XXXX
    RECORD ID IS XXXX
    LOCATION MODE IS XXXX
    WITHIN XXXX AREA FROM XXXX THRU XXXX
SET DESCRIPTION
    SET NAME IS XXXX
    ORDER IS XXXX
    MODE IS XXXX
    MEMBER IS XXXX
    .
    .
    .
```

Figure 5.10

Using a Data Definition Language to Define a Schema

Another important step in creating a database is to establish a **data dictionary**, a detailed description of all data used in the database. The data dictionary contains the following data:

- Name of the data item
- Aliases or other names that may be used to describe the item
- Range of values that can be used
- Type of data (such as alphanumeric or numeric)
- Amount of storage needed for the item
- Notation of the person responsible for updating it and the various users who can access it
- List of reports that use the data item

data dictionary
A detailed description of all the data used in the database.

A data dictionary can also include a description of data flows, the way records are organized, and the data-processing requirements. Figure 5.11 shows a typical data dictionary entry.

Figure 5.11

A Typical Data Dictionary Entry

NORTHWESTERN MANUFACTURING

PREPARED BY:	D. BORDWELL
DATE:	04 AUGUST 2007
APPROVED BY:	J. EDWARDS
DATE:	13 OCTOBER 2007
VERSION:	3.1
PAGE:	1 OF 1
DATA ELEMENT NAME:	PARTNO
DESCRIPTION:	INVENTORY PART NUMBER
OTHER NAMES:	PTNO
VALUE RANGE:	100 TO 5000
DATA TYPE:	NUMERIC
POSITIONS:	4 POSITIONS OR COLUMNS

For example, the information in a data dictionary for the part number of an inventory item can include the following data:

- Name of the person who made the data dictionary entry (D. Bordwell)
- Date the entry was made (August 4, 2007)
- Name of the person who approved the entry (J. Edwards)
- Approval date (October 13, 2007)
- Version number (3.1)
- Number of pages used for the entry (1)
- Part name (PARTNO)
- Part names that might be used (PTNO)
- Range of values (part numbers can range from 100 to 5,000)
- Type of data (numeric)
- Storage required (four positions are required for the part number)

A data dictionary is valuable in maintaining an efficient database that stores reliable information with no redundancy, and makes it easy to modify the database when necessary. Data dictionaries also help computer and system programmers who require a detailed description of data elements stored in a database to create the code to access the data.

Storing and Retrieving Data

One function of a DBMS is to be an interface between an application program and the database. When an application program needs data, it requests the data through the DBMS. Suppose that to calculate the total price of a new car, an auto dealer pricing program needs price data on the engine option—six cylinders instead of the standard four cylinders. The application program requests this data from the DBMS. In doing so, the application program follows a logical access path. Next, the DBMS, working with various system programs, accesses a storage device, such as disk drives, where the data is stored. When the DBMS goes to this storage device to retrieve the data, it follows a path to the physical location (physical access path) where the price of this option is stored. In the pricing example, the DBMS might go to a disk drive to retrieve the price data for six-cylinder engines. This relationship is shown in Figure 5.12.

This same process is used if a user wants to get information from the database. First, the user requests the data from the DBMS. For example, a user might give a command, such as LIST ALL OPTIONS FOR WHICH PRICE IS GREATER THAN 200 DOLLARS. This is the logical access path (LAP). Then, the DBMS might go to the options price section of a disk to get the information for the user. This is the physical access path (PAP).

Two or more people or programs attempting to access the same record in the same database at the same time can cause a problem. For example, an inventory control program might attempt to reduce the inventory level for a product by ten units because ten units were just shipped to a customer. At the same time, a purchasing program might attempt to increase

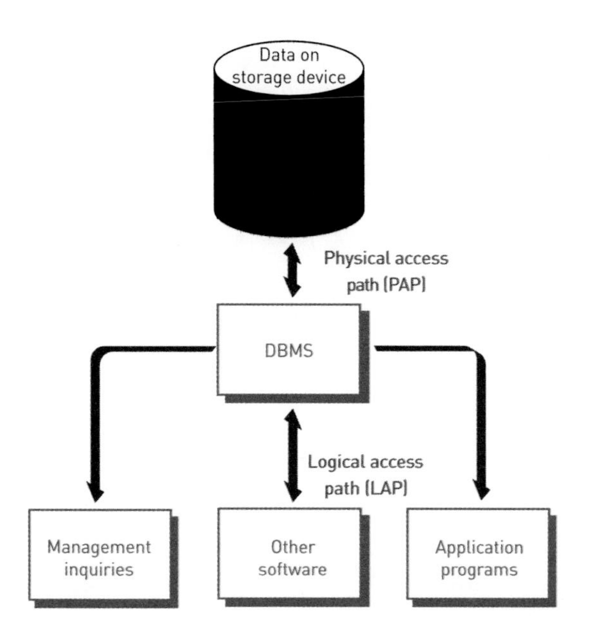

Figure 5.12

Logical and Physical Access Paths

the inventory level for the same product by 200 units because more inventory was just received. Without proper database control, one of the inventory updates might be incorrect, resulting in an inaccurate inventory level for the product. **Concurrency control** can be used to avoid this potential problem. One approach is to lock out all other application programs from access to a record if the record is being updated or used by another program.

concurrency control
A method of dealing with a situation in which two or more people need to access the same record in a database at the same time.

Manipulating Data and Generating Reports

After a DBMS has been installed, employees, managers, and consumers can use it to review reports and obtain important information. For example, the Food Allergen and Consumer Protection Act, effective in 2006, requires that food manufacturing companies generate reports on the ingredients, formulas, and food preparation techniques for the public. Using a DBMS, a company can easily manage this requirement.

Some databases use *Query-by-Example (QBE)*, which is a visual approach to developing database queries or requests. Like Windows and other GUI operating systems, you can perform queries and other database tasks by opening windows and clicking the data or features you want (see Figure 5.13).

In other cases, database commands can be used in a programming language. For example, C++ commands can be used in simple programs that will access or manipulate certain pieces of data in the database. Here's another example of a DBMS query: SELECT * FROM EMPLOYEE WHERE JOB_CLASSIFICATION = "C2". The * tells the program to include all columns from the EMPLOYEE table. In general, the commands that are used to manipulate the database are part of the **data manipulation language** (DML). This specific language, provided with the DBMS, allows managers and other database users to access, modify, and make queries about data contained in the database to generate reports. Again, the application programs go through schemas and the DBMS before actually getting to the physically stored data on a device such as a disk.

data manipulation language (DML)
The commands that are used to manipulate the data in a database.

In the 1970s, D. D. Chamberlain and others at the IBM Research Laboratory in San Jose, California, developed a standardized data manipulation language called *Structured Query Language (SQL*, pronounced like *sequel*). The EMPLOYEE query shown earlier is written in SQL. In 1986, the American National Standards Institute (ANSI) adopted SQL as the standard query language for relational databases. Since ANSI's acceptance of SQL, interest in making SQL an integral part of relational databases on both mainframe and personal computers has increased. SQL has many built-in functions, such as average (AVG), the largest value (MAX), the smallest value (MIN), and others. Table 5.6 contains examples of SQL commands.

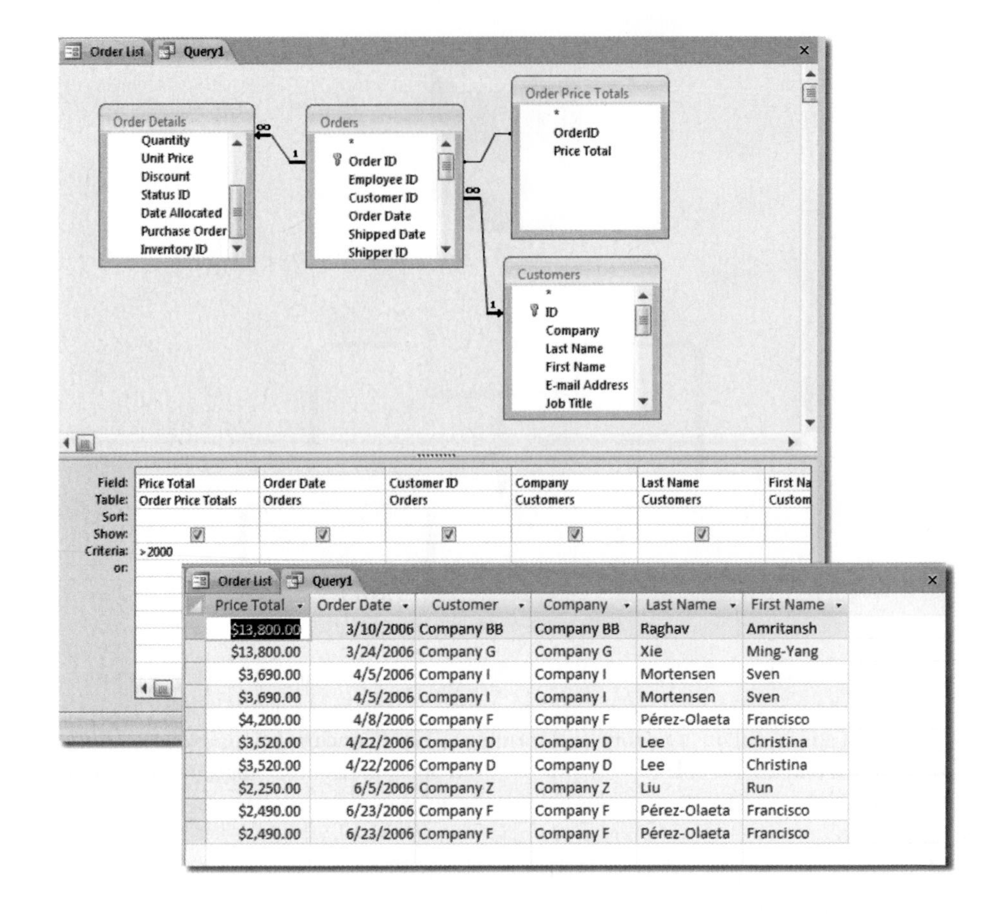

Figure 5.13

Query by Example

Some databases use Query-by-Example (QBE) to generate reports and information.

Table 5.6

Examples of SQL Commands

SQL Command	Description
SELECT ClientName, Debt FROM Client WHERE Debt > 1000	This query displays all clients (ClientName) and the amount they owe the company (Debt) from a database table called Client for clients who owe the company more than $1,000 (WHERE Debt > 1000).
SELECT ClientName, ClientNum, OrderNum FROM Client, Order WHERE Client.ClientNum=Order.ClientNum	This command is an example of a join command that combines data from two tables: the client table and the order table (FROM Client, Order). The command creates a new table with the client name, client number, and order number (SELECT ClientName, ClientNum, OrderNum). Both tables include the client number, which allows them to be joined. This is indicated in the WHERE clause, which states that the client number in the client table is the same as (equal to) the client number in the order table (WHERE Client.Client Num= Order.ClientNum).
GRANT INSERT ON Client to Guthrie	This command is an example of a security command. It allows Bob Guthrie to insert new values or rows into the Client table.

SQL lets programmers learn one powerful query language and use it on systems ranging from PCs to the largest mainframe computers (see Figure 5.14). Programmers and database users also find SQL valuable because SQL statements can be embedded into many programming languages, such as the widely used C++ and COBOL languages. Because SQL uses standardized and simplified procedures for retrieving, storing, and manipulating data in a database system, the popular database query language can be easy to understand and use.

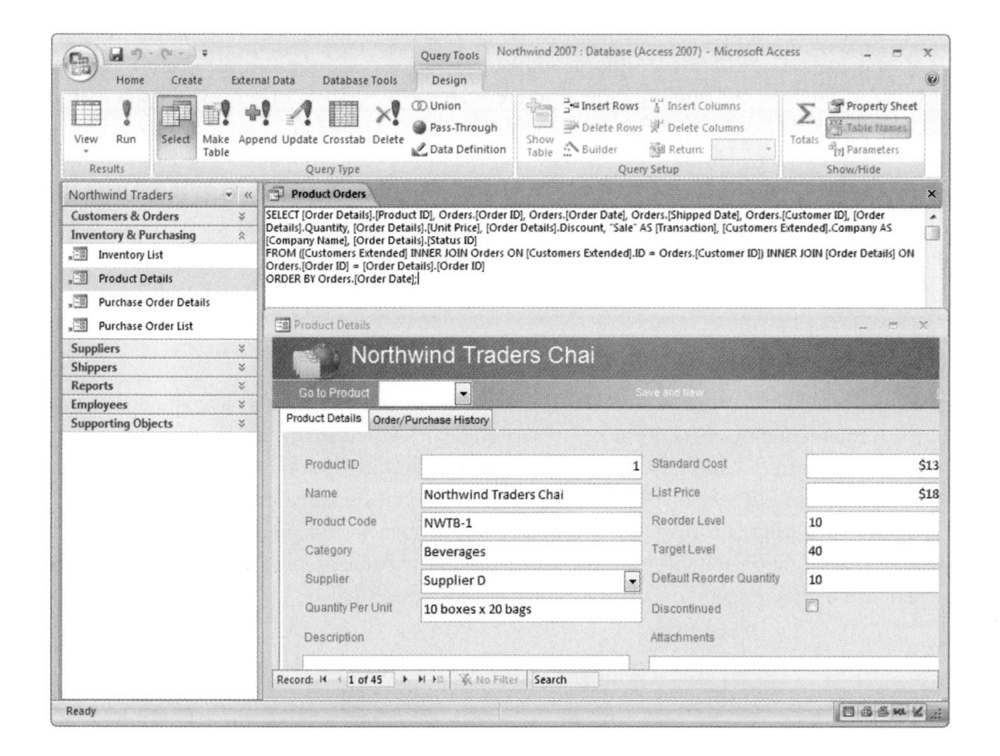

Figure 5.14

Structured Query Language

Structured Query Language (SQL) has become an integral part of most relational databases, as shown by this screen from Microsoft Access 2007.

After a database has been set up and loaded with data, it can produce desired reports, documents, and other outputs (see Figure 5.15). These outputs usually appear in screen displays or hard-copy printouts. The output-control features of a database program allow you to select the records and fields to appear in reports. You can also make calculations specifically for the report by manipulating database fields. Formatting controls and organization options (such as report headings) help you to customize reports and create flexible, convenient, and powerful information-handling tools.

Monthly Sales Report

Sunday, July 29, 2007 5:36:07 PM

June, 2006

Product	Sales
Northwind Traders Boysenberry Spread	$2,250.00
Northwind Traders Dried Apples	$1,590.00
Northwind Traders Fruit Cocktail	$1,560.00
Northwind Traders Chocolate	$1,020.00
Northwind Traders Dried Pears	$900.00
Northwind Traders Cajun Seasoning	$660.00
Northwind Traders Coffee	$230.00
Northwind Traders Clam Chowder	$96.50
June Sales Total	$8,306.50

Page 1 of 1

Figure 5.15

Database Output

A database application offers sophisticated formatting and organization options to produce the right information in the right format.

A DBMS can produce a wide variety of documents, reports, and other output that can help organizations achieve their goals. The most common reports select and organize data to present summary information about some aspect of company operations. For example, accounting reports often summarize financial data such as current and past-due accounts. Many companies base their routine operating decisions on regular status reports that show the progress of specific orders toward completion and delivery.

Databases can also provide support to help executives and other people make better decisions. A database by Intellifit, for example, can be used to help shoppers make better decisions and get clothes that fit when shopping online. The database contains true sizes of apparel from various clothing companies that do business on the Web. The process starts when a customer's body is scanned into a database at one of the company's locations, typically in a shopping mall. About 200,000 measurements are taken to construct a 3-D image of the person's body shape. The database then compares the actual body dimensions with sizes given by Web-based clothing stores to get an excellent fit.[14]

Database Administration

Database systems require a skilled DBA. A DBA is expected to have a clear understanding of the fundamental business of the organization, be proficient in the use of selected database management systems, and stay abreast of emerging technologies and new design approaches. The role of the DBA is to plan, design, create, operate, secure, monitor, and maintain databases. Typically, a DBA has a degree in computer science or management information systems and some on-the-job training with a particular database product or more extensive experience with a range of database products. See Figure 5.16.

Figure 5.16

Database Administrator

The role of the database administrator (DBA) is to plan, design, create, operate, secure, monitor, and maintain databases.

(Source: BananaStock / Alamy.)

The DBA works with users to decide the content of the database—to determine exactly what entities are of interest and what attributes are to be recorded about those entities. Thus, personnel outside of IS must have some idea of what the DBA does and why this function is important. The DBA can play a crucial role in the development of effective information systems to benefit the organization, employees, and managers.

The DBA also works with programmers as they build applications to ensure that their programs comply with database management system standards and conventions. After the database is built and operating, the DBA monitors operations logs for security violations. Database performance is also monitored to ensure that the system's response time meets users' needs and that it operates efficiently. If there is a problem, the DBA attempts to correct it before it becomes serious.

data administrator
A nontechnical position responsible for defining and implementing consistent principles for a variety of data issues.

Some organizations have also created a position called the *data administrator*, a nontechnical, but important role that ensures that data is managed as an important organizational resource. The **data administrator** is responsible for defining and implementing consistent principles for a variety of data issues, including setting data standards and data definitions that apply across all the databases in an organization. For example, the data administrator would ensure that a term such as "customer" is defined and treated consistently in all corporate databases. This person also works with business managers to identify who should have read or update access to certain databases and to selected attributes within those databases. This information is then communicated to the database administrator for implementation. The data administrator can be a high-level position reporting to top-level managers.

Popular Database Management Systems

Some popular DBMSs for single users include Microsoft Access and FileMaker Pro. The complete database management software market encompasses software used by professional programmers that runs on midrange, mainframe, and supercomputers. The entire market, including IBM, Oracle, and Microsoft, generates billions of dollars per year in revenue. Although Microsoft rules in the desktop PC software market, its share of database software on larger computers is small.

Like other software products, a number of open-source database systems are available, including PostgreSQL and MySQL. Open-source software was described in Chapter 4. In addition, many traditional database programs are now available on open-source operating systems. The popular DB2 relational database from IBM, for example, is available on the Linux operating system. The Sybase IQ database and other databases are also available on the Linux operating system.

A new form of database system is emerging that some refer to as *Database as a Service* (*DaaS*) and others as Database 2.0. DaaS is similar to software as a service (SaaS). Recall that a SaaS system is one in which the software is stored on a service provider's servers and accessed by the client company over a network. In DaaS, the database is stored on a service provider's servers and accessed by the client over a network, typically the Internet. In DaaS, database administration is provided by the service provider. SaaS and DaaS are both part of the larger cloud computing trend. Recall from Chapter 3 that cloud computing uses a giant cluster of computers that serves as a host to run applications that require high-performance computing. In cloud computing, all information systems and data are maintained and managed by service providers and delivered over the Internet. Businesses and individuals are freed from having to install, service, maintain, upgrade, and safeguard their systems.

More than a dozen companies are moving in the DaaS direction. They include Google, Microsoft, Intuit, Serran Tech, MyOwnDB, and Trackvia.[15] XM Radio, Google, JetBlue Airways, Bank of America, Southwest Airlines, and others use QuickBase from service provider Intuit to manage their databases out of house.[16] JetBlue, for example, uses a DaaS from Intuit to organize and manage IT projects.[17] Because the database and DBMS are available from any Internet connection, those involved in managing and implementing systems development projects can record their progress and check on others' progress from any location.

Special-Purpose Database Systems

In addition to the popular database management systems just discussed, some specialized database packages are used for specific purposes or in specific industries. For example, the Israeli Holocaust Database (*www.yadvashem.org*) is a special-purpose database available through the Internet and contains information on about three million people in 14 languages. A unique special-purpose DBMS for biologists called Morphbank (*www.morphbank.net*) allows researchers from around the world to continually update and expand a library of over 96,000 biological images to share with the scientific community and the public. The iTunes store music and video catalog is a special-purpose database system. When you search for your favorite artist, you are querying the database.

Selecting a Database Management System

The database administrator often selects the best database management system for an organization. The process begins by analyzing database needs and characteristics. The information needs of the organization affect the type of data that is collected and the type of database management system that is used. Important characteristics of databases include the following:

- **Database size.** The number of records or files in the database
- **Database cost.** The purchase or lease costs of the database
- **Concurrent users.** The number of people who need to use the database at the same time (the number of concurrent users)
- **Performance.** How fast the database is able to update records
- **Integration.** The ability to be integrated with other applications and databases
- **Vendor.** The reputation and financial stability of the database vendor

The Web-based Morphbank database allows scientists from around the world to upload and share biological and microscopic photographs and descriptions that support research in many areas.

(Source: *www.morphbank.net*)

For many organizations, database size doubles about every year or two. With the increasing use of digital media—images, video, and audio—data storage demands are growing exponentially. In fact, the volume of data being created has surpassed the world's available storage capacity.[18] The growing need for data storage has not escaped the notice of large technology companies such as Google and Microsoft, who are buying hundreds of acres of land and building huge data centers to support the world's data storage needs.[19] Meanwhile, many businesses and government agencies are working to consolidate data dispersed across the organization into smaller, more efficient centralized systems.

Using Databases with Other Software

Database management systems are often used with other software or the Internet. A DBMS can act as a front-end application or a back-end application. A *front-end application* is one that directly interacts with people or users. Marketing researchers often use a database as a front end to a statistical analysis program. The researchers enter the results of market questionnaires or surveys into a database. The data is then transferred to a statistical analysis program to determine the potential for a new product or the effectiveness of an advertising campaign. A *back-end application* interacts with other programs or applications; it only indirectly interacts with people or users. When people request information from a Web site, the Web site can interact with a database (the back end) that supplies the desired information. For example, you can connect to a university Web site to find out whether the university's library has a book you want to read. The Web site then interacts with a database that contains a catalog of library books and articles to determine whether the book you want is available.

DATABASE APPLICATIONS

Today's database applications manipulate the content of a database to produce useful information. Common manipulations are searching, filtering, synthesizing, and assimilating the data contained in a database, using a number of database applications. These applications allow users to link the company databases to the Internet, set up data warehouses and marts, use databases for strategic business intelligence, place data at different locations, use online processing and open connectivity standards for increased productivity, develop databases with the object-oriented approach, and search for and use unstructured data, such as graphics, audio, and video.

Linking the Company Database to the Internet

Linking databases to the Internet is one reason the Internet is so popular. A large percentage of corporate databases are accessed over the Internet through a standard Web browser. Being able to access bank account data, student transcripts, credit card bills, product catalogs, and a host of other data online is convenient for individual users, and increases effectiveness and efficiency for businesses and organizations. Amazon.com, Apple's iTunes store, eBay, and others have made billions of dollars by combining databases, the Internet, and smart business models.

As discussed in the Ethical and Societal Issues sidebar, Google is rolling out a DBMS that will provide patients and physicians with one storage location for all medical records, accessed through a Web browser.[20] Access to private medical information over the public Web has some privacy advocates concerned. However, the convenience that the system offers by dramatically reducing the number of paper forms to fill out and store, along with the reduction of clerical errors through streamlined data management procedures, has most in the field supporting the move to a centralized system. Google protects patient records with encryption and authentication technologies.

Developing a seamless integration of traditional databases with the Internet is often called a *semantic Web*. A semantic Web allows people to access and manipulate a number of traditional databases at the same time through the Internet. The World Wide Web Consortium has established standards for a semantic Web in hopes of some day evolving the Web into one big database that is easy to manage and traverse. Yahoo has recently announced its commitment to complying with the standards for a semantic Web.[21]

Although the semantic Web standards have not been embraced by all businesses, many software vendors—including IBM, Oracle, Microsoft, Macromedia, and Inline Internet Systems—are incorporating the Internet into their products. Such databases allow companies to create an Internet-accessible catalog, which is a database of items, descriptions, and prices. As evidenced by the Web, most companies are using these tools to take their business online.

In addition to the Internet, organizations are gaining access to databases through networks to find good prices and reliable service. Connecting databases to corporate Web sites and networks can lead to potential problems, however. A recent study found that nearly half a million database servers were vulnerable to attack over the Internet due to the lack of proper security measures.[22]

Data Warehouses, Data Marts, and Data Mining

The raw data necessary to make sound business decisions is stored in a variety of locations and formats. This data is initially captured, stored, and managed by transaction processing systems that are designed to support the day-to-day operations of the organization. For decades, organizations have collected operational, sales, and financial data with their online transaction processing (OLTP) systems. The data can be used to support decision making using data warehouses, data marts, and data mining.

Data Warehouses

data warehouse
A database that collects business information from many sources in the enterprise, covering all aspects of the company's processes, products, and customers.

A **data warehouse** is a database that holds business information from many sources in the enterprise, covering all aspects of the company's processes, products, and customers. The data warehouse provides business users with a multidimensional view of the data they need to analyze business conditions. Data warehouses allow managers to ***drill down*** to get more detail or ***roll up*** to take detailed data and generate aggregate or summary reports. A data warehouse is designed specifically to support management decision making, not to meet the needs of transaction processing systems. A data warehouse stores historical data that has been extracted from operational systems and external data sources (see Figure 5.17). This operational and external data is "cleaned up" to remove inconsistencies and integrated to create a new information database that is more suitable for business analysis.

Figure 5.17

Elements of a Data Warehouse

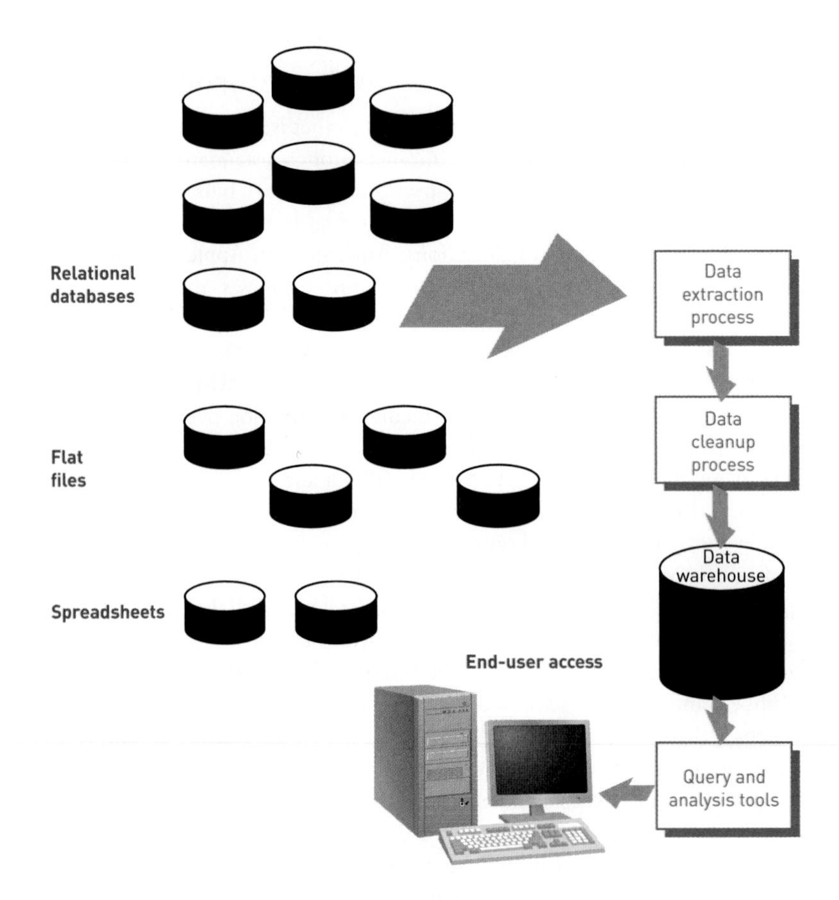

Data warehouses typically start out as very large databases, containing millions and even hundreds of millions of data records. As this data is collected from the various production systems, a historical database is built that business analysts can use. To keep it fresh and accurate, the data warehouse receives regular updates. Old data that is no longer needed is purged from the data warehouse. Updating the data warehouse must be fast, efficient, and automated, or the ultimate value of the data warehouse is sacrificed. It is common for a data warehouse to contain from three to ten years of current and historical data. Data-cleaning tools can merge data from many sources into one database, automate data collection and verification, delete unwanted data, and maintain data in a database management system. Data warehouses can also get data from unique sources. Oracle's Warehouse Management software, for example, can accept information from Radio Frequency Identification (RFID) technology, which is being used to tag products as they are shipped or moved from one location to another. Instead of recalling hundreds of thousands of cars because of a possible defective part, automotive companies could determine exactly which cars had the defective parts and only recall the 10,000 cars with the bad parts using RFID. The savings would be huge.

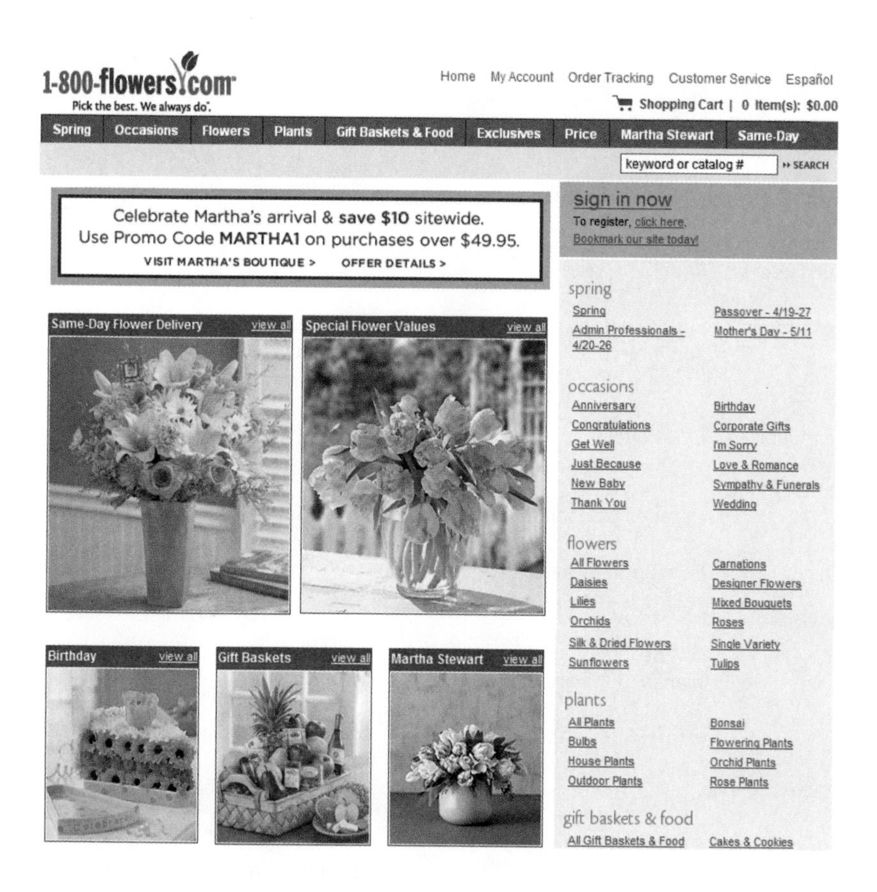

1-800-flowers.com uses a data warehouse to reference customer historical data. DBMS software accessed over the corporate intranet gives marketing professionals the information they need to determine customer interests based on past interactions.[23]

The primary advantage of data warehousing is the ability to relate data in innovative ways. However, a data warehouse can be extremely difficult to establish, with the typical cost exceeding $2 million. Table 5.7 compares online transaction processing (OLTP) and data warehousing.

Data Marts

A **data mart** is a subset of a data warehouse. Data marts bring the data warehouse concept—online analysis of sales, inventory, and other vital business data that has been gathered from transaction processing systems—to small and medium-sized businesses and to departments within larger companies. Rather than store all enterprise data in one monolithic database, data marts contain a subset of the data for a single aspect of a company's business—for example, finance, inventory, or personnel. In fact, a specific area in the data mart might contain more detailed data than the data warehouse would provide.

data mart
A subset of a data warehouse.

Data marts are most useful for smaller groups who want to access detailed data. A warehouse contains summary data that can be used by an entire company. Because data marts typically contain tens of gigabytes of data, as opposed to the hundreds of gigabytes in data warehouses, they can be deployed on less powerful hardware with smaller secondary storage devices, delivering significant savings to an organization. Although any database software can be used to set up a data mart, some vendors deliver specialized software designed and priced specifically for data marts. Already, companies such as Sybase, Software AG, Microsoft, and others have announced products and services that make it easier and cheaper to deploy these scaled-down data warehouses. The selling point: Data marts put targeted business information into the hands of more decision makers. For example, the Defense Acquisition University (DAU), which is responsible for continuing education and career management for employees of the U.S. Department of Defense, uses data marts to provide administrators, instructors, and staff with domain-specific information.[24] A data warehouse is used to combine information from more than 50 disconnected sources, and the DBMS then organizes the information into area-specific data marts, which produce reports accessible through an online dashboard application. The system is estimated to save DAU personnel three to five years of labor.

Characteristic	OLTP Database	Data Warehousing
Purpose	Support transaction processing	Support decision making
Source of data	Business transactions	Multiple files, databases—data internal and external to the firm
Data access allowed users	Read and write	Read only
Primary data access mode	Simple database update and query	Simple and complex database queries with increasing use of data mining to recognize patterns in the data
Primary database model employed	Relational	Relational
Level of detail	Detailed transactions	Often summarized data
Availability of historical data	Very limited—typically a few weeks or months	Multiple years
Update process	Online, ongoing process as transactions are captured	Periodic process, once per week or once per month
Ease of process	Routine and easy	Complex, must combine data from many sources; data must go through a data cleanup process
Data integrity issues	Each transaction must be closely edited	Major effort to "clean" and integrate data from multiple sources

Table 5.7

Comparison of OLTP and Data Warehousing

Data Mining

Data mining is an information-analysis tool that involves the automated discovery of patterns and relationships in a data warehouse. Like gold mining, data mining sifts through mountains of data to find a few nuggets of valuable information. The University of Maryland has developed a data-mining technique to "forecast terrorist behavior based on past actions."[25] The system uses a real-time data extraction tool called T-REX to scour an average of 128,000 articles a day and forecast future activities of over 110 terrorist groups.

data mining
An information-analysis tool that involves the automated discovery of patterns and relationships in a data warehouse.

Data mining's objective is to extract patterns, trends, and rules from data warehouses to evaluate (i.e., predict or score) proposed business strategies, which will improve competitiveness, increase profits, and transform business processes. It is used extensively in marketing to improve customer retention; cross-selling opportunities; campaign management; market, channel, and pricing analysis; and customer segmentation analysis (especially one-to-one marketing). In short, data-mining tools help users find answers to questions they haven't thought to ask.

E-commerce presents another major opportunity for effective use of data mining. Attracting customers to Web sites is tough; keeping them can be next to impossible. For example, when retail Web sites launch deep-discount sales, they cannot easily determine how many first-time customers are likely to come back and buy again. Nor do they have a way of understanding which customers acquired during the sale are price sensitive and more likely to jump on future sales. As a result, companies are gathering data on user traffic through their Web sites and storing the data in databases. This data is then analyzed using data-mining techniques to personalize the Web site and develop sales promotions targeted at specific customers.

predictive analysis
A form of data mining that combines historical data with assumptions about future conditions to predict outcomes of events, such as future product sales or the probability that a customer will default on a loan.

Predictive analysis is a form of data mining that combines historical data with assumptions about future conditions to predict outcomes of events, such as future product sales or the probability that a customer will default on a loan. Retailers use predictive analysis to upgrade occasional customers into frequent purchasers by predicting what products they will buy if offered an appropriate incentive. Genalytics, Magnify, NCR Teradata, SAS Institute,

MySpace.com mines the data of all of its members to determine which ads should be displayed for each member to attract the maximum attention and hits.[26]

Sightward, SPSS, and Quadstone have developed predictive analysis tools. Predictive analysis software can be used to analyze a company's customer list and a year's worth of sales data to find new market segments that could be profitable.

The City of Richmond Police Department uses predictive analysis to predict "when and where crimes were likely to occur, so officers can be on hand to prevent their occurrence."[27]

(Source: Courtesy of Mitch Kezar.)

Traditional DBMS vendors are well aware of the great potential of data mining. Thus, companies such as Oracle, Sybase, Tandem, and Red Brick Systems are all incorporating data-mining functionality into their products. Table 5.8 summarizes a few of the most frequent applications for data mining.

Application	Description
Branding and positioning of products and services	Enable the strategist to visualize the different positions of competitors in a given market using performance (or other) data on dozens of key features of the product and then to condense all that data into a perceptual map of only two or three dimensions.
Customer churn	Predict current customers who are likely to switch to a competitor.
Direct marketing	Identify prospects most likely to respond to a direct marketing campaign (such as a direct mailing).
Fraud detection	Highlight transactions most likely to be deceptive or illegal.
Market basket analysis	Identify products and services that are most commonly purchased at the same time (e.g., nail polish and lipstick).
Market segmentation	Group customers based on who they are or on what they prefer.
Trend analysis	Analyze how key variables (e.g., sales, spending, promotions) vary over time.

Table 5.8

Common Data-Mining Applications

Business Intelligence

The use of databases for business-intelligence purposes is closely linked to the concept of data mining. **Business intelligence (BI)** involves gathering enough of the right information in a timely manner and usable form and analyzing it so that it can have a positive effect on business strategy, tactics, or operations. IMS Health, for example, provides a BI system designed to assist businesses in the pharmaceutical industry with custom marketing to physicians, pharmacists, nurses, consumers, government agencies, and nonprofit healthcare organizations.[28] Business intelligence turns data into useful information that is then distributed throughout an enterprise. It provides insight into the causes of problems, and when implemented can improve business operations and sometimes even save lives. For example, BI software at the Sahlgrenska University Hospital in Gothenburg, Sweden, has helped neurosurgeons save lives by identifying complications in patient conditions after cranial surgery.[29] The Information Systems at Work box shows how business intelligence is used in the utilities industry.

business intelligence
The process of gathering enough of the right information in a timely manner and usable form and analyzing it to have a positive impact on business strategy, tactics, or operations.

Competitive intelligence is one aspect of business intelligence and is limited to information about competitors and the ways that knowledge affects strategy, tactics, and operations. Competitive intelligence is a critical part of a company's ability to see and respond quickly and appropriately to the changing marketplace. Competitive intelligence is not espionage—the use of illegal means to gather information. In fact, almost all the information a competitive-intelligence professional needs can be collected by examining published information sources, conducting interviews, and using other legal, ethical methods. Using a variety of analytical tools, a skilled competitive-intelligence professional can by deduction fill the gaps in information already gathered.

competitive intelligence
One aspect of business intelligence limited to information about competitors and the ways that knowledge affects strategy, tactics, and operations.

The term **counterintelligence** describes the steps an organization takes to protect information sought by "hostile" intelligence gatherers. One of the most effective counterintelligence measures is to define "trade secret" information relevant to the company and control its dissemination.

counterintelligence
The steps an organization takes to protect information sought by "hostile" intelligence gatherers.

INFORMATION SYSTEMS @ WORK

Yangtze Power Harnesses the Power

Perhaps you've heard of the Yangtze River in China, and the enormous Three Gorges Dam being erected to harness the river's force for hydroelectric power. Due to be completed in 2011, the Three Gorges Dam will generate 22,500 megawatts of electricity, more than any other hydroelectric facility in the world. The company that will operate the dam is Yangtze Power, China's largest publicly listed utility company.

For years, Yangtze Power has managed the Gezhouba Power Station and six commissioned generating units. It has maintained business data in five databases, supporting its five divisions: Power Generation Management, Finance, Human Resources, Contract Management, and Safety and Control Management. Keeping data in siloed systems—separate, unconnected systems—limited information transfer through the enterprise. If a manager from Human Resources wanted to evaluate data from Contract Management, he would have to e-mail someone in that department to have a report generated and transferred. As Yangtze Power looked ahead to growth and the addition of the world's largest hydroelectric power generator, the company knew that its information would need to flow more freely through the enterprise in order for it to make the best business decisions in a timely fashion.

After evaluating products from Business Objects, Cognos, Informatica, MicroStrategy, and Oracle, Yangtze Power decided to go with Oracle to design one centralized database for all of its information because it was the only company that could provide one integrated system.

In March 2007, Yangtze Power's technology team worked with Oracle to develop a needs analysis and begin data preparation. Requirements were defined to cover six major areas of the business, including 65 performance indices and 370 reports. Through extensive preparation and testing, the system was up and running by November 2007.

Oracle's business intelligence tools allow senior managers to analyze performance on a daily basis, highlight areas for improvement, and monitor the results of business strategies. Each morning, reports on the previous day's critical activities are waiting on managers' desks. The new database stores three years of data, so that managers can draw on historical data when analyzing business performance. Communication between departments has improved, since everyone accesses the same data from a central system, and reports can easily be generated tailored to meet any business need.

Oracle's BI tools are used to create customized reports and charts including pie charts, broken curve diagrams, histograms, and radar maps. Being able to visualize data and trends in data enables a deeper analysis of the organization's business performance.

Yangtze Power has gained control over its flow of information through the enterprise. Now it is working to gain similar control over the raging waters of the Yangtze River.

Discussion Questions

1. What was wrong with Yangtze Power's previous database system, and how was it affecting the business?
2. What solution did Yangtze opt for, and how did it improve business?

Critical Thinking Questions

1. How does a centralized database improve communications within an organization?
2. In what situations might one centralized database not be practical for an enterprise?

Sources: Oracle Success Stories, "Yangtze Power Improves Business Intelligence with Integrated Database and Analysis Tools," 2008, *www.oracle.com/customers/snapshots/yangtze-power-case-study.pdf*, Yangtze River Web site, *www.yangtzeriver.org*, accessed April 2, 2008, Oracle Database and BI Tools, *www.oracle.com/database*, accessed April 2, 2008.

Distributed Databases

distributed database
A database in which the data can be spread across several smaller databases connected via telecommunications devices.

Distributed processing involves placing processing units at different locations and linking them via telecommunications equipment. A **distributed database**—a database in which the data can be spread across several smaller databases connected through telecommunications devices—works on much the same principle. A user in the Milwaukee branch of a clothing manufacturer, for example, might make a request for data that is physically located at corporate headquarters in Milan, Italy. The user does not have to know where the data is physically stored (see Figure 5.18).

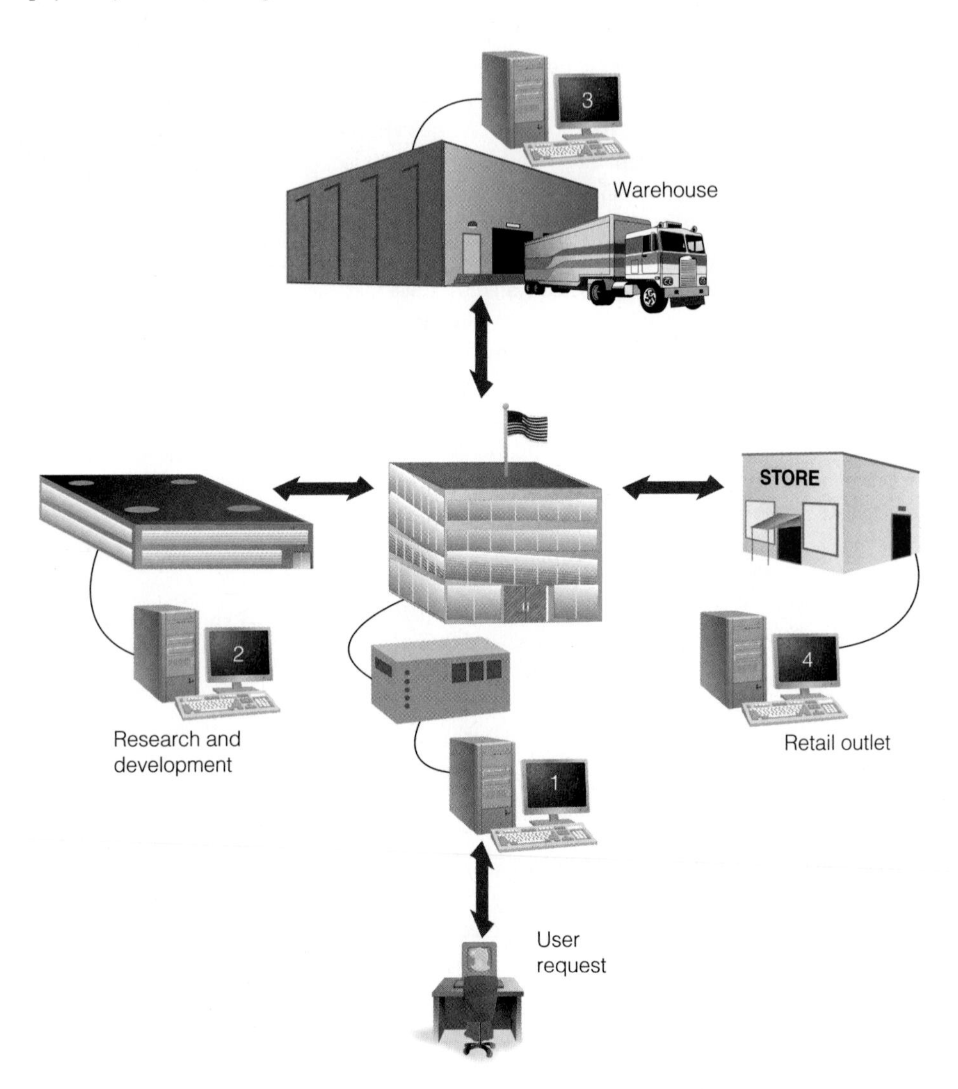

Figure 5.18

The Use of a Distributed Database

For a clothing manufacturer, computers might be located at corporate headquarters, in the research and development center, in the warehouse, and in a company-owned retail store. Telecommunications systems link the computers so that users at all locations can access the same distributed database no matter where the data is actually stored.

Distributed databases give corporations and other organizations more flexibility in how databases are organized and used. Local offices can create, manage, and use their own databases, and people at other offices can access and share the data in the local databases. Giving local sites more direct access to frequently used data can improve organizational effectiveness and efficiency significantly. The New York City Police Department, for example, has thousands of officers searching for information located on servers in offices around the city.

Despite its advantages, distributed processing creates additional challenges in integrating different databases (information integration), maintaining data security, accuracy, timeliness, and conformance to standards. Distributed databases allow more users direct access at different sites; thus, controlling who accesses and changes data is sometimes difficult. Also, because distributed databases rely on telecommunications lines to transport data, access to data can be slower.

To reduce telecommunications costs, some organizations build a replicated database. A **replicated database** holds a duplicate set of frequently used data. The company sends a copy of important data to each distributed processing location when needed or at predetermined times. Each site sends the changed data back to update the main database on an update cycle that meets the needs of the organization. This process, often called *data synchronization*, is used to make sure that replicated databases are accurate, up to date, and consistent with each other. A railroad, for example, can use a replicated database to increase punctuality, safety, and reliability. The primary database can hold data on fares, routings, and other essential information. The data can be continually replicated and downloaded on a read-only basis from the master database to hundreds of remote servers across the country. The remote locations can send back the latest figures on ticket sales and reservations to the main database.

replicated database
A database that holds a duplicate set of frequently used data.

Online Analytical Processing (OLAP)

For nearly two decades, multidimensional databases and their analytical information display systems have provided flashy sales presentations and trade show demonstrations. All you have to do is ask where a certain product is selling well, for example, and a colorful table showing sales performance by region, product type, and time frame appears on the screen. Called **online analytical processing (OLAP)**, these programs are now being used to store and deliver data warehouse information efficiently. The leading OLAP software vendors include Microsoft, Cognos, SAP, Business Objects, MicroStrategy, Applix, Infor, and Oracle. Lufthansa Cargo depends on OLAP to deliver up-to-the-minute company statistics that help the company compete in the growing global air-freight market.[30] The market is growing by six percent annually, and competitors are emerging all around the world to get a piece of the action. Lufthansa Cargo uses OLAP to analyze its data to provide the fastest service to its customers and the lowest rates.

online analytical processing (OLAP)
Software that allows users to explore data from a number of perspectives.

The value of data ultimately lies in the decisions it enables. Powerful information-analysis tools in areas such as OLAP and data mining, when incorporated into a data warehousing architecture, bring market conditions into sharper focus and help organizations deliver greater competitive value. OLAP provides top-down, query-driven data analysis; data mining provides bottom-up, discovery-driven analysis. OLAP requires repetitive testing of user-originated theories; data mining requires no assumptions and instead identifies facts and conclusions based on patterns discovered. OLAP, or multidimensional analysis, requires a great deal of human ingenuity and interaction with the database to find information in the database. A user of a data-mining tool does not need to figure out what questions to ask; instead, the approach is, "Here's the data, tell me what interesting patterns emerge." For example, a data-mining tool in a credit card company's customer database can construct a profile of fraudulent activity from historical information. Then, this profile can be applied to all incoming transaction data to identify and stop fraudulent behavior, which might otherwise go undetected. Table 5.9 compares the OLAP and data-mining approaches to data analysis.

Table 5.9
Comparison of OLAP and Data Mining

Characteristic	OLAP	Data Mining
Purpose	Supports data analysis and decision making	Supports data analysis and decision making
Type of analysis supported	Top-down, query-driven data analysis	Bottom-up, discovery-driven data analysis
Skills required of user	Must be very knowledgeable of the data and its business context	Must trust in data-mining tools to uncover valid and worthwhile hypotheses

Object-Relational Database Management Systems

object-oriented database
A database that stores both data and its processing instructions.

An **object-oriented database** uses the same overall approach of objected-oriented programming that was discussed in Chapter 4. With this approach, both the data and the processing instructions are stored in the database. For example, an object-oriented database could store monthly expenses and the instructions needed to compute a monthly budget from those expenses. A traditional DBMS might only store the monthly expenses. The King County Metro Transit system in the state of Washington uses an object-oriented database in a system supplied by German vendor Init to manage the routing and accounting of its bus line.[31] Object-oriented databases are useful when a database contains complex data that needs to be processed quickly and efficiently.

In an object-oriented database, a *method* is a procedure or action. A sales tax method, for example, could be the procedure to compute the appropriate sales tax for an order or sale—for example, multiplying the total amount of an order by five percent, if that is the local sales tax. A *message* is a request to execute or run a method. For example, a sales clerk could issue a message to the object-oriented database to compute sales tax for a new order. Many object-oriented databases have their own query language, called *object query language (OQL)*, which is similar to SQL, discussed previously.

object-oriented database management system (OODBMS)
A group of programs that manipulate an object-oriented database and provide a user interface and connections to other application programs.

An object-oriented database uses an **object-oriented database management system (OODBMS)** to provide a user interface and connections to other programs. Computer vendors who sell or lease OODBMSs include Versant and Objectivity. Many organizations are selecting object-oriented databases for their processing power. Versant's OODBMS, for example, is being used by companies in the telecommunications, defense, online gaming, and healthcare industries, and by government agencies. The *Object Data Standard* is a design standard created by the *Object Database Management Group (www.odmg.org)* for developing object-oriented database systems.

object-relational database management system (ORDBMS)
A DBMS capable of manipulating audio, video, and graphical data.

An **object-relational database management system (ORDBMS)** provides a complete set of relational database capabilities plus the ability for third parties to add new data types and operations to the database. These new data types can be audio, images, unstructured text, spatial, or time series data that require new indexing, optimization, and retrieval features. Each of the vendors offering ORDBMS facilities provides a set of application programming interfaces to allow users to attach external data definitions and methods associated with those definitions to the database system. They are essentially offering a standard socket into which users can plug special instructions. DataBlades, Cartridges, and Extenders are the names applied by Oracle and IBM to describe the plug-ins to their respective products. Other plug-ins serve as interfaces to Web servers.

Visual, Audio, and Other Database Systems

In addition to raw data, organizations are increasingly finding a need to store large amounts of visual and audio signals in an organized fashion. Credit card companies, for example, enter pictures of charge slips into an image database using a scanner. The images can be stored in the database and later sorted by customer name, printed, and sent to customers along with their monthly statements. Image databases are also used by physicians to store x-rays and transmit them to clinics away from the main hospital. Financial services, insurance companies, and government branches are using image databases to store vital records and replace paper documents. Drug companies often need to analyze many visual images from laboratories. Chesapeake Energy maintains a database filled with scanned images of terrain and drilling locations.[32] Visual databases can be stored in some object-relational databases or special-purpose database systems. Many relational databases can also store graphic content.

Combining and analyzing data from different databases is an increasingly important challenge. Global businesses, for example, sometimes need to analyze sales and accounting data stored around the world in different database systems. Companies such as IBM are developing *virtual database systems* to allow different databases to work together as a unified database system. Banc of America Securities Prime Brokerage, for example, turned to database virtualization to address management and performance problems. Since its implementation, the virtual database system has reduced storage administration by 95 percent and decreased the need for more storage capacity by 50 percent.[33]

In addition to visual, audio, and virtual databases, other special-purpose database systems meet particular business needs. *Spatial data technology* involves using a database to store and access data according to the locations it describes and to permit spatial queries and analysis. MapInfo software from Pitney Bowes allows businesses such as Home Depot, Sonic Restaurants, CVS Corporation, and Chico's to choose the optimal location for new stores and restaurants based on geospatial demographics.[34] The software provides information about local competition, populations, and traffic patterns to predict how a business will fare in a particular location. Builders and insurance companies use spatial data to make decisions related to natural hazards. Spatial data can even be used to improve financial risk management with information stored by investment type, currency type, interest rates, and time.

Spatial data technology is used by NASA to store data from satellites and Earth stations. Location-specific information can be accessed and compared.

(Source: Courtesy of NASA.)

SUMMARY

Principle

Data management and modeling are key aspects of organizing data and information.

Data is one of the most valuable resources that a firm possesses. It is organized into a hierarchy that builds from the smallest element to the largest. The smallest element is the bit, a binary digit. A byte (a character such as a letter or numeric digit) is made up of eight bits. A group of characters, such as a name or number, is called a field (an object). A collection of related fields is a record; a collection of related records is called a file. The database, at the top of the hierarchy, is an integrated collection of records and files.

An entity is a generalized class of objects for which data is collected, stored, and maintained. An attribute is a characteristic of an entity. Specific values of attributes—called data items—can be found in the fields of the record describing an entity. A data key is a field within a record that is used to identify the record. A primary key uniquely identifies a record, while a secondary key is a field in a record that does not uniquely identify the record.

Traditional file-oriented applications are often characterized by program-data dependence, meaning that they have data organized in a manner that cannot be read by other programs. To address problems of traditional file-based data management, the database approach was developed. Benefits of this approach include reduced data redundancy, improved data consistency and integrity, easier modification and updating, data and program independence, standardization of data access, and more-efficient program development.

One of the tools that database designers use to show the relationships among data is a data model. A data model is a map or diagram of entities and their relationships. Enterprise data modeling involves analyzing the data and information needs of an entire organization. Entity-relationship (ER) diagrams can be employed to show the relationships between entities in the organization.

The relational model places data in two-dimensional tables. Tables can be linked by common data elements, which are used to access data when the database is queried. Each row represents a record. Columns of the tables are called attributes, and allowable values for these attributes are called the domain. Basic data manipulations include selecting, projecting, and joining. The relational model is easier to control, more flexible, and more intuitive than the other models because it organizes data in tables.

Principle

A well-designed and well-managed database is an extremely valuable tool in supporting decision making.

A DBMS is a group of programs used as an interface between a database and its users and other application programs. When an application program requests data from the database, it follows a logical access path. The actual retrieval of the data follows a physical access path. Records can be considered in the same way: A logical record is what the record contains; a physical record is where the record is stored on storage devices. Schemas are used to describe the entire database, its record types, and their relationships to the DBMS.

A DBMS provides four basic functions: providing user views, creating and modifying the database, storing and retrieving data, and manipulating data and generating reports. Schemas are entered into the computer via a data definition language, which describes the data and relationships in a specific database. Another tool used in database management is the data dictionary, which contains detailed descriptions of all data in the database.

After a DBMS has been installed, the database can be accessed, modified, and queried via a data manipulation language. A more specialized data manipulation language is the query language, the most common being Structured Query Language (SQL). SQL is used in several popular database packages today and can be installed on PCs and mainframes.

Popular single-user DBMSs include Corel Paradox and Microsoft Access. IBM, Oracle, and Microsoft are the leading DBMS vendors. Database as a Service (DaaS), or Database 2.0, is a new form of database service in which clients lease use of a database on a service provider's site.

Selecting a DBMS begins by analyzing the information needs of the organization. Important characteristics of databases include the size of the database, the number of concurrent users, its performance, the ability of the DBMS to be integrated with other systems, the features of the DBMS, the vendor considerations, and the cost of the database management system.

Principle

The number and types of database applications will continue to evolve and yield real business benefits.

Traditional online transaction processing (OLTP) systems put data into databases very quickly, reliably, and efficiently, but they do not support the types of data analysis that today's businesses and organizations require. To address this need, organizations are building data warehouses, which are relational database management systems specifically designed to support management decision making. Data marts are subdivisions of data warehouses, which are commonly devoted to specific purposes or functional business areas.

Data mining, which is the automated discovery of patterns and relationships in a data warehouse, is emerging as a practical approach to generating hypotheses about the patterns and anomalies in the data that can be used to predict future behavior.

Predictive analysis is a form of data mining that combines historical data with assumptions about future conditions to forecast outcomes of events such as future product sales or the probability that a customer will default on a loan.

Business intelligence is the process of getting enough of the right information in a timely manner and usable form and analyzing it so that it can have a positive effect on business strategy, tactics, or operations. Competitive intelligence is one aspect of business intelligence limited to information about competitors and the ways that information affects strategy, tactics, and operations. Competitive intelligence is not espionage—the use of illegal means to gather information. Counterintelligence describes the steps an organization takes to protect information sought by "hostile" intelligence gatherers.

With the increased use of telecommunications and networks, distributed databases, which allow multiple users and different sites access to data that may be stored in different physical locations, are gaining in popularity. To reduce telecommunications costs, some organizations build replicated databases, which hold a duplicate set of frequently used data.

Multidimensional databases and online analytical processing (OLAP) programs are being used to store data and allow users to explore the data from a number of different perspectives.

An object-oriented database uses the same overall approach of objected-oriented programming, first discussed in Chapter 4. With this approach, both the data and the processing instructions are stored in the database. An object-relational database management system (ORDBMS) provides a complete set of relational database capabilities, plus the ability for third parties to add new data types and operations to the database. These new data types can be audio, video, and graphical data that require new indexing, optimization, and retrieval features.

In addition to raw data, organizations are increasingly finding a need to store large amounts of visual and audio signals in an organized fashion. A number of special-purpose database systems are also being used.

CHAPTER 5: SELF-ASSESSMENT TEST

Data management and modeling are key aspects of organizing data and information.

1. A group of programs that manipulate the database and provide an interface between the database and the user of the database and other application programs is called a(n) ________________.
   a. GUI
   b. operating system
   c. DBMS
   d. productivity software
2. A(n) ________________ is a skilled and trained IS professional who directs all activities related to an organization's database.
3. Data redundancy is a desirable quality in a database. True or False?
4. A(n) ________________ is a field or set of fields that uniquely identifies a database record.
   a. attribute
   b. data item
   c. key
   d. primary key
5. A(n) ________________ uses basic graphical symbols to show the organization of and relationships between data.
6. What database model places data in two-dimensional tables?
   a. relational
   b. network
   c. normalized
   d. hierarchical

A well-designed and well-managed database is an extremely valuable tool in supporting decision making.

7. ________________ involves combining two or more database tables.
8. After data has been placed into a relational database, users can make inquiries and analyze data. Basic data manipulations include selecting, projecting, and optimizing. True or False?
9. Because the DBMS is responsible for providing access to a database, one of the first steps in installing and using a database involves telling the DBMS the logical and physical structure of the data and relationships among the data in the database. This description of an entire database is called a(n) ________________.

10. The commands used to access and report information from the database are part of the ______________.
    a. data definition language
    b. data manipulation language
    c. data normalization language
    d. schema
11. Access is a popular DBMS for ______________.
    a. personal computers
    b. graphics workstations
    c. mainframe computers
    d. supercomputers
12. A new trend in database management, known as Database as a Service, places the responsibility of storing and managing a database on a service provider. True or False?

The number and types of database applications will continue to evolve and yield real business benefits.

13. A(n) ______________ holds business information from many sources in the enterprise, covering all aspects of the company's processes, products, and customers.
14. An information-analysis tool that involves the automated discovery of patterns and relationships in a data warehouse is called ______________.
    a. a data mart
    b. data mining
    c. predictive analysis
    d. business intelligence
15. ______________ allows users to predict the future based on database information from the past and present.

CHAPTER 5: SELF-ASSESSMENT TEST ANSWERS

(1) c (2) database administrator (3) False (4) d (5) entity-relationship diagram (6) a (7) Joining (8) False (9) schema (10) b (11) a (12) True (13) data warehouse (14) b (15) Predictive analysis

REVIEW QUESTIONS

1. What is an attribute? How is it related to an entity?
2. Define the term *database*. How is it different from a database management system?
3. What is the hierarchy of data in a database?
4. What is a flat file?
5. What is the purpose of a primary key? How can it be useful in controlling data redundancy?
6. What is the purpose of data cleanup?
7. What are the advantages of the database approach?
8. What is data modeling? What is its purpose? Briefly describe three commonly used data models.
9. What is a database schema, and what is its purpose?
10. How can a data dictionary be useful to database administrators and DBMS software engineers?
11. Identify important characteristics in selecting a database management system.
12. What is the difference between a data definition language (DDL) and a data manipulation language (DML)?
13. What is the difference between projecting and joining?
14. What is a distributed database system?
15. What is a data warehouse, and how is it different from a traditional database used to support OLTP?
16. What is meant by the "front end" and the "back end" of a DBMS?
17. What is data mining? What is OLAP? How are they different?
18. What is an ORDBMS? What kind of data can it handle?
19. What is business intelligence? How is it used?
20. In what circumstances might a database administrator consider using an object-oriented database?

DISCUSSION QUESTIONS

1. You have been selected to represent the student body on a project to develop a new student database for your school. What actions might you take to fulfill this responsibility to ensure that the project meets the needs of students and is successful?
2. Your company wants to increase revenues from its existing customers. How can data mining be used to accomplish this objective?
3. You are going to design a database for your cooking club to track its recipes. Identify the database characteristics

most important to you in choosing a DBMS. Which of the database management systems described in this chapter would you choose? Why? Is it important for you to know what sort of computer the database will run on? Why or why not?

4. Make a list of the databases in which data about you exists. How is the data in each database captured? Who updates each database and how often? Is it possible for you to request a printout of the contents of your data record from each database? What data privacy concerns do you have?
5. If you were the database administrator for the iTunes store, how might you use predictive analysis to determine which artists and movies will sell most next year?
6. You are the vice president of information technology for a large, multinational consumer packaged goods company (such as Procter & Gamble or Unilever). You must make a presentation to persuade the board of directors to invest $5 million to establish a competitive-intelligence organization—including people, data-gathering services, and software tools. What key points do you need to make in favor of this investment? What arguments can you anticipate that others might make?
7. Briefly describe how visual and audio databases can be used by companies today.
8. Identity theft, where people steal your personal information, continues to be a threat. Assume that you are the database administrator for a corporation with a large database. What steps would you implement to help prevent people from stealing personal information from the corporate database?
9. What roles do databases play in your favorite online activities and Web sites?

PROBLEM-SOLVING EXERCISES

1. Develop a simple data model for the music you have on your MP3 player or in your CD collection, where each row is a song. For each row, what attributes should you capture? What will be the unique key for the records in your database? Describe how you might use the database.
2. A video movie rental store is using a relational database to store information on movie rentals to answer customer questions. Each entry in the database contains the following items: Movie ID No. (primary key), Movie Title, Year Made, Movie Type, MPAA Rating, Number of Copies on Hand, and Quantity Owned. Movie types are comedy, family, drama, horror, science fiction, and western. MPAA ratings are G, PG, PG-13, R, NC-17, and NR (not rated). Use a single-user database management system to build a data-entry screen to enter this data. Build a small database with at least ten entries.
3. To improve service to their customers, the salespeople at the video rental store have proposed a list of changes being considered for the database in the previous exercise. From this list, choose two database modifications and modify the data-entry screen to capture and store this new information.

   Proposed changes:

   a. Add the date that the movie was first available to help locate the newest releases.
   b. Add the director's name.
   c. Add the names of three primary actors in the movie.
   d. Add a rating of one, two, three, or four stars.
   e. Add the number of Academy Award nominations.

4. Your school maintains information about students in several interconnected database files. The student_contact file contains student contact information. The student_grades file contains student grade records, and the student_financial file contains financial records including tuition and student loans. Draw a diagram of the fields these three files might contain, which field is a primary key in each file, and which fields serve to relate one file to another. Use Figure 5.7 as a guide.

TEAM ACTIVITIES

1. In a group of three or four classmates, communicate with the person at your school that supervises information systems. Find out how many databases are used by your school and for what purpose. Also find out what policies and procedures are in place to protect the data stored from identity thieves and other threats.
2. As a team of three or four classmates, interview business managers from three different businesses that use databases to help them in their work. What data entities and data attributes are contained in each database? How do they access the database to perform analysis? Have they received training in any query or reporting tools? What do they like about their database and what could be improved? Do any of them use data-mining or OLAP techniques? Weighing the information obtained, select one of these databases as

being most strategic for the firm and briefly present your selection and the rationale for the selection to the class.

3. Imagine that you and your classmates are a research team developing an improved process for evaluating auto loan applicants. The goal of the research is to predict which applicants will become delinquent or forfeit their loan. Those who score well on the application will be accepted; those who score exceptionally well will be considered for lower-rate loans. Prepare a brief report for your instructor addressing these questions:

a. What data do you need for each loan applicant?
b. What data might you need that is not typically requested on a loan application form?
c. Where might you get this data?
d. Take a first cut at designing a database for this application. Using the chapter material on designing a database, show the logical structure of the relational tables for this proposed database. In your design, include the data attributes you believe are necessary for this database, and show the primary keys in your tables. Keep the size of the fields and tables as small as possible to minimize required disk drive storage space. Fill in the database tables with the sample data for demonstration purposes (ten records). After your design is complete, implement it using a relational DBMS.

WEB EXERCISES

1. Use a Web search engine to find information on specific products for one of the following topics: business intelligence, object-oriented databases, or database as a service. Write a brief report describing what you found, including a description of the database products and the companies that developed them.
2. List your five favorite Web sites. Consider the services that they provide. For each site, suggest how one or more databases might be used on the back end to supply information to visitors.

CAREER EXERCISES

1. What type of data is stored by businesses in a professional field that interests you? How many databases might be used to store that data? How would the data be organized within each database?
2. How could you use business intelligence (BI) to do a better job at work? Give some specific examples of how BI can give you a competitive advantage.

CASE STUDIES

Case One

The Getty Vocabularies

J. Paul Getty was an American industrialist who made his fortune in the oil business. He made his first million at age 25 in 1916, and later became the world's first billionaire. Getty viewed art as a 'civilizing influence in society, and strongly believed in making art available to the public for its education and enjoyment.' To that end, he created an art museum in Los Angeles, California, and established the J. Paul Getty Trust, commonly referred to as the Getty.

The Getty includes four branches: the Getty Museum, a research institute, a conservation institute, and a foundation. In the 1980s, the Getty discovered a need within the art research community. Researchers lacked a common vocabulary with which to discuss art and artists' work. Establishing a scientific vocabulary with which to describe artwork, style, and technique would allow the study and appreciation of artwork to flourish. To meet this need, the Getty created and published the *Art and Architecture Thesaurus* (AAT) in 1990. The three-volume tome, which includes a thesaurus of geographic names and the Union List of Artist Names, has become a priceless resource for art historical research. It provides tools, standards, and best practices for documenting works of art, just as the Library of Congress provides a standard cataloging tool for libraries.

However, the massive AAT is difficult to search and is expensive to edit and update. Recognizing that a digital version of the resource would provide many benefits, the Getty recently began porting the AAT and associated volumes into a database that can be electronically searched and edited over the Web. To do so, the Getty had to first select a database technology in which to house the information, and a DBMS for use in searching and editing the contents.

One challenge of building an online AAT was that the various components of the resource were stored using different proprietary technologies. The first task was to collect them into one common technology, which required a custom-designed system. Technicians within the Getty opted to use Oracle databases and a product called PowerBuilder from Sybase, Inc., for the user interface. Custom coding was done in Perl and SQR programming languages to merge the components into a cohesive system. The result is a system called the Vocabulary Coordination System (VCS). The VCS is used to collect, analyze, edit, merge, and distribute the terminology managed by the Getty vocabularies. A special Web-based interface was developed that made searching the volumes easy enough for anyone to manage. You can try it yourself at *www.getty.edu/research/conducting_research/vocabularies*.

The resulting system was so impressive that it won the Getty the Computerworld Honors Award in Media, Arts & Entertainment for innovative use of technology. The system makes it easy for scholars to update information in the vocabularies, and for everyone from school children to professional art historians to research and learn about art and art history. The Getty online vocabularies are an ideal realization of J. Paul Getty's original philosophy of promoting human civility through cultural awareness, creativity, and aesthetic enjoyment.

Discussion Questions

1. What purpose do the Getty vocabularies serve, and how are they supported through database technology?
2. How does using the Web as a front end to this database further support J. Paul Getty's vision?

Critical Thinking Questions

1. What concerns do you think the designers of the database had when making this valuable resource available online to the general public?
2. Why did the database designers need to use custom-designed code to collect the original data?

Sources: Pratt, Mary K., "The Getty makes art accessible with online database." *Computerworld*, March 10, 2008, *www.computerworld.com/action/article.do?command=viewArticleBasic&taxonomyName=Databases&articleId=310236&taxonomyId=173&intsrc=kc_li_story*; Staff, "The Computerworld Honors Program: Web-Based Global Art Resources: The Getty Vocabularies," *Computerworld*, 2007, *www.cwhonors.org/viewCaseStudy.asp?NominationID=112*; The Getty Web site, *www.getty.edu*, accessed April 1, 2008.

Case Two

ETAI Manages Auto Parts Overload with Open-Source Database

If you need a hard-to-find automobile part for a European import, you could probably find it in a catalog published by the ETAI Group in France. The ETAI catalog includes over 30 million parts for over 50,000 European car models manufactured during the past 15 years. The catalog is updated 100 times each year to stay current with the latest models.

While maintaining an average auto parts catalog might not seem a daunting task, this one is an exception. ETAI collects auto parts information from nine databases provided by parts manufacturers. Each database uses a unique design with different formats for parts numbers and varying amounts and types of fields for each part record. Over many years, ETAI had developed a system for collating the data using a variety of programming languages and platforms. The entire process required 15 steps and two to three weeks. It was so complicated that if ETAI's database administrator were to leave, his replacement would have a difficult time learning how the complicated system worked.

Philippe Bobo, the director of software and information systems at ETAI, knew it was time to improve the system. He and his team tested products from a variety of vendors over a five-week period, and eventually decided to work with Talend Open Data Solutions, based in Los Altos, California. Talend specializes in open-source database management systems that integrate data from various types of systems into a single target system—exactly what ETAI needed.

Talend designed a system for ETAI using a single standard programming language that queries the nine auto parts databases and streams the results into one data warehouse. It then cleans the data and standardizes it for output to a catalog format. The 15-step, three-week process is now reduced to one step and two days.

Philippe likes the open-source nature of Talend's solution because it makes it possible for his own software engineers to work with and adjust the software over time to accommodate new needs in the system. Updating the DBMS has reduced labor costs and production time, and made it possible for ETAI to expand into other types of catalogs and service manuals.

Discussion Questions

1. What challenges did ETAI face that made creating their catalog a three-week-long ordeal?
2. How did the solution provided by Talend reduce the job time by 90 percent?

Critical Thinking Questions

1. What benefits were provided by the open-source solution?
2. Why couldn't ETAI standardize the data formats in the nine databases?

Sources: Weiss, Todd R., "ETAI avoids data traffic jam with open source," *Computerworld*, December 17, 2007, *www.computerworld.com/action/article.do?command=viewArticleBasic&articleId=90531 61&intsrc=news_list*; Weiss, Todd R., "ETAI Rides Open Source to Ease Data Traffic Jam," *Computerworld*, December 31, 2007, *www.computerworld.com/action/article.do?command=viewArticleBasic&articleId=30982 1*; Talend Open Data Solutions Web site, *www.talend.com*, accessed March 31, 2008; ETAI Web site, *www.etai.fr/g_instit/atout.htm*, accessed March 31, 2008.

Questions for Web Case

See the Web site for this book to read about the Whitmann Price Consulting case for this chapter. Following are questions concerning this Web case.

Whitmann Price Consulting: Database Systems and Business Intelligence

Discussion Questions

1. How will Whitmann Price consultants and the company itself benefit from their ability to call up corporate information in an instant anywhere and at any time?
2. Why will the database itself not require a change to support the new advanced mobile communications and information system?

Critical Thinking Questions

1. The Web has acted as a convenient standard for accessing all types of information from various types of computing platforms. How will this benefit the systems developers of Whitmann Price in developing forms and reports for the new mobile system?
2. What are the suggested limitations of using a BlackBerry device for accessing and interacting with corporate data?

NOTES

Sources for opening vignette: Havenstein, Heather, "Wal-Mart CTO details HP data warehouse move," *ITWorld Canada*, August 3, 2007, *www.itworldcanada.com/a/Enterprise-Business-Applications/efb96e0a-18de-47e6-ac61-ddab5cc55b5b.html*; Wal-Mart Corporate Fact Sheet, accessed March 30, 2008; Hayes Weier, Mary, "Wal-Mart Speaks Out On HP Neoview Decision," *Information Week*, August 3, 2007, *www.informationweek.com/management/showArticle.jhtml?articleID=201203010, www.walmartstores.com/media/factsheets/fs_2230.pdf*; HP Neoview Enterprise Data Warehouse Web site, *http://h20331.www2.hp.com/enterprise/cache/414444-0-0-225-121.html*, accessed March 30, 3008.

1 Wailgum, Thomas, "Hollywood agency updates systems to woo talent," *Computerworld*, January 19, 2007, *www.computerworld.com/action/article.do?command=viewArticleBasic&taxonomyName=Business_Intelligence&a rticleId=9008545&taxonomyId=9&intsrc=kc_li_story*.

2 Havenstein, Heather, "City of Albuquerque puts BI capabilities into residents' hands," *Computerworld*, September 17, 2007, *www.computerworld.com/action/article.do?command=viewArticleBasic&taxonomyName=Data_Mining&articleId = 301748&taxonomyId=54&intsrc=kc_li_story*.

3 Vijayan, Jaikumar, "Harvard grad students hit in computer intrusion," *Computerworld*, March 13, 2008, *www.computerworld.com/action/article.do?command=viewArticleBasic&articleId=9068221&source=rss_news10*.

4 Staff, "Bacs database fault leaves 400,000 without pay," *Computerworld UK*, March 30, 2007, *www.itworldcanada.com/a/Information-Architecture/94cb5e9e-79c5-4f29-a909-c7025be0d0b4.html*.

5 Mearian, Lucas, "Study: Digital universe and its impact bigger than we thought," *Computerworld*, March 11, 2008, *www.computerworld.com/action/article.do?command=viewArticleBasic&articleId=9067639&source=rss_news10*.

6 Koman, Richard, "Exploding Digital Data Growth Is a Challenge for IT," *Top Tech News*, March 11, 2008, *www.toptechnews.com/story.xhtml?story_id=58752*.

7 Nakashima, Ellen, "FBI Prepares Vast Database Of Biometrics," *Washington Post*, December 22, 2007, *www.washingtonpost.com/wp-dyn/content/article/2007/12/21/AR2007122102544_pf.html*.

8 Kolbasuk McGee, Marianne, "Wal-Mart Requires In-Store Clinics To Use E-Health Records System," *Information Week*, February 16, 2008, *www.informationweek.com/story/showArticle.jhtml?articleID=206504257&cid=RSSfeed_IWK_All*.

9 Pratt, Mary, "Steven Barlow: Master of Data Warehousing," *Computerworld*, July 9, 2007, *www.computerworld.com/action/article.do?command=viewArticleBasic&taxonomyName=data_warehousing&arti cleId=297032&taxonomyId=55&intsrc=kc_feat*.

10 Mullins, Craig, "The Database Report - July 2007," *TDNA*, July 10, 2007, *www.tdan.com/view-featured-columns/5603*.

11 Fonseca, Brian, "Baseball Hall of Fame on deck for archive format change?" *Computerworld*, July 24, 2007, *www.computerworld.com/action/article.do?command=viewArticleBasic&taxonomyName=servers_and_data_cente r&articleId=9027849&taxonomyId=154&intsrc=kc_top*.

12 Microsoft Staff, "Microsoft Transforms Management Training into an Interactive, On-the-Job Experience," Microsoft Case Studies, August 30, 2007, *www.microsoft.com/casestudies/casestudy.aspx?casestudyid=4000000613*.

13 Oracle Staff, "Lighting Manufacturer Surya Roshni Streamlines Supply Chain with Bright Results," Oracle Customer Snapshot, 2008, *www.oracle.com/customers/snapshots/surya-roshni-snapshot.pdf*.

14 Staff, "INTELLIFIT Moves From Virtual Fitting (match-to-order) to True Mass Customization: Custom-made jeans with a high-tech twist," *Mass Customization & Open Innovation News*, February 15, 2008, *http://mass-customization.blogs.com/mass_customization_open_i/2008/02/intellifit-move.html*.

15 IT Redux Web site, accessed March 23, 2008, *http://itredux.com/office-20/database/?family=Database*.

16 Lai, Eric, "Cloud database vendors: What, us worry about Microsoft?" *Computerworld*, March 12, 2008, *www.computerworld.com/action/article.do?command=viewArticleBasic&articleId=9067979&pageNumber=1*.

17 Intuit Staff, "Jet Blue – The Challenge," Intuit Case Studies, Accessed April 10, 2008, http://quickbase.intuit.com/customers/jetblue_v4.pdf

18 Arellano, Nestor, "Information created in 2007 will exceed storage capacity says IDC," *ITWorld Canada*, March 23, 2008, *www.itworldcanada.com/a/Information-Architecture/844a38f1-ccaf-445f-867f-498f041587c1.html*.

19 Gittlen, Sandra, "Data center land grab: How to get ready for the rush," *Computerworld*, March 12, 2007, *www.computerworld.com/action/article.do?command=viewArticleBasic&articleId=9012963*.

20 AP Staff, "Google ventures into health records biz," CNN.com, February 21, 2008, *www.cnn.com/2008/TECH/02/21/google.records.ap*.

21 Dixon, Guy, "Yahoo embraces semantic Web standards," vnunet.com, March 18, 2008, *www.vnunet.com/vnunet/news/2212249/yahoo-embraces-semantic-web*.

22 McMillan, Robert, "Researcher: Half a million database servers have no firewall," *Computerworld*, November 14, 2007, *www.computerworld.com/action/article.do?command=viewArticleBasic&articleId=9046821&source=rss_news10*.

23 SAS Staff, "1-800-FLOWERS.COM Gathers a Bouquet of CRM Capabilities," SAS Success Stories, *www.sas.com/success/1800flowers_IT.html*, accessed March 23, 2008.

24 Computerworld Staff, "The Computerworld Honors Program: Defense Acquisition University," *Computerworld*, 2007, *www.cwhonors.org/viewCaseStudy.asp?NominationID=313*.

25 Vijayan, Jaikumar, "Univ. of Md. launches data mining portal for counter-terrorism research," *Computerworld*, February 26, 2008, *www.computerworld.com/action/article.do?command=viewArticleBasic&articleId=9064938*.

26 Stone, Brad, "MySpace to Discuss Effort to Customize Ads," *New York Times*, September 18, 2007, *www.nytimes.com/2007/09/18/technology/18myspace.html?_r=1&oref=slogin*.

27 Computerworld Staff, "The Computerworld Honors Program: City of Richmond Police Department," *Computerworld*, 2007, *www.cwhonors.org/viewCaseStudy.asp?NominationID=303*.

28 Nickum, Chris, "The BI Prescription," *OptimizeMag.com*, April 2007, Page 45

29 Computerworld Staff, "The Computerworld Honors Program: QlikTech International," *Computerworld*, 2007, *www.cwhonors.org/viewCaseStudy.asp?NominationID=155*.

30 Cognos Staff, "Lufthansa Cargo," Lufthansa Cargo Case Study, Accessed April 10, 2008, http://www.cognos.com/pdfs/success_stories/ss_lufthansa-cargo.pdf

31 Staff, "init wins contract worth 25 mill. US Dollar from Seattle," Init News Release, March 13, 2007, *www.init-ka.de/en_news/PR_AH_2007/AH_070313_Seattle_en.php*.

32 Babcock, Charles, "Oracle Touts Virtualization, Apps – and 11g Of Course," *Information Week*, November 19, 2007, page 30.

33 Fonseca, Brian, "Virtualization Cutting Storage Costs for Some Large Firms," *Computerworld*, April 23, 2007, *www.computerworld.com/action/article.do?command=viewArticleBasic&taxonomyId=154&articleId=290230&int src=hm_topic*.

34 Staff, "MapInfo Customer Testimonials," MapInfo Web site, *www.mapinfo.com/location/integration?txtTopNav=836a2545d8a37f00dev-vcm100001a031dc7____&txtLeftNav=6e5819cd57f47f00dev-vcm100001a031dc7____&txtExtNav=391495ac2a771110Vgn-VCM1000002102ldc7____*, accessed March 23, 2008.

CHAPTER • 6 •

Telecommunications and Networks

PRINCIPLES	LEARNING OBJECTIVES
▪ A telecommunications system and network have many fundamental components.	▪ Identify and describe the fundamental components of a telecommunications system. ▪ Identify two broad categories of telecommunications media and their associated characteristics. ▪ Identify several telecommunications hardware devices and discuss their functions.
▪ Telecommunications, networks, and their associated applications are essential to organizational success.	▪ Describe the benefits associated with the use of a network. ▪ Name three distributed processing alternatives and discuss their basic features. ▪ List and describe several telecommunications applications that organizations benefit from today.

Information Systems in the Global Economy Deloitte, Milan, Italy

Unified Communications for Financial Management

Deloitte Touche Tohmatsu, also known as Deloitte & Touche, or just Deloitte, is one of the world's largest financial services firms. Established more than 150 years ago by William Welch Deloitte, the company grew from a small accountancy office in London to a global network of member firms employing 150,000 professionals in 140 countries with $20 billion in total revenues.

Today's demanding and competitive business environment requires professionals to extract the most value out of every minute of the workday. This means staying connected to corporate networks, accessing information, and communicating with colleagues from all locations: at the office, in a warehouse, in conference rooms, at airports, in vehicles, and at home. Businesses such as Deloitte realize that the quality of telecommunications systems and the services they deliver can dramatically affect a company's success.

Recently, Deloitte in Milan, Italy, conducted a major reorganization to consolidate its five branches into one complex. The consolidation was intended to improve communications and service while reducing costs. The reorganization presented the opportunity to replace Deloitte's telecommunications systems with the most current technologies.

Deloitte decided to invest in a unified communications system that provides voice, video, and data communications over one network. A unified communications system can help companies save money because it can eliminate the need to set up multiple networks for differing uses. However, a unified system can do more than reduce redundancy: A company can design software and hardware to integrate various forms of telecommunications into one powerful and easy-to-use system that is accessible anywhere anytime from a variety of devices.

The unified communications system that Deloitte purchased from Cisco included several interconnected services, including the following:

- About 1,200 unified communications lines to stream voice, video, and data to computers and videophones
- Videoconferencing services using Internet Protocol (IP)
- Television broadcasting service over IP
- Call management system
- Local area network management system
- Wi-Fi wireless networking in all buildings in the complex

Deloitte uses the television broadcasting capabilities of the new system to broadcast regular messages from the CEO. It uses the videophone service to improve communications between employees and to support conferences and meetings while team members are away from the office.

The system also allows Deloitte to make better use of its office space. In most businesses, many offices and desks are often unused throughout the day due to employees that are traveling, in meetings, or otherwise out of the office. Because everything Deloitte employees need is delivered over the network, and the network is available anywhere, employees can use any available space. The Hot-Desks service, provided through the network, keeps track of which workstations, desks, and conference rooms are in use and can direct employees to unused space.

Deloitte provides notebook PCs with integrated wireless networking capabilities to all employees, making it possible for work to take place in any area of the complex through the Wi-Fi network. The notebooks include a SoftPhone application that employees use to access telephone and videophone services over the local network or over the Internet while traveling. If an employee receives a phone call at the office, he or she can answer the call on the notebook PC from any location. This means Deloitte employees can take the office with them wherever they go.

Deloitte's unified communications system also includes call center management that routes incoming calls to the proper party. The new system is easily managed and controlled through one central interface. This gives the company greater control over its telecommunications at a much lower cost. Deloitte professionals are experiencing a notable increase in collaborative work and improvements to the entire organizational structure. Deloitte provides an ideal example of the important role telecommunications plays in the success of a business.

As you read this chapter, consider the following:

- What services do new telecommunications and network technologies offer to assist individuals and organizations in being more effective?
- What role does telecommunications play in connecting organizations and growing the global economy?

Why Learn About Telecommunications and Networks?

Effective communication is essential to the success of every major human undertaking, from building great cities to waging war to running a modern organization. Today we use electronic messaging and networking to shrink the world and enable people everywhere to communicate and interact effectively without requiring face-to-face meetings. Regardless of your chosen major or future career field, you will need the communications capabilities provided by telecommunications and networks, especially if your work involves the supply chain. Among all business functions, supply chain management might use telecommunications and networks the most because it requires cooperation and communications among workers in inbound logistics, warehouse and storage, production, finished product storage, outbound logistics, and most important, with customers, suppliers, and shippers. All members of the supply chain must work together effectively to increase the value perceived by the customer, so partners must communicate well. Other employees in human resources, finance, research and development, marketing, and sales positions must also use communications technology to communicate with people inside and outside the organization. To be a successful member of any organization, you must be able to take advantage of the capabilities that these technologies offer you. This chapter begins by discussing the importance of effective communications.

In today's high-speed global business world, organizations need always-on, always-connected computing for traveling employees and for network connections to their key business partners and customers. As we saw in the opening vignette, forward-thinking companies such as Deloitte hope to save billions of dollars, reduce time to market, and enable collaboration with their business partners by using telecommunications systems. Here are just a few additional examples of organizations using telecommunications and networks to move ahead.

- Wal-Mart, the world's largest retailer with $345 billion in sales, plans to include RFID tags on products in its 4,068 North American stores to improve inventory accuracy, thereby reducing untracked sales and all but eliminating lost or missing merchandise. The net result is a savings of $287 million per year. Telecommunications between the RFID chips and scanners on forklift trucks and between the trucks and in-store computers is an essential component of achieving these savings.[1]
- Procter & Gamble (P&G) has implemented 13 of its planned 40 Video Collaboration Studios using the Cisco TelePresence system to empower the P&G community of more

than 138,000 employees working in over 80 countries worldwide. The Studios foster a high degree of communication and collaboration without requiring members of a team to physically travel to meet. Use of this telecommunications technology is credited with helping P&G to bring its products to market faster and compete more effectively. Aflac, BT, McKesson, SAP, Verizon, and more than 100 other Cisco customers are also experimenting with the use of this technology.[2]

- Home appliance manufacturers are adding telecommunications capabilities to their products to make them more appealing and useful. Whirlpool is testing the concept of making refrigerators a central information hub by providing removable digital photo frames capable of displaying digital images and providing news and weather updates with its most advanced models. Future plans call for refrigerators that can play music from MP3 players or satellite radios. Not to be outdone, LG Electronics plans to offer refrigerators with a 15-inch LCD HDTV for TV and video playback.[3]
- Thousands of companies are employing Webcasts to inform and educate potential customers about their products and services.
- High technology companies such as Boeing use a wide range of telecommunications technologies to support their business and collaborate with people from inside and outside the company. Boeing has created the LabNet to connect the various Boeing Labs and customers that test concepts and features under development. The LabNet enables all participants to visualize live, simulated, and computer-generated fighter jets as they demonstrate their performance under various test scenarios.[4]

AN OVERVIEW OF TELECOMMUNICATIONS

Telecommunications refers to the electronic transmission of signals for communications, by means such as telephone, radio, and television. Telecommunications is creating profound changes in business because it lessens the barriers of time and distance. Advances in telecommunications technology allow us to communicate rapidly with business partners, clients, and coworkers almost anywhere in the world. Telecommunications also reduces the amount of time needed to transmit information that can drive and conclude business actions. Telecommunications not only is changing the way organizations operate, but the nature of commerce itself. As networks connect to one another and transmit information more freely, a competitive marketplace demands excellent quality and service from all organizations.

Figure 6.1 shows a general model of telecommunications. The model starts with a sending unit (1) such as a person, a computer system, a terminal, or another device that originates the message. The sending unit transmits a signal (2) to a telecommunications device (3). The telecommunications device—a hardware component that facilitates electronic communication—performs many tasks, which can include converting the signal into a different form or from one type to another. The telecommunications device then sends the signal through a medium (4). A **telecommunications medium** is any material substance that carries an electronic signal to support communications between a sending and receiving device. Another telecommunications device (5) connected to the receiving device (6) receives the signal. The process can be reversed, and the receiving unit (6) can send a message to the original sending unit (1). An important characteristic of telecommunications is the speed at which information is transmitted, which is measured in bits per second (bps). Common speeds are in the range of thousands of bits per second (Kbps) to millions of bits per second (Mbps) and even billions of bits per second (Gbps).

telecommunications medium
Any material substance that carries an electronic signal and serves as an interface between a sending device and a receiving device.

A **telecommunications protocol** defines the set of rules that governs the exchange of information over a communications medium. The goal is to ensure fast, efficient, error-free communications and to enable hardware, software, and equipment manufacturers and service providers to build products that interoperate effectively. The *Institute of Electrical and Electronics Engineers (IEEE)* is a leading standards-setting organization whose IEEE 802 network standards are the basis for many telecommunications devices and services. The *International Telecommunication Union (ITU)* is a specialized agency of the United Nations with

telecommunications protocol
A set of rules that governs the exchange of information over a communications medium.

Figure 6.1

Elements of a Telecommunications System

Telecommunications devices relay signals between computer systems and transmission media.

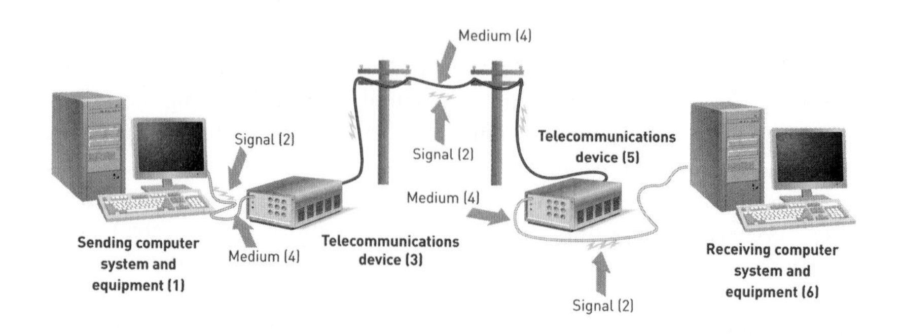

headquarters in Geneva, Switzerland. The international standards produced by the ITU are known as Recommendations and carry a high degree of formal international recognition.

synchronous communications
A form of communications where the receiver gets the message instantaneously, when it is sent.

asynchronous communications
A form of communications where the receiver gets the message after some delay—sometimes hours or days after the message is sent.

Communications between two people can occur synchronously or asynchronously. With **synchronous communications**, the receiver gets the message instantaneously, when it is sent. Voice and phone communications are examples of synchronous communications. With **asynchronous communications**, the receiver gets the message after some delay—sometimes hours or days after the message is sent. Sending a letter through the post office or e-mail over the Internet are examples of asynchronous communications. Both types of communications are important in business.

Using telecommunications can help businesses solve problems, coordinate activities, and capitalize on opportunities. To use telecommunications effectively, you must carefully analyze telecommunications media and devices.

Telecommunications technology enables business people to communicate with coworkers and clients from remote locations.

(Source: © BananaStock / Alamy.)

Basic Telecommunications Channel Characteristics

simplex channel
A communications channel that can transmit data in only one direction.

half-duplex channel
A communications channel that can transmit data in either direction, but not simultaneously.

full-duplex channel
A communications channel that permits data transmission in both directions at the same time, so a full-duplex channel is like two simplex channels.

The transmission medium carries messages from the source of the message to its receivers. A transmission medium can be divided into one or more telecommunications channels, each capable of carrying a message. Telecommunications channels can be classified as simplex, half-duplex, or full-duplex.

A **simplex channel** can transmit data in only one direction and is seldom used for business telecommunications. Doorbells and the radio operate using a simplex channel. A **half-duplex channel** can transmit data in either direction, but not simultaneously. For example, A can begin transmitting to B over a half-duplex line, but B must wait until A is finished to transmit back to A. Personal computers are usually connected to a remote computer over a half-duplex channel. A **full-duplex channel** permits data transmission in both directions at the same time, so a full-duplex channel is like two simplex channels. Private leased lines or two standard phone lines are required for full-duplex transmission.

Channel Bandwidth

In addition to the direction of data flow supported by a telecommunications channel, you must consider the speed at which data can be transmitted. Telecommunications **channel bandwidth** refers to the rate at which data is exchanged, usually measured in bits per second (bps)—the broader the bandwidth, the more information can be exchanged at one time. **Broadband communications** is a relative term but generally means a telecommunications system that can exchange data very quickly. For example, for wireless networks, broadband lets you send and receive data at a rate greater than 1.5 Mbps.

channel bandwidth
The rate at which data is exchanged over a telecommunications channel, usually measured in bits per second (bps).

broadband communications
A telecommunications system in which a very high rate of data exchange is possible.

Telecommunications professionals consider the capacity of the channel when they recommend transmission media for a business. In general, today's organizations need more bandwidth for increased transmission speed to carry out their daily functions. Another key consideration is the type of telecommunications media to use.

Telecommunications Media

Each telecommunications media type can be evaluated according to characteristics such as cost, capacity, and speed. In designing a telecommunications system, the transmission media selected depends on the amount of information to be exchanged, the speed at which data must be exchanged, the level of concern for data privacy, whether the users are stationary or mobile, and many other business requirements. The transmission media are selected to support the goals of the information and organizational systems at the lowest cost, but still allow for possible modifications should your business requirements change. Transmission media can be divided into two broad categories: *guided transmission media,* in which telecommunications signals are guided along a solid medium, and *wireless,* in which the telecommunications signal is broadcast over airwaves as a form of electromagnetic radiation.

Guided Transmission Media Types

Guided transmission media are available in many types. Table 6.1 summarizes the guided media types by physical media type. These guided transmission media types are discussed in the sections following the table.

Table 6.1
Guided Transmission Media Types

Media Type	Description	Advantages	Disadvantages
Twisted-pair wire	Twisted pairs of copper wire, shielded or unshielded	Used for telephone service; widely available	Transmission speed and distance limitations
Coaxial cable	Inner conductor wire surrounded by insulation	Cleaner and faster data transmission than twisted-pair wire	More expensive than twisted-pair wire
Fiber-optic cable	Many extremely thin strands of glass bound together in a sheathing; uses light beams to transmit signals	Diameter of cable is much smaller than coaxial; less distortion of signal; capable of high transmission rates	Expensive to purchase and install
Broadband over power lines	Data is transmitted over standard high-voltage power lines	Can provide Internet service to rural areas where cable and phone service may be nonexistent	Can be expensive and may interfere with ham radios and police and fire communications

Twisted-Pair Wire

Twisted-pair wire contains two or more twisted pairs of wire, usually copper (see Figure 6.2). Proper twisting of the wire keeps the signal from "bleeding" into the next pair and creating electrical interference. Because the twisted-pair wires are insulated, they can be placed close together and packaged in one group. Hundreds of wire pairs can be grouped into one large wire cable.

Twisted-pair wires are classified by category (Category 1, 2, 3, 4, 5, 5E, and 6). The lower categories are used primarily in homes. Higher categories are used in networks and can carry data at higher speeds. For example, 10 Gigabit Ethernet is a standard for transmitting data in full-duplex mode at the speed of 10 billion bits per second for limited distances over category 5 or 6 twisted-pair wire. The 10 Gigabit Ethernet cable can be used for the

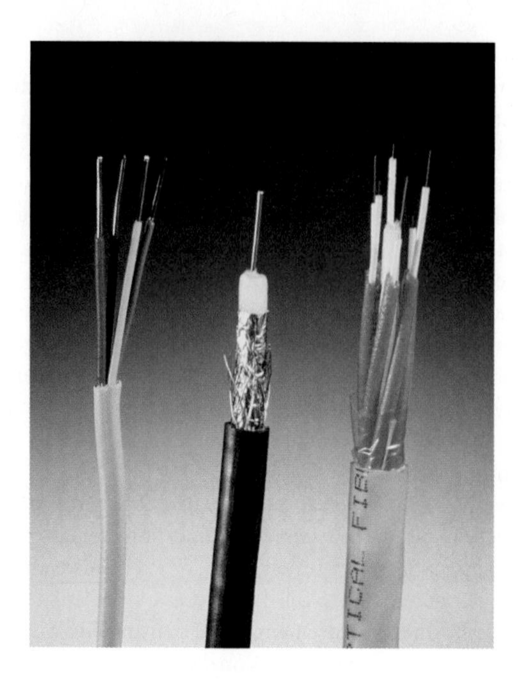

Figure 6.2

Types of Guided Transmission Media

Twisted-pair wire (left), coaxial cable (middle), fiber-optic cable (right)

(Source: © Greg Pease/Getty Images.)

high-speed links that connect groups of computers or to move data stored in large databases on large computers to stand-alone storage devices.

The Niagara Falls Bridge Commission (NFBC) is a joint United States and Canadian agency that monitors three border crossings spanning the Niagara River between western New York State and southern Ontario. NFBC also operates a fourth site that processes more than seven million border crossings per year. The NFBC relies on a 10 Gigabit Ethernet network to carry video and other data from border locations to its operations center in Lewiston, New York. Video data from 170 cameras is used to support monitoring for unusual or suspicious activity along the border as well as to manage the flow of traffic.[5]

Coaxial Cable

Figure 6.2 (middle) also shows a typical coaxial cable, similar to that used in cable television installations. When used for data transmission, coaxial cable falls in the middle of the guided transmission media in terms of cost and performance. The cable itself is more expensive than twisted-pair wire but less than fiber-optic cable (discussed next). However, the cost of installation and other necessary communications equipment makes it difficult to compare the total costs of each medium. Coaxial cable offers cleaner and crisper data transmission (less noise) than twisted-pair wire. It also offers a higher data transmission rate.

Cable companies are aggressively courting customers for telephone service, enticing them away from the phone companies by bundling Internet and phone services along with TV. For example, Comcast provides new movies on demand the same day as the DVD release to its more than 24 million subscribers through its Project Infinity.[6]

Fiber-Optic Cable

Fiber-optic cable, consisting of many extremely thin strands of glass or plastic bound together in a sheathing (also known as a jacket), transmits signals with light beams (see Figure 6.2, right). These high-intensity light beams are generated by lasers and are conducted along the transparent fibers. These fibers have a thin coating, called *cladding*, which effectively works like a mirror, preventing the light from leaking out of the fiber. The much smaller diameter of fiber-optic cable makes it ideal when there is no room for bulky copper wires—for example, in crowded conduits, which can be pipes or spaces carrying both electrical and communications wires. Fiber-optic cable and associated telecommunications devices are more expensive to purchase and install than their twisted-pair wire counterparts, although the cost is decreasing.

Verizon has been building a fiber-optic network since 2004 at a budgeted cost of $18 billion. This has required Verizon crews to remove traditional twisted-pair wires used to carry phone calls and replace them with hair-thin strands of optical fiber in thousands of towns and cities. When complete, the Verizon Fiber Optic Service (FiOS) network will take fiber directly to subscribers' homes and provide downstream connection speeds (the speed that data is transmitted to your computer) of 5 Mbps to 50 Mbps. The network will be used to deliver high-speed Internet connection, telephone service, and TV including video on demand. Verizon hopes that the new infrastructure will enable it to win customers from the cable companies.[7]

Broadband over Power Lines

Many utilities, cities, and organizations are experimenting with *broadband over power lines (BPL)* to provide Internet access to homes and businesses over standard high-voltage power lines. This form of BPL is called *access BPL*. A system called *in-premise BPL* can be used to create a local area network using the building's wiring. A potential problem with BPL is that transmitting data over unshielded power lines can interfere with both amateur (ham) radio broadcasts and police and fire radios. However, BPL can provide Internet service in rural areas where broadband access has lagged because electricity is more prevalent in homes than cable or even telephone lines.

The U.S. Transportation Security Administration is testing the use of in-premise BPL at selected airports to connect airport passenger and other screening systems, cameras at ticket counters, and passport readers.[8] To access the Internet, BPL users connect their computer to a special hardware device that plugs into any electrical wall socket. Comtrend Corporation offers a PowerGrid 904 adapter that enables data transmission speeds of up to 400 Mbps.[9]

Wireless Communications Options

Wireless communications coupled with the Internet is revolutionizing how and where we gather and share information, collaborate in teams, listen to music or watch video, and stay in touch with our families and coworkers while on the road. With wireless capability, a coffee shop can become our living room and the bleachers at a ball park can become our office. The many advantages and freedom provided by wireless communications are causing many organizations to consider moving to an all-wireless environment. Shopanista, a shuttle for shoppers in Los Angeles, California, made the decision to move to wireless after tiring of the hassles of moving wired devices.[10]

Wireless transmission involves the broadcast of communications in one of three frequency ranges: radio, microwave, or infrared frequencies, as shown in Table 6.2. In some cases, the use of wireless communications is regulated and the signal must be broadcast within a specific frequency range to avoid interference with other wireless transmissions. For example, radio and TV stations must gain approval to use a certain frequency to broadcast their signals. In those cases where wireless communications are not regulated, there is a high potential for interference between signals.

Table 6.2

Frequency Ranges Used for Wireless Communications

Technology	Description	Advantages	Disadvantages
Radio frequency range	Operates in the 3KHz–300 MHz range	Supports mobile users; costs are dropping	Signal highly susceptible to interception
Microwave—terrestrial and satellite frequency range	High-frequency radio signal (300 MHz–300 GHz) sent through atmosphere and space (often involves communications satellites)	Avoids cost and effort to lay cable or wires; capable of high-speed transmission	Must have unobstructed line of sight between sender and receiver; signal highly susceptible to interception
Infrared frequency range	Signals in the 300 GHz–400 THz frequency range sent through air as light waves	Lets you move, remove, and install devices without expensive wiring	Must have unobstructed line of sight between sender and receiver; transmission effective only for short distances

With the spread of wireless network technology to support devices such as PDAs, mobile computers, and cell phones, the telecommunications industry needed new protocols to define how these hardware devices and their associated software would interoperate on the networks provided by telecommunications carriers. Today more than 70 active groups set standards at a regional, national, and global level resulting in a dizzying array of communications standards and options.[11] Some of the more widely used wireless communications options are discussed next.

Short Range Wireless Options

Many wireless solutions provide communications over very short distances including near field communications, Bluetooth, ultra wideband, infrared transmission, and Zigbee.

Near Field Communication (NFC)

Near Field Communication (NFC)
A very short-range wireless connectivity technology designed for cell phones and credit cards.

Near Field Communication (NFC) is a very short-range wireless connectivity technology designed for cell phones and credit cards. With NFC, consumers can wave their credit cards or even cell phones within a few inches of point-of-sale terminals to pay for purchases. Consumers are using the technology in Germany and Austria, and pilot projects are being conducted in London, Singapore, the Netherlands, and Finland. In the United States, MasterCard and Visa are testing devices with embedded NFC and are looking for partners to explore the widespread use of NFC technology in phones and credit cards.[12]

Bluetooth

Bluetooth
A wireless communications specification that describes how cell phones, computers, faxes, personal digital assistants, printers, and other electronic devices can be interconnected over distances of 10–30 feet at a rate of about 2 Mbps.

Bluetooth is a wireless communications specification that describes how cell phones, computers, personal digital assistants, printers, and other electronic devices can be interconnected over distances of 10–30 feet at a rate of about 2 Mbps. Bluetooth enables users of multifunctional devices to synchronize with information in a desktop computer, send or receive faxes, print, and, in general, coordinate all mobile and fixed computer devices. The Bluetooth technology is named after the tenth century Danish King Harald Blatand, or Harold Bluetooth in English. He had been instrumental in uniting warring factions in parts of what is now Norway, Sweden, and Denmark—just as the technology named after him is designed to allow collaboration between differing devices such as computers, phones, and other electronic devices.

All types of businesses find Bluetooth technology helpful. For example, MedicMate is a mobile software developer with an application designed for medical professionals who can't carry patient files from one patient or facility to another. The application is also intended for institutions who want to provide electronic patient or pharmaceutical data wirelessly, but have not implemented broadband infrastructure. The application runs on any mobile handset with a touch screen and the Windows Mobile 2003 operating system. With it, the user can create or display patient data, problem lists, sticky notes, and patient alarms. The application works with infrared or Bluetooth technology to send and receive data wirelessly.[13]

Ultra Wideband (UWB)

ultra wideband (UWB)
A wireless communications technology that transmits large amounts of digital data over short distances of up to 30 feet using a wide spectrum of frequency bands and very low power.

Ultra wideband (UWB) is a wireless communications technology that transmits large amounts of digital data over short distances of up to 30 feet using a wide spectrum of frequency bands and very low power. Ultra wideband has the potential to replace Bluetooth's 2 Mbps transmission speed with 400 Mbps rates for wirelessly connecting printers and other devices to desktop computers or enabling completely wireless home multimedia networks.[14] The manufacturers of electronic entertainment devices are particularly interested in the use of UWB. With UWB, a digital camcorder could play a just-recorded video on an HDTV without anyone having to fiddle with wires. A portable MP3 player could stream audio to high-quality surround-sound speakers anywhere in the room. A mobile computer user could wirelessly connect to a digital projector in a conference room to deliver a presentation.

Infrared Transmission

infrared transmission
A wireless communications technology that operates at a frequency of 300 GHz and above that requires line-of-sight transmission and operates over short distances—such as a few yards.

Infrared transmission sends signals at a frequency of 300 GHz and above. Infrared transmission requires line-of-sight transmission and short distances—such as a few yards. Infrared transmission allows handheld computers to transmit data and information to larger computers within the same room and to connect a display screen, printer, and mouse to a computer.

The Apple Remote is a remote control device made for use with Apple infrared products. It has six buttons: Menu, Play/Pause, Volume Up, Volume Down, Previous/Rewind, and Next/Fast-Forward. The new Mac Mini features an infrared port designed to work with the Apple Remote and support Front Row, a multimedia application that allows users to access shared iTunes and iPhoto libraries and video throughout their homes.

Zigbee

Zigbee is a form of wireless communications frequently used in security systems and heating and cooling control systems. Zigbee is a relatively low-cost technology and requires little power, which allows longer life with smaller batteries.

Energy Optimizers, Ltd, a company in the United Kingdom, has developed a plug-in electricity meter called the Plogg that can monitor the energy usage of appliances. The device uses the Zigbee protocol to collect data from refrigerators, air conditioners, and other appliances and relay it to a central server via the Internet. Should an appliance be left on after hours, the Plogg can alert someone or turn the device down or off.[15]

Medium Range Wireless Options

Wi-Fi is a wireless telecommunications technology brand owned by the Wi-Fi Alliance, which consists of about 300 technology companies including AT&T, Dell, Microsoft, Nokia, and Qualcomm. The alliance exists to improve the interoperability of wireless local area network products based on the IEEE 802.11 series of telecommunications standards.

With a Wi-Fi wireless network, the user's computer, smartphone, or personal digital assistant has a wireless adapter that translates data into a radio signal and transmits it using an antenna. A wireless access point, which consists of a transmitter with an antenna, receives the signal and decodes it. The access point then sends the information to the Internet over a wired connection (see Figure 6.3). When receiving data, the wireless access point takes the information from the Internet, translates it into a radio signal, and sends it to the device's wireless adapter. These devices typically come with built-in wireless transmitters and software to enable them to alert the user to the existence of a Wi-Fi network. The area covered by one or more interconnected wireless access points is called a "hot spot." Current Wi-Fi access points have a maximum range of about 300 feet outdoors and 100 feet within a dry-walled building. Wi-Fi has proven so popular that hot spots are popping up in places such as airports, coffee shops, college campuses, libraries, and restaurants.

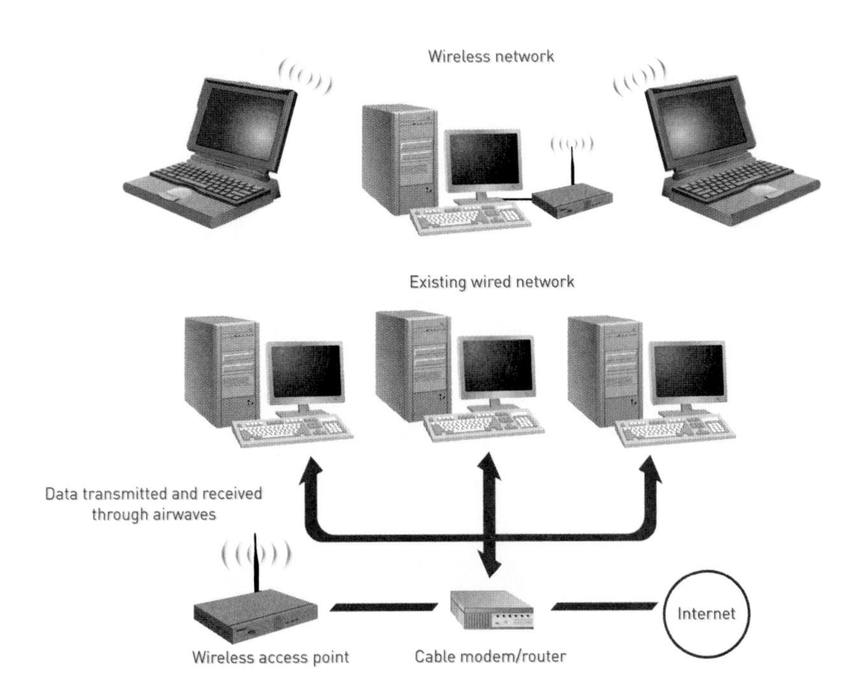

Figure 6.3
Wi-Fi Network

Over 100 U.S. city governments have implemented municipal Wi-Fi networks for use by meter readers and other municipal workers and to partially subsidize Internet access to their citizens and visitors. Supporters of the networks believe that the presence of such networks stimulates economic development by attracting new businesses. Critics doubt the long-term viability of municipal Wi-Fi networks because the technology cannot easily handle rapidly increasing numbers of users. Also, because municipal Wi-Fi networks use an unlicensed bandwidth available to any user and they operate at up to 30 times the power of existing home and business Wi-Fi networks, critics claim interference is inevitable with these networks. Competing Internet service providers (cable, telephone, and satellite, for example) complain that municipal Wi-Fi networks are subsidized to such an extent that they have an unfair competitive cost advantage.

Municipal Wi-Fi network projects for San Francisco, Chicago, Philadelphia, St. Louis, and Houston were delayed or cancelled as the costs of the projects became too high for both the cities involved and the builders of the networks.[16] Many of the other existing municipal Wi-Fi networks have failed to achieve their goals for number of subscribers.[17] On the other hand, when Denver International Airport switched its public Wi-Fi offering from paid ($7.95 per day) to advertising supported, the number of users increased ten-fold with some 8,000 connections to the network each day.[18]

Several airlines plan to offer passengers with Wi-Fi enabled devices access to the Internet, e-mail, and stored in-flight entertainment.[19] Surveys have shown that 80 percent of business travelers and over 50 percent of leisure travelers want onboard Internet access. Southwest spokeswoman Whitney Eichinger says, "We hope that the Internet will be expected on airplanes just as it's expected in a hotel or coffee shop."[20]

Wide Area Wireless Network Options

Many solutions provide wide area network options including satellite and terrestrial microwave transmission, wireless mesh, 3G, 4G, and WiMAX.

Microwave Transmission

Microwave is a high-frequency (300 MHz–300 GHz) signal sent through the air (see Figure 6.4). Terrestrial (Earth-bound) microwaves are transmitted by line-of-sight devices, so that the line of sight between the transmitter and receiver must be unobstructed. Typically, microwave stations are placed in a series—one station receives a signal, amplifies it, and retransmits it to the next microwave transmission tower. Such stations can be located roughly 30 miles apart before the curvature of the Earth makes it impossible for the towers to "see one another." Microwave signals can carry thousands of channels at the same time.

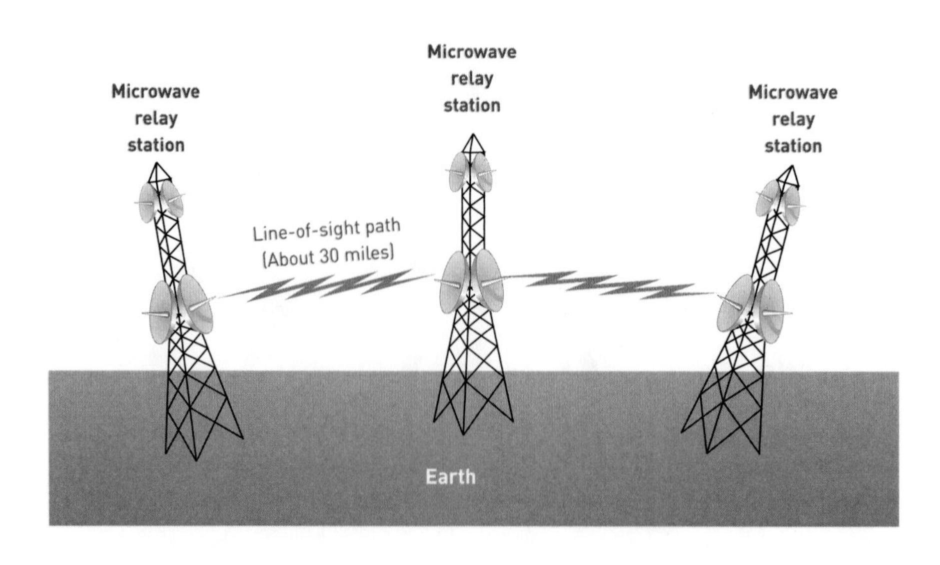

Figure 6.4

Microwave Communications

Because they are line-of-sight transmission devices, microwave dishes are frequently placed in relatively high locations, such as atop mountains, towers, or tall buildings.

A communications satellite also operates in the microwave frequency range (see Figure 6.5). The satellite receives the signal from the Earth station, amplifies the relatively weak signal, and then rebroadcasts it at a different frequency. The advantage of satellite communications is that it can receive and broadcast over large geographic regions. Such problems as the curvature of the Earth, mountains, and other structures that block the line-of-sight microwave transmission make satellites an attractive alternative. Geostationary, low earth orbit, and small mobile satellite stations are the most common forms of satellite communications.

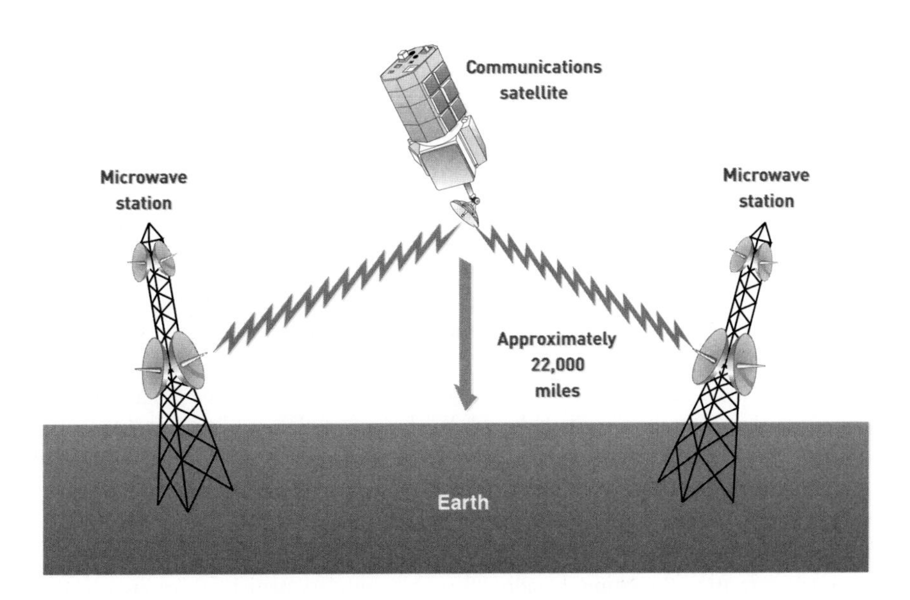

Figure 6.5

Satellite Transmission

Communications satellites are relay stations that receive signals from one Earth station and rebroadcast them to another.

A *geostationary satellite* orbits the Earth directly over the equator, approximately 22,300 miles above the Earth so that it appears stationary. The U.S. National Weather Service relies on the Geostationary Operational Environmental Satellite program for weather imagery and quantitative data to support weather forecasting, severe storm tracking, and meteorological research.

A *low earth orbit (LEO) satellite* system employs many satellites, each in an orbit at an altitude of less than 1,000 miles. The satellites are spaced so that, from any point on the Earth at any time, at least one satellite is on a line of sight.

A *very small aperture terminal (VSAT)* is a satellite ground station with a dish antenna smaller than 3 meters in diameter. News organizations employ VSAT dishes that run on battery power to quickly establish communications and transmit news stories from remote locations. Many people are also investing in VSAT technology in their homes to receive TV and send and receive computer communications.

Kerr-McGee is the largest independent oil producer and leaseholder in the deep waters of the Gulf of Mexico. The firm implemented a VSAT-based communications system to enable drilling rig crews to communicate reliably with each other, the main office, and friends and family back home. The network can also be used to share data and video of underwater pipelines with technical experts at headquarters and regional offices so that faster and better operational decisions can be made.[21]

Wireless Mesh

Wireless mesh uses multiple Wi-Fi access points to link a series of interconnected local area networks to form a wide area network capable of serving a large campus or entire city. Communications are routed among network nodes by allowing for continuous connections and reconfiguration around blocked paths by “hopping” from node to node until a connection can be established. Mesh networks are very robust: If one node fails, all the other nodes can still communicate with each other, directly or through one or more intermediate nodes.

wireless mesh

A way to route communications between network nodes (computers or other devices) by allowing for continuous connections and reconfiguration around blocked paths by “hopping” from node to node until a connection can be established.

The city of Tempe, Arizona implemented a mesh network to provide broadband wireless access for residents, visitors, students, and mobile workers on their laptop, PDA, or smartphone. The network has not been widely accepted—it has 1,000 outdoor access points but only 500 subscribers.[22]

3G Wireless Communications

The International Telecommunications Union (ITU) established a single standard for cellular networks in 1999. The goal was to standardize future digital wireless communications and allow global roaming with a single handset. Called IMT-2000, now referred to as 3G, this standard provides for faster transmission speeds in the range of 2–4 Mbps. Originally, 3G was supposed to be a single, unified, worldwide standard, but the 3G standards effort split into several different standards. One standard is the Universal Mobile Telephone System (UMTS), which is the preferred solution for European countries that use Global System for Mobile (GSM) communications. GSM is the *de facto* wireless telephone standard in Europe with more than 120 million users worldwide in 120 countries. Another 3G-based standard is Code-Division Multiple Access (CDMA), which is used in Australia, Canada, China, India, Israel, Mexico, South Korea, the United States, and Venezuela. The wide variety of 3G cellular communications protocols can support many business applications. The challenge is to enable these protocols to intercommunicate and support fast, reliable, global wireless communications.

3G wireless communication is useful for business travelers, people on the go, and people who need to get or stay connected. Although Wi-Fi is an option, 3G is preferable to mobile users concerned about the availability, cost, and security associated with the use of public Wi-Fi networks.

4G Wireless Communications

4G stands for fourth-generation broadband mobile wireless, which is expected to deliver more advanced versions of enhanced multimedia, smooth streaming video, universal access, portability across all types of devices, and eventually, worldwide roaming capability. 4G will also provide increased data transmission rates in the 20–40 Mbps range.

Pine Cellular, Pine Telephone, and Choctaw Electric are deploying Nortel 4G technology to provide homes and businesses in southeastern Oklahoma with reliable, wireless high-speed Internet. The network will provide low-cost, broad coverage and deliver wireless services to rural areas where construction of a wired network is less economical. The 4G services will be provided at no charge to the local police and fire departments and public schools.[23]

Worldwide Interoperability for Microwave Access (WiMAX)

Worldwide Interoperability for Microwave Access (WiMAX)
The common name for a set of IEEE 802.16 wireless metropolitan area network standards that support different types of communications access.

Worldwide Interoperability for Microwave Access (WiMAX) is the common name for a set of IEEE 802.16 wireless metropolitan area network standards that support various types of communications access. In many respects, WiMAX operates like Wi-Fi, only over greater distances and at faster transmission speeds. A WiMAX tower connects directly to the Internet via a high-bandwidth, wired connection. A WiMAX tower can also communicate with another WiMAX tower using a line-of-sight, microwave link. The distance between the WiMAX tower and an antenna can be as great as 30 miles. WiMAX can support data communications at a rate of 70 Mbps. Fewer WiMAX base stations are required to cover the same geographical area than when Wi-Fi technology is used. Mobile WiMAX refers to systems built based on the 802.16e standard and provides both fixed and mobile access over the same network infrastructure. Fixed WiMAX is based on the 802.16-2004 standard designed to deliver communications to homes and offices, but it cannot support mobile users. WiMAX is considered a 4G service.

In mid-2008, Sprint Nextel combined its wireless broadband unit with Clearwire to create a new communications company whose goal is to build the first national WiMAX network bringing coverage to 120 million people by the end of 2010. AT&T and Verizon Wireless have chosen a different direction and plan to upgrade their wireless networks with a future technology called Long Term Evolution.[24] To supply the necessary phones, computer chips, and other equipment, Sprint is working with Intel, Motorola, Nokia, and Samsung to provide WiMAX-capable PC cards, gaming devices, laptops, cameras, and even phones. Sprint also plans to implement a business model much different from the typical cellular model. Sprint's strategy is to allow any WiMAX-compliant device to run on its network. Users will be able to buy such devices at a variety of retail stores and will not be required to sign a contract with Sprint to use the network.[25]

Several countries are actively rolling out WiMAX networks. Aircel Business Solutions, one of the largest telecom groups in India, has completed a WiMAX network that covers 44 cities in India.[26] SB Broadband in Sweden is building a WiMAX network to cover nine cities in southwest Sweden. The Telmex WiMAX network in Chile covers 98 percent of the population. The general manager of Telmex, Eduardo Diaz Corona, stated, "We all know that in the world today, being connected [to the Internet] is not a luxury but a need. The technology is no longer for the privileged few."[27]

Most telecommunications experts agree that WiMAX is an attractive option for developing countries with little or no wireless telephone infrastructure. However, it is not clear whether WiMAX will be as successful in developed countries such as the United States, where regular broadband is plentiful and cheap and 3G wireless networks already cover most major metropolitan areas.

WiMAX is a key component of Intel's broadband wireless strategy to deliver innovative mobile platforms for "anytime, anywhere" Internet access. Intel has placed a large bet on the success of WiMAX and hopes to have 1.3 billion people using WiMAX to connect to the Internet by 2012.[28] Higher-end notebook computers will have WiMAX technology that uses a chip called Rosedale. WiMAX cards that plug into a slot in the computer will also be available.[29]

Future Wireless Communications Developments

In 1997, Congress enacted a law requiring that U.S. television stations move to all-digital broadcasts and abandon the analog spectrum available for analog signals in the 700 MHz frequency band used to carry UHF stations 52 to 69. A **digital signal** represents bits, whereas an **analog signal** is a variable signal continuous in both time and amplitude so that any small fluctuations in the signal are meaningful. In 2006, the government set a deadline of February 18, 2009 for all stations to cease analog broadcasting.[30] This shift to digital broadcasting might have a long-term benefit to the public as it frees up portions of the 700 MHz frequency band so it can be reallocated for other purposes.

digital signal
A signal that represents bits.

analog signal
A variable signal continuous in both time and amplitude so that any small fluctuations in the signal are meaningful.

At the 700 MHz frequency, signals travel about four times farther than the signals at the higher frequencies used by Wi-Fi and WiMAX. Also, the lower frequency penetrates walls more effectively. Thus, it is hoped that the reuse of this spectrum by wireless communications providers will enable easier and less expensive deployment of broadband wireless networks, resulting in newer, less expensive, and more widespread high-speed networks for businesses and consumers.[31] A portion of the 700 MHz spectrum is set aside for use by public safety agencies such as police and fire departments.[32] Communications problems during the September 11, 2001, terrorist attacks on the U.S. and later disasters such as Hurricane Katrina illustrated the need for a national voice and data network for public safety agencies.

In another interesting development, Google formed a 34-member alliance consisting of hardware, software, and telecommunications companies to develop Android, a software development platform for mobile phones based on the Linux operating system that will likely compete with Apple's popular iPhone,[33] a combination mobile phone, widescreen iPod, and Internet access device to support e-mail and Web browsing. The iPhone can connect to the Internet either via Wi-Fi or AT&T's Edge data network (considered to be a 2.5G network, not quite as fast as a 3G network).

Apple iPhone

The iPhone is a combination mobile phone, widescreen iPod, and Internet access device.

(Source: Courtesy of Apple Computer.)

NETWORKS AND DISTRIBUTED PROCESSING

computer network
The communications media, devices, and software needed to connect two or more computer systems or devices.

A **computer network** consists of communications media, devices, and software needed to connect two or more computer systems or devices. The computers and devices on the networks are also called *network nodes*. After they are connected, the nodes can share data, information, and processing jobs. Increasingly, businesses are linking computers in networks to streamline work processes and enable employees to collaborate on projects. If a company uses networks effectively, it can grow into an agile, powerful, and creative organization, giving it a long-term competitive advantage. Organizations can use networks to share hardware, programs, and databases. Networks can transmit and receive information to improve organizational effectiveness and efficiency. They enable geographically separated workgroups to share documents and opinions, which fosters teamwork, innovative ideas, and new business strategies.

Network Types

Depending on the physical distance between nodes on a network and the communications and services it provides, networks can be classified as personal area, local area, metropolitan area, or wide area.

Personal Area Networks

personal area network (PAN)
A network that supports the interconnection of information technology within a range of 33 feet or so.

A **personal area network** (PAN) is a wireless network that connects information technology devices within a range of 33 feet or so. One device serves as the controller during wireless PAN initialization, and this controller device mediates communication within the PAN. The controller broadcasts a beacon that synchronizes all devices and allocates time slots for the devices. With a PAN, you can connect a laptop, digital camera, and portable printer without physical cables. You can download digital image data from the camera to the laptop and then print it on a high-quality printer—all wirelessly.

Ford and Microsoft collaborated to develop the Sync service for in-car communications and entertainment. The Sync service creates a wireless connection to cell phones and MP3 players. Sync enables car occupants to place "hands free" cell phone calls using voice commands. Users can also request specified songs from a connected media player using voice commands.[34]

Local Area Networks

local area network (LAN)
A network that connects computer systems and devices within a small area, such as an office, home, or several floors in a building.

A network that connects computer systems and devices within a small area, such as an office, home, or several floors in a building is a **local area network** (LAN). Typically, LANs are wired into office buildings and factories (see Figure 6.6). Although LANs often use unshielded twisted-pair wire, other media—including fiber-optic cable—is also popular. Increasingly, LANs are using some form of wireless communications. You can build LANs to connect personal computers, laptop computers, or powerful mainframe computers.

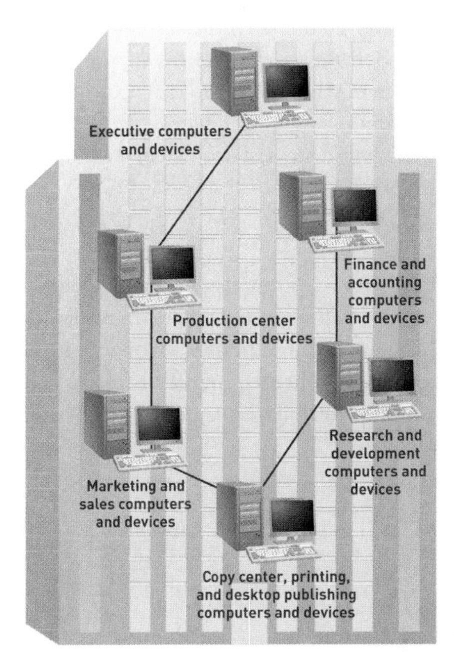

Figure 6.6

A Typical LAN

All network users within an office building can connect to other users' devices for rapid communication. For instance, a user in research and development could send a document from her computer to be printed at a printer in the desktop publishing center.

A basic type of LAN is a simple peer-to-peer network that a small business might use to share files and hardware devices such as printers. In a peer-to-peer network, you set up each computer as an independent computer, but let other computers access specific files on its hard drive or share its printer. These types of networks have no server. Instead, each computer is connected to the next machine. Examples of peer-to-peer networks include Windows for Workgroups, Windows NT, Windows 2000, and AppleShare. Performance of the computers on a peer-to-peer network is usually slower because one computer is actually sharing the resources of another computer.

With more people working at home, connecting home computing devices and equipment into a unified network is on the rise. Small businesses are also connecting their systems and equipment. A home or small business can connect network resources, computers, printers, scanners, and other devices. A person working on one computer, for example, can use data and programs stored on another computer's hard disk. In addition, several computers on the network can share a single printer. To make home and small business networking a reality, many companies are offering networking standards, devices, and procedures.

Disneyland's House of the Future in Tomorrowland features information technology designed to enhance everyday living. A LAN that senses the presence of people throughout the home is the key to making this all work. The lights, temperature, and even the paintings on the wall adjust to preset personal preferences as people enter and leave the rooms. If someone clicks on the remote, the network dims the lights, shuts off any music, and draws the shades in preparation for the TV to turn on. The network also enables people to easily transfer music, photos, and videos among computers and TVs throughout the home.[35]

Metropolitan Area Networks

A **metropolitan area network** (MAN) is a telecommunications network that connects users and their computers in a geographical area that spans a campus or city. Most MANs have a range of roughly 30 to 90 miles. For example, a MAN might redefine the many networks within a city into a single larger network or connect several LANs into a single campus LAN.

metropolitan area network (MAN)
A telecommunications network that connects users and their devices in a geographical area that spans a campus or city.

The Miami-Dade Police Department consists of 3,000 officers and 1,500 civilians who serve and protect more than two million citizens over a 2,100 square mile area. The department implemented a MAN to enable its officers to gain easy access to the data they need while staying mobile on the streets rather than behind a desk. Officers in cruisers connect to hot spots in station parking lots to gain access to the network. Here they can download reports and access local and national databases for fingerprints, mug shots, and other information about suspects. Officers can also participate in pretrial meetings via videoconferencing at their district stations and save the hours required to go downtown and meet face-to-face with prosecutors and others. Bob Reyes, the systems support manager, says: "We're in the midst of a five-year plan. Much of what that involves is putting computers into the cars and providing our officers with timely upgrades, patches, and virus controls through wireless connectivity when they enter their district station areas without having to go inside or wait in line to plug into the network."[36]

Wide area networks

wide area network (WAN)
A telecommunications network that ties together large geographic regions.

A **wide area network** (WAN) is a telecommunications network that connects large geographic regions. A WAN might be privately owned or rented and includes public (shared users) networks. When you make a long-distance phone call or access the Internet, you are using a WAN. WANs usually consist of computer equipment owned by the user, together with data communications equipment and telecommunications links provided by various carriers and service providers (see Figure 6.7).

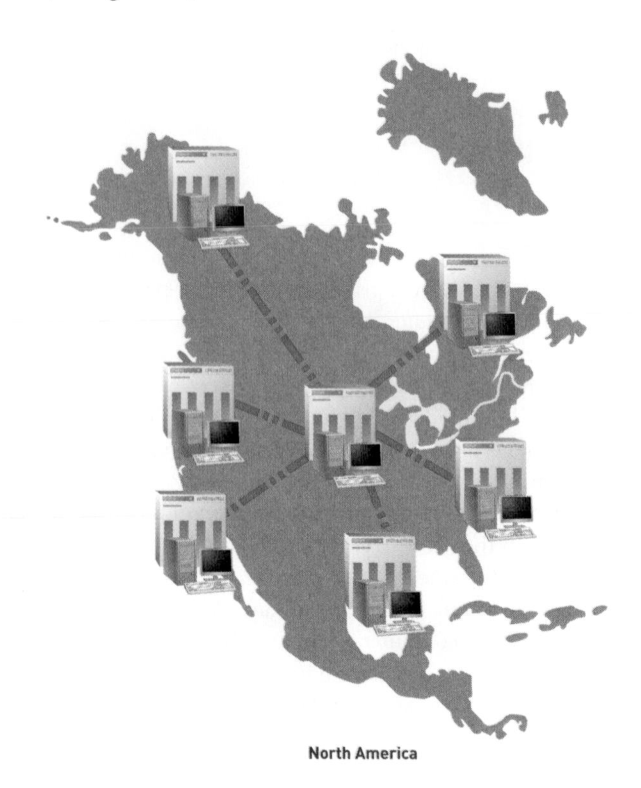

Figure 6.7

A Wide Area Network

WANs are the basic long-distance networks used around the world. The actual connections between sites, or nodes (shown by dashed lines), might be any combination of guided and wireless media. When you make a long-distance telephone call or access the Internet, you are using a WAN.

WANs often provide communications across national borders, which involves national and international laws regulating the electronic flow of data across international boundaries, often called *transborder data flow*. Many countries, including those in the European Union, have strict laws limiting the use of telecommunications and databases, making normal business transactions such as payroll costly, slow, or even impossible.

Basic Processing Alternatives

centralized processing
Processing alternative in which all processing occurs at a single location or facility.

When an organization needs to use two or more computer systems, it can implement one of three basic processing alternatives: centralized, decentralized, or distributed. With **centralized processing**, all processing occurs in a single location or facility. This approach offers the highest degree of control because a single centrally managed computer performs all data processing. The Ticketmaster reservation service is an example of a centralized system. One central computer with a database stores information about all events and records the purchases of seats. Ticket clerks at various ticket selling locations can enter order data and print the results, or customers can place orders directly over the Internet.

With **decentralized processing**, processing devices are placed at various remote locations. Each processing device is isolated and does not communicate with any other processing device. Decentralized systems are suitable for companies that have independent operating units, such as 7-Eleven, where each of its 5,800 U.S. stores is managed to meet local retail conditions. Each store has a computer that runs over 50 business applications such as cash register operations, gasoline pump monitoring, and merchandising.

decentralized processing
Processing alternative in which processing devices are placed at various remote locations.

With **distributed processing**, processing devices are placed at remote locations but are connected to each other via a network. One benefit of distributed processing is that managers can allocate data to the locations that can process it most efficiently. Kroger operates over 2,400 supermarkets, each with its own computer to support store operations such as customer checkout and inventory management. These computers are connected to a network so that sales data gathered by each store's computer can be sent to a huge data repository on a mainframe computer for efficient analysis by marketing analysts and product supply chain managers.

distributed processing
Processing alternative in which computers are placed at remote locations but are connected to each other via a network.

The September 11, 2001, terrorist attacks and the current relatively high level of natural disasters such as Hurricane Katrina sparked many companies to distribute their workers, operations, and systems much more widely, a reversal of the previous trend toward centralization. The goal is to minimize the consequences of a catastrophic event at one location while ensuring uninterrupted systems availability.

File Server Systems

Users can share data through file server computing, which allows authorized users to download entire files from certain computers designated as file servers. After downloading data to a local computer, a user can analyze, manipulate, format, and display data from the file (see Figure 6.8).

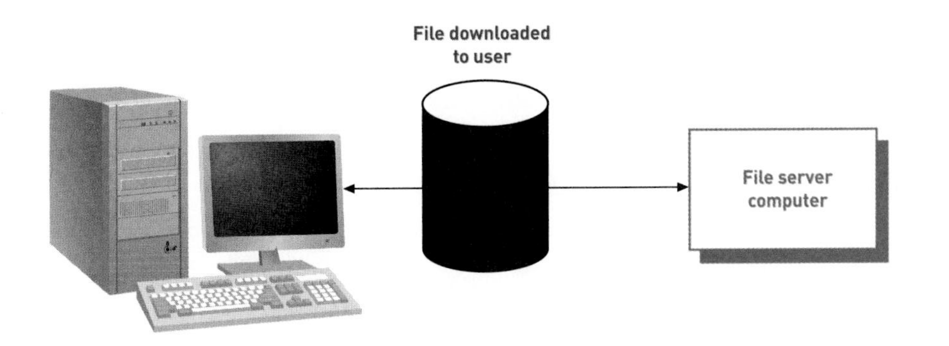

Figure 6.8

File Server Connection

The file server sends the user the entire file that contains the data requested. The user can then analyze, manipulate, format, and display the downloaded data with a program that runs on the user's personal computer.

Client/Server Systems

In **client/server** architecture, multiple computer platforms are dedicated to special functions such as database management, printing, communications, and program execution. These platforms are called *servers*. Each server is accessible by all computers on the network. Servers can be computers of all sizes; they store both application programs and data files and are equipped with operating system software to manage the activities of the network. The server distributes programs and data to the other computers (clients) on the network as they request them. An application server holds the programs and data files for a particular application, such as an inventory database. The client or the server can do the processing.

client/server
An architecture in which multiple computer platforms are dedicated to special functions such as database management, printing, communications, and program execution.

A client is any computer (often a user's personal computer) that sends messages requesting services from the servers on the network. A client can converse with many servers concurrently. For example, a user at a personal computer initiates a request to extract data that resides in a database somewhere on the network. A data request server intercepts the request and determines on which database server the data resides. The server then formats the user's request into a message that the database server will understand. When it receives the message, the database server extracts and formats the requested data and sends the results to the client. The database server sends only the data that satisfies a specific query—not the entire file (see Figure 6.9). As with the file server approach, when the downloaded data is on the user's machine, it can then be analyzed, manipulated, formatted, and displayed by a program that runs on the user's personal computer.

Client
Server
Server
Client
Server

Figure 6.9

Client/Server Connection

Multiple computer platforms, called *servers*, are dedicated to special functions. Each server is accessible by all computers on the network. The client requests services from the servers, provides a user interface, and presents results to the user.

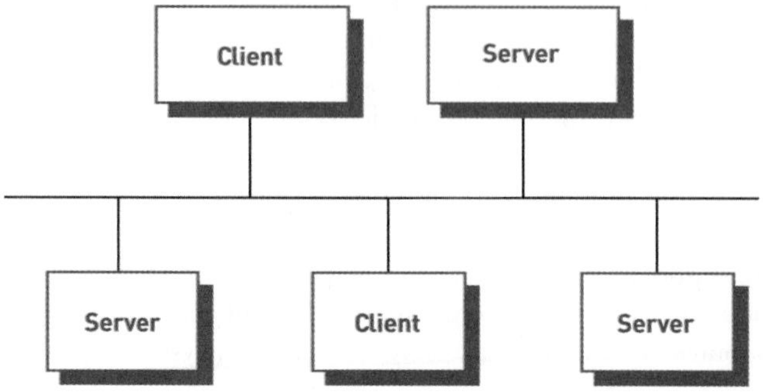

Table 6.3 lists the advantages and disadvantages of client/server architecture.

Table 6.3

Advantages and Disadvantages of Client/Server Architecture

Advantages	Disadvantages
Moving applications from mainframe computers and terminal-to-host architecture to client/server architecture can yield significant savings in hardware and software support costs.	Moving to client/server architecture is a major two- to five-year conversion process.
Minimizes traffic on the network because only the data needed to satisfy a user query is moved from the database to the client device.	Controlling the client/server environment to prevent unauthorized use, invasion of privacy, and viruses is difficult.
Security mechanisms can be implemented directly on the database server through the use of stored procedures.	Using client/server architecture leads to a multivendor environment with problems that are difficult to identify and isolate to the appropriate vendor.

Telecommunications Hardware

Networks require various telecommunications hardware devices to operate including modems, multiplexers, front-end processors, private branch exchanges, switches, bridges, routers, and gateways.

Modems

At each stage of the communications process, transmission media of differing types and capacities may be used. If you use an analog telephone line to transfer data, it can only accommodate an analog signal. Because a computer generates a digital signal represented by bits, you need a special device to convert the digital signal to an analog signal, and vice versa (see Figure 6.10). Translating data from digital to analog is called *modulation*, and translating data from analog to digital is called *demodulation*. Thus, these devices are modulation/demodulation devices, or **modems**. Penril/Bay Networks, Hayes, Microcom, Motorola, and U.S. Robotics are modem manufacturers.

modem
A telecommunications hardware device that converts (modulates and demodulates) communications signals so they can be transmitted over the communication media.

Modems can dial telephone numbers, originate message sending, and answer incoming calls and messages. Modems can also perform tests and checks on how well they are operating. Some modems can vary their transmission rates based on detected error rates and other conditions. Cellular modems in laptop personal computers allow people on the go to connect to wireless networks and communicate with other users and computers.

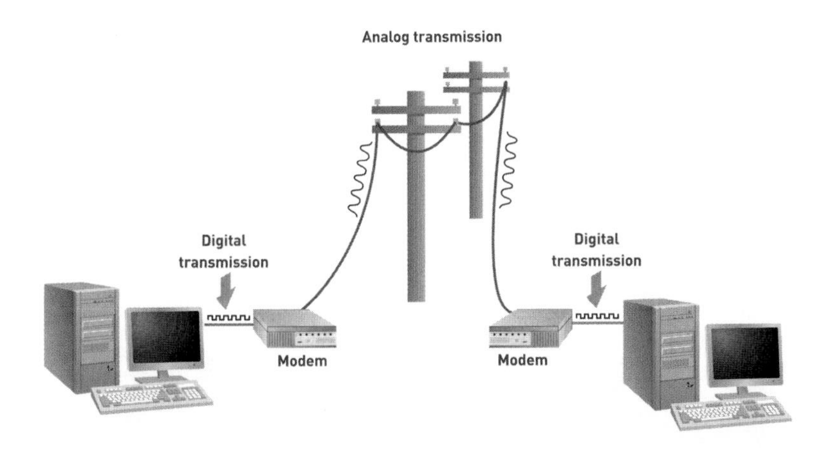

Figure 6.10

How a Modem Works

Digital signals are modulated into analog signals, which can be carried over existing phone lines. The analog signals are then demodulated back into digital signals by the receiving modem.

With a cellular modem, you can connect to other computers while in your car, on a boat, or in any area that has cellular transmission service. You can use PC memory card expansion slots for standardized credit card-sized PC modem cards, which work like standard modems. PC modems are becoming increasingly popular with notebook and portable computer users.

A cable modem can deliver network and Internet access at up to 10Mbps.

(Source: Courtesy of D-Link Systems, Inc.)

Cable company network subscribers use a cable modem, which has a low initial cost and can transmit at speeds up to 10 Mbps. The cable modem is always on, so you can be connected to the Internet around the clock. Digital subscriber line (DSL) is a family of services that provides high-speed digital data communications service over the wires of the local telephone company. Subscribers employ a DSL modem to connect their computers to this service.

Multiplexers

A **multiplexer** is a device that combines data from multiple data sources into a single output signal that carries multiple channels, thus reducing the number of communications links needed and, therefore, lowering telecommunications costs (see Figure 6.11). Multiplexing is commonly used on long-distance phone lines, combining many individual phone calls onto a single long-distance line without affecting the speed or quality of an individual call. At the receiving end, a demultiplexer chooses the correct destination from the many possible destinations and routes each individual call to its correct destination.

multiplexer
A device that combines data from multiple data sources into a single output signal that carries multiple channels, thus reducing the number of communications links needed and therefore, lowering telecommunications costs.

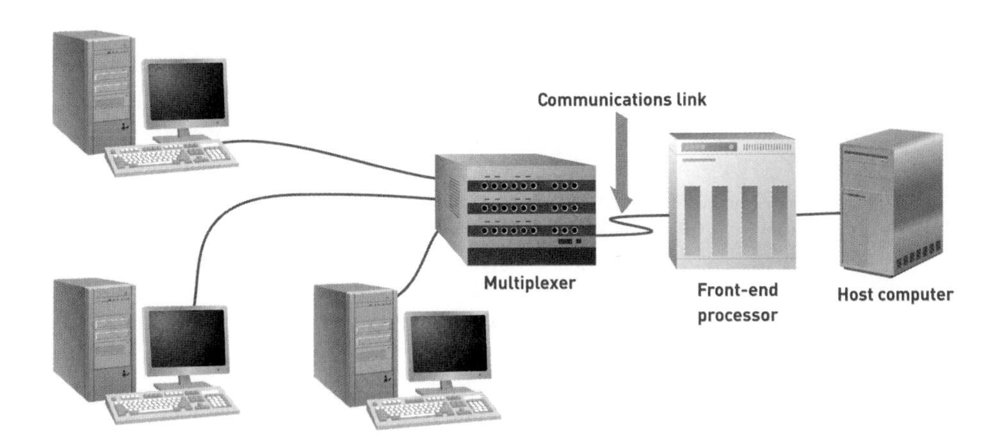

Figure 6.11

Use of a Multiplexer to Consolidate Data Communications onto a Single Communications Link

U.S. Bancorp is a financial services holding company that provides a wide range of services to banks, financial institutions, and government institutions. Its retail division is U.S. Bank, the sixth largest bank in the U.S. The firm uses multiplexers to create a nationwide fiber-optic network to connect its various communications centers.[37]

Telecommunications networks require state-of-the-art computer software technology to continuously monitor the flow of voice, data, and image transmission over billions of circuit-miles worldwide.

(Source: © Roger Tully/Getty Images.)

front-end processor
A special-purpose computer that manages communications to and from a computer system serving hundreds or even thousands of users.

private branch exchange (PBX)
A telephone switching exchange that serves a single organization.

switch
A telecommunications device that uses the physical device address in each incoming message on the network to determine to which output port it should forward the message to reach another device on the same network.

bridge
A telecommunications device that connects one LAN to another LAN using the same telecommunications protocol.

router
A telecommunications device that forwards data packets across two or more distinct networks toward their destinations, through a process known as routing.

gateway
A telecommunications device that serves as an entrance to another network.

network operating system (NOS)
Systems software that controls the computer systems and devices on a network and allows them to communicate with each other.

Front-End Processors

Front-end processors are special-purpose computers that manage communications to and from a computer system serving hundreds or even thousands of users. They poll user devices to see if they have messages to send; facilitate efficient, error-free communications; perform message and transaction switching; multiplexing; transaction security; and end-to-end transaction management and reporting—important functions needed to support mission critical transaction environments such as banking, point-of-sale, and healthcare applications. By performing this work, the front-end processor relieves the primary computer system of much of the overhead processing associated with telecommunications.

Private Branch Exchange (PBX)

A **private branch exchange** (PBX) is a telephone switching exchange that serves a single organization. It enables users to share a certain number of outside lines (trunk lines) to make telephone calls to people outside the organization. This sharing reduces the number of trunk lines required, which reduces the organization's telephone expense. With a PBX, you typically need to dial three or four digits to reach anyone else within the organization. The PBX can also provide many other functions such as voice mail, voice paging, three-way calling, call transfer, and call waiting. Centrex is a form of PBX with all switching occurring at the local telephone office instead of on the organization's premises.

Switches, Bridges, Routers, and Gateways

Telecommunications hardware devices switch messages from one network to another at high speeds. A **switch** uses the physical device address in each incoming message on the network to determine to which output port it should forward the message to reach another device on the same network. A **bridge** connects one LAN to another LAN that uses the same telecommunications protocol. A **router** forwards data packets across two or more distinct networks toward their destinations through a process known as routing. Often, an Internet service provider (ISP) installs a router in a subscriber's home that connects the ISP's network to the network within the home. A **gateway** is a network device that serves as an entrance to another network.

Telecommunications Software

A **network operating system** (NOS) is systems software that controls the computer systems and devices on a network and allows them to communicate with each other. The NOS performs the similar functions for the network as operating system software does for a computer, such as memory and task management and coordination of hardware. When network equipment (such as printers, plotters, and disk drives) is required, the NOS makes sure that these resources are used correctly. Novell NetWare, Windows 2000, Windows 2003, and Windows 2008 are common network operating systems.

MySpace, the popular social networking Web site that offers an interactive, user-submitted network of friends, personal profiles, blogs, photos, music, and videos internationally, was one of the first very busy Web sites to adopt the use of Windows Server 2008.

Because companies use networks to communicate with customers, business partners, and employees, network outages or slow performance can mean a loss of business. Network management includes a wide range of technologies and processes that monitor the network and help identify and address problems before they can create a serious impact.

Software tools and utilities are available for managing networks. With **network-management software**, a manager on a networked personal computer can monitor the use of individual computers and shared hardware (such as printers), scan for viruses, and ensure compliance with software licenses. Network-management software also simplifies the process of updating files and programs on computers on the network—a manager can make changes through a communications server instead of having to visit each individual computer. In addition, network-management software protects software from being copied, modified, or downloaded illegally and performs error control to locate telecommunications errors and potential network problems. Some of the many benefits of network-management software include fewer hours spent on routine tasks (such as installing new software), faster response to problems, and greater overall network control.

network-management software
Software that enables a manager on a networked desktop to monitor the use of individual computers and shared hardware (such as printers), scan for viruses, and ensure compliance with software licenses.

Today, most IS organizations use network management software to ensure that their network remains up and running and that every network component and application is performing acceptably. The software enables IS staff to identify and resolve fault and performance issues before they affect customers and service. The latest network-management technology even incorporates automatic fixes—the network-management system identifies a problem, notifies the IS manager, and automatically corrects the problem before anyone outside the IS department notices it.

T-Mobile Austria GmbH is a subsidiary of T-Mobile International and serves about one-third of all mobile users in Austria. Its infrastructure is highly diverse and includes a mix of hardware from Alcatel, Cisco, Ericsson, Hewlett-Packard, and Siemens using the Microsoft Windows NT and 2000, Solaris, HP-UX, and Linux operating systems. This collection of systems, hardware, and applications requires constant monitoring to detect potential device failures or system bottlenecks before they can generate customer complaints or service failures. "Tivoli Netcool service monitors our Internet services, our mobile radio networks, and most importantly, provides round-the-clock management of our host and server devices. This ensures that important applications will never fail without being noticed," says Dr. Sabine Ringhofer, Senior Manager, Network Operations, T-Mobile Austria.[38]

Securing Data Transmission

The interception of confidential information by unauthorized individuals can cause a compromise of private information about employees or customers, reveal marketing or new product development plans, or cause organizational embarrassment. Organizations with widespread operations need a way to maintain the security of communications with employees and business partners, wherever their facilities are located.

Guided media networks have an inherently secure feature; only devices physically attached to the network can access the data. Wireless networks, on the other hand, are surprisingly often configured by default to allow access to any device that attempts to "listen to" broadcast communications. Action must be taken to override the defaults.

Encryption of data is one approach taken to protect the security of communications over both wired and wireless networks. **Encryption** is the process of converting an original message into a form that can only be understood by the intended receiver. A key is a variable value that is applied (using an algorithm) to a set of unencrypted text to produce encrypted text or to decrypt encrypted text (see Figure 6.12). The key is chosen from one of a large number of possible encryption keys. The longer the key, the greater the number of possible encryption keys. An encryption protocol based on a 56-bit key, for example, has 2^{56} different possible keys while one based on a 128-bit key has 2^{128} different possible keys. Of course, it is essential

encryption
The process of converting an original message into a form that can only be understood by the intended receiver.

that the key be kept secret from possible interceptors. A hacker who obtains the key by whatever means can recover the original message from the encrypted data.

Figure 6.12

The Encryption Process

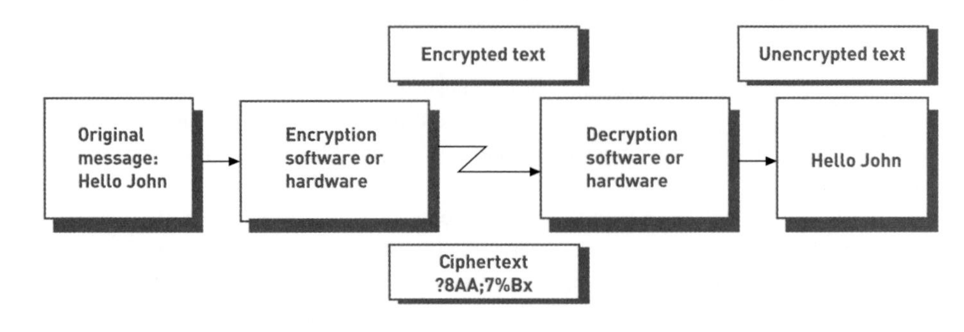

Encryption methods rely on the limitations of computing power for their security—if breaking a code requires too much computing power, even the most determined hacker cannot be successful. In an alarming breakthrough, two research groups in Australia and China working independently have built laser-based quantum computers that can implement Shor's algorithm, a mathematical routine capable of cracking modern encryption methods. This development raises the potential that others could build such computers and break the codes that protect our national security secrets, banking transactions, and business data.[39]

Securing Wireless Networks

Wired equivalent privacy (WEP)
An early attempt at securing wireless communications based on encryption using a 64- or 128-bit key that is not difficult for hackers to crack.

Wi-Fi Protected Access (WPA)
A security protocol that offers significantly improved protection over WEP.

WEP and WPA are the two main approaches to securing wireless networks such as Wi-Fi and WiMAX. **Wired equivalent privacy (WEP)** used encryption based on 64-bit key, which has been upgraded to a 128-bit key. WEP represents an early attempt at securing wireless communications and is not difficult for hackers to crack. Most wireless networks now employ the **Wi-Fi Protected Access (WPA)** security protocol that offers significantly improved protection over WEP.

The following steps, while not foolproof, help safeguard a wireless network:

- Connect to the router and change the default logon (admin) and password (password) for the router. These defaults are widely known by hackers.
- Create a service set identifier (SSID). This is a 32-character unique identifier attached to the header portion of packets sent over a wireless network that differentiates one network from another. All access points and devices attempting to connect to the network must use the same SSID.
- Configure the security to WEP or WPA, preferably WPA if all devices connected to the network are WPA compatible. Surprisingly, many routers are shipped with encryption turned off.
- Disable SSID broadcasting. By default, wireless routers broadcast a message communicating the SSID so wireless devices within range (such as a laptop) can identify and connect to the wireless network. If a device doesn't know the wireless network's SSID, it cannot connect. Disabling the broadcasting of the SSID will discourage all but the most determined and knowledgeable hackers.
- Configure each wireless computer on the network to access the network by setting the security to WEP or WPA and entering the same password entered to the router.

War driving involves hackers driving around with a laptop and antenna trying to detect insecure wireless access points. Once connected to such a network, the hacker can gather enough traffic to analyze and crack the encryption. On a WEP-encrypted network and with 85,000 packets to analyze, there is about a 95 percent probability that the hacker can crack the code in less than two minutes using the program aircrack-ptw running on an ordinary Pentium personal computer.[40] This approach was probably used to swipe data on some 45 million credit and debit card customers of TJX, the parent company of T.J. Maxx, Marshalls, Winners, Home Goods, and other retailers.[41]

Other Encryption Methods

Data Encryption Standard (DES) is an early data encryption standard developed in the 1970s that uses a 56-bit private key algorithm. Today's computers can crack the DES code in a matter of minutes.[42] As a result, the Triple-DES algorithm was developed. This algorithm encrypts the data with one 56-bit key and then encrypts it a second time with a different 56-bit key. The result is encrypted a third time using the original 56-bit key.[43]

Data Encryption Standard (DES)
An early data encryption standard developed in the 1970s that uses a 56-bit private key algorithm.

State and federal regulatory requirements do not allow banks to use wireless communications without an approved encryption system to protect communications. ERF Wireless is a communications service provider whose broadband wireless service called BankNet meets these requirements for several financial institutions in Texas, Missouri, and Louisiana. The system is based on use of Triple-DES encryption and can transmit data at 10 Mbps.[44]

Advanced Encryption Standard (AES) is an extremely strong data encryption standard sponsored by the National Institute of Standards and Technology based on a key size of 128 bits, 192 bits, or 256 bits. It replaces DES and can encrypt data much faster than Triple-DES. It is used to send and receive unclassified material by U.S. government agencies and may eventually become the encryption standard for commercial transactions in the private sector. If a computer could crack a DES key in one second, it would take that machine approximately 149 trillion years to crack a 128-bit AES key. To put that into perspective, the universe is believed to be less than 20 billion years old.

Advanced Encryption Standard (AES)
An extremely strong data encryption standard sponsored by the National Institute of Standards and Technology based on a key size of 128 bits, 192 bits, or 256 bits.

Encryption for the U.S. military and other classified communications is handled by other secret algorithms.

Virtual Private Network (VPN)

The use of a virtual private network is another means used to secure the transmission of communications. A **virtual private network (VPN)** is a private network that uses a public network (usually the Internet) to connect multiple remote locations. A VPN provides network connectivity over a potentially long physical distance and thus can be considered a form of wide area network. VPNs support secure, encrypted connections between a company's private network and remote users through a third-party service provider. Telecommuters, salespeople, and frequent travelers find the use of a VPN to be a safe, reliable, low-cost way to connect to the corporate intranet.

virtual private network (VPN)
A private network that uses a public network (usually the Internet) to connect multiple remote locations.

ROI, the Dutch Institute for Public Administration, provides training programs for government agencies. Each year its 400-plus employees train thousands of civil servants in the Netherlands and throughout Europe. Much of the information customers share with ROI is highly confidential so that IT security is critical. The bulk of ROI employees do not work at the headquarters in The Hague. To access data stored there, they use a VPN to communicate with each other and to share customer information confidentially.[45]

Often users are provided with a security token that displays a constantly changing password to log onto the VPN. This solution avoids the problem of users forgetting their password while providing added security through use of a password constantly changing every 30–60 seconds. Technological and Commercial joint-stock Bank (Techcombank) is one of Vietnam's largest and fastest growing banks with 110 branches and offices in 20 provinces and cities. Over 100,000 customers access its services online via a secure Web site. The customer's identity is verified using two parameters: the customer-created password and a one-time six-digit secure password generated by an RSA SecurID security token. This two-factor security solution is considered to be an extremely secure and strong approach to Internet banking consistent with global best practices.[46]

TELECOMMUNICATIONS SERVICES AND NETWORK APPLICATIONS

Telecommunications and networks are a vital part of today's information systems. In fact, it is hard to imagine how organizations could function without them. For example, when a business needs to develop an accurate monthly production forecast, a manager simply downloads sales forecast data gathered directly from customer databases. Telecommunications provides the network link, allowing the manager to access the data quickly and generate the production report, which supports the company's objective of better financial planning. This section looks at some of the more significant telecommunications services and network applications.

Cellular Phone Services

Cellular phones operate using radio waves to provide two-way communications. The cell phone has become ubiquitous and is an essential part of life in the twenty-first century. It is estimated that approximately 170 million landlines were in use in the U.S. at the end of 2007, compared to 250 million cell phones—including residential and business cell phones.[47] In the United States, the largest wireless license holders run mobile networks that are in near-constant transition; these companies include AT&T, Nextel, Sprint PCS, T-Mobile, and Verizon.

With cellular transmission, a local area such as a city is divided into cells. As a person with a cellular device such as a mobile phone moves from one cell to another, the cellular system passes the phone connection from one cell to another (see Figure 6.13). The signals from the cells are transmitted to a receiver and integrated into the regular phone system. Cellular phone users can thus connect to anyone who has access to regular phone service, such as a child at home or a business associate in London. They can also contact other cellular phone users. Because cellular transmission uses radio waves, people with special receivers can listen to cellular phone conversations, so they are not secure.

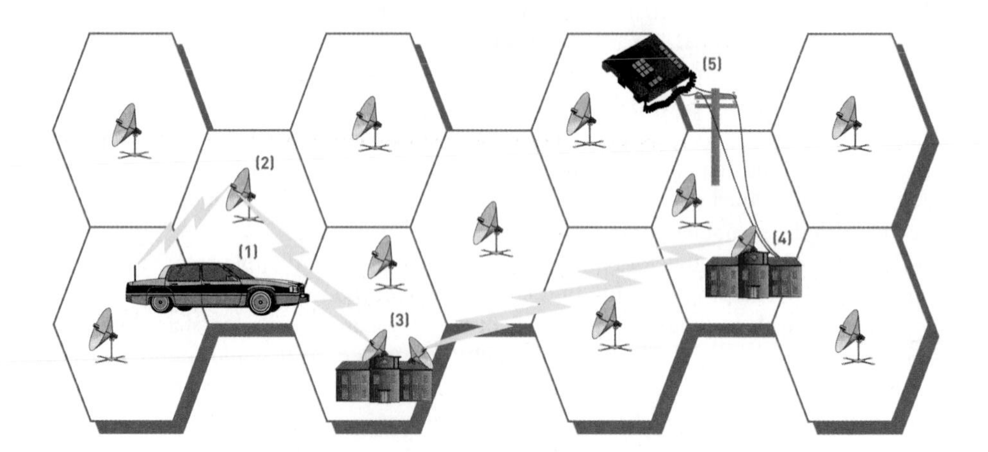

Figure 6.13

A Typical Cellular Transmission Scenario

Using a cellular car phone, the caller dials the number (1). The signal is sent from the car's antenna to the low-powered cellular antenna located in that cell (2). The signal is sent to the regional cellular phone switching office, also called the *mobile telephone switching office (MTSO)* (3). The signal is switched to the local telephone company switching station closest to the call destination (4). Now integrated into the regular phone system, the call is switched to the number originally dialed (5), all without the need for operator assistance.

Increasingly, workers rely on their mobile phones as their primary business phones. However, they frequently encounter problems with poor in-building coverage and find it difficult to place calls or conduct an extended conversation. A picocell is a miniature cellular base station designed to serve a very small area such as part of a floor inside a building. Many communications companies now offer picocell solutions to boost cell phone signals (e.g., the Spotwave Zen)[48] or enable the cell phone to operate over other wireless networks (e.g., RadioFrame Picocell and Femtocell Base Stations), thus guaranteeing a strong, reliable cell signal.[49] Picocells are being installed to provide service on aircraft registered in European, Asian, and Middle Eastern countries to enable passengers to place calls and text messages. Picocells are also enabling aircraft crew to update approach charts and access management networks while parked at the gate.[50] The Ethical and Societal Issues sidebar describes how data about cell phone usage can help direct drivers to avoid congested highways.

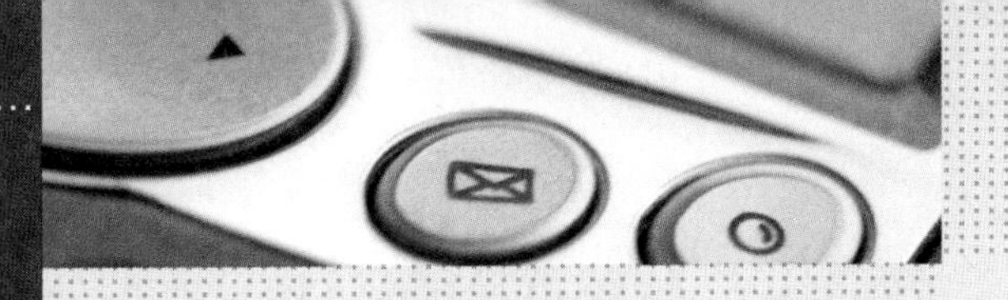

ETHICAL AND SOCIETAL ISSUES

Bangalore Clears Congestion with Telecommunications

Bangalore is India's third most populous city and the hub of the country's information services industry; in fact, Bangalore is known as the Silicon Valley of India. Bangalore is host to many United States companies that outsource portions of their workload. The crowded and congested Bangalore area is growing in population at exponential rates. The economy is booming as well. With many people moving to Bangalore, and with many residents able to afford a car for the first time, it is estimated that 700 new vehicles are being added to Bangalore's crowded streets every day. Due to a chaotic traffic management system, the traffic in many areas of Bangalore grinds to a halt for hours each day.

Recently, the commissioner of police and traffic in Bangalore partnered with Indian cell phone giant Bharti Airtel and the geographic information systems company Mapunity to harness the power of telecommunications and apply it to Bangalore's traffic problems. The collaboration benefitted all partners.

Bharti Airtel, or Airtel for short, is India's largest global system for mobile communications (GSM) cell phone carrier with 62 million subscribers. Airtel was experiencing a high rate of dropped calls in areas of Bangalore where traffic congestion was the most dense. Customers were not happy. Imagine being stuck in traffic for hours with no cell phone service. To resolve the problem, Airtel constructed small cell towers to service only the areas of highway where traffic was the worst. The investment paid off. Hardly any calls were dropped during peak rush hour.

Bangalore's police and traffic commissioner, M.N. Reddi, had a brilliant idea. He asked Airtel to provide his agency with streaming information on how many cell phones were connected to the network along Bangalore highways. Airtel could easily provide this information because of its recently installed mini towers. Reddi discovered that the number of cell phones in a given area was a perfect indicator of the number of cars on the highway and the level of traffic congestion.

Reddi contacted Mapunity, which uses Google Maps and geographic information systems software from the N.S. Raghavan Center for Entrepreneurial Learning (NSRCEL), to build a system that employs cell phone usage data to inform police and commuters of traffic conditions in real time. The system, called Bangalore Transport Information System (BTIS), can be accessed on the Web at *www.btis.in* or through cell phone text messaging.

Having a finger on the pulse of Bangalore traffic has allowed BTIS to offer several services to commuters. At the Web site, commuters can view traffic patterns on a map and plan their commute to follow the least congested roads. A commuter can also provide the site with a starting point and destination and allow the software to suggest the fastest route—one that is most direct and also least congested. The Web site also provides a tool to arrange for carpooling. Cell phone users can send a text message to the system using different codes for different areas and receive a message outlining heavy traffic areas to avoid. Reddi is also installing plasma screens in the lobbies of Bangalore hotels and large displays in technology parks that show commuters areas to avoid as they take off on the road.

After two weeks of operation, the free service was getting around 4,000 text message requests and 2,000 people visiting the site each day—numbers that are growing by five percent daily. The system is being expanded to serve other cities in India.

Discussion Questions

1. What people, companies, and organizations were affected by the traffic congestion in Bangalore? In what way?
2. What telecommunications technologies are used by the BTIS?

Critical Thinking Questions

1. How did the solution provided by the partnership between the Bangalore traffic commissioner and Airtel save the city and the company money? Consider the cost of a solution had these two entities not become partners.
2. What privacy issues does Airtel raise by providing customer usage data to the Bangalore police? How can these issues be addressed?

Sources: Goswami, Kanika, "Traffic Problem Finds Cell Phone Solution," *Computerworld*, August 30, 2007, *www.computerworld.com/action/article.do?command=viewArticleBasic&articleId=9033 738&pageNumber=1*; BangaloreTransport Information System, *www.btis.in*, accessed April 26, 2008; Airtel Web site, *www.airtel.in*, accessed April 26, 2008; Mapunity Web site, www.mapunity.com, accessed April 26, 2008.

Digital Subscriber Line (DSL) Service

digital subscriber line (DSL)
A telecommunications service that delivers high-speed Internet access to homes and small businesses over the existing phone lines of the local telephone network.

A **digital subscriber line** (DSL) is a telecommunications service that delivers high-speed Internet access to homes and small businesses over the existing phone lines of the local telephone network (see Figure 6.14). Most home and small business users are connected to an *asymmetric DSL (ADSL)* line designed to provide a connection speed from the Internet to the user (download speed) that is three to four times faster than the connection from the user back to the Internet (upload speed). ADSL does not require an additional phone line and yet provides "always-on" Internet access. A drawback of ADSL is that the farther the subscriber is from the local telephone office, the poorer the signal quality and the slower the transmission speed. ADSL provides a dedicated connection from each user to the phone company's local office, so the performance does not decrease as new users are added. Cable modem users generally share a network loop that runs through a neighborhood so that adding users means lowering the actual transmission speeds. *Symmetric DSL (SDSL)* is used mainly by small businesses and does not allow you to use the phone at the same time, but the speed of receiving and sending data is the same.

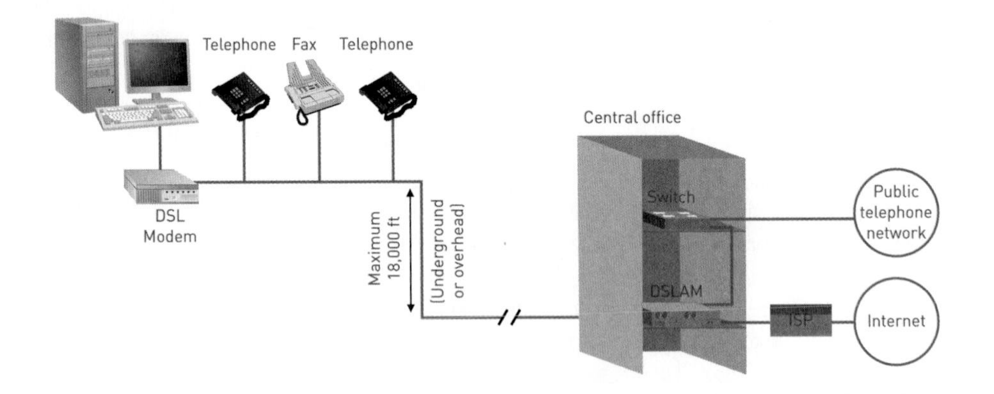

Figure 6.14

Digital Subscriber Line (DSL)

At the local telephone company's central office, a DSL Access Multiplexer (DSLAM) takes connections from many customers and multiplexes them onto a single, high-capacity connection to the Internet. Subscriber phone calls can be routed through a switch at the local telephone central office to the public telephone network.

Voice over Internet Protocol (VoIP) Services

Voice over Internet Protocol (VoIP)
A collection of technologies and communications protocols that enables your voice to be converted into packets of data that can be sent over a data network such as the Internet, a WAN or LAN.

Voice over Internet Protocol (VoIP) is a collection of technologies and communications protocols that enables voice conversations to be converted into packets of data that can be sent over a data network such as the Internet, a WAN, or a LAN. You can use VoIP to make a call directly from a computer equipped with appropriate software and a microphone, a special VoIP phone, or an ordinary phone connected to an analog telephone adapter that converts the analog voice signal into data packets (see Figure 6.15).

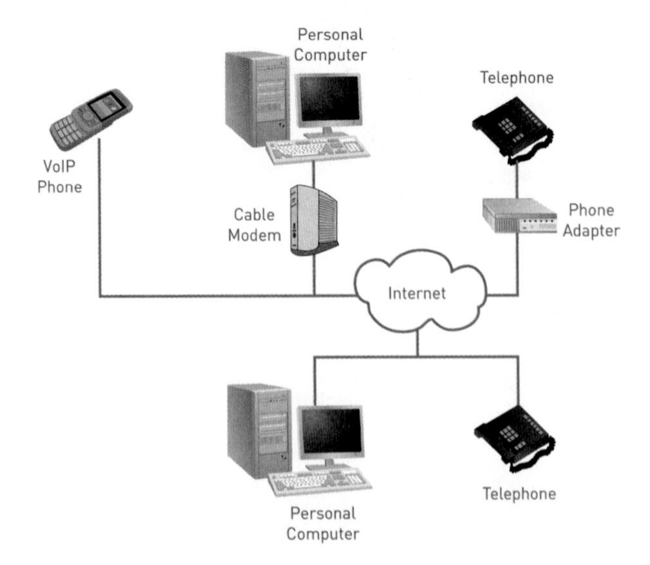

Figure 6.15

VoIP Options

Some cell phones, such as the BlackBerry Curve 8320, can send and receive data and make VoIP phone calls over wireless Wi-Fi 802.11 b/g networks. When you exit the hot spot, the phone switches the call seamlessly to the cellular network. Being able to make phone calls over Wi-Fi is a great option in areas where the cellular service is spotty.[51]

Some of the advantages associated with the use of VoIP include:

- For corporate customers, some telecommunications-related cost savings are possible by using a single data network to carry both voice and data.
- VoIP enables online retailers to provide "click to talk" customer service—online customers needing additional help can click a hyperlink to create a VoIP phone connection to a live customer service representative.
- Some VoIP providers permit users to call anywhere in the world at a low cost per minute.

Some of the disadvantages of VoIP include:

- VoIP users may not be able to place calls if power is lost either locally or at the broadband carrier itself because broadband modems and other VoIP equipment depend on electricity from the power company. Conventional phones can continue to operate during a local power failure because they receive their power from the telephone company's local office, which operates on a separate power supply.
- Some VoIP calls might have lower quality than conventional phone calls if data packets are lost or delayed at any point in the network between VoIP users.
- Support for the sending of faxes over VoIP is still limited.
- Not all VoIP services connect directly to emergency services through 911.
- Not all VoIP providers offer directory assistance or white page listings.

Merrill Lynch, the global financial management and advisory company, implemented an advanced VoIP voice trading system. Over 4,000 IQ/MAX next-generation trading desktops were installed on its two largest trading floors in New York and London plus other trading floors around the globe. Tony Kerrison, CTO, Global Technology Infrastructure at Merrill Lynch, said, "... what matters most to our traders is efficiency, dependability, and user friendliness. The outstanding VoIP technology IQ/MAX delivers will provide our traders with highly advanced, reliable, and collaborative trading floor communications."[52]

Linking Personal Computers to Mainframes and Networks

One of the most basic ways that telecommunications connect users to information systems is by connecting personal computers to mainframe computers so that data can be downloaded or uploaded. For example, a user can download a data file or document file from a database to a personal computer. Some telecommunications software programs instruct the computer to connect to another computer on the network, download or send information, and then disconnect from the telecommunications line. These programs are called *unattended systems* because they perform the functions automatically, without user intervention.

Voice Mail

With **voice mail**, users can send, receive, and store verbal messages for and from other people around the world. Some voice mail systems assign a code to a group of people. Suppose the code 100 stands for the 238 sales representatives in a company. If anyone calls the voice mail system, enters the number 100, and leaves a message, all 238 sales representatives receive the same message. Call management systems can be linked to corporate e-mail and instant messaging systems. Calls to employees can be generated from instant messages or converted into e-mail messages to ensure quicker access and response.

voice mail
Technology that enables users to send, receive, and store verbal messages for and from other people around the world.

Reverse 911 Service

Reverse 911 service is a communications solution that delivers emergency notifications to users in a selected geographical area. The technology employs databases of phone numbers and contact information to send over 250,000 voice or text messages via phone, pager, cell phone, or e-mail per hour. A series of messages was sent to some 500,000 residents of San Diego using reverse 911 services when disastrous wildfires swept the area in October 2007.[53]

reverse 911 service
A communications solution that delivers emergency notifications to users in a selected geographical area.

Voice-to-Text Services

Voice mail is more difficult to manage than e-mail because you must deal with messages one-by-one without knowing who has called you and without being able to prioritize the messages. In recognition of these shortcomings of voice mail, several services (e.g., Jott, SpinVox, GotVoice, and SimulScribe) are now available to convert speech to text so that you can manage voice mails more effectively. If you subscribe to one of these **voice-to-text services**, your voice mail no longer reaches your phone service provider's voice mail service. Instead, it is rerouted to the voice-to-text service, translated into text, and then sent to your regular e-mail account or a special account for converted e-mail messages. You can also temporarily disable the voice-to-text service and receive voice messages.[54]

voice-to-text service
A service that captures voice mail messages, converts them to text, and sends them to an e-mail account.

Home and Small Business Networks

Small businesses and many families own more than one computer and look for ways to set up a simple network to share printers or an Internet connection; access shared files such as photos, MP3 audio files, spreadsheets, and documents on different machines; play games that support multiple concurrent players; and send the output of network-connected devices, such as a security camera or DVD player, to a computer.

One simple solution is to establish a wireless network that covers your home or small business. To do so, you can buy an 802.11*n* access point, connect it to your cable modem or DSL modem, and then use it to communicate with all your devices. For less than $100, you can purchase a combined router, firewall, Ethernet hub, and wireless hub in one small device. Computers in your network connect to this box with a wireless card, which is connected by cable or DSL modem to the Internet. This enables each computer in the network to access the Internet. The firewall filters the information coming from the Internet into your network. You can configure it to reject information from offensive Web sites or potential hackers. The router can also encrypt all wireless communications to keep your network secure.

In addition, you can configure your computers to share printers and files. Windows Vista includes a Network and Sharing Center that helps with network configuration. Some of the basic configuration steps include assigning each computer to a workgroup and giving it a name, identifying the files you want to share (placing an optional password on some files), and identifying the printers you want to share.

Electronic Document Distribution

Electronic document distribution lets you send and receive documents in a digital form without printing them (although printing is possible). It is much faster to distribute electronic documents via networks than to mail printed forms. Viewing documents on screen instead of printing them also saves paper and document storage space. Accessing and retrieving electronic documents is also much faster.

electronic document distribution
A process that enables the sending and receiving of documents in a digital form without being printed (although printing is possible).

An issue in the use of some paper documents (e.g., college transcripts and various government documents) is the requirement to verify their authenticity. Hewlett-Packard has developed technology at its lab in Bangalore, India, that enables authentication of paper documents based on a two-dimensional bar code printed on the back of the document. The bar code is read by a scanner and verified online in a database accessible at a Web site established by the issuing authority. Bangalore now issues bar-coded transcripts.[55]

Call Centers

A call center is a physical location where an organization handles customer and other telephone calls, usually with some amount of computer automation. Call centers are used by customer service organizations, telemarketing companies, computer product help desks, charitable and political campaign organizations, and any organization that uses the telephone to sell or support products and services. An automatic call distributor (ACD) is a telephone facility that manages incoming calls, handling them based on the number called and an associated database of instructions. Call centers frequently employ an ACD to validate callers, place outgoing calls, forward calls to the right party, allow callers to record messages, gather usage statistics, balance the workload of support personnel, and provide other services.

The National Do Not Call Registry was set up in 2003 by the U.S. Federal Trade Commission. Telemarketers who call numbers on the list face penalties of up to $11,000 per call, as well as possible consumer lawsuits. Over 60 million consumers have signed onto the list by logging on to the site *www.donotcall.gov*. Although the registry has greatly reduced the number of unwanted calls to consumers, it has created several compliance-related issues for direct marketing companies.

Offshore call centers that provide technical support services are a fact of life for many technology vendors and their customers. However, vendors and users agree that support operations have to balance their desire to reduce labor costs with customer satisfaction considerations. In a cost-cutting move, Barclays PLC's Barclaycard credit card unit closed its customer call center in Manchester, England and transferred 630 of the jobs to Mumbai and Delhi, India. Barclaycard CEO Anthony Jenkins stated: "Barclaycard's business is becoming more global, and to stay successful, we must change how we operate to reflect this." Interestingly, rival financial services firm Lloyds TSB decided to close its call center in Mumbai, India, and enable customers to contact representatives at local branches directly.[56]

Offshore call centers provide technical support services for many technology vendors and their customers.

(Source: © STR/AFP/Getty Images.)

Telecommuting and Virtual Workers and Workgroups

telecommuting
A work arrangement whereby employees work away from the office using personal computers and networks to communicate via e-mail with other workers and to pick up and deliver results.

Employees are increasingly performing work away from the traditional office setting. Many enterprises have adopted policies for **telecommuting** so that employees can work away from the office using computing devices and networks. This means workers can be more effective and companies can save money on office and parking space and office equipment. According to a survey by CDW Corporation, 44 percent of respondents who work for the federal government have the option to telecommute while only 15 percent of private sector employees can.[57]

Telecommuting is popular among workers for several reasons. Parents find that eliminating the daily commute helps balance family and work responsibilities. Qualified workers who otherwise might be unable to participate in the normal workforce (e.g., those who are physically challenged or who live in rural areas too far from the city office to commute regularly) can use telecommuting to become productive workers. When gas prices soar, telecommuting can help workers reduce significant expenses. Extensive use of telecommuting can lead to decreased need for office space, potentially saving a large company millions of dollars. Corporations are also being encouraged by public policy to try telecommuting as a means of reducing traffic congestion, oil consumption, and air pollution. Large companies view telecommuting as a means to distribute their workforce and reduce the impact of a disaster at a central facility.

Some types of jobs are better suited for telecommuting than others, including jobs held by salespeople, secretaries, real estate agents, computer programmers, and legal assistants, to name a few. Telecommuting also requires a special personality type to be effective. Telecommuters need to be strongly self-motivated, organized, focused on their tasks with minimal supervision, and have a low need for social interaction. Jobs unsuitable for telecommuting include those that require frequent face-to-face interaction, need much supervision, and have many short-term deadlines. Employees who choose to work at home must be able to work independently, manage their time well, and balance work and home life.

Cable TV provider Cox Communications employs about 22,000 people, of which roughly 10 percent are teleworkers who provide customer service to Cox's six million subscribers. Cox estimates a net savings of roughly $3,400 per year per call agent in reduced office space, energy, and the cost of employee parking. Cox requires teleworkers to come to the regional office every couple of weeks for meetings and employs videoconferencing as well. Josh Nelson, vice president of information and technology for Cox, states: "Working from home doesn't mean that they want to be alone. We don't want to lose our company culture, we don't want to lose the connection with our employees."[58]

Videoconferencing

videoconferencing
A telecommunications system that combines video and phone call capabilities with data or document conferencing.

Videoconferencing enables people to hold a conference by combining voice, video, and audio transmission. Videoconferencing reduces travel expenses and time, and increases managerial effectiveness through faster response to problems, access to more people, and less duplication of effort by geographically dispersed sites. Almost all videoconferencing systems combine video and phone call capabilities with data or document conferencing (see Figure 6.16). You can see the other person's face, view the same documents, and swap notes and drawings. With some systems, callers can change live documents in real time. Many businesses find that the document- and application-sharing feature of the videoconference enhances group productivity and efficiency. It also fosters teamwork and can save corporate travel time and expense.

Figure 6.16

Videoconferencing

Videoconferencing allows participants to conduct long-distance meetings "face to face" while eliminating the need for costly travel.

(Source: © Comstock Images/ Alamy.)

Group videoconferencing is used daily in a variety of businesses as an easy way to connect work teams. Members of a team meet in a specially prepared videoconference room equipped with sound-sensitive cameras that automatically focus on the person speaking, large TV-like monitors for viewing the participants at the remote location, and high-quality speakers and microphones. Videoconferencing costs have declined steadily, while video quality and synchronization of audio to data—once weak points for the technology—have improved.

Jeffrey Marshall, senior vice president and chief information officer at retailer Kohl's, employs videoconferencing to interact with job candidates that have passed an initial screening by its in-house recruiters.[59] Another example is Google's Earth Outreach, a program to help nonprofit organizations communicate their mission and advocate their work. Google held a videoconference to help kick off the program that included an appearance by noted wildlife advocate Jane Goodall, who participated via videoconference from London.[60] Read the Information Systems @ Work special feature to learn how one company makes effective use of videoconferencing (telepresence) technology.

INFORMATION SYSTEMS @ WORK

Telepresence Eliminates Travel and Saves Valuable Human Resources

Derek Chan, head of digital operations at DreamWorks Studios, is ecstatic. DreamWorks used to release an animated film every 18 to 36 months. "Now we're doing a show in May, another in November, and then May. We're reaching a scale no one else has been able to do. When we ended up building these systems, it changed the landscape for us," says Chan. The systems Chan is using are telepresence systems.

DreamWorks Animation SKG, creators of many popular films including "Shrek," "Madagascar," "Chicken Run," and "Bee Movie," is well known for its high-quality 3-D animation. Creating these movies takes the combined effort of many top animators located around the world. For DreamWorks, a great deal of collaboration takes place between its home studios in Glendale, California, and its subsidiary, Pacific Data Images, 400 miles north in Redwood City.

The long distance between sites was causing important DreamWorks executives, artists, and directors to waste time traveling rather than creating. DreamWorks tried a variety of network conferencing systems, but none provided a smooth stream of communication—in-person visits were much more productive. DreamWorks partnered with Hewlett-Packard (HP) to create a videoconferencing system that allows people around the world to communicate as though they were sitting around a conference table. The result is an HP product called HP Halo, a telepresence and videoconferencing system. The technology has proven successful and similar systems are being offered by other vendors under the general title of telepresence systems.

Telepresence participants sit at a long, one-sided conference table facing a wall covered with large displays. The room is equipped with unobtrusive video cameras and a high-resolution document camera. When connected to the telepresence studio at another location, the displays show the other meeting participants seated as if across the table. A large display above the participants shows documents that people want to share using the document camera or directly from a PC. Microphones and speakers allow participants to converse in a natural voice. People outside the telepresence environment can dial into the system to join in on audio.

Network connections are the most important component of the telepresence system. DreamWorks' HP Halo system provides a dedicated high-bandwidth network line between its Glendale studios and its Redwood City offices. Users describe the so-called tele-immersive environment as being stunningly lifelike. It's as though the participants are physically together.

For DreamWorks, this means a fundamental change in the way it does business. Teams can work together by sharing and discussing documents, images, and video, while cutting the time and cost of travel. Animators and producers use the system to collaborate from disperse geographic locations, developing storyboards, reviewing artwork, and adjusting character designs in real time. The document camera allows artists to sketch ideas to share with the group. The network is fast enough to transfer video clips from motion pictures while it is transferring live action video of participants.

Using HP Halo, DreamWorks became the first company to release two animated films in one year. Executives that previously traveled overseas once every three weeks now travel once every three months. Trips between Glendale and Redwood City have been reduced by as much as 80 percent.

Discussion Questions

1. What network considerations are involved when introducing a telepresence system? How might DreamWorks' requirements vary from a retail company such as Home Depot?
2. How did telepresence allow DreamWorks employees to be more productive and efficient?

Critical Thinking Questions

1. Although some workforces are becoming more mobile, others such as DreamWorks find it more effective to stay put. What types of business activities are best carried out through travel? What kinds of professionals benefit from avoiding travel? How do telecommunications assist both?
2. In your chosen career, do you anticipate a lot of travel or a little? What types of activities will you perform that require telecommunications?

Sources: King, Julia, "Premier 100 IT Leaders 2008," *Computerworld*, December 10, 2007, *www.computerworld.com/action/article.do?command=viewArticleBasic&articleId=305908&pageNumber=1*; HP Staff, "DreamWorks speeds film production with HP Halo Collaboration Studio," HP Case Study, *http://h20219.www2.hp.com/enterprise/downloads/Case%20Study_DreamWorks%20hi-res_7_17_07.pdf*, accessed April 28, 2008; HP Halo Web site, *http://h20219.www2.hp.com/enterprise/cache/570006-0-0-0-121.html*, accessed April 28, 2008.

Electronic Data Interchange

electronic data interchange (EDI)
An intercompany, application-to-application communication of data in a standard format, permitting the recipient to perform a standard business transaction, such as processing purchase orders.

Electronic data interchange (EDI) is a way to communicate data from one company to another and from one application to another in a standard format, permitting the recipient to perform a standard business transaction, such as processing purchase orders. Connecting corporate computers among organizations is the idea behind EDI, which uses network systems and follows standards and procedures that can process output from one system directly as input to other systems, without human intervention. EDI can link the computers of customers, manufacturers, and suppliers (see Figure 6.17). This technology eliminates the need for paper documents and substantially cuts down on costly errors. Customer orders and inquiries are transmitted from the customer's computer to the manufacturer's computer. The manufacturer's computer determines when new supplies are needed and can place orders by connecting with the supplier's computer.

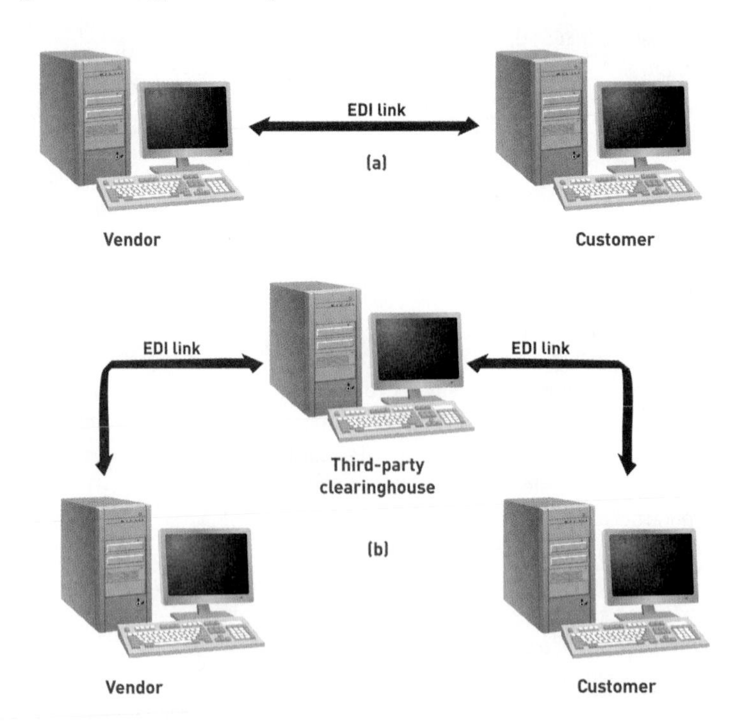

Figure 6.17

Two Approaches to Electronic Data Interchange

Many organizations now insist that their suppliers use EDI systems. Often, the vendor and customer (a) have a direct EDI connection; or (b) the link is provided by a third-party clearinghouse that converts data and performs other services for the participants.

Among other products, Johnson Controls provides automotive interiors and batteries for automobiles and hybrid electric vehicles. The firm employs a third-party vendor, Covisint, to handle EDI transactions with its automotive customers and suppliers to enable collaboration on scheduling, shipping, and orders. Sue Kemp, global vice president and general manager of information technology at Johnson Controls, believes her firm has reduced its EDI-related costs by 10 percent because: "Before, we had to build a highway to everyone else. But now we pretty much have outsourced our whole EDI to them."[61]

Public Network Services

public network services
Systems that give personal computer users access to vast databases and other services, usually for an initial fee plus usage fees.

Public network services give personal computer users access to vast databases, the Internet, and other services, usually for an initial fee plus usage fees. Public network services allow customers to book airline reservations, check weather forecasts, find information on TV programs, analyze stock prices and investment information, communicate with others on the network, play games, and receive articles and government publications. Fees are based on the services used, and can range from under $15 to over $500 per month. Providers of public network services include Microsoft, America Online, and Prodigy. These companies provide an array of services, including news, electronic mail, and investment information. AOL is the largest provider of public network services.

Electronic Funds Transfer

Electronic funds transfer (EFT) is a system of transferring money from one bank account directly to another without any paper money changing hands. It is used for both credit transfers, such as payroll payments, and for debit transfers, such as mortgage payments. The benefits of EFT include reduced administrative costs, increased efficiency, simplified bookkeeping, and greater security. One of the most widely used EFT programs is direct deposit, which deposits employee payroll checks directly into the designated bank accounts. The two primary components of EFT, wire transfer and automated clearing house, are summarized in Table 6.4.

electronic funds transfer (EFT)
A system of transferring money from one bank account directly to another without any paper money changing hands.

Table 6.4

Comparison of ACH Payments and Wire Transfers

	ACH Payments	Wire Transfers
When does payment clear?	Overnight	Immediately
Can payment be canceled?	Yes	No
Is there a guarantee of sufficient funds?	No	Yes
What is the approximate cost per transaction?	$.25	$10–$40

Distance Learning

With over 250,000 students, the University of Phoenix is the largest for-profit school in the U.S. The school offers both distance learning and campus-based educational programs. In the distance learning program, students complete coursework through electronic forums. They can download lectures and assignments from a student Web site and study them at a time and place most convenient to them. They can develop teamwork skills through collaboration on team learning assignments.[62]

Often called **distance learning** or cyberclasses, such electronic classes are likely to be the wave of the future. With distance learning software and systems, instructors can easily create course home pages on the Internet. Students can access the course syllabus and instructor notes on the Web page. Electronic mailing lists allow students and the instructor to e-mail one another for homework assignments, questions, or comments about material presented in the course. It is also possible to form chat groups so that students can work together as a "virtual team," which meets electronically to complete a group project.

distance learning
The use of telecommunications to extend the classroom.

Shared Workspace

Collaboration is much easier when all participants can be assembled in one place at the same time. However, in today's fast-paced environment with people resources spread worldwide, it is seldom possible to create this ideal situation. A **shared workspace** is a common work area where authorized project members and colleagues can share documents, issues, models, schedules, spreadsheets, photos, and all forms of information to keep each other current on the status of projects or topics of common interest. This reduces time and geographical barriers and makes it easier to collaboratively create, organize, share, and manage information.

shared workspace
A common work area where authorized project members and colleagues can share documents, issues, models, schedules, spreadsheets, photos, and all forms of information to keep each other current on the status of projects or topics of common interest.

The U.S. Marine Corps migrated many of their administrative tasks that involved the e-mail exchange of Microsoft Office Word and Excel documents to a shared workspace to collect the necessary information and make it available to those who need to see it. In addition, the use of the shared workspace has considerably reduced the need for Marines to leave Quantico for meetings at the Pentagon, thus decreasing travel costs and time away from the base.[63]

Unified Communications

unified communications
A technology solution that provides a simple and consistent user experience across all types of communications such as instant messaging, fixed and mobile phone, e-mail, voice mail, and Web conferencing.

Unified communications provides a simple and consistent user experience across all types of communications such as instant messaging, fixed and mobile phone, e-mail, voice mail, and Web conferencing. The concept of *presence* (knowing where one's desired communication participants are and if they are available at this instant) is a key component of unified communications. The goal is to reduce the time required to make decisions and communicate results, thus greatly improving productivity.

All of the ways that unified communications can be implemented rely on fast, reliable communications networks. Typically, users have a device capable of supporting the various forms of communications (e.g., laptop with microphone and video camera or a smartphone) that is loaded with software that supports unified communications. The users' devices also connect to a server that keeps track of the presence of each user.

World-renowned electric guitar maker Gibson Guitar has over 4,000 workers spread across three continents. "It was challenging to get everyone together and make decisions. There were phone calls being made and e-mails sent with 24-hour response time," according to Gibson Guitar's Director of IT Kathy Benner. The firm became an early adopter of unified communications to link its business operations, speed up decision making, and cut costs.[64]

Global Positioning System Applications

The Global Positioning System (GPS) is a fully functional global navigation satellite system employing over two dozen satellites in orbit at roughly 12,500 miles above the Earth. The satellites transmit microwave signals so a GPS receiver can precisely determine their location, speed, direction, and time.[65]

To determine its position, a GPS receiver receives the signals from three or more GPS satellites and determines its exact distance from each satellite. It then uses these distances to triangulate its precise location in terms of latitude, longitude, and altitude. A GPS receiver must have a clear line of sight to the satellite to operate, so dense tree cover and buildings can keep it from operating.

GPS tracking technology has become the standard by which fleet managers monitor the movement of their cars, trucks, and vehicles. GPS tracking quickly exposes inefficient routing practices, wasted time on the job, and speeding. Even small fleet operators can achieve significant benefits from the use of GPS tracking. Amherst Alarm installs, services, and monitors alarm systems in western New York. The company uses GPS technology to track its 12 vehicles so that it can locate its technicians throughout the day. This ensures efficiency on jobs and helps to route the closest available technician to the next service location.[66] In another example, the Boston school system placed GPS devices on its buses to ensure that it always knows where each school bus is.[67]

Computer-based navigation systems are also based on GPS technology. These systems come in all shapes and sizes and with varying capabilities—from PC-based systems installed in automobiles for guiding you across the country to handheld units you carry while hiking through a national forest. All systems need a GPS antenna to receive satellite signals to pinpoint your location. On most of these systems, your location is superimposed on a map stored on CDs or a DVD. Portable systems can be moved from one car to another or carried in your knapsack. Some systems come with dynamic rerouting capability where the path recommended depends on weather and road conditions continually transmitted to a receiver in your car connected to a satellite radio system.

Most new cell phones include an internal GPS chip that can provide navigation features if the user activates the associated software and agrees to pay the additional fees. Verizon customers can use Verizon VZ Navigator software, and AT&T, Sprint, T-Mobile, and Verizon Wireless customers can use TeleNav's GPS Navigator.[68] Some employers use GPS-enabled phones to track their employees' locations. The Whereifone locator phone provides GPS coordinates and can dial emergency phone numbers.[69] Parents and caregivers can track the phone's location by phone or online and can receive notification if it leaves a designated safe area. The Shroud is an online game that lets players use GPS-enabled phones to enhance the virtual reality of the game.

Specialized Systems and Services

With millions of personal computers in businesses across the country, interest in specialized and regional information services is increasing. Specialized services, which can be expensive, include professional legal, patent, and technical information. For example, investment companies can use systems such as Quotron and Shark to get up-to-the-minute information on stocks, bonds, and other investments.

Another example of a specialized communications service is the Nike+iPod Sports Kit, the result of a collaboration effort between Nike and Apple. The product provides a useful device for serious runners who want to monitor their performance and listen to their favorite music while training. The Nike+iPod Sports Kit consists of a small accelerometer to measure the distance and pace of the athlete. The accelerometer and its transmitter can be mounted under the inner sole of the Nike+ model shoe, which includes a special pocket in which to place the device. The accelerometer communicates wirelessly with a receiver plugged into an iPod nano to enable the athlete to view the walk or run history. The Nike+iPod Sports Kit can store training information including the length of the workout, the distance covered, the pace, and number of calories burned. This information can be displayed on the screen of the iPod or broadcast through the headphones.[70]

Organizations are finding many useful applications of telecommunications systems to help them dramatically improve their products and services and enhance their ability to collaborate among customers, suppliers, and business partners.

SUMMARY

Principle

A telecommunications system and network have many fundamental components.

Telecommunications and networks are creating profound changes in business because they remove the barriers of time and distance.

The effective use of networks can turn a company into an agile, powerful, and creative organization, giving it a long-term competitive advantage. Networks let users share hardware, programs, and databases across the organization. They can transmit and receive information to improve organizational effectiveness and efficiency. They enable geographically separated workgroups to share documents and opinions, which fosters teamwork, innovative ideas, and new business strategies.

In a telecommunications system, the sending unit transmits a signal to a telecommunications device, which performs a number of functions such as converting the signal into a different form or from one type to another. The telecommunications device then sends the signal through a medium that carries the electronic signal. The signal is received by another telecommunications device that is connected to the receiving computer.

Communications can be classified as synchronous or asynchronous.

A transmission medium can be divided into one or more communications channels, each capable of carrying a message. Telecommunications channels can be classified as simplex, half-duplex, or full-duplex.

Telecommunications protocols define the set of rules that governs the exchange of information over a telecommunications channel to ensure fast, efficient, error-free communications and to enable hardware, software, and equipment manufacturers and service providers to build products that interoperate effectively. There is a myriad of telecommunications protocols, including international, national, and regional standards.

Channel bandwidth refers to the rate at which data is exchanged, usually expressed in bits per second.

Principle

Telecommunications, networks, and their associated applications are essential to organizational success.

The telecommunications media that physically connect data communications devices can be divided into two broad categories: guided transmission media and wireless media. Guided transmission media include twisted-pair wire cable, coaxial cable, fiber-optic cable, and broadband over power lines. Wireless transmission involves the broadcast of communications in one of three frequency ranges: radio, microwave, or infrared.

Wireless communications solutions for very short distances include near field communications, Bluetooth, ultra wideband, infrared transmission, and Zigbee. Wi-Fi is a popular wireless communications solution for medium range distances. Wireless communications solutions for long distances include satellite and terrestrial microwave transmission, wireless mesh, 3G, 4G, and WiMAX.

Reallocating frequency in the 700 MHz spectrum might lead to new and more effective wireless solutions as well as a national voice and data network for public safety agencies.

The geographic area covered by a network determines whether it is called a personal area network (PAN), local area network (LAN), metropolitan area network (MAN), or wide area network (WAN).

The electronic flow of data across international and global boundaries is often called transborder data flow.

When an organization needs to use two or more computer systems, it can follow one of three basic data-processing strategies: centralized (all processing at a single location, high degree of control), decentralized (multiple processors that do not communicate with one another), or distributed (multiple processors that communicate with each other). Distributed processing minimizes the consequences of a catastrophic event at one location while ensuring uninterrupted systems availability.

A client/server system is a network that connects a user's computer (a client) to one or more host computers (servers). A client is often a PC that requests services from the server, shares processing tasks with the server, and displays the results.

Numerous telecommunications devices commonly employed include modem, multiplexer, front-end processor, PBX, switches, bridges, routers, and gateways.

Telecommunications software performs important functions, such as error checking and message formatting. A network operating system controls the computer systems and devices on a network, allowing them to communicate with one another. Network-management software enables a manager to monitor the use of individual computers and shared hardware, scan for viruses, and ensure compliance with software licenses.

The interception of confidential information by unauthorized parties is a major concern for organizations. Encryption of data and the use of virtual private networks are two common solutions to this problem. Special measures must be taken to secure wireless networks.

The wide range of telecommunications and network applications includes cellular phone services, digital subscriber line (DSL), VoIP, linking personal computer to mainframes, voice mail, reverse 911 service, voice-to-text services, home and small business networks, electronic document distribution, call centers, telecommuting, videoconferencing, electronic data interchange, public network services, electronic funds transfer, distance learning, shared workspaces, unified communications, global positioning system applications, and specialized systems and services.

CHAPTER 6: SELF-ASSESSMENT TEST

A telecommunications system and network have many fundamental components.

1. Videoconferencing is an example of asynchronous communications. True or False?
2. Gbps stands for billions of bytes per second. True or False?
3. A simplex channel can transmit data in only one direction and is seldom used for business communications. True or False?
4. Two broad categories of transmission media are ______________.
   a. guided and wireless
   b. shielded and unshielded
   c. twisted and untwisted
   d. infrared and microwave

Telecommunications, networks, and their associated applications are essential to organizational success.

5. Verizon has been replacing its traditional twisted-wire pair network with an all ________ _______ network in hopes of winning new customers from the cable companies.
6. Many wireless communications options are available. True or False?
7. Which of the following is a telecommunications service that delivers high-speed Internet access to homes and small businesses over existing phone lines?
   a. BPL
   b. DSL
   c. Wi-Fi
   d. Ethernet
8. A device that encodes data from two or more devices onto a single telecommunications channel is called a(n) ____________.
9. A(n) ________ is a network that can connect technology devices within a range of 33 feet or so.
10. ______ _________is a way to route communications between network nodes by allowing for continuous connections and reconfiguration around blocked paths by "hopping" from node to node until a connection is established.
11. Telecommuting enables organizations to save money because less office space, parking space, and office equipment is required. True or False?
12. ______ is a system of transferring money from one bank account directly to another without any paper money changing hands.

CHAPTER 6: SELF-ASSESSMENT TEST ANSWERS

(1) False (2) False (3) True (4) a (5) fiber-optic (6) True (7) b (8) multiplexer (9) personal area network, or PAN (10) mesh networking (11) True (12) Electronic funds transfer

REVIEW QUESTIONS

1. What is the difference between synchronous and asynchronous communications?
2. Describe the elements and steps involved in the telecommunications process.
3. What is a telecommunications protocol?
4. What are the names of the three primary frequency ranges employed in wireless communications?
5. Define the term *computer network*.
6. What advantages and disadvantages are associated with the use of client/server computing?
7. What is the Bluetooth telecommunications specification? What capabilities does it provide?
8. What is the difference between near field communication and ultra wideband?
9. What is the difference between Wi-Fi and WiMAX communications?
10. What role do the bridge, router, gateway, and switch play in a network?
11. Describe a local area network and its various components.
12. What is a metropolitan area network?
13. What is EDI? Why are companies using it?
14. Identify two approaches to securing the transmission of confidential data.
15. What is meant by the term *unified communications*?

DISCUSSION QUESTIONS

1. How might you use a local area network in your home? What devices could connect to such a network?
2. Why is an organization that employs centralized processing likely to have a different management decision-making philosophy than an organization that employs distributed processing?
3. What are the pros and cons of public Wi-Fi access?
4. Briefly discuss the pros and cons of e-mail versus voice mail. Under what circumstances would you use one and not the other?
5. What is a shared workspace? Describe how you might use a shared workspace if you were the manager of an important global project for your organization.
6. Develop a set of criteria to use to determine if a given job position is a good candidate for telecommuting.
7. Do you think that this course is a good candidate for a distance learning course? Why or why not?
8. Why do you think so many wireless communications protocols have been developed? Will the number of protocols increase or decrease over time?

PROBLEM-SOLVING EXERCISES

1. You have been hired as a telecommunications consultant to help an organization assess the benefits and potential cost savings associated with replacing an existing wired LAN with a wireless Wi-Fi network for an organization of 450 people located in a three-story building. You have determined that the cost to remove the current LAN and replace it with a new Wi-Fi network is $150,000. The cost of moving an existing LAN jack or installing a new one for the old LAN was $125 per change. The number of moves and installs averaged 25 per month. Develop a spreadsheet to analyze the costs and savings over a five-year period. Write a recommendation to management based on your findings and any other factors that might support or not support the installation of a Wi-Fi network.
2. As the CIO of a hospital, you are convinced that installing a wireless network and portable computers is a necessary step to reduce costs and improve patient care. Use PowerPoint or similar software to make a convincing presentation to management for adopting such a program. Your presentation must address points such as defining the benefits and potential issues that make such a program a success.

TEAM ACTIVITIES

1. With a group of your classmates, develop a proposal to install videoconferencing equipment in one of your school's classrooms so that students can view lectures at a distant videoconferencing facility. What sort of equipment is required, who provides this equipment, and what does it cost to install and operate?
2. Form a team to identify the public locations (such as an airport, public library, or café) in your area where wireless LAN connections are available. Visit at least two locations and write a brief paragraph discussing your experience at each location trying to connect to the Internet.

WEB EXERCISES

1. Do research on the Web to identify the latest 4G communications developments. Which 4G communications option seems to be the most broadly used? Write a short report on what you found.
2. Go online to identify and document an interesting use of encryption or VPN at a commercial organization.

CAREER EXERCISES

1. Consider a future job position in which you are familiar through work experience, coursework, or a study of industry performance. How might you employ some of the telecommunications and network applications described in this chapter in this future role?
2. One of the many online job-search companies includes Monster.com. Investigate one or more of these companies and research the positions available in the telecommunications industry, including the Internet. You might be asked to summarize your findings for your class in a written or verbal report.

CASE STUDIES

Case One

Latest Telecom Technologies Feed Crucial Information to Physicians in CHA

Cambridge Health Alliance (CHA) serves the residents of Cambridge, Somerville, and Boston's Metro-North in Massachusetts. CHA includes three hospital campuses and more than 20 primary care and specialty practices—over 5,000 professionals in all.

Physicians' ability to access information and medical images such as MRIs quickly and at any location within CHA properties is essential for helping patients and saving lives. In most of today's state-of-the-art medical facilities, medical information and images are delivered to desktop, notebook, and handheld PCs through local area networks, both wired and wirelessly, using systems like CHA's Picture Archive and Communications Systems (PACS) and Computerized Physician Order Entry (CPOE). The most current technologies, such as secure wireless local area networks and Voice over IP (VoIP), provide faster communications services and cost savings.

CHA is rapidly approaching a completely digital environment. Today, nurses and physicians take notes and write orders on handheld and tablet PCs wirelessly connected to the CHA local area network. CHA's three hospitals and administrative offices are connected over a dedicated SONET ring network. SONET is a fast, fiber optic-based network that can handle layers of data and voice communications simultaneously. Medical communications require the most powerful of networks to transfer large amounts of high-resolution images. CHA can make connections and receive responses in less than one millisecond for critical applications.

CHA employs VoIP to provide voice communications services over its data network. The VoIP network helps CHA resolve problems that it faces with linguistics. Around Boston, 15 primary and 43 secondary languages are spoken. CHA needs to communicate with its patients in all of these languages. It can do so by using its VoIP system to connect interpreters with patients and healthcare workers. CHA even provides videoconferencing for more effective communication.

CHA also uses wireless VoIP handsets to allow its employees to maintain voice communications with each other and on-site patients over its local network. Some CHA physicians clip Vocera communications badges to their lapel so they can speak the name of the person they want to contact, and Vocera makes the connection. The Vocera badges use VoIP over the local wireless network. They can also track the location of people around the facility. The VoIP technology makes physicians and staff much easier to reach.

The CHA network also saves the organization money. Rather than needing specialists on-site at all locations, one specialist can service all locations over the network. For example, a radiological specialist can view and evaluate medical images submitted over the network and provide an evaluation in minutes.

CHA's new network is much easier to administer. While it is often estimated that organizations spend 70 percent of IS resources supporting legacy systems and 30 percent on new innovations, Don Peterson, CHA manager of network engineering, calculates 50 percent for each for CHA. "As network manager, what I like best about this network is that it allows us to spend more time moving forward and less time worrying about what's already installed," says Peterson.

Discussion Questions

1. Why are high-performance telecommunications technologies important to the medical profession?
2. How does CHA use VoIP to provide better service to patients?

Critical Thinking Questions

1. What would be the consequence of complete network failure to a medical organization like CHA? How could this be prevented?
2. What issues of privacy arise in the medical setting regarding wireless networking?

Sources: Nortel Staff, "IT: A Critical Component of Cutting-Edge Healthcare," Nortel Case Study, *www.nortel.com/corporate/success/ss_stories/collateral/nn120660.pdf*, accessed April 26, 2008; Cambridge Health Alliance Web site, *www.cha.harvard.edu*, accessed April 26, 2008.

Case Two

Del Monte Provides Secure Connections for Telecommuters

San Francisco-based Del Monte Foods is one of America's largest and most well-known food companies and the second largest pet foods company. It generated approximately $3.4 billion in net sales in 2007 through its numerous brands, which include Del Monte, StarKist, Contadina, Milkbone, 9Lives, Meow Mix, and Nature's Recipe.

Del Monte depends on telecommunications networks to supply its 7,800 full-time employees with access to information systems such as the corporate enterprise resource planning (ERP), data warehouse, and customer relationship management (CRM) applications. An increasing number of Del Monte employees work from home offices or remote sales offices. Del Monte needed a system that employees could use to access the corporate network so that they could work as effectively as employees in the corporate offices.

The challenge with providing access to corporate networks to people outside the network is security. By opening connections over the Internet, a company makes its network more vulnerable to hackers. Del Monte wanted to provide access to its corporate data and services to employees outside the office and to select business partners and other third parties without putting its network at risk. Del Monte needed a secure intranet and extranet.

Del Monte worked with telecommunications professionals to set up a secure Web site that employees and partners could access from any Internet connection. The Web site uses VPN authentication and Cisco's Secure Access Control Server to keep hackers out and allow authorized users in. Once logged on, the user can access only the portions of the network and data that they have been authorized to access. For example, a sales representative may need access to Del Monte's data warehouse to track an order, while an accountant may be provided with access to the ERP.

Del Monte supplies an even more secure connection for employees that work from home. These employees are given a Cisco ASA 5500 Series Adaptive Security Appliance, a network device that provides a firewall and intrusion prevention system (IPS) to keep hackers out, and virtual private networking (VPN) to encrypt and safeguard data flowing over the network. Telecommuters connect their PCs and telephones to the device to enjoy the same quality of network service as those in corporate headquarters.

The Cisco device allows professionals to receive business phone calls at home, while referencing data acquired from the corporate information systems. Del Monte calls it an "office in a box." Telecommuters can even use the system to attend meetings through videoconferencing software.

Using Del Monte's secure extranet and secure home office system, employees at home offices and remote locations can communicate with employees anywhere to collaborate more effectively over one integrated network for voice, video, and data. The system saves Del Monte money by allowing the company to remove expensive T1 lines from remote offices and replace them with high-speed Internet connections. The system is also easy to manage and expand. Del Monte could easily add new security features as they are needed.

Network engineers at Del Monte are currently working on using the service for disaster recovery. If Del Monte's corporate offices were to experience a fire, earthquake, or another natural disaster, the company could continue operations by using the extranet to allow all employees to access network resources from home.

Discussion Questions

1. What is a primary concern of making a private network available to employees who are outside the office? Why?
2. What technologies did Del Monte employ to address this primary concern?

Critical Thinking Questions

1. What benefits do Del Monte and its employees enjoy by providing extranet access to the Del Monte network?
2. If you could choose whether to work at home or in a corporate office, which would you choose and why?

Sources: Cisco Staff, "Food Manufacturer Extends its Workplace with Secure Remote Access," Cisco Success Story, *www.cisco.com/en/US/prod/collateral/vpndevc/ps6032/ps6094/ps6120/case_study_c36-464676_v1.pdf*, accessed April 28, 2008; Del Monte Web site, *www.delmonte.com*, accessed April 28, 2008.

Questions for Web Case

See the Web site for this book to read about the Whitmann Price Consulting case for this chapter. Following are questions concerning this Web case.

Whitmann Price Consulting: Telecommunications and Networks

Discussion Questions

1. What role does bandwidth play in the successful delivery of the Advanced Mobile Communications and Information System?
2. When does functionality transform the standard BlackBerry device into an Advanced Mobile Communications and Information System?

Critical Thinking Questions

1. Describe three telecommunications and network technologies used to connect the BlackBerry with other devices.
2. At this stage in the process, what actions might Sandra and Josh take to reduce the overall costs of the Advanced Mobile Communications and Information System?

NOTES

Sources for the opening vignette: Cisco Staff, "Convergence—Under One Roof," Cisco Case Study, *www.cisco.com/en/US/prod/collateral/voicesw/ps6788/vcallcon/ps556/prod_case_study_ deloittel.pdf*, accessed April 28, 2008; Deloitte Web site, *www.deloitte.com*, accessed April 28, 2008; Cisco Voice and Unified Communications Web site, *www.cisco.com/en/US/products/sw/voicesw*, accessed April 28, 2008.

1 Nystedt, Dan, "Wal-Mart Eyes $287 Million Benefit from RFID," *Network World*, October 12, 2007.
2 Reese, Brad, "Procter & Gamble CIO Filippo Passerini Deploys Expensive Rollout of Cisco Telepresence System," *Network World*, December 30, 2007.
3 Shah, Adam, "Digital Gear—CES—Home Appliances Get Digital Photo Frames," *Network World*, December 26, 2007.
4 Desmond, Paul, "Building Virtual Worlds at Boeing," *Network World*, October 29, 2007.
5 Duffy, Jim, "10 Gigabit Ethernet Secures Border at Niagara Falls," *Network World*, January 21, 2008.
6 Gruenwedel, Erik, "Comcast Expands Video-on-Demand Slate," Reuters, February 5, 2008.
7 Reardon, Marguerite, "Verizon's Fiber-Optic Payoff," *CNET News*, June 21, 2007
8 Hoover, Nicholas, J., "TSA to Test Broadband Over Powerline Technology," *InformationWeek*, January 6, 2007.
9 Staff, "BPL Network Adapter Speeds of 400 Mbps Achieved by Comtrend," *Broadband Focus.com*, November 15, 2007.
10 Brown, Damon, "Letting Go of Your Landline, *Technology, Inc.*, January 2007.
11 "Standard Setting Organization and Standards List—Wireless and Mobile," *consortiuminfo.org/links/wireless*, accessed March 30, 2008.
12 Malykhina, Elena, "Deployments of Contactless Payment Systems Slower than Expected," *InformationWeek*, January 10, 2008.
13 Sweeney, Terry, "MedicMate Mobilizes Bedside Manner," *InformationWeek*, April 10, 2008.
14 Griffith, Eric, "In 2008, UWB Takes the World," *Ultrawideband Planet.com*, June 18, 2007.
15 Hamblen, Matt, "Plogg Uses Zigbee to Monitor Electricity," *Computerworld*, March 18, 2008.
16 Wailgum, Thomas, "Can Muni Wi-Fi Be Saved?" *Network World*, October 2, 2007.
17 Cox, John, "Municipal Wi-Fi vs. 3G," *Network World*, October 26, 2007.
18 Lawson, Stephen, "Denver Airport Goes Fast and Free on Wi-Fi," *Network World*, December 10, 2007.
19 Reed, Brad, "How Four Airlines Plan to Connect Fliers to the Web," *Network World*, December 7, 2007.
20 Terdiman, Daniel, "In-Flight Internet: Grounded for Life?" *CNET News*, January 25, 2008.
21 Oil and Gas Case Study, CapRock Web site, *www.caprock.com*, accessed February 6, 2008.
22 Lawson, Stephen, "Year End—New Mobile Approaches Got a Reality Check," *Network World*, December 18, 2007.
23 "Pine Cellular and Choctaw Electric Cooperative Will Use Nortel 4G WiMAX to Bring Broadband to Areas of Southeastern Oklahoma," *Fierce Wireless*, September 27, 2007.
24 Bloomberg News, "Sprint and Clearwire to Build WiMAX Network," *International Herald Tribune*, May 7, 2008.
25 Gohring, Nancy, "CES—Sprint Insists Its WiMAX Network Is on Track," *Network World*, January 8, 2008.
26 "Aircel Expands WiMAX in India," *WiMAX Day*, March 14, 2007.
27 "Telmex Covers Chile with WiMAX," *WiMAX Day*, January 17, 2008.
28 "Intel Hopes for 1.3 Billion WiMAX Users," *WiMAX Day*, May 14, 2007.
29 "Intel: WiMAX to Take Off," *PC World*, January 8, 2008.
30 Malik, Om, "700 MHz Explained in 10 Steps," *GigaOM*, March 14, 2007.
31 Johnson, Johna Till, "FCC Spectrum Auction: What's Google on About?" *Network World*, July 26, 2007.
32 Gross, Grant, "Debate Heats Up Over Open-Access Spectrum Rules," *Network World*, July 10, 2007.
33 Gardner, David W., "Google's Android Builds on Past Features, Hints at Future Enhancements," *InformationWeek*, November 5, 2007.
34 Massy, Kevin, "Ford and Microsoft in Sync for In-Car Infotainment," *CNET News*, January 7, 2007.
35 Flaccus, Gillian, "Disneyland to Open House of the Future 2.0," *Tech News World*, February 13, 2008.
36 "Case Study: Miami-Dade Police Department," *www.nortel.com*, accessed February 15, 2008.
37 Gardner, David W., "Qwest Expands Its Nationwide Fiber Optic Network," *InformationWeek*, July 16 2007.
38 "T-Mobile Austria Looks to IBM Tivoli Netcool Software for End-to-End Visibility of Its Wireless Networks," March 27, 2007, *www-01.ibm.com/software/success*.
39 Das, Saswato, "Quantum Threat to Our Secret Data," *New Scientist Tech*, September 13, 2007.
40 Bangeman, Eric, "New Attack Cracks WEP Code in Record Time," *ars technical*, April 4, 2007.
41 Greenemeier, Larry, "T.J. Maxx Data Theft Likely Due to Wireless 'Wardriving'," *InformationWeek*, May 8, 2007.
42 Bridis, Ted, "Code-Breakers Crack Government Approved Encryption Standard," Associated Press, July 17, 1998.
43 "DES," *Network World*, *www.networkworld.com/community/node/16520*, accessed February 14, 2008.
44 Messmer, Ellen, "Encrypted Wireless Service Targets Financial Institutions," *Network World*, February 8, 2008.
45 "ROI: Easy, Integrated Solution Gives Workers Remote Access, Improved Productivity," Microsoft Forefront, July 2007, *http://whitepapers.zdnet.com/casestudy.aspx?&cname=LAN+-+WAN&docid=332964*.
46 "Case Study: Techcombank", RSA Web site, *www.rsa.com*, accessed February 20, 2008.
47 Sarkar, Dibya, "Cell Phone Spending Surpasses Landlines," *newsvine.com*, December 18, 2007.
48 Spotwave Web site, *www.spotwave.com*, accessed February 10, 2008.
49 "RadioFrame Provides Picocells for Orange's Onsite Service," *cellular-news*, January 23, 2008.
50 "Airlines Will Install over 4,000 Picocells for Passenger GSM and Wi-Fi Services between 2008 and 2011, according to Freesky Research," WebWire, January 30, 2008.
51 Cassavoy, Liane, "RIM BlackBerry Curve 8320," *PC World*, October 25, 2007.
52 Shifrin, Tash, "Merrill Lynch Picks VoIP-Powered Trading Systems, *Computerworld*, November 15, 2007.
53 Thibodeau, Patrick, "San Diego Puts Web Collaboration, Reverse 911 Systems to Use in Wildfire Battle," *Computerworld*, October 24, 2007.
54 DeJean, David, "Voice Mail Driving You Crazy? Get It in Writing," *InformationWeek*, October 18, 2007.
55 Ribeiro, John, "HP's India Lab Secures Paper Documents," *Network World*, July 13, 2007.
56 Bucken, Mike, "Global Dispatches: Barclays Unit Shifting Call Center Jobs to India," *Computerworld*, March 19, 2007.
57 Jones, K. C., "Federal Government Beats Private Sector in Telecommuting," *InformationWeek*, March 22, 2007.
58 Mullins, Robert, "How Cox Communications Joined the Teleworking Revolution," *PC World*, November 15, 2007.

59 Howze, Jane, "The Hiring Manager Interviews: Kohl's CIO Jeff Marshall Hires Candidates Who Embrace Change," *CIO*, October 17, 2007.
60 Montalbano, Elizabeth, "Google Sets Sights on Nonprofits with Outreach Program," *CIO*, June 26, 2007.
61 Messmer, Ellen, "Dot-com Survivor Covisint finds B2B Niche," *Network World*, November 1, 2007.
62 "About Us," University of Phoenix Web site, *www.uopxonline.com/aboutus.asp*, accessed February 21, 2008.
63 Lai, Eric, "Marines Deploy SharePoint to Improve Administration," *Computerworld*, June 15, 2007.
64 Malykhina, Elena, "Guitar Legend Riffs on Microsoft's Communications Server," *InformationWeek*, October 5, 2007.
65 Global Positioning System, *Wikipedia*, accessed February 4, 2008.
66 "Case Study: Amherst Alarm," Sage Quest Web site, *sage-quest.com*, accessed January 28, 2008.
67 "GPS Helps Cities Save on Gas and Catch Employees Goofing Off," *KOMOTV.com*, November 16, 2007.
68 McLaughlin, Molly K., "Cheap GPS Cell Phones," *PC Magazine*, August 8, 2007.
69 Whereifone Web site, *www.wherify.com/wherifone*, accessed February 6, 2008.
70 Apple Nike + iPod Sport Kit product description, *amazon.com*, accessed February 16, 2008.

CHAPTER

• 7 •

The Internet, Intranets, and Extranets

PRINCIPLES	LEARNING OBJECTIVES
▪ The Internet is like many other technologies—it provides a wide range of services, some of which are effective and practical for use today, others that are still evolving, and still others that will fade away from lack of use.	▪ Briefly describe how the Internet works, including alternatives for connecting to it and the role of Internet service providers.
▪ Originally developed as a document-management system, the World Wide Web is a hyperlink-based system that is easy to use for personal and business applications.	▪ Describe the World Wide Web and how it works. ▪ Explain the use of Web browsers, search engines, and other Web tools. ▪ Identify and briefly describe the applications associated with the Internet and the Web.
▪ Because use of the Internet and World Wide Web is becoming universal in the business environment, management, service and speed, privacy, and security issues must continually be addressed and resolved.	▪ Identify who is using the Web to conduct business and discuss some of the pros and cons of Web commerce. ▪ Outline a process for creating Web content. ▪ Describe Java and discuss its potential impact on the software world. ▪ Define the terms *intranet* and *extranet* and discuss how organizations are using them. ▪ Identify several issues associated with the use of networks.

Information Systems in the Global Economy Lamborghini, Italy

Web Portal and Online Collaboration Shrink Distances

The Internet has enabled thousands of businesses to extend their reach beyond borders to become global competitors. For a global business to succeed, all of its offices and personnel spread around the world must stay in sync. This is typically accomplished by being in close communications with headquarters. The Internet and Web allow employees separated by thousands of miles to operate as though they were seated across the table from each other. Automobile manufacturer Lamborghini, with headquarters in Sant'Agata Bolognese, Italy, recently discovered the benefits of employing the latest technologies to shrink the distances between its more than 100 dealerships scattered around the world.

The famous manufacturer of elite sports and racing cars had anything but elite communications between dealerships and headquarters. Until recently, Lamborghini dealers relied on telephone, e-mail, and snail mail to order new cars and spare parts and to learn about the latest marketing programs and business procedures. The multiple forms of inefficient communications were difficult to manage at headquarters and allowed important tasks to fall through the cracks. Executives at the company realized that their problems could be remedied by more efficient use of Internet and Web technologies.

Lamborghini decided to create a Web portal—a custom-designed Web page that would provide dealerships with direct access to corporate databases and information systems over the Internet, resources that were previously only accessible by employees at headquarters. Secured by password authentication and encryption, dealers could log on to the Web portal to check inventory, place orders for cars and parts, read daily announcements, and access procedural instructions. The new portal dramatically cuts down on paperwork and manual processing of orders at headquarters. The portal has also reduced errors by eliminating the two-stage system that used to require headquarters staff to interpret and enter order data into the system.

Lamborghini executives were so impressed by the savings and improvements offered by the new Web portal that they looked for ways to further enhance the system. Observing the benefits of online social networks and Web 2.0 technologies, Lamborghini looked for ways to implement these technologies into its own Web portal system. They added social features that allowed dealers and personnel to post announcements and requests that others fielded on the network. For example, if a dealership needed a part for a vintage car that Lamborghini no longer manufactures, this system could find someone within the network that had the part. Lamborghini also added a chat utility and file sharing features to allow dealers around the world to collaborate on promotions and sales tactics.

Ultimately, Lamborghini created an intranet—a secure private network accessible over the Internet—so that dealers could have expanded access to the information systems and data stored on servers at headquarters. Using this system, the Lamborghini dealer in Orange County, California, could access the same information, systems, databases, and services as the vice president of sales in Sant'Agata Bolognese, Italy. The dealer could also develop relationships with the other hundred dealers around the world for more effective business practices.

Finally, Lamborghini developed a state-of-the-art Web site that provides an experience more like a top-rate motion picture or video game than an automobile Web site. The site uses the latest technologies to thoroughly impress visitors and reflect the high quality of the corporation and its products.

As you read this chapter, consider the following:

- What unique features of the Internet and Web make them popular choices for many business communication applications?
- In what ways do people use the Internet and Web to improve their quality of life? How do businesses use these technologies to improve the bottom line?

Why Learn About the Internet?

To say that the Internet has had a big effect on organizations of all types and sizes would be a huge understatement. Since the early 1990s, when the Internet was first used for commercial purposes, it has affected all aspects of business. Businesses use the Internet to sell and advertise their products and services, reaching out to new and existing customers. If you are undecided about a career, you can use the Internet to investigate career opportunities and salaries using sites such as *www.monster.com* and HotJobs at *www.yahoo.com*. Most companies have Internet sites that list job opportunities, descriptions, qualifications, salaries, and benefits. If you have a job, you probably use the Internet daily to communicate with coworkers and your boss. People working in every field and at every level use the Internet in their jobs. Purchasing agents use the Internet to save millions of dollars in supplies every year. Travel and events-management agents use the Internet to find the best deals on travel and accommodations. Automotive engineers use the Internet to work with other engineers around the world developing designs and specifications for new automobiles and trucks. Property managers use the Internet to find the best prices and opportunities for commercial and residential real estate. Whatever your career, you will probably use the Internet daily. This chapter starts by exploring how the Internet works and then investigates the many exciting opportunities for using the Internet to help you achieve your goals.

Internet
A collection of interconnected networks, all freely exchanging information.

The Internet is the world's largest computer network. Actually, the **Internet** is a collection of interconnected networks, all freely exchanging information (see Figure 7.1). Research firms, colleges, and universities have long been part of the Internet, and businesses, high schools, elementary schools, and other organizations have joined it as well. Nobody knows exactly how big the Internet is because it is a collection of separately run, smaller computer networks. There is no single place where all the connections are registered. Figure 7.2 shows the staggering growth of the Internet, as measured by the number of Internet host sites or domain names. Domain names are discussed later in the chapter.

Figure 7.1
Routing Messages over the Internet

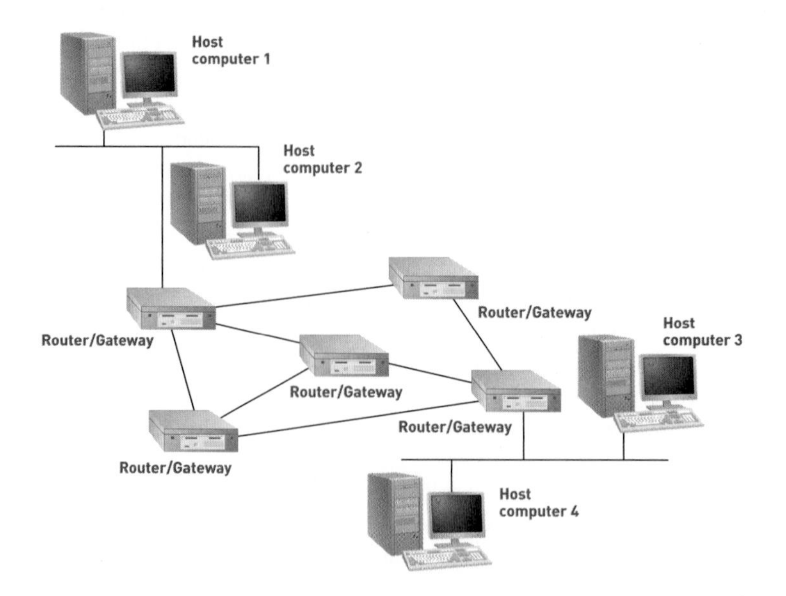

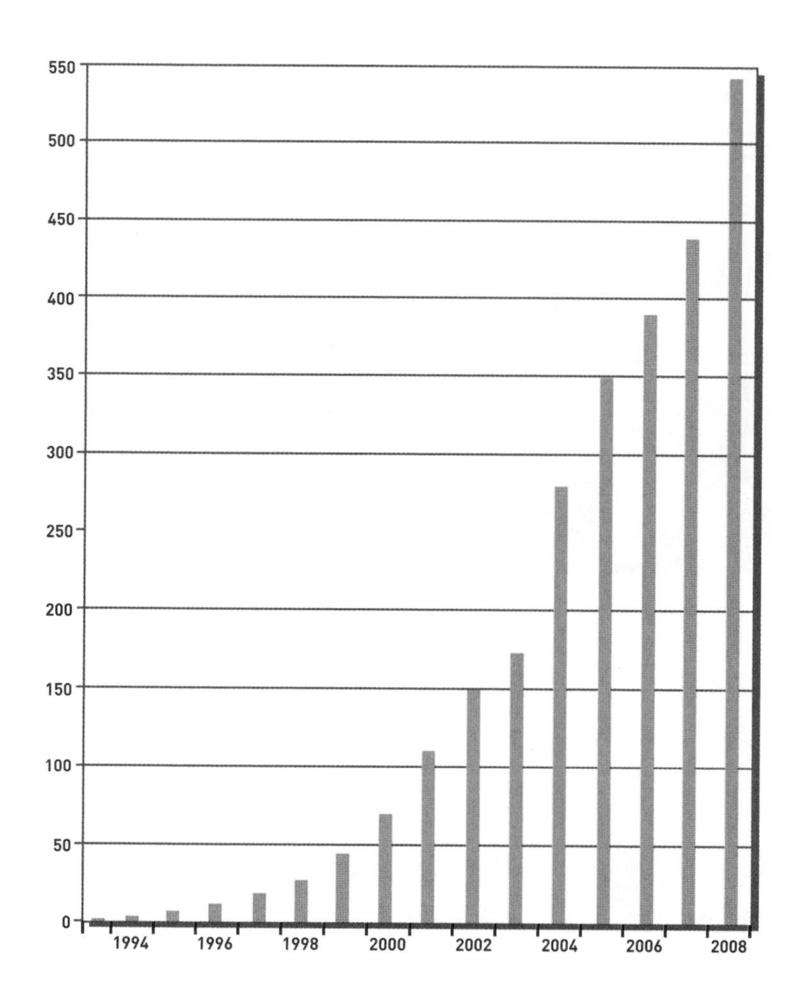

Figure 7.2

Internet Growth: Number of Internet Domain Names

(Source: Data from "The Internet Domain Survey," *www.isc.org*.)

USE AND FUNCTIONING OF THE INTERNET

The Internet is truly international in scope, with users on every continent—including Antarctica. More than 215 million people in the United States (71.4 percent of the population) use the Internet. Although the United States has high Internet penetration among its population, it does not constitute the majority of people online. Of all the people using the Internet, citizens of Asian countries make up 39 percent, Europeans 26 percent, and North Americans only 18 percent.[1] The Internet is expanding around the globe but at differing rates for each country. For example, most Internet usage in South Korea is through high-speed broadband connections, and over 71 percent of the population is online. In North Korea, however, Internet use and other civil liberties are restricted by the government. Several people and organizations are working to provide Internet access to developing countries.[2] More than 1.3 billion people use the Internet around the world, and as Figure 7.2 shows, if the rate of growth continues, the number of users will surge to more than 2 billion in a few years.

The ancestor of the Internet was the **ARPANET**, a project started by the U.S. Department of Defense (DoD) in 1969. The ARPANET was both an experiment in reliable networking and a means to link DoD and military research contractors, including many universities doing military-funded research. (*ARPA* stands for the Advanced Research Projects Agency, the branch of the DoD in charge of awarding grant money. The agency is now known as DARPA—the added *D* is for *Defense*.) The ARPANET was highly successful, and every university in the country wanted to use it. This wildfire growth made it difficult to manage the ARPANET, particularly its large and rapidly growing number of university sites. So, the ARPANET was broken into two networks: MILNET, which included all military sites, and a new, smaller ARPANET, which included all the nonmilitary sites. The two networks

ARPANET
A project started by the U.S. Department of Defense (DoD) in 1969 as both an experiment in reliable networking and a means to link DoD and military research contractors, including many universities doing military-funded research.

China has over 210 million Internet users online, which is only 16 percent of its population.

(Source: Courtesy of Reuters/STR / Landov.)

Internet Protocol (IP)
A communication standard that enables traffic to be routed from one network to another as needed.

remained connected, however, through use of the **Internet Protocol (IP)**, which enables traffic to be routed from one network to another as needed. All the networks connected to the Internet speak IP, so they all can exchange messages. Table 7.1 outlines a brief history of the Internet. Katie Hafner's book, *Where Wizards Stay Up Late: The Origins of the Internet,* provides a more detailed description of the history of the Internet.[3]

Table 7.1

A Brief History of the Internet

Event	Date
ARPANET is created	1969
TCP/IP becomes the protocol for ARPANET	1982
Domain Name System (DNS) is created	1984
Tim Berners-Lee creates the World Wide Web	1991
Commercial Internet Exchange (CIX) Association is established to allow businesses to connect to the Internet	1991

Today, people, universities, and companies are attempting to make the Internet faster and easier to use. To speed Internet access, a group of corporations and universities called the University Corporation for Advanced Internet Development (UCAID) is working on a faster, alternative Internet. Called Internet2 (I2), Next Generation Internet (NGI), or Abilene, depending on the universities or corporations involved, the new Internet offers the potential of faster Internet speeds, up to 2 Gbps per second or more.[4] An offshoot of Internet2 that some call Internet3, which is officially named the *National LambdaRail* (NLR), is a cross-country, high-speed (10 Gbps), fiber-optic network dedicated to research in high-speed networking applications.[5] The NLR provides a unique national networking infrastructure to advance networking research and next-generation network-based applications in science, engineering, and medicine. This new high-speed fiber-optic network will support the ever-increasing need of scientists to gather, transfer, and analyze massive amounts of scientific data.

How the Internet Works

The Internet transmits data from one computer (called a *host*) to another (see Figure 7.1). If the receiving computer is on a network to which the first computer is directly connected, it can send the message directly. If the receiving and sending computers are not directly connected to the same network, the sending computer relays the message to another computer that can forward it. The message is typically sent through one or more routers (see Chapter 6) to reach its destination. It is not unusual for a message to pass through a dozen or more routers on its way from one part of the Internet to another.

The various networks that are linked to form the Internet work much the same way—they pass data around in chunks called *packets*, each of which carries the addresses of its sender and its receiver along with other technical information. The set of conventions used to pass packets from one host to another is the IP. Many other protocols are used in connection with IP. The best known is the **Transmission Control Protocol** (**TCP**). Many people use TCP/IP as an abbreviation for the combination of TCP and IP used by most Internet applications. After a network following these standards links to a **backbone**—one of the Internet's high-speed, long-distance communications links—it becomes part of the worldwide Internet community.

Transmission Control Protocol (TCP)
The widely used Transport-layer protocol that most Internet applications use with IP.

backbone
One of the Internet's high-speed, long-distance communications links.

Each computer on the Internet has an assigned address called its **Uniform Resource Locator** (**URL**) to identify it to other hosts. The URL gives those who provide information over the Internet a standard way to designate where Internet resources such as servers and documents are located. Consider the URL for Course Technology, *http://www.course.com.*

Uniform Resource Locator (URL)
An assigned address on the Internet for each computer.

The "http" specifies the access method and tells your software to access a file using the Hypertext Transport Protocol. This is the primary method for interacting with the Internet. In many cases, you don't need to include http:// in a URL because it is the default protocol. Thus, *http://www.course.com* can be abbreviated as *www.course.com.*

The "www" part of the address signifies that the address is associated with the World Wide Web service, discussed later. The "course.com" part of the address is the domain name that identifies the Internet host site. Domain names must adhere to strict rules. They always have at least two parts, with each part separated by a dot (period). For some Internet addresses, the far right part of the domain name is the country code (such as au for Australia, ca for Canada, dk for Denmark, fr for France, and jp for Japan). Many Internet addresses have a code denoting affiliation categories. (Table 7.2 contains a few popular categories.) The far left part of the domain name identifies the host network or host provider, which might be the name of a university or business. Other countries outside the United States use different top-level domain affiliations than the ones described in the table.

Table 7.2

U.S. Top-Level Domain Affiliations

Affiliation ID	Affiliation
com	Business sites
edu	Educational sites
gov	Government sites
net	Networking sites
org	Nonprofit organization sites

The Internet Corporation for Assigned Names and Numbers (ICANN) is responsible for managing IP addresses and Internet domain names. One of its primary concerns is to make sure that each domain name represents only one individual or entity——the one that legally registers it. For example, if your teacher wanted to use *www.course.com* for a course Web site, he or she would soon discover that that domain name has already been registered by Course Technology and is not available. ICANN uses companies called *accredited domain name registrars* to handle the business of registering domain names. For example, you can visit *godaddy.com*, an accredited registrar, to find out if a particular name has already been registered, and if not, register it for around $10 per year.

Millions of domain names have been registered. Some people, called *cyber-squatters*, have registered domain names in the hope of selling the names to corporations or people. The domain name Business.com, for example, sold for $7.5 million. In one case, a federal judge ordered the former owner of one Web site to pay the person who originally registered the domain name $40 million in compensatory damages and an additional $25 million in punitive damages. But some companies are fighting back, suing people who register domain names only to sell them to companies. Today, the ICANN has the authority to resolve domain-name disputes. Under new rules, if an address is found to be "confusingly similar" to a registered trademark, the owner of the domain name has no legitimate interest in the name. The rule was designed in part to prevent cyber-squatters.

Accessing the Internet

Although you can connect to the Internet in numerous ways (see Figure 7.3), Internet access is not distributed evenly throughout the world. Which access method you choose is determined by the size and capability of your organization or system.

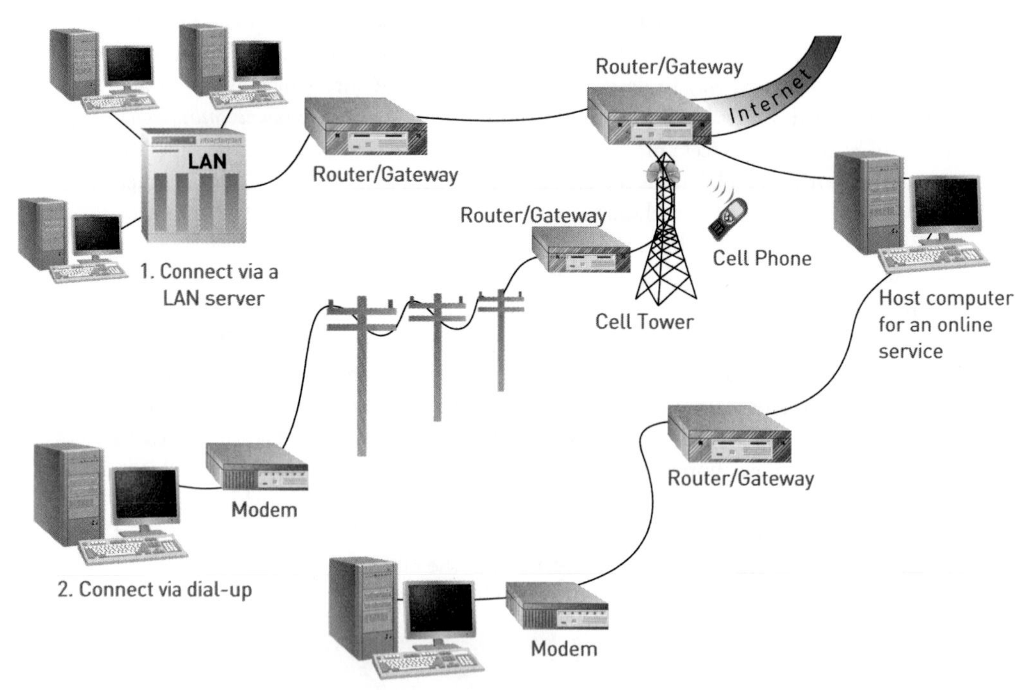

Figure 7.3

Several Ways to Access the Internet

Users can access the Internet in several ways, including using a LAN server, dialing into a server using the telephone lines, using a high-speed service, or accessing the Internet over a wireless network.

Connect via LAN Server

This approach is used by businesses and organizations that manage a local area network (LAN). By connecting a server on the LAN to the Internet using a router, all users on the LAN are provided access to the Internet. Business LAN servers are typically connected to the Internet at very fast data rates, sometimes in the hundreds of Mbps. In addition, you can share the higher cost of this service among several dozen LAN users to allow a reasonable cost per user.

Connect via Dial-up

Connecting to the Internet through a dial-up connection requires a modem that allows the computer to use standard phone lines. The modem then contacts a server managed by the Internet service provider (ISP). Dial-up connections use TCP/IP protocol software plus Serial Line Internet Protocol (SLIP) or Point-to-Point Protocol (PPP) software. SLIP and PPP are two communications protocols that transmit packets over telephone lines, allowing dial-up access to the Internet. After the connection is made, you are on the Internet and can access any of its resources. Dial-up is considered the slowest of connections because it is restricted by the 56 Kbps limitation of traditional phone line service. A dial-up connection also ties up the phone line so that it is unavailable for voice calls.

Connect via High-Speed Service

Several "high-speed" Internet services are available for home and business. They include cable modem connections from cable television companies, DSL connections from phone companies, and satellite connections from satellite television companies. These technologies were discussed in Chapter 6. High-speed services provide data transfer rates between 1 and 7 Mbps. Unlike dial-up, high-speed services provide "always connected" service that does not tie up the phone line.

Connect Wirelessly

In addition to connecting to the Internet through wired systems such as phone lines and television cables, wireless Internet access is very popular. Thousands of public Wi-Fi services are available in coffee shops, airports, hotels, and elsewhere, where Internet access is provided free, for an hourly rate, or for a monthly subscription fee. Wi-Fi is even making its way into aircraft, allowing business travelers to be productive during air travel by accessing e-mail and corporate networks.[6]

Cell phone carriers also provide Internet access for handsets or notebooks equipped with connect cards. New 3G mobile phone services rival wired high-speed connections enjoyed at home and work. Sprint, Verizon, AT&T, and other popular carriers are working to bring 4G service to subscribers soon. 4G cell phone service will compete strongly against today's wired services. Wireless devices also require specific protocols and approaches to connect. For example, *wireless application protocol (WAP)* is used to connect cell phones and other devices to the Internet. See Figure 7.4.

Figure 7.4

Connecting Devices to Wireless Networks

Notebook computers can use a connect card to take advantage of cell phone carrier data services.

(Source: Courtesy of vario images GmbH & Co.KG / Alamy.)

When Apple introduced the iPhone, one of its slogans was the "Internet in your pocket." The iPhone serves to prove the popularity of, and potential for, Internet services over a handset.[7] Intel picked up on Apple's slogan and added it to its own marketing campaign for its processor called the Atom, which is designed to bring the Internet to more mobile devices.[8]

Internet Service Providers

An **Internet service provider** (**ISP**) is any company that provides Internet access to people and organizations. Some ISPs such as America Online (AOL) and Microsoft Network (MSN) offer extended information services through software installed on the subscriber's PC. Many others simply offer a connection to the Internet that subscribers use with a Web browser and other Internet software to access services. Thousands of organizations serve as ISPs, ranging from universities making unused communications line capacity available to students and faculty to major communications giants such as AT&T and Verizon. To use this type of

Internet service provider (ISP)
Any company that provides Internet access to people or organizations.

connection, you must have an account with the service provider and software that allows a direct link via TCP/IP.

For a dial-up connection, ISPs typically charge a monthly fee that can range from $10 to $30 for unlimited Internet access. The fee normally includes e-mail. Many ISPs and online services offer broadband Internet access through DSLs, cable, or satellite transmission. Broadband users pay between $30 and $60 per month for unlimited service. Broadband rates differ based on the speed of the connection. Some businesses and universities use the very fast T1 or T3 lines to connect to the Internet. T1 and T3 support high data rates, but have additional value over DSL and cable because they can send many signals simultaneously. Table 7.3 compares the speed of modem, DSL, cable, and T1 Internet connections to perform basic tasks. This table uses advertised connection speeds; your performance will be slower. These technologies were discussed in Chapter 6.

Table 7.3

Approximate Times to Perform Basic Tasks at Advertised Connection Speeds

Task	Modem (56Kbps)	T1 (1.4 Mbps)	DSL (3 Mbps)	Cable (7 Mbps)	T3 (44 Mbps)
Send 20-page term paper (500 KB)	9 seconds	0.36 seconds	0.17 seconds	.07 seconds	.01 seconds
Send a four-minute song as an MP3 file (4.5 MB)	80 seconds	3.2 seconds	1.5 seconds	.64 seconds	0.1 seconds
Send a full-length motion picture as a compressed file (1.4 GB)	About 7 hours	About 16 minutes	7.8 minutes	3.3 minutes	.53 seconds

Some ISPs are experimenting with low-fee or no-fee Internet access, though strings are attached to the no-fee offers in most cases. Some free ISPs require that customers provide detailed demographic and personal information. In other cases, customers receive extra advertising when using the Web. For example, a *pop-up ad* is a window that is displayed when someone visits a Web site. It opens and advertises a product or service. Some e-commerce retailers have posted ads that resemble computer-warning messages and have been sued for deceptive advertising. A *banner ad* appears as a banner or advertising element within a Web site's layout, which you can ignore or click to go to the advertiser's Web site.

ETHICAL AND SOCIETAL ISSUES

Comcast, Packet Shaping, and Net Neutrality

In 2008, the U.S. Federal Communications Commission (FCC) required the telecommunications and cable TV giant Comcast to testify about its practice of packet shaping. Packet shaping is a technique that some Internet service providers (ISPs) use to control the volume of network traffic to optimize or guarantee performance. It involves filtering Internet traffic, which travels the Internet in data packets, and slowing down some types of packets in order to speed up others.

Some Comcast customers noticed that when they attempted to upload files on Bit Torrent, a popular file-sharing utility, the transfers were extremely slow and often stopped all together. Reporters with the Associated Press ran some nationwide tests and determined that Comcast was indeed filtering Internet packets, determining which packets were uploading files with Bit Torrent, and then shutting down those packets. When the news was released, it made front page headlines, and many Internet users were enraged.

Why were they enraged? The people who designed the Internet and Web intended it to be run with neutrality. Network neutrality refers to a principle applied to Internet services whereby all data is delivered to all users with equal priority. Most importantly, many believe that managers of Internet traffic do not have the right to examine what is in packets. To do so would be an invasion of privacy. A process called "deep packet inspection" allows automated systems to analyze the contents of every packet flowing through the ISP and take actions based on a rule. For example, packets with the words "bomb instructions" might be detained and the owner tracked down. But would such an action be in accordance with the privacy guaranteed to U.S. citizens in the Constitution?

Some organizations do not support net neutrality. Comcast claims that 80 percent of its bandwidth is taken up by 10 percent of its users involved in illegal file sharing, with many of the files being huge video files. If this were true, and if it were affecting the overall performance of the Internet, they might have a case for packet shaping.

Law enforcement would also like the ability to filter packets to find criminal activity. FBI director Robert Mueller has indicated that he would like the House of Representatives to grant permission to filter all Internet traffic to detect illegal activities. The Recording Industry Association of America and the Motion Picture Association of America would like ISPs to filter out packets that are involved in illegal file sharing and copyright violation. However, the ISPs are not eager to become the Internet's traffic cops. When false arrests are made, as they inevitably would be, the ISP would be liable.

After hearing Comcast's case and the voices of those supporting net neutrality, the FCC stated that it plans to regulate how and when ISPs can practice packet shaping. Comcast has promised to change its packet-shaping practices. Net neutrality is likely to be a hot topic in federal courts in coming years.

Discussion Questions

1. What is packet shaping and do ISPs use it?
2. What are the principles of network neutrality?

Critical Thinking Questions

1. Do you favor network neutrality or packet shaping? Why?
2. Is it necessary for the U.S. government to intervene and regulate ISPs? What laws would you like to see enacted?

Sources: Svensson, Peter, "Comcast blocks some Internet traffic," Associated Press/MSNBC, *www.msnbc.msn.com/id/21376597*; Kumar, Vishesh, "Comcast, BitTorrent to Work Together on Network Traffic," *The Wall Street Journal*, March 27, 2008, *http://online.wsj.com/article/SB120658178504567453.html*; Stokes, Jon, "FBI wants to move hunt for criminals into Internet backbone," *Ars Technica*, April 24, 2008, *http://arstechnica.com/news.ars/post/20080424-fbi-wants-to-move-hunt-for-criminals-into-internet-backbone.html*.

THE WORLD WIDE WEB

World Wide Web
A collection of tens of millions of server computers that work together as one in an Internet service using hyperlink technology to provide information to billions of users.

hyperlink
Highlighted text or graphics in a Web document that, when clicked, opens a new Web page or section of the same page containing related content.

Web browser
Web client software such as Internet Explorer, Firefox, and Safari used to view Web pages.

The World Wide Web was developed by Tim Berners-Lee at CERN, the European Organization for Nuclear Research in Geneva, Switzerland. He originally conceived of it as an internal document-management system. From this modest beginning, the **World Wide Web** (the Web, WWW, or W3) has grown to a collection of tens of millions of server computers that work together as one in an Internet service using hyperlink technology to provide information to billions of users. These computers, called *Web servers*, are scattered all over the world and contain every imaginable type of data. Web users use **hyperlinks**, highlighted text or graphics in a Web document, that, when clicked, open a new Web page or section of the same page containing related content. Thanks to the high-speed Internet circuits connecting them and hyperlink technology, users can jump between Web pages and servers effortlessly—creating the illusion of using one big computer. Because of the vast amount of information available on the Web and the wide variety of media, the Web has become the most popular means of information access in the world today.

In short, the Web is a hyperlink-based system that uses the client/server model. It organizes Internet resources throughout the world into a series of linked files, called pages, accessed and viewed using Web client software called a **Web browser**. Internet Explorer, Firefox, and Safari are three popular Web browsers. See Figure 7.5. A collection of pages on one particular topic, accessed under one Web domain, is called a Web site. The Web was originally designed to support formatted text and pictures on a page. It has evolved to support many more types of information and communication including user interactivity, animation, and video. Web *plug-ins* help provide additional features to standard Web sites. Adobe Flash and Real Player are examples of Web plug-ins.

Figure 7.5

Mozilla Firefox

Web browsers such as Firefox let you access Internet resources throughout the world using a series of linked Web pages.

A *Web portal* is an entry point or doorway to the Internet. Web portals include AOL, MSN, Google, Yahoo!, and others. For example, some people use Yahoo.com as their Web portal, which means they have set Yahoo! as the starting point when they open their browsers. When they enter the Internet, the Yahoo! Web site appears. Users can often customize Web portals and choose from a variety of widgets—small useful applications and services—to add to the page.[9] Web browser settings use the term *home page* to refer to your starting point. This setting can apply to any Web page you prefer. A *corporate Web portal* refers to the company's Internet site, which is a gateway or entry point to corporate data and resources.

Hypertext Markup Language (HTML) is the standard page description language for Web pages. One way to think about HTML is as a set of highlighter pens that you use to mark up plain text to make it a Web page—red for the headings, yellow for bold, and so on. The **HTML tags** let the browser know how to format the text: as a heading, list, or body text, for example. Users "mark up" a page by placing HTML tags before and after a word or words. For example, to turn a sentence into a heading, you place the <h1> tag at the start of the sentence. At the end of the sentence, you place the closing tag </h1>. When you view this page in your browser, the sentence will be displayed as a heading. HTML also provides tags to import objects stored in files, such as photos, audio, and movies, into a Web page. In short, a Web page is made up of three components: text, tags, and references to files. The text is your message, the tags are codes that mark the way words will be displayed, and the references to files insert photos and media into the Web page at specific locations. All HTML tags are enclosed in a set of angle brackets (< and >), such as <h2>. The closing tag has a forward slash in it, such as </b> for closing bold. Consider the following text and tags:

Hypertext Markup Language (HTML)
The standard page description language for Web pages.

HTML tags
Codes that let the Web browser know how to format text—as a heading, as a list, or as body text—and whether images, sound, and other elements should be inserted.

```
<h1 align="center">Principles of Information Systems</h1>
```

This HTML code centers Principles of Information Systems as a major, or level 1, heading. The "h1" in the HTML code indicates a first-level heading. On some browsers, the heading might be 14-point type with a Times Roman font. On other browsers, it might be a larger 18-point type in a different font. Figure 7.6 shows a simple document and its corresponding HTML tags. Notice the <html> tag at the top indicating the beginning of the HTML code. The <title> indicates the beginning of the title: "Course Technology—Leading the Way in IT Publishing." The </title> tag indicates the end of the title.

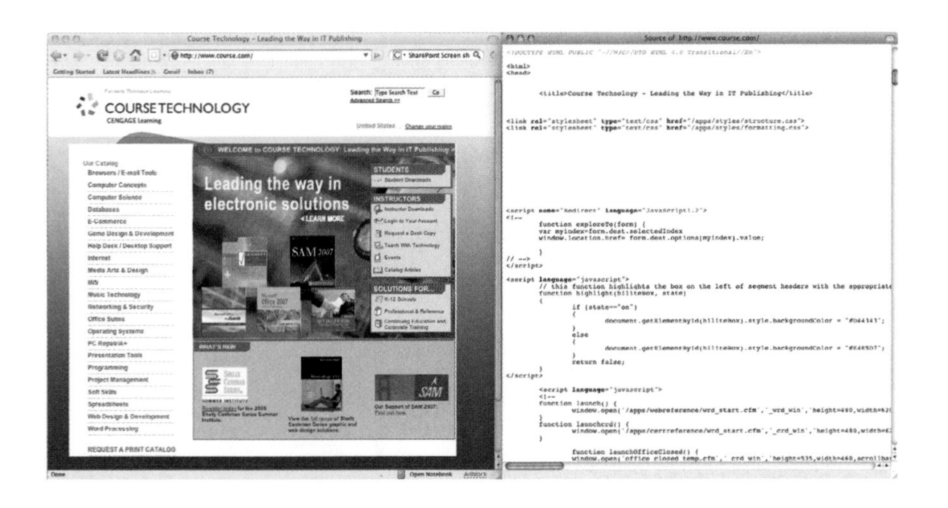

Figure 7.6

Sample Hypertext Markup Language

The window on the left is a Web document, and the window on the right shows the corresponding HTML tags.

Some newer Web standards are gaining in popularity, including Extensible Markup Language (XML), Extensible Hypertext Markup Language (XHTML), Cascading Style Sheets (CSS), Dynamic HTML (DHMTL), and Wireless Markup Language (WML), which can display Web pages on small screens such as smartphones and PDAs. XHTML is a combination of XML and HTML that has been approved by the World Wide Web Consortium (W3C).

Extensible Markup Language (XML) is a markup language for Web documents containing structured information, including words and pictures. XML does not have a predefined tag set. With HTML, for example, the <h1> tag always means a first-level heading. The content and formatting are contained in the same HTML document. XML Web documents contain the content of a Web page. The formatting of the content is contained in a style sheet. A few typical instructions in XML follow:

Extensible Markup Language (XML)
The markup language for Web documents containing structured information, including words, pictures, and other elements.

<chapter>Hardware
<topic>Input Devices
<topic>Processing and Storage Devices
<topic>Output Devices

Cascading Style Sheet (CSS)
A file or portion of an HTML file that defines the visual appearance of content in a Web page.

A **Cascading Style Sheet (CSS)** is a file or portion of an HTML file that defines the visual appearance of content in a Web page. Using CSS is convenient because you only need to define the technical details of the page's appearance once, rather than in each HTML tag. For example, the visual appearance of the preceding XML content may be contained in the following style sheet. This style sheet specifies that the chapter title "Hardware" is displayed on the Web page in a large Arial font (18 points). "Hardware" will also appear in bold blue text. The "Input Devices" title will appear in a smaller Arial font (12 points) and italic red text.

chapter: (font-size: 18pt; color: blue; font-weight: bold; display: block; font-family: Arial; margin-top: 10pt; margin-left: 5pt)
topic: (font-size: 12pt; color: red; font-style: italic; display: block; font-family: Arial; margin-left: 12pt)

Many new Web sites being developed use CSS to define the visual design and layout of Web pages, XML to define the content, and XHTML to join the content (XML) with the design (CSS). See Figure 7.7. This modular approach to Web design allows you to change the visual design without affecting the content, or to change the content without affecting the visual design.

Figure 7.7

XML, CSS, and XHTML

Today's Web sites are created using XML to define content, CSS to define the visual style, and XHTML to put it all together.

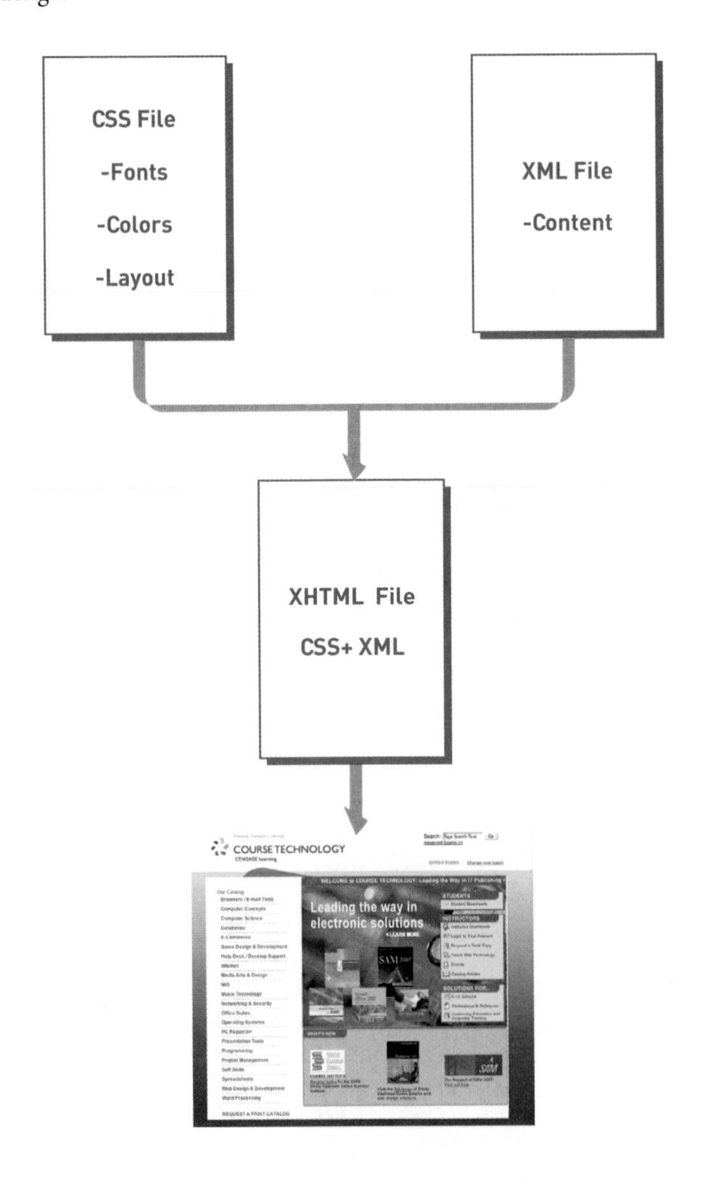

Web 2.0 and the Social Web

Over the past few years, the Web has been evolving from a one-directional resource where users only obtain information to a two-directional resource where users obtain and contribute information. Consider Web sites such as YouTube, Wikipedia, and MySpace as examples. The Web has also grown in power to support full-blown software applications, such as Google Docs, and is becoming a computing platform on its own. These two major trends in how the Web is used and perceived have created dramatic changes on the Web, so that the new form of the Web has earned the title of **Web 2.0.**[10]

Web 2.0
The Web as a computing platform that supports software applications and the sharing of information between users.

The original Web, now referred to as Web 1.0, provided a platform for technology-savvy developers and the businesses and organizations that hired them to publish information for the general public to view. The introduction of user-generated content supported by Wikipedia, blogging, and podcasting made it clear that those using the Web were also interested in contributing to its content. This led to the development of Web sites with the sole purpose of supporting user-generated content and user feedback.

Web sites such as YouTube and Flickr allow users to share video and photos with other people, groups, and the world. With social networking Web sites such as Facebook and MySpace, users can post information about their interests and find like-minded people. Using microblogging sites such as Twitter and Jaiku, people can post thoughts and ideas throughout the day for friends to read. Social bookmarking sites such as Digg and del.icio.us allow users to pool their votes to determine what online news stories and Web pages are most interesting each moment of the day. Similarly, Epinions and many retail Web sites allow consumers to voice their opinions about products. All of these popular Web sites serve as examples of how the Web has transformed to become the town square where people share information, ideas, and opinions; meet with friends; and make new acquaintances.

The introduction of powerful Web-delivered applications such as Google Docs, Adobe Photoshop Express, Xcerion Web-based OS, and Microsoft Maps have elevated the Web from an online library to a platform for computing.[11] Many of the computer activities traditionally provided through software installed on a PC can now be carried out using rich Internet applications (RIAs) in a Web browser without installing any software. A **rich Internet application** is software that has the functionality and complexity of traditional application software, but runs in a Web browser and does not require local installation. RIAs are the result of continuously improving programming languages and platforms designed for the Web.

rich Internet application
Software that has the functionality and complexity of traditional application software, but does not require local installation and runs in a Web browser.

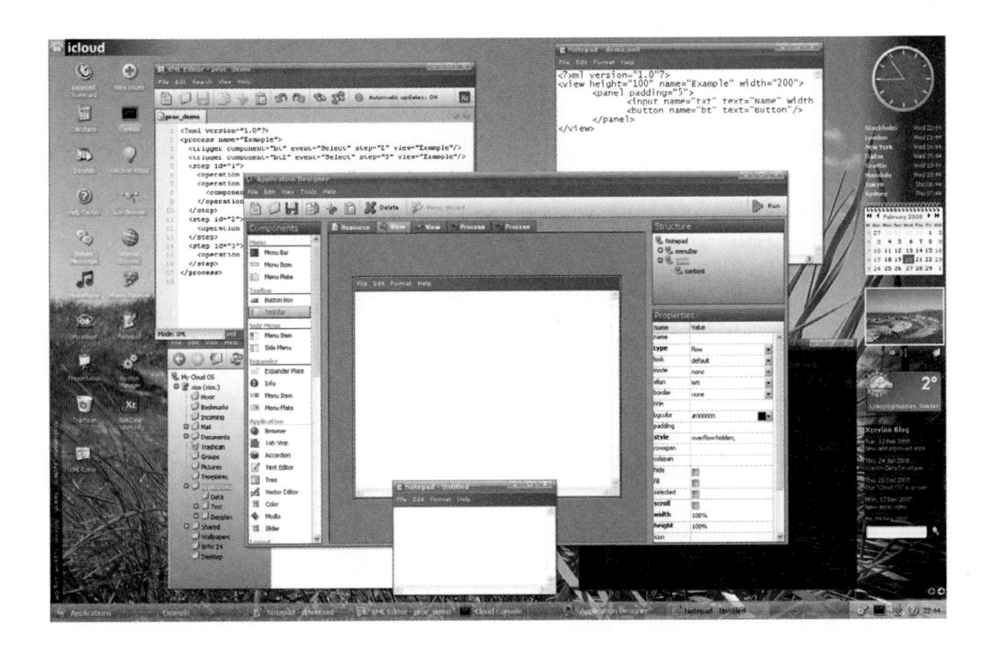

iCloud from Xcerion is a Web-based operating system that runs in a browser window.

(Source: Courtesy of Xcerion AB.)

Web Programming Languages

Java
An object-oriented programming language from Sun Microsystems based on C++ that allows small programs (applets) to be embedded within an HTML document.

Several programming languages are key to the Web. **Java**, for example, is an object-oriented programming language from Sun Microsystems based on the C++ programming language, which allows small programs, called *applets*, to be embedded within an HTML document. When the user clicks the appropriate part of an HTML page to retrieve an applet from a Web server, the applet is downloaded onto the client workstation, where it begins executing. Unlike other programs, Java software can run on any type of computer. Programmers use Java to make Web pages come alive, adding splashy graphics, animation, and real-time updates.

In addition to Java, companies use a variety of other programming languages and tools to develop Web sites. Software services delivered over the Web may run on the Web server, delivering the results of the processing to the user, or may run directly on the client—the user's PC. These two categories are commonly referred to as client-side and server-side software. JavaScript, VBScript, and ActiveX (used with Internet Explorer) are Internet languages used to develop Web pages and perform important functions, such as accepting user input. *Asynchronous JavaScript and XML (AJAX)* has become a popular programming language for developing RIAs. Programs built with AJAX run smoothly on the client PC, occasionally exchanging messages with the server.

Hypertext Preprocessor, or *PHP*, is an open-source programming language. PHP code or instructions can be embedded directly into HTML code. Unlike some other Internet languages, PHP can run on a Web server, with the results being transferred to a client computer. PHP can be used on a variety of operating systems, including Microsoft Windows, Macintosh OS X, HP-UX, and others. It can also be used with a variety of database management systems, such as DB2, Oracle, Informix, MySQL, and many others. These characteristics—running on different operating systems and database management systems, and being an open-source language—make PHP popular with many Web developers. Perl is another popular server-side programming language.

Adobe Flash and *Microsoft Silverlight* provide development environments for creating rich Web animation and interactive media. Both Flash and Silverlight require the user to install a browser plug-in to run. Flash has become so common that popular browsers include it as a standard feature. Microsoft Silverlight is a relatively new technology that is working to become a competitor to Flash. Any Web site that you visit that provides sophisticated animation and interaction is probably created with Flash or Silverlight. Such pages often take longer to load than standard HTML pages.

Developing Web Content

The art of Web design involves working within the technical limitations of the Web and using a set of tools to make appealing designs. Popular tools for creating Web pages and managing Web sites include Adobe Dreamweaver, Microsoft Expression Web, and Nvu. (See Figure 7.8.) Today's Web development applications allow the user to create Web sites using software that resembles a word processor. The software includes features that allow the developer to work directly with the HTML code or use autogenerated code. Web development software also helps the designer keep track of all files in a Web site and the hyperlinks that connect them.

After you create Web pages, your next step is to place, or publish, the content on a Web server. Popular publishing options include using ISPs, free sites, and Web hosting. Web hosting services provide space on their Web servers for people and businesses that don't have the financial resources, time, or skills to host their own Web sites. A Web host can charge $15 or more per month, depending on services. Some Web hosting sites include domain name registration, Web authoring software, and activity reporting and monitoring of the Web site. Some ISPs also provide limited Web space, typically 1 to 6 MB, as part of their monthly fee. If more disk space is needed, additional fees are charged. Free sites offer limited space for a Web site. In return, free sites often require the user to view advertising or agree to other terms and conditions.

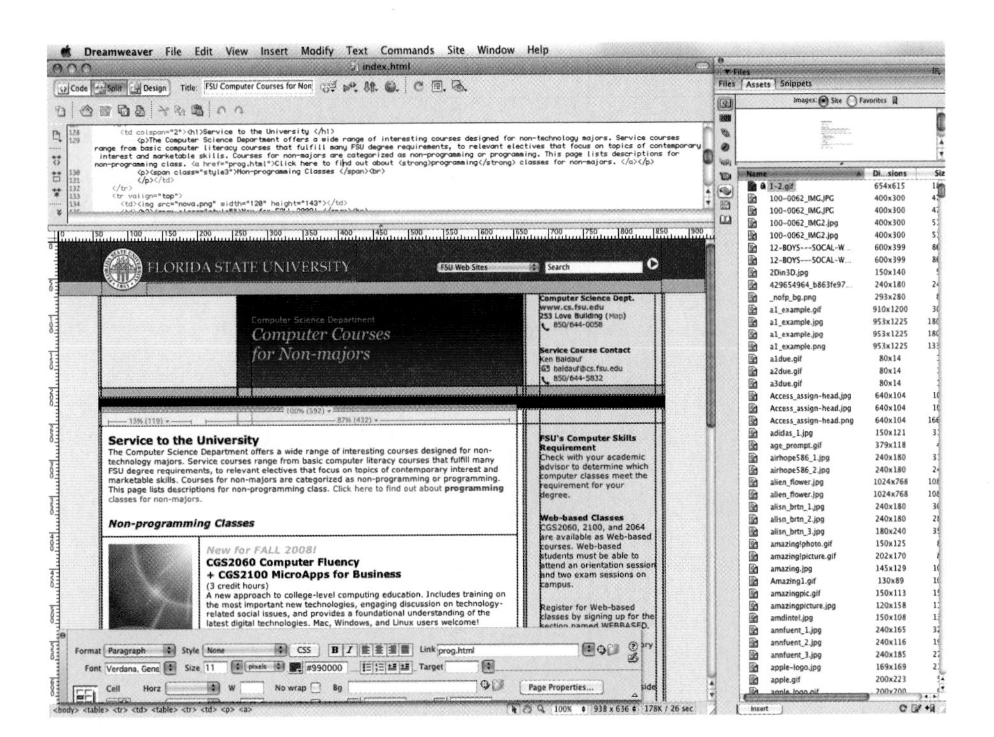

Figure 7.8

Creating Web Pages

Adobe Dreamweaver makes Web design nearly as easy as using a word processor.

Some Web developers are creating programs and procedures to combine two or more Web applications into a new service, called a *mash-up*. A mash-up is named for the process of mixing two or more hip-hop songs into one song. A Web site that provides crime information, for example, can be mashed up with a mapping Web site to produce a Web site with crime information placed on top of a map of a metropolitan area. People are becoming creative in how they mash up several Web sites into new ones. Google and Yahoo both provide developers with tools for creating mash-ups. Google's is called the Mashup Editor (*editor.googlemashups.com*), and Yahoo's is called Pipes (*pipes.yahoo.com*).[12]

Some corporations maintain and manage their large enterprise Web sites using a *content management system (CMS)*. A CMS consists of both software and support. Companies that provide a CMS can charge from $15,000 to more than $500,000 annually, depending on the complexity of the Web site being maintained and the services being performed. Leading CMS vendors include BroadVision, EBT, FileNet, and Vignette. Many of these products are popular because they take a newer approach to developing and maintaining Web content called Web services, discussed next.

Many products make it easy to develop Web content and interconnect Web services, discussed in the next section. Microsoft, for example, provides a development and Web services platform called .NET. The .NET platform allows developers to use different programming languages to create and run programs, including those for the Web. The .NET platform also includes a rich library of programming code to help build XML Web applications. Other popular Web development platforms include Sun JavaServer Pages, Microsoft ASP, and Adobe Cold Fusion. See Figure 7.9.

Web Services

Web services consist of standards and tools that streamline and simplify communication among Web sites, promising to revolutionize the way we develop and use the Web for business and personal purposes. Internet companies, including Amazon, eBay, and Google, are now using Web services. Amazon, for example, has developed Amazon Web Services (AWS) to make the contents of its huge online catalog available to other Web sites or software applications. Mitsubishi Motors of North America uses Web services to link about 700 automotive dealers on the Internet.

Web services
Standards and tools that streamline and simplify communication among Web sites for business and personal purposes.

The key to Web services is XML. Just as HTML was developed as a standard for formatting Web content into Web pages, XML is used within a Web page to describe and

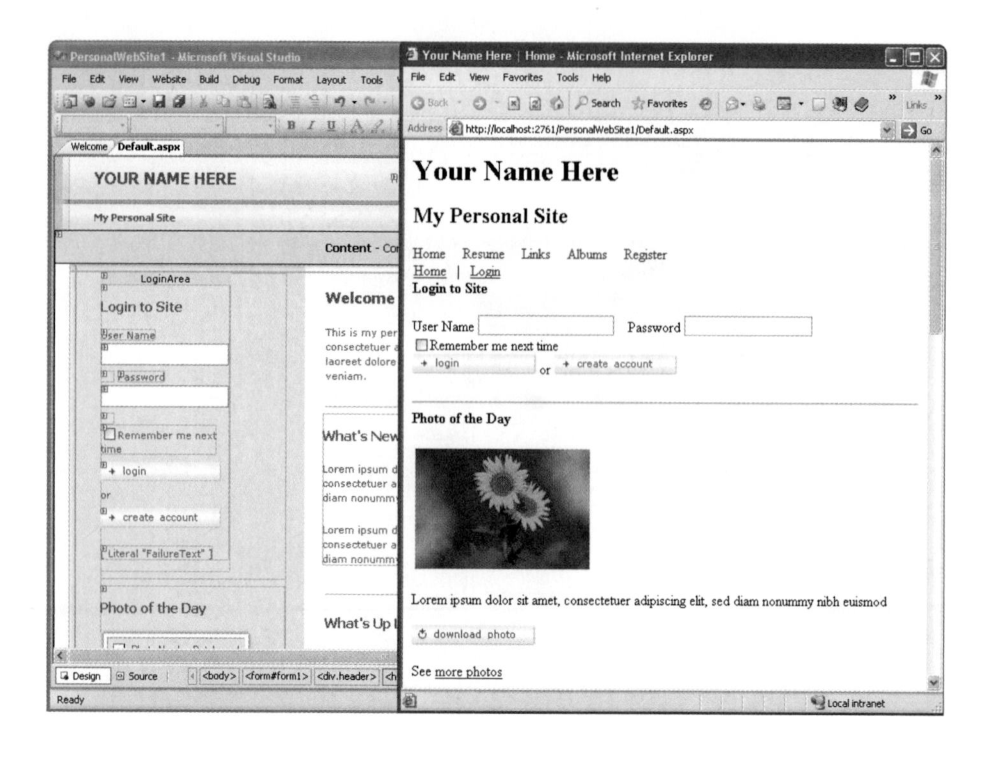

Figure 7.9

Developing a Web Page Using Microsoft Visual Studio

The window on the left shows the Web page being developed using Microsoft Visual Studio, part of the .NET Web development platform. The window on the right shows the Web page in a browser.

transfer data between Web service applications. XML is easy to read and has wide industry support. In addition to XML, three other components are used in Web service applications:

1. SOAP (Simple Object Access Protocol) is a specification that defines the XML format for messages. SOAP allows businesses, their suppliers, and their customers to communicate with each other. It provides a set of rules that makes it easier to move information and data over the Internet.
2. WSDL (Web Services Description Language) provides a way for a Web service application to describe its interfaces in enough detail to allow a user to build a client application to talk to it. In other words, it allows one software component to connect to and work with another software component on the Internet.
3. UDDI (Universal Discovery Description and Integration) is used to register Web service applications with an Internet directory so that potential users can easily find them and carry out transactions over the Web.

INTERNET AND WEB APPLICATIONS

The types of Internet and Web applications available are vast and ever expanding. Here are some examples.

Search Engines and Web Research

search engine
A valuable tool that enables you to find information on the Web by specifying words that are key to a topic of interest, known as keywords.

A search engine is a valuable tool that enables you to find information on the Web by specifying words that are key to a topic of interest—known as keywords. You can also use operators such as OR and NOT for more precise search results.[13] Table 7.4 provides examples of the use of operators in Google searches as listed on Google's help page (*www.google.com/help/cheatsheet.html*).

Search engines provide their services for free, which leaves many wondering how they make money. Search engines make money from companies that place advertisements with the search results. Search companies learn about visitors' interests by the topics for which they search, and can target ads to each user based on those interests. Google, Microsoft, and Yahoo! are leaders in the search engine business.

Table 7.4

Using Operators in Google Web Searches

Keywords and Operator Entered	Search Engine Interpretation
vacation Hawaii	The words "vacation" and "Hawaii"
Maui OR Hawaii	Either the word "Maui" or the word "Hawaii"
"To each his own"	The exact phrase "To each his own"
virus -computer	The word virus, but not the word computer
Star Wars Episode +I	The movie title "Star Wars Episode", including the Roman numeral I
~auto loan	Loan information for both the word "auto" and its synonyms, such as "truck" and "car"
define:computer	Definitions of the word "computer" from around the Web
red * blue	The words "red" and "blue" separated by one or more words

Search engines scour the Web with bots (automated programs) called spiders that follow all Web links in an attempt to catalog every Web page by topic. The process is called Web crawling, and due to the ever-changing nature of the Web, it is a job that never ends. Google maintains over four billion indexed Web pages in a database on 30 clusters of up to 2,000 computers, each totaling over 30 petabytes of data.

One of the challenges of Web crawling is determining which of the words on any given Web page describe its topic. Different search engines use different methods. Methods include counting word occurrences within the Web page, evaluating nouns and verbs in the page's title and subtitle, using keywords provided by the page's author in a meta tag, and evaluating the words used in links to the page from other pages. Once the search engine has a reasonable idea of a page's topic, it records the URL, page title, and associated information and keywords in a database.

After building the search database, the next challenge facing a search engine is to determine which of the hundreds or thousands of Web pages associated with a particular keyword are most useful. The method of ranking Web pages from most relevant to least differs from search engine to search engine. Google uses a popularity contest approach. Web pages that are referenced from other Web pages are ranked higher than those that are not. Each reference is considered a vote for the referenced page. The more votes a Web page gets, the higher its rank. References from higher-ranked pages weigh more heavily than those from lower-ranked pages.

Today's heated competition in the search engine market is pressing the big players to expand their services. Table 7.5 lists some of the newer search engine services available and being developed.

Search engines have become important to businesses as a tool to drive visitors to the business' Web site. Many businesses invest in search engine optimization (SEO)—a process for driving traffic to a Web site by using techniques that improve the site's ranking in search results. SEO is based on the understanding that Web page links listed on the first page of search results, as high on the list as possible, have a far greater chance of being clicked. SEO professionals study the algorithms employed by search engines, altering the Web page contents and other variables to improve the page's chance of being ranked number one.[14]

SEO has become a valuable marketing tool. Tax software company TaxEngine.com hired an SEO company to improve its visibility and allow it to compete with larger companies such as H&R Block and TurboTax. The SEO approach saved the company 50 percent in marketing costs over the previous year while increasing its business. The SEO optimized the company's hit results for over 100 highly competitive search keywords.[15]

In addition to search engines, you can use other Internet sites to research information. For example, *www.findarticles.com* contains millions of articles on a variety of topics, including business and finance, health and fitness, sports, and reference and education. You

Table 7.5
Search Engine Services

Service	What it Does
Alerts	Receive news and search results via e-mail
Answers	Ask a question, set a price, get an answer
Catalogs	Search and browse mail-order catalogs
Desktop Search for Enterprise	Search your company's network
Images	Search for images on the Web
Local	Find local businesses and services
Maps	View maps and get directions
Mobile	Search the Web from your cell phone
News	Search thousands of news stories
Personalized Search Page	Customize your search page with current news and weather
Print	Search the full text of books
Ride Finder	Find a taxi, limousine, or shuttle using real-time position of vehicles
Scholar; University Search	Search through journal articles, abstracts, and other scholarly literature; search a specific school's Web site
Search by Location	Filter results by geographic location
Search History	Maintain a history of past searches and the Web sites that produced results
Search Toolbar	Access search from the toolbar of your browser or operating system taskbar
Shopping	Find the best deal on consumer products
Video	Search recent TV programs online

can use many news organizations, including CNN (*www.cnn.com*) and Fox News (*www.foxnews.com*), to access current information on a variety of topics. Some Web sites maintain versions in different languages, especially for research purposes. Others offer programs such as Flexnet that combine online learning with face-to-face instruction. Wikipedia, an encyclopedia with 1.9 million English-language entries created and edited by tens of thousands of contributors, is another example of a Web site that can be used to research information.[16] (See Figure 7.10.) In Hawaiian, *wiki* means quick, so a wikipedia provides quick access to information. The Web site is both open source and open editing, which means that people can add or edit entries in the encyclopedia at any time. Because thousands of people are monitoring Wikipedia, the Web-based encyclopedia is self-regulating. Incorrect, outdated, or offensive material is usually removed, although people with an unobjective point of view have distorted information on Wikipedia intentionally. Jimmy Wales, the founder of Wikipedia, would like to expand the wiki concept into books, manuals, and quotations. Some think that the approach of Wikipedia can be used to allow people to collaborate on important projects. Squidoo (*www.squidoo.com*) is a Web site you can use to find information about a person's view of a particular topic, often called a "lens." You can find lenses on a wide variety of topics, including the arts, computers and technology, education, health, movies, music, news, and much more.

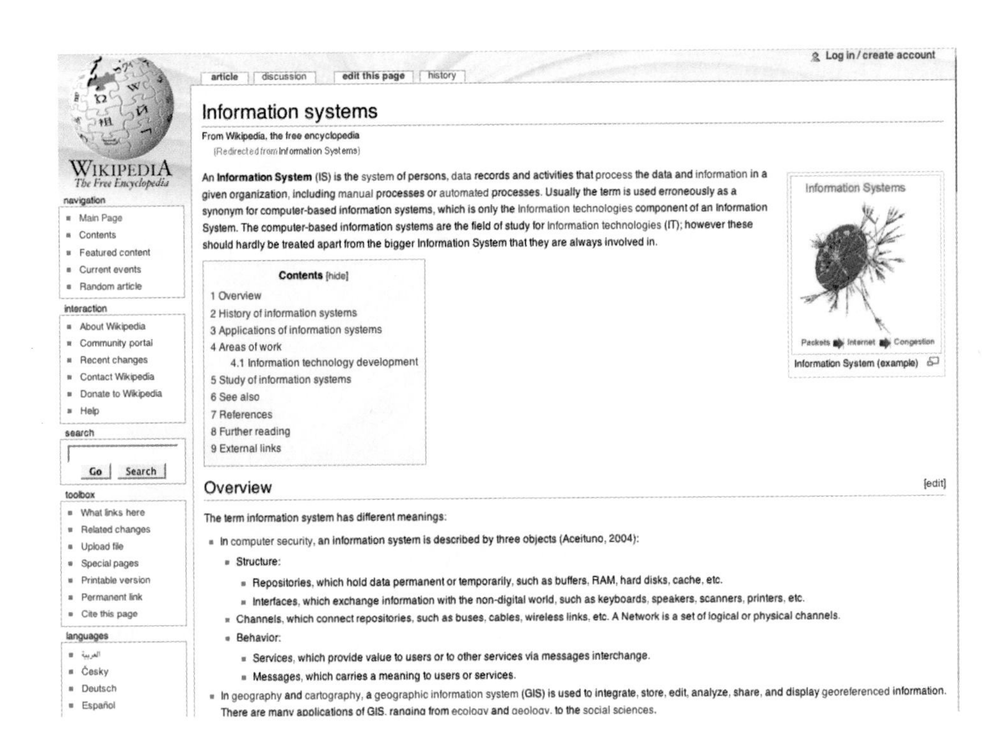

Figure 7.10

Wikipedia

Wikipedia captures the knowledge of tens of thousands of experts.

Business Uses of the Web

In 1991, the Commercial Internet Exchange (CIX) Association was established to allow businesses to connect to the Internet. Since then, firms have been using the Internet for a number of applications, as discussed in this section.

E-mail, Instant Messaging, and Video Chat

E-mail is no longer limited to simple text messages. Depending on your hardware and software and the hardware and software of your recipient, you can embed images, sound, and video in a message and attach any kind of file. The authors of this book, for example, attached chapter files to e-mail messages that were sent to editors and reviewers for feedback.

Many people use online e-mail services such as Hotmail, MSN, and Gmail to access e-mail. Online e-mail services store messages on the server, so users need to be connected to the Internet to view, send, and manage e-mail. Other people prefer to use software such as Microsoft Outlook, Apple Mail, or Mozilla Thunderbird, which retrieve e-mail from the server and deliver it to the user's PC. Post office protocol (POP) is used to transfer messages from e-mail servers to your PC. E-mail software typically includes more information management features than online e-mail services, and lets you save your e-mail on your own PC, making it easier to manage and organize messages and to keep the messages private and secure. Another protocol called Internet message access protocol (IMAP) allows you to view e-mail using Outlook or other e-mail software without downloading and storing the messages locally. Some users prefer this method because it allows them to view messages from any Internet-connected PC.

Business users that access e-mail from smartphones such as the BlackBerry take advantage of a technology called Push e-mail. Push e-mail uses corporate server software that transfers, or pushes, e-mail out to the handset as soon as it arrives at the corporate e-mail server. To the BlackBerry user, it appears as though e-mail is delivered directly to the handset. Push e-mail allows the user to view e-mail from any mobile or desktop device connected to the corporate server. This arrangement allows users flexibility in where, when, and how they access and manage e-mail.[17]

Since text-based communications lack the benefit of facial expression, voice inflection, and body language, users have developed methods of expressing emotion through typed characters. Text messaging has also led to abbreviations for common expressions that save typing time. Table 7.6 lists some expressions and abbreviations frequently used in personal

BlackBerry users have instant access to e-mail sent to their business account.

(Source: Courtesy of Marvin Woodyatt/Photoshot /Landov.)

e-mail messages, text messaging, and other forms of text communications. These abbreviations are normally not appropriate for business correspondence.

Table 7.6

Some Common Abbreviations Used in Personal E-Mail

Expressions	Abbreviations
;-) Smile with a wink	AAMOF—As a matter of fact
;-(Frown with a wink	AFAIK—As far as I know
:-# My lips are sealed	BTW—By the way
:-D Laughing	CUL8R—See you later
:-0 Shocked	F2F—Face to face
:-] Blockhead	LOL—Laughing out loud
:-@ Screaming	OIC—Oh, I see
:-& Tongue-tied	TIA—Thanks in advance
%-) Brain-dead	TTFN—Ta-Ta for now

Some companies use bulk e-mail to send legitimate and important information to sales representatives, customers, and suppliers around the world. With its popularity and ease of use, however, some people feel they are drowning in too much e-mail.[18] Over a trillion e-mail messages are sent from businesses in North America each year. This staggering number is up from 40 billion e-mail messages in 1995. Many messages are copies sent to long lists of corporate users. Users are taking steps to cope with and reduce the mountain of e-mail. Some companies have banned the use of copying others on e-mails unless it is critical. Some e-mail services scan for possible junk or bulk mail, called *spam*, and delete it or place it in a separate file. More than half of all e-mail can be considered spam. Some business executives receive 300 or more spam e-mails in their corporate mailboxes every morning. Mukesh Lulla, president of TeamF1, a networking and security-software company, receives 300 to 400 e-mail messages daily, *not including* spam.[19] While spam-filtering software can prevent or discard unwanted messages, other software products can help users sort and answer large amounts of legitimate e-mail. For example, software from ClearContext, Seriosity, and Xobni rank and sort messages based on sender, content, and context, allowing individuals to focus on the most urgent and important messages first.

instant messaging
A method that allows two or more people to communicate online using the Internet.

Instant messaging is online, real-time communication between two or more people who are connected to the Internet. With instant messaging, two or more windows or panes open, with each one displaying text a person is typing. Because the typing is displayed in real time, instant messaging is like talking to someone using the keyboard. See Figure 7.11.

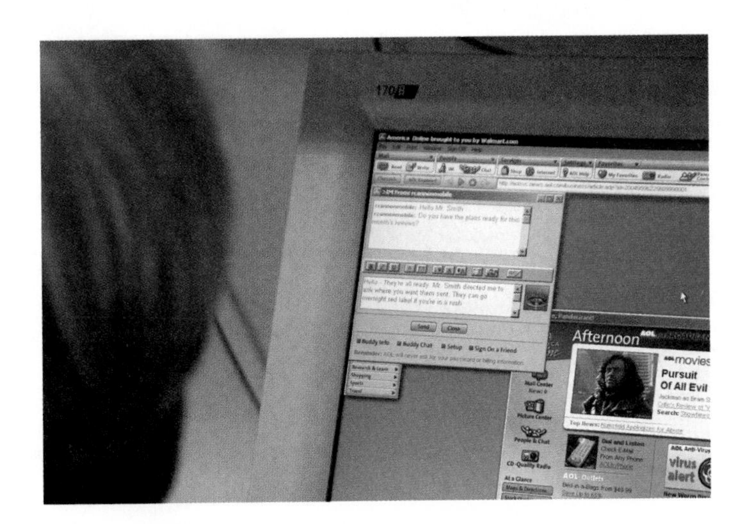

Figure 7.11

Instant Messaging

Instant messaging lets you converse with another Internet user by exchanging messages instantaneously.

Many companies offer instant messaging, including America Online, Yahoo!, and Microsoft. America Online is one of the leaders in instant messaging, with millions of people using AOL Instant Messenger (AIM) and its client program ICQ. In addition to being able to type messages on a keyboard and have the information instantly displayed on the other person's screen, some instant messaging programs are allowing voice communication or connection to cell phones. Today, instant messaging can be delivered over the Internet, through the cell phone services, and via other telecommunications services first discussed in Chapter 6.

As more people are connecting to the Internet over broadband connections, increasing numbers of users are turning to video chat. Services such as Apple iChat and Skype provide computer-to-computer video chat so users can speak to each other face-to-face. Some video chat services support conference calling as well.

Career Information and Job Searching

The Internet is an excellent source of job-related information. People looking for their first job or seeking information about new job opportunities can find a wealth of information. Search engines can be a good starting point for searching for specific companies or industries. You can use a directory on Yahoo's home page, for example, to explore industries and careers. Most medium and large companies have Internet sites that list open positions, salaries, benefits, and people to contact for further information. The IBM Web site, *www.ibm.com*, has a link to "Jobs at IBM." When you click this link, you can find information on jobs with IBM around the world. Some Internet sites specialize in certain careers or industries. The site *www.directmarketingcareers.com* lists direct marketing jobs and careers. Some sites can help you develop a résumé and find a good job. They can also help you develop an effective cover letter for a résumé, prepare for a job interview, negotiate an employment contract, and more. In addition, several Internet sites specialize in helping you find job information and even apply for jobs online, including *www.monster.com*, *www.hotjobs.com*, and *www.careerbuilder.com*. You must be careful when applying for jobs online, however. Some bogus companies or Web sites will steal your identity by asking for personal information. People eager to get a job often give their Social Security number, birth date, and other personal information. The result can be no job, large bills on your credit card, and ruined credit.

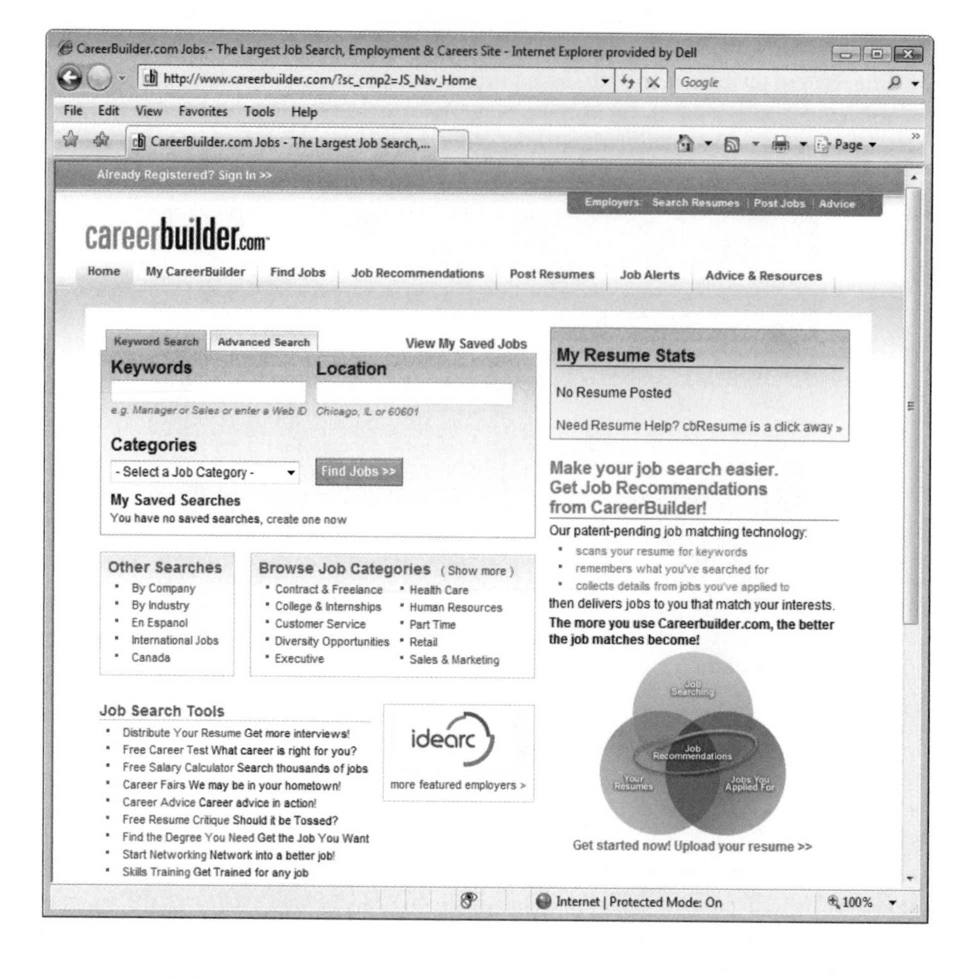

Several Internet sites specialize in helping people get job information and even apply for jobs online.

Telnet, SSH, and FTP

Telnet is a network protocol that enables users to log on to networks remotely over the Internet. Telnet software uses a command-line interface that allows the user to work on a remote server directly. Since Telnet is not secured with encryption, most users are switching to *secure shell (SSH)*, which provides Telnet functionality through a more secure connection.

File Transfer Protocol (FTP)
A protocol that describes a file transfer process between a host and a remote computer and allows users to copy files from one computer to another.

File Transfer Protocol (FTP) is a protocol that supports file transfers between a host and a remote computer. Using FTP, users can copy files from one computer to another. For example, companies use it to transfer vast amounts of business transactional data to the computers of its customers and suppliers. You can also use FTP to gain access to a wealth of free software on the Internet and to upload or download content to a Web site. The authors and editors of this book used an FTP site provided by the publisher, Course Technology, to share and transfer important files during the publication process. Chapter files and artwork, for example, were uploaded to a Course Technology Web site and downloaded by authors and editors to review. Like Telnet, FTP connections are not encrypted, and are therefore not secure. Many users are switching to secure FTP (SFTP) for more secure file transfers.

Web Log (Blog), Video Log (Vlog), and Podcasting

Web log (blog)
A Web site that people can create and use to write about their observations, experiences, and feelings on a wide range of topics.

A **Web log**, typically called a **blog**, is a Web site that people can create and use to write about their observations, experiences, and opinions on a wide range of topics.[20] The community of blogs and bloggers is often called the *blogosphere.* A *blogger* is a person who creates a blog, while *blogging* refers to the process of placing entries on a blog site. A blog is like a journal. When people post information to a blog, it is placed at the top of the blog page. Blogs can include links to external information and an area for comments submitted by visitors. Video content can also be placed on the Internet using the same approach as a blog. This is often called a *video log* or *vlog*. Blogs are easy to post, but they can cause problems when people tell or share too much. People have been fired for blogging about work, and the daughter of a politician embarrassed her father when she made personal confessions on a blog.

Blog sites, such as *www.blogger.com* and *www.blogcatalog.com*, can include information and tools to help people create and use Web logs. To set up a blog, you can go to the Web site of a blog service provider, such as *www.livejournal.com*, create a username and password, select a theme, choose a URL, follow any other instructions, and start making your first entry. Blog search engines include Technorati and Blogdigger. You can also use Google to locate a blog.

A corporate blog can be useful for communicating with customers, partners, and employees. However, companies and their employees need to be cautious about the legal risks of blogging.[21] Blogging can expose a corporation and its employees to charges of defamation, copyright and trademark infringement, invasion of privacy, and revealing corporate secrets.

A *podcast* is an audio broadcast over the Internet. The name podcast comes from the word *iPod*, Apple's portable music player, and the word *broadcasting*. A podcast is an audio blog, like a personal radio station on the Internet, and extends blogging by adding audio messages. Using PCs, recording software, and microphones, you can record audio messages and place them on the Internet. You can then listen to the podcasts on your PC or download the audio material to a digital audio player, such as Apple's iPod. You can also use podcasting to listen to TV programs, your favorite radio personalities, music, and messages from your friends and family at any time and place. Finding good podcasts, however, can be challenging. Apple's iTunes provides free access to tens of thousands of podcasts sorted by topic and searchable by keyword. After you find a podcast, you can download it to a PC (Windows or Mac) and to an MP3 music player such as the iPod for future listening.

Figure 7.12

iTunes Podcasts

iTunes provides free access to tens of thousands of podcasts.

People and corporations can use podcasts to listen to audio material, increase revenues, or advertise products and services. You can listen to podcasts of radio programs, including some programs from National Public Radio (NPR), while you are driving, walking, making a meal, or most other activities. Clear Channel Communications, a radio broadcasting corporation, sells memberships to podcasts of popular radio shows and personalities, including Rush Limbaugh, NPR programs, and others. ABC News uses podcasts to allow people to listen to some TV programs, such as Nightline. Colleges and universities often use blogs and podcasts to deliver course material to students.

Many blogs, vlogs, and podcasts offer automatic updates to a PC using a technology called Really Simple Syndication (RSS). RSS is a collection of Web technologies that allow users to subscribe to Web content that is frequently updated. With RSS, you can receive a blog update without actually visiting the blog Web site. You can also use RSS to receive other updates on the Internet from news Web sites and podcasts. Software used to subscribe to RSS feeds is called *aggregator software.* Google Reader is a popular aggregator for subscribing to blogs.

Usenet and Newsgroups

Usenet is an older technology that uses e-mail to provide a centralized news service. Topic areas in Usenet are called newsgroups. A newsgroup is essentially an online discussion group that focuses on a particular topic. Newsgroups are organized into various hierarchies by general topic, and each topic can contain many subtopics. Table 7.7 provides some examples. Usenet is actually a protocol that describes how groups of messages can be stored on and sent between computers. Following the Usenet protocol, e-mail messages are sent to a host computer that acts as a Usenet server. This server gathers information about a single topic into a central place for messages. A user sends e-mail to the server, which stores the messages. The user can then log on to the server to read these messages or have software on the computer log on and automatically download the latest messages to be read at leisure. Thus, Usenet forms a virtual forum for the electronic community, and this forum is divided into newsgroups. Blogging, RSS, and social networking sites have drawn many users away from Usenet forums.

Table 7.7

Selected Usenet Newsgroups

Newsgroup Address	Description
alt.airline	Current schedules of various airlines
alt.books	Index of book reviews
alt.current-events.net-abuse.spam	Reports of e-mail and newsgroup spam abuse
alt.fan	Fans of various performers, artists, and others
alt.politics	Index of political discussions
alt.sports.baseball	Major league baseball
biz.ecommerce	Internet retailers
misc.legal	Miscellaneous discussions of legal matters
news.software	Usenet software
rec.backcountry	Activities in the great outdoors
rec.food.restaurants	Discussion of dining out

Chat Rooms

chat room
A facility that enables two or more people to engage in interactive "conversations" over the Internet.

A **chat room** is a facility that enables two or more people to engage in interactive "conversations" over the Internet. When you participate in a chat room, dozens of people might be participating from around the world. Multiperson chats are usually organized around specific topics, and participants often adopt nicknames to maintain anonymity. One form of chat room, Internet Relay Chat (IRC), requires participants to type their conversation rather than speak. Voice chat is also an option, but you must have a microphone, sound card, speakers, a fast modem or broadband, and voice-chat software compatible with the other participants'. Most of the functionality of chat is available in instant messaging software.

Internet Phone and Videoconferencing Services

Internet phone service enables you to communicate with others around the world. This service is relatively inexpensive and can make sense for international calls. With some services, you can use the Internet to call someone who is using a standard phone. You can also keep your phone number when you move to another location. According to one Internet phone user who moved from Madison, Wisconsin, to California, "I was so happy about that. Nothing changed for my customers. For all they knew I was still in Madison." Cost is often a big factor for those using Internet phones—a call can be as low as 1 cent per minute for calls within the United States. Low rates are also available for calling outside the United States. In addition, voice mail and fax capabilities are available. Some cable TV companies, for example, are offering cable TV, phone service, and caller ID for under $40 a month. Skype offers free and low-priced Internet phone and video phone service from any Internet connected computer.

Figure 7.13

Wi-Fi Phone

The BlackBerry Curve 8320 smartphone, combined with T-Mobile's HotSpot @ Home service, uses the Internet for calls while at home and the cell network while away.

(Source: Courtesy of Research In Motion.)

Using *voice-over-IP (VoIP)* technology, as described in Chapter 6, network managers can route phone calls and fax transmissions over the same network they use for data—which means no more separate phone bills. See Figure 7.13. Gateways installed at both ends of the communications link convert voice to IP packets and back. With the advent of widespread, low-cost Internet telephony services, traditional long-distance providers are being pushed to either respond in kind or trim their own long-distance rates. VoIP (pronounced *voyp*) is growing rapidly.

Although the technology for VoIP has existed for decades, the widespread use of VoIP is just beginning. Today, many companies offer Web phone service using VoIP, including Vonage, AT&T, Comcast, Verizon, AOL, Packet8, Callserve, Net2Phone, and WebPhone. Even so, there are obstacles to using VoIP. Some service providers, for example, might have trouble connecting their customers to emergency 911 service. In 2005, the Federal Communications Commission (FCC) issued an order that Web phone services must notify their customers of the potential problems with making emergency 911 calls.

Internet videoconferencing, which supports both voice and visual communications, is another important Internet application. Microsoft's NetMeeting, a utility within Windows XP, is an inexpensive and easy way for people to meet and communicate on the Web. Windows Vista offers Windows Meeting Space to provide the same service. The Internet can also be used to broadcast group meetings, such as sales seminars, using presentation software and videoconferencing equipment. These Internet presentations are often called Webcasts or Webinars. WebEx and GoToMeeting are two popular Web conferencing tools. The ideal video product will support multipoint conferencing, in which users appear simultaneously on multiple screens. Hewlett-Packard (HP) has produced such a system called Halo (see Figure 7.14). When using Halo, it appears as though you are speaking with a number of people across a table, though those people may actually be located around the world.

Social Networks

Social networking Web sites provide Web-based tools for users to share information about themselves with people on the Web and to find, meet, and converse with other members. The most popular social networking sites are MySpace and Facebook. Both sites provide members with a personal Web page and allow them to post photos and information about themselves (see Figure 7.15). Social networking sites allow members to send messages to each

Figure 7.14

Halo Collaboration Meeting Room

HP's Halo telepresence system allows people at various locations to meet as though they were gathered around a table.

(Source: Hewlett-Packard Web site, *www.hp.com*.)

other and post comments on each other's pages. Members accumulate friends through invitation. Special interest groups can be created and joined as well. Social networking Web sites also provide tools to search for people with similar interests. The power of social networks is now being harnessed for business purposes.[22] Many businesses are using the information posted in member profiles to find potential clients. Linked In is a social network that allows professionals to find others who work in the same field, applying social networking techniques for business networking.

Figure 7.15

Social Networking Web Sites

Facebook is a social networking site that provides members with Web pages to post photos and information about themselves.

Facebook provides an application development platform so that technically proficient members can create applications to run within Facebook. This has led to hundreds of tools that Facebook users can add to their pages. For example, Facebook has tools to connect to people with similar music tastes, to see your daily horoscope, to share videos, to find "Mr. or Ms. Right," to express your mood, and many more.

The U.S. intelligence community is adopting social networking to share information among operatives and analysts.[23] A-Space will be a private online social network designed for intelligence professionals to communicate online. It includes blogs, a searchable database, libraries of reports, collaborative word processing, and other useful tools to allow those in the field to quickly exchange and access information.

Another social networking site called Twitter (*www.twitter.com*) allows members to report on what they are doing throughout the day. Referred to as a microblogging service, Twitter allows users to send short text updates (up to 140 characters long) from cell phones or the Internet to their Twitter page to let others know what they are doing.[24] Twitter updates can be forwarded to MySpace or Facebook Web sites.

Media Sharing

Media-sharing Web sites such as YouTube for video sharing and Flickr for photo sharing provide methods for members to store and share digital media files on the Web. YouTube allows members to post homemade video content in categories such as comedy, entertainment, film and animation, how-to, news, people, pets, sports, and travel. As mentioned earlier, Flickr allows members to upload photos to their own personal online photo album and choose photos to share with the community.

What makes these media-sharing sites part of Web 2.0 is their focus on community. Both Flickr and YouTube provide ways for members to comment on the media. YouTube allows visitors to e-mail favorite video clips to friends. Both sites provide methods for visitors to view the most popular media or search on a particular topic.

Flickr uses a methodology of ranking content that has become popular with many Web 2.0 sites. Formally called a folksonomy or collaborative tagging, Flickr allows users to associate descriptive tags with photos. For example, you might tag a photo of your pet Weimaraner at the beach with "Dog," "Pet," "Weimaraner," and "Beach." Using associated tags, Flickr can easily group common photos together and gather statistics on photos. Flickr uses this information to create a tag cloud—a diagram of keyword links with the size of each word representing the number of photos that use that tag (see Figure 7.16). Smugmug is a photo sharing site that supports higher-quality images than most such sites and allows members to sell their photos to others for profit[25].

Figure 7.16

Flickr's Tag Cloud

Flickr's tag cloud uses font size to indicate which tags have the most photos associated with them; users click a tag to see the associated photos.

(Source: Flickr Web site, *www.flickr.com*.)

Social Bookmarking

Social bookmarking sites are another example of Web 2.0. These sites provide a way for Web users to store, classify, share, and search Web bookmarks—also referred to as favorites. The typical purpose of social bookmarking sites is to provide a view of the most popular Web sites, videos, blog articles, or other Web content at any given moment. Often social bookmarking sites include Web browser add-ons (extensions) that provide a button on the toolbar for recommending Web content. For example, del.icio.us is a social bookmarking Web site that provides a "what's hot right now" button. When you sign up for del.icio.us, you can download software to install on your computer that provides two buttons on your browser toolbar. When you find a page you want to bookmark, you click the Tag button on the toolbar to store the link in your bookmark list on http://del.icio.us. Pages you bookmark are tallied with other users' bookmarks to determine the most popular pages on the Web at any given moment.

Digg is another popular social bookmarking site dedicated to news. Many online news services provide "Digg this" buttons on articles so that readers can bookmark the article. At *www.digg.com* you can see the most popular news articles of the moment listed sequentially, with the articles that accumulated the most "digs" listed first. Digg also provides links to the most popular videos and podcasts.

Content Streaming

content streaming
A method for transferring multimedia files over the Internet so that the data stream of voice and pictures plays more or less continuously without a break, or very few of them; enables users to browse large files in real time.

Content streaming is a method for transferring multimedia files, radio broadcasts, and other content over the Internet so that the data stream of voice and pictures plays more or less continuously, without a break, or with very few of them. It also enables users to browse large files in real time. For example, rather than wait for an entire 5 MB video clip to download before they can play it, users can begin viewing a streamed video as it is being received. Content streaming works best when the transmission of a file can keep up with the playback of the file.

Shopping on the Web

Shopping on the Web for books, clothes, cars, medications, and even medical advice can be convenient, easy, and cost effective. Amazon.com, for example, sells short stories by popular authors for 49 cents per story. The service, called Amazon Shorts, has stories that vary in length from 2,000 to 10,000 words by authors such as Danielle Steel, Terry Brooks, and others. The company also sells traditional books and other consumer products. To add to their other conveniences, many Web sites offer free shipping and pickup for returned items that don't fit or otherwise meet a customer's needs.

bot
A software tool that searches the Web for information such as products and prices.

Increasingly, people are using bots to help them search for information or shop on the Internet. A **bot**, also called an *intelligent agent,* is a software tool that searches the Web for information, products, or prices. A bot, short for *robot,* can find the best prices or features from multiple Web sites. Shopping.com uses bots to identify the best prices on merchandise.

Web Auctions

Web auction
An Internet site that matches buyers and sellers.

A **Web auction** is a way to connect buyers and sellers. Web auction sites are a place where businesses are growing their markets or reaching customers for a low cost per transaction. Web auctions are transforming the customer-supplier relationship.

One of the most popular auction sites is eBay, which often has millions of auctions occurring at the same time. The eBay site is easy to use and includes thousands of products and services in many categories. eBay remains a good way to get rid of things you don't need or find bargains on things you do need. eBay drop-off stores allow people who are inexperienced with Internet auctions or too busy to develop their own listings to sell items on the popular Web site. In addition to eBay, you can find a number of other auction sites on the Web. Traditional companies are even starting their own auction sites.

Although auction Web sites are excellent for matching buyers and sellers, they can present problems. Auction sites cannot always determine whether the people and companies listing products and services are legitimate. In addition, some Web sites have had illegal or questionable items offered. Many Web sites have an aggressive fraud investigation system to prevent and help prosecute fraudulent use of their sites. Even with these potential problems, the use of Web auction sites is expected to continue to grow rapidly.

Music, Radio, Video, and TV on the Internet

Music, radio, and video are hot growth areas on the Internet. Audio and video programs can be played on the Internet, or files can be downloaded for later use. Using music players and music formats such as MP3, discussed in Chapter 3, you can download music from the

eBay is a popular auction Web site.

Internet and listen to it anywhere using small, portable digital audio players. Subscribers to a music service such as Napster or Rhapsody can download an unlimited number of songs from the site, as long as they pay the annual subscription service fee. Pay-per-song services such as those offered by Apple iTunes and Amazon allow users to purchase songs for around $0.99 per song.[26] A music service called SpiralFrog supports free music downloads for users willing to provide marketing information and watch ads.[27]

Radio broadcasts are now available on the Internet. Entire audio books can also be downloaded for later listening using services such as Audible.com. Audible provides a subscription service that allows users to download one or more books per month that can then be played using PC media software and portable digital music players.

Video and TV are also becoming available on the Internet. One way to put TV programming on the Internet is to use the Internet Protocol Television (IPTV) protocol. With the potential of offering an almost unlimited number of programs, IPTV can serve a vast array of programming on specialty areas, such as yoga, vegetarian food, unusual sporting events, and news from a city or region of a state. Google offers a service that allows people to download selected television shows, movies, and other video. Some episodes of television shows cost users $1.99 to download. According to Jennifer Feikin, director of Google Video, "It's the biggest marketplace of content that was previously offline and is now brought online." Devices such as Apple's video-enabled iPod can be used to view the video content. Cell phones are being designed to receive and display Internet television and video.[28]

Other Web sites providing television content include Joost (*joost.com*), which offers over 20,000 free TV shows on over 400 channels.[29] Hulu streams popular television programming from broadcast and cable networks at *www.hulu.com*.[30]

A number of innovative devices let you record TV programs and view them at any time and place.[31] A California company called Sling Media (*www.slingmedia.com*), for example, offers a device that can broadcast any TV program coming into your home to a broadband Internet-connected PC.[32] Once on the Internet, you can watch the TV program at any time and place that has broadband Internet service. The device, called a Slingbox, costs about $250 and doesn't require monthly service fees.

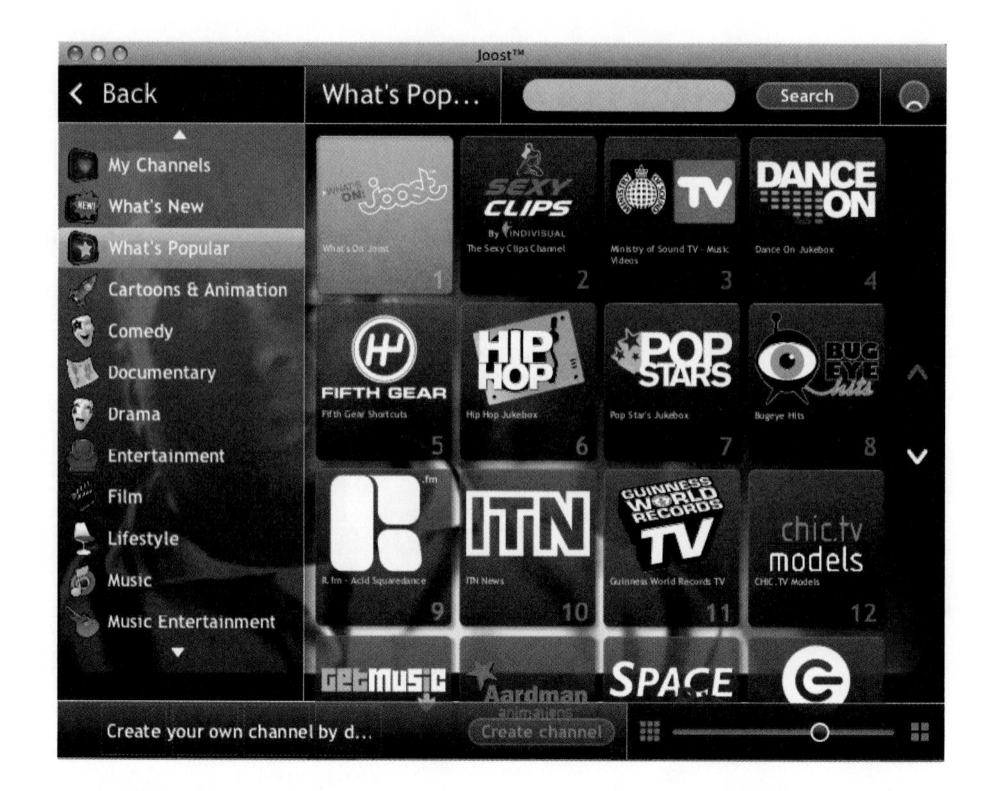

Internet TV service Joost offers 20,000 free TV shows on over 400 channels.

E-Books and Audio Books

Digital books, both in text and audio form, are growing in popularity thanks to appealing devices and services. Amazon's e-book reader called the Kindle jump-started the e-book market with its small form factor and high capacity.[33] The Kindle wirelessly connects to Amazon's e-book service using free wireless service from Sprint to download books, blogs, newspapers, and periodicals. It can store about 200 books.

Audio books have become more popular due to the popularity of the iPod and services like audible.com. Audio books are either read by a narrator or performed by actors. They may be abridged (consolidated and edited for audio format), or unabridged (read word for word from the book). Audio book services may allow you to purchase books individually, or sign up for a membership and receive a new book each month. Audio books can be transferred from PC to a portable device such as an iPod or Kindle.

Office on the Web

Having an Internet office with access to files and information can be critical for people who travel frequently or work at home. An Internet office is a Web site that contains files, phone numbers, e-mail addresses, an appointment calendar, and more. Using a standard Web browser, you can access important business information. An Internet office allows your desktop computer, phone books, appointment schedulers, and other important information to be with you wherever you are.

Many services and software products offer remote access to your files and programs over the Internet on the Web. As mentioned earlier, Microsoft and Google both support online document storage and sharing. Both companies and others also provide ways to access contact lists and calendar software and data online. For example, 37signals.com provides online project management, contact management, calendar, and group chat applications. Microsoft SharePoint provides businesses with collaborative workspaces and social computing tools to allow people at different locations to work on projects together. (See Figure 7.17.) Over time, increasing amounts of software, services, and storage will be available online through your Web browser.

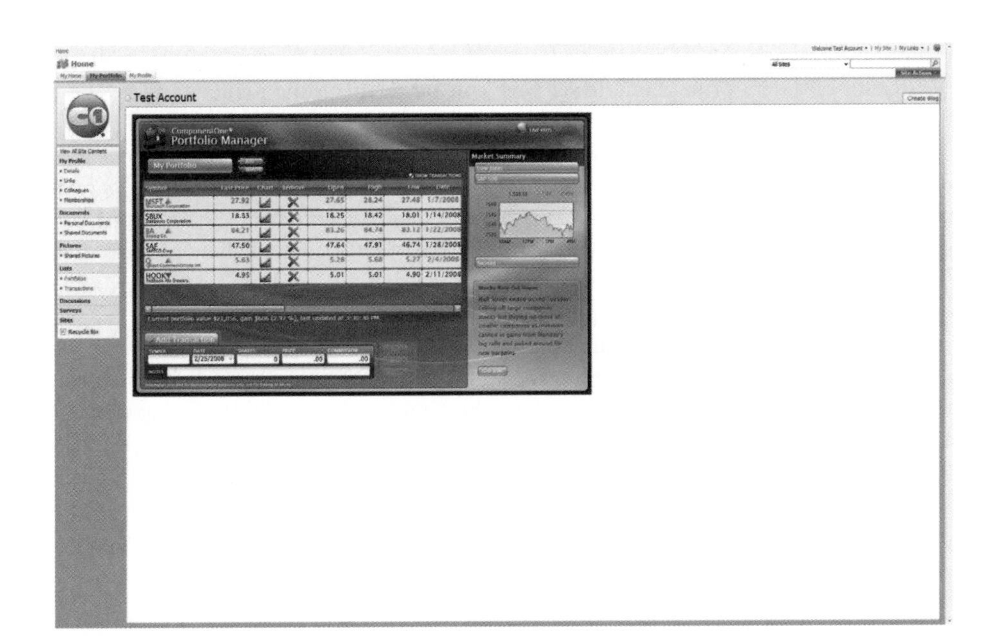

Figure 7.17

Microsoft SharePoint

Microsoft SharePoint provides a Web portal for sharing documents and information between organization members.

(Source: Courtesy of Microsoft Corporation.)

Internet Sites in Three Dimensions

Some Web sites offer three-dimensional views of places and products. For example, a 3-D Internet auto showroom allows people to select different views of a car, simulating the experience of walking around in a real auto showroom. (See Figure 7.18.) When looking at a 3-D real estate site on the Web, people can tour the property, go into different rooms, look at the kitchen appliances, and even take a virtual walk in the garden. Map Web sites and Internet-powered software like Windows Live Search and Google Earth provide views of cities and locations in a 3-D environment.[34] Second Life provides an entire 3-D virtual world for users to explore using avatars—characters within the world.[35][36]

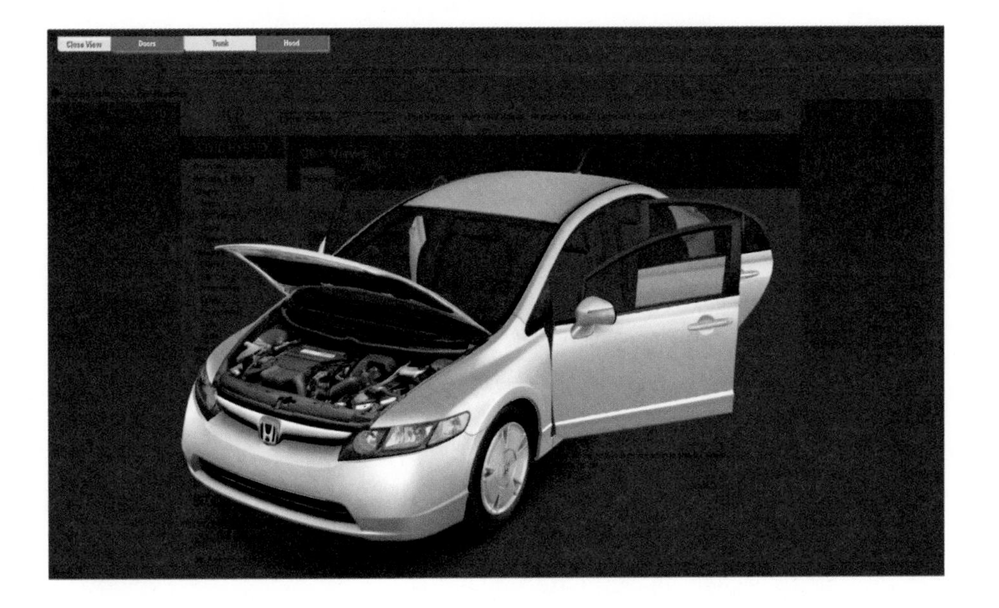

Figure 7.18

3-D Honda Hybrid

3-D graphics technologies allow consumers to examine merchandise from all angles.

(Source: American Honda Web site, *www.honda.com*.)

Other Internet Services and Applications

Other Internet services are constantly emerging. A vast amount of information is available over the Internet from libraries. Many articles that served as the basis of the sidebars, cases, and examples used throughout this book were obtained from university libraries online. Movies can be ordered and even delivered over the Internet. The Internet can provide critical information during times of disaster or terrorism. During a medical emergency, critical medical information can be transmitted over the Internet. People wanting to consolidate their credit card debt or to obtain lower payments on their existing home mortgages have turned to sites such as Quicken Loan, E-Loan, and LendingTree for help.

The Internet can also be used to translate words, sentences, or complete documents from one language into another. For example, Babel Fish Translation (*www.world.altavista.com*) and Free Translation (*www.freetranslation.com*) can translate a block of text from one language into another. Some search engines can translate Web pages and allow you to search for Web sites published in certain languages or countries. Clicking Language Tools on the home page of Google, for example, provides these capabilities.

The Internet also facilitates distance learning, which has dramatically increased in the last several years. Many colleges and universities now allow students to take courses without visiting campus. In fact, you might be taking this course online. MIT is offering all of its 1,800 courses free online.[37] Businesses are also taking advantage of distance learning through the Internet. Video cameras can be attached to computers and connected to the Internet. These Internet cameras can be used to conduct job interviews, hold group meetings with people around the world, monitor young children at daycare centers, check rental properties and second homes from a distance, and more.

INFORMATION SYSTEMS @ WORK

Chevron Takes to the Clouds

The importance of the Internet and Web is illustrated in the many applications discussed in this chapter. However, their importance may far exceed the sum of these applications. The Internet and the Web are quickly becoming equivalent with computing, especially with the rise of Web 2.0 technologies, which provides a platform for computing. This is evidenced by online applications such as Google Docs and Adobe Photoshop Express, and extends to the way businesses are managing their information systems. Take Chevron as an example.

Chevron Corporation is based in San Ramon, California and is one of the world's leading energy companies. It employs more than 55,000 people in 180 countries to produce and transport crude oil and natural gas and to refine, market, and distribute fuels and energy products. Chevron works with many suppliers, each of which provides the company with continuously updated editions of their products and services catalogs. Chevron negotiates the price of the items and services that they purchase from the suppliers. Managing supplier catalogs and negotiating prices while staying within a budget has been such a chore that Chevron, like most other large enterprises, decided to pay an outside information systems company millions of dollars annually to manage the responsibility.

The complicated process of negotiating prices with suppliers became unwieldy for Chevron and its catalog management services provider. Most negotiating was done through e-mail with spreadsheet attachments. The company hired by Chevron to manage the process acted as the middleman. Eventually, it became apparent that the system was ineffective and costly. Chevron searched for an alternative and found it in a new service from Ketera Technologies.

Ketera Technologies provides on-demand services that the company calls "spend management solutions." According to its Web site (*www.ketera.com*), Ketera provides "applications and services needed to control and reduce corporate spending at a low cost of ownership." The application and services that Ketera provides run on Ketera servers maintained by Ketera staff and delivered to clients over the Internet through a Web browser.

Using Ketera's services over the Web, Chevron suppliers can change prices, which Chevron executives then approve or adjust to stay in budget. Ketera's service provides numerous tools that Chevron can use to analyze and control its spending. Most importantly, since the service is provided on the Web, all involved parties have convenient access to a centralized system, allowing Chevron to negotiate with all suppliers from one location and using one tool.

Ketera is one of many software companies providing software services over the Internet. Recall from previous chapters that this method of delivering software is referred to as software as a service, or SaaS. Increasing numbers of businesses of all sizes are turning to SaaS. One compelling reason involves accounting. To stay profitable in a challenging economy, many companies are reducing capital expenditures. Software purchases are classified as capital expenditures, while Web- delivered software services are not. Web-delivered software is sold on a subscription basis, and classified by accountants as a maintenance expense, an area of the budget that is not being cut as much as capital expenses.

Internet security has also developed to the point where companies are more confident that the data being managed will remain private and secure as it travels the Internet. SaaS vendors are providing clients with adequate assurances that the data will remain safe in their hands.

Industry analysts expect that by 2011 the worldwide market for SaaS will grow from $6.3 billion to $19.3 billion. This estimate and the rapid increase of offerings of Web-based software for both personal and business use indicate a strong migration from private ownership of software and information systems to a model in which software and information systems are supplied as subscription services delivered over the Internet. This is what defines cloud computing. Cloud computing will make the Internet and Web more important to businesses and society than ever.

Discussion Questions

1. What benefits and dangers are presented by SaaS and cloud computing?
2. Name two reasons companies are turning to SaaS.

Critical Thinking Questions

1. Do you think SaaS suffers from any limitations because it is delivered over the Internet? What are they?
2. How might the increasing number of software services and data being delivered over the Internet affect its infrastructure?

Sources: Vara, Vauhini, "Web-Based Software Services Take Hold," *The Wall Street Journal*, May 15, 2007, Page B3; Baker, Stephen, "Google and the Wisdom of Clouds," *BusinessWeek*, December 24, 2007, Pages 49–55; Ketera Web site, *www.ketera.com*, accessed May 6, 2008.

INTRANETS AND EXTRANETS

intranet
An internal corporate network built using Internet and World Wide Web standards and technologies; used by employees to gain access to corporate information.

An **intranet** is an internal corporate network built using Internet and World Wide Web standards and technologies. Employees of an organization use it to gain access to corporate information. After getting their feet wet with public Web sites that promote company products and services, corporations are seizing the Web as a swift way to streamline—even transform—their organizations. These private networks use the infrastructure and standards of the Internet and the World Wide Web. Using an intranet offers one considerable advantage: Many people are already familiar with Internet technology, so they need little training to make effective use of their corporate intranet.

An intranet is an inexpensive yet powerful alternative to other forms of internal communication, including conventional computer networks. One of an intranet's most obvious virtues is its ability to reduce the need for paper. Because Web browsers run on any type of computer, the same electronic information can be viewed by any employee. That means that all sorts of documents (such as internal phone books, procedure manuals, training manuals, and requisition forms) can be inexpensively converted to electronic form on the Web and be constantly updated. An intranet provides employees with an easy and intuitive approach to accessing information that was previously difficult to obtain. For example, it is an ideal solution to providing information to a mobile salesforce that needs access to rapidly changing information.

extranet
A network based on Web technologies that links selected resources of a company's intranet with its customers, suppliers, or other business partners.

A growing number of companies offer limited access to their private corporate network for selected customers and suppliers. Such networks are referred to as extranets; they connect people who are external to the company. An **extranet** is a network that links selected resources of the intranet of a company with its customers, suppliers, or other business partners. Again, an extranet is built around Web technologies.

Security and performance concerns are different for an extranet than for a Web site or network-based intranet. User authentication and privacy are critical on an extranet so that information is protected. Obviously, the network must perform well to provide quick response to customers and suppliers. Table 7.8 summarizes the differences between users of the Internet, intranets, and extranets.

Table 7.8
Summary of Internet, Intranet, and Extranet Users

Type	Users	Need User ID and Password?
Internet	Anyone	No
Intranet	Employees	Yes
Extranet	Business partners	Yes

virtual private network (VPN)
A secure connection between two points on the Internet.

tunneling
The process by which VPNs transfer information by encapsulating traffic in IP packets over the Internet.

Secure intranet and extranet access applications usually require the use of a virtual private network (VPN). A **virtual private network (VPN)** is a secure connection between two points on the Internet. VPNs transfer information by encapsulating traffic in IP packets and sending the packets over the Internet, a practice called **tunneling**. Most VPNs are built and run by ISPs. Companies that use a VPN from an ISP have essentially outsourced their networks to save money on wide area network equipment and personnel.

NET ISSUES

The topics raised in this chapter apply not only to the Internet and intranets but also to LANs, private WANS, and every type of network. Control, access, hardware, and security problems affect all networks, so you should be familiar with the following issues:

- **Management issues.** Although the Internet is a huge, global network, it is managed at the local level; no centralized governing body controls the Internet. Preventing attacks is always an important management issue. Increasingly, states are proposing legislation to help collect sales tax from Internet sales.
- **Service and speed issues.** The growth in Internet traffic continues to be significant. Traffic volume on company intranets is growing even faster than the Internet. Companies setting up an Internet or intranet Web site often underestimate the amount of computing power and communications capacity they need to serve all the "hits" (requests for pages) they get from Web cruisers.
- **Privacy, fraud, security, and unauthorized Internet sites.** As use of the Internet grows, privacy, fraud, and security issues become even more important. People and companies are reluctant to embrace the Internet unless these issues are successfully addressed. Unauthorized and unwanted Internet sites are also problems some companies face. A competitor or an unhappy employee can create an Internet site with an address similar to a company's. When someone searches for information about the company, he or she might find an unauthorized site instead. While the business use of the Web has soared, online scams have put the brakes on some Internet commerce. Many Internet users have cut back on their Internet shopping and banking because of potential Internet scams and concerns about privacy and identity theft. In a business setting, the Web can also be a distraction to doing productive work. Although many businesses block certain Web sites at work, others monitor Internet usage. Workers have been fired for inappropriate or personal use of the Internet while on the job.

SUMMARY

Principle

The Internet is like many other technologies—it provides a wide range of services, some of which are effective and practical for use today, others that are still evolving, and still others that will fade away from lack of use.

The Internet started with ARPANET, a project sponsored by the U.S. Department of Defense (DoD). Today, the Internet is the world's largest computer network. Actually, it is a collection of interconnected networks, all freely exchanging information. The Internet transmits data from one computer (called a host) to another. The set of conventions used to pass packets from one host to another is known as the Internet Protocol (IP). Many other protocols are used with IP. The best known is the Transmission Control Protocol (TCP). TCP is so widely used that many people refer to the Internet protocol as TCP/IP, the combination of TCP and IP used by most Internet applications. Each computer on the Internet has an assigned address to identify it from other hosts, called its Uniform Resource Locator (URL).

People can connect to the Internet in several ways: via a LAN whose server is an Internet host, or via a dial-up connection, high-speed service, or wireless service. An Internet service provider is any company that provides access to the Internet. To use this type of connection, you must have an account with the service provider and software that allows a direct link via TCP/IP.

Principle

Originally developed as a document-management system, the World Wide Web is a hyperlink-based system that is easy to use for personal and business applications.

The Web is a collection of tens of millions of servers that work together as one in an Internet service providing information via hyperlink technology to billions of users worldwide. Thanks to the high-speed Internet circuits connecting them and hyperlink technology, users can jump between Web pages and servers effortlessly—creating the illusion of using one big computer. Because of its ability to handle multimedia objects and hypertext links between distributed objects, the Web is emerging as the most popular means of information access on the Internet today.

As a hyperlink-based system that uses the client/server model, the Web organizes Internet resources throughout the world into a series of linked files, called pages, accessed and viewed using Web client software, called a Web browser. Internet Explorer, Firefox, and Safari are three popular Web browsers. A collection of pages on one particular topic, accessed under one Web domain, is called a Web site.

Hypertext Markup Language (HTML) is the standard page description language for Web pages. The HTML tags let the browser know how to format the text: as a heading, as a list, or as body text, for example. HTML also indicates where images, sound, and other elements should be inserted. Some newer Web standards are gaining in popularity, including Extensible Markup Language (XML), Extensible Hypertext Markup Language (XHTML), Cascading Style Sheets (CSS), Dynamic HTML (DHMTL), and Wireless Markup Language (WML).

Web 2.0 refers to the Web as a computing platform that supports software applications and the sharing of information between users. Over the past few years, the Web has been changing from a one-directional resource where users find information to a two-directional resource where users find and share information. The Web has also grown in power to support complete software applications and is becoming a computing platform on its own. A rich Internet application (RIA) is software that has the functionality and complexity of traditional application software, but runs in a Web browser and does not require local installation. Java is an object-oriented programming language from Sun Microsystems based on the C++ programming language, which allows small programs, called applets, to be embedded within an HTML document.

Principle

Because use of the Internet and the World Wide Web is becoming universal in the business environment, management, service and speed, privacy, and security issues must continually be addressed and resolved.

Internet and Web applications include Web browsers; e-mail; career information and job searching; Telnet; FTP; Web logs (blogs); podcasts; Usenet and newsgroups; chat rooms; Internet phone; Internet video; content streaming; instant messaging; shopping on the Web; Web auctions; music, radio, and video; office on the Web; 3-D Internet sites; free software; and other applications.

You use a search engine to find information on the Web by specifying words that are key to a topic of interest, known as keywords. Search engines scour the Web with bots (automated programs) called spiders that follow all Web links in an attempt to catalog every Web page by topic.

You use e-mail to send messages. Various technologies are available for accessing and managing e-mail including online e-mail services, POP, and IMAP. The Internet also offers a vast amount of career and job search information.

Telnet and SSH enable you to log on to remote computers. You use FTP to transfer a file from another computer to your computer or vice versa. Web logs (blogs) are Internet sites that people and organizations can create and use to write about their observations, experiences, and opinions on a wide range of topics. A podcast is an audio broadcast over the Internet. Usenet supports newsgroups, which are online discussion groups focused on a particular topic. Chat rooms let you talk to dozens of people at one time, who can be located all over the world. You can also use Internet phone service to communicate with others around the world. Internet video enables people to conduct virtual meetings.

Online social networks provide Web-based tools for users to share information about themselves with others on the Web and to find, meet, and converse with other members. Media-sharing Web sites such as YouTube for video sharing and Flickr for photo sharing provide methods for members to store and share digital media files on the Web. Social bookmarking sites let Web users store, classify, share, and search Web bookmarks—also referred to as favorites. Content streaming is a method of transferring multimedia files over the Internet so that the data stream of voice and pictures plays continuously. Instant messaging allows people to communicate in real time using the Internet. Shopping on the Web is popular for a host of items and services. Web auctions match people looking for products and services with people selling these products and services. You can also use the Web to download and play music, listen to radio, and view video programs. With office on the Web, you can store important files and information on the Internet. When telecommuting or traveling, you can download these files and information or send them to other people. Some Internet sites are three-dimensional, allowing you to manipulate the site to see different views of products and images on the Internet. A wealth of free software and services is available through the Internet. Some of the free information, however, might be misleading or even false. Other Internet services include information about space exploration, fast information transfer, obtaining a home loan, and distance learning.

An intranet is an internal corporate network built using Internet and World Wide Web standards and products. Because Web browsers run on any type of computer, the same electronic information can be viewed by any employee. That means that all sorts of documents can be converted to electronic form on the Web and constantly be updated.

An extranet is a network that links selected resources of the intranet of a company with its customers, suppliers, or other business partners. It is also built around Web technologies. Security and performance concerns are different for an extranet than for a Web site or network-based intranet. User authentication and privacy are critical on an extranet. Obviously, the network must perform well to provide quick response to customers and suppliers.

Management issues, service, and speed affect all networks. No centralized governing body controls the Internet. Also, because the amount of Internet traffic is so large, service bottlenecks often occur. Privacy, fraud, and security issues must continually be addressed and resolved.

CHAPTER 7: SELF-ASSESSMENT TEST

The Internet is like many other new technologies—it provides a wide range of services, some of which are effective and practical for use today, others that are still evolving, and still others that will fade away from lack of use.

1. The ________________ was the ancestor of the Internet and was developed by the U.S. Department of Defense.
2. On the Internet, what enables traffic to flow from one network to another?
   a. Internet Protocol
   b. ARPANET
   c. Uniform Resource Locator
   d. LAN server
3. Each computer on the Internet has an address called the *Transmission Control Protocol*. True or False?
4. What organization is responsible for managing Internet addresses?
   a. Internet Corporation for Assigned Names and Numbers (ICANN)
   b. Internet Society (ISOC)
   c. Defense Advanced Research Projects Agency (DARPA)
   d. America Online (AOL)
5. A(n) ________________ is a company that provides people and organizations with access to the Internet.

Originally developed as a document-management system, the World Wide Web is a hyperlink-based system that is easy to use for personal and business applications.

6. A podcast is an online Web site that people can create and use to write about their observations, experiences, and opinions on a wide range of topics. True or False?
7. Which technology helps you easily specify the visual appearance of Web pages in a Web site?
   a. HTML
   b. XHTML
   c. XML
   d. CSS

8. ____________ refers to the Web as a computing platform that supports software applications and the sharing of information between users.

9. What is the standard page description language for Web pages?
   a. Home Page Language
   b. Hypermedia Language
   c. Java
   d. Hypertext Markup Language (HTML)

Because use of the Internet and the World Wide Web is becoming more universal in the business environment, management, service and speed, privacy, and security issues must continually be addressed and resolved.

10. Digg and del.icio.us are examples of ________________ Web sites.
    a. media sharing
    b. social network
    c. social bookmarking
    d. content streaming

11 A(n) __________________ is a network based on Web technology that links customers, suppliers, and others to the company.

12. An intranet is an internal corporate network built using Internet and World Wide Web standards and products. True or False?

CHAPTER 7: SELF-ASSESSMENT TEST ANSWERS

(1) ARPANET (2) a (3) False (4) a (5) Internet service provider (ISP) (6) False (7) d (8) Web 2.0 (9) d (10) c (11) extranet (12) True

REVIEW QUESTIONS

1. What is the Internet? Who uses it and why?
2. What is ARPANET?
3. Identify the features of the Internet that make it unlikely to stop working from a single point of failure. Why do you think the Internet has such a high degree of redundancy?
4. Explain the naming conventions used to identify Internet host computers.
5. What is a Web browser? Provide two examples.
6. Briefly describe three different ways to connect to the Internet. What are the advantages and disadvantages of each approach?
7. What is an Internet service provider? What services do they provide?
8. What are the advantages and disadvantages of e-mail?
9. What benefit does IMAP e-mail have over POP?
10. What is a podcast?
11. For what are Telnet and FTP used?
12. What is an Internet chat room?
13. What is content streaming?
14. What is instant messaging?
15. What is the Web? Is it another network like the Internet or a service that runs on the Internet?
16. What is a URL and how is it used?
17. What is an intranet? Provide three examples of the use of an intranet.
18. What is an extranet? How is it different from an intranet?
19. Describe at least three important Internet issues.

DISCUSSION QUESTIONS

1. Instant messaging is being widely used today. Describe how this technology could be used in a business setting. Are there any drawbacks or limitations to using instant messaging in a business setting?
2. Your company is about to develop a new Web site. Describe how you could use Web services for your site.
3. Why is it important to have an organization that manages IP addresses and domain names?
4. Describe how a company could use a blog and podcasting.
5. Briefly describe how the Internet phone service operates. Discuss the potential impact that this service could have on traditional telephone services and carriers.
6. Why is XML an important technology?
7. How do XHTML, CSS, and XML work together to create a Web page?
8. Identify three companies with which you are familiar that are using the Web to conduct business. Describe their use of the Web.
9. What is Voice over IP (VoIP), and how could it be used in a business setting?
10. What are the defining characteristics of a Web 2.0 site?
11. One of the key issues associated with the development of a Web site is getting people to visit it. If you were developing a Web site, how would you inform others about it

and make it interesting enough that they would return and tell others about it?

12. Downloading music, radio, and video programs from the Internet is easier than in the past, but some companies are still worried that people will illegally obtain copies of this programming without paying the artists and producers royalties. If you were an artist or producer, what would you do?
13. How could you use the Internet if you were a traveling salesperson?
14. Briefly summarize the differences in how the Internet, a company intranet, and an extranet are accessed and used.

PROBLEM-SOLVING EXERCISES

1. Do research on the Web to find several popular Web auction sites. After researching these sites, use a word processor to write a report on the advantages and potential problems of using a Web auction site to purchase a product or service. Also discuss the advantages and potential problems of selling a product or service on a Web auction site. How could you prevent scams on an auction Web site?
2. Develop a brief proposal for creating a business Web site. How could you use Web services to make creating and maintaining the Web site easier and less expensive? Develop a simple spreadsheet to analyze the income you need to cover your Web site and other business expenses.
3. Think of a business that you might like to establish. Use a word processor to define the business in terms of what product(s) or service(s) it provides, where it is located, and its name. Go to *www.godaddy.com* and find an appropriate domain name for your business that is not yet taken. Write a paragraph about your experience finding a name, and why you chose the name that you did.
4. You have been hired to research the use of a blog for a company. Develop a brief report on the advantages and disadvantages of using a blog to advertise corporate products and services. Using a graphics program, prepare a slide show to help you make a verbal presentation.

TEAM ACTIVITIES

1. With your teammates, identify a company that is making effective use of a company extranet. Find out all you can about its extranet. Try to speak with one or more of the customers or suppliers who use the extranet and ask what benefits it provides from their perspective.
2. Your group will use Web 2.0 sites to organize a social gathering. First choose a group name based on what type of social event you are planning. This could be an actual event that group members will attend such as "Pizza Extravaganza." Use Facebook to create a group page and use it to communicate with group members. Use the group page to establish who will be the group leader. Each member should use Google Calendar to post his or her activities for the week the event is to take place. Share your calendars with everyone in the group. The group leader should examine everyone's calendar to determine a date and time when everyone is available for the event. Create the event and invite the other group members using Google Calendar and Gmail. The leader should create a document using Google Docs that lists details of the event—the title, the purpose, activities on the agenda, food that will be available, the responsibilities of those attending, etc. Share the document with group members. Group members should share their ideas by editing the document. The group leader should judiciously decide which edits to keep and which to reject. Present your instructor with information to join your Facebook group and to view your calendars and Google doc. Write a summary of your experiences with this exercise.
3. Have each team member use a different search engine to find information about podcasting. Meet as a team and decide which search engine was the best for this task. Write a brief report to your instructor summarizing your findings.

WEB EXERCISES

1. This chapter covers a number of powerful Internet tools, including Internet phones, search engines, browsers, e-mail, newsgroups, Java, and intranets. Pick one of these topics and find more information on the Internet. You might be asked to develop a report or send an e-mail message to your instructor about what you found.

2. The Internet can be a powerful source of information about various industries and organizations. Locate several industry or organization Web sites. Which Web site is the best designed? Which one provides the most amount of information?
3. Research some of the potential disadvantages of using the Internet, such as privacy, fraud, or unauthorized Web sites. Write a brief report on what you found.
4. Set up an account on *www.twitter.com* and invite a few friends to join. Use Twitter to send messages to your friends on their cell phones, keeping everyone posted on what you are doing throughout the day. Write a review of the service to submit to your instructor.

CAREER EXERCISES

1. Use the Internet to explore starting salaries, benefits, and job descriptions for a career in developing or managing a Web site. Monster.com is a good place to start.
2. Describe how the Internet can be used on the job for two careers that interest you.

CASE STUDIES

Case One

The Best Online Brick-and-Mortar Retailer

Guess which brick-and-mortar retail business—that is, a business with a physical store—attracts the most customers to its Web site. Wal-Mart? Target? Best Buy? A recent study by Nielsen NetRatings revealed that J.C. Penney attracts more shoppers to its Web site than any other brick-and-mortar retailer. About as many people visit jcpenney.com as visit Amazon.com or eBay. For J.C. Penney, that's over 300,000 unique paying customers per month. What's their secret?

J.C. Penney knows how to create synergy between different avenues of sales. Synergy occurs when separate entities combine to create a greater effect than the sum of their separate effects. A common analogy is a peanut butter and jelly sandwich, which tastes better because of combined flavors. J.C. Penney's peanut butter has been its more than 1,000 department stores, and its jelly is its catalog business, the nation's largest. Through these two sales vehicles, J.C. Penney can provide the merchandise customers desire when they desire it.

The synergy between J.C. Penney's catalog and store occurs by each supporting the other to meet customers' needs. If an item is unavailable to a customer in the store, the customer is directed to the catalog desk, where he or she can browse through three times the amount of merchandise as is available in stock. By delivering catalogs to tens of thousands of households, J.C. Penney reaches customers that might not otherwise visit their stores.

Moving online was natural for this company because it had a long history of experience selling to customers remotely through its printed catalog. The Web provided a more powerful catalog for the retailer, one that reaches millions of potential customers. J.C. Penney integrated its Web presence with its in-store and catalog sales to create more synergy and more retail power.

At the turn of the millennium, J.C. Penney's stockholders were concerned about the future of the company. In the late 1990s, Penney's catalog revenues peaked at about $4 billion and started to decline. Catalog sales continued declining over time until in 2006 they reached $1.7 billion. In that same period, J.C. Penney's online sales increased to $1.5 billion in 2007. The total revenue for J.C. Penney in 2007 was $19.9 billion. While the catalog sales have continued to decline, the combined catalog and Internet sales as well as total sales for the business have steadily increased over the past four years.

This indicates that the synergy between Internet and in-store sales is strong. JCPenney.com is working to lure customers into the brick-and-mortar stores. Like the catalog, JCPenney.com lists three times as much merchandise as is stocked in the stores. Computer terminals are provided at Penney's 35,000 check-out registers to allow in-store customers to shop online for items that they could not find in the store. Listing so many items online provided J.C. Penney with a low-cost mechanism for selling slow-moving items. Online customers can check the availability of items in local stores, allowing them to find what they like from the comfort of their own home and pick it up locally the same day—without incurring shipping charges.

Penney's online sales accounted for 6 percent of total sales compared with 4 percent for Sears, and only 1 percent for Wal-Mart. Plus, Penney's online customers are considerably younger than its in-store customers, enabling the company to reach out to the next generation.

Discussion Questions

1. What methods does J.C. Penney use to create synergy between its Web site and brick-and-mortar store?
2. Why was J.C. Penney more adept at moving to the Web than other retail businesses?

Critical Thinking Questions

1. If J.C. Penney's online sales account for only 6 percent of total sales, why is it considered so valuable?
2. What other ways might J.C. Penney take advantage of its Web site to boost its total sales?

Sources: Berner, Robert, "J.C. Penney Gets the Net," *BusinessWeek*, May 7, 2007, page 70; J.C. Penney Corporate Web site, *www.jcpenney.net*, accessed May 7, 2008.

Case Two

Procter & Gamble Implement Enterprise 2.0

Procter & Gamble (P&G) owns a large portfolio of familiar brands such as Pampers, Tide, Bounty, Folgers, Pringles, Charmin, and Crest. P&G operates in more than 80 countries worldwide, with net sales increasing continuously over the past ten years to over $76 billion in 2007.

Procter & Gamble's CEO, A.G. Lafley, believes in communication and collaboration. He is pushing P&G IT Innovation Manager Joe Schueller to find more effective and innovative ways for P&G's 138,000 employees to collaborate online. Naturally, Schueller looked immediately to Web 2.0 technologies for ideas. When applied to an enterprise, Web 2.0 technologies are referred to as Enterprise 2.0.

Schueller is not a fan of e-mail. He sees it as a barrier to employees' use of more effective means of communication. Replying to all recipients of a message ends up wasting the time of people who do not need to receive, read, and respond to the message. Instead Schueller has equipped P&G employees with easy access to a corporate blog. For some types of group communications, Schueller finds blogs the ideal tool. Information is not forced on people. Those interested can follow the blog and post comments to add to the dialog.

Schueller is harnessing the power of the wiki as a content and knowledge management system. Members of the organization who have valuable knowledge about P&G topics can post articles and advice. That helps corporate knowledge stay within the company, even when knowledgeable employees leave.

P&G banked on Microsoft products to provide most of its Enterprise 2.0 functionality. Microsoft Live Communications Server provides instant messaging, unified communications, and presence—the ability to access communications services from any location. Live Meeting provides Web conferencing, and SharePoint provides a platform for content management and collaboration. Roughly 80,000 P&G employees use corporate instant messaging tools.

Besides using Microsoft products, P&G also uses software and tools from other vendors for its Enterprise 2.0 investments. For example, P&G uses a product from Connectbeam that works with Google search tools to allow employees to share bookmarks and tag articles, pages, and documents with descriptive words to make information easier to find. P&G has launched a corporate social networking site so that employees can let others know who they are and in which areas of corporate activities they are involved. The goal is to encourage employees to easily find others with expert knowledge. All of these Enterprise 2.0 applications are accessed through a unified portal that also includes RSS feeds of business news.

P&G is serving as inspiration to other companies who are developing an interest in Enterprise 2.0. Information systems departments see Web 2.0 technologies as a chance to provide real value to the organization. Bank of America, Boeing, the CIA, FedEx, Morgan Stanley, and Pfizer are examining Schueller's example. Motorola has also invested in Enterprise 2.0, with an intranet that includes 4,400 blogs and 4,200 wiki pages.

Discussion Questions

1. What qualities of Web 2.0 applications are appealing for enterprise use?
2. Why might a company not want to use Web 2.0 applications?

Critical Thinking Questions

1. How can each of the five Enterprise 2.0 applications used by P&G help its employees be more effective and efficient?
2. Compare and contrast e-mail, IM, and blogs as tools for effective communications.

Sources: Hoover, Nicholas, "Beyond E-Mail," *Information Week*, June 25, 2007, pages 29-30; Procter and Gamble Corporate Web site, *www.pg.com/en_US/index.jhtml*, accessed May 7, 2008.

Questions for Web Case

See the Web site for this book to read about the Whitmann Price Consulting case for this chapter. Following are questions concerning this Web case.

Whitmann Price Consulting: The Internet, Intranets, and Extranets

Discussion Questions

1. Why do you think it is easiest and most economical to develop custom-designed applications using Web standards?
2. What additional security concerns arise when providing access to private information over a wireless public network?

Critical Thinking Questions

1. In what situations might Whitmann Price consider developing an extranet?
2. After the Advanced Mobile Communications and Information System is up and running, what would happen if the organization decided to switch to Palm Treo devices to replace the BlackBerries? The Palm Treo has many of the same features as a BlackBerry, including a Web browser. Do you think Whitmann Price custom-designed applications would still work on the new devices? Why or why not?

NOTES

Sources for the opening vignette: IBM Staff, "Lamborghini accelerates time-to-value with IBM Lotus and WebSphere technologies," IBM Success Stories, February 19, 2008, *www-01.ibm.com/software/success/cssdb.nsf/CS/STRD- 7BYLVZ?OpenDocument*; Lamborghini Web site, *www.lamborghini.com*, accessed May 6, 2008.

1 Internet Usage World Stats Web site, *www.internetworldstats.com*, accessed April 1, 2008.
2 Zhao, Michael, "60-Mile Wi-Fi," *Forbes*, April 9, 2007, pages 76-78.
3 Hafner, Katie, *Where Wizards Stay Up Late: The Origins of the Internet*, Touchstone, New York, 1996.
4 Internet2 Web site, *www.internet2.edu*, accessed April 1, 2008.
5 National LambdaRail Web site, *www.nlr.net*, accessed April 1, 2008.
6 Swibel, Matthew, "Fly the Connected Skies," *Forbes*, November 26, 2007, page 56.
7 Worthen, Ben, "Web Surfing on iPhone Erases Doubts of Mobile Devices' Future Online Role," *Wall Street Journal*, December 11, 2007, Business Technology section, page 84.
8 Case, Loyd, "Intel Launches Low-Power Atom Processor," *PC Magazine*, April 2, 2008, *www.pcmag.com/article2/0,2817,2280892,00.asp*.
9 Mossberg, Walter, "Desktop Modules Help to Personalize Data, Cut Through Clutter," *The Wall Street Journal*, February 1, 2007, Marketplace section, page B1.
10 Neville, Jeffrey, "Web 2.0's Wild Blue Yonder," *Information Week*, January 8, 2007, page 45.
11 Larkin, Erik, "Back Up Your Files Online Without Even Trying," *PC World*, May 2007, page 112.
12 Claburn, Thomas, "Mashups Made Easy," *InformationWeek*, February 12, 2007, page 14.
13 Spanbauer, Scott, "Advanced Google: Search Faster, Find More," *PC World*, February 2008, pages 128-130.
14 Delaney, Kevin, "How Search-Engine Rules Cause Sites to Go Missing," *The Wall Street Journal*, page B1.
15 SEO staff, "Case Study: Tax Engine," SEO Case Studies, *www.seo.com/clients/tax-engine-seo-case-study*,accessed May 4, 2008.
16 Gomes, Lee, "Forget the Articles, Best Wikipedia Read Is Its Discussions," *Wall Street Journal*, August 15, 2007, Marketplace section, page B1.
17 Vascellaro, Jessica, "RIM Upgrades Email for BlackBerry Users," *Wall Street Journal*, January 23, 2008, Technology section, page B5.
18 Hoover, Nicholas, "More E-Mail, More Problems," *InformationWeek*, January 22, 2007, pages 43-47.
19 Buckman, Rebecca, "Email's Friendly Fire," *Wall Street Journal*, November 27, 2007, Marketplace section, page B1.
20 Lyons, Daniel, "Easy Blogging," *Forbes*, April 9, 2007, pages 56–57.
21 Savell, Lawrence, "Blogger Beware!" *Computerworld*, September 24, 2007, pages 32–34.
22 Enrich, David, "Turning an Online Community into a Business," *The Wall Street Journal*, page B8.
23 Heher, Ashley, "Social-networking site to keep U.S. spies in touch," *Rocky Mountain News*, September 10, 2007, Business section, page 6.
24 Gomes, Lee, "Web Is Now So Filled with Idle Chat, It's Almost Like Phoning," *The Wall Street Journal*, July 11, 2007, page B1.
25 Armstrong, Larry, "An Idea That Really Clicked," *Business Week*, January 10, 2008, *www.businessweek.com/magazine/content/08_03/b4067202218875.htm?chan=magazine+channel_in+depth+--+second+careers*.
26 Smith, Ethan and Vara, Vauhini, "Music Service from Amazon Takes on iTunes," *Wall Street Journal*, May 17, 2007, Personal Journal section, page D1.
27 Coyle, Jake, "Site allows free music downloads," *Rocky Mountain News*, September 24, 2007, Business section page 4.
28 Yuan, Li, "Cellphone Video Gets on the Beam," *The Wall Street Journal*, January 4, 2007, Technology Journal section, page B3.
29 Karnitschnig, Matthew, "Viacom Charts New Course Online," *The Wall Street Journal*, February 20, 2007, page A3.
30 Hardy, Quentin, "Better Than YouTube," *Forbes*, May 21, 2007, page 72.
31 Grant, Peter, "Find It on the Web, Watch It on TV," *The Wall Street Journal*, January 3, 2007, Marketplace section, page B1.
32 Cassavoy, Liane, "Slingbox Keeps You in Touch With Your TV," *PCWorld*, January 2007, page 78.
33 Perenson, Melissa, "Amazon Kindles Interest in E-Books," *PC World*, February 2008, page 64.
34 O'Reilly, Dennis, "Windows Live Search Gains 3D Map Views," *PCWorld*, February, 2007, page 76.
35 Kirkpatrick, David, "It's Not a Game," *Fortune*, February 5, 2007, pages 56–62.
36 Tynan, Dan, "Traveling the Web's Third Dimension," *PCWorld*, July 2007, page 49.
37 Reuters staff, "MIT to offer its courses free online by year end," Reuters, March 9, 2007, *www.reuters.com/article/domesticNews/idUSN0927676520070310*.

PART

3

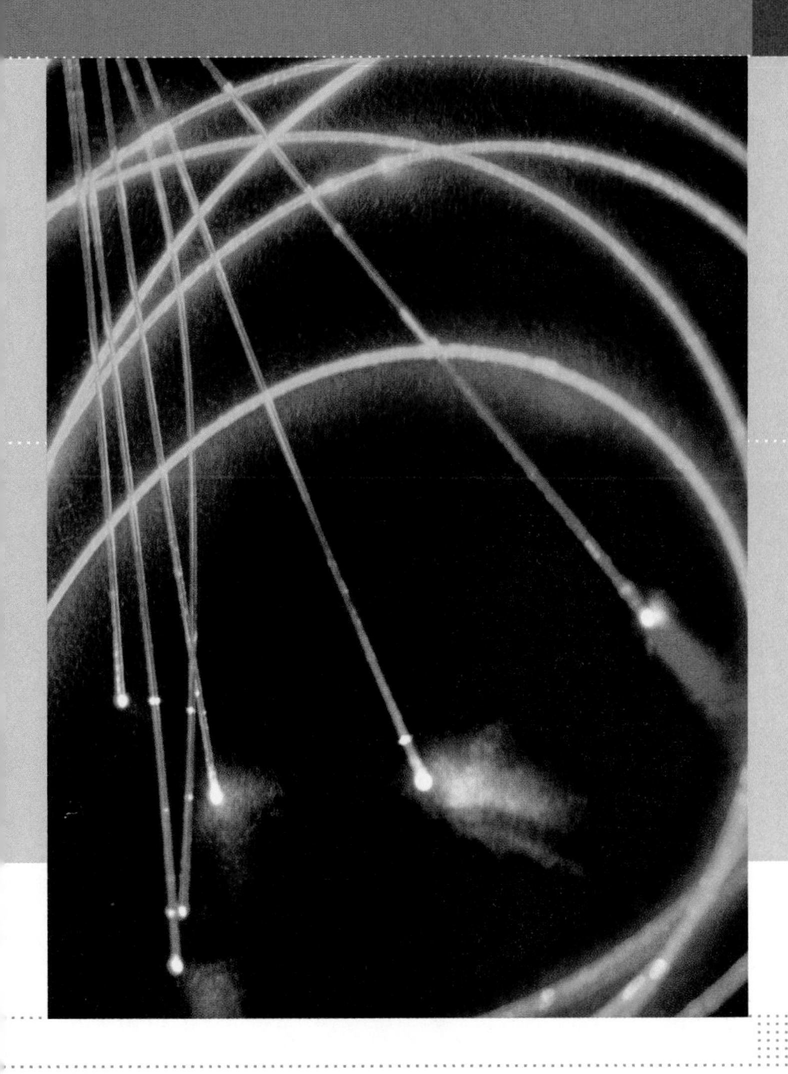

Business Information Systems

CHAPTER • 8 • Electronic and Mobile Commerce

PRINCIPLES	LEARNING OBJECTIVES
▪ Electronic commerce and mobile commerce are evolving, providing new ways of conducting business that present both opportunities for improvement and potential problems.	▪ Describe the current status of various forms of e-commerce, including B2B, B2C, C2C, and e-Government. ▪ Outline a multistage purchasing model that describes how e-commerce works. ▪ Define m-commerce and identify some of its unique challenges.
▪ E-commerce and m-commerce can be used in many innovative ways to improve the operations of an organization.	▪ Identify several e-commerce and m-commerce applications. ▪ Identify several advantages associated with the use of e-commerce and m-commerce.
▪ Although e-commerce and m-commerce offer many advantages, users must be aware of and protect themselves from many threats associated with use of this technology.	▪ Identify the major issues that represent significant threats to the continued growth of e-commerce and m-commerce.
▪ Organizations must define and execute a strategy to be successful in e-commerce and m-commerce.	▪ Outline the key components of a successful e-commerce and m-commerce strategy.
▪ E-commerce and m-commerce require the careful planning and integration of a number of technology infrastructure components.	▪ Identify the key components of technology infrastructure that must be in place for e-commerce and m-commerce to work. ▪ Discuss the key features of the electronic payment systems needed to support e-commerce and m-commerce.

Information Systems in the Global Economy Staples, United States

Staples Upgrades E-Commerce System to Increase Conversion Rate

Staples Inc. created the first office supply superstore in 1986 and has grown to over 2,000 stores in 22 countries. Staples has a reputation for using technology and information systems, both in the store and on the Internet, to provide customers with easy access to the office supplies that they need. After investing heavily in online sales, Staples has become the second largest Internet retailer after Amazon.com. Staples e-commerce sales total over $5 billion annually, nearly one-third of its total sales.

Staples e-commerce sales include selling online to independent consumers, called business-to-consumer e-commerce, or B2C, and selling to businesses at special bulk rates, called business-to-business e-commerce or B2B. Staples provides two Web sites to cater to its two types of customers: Staples.com for B2C home office and small businesses and StaplesLink.com for B2B larger businesses. According to IBM, who works with Staples in developing their e-commerce technologies, both e-commerce channels figure prominently in the company's long-term growth strategy.

Recently Staples decided to invest in its B2C site so it could better support the rapidly changing business strategies that make Staples a market leader. Staples also needed its Web site to accommodate surges in customer volume without any loss in performance. Staples knows that reliability and performance are foundational requirements for an e-commerce Web site to succeed. The ability to execute online business initiatives quickly gives a company an advantage over competitors. Due to complexity and functional limitations in the information systems, Staples.com was falling short of these requirements.

Staples worked with consultants from IBM to upgrade its Staples.com hardware, software, and overall information systems. Powerful new Web servers were installed that were more efficient and scalable so that additional power could be added as needed. IBM WebSphere Commerce software was a key component in creating a new e-commerce system that is stable and can manage customer transaction data more efficiently. The new system works seamlessly with Staples back-end systems for unified database management.

Staples views its new e-commerce system as a "foundation of a new way of interacting with its customers," according to an IBM case study. The company is using the system to create a unique online shopping experience for its customers, which is a central reason it now leads in the market. The new Staples.com provides a personalized and custom-designed online environment for its customers. Staples believes that allowing customers to quickly find items that suit their unique needs is crucial for customer retention. This is the philosophy behind Staples "easy" marketing strategy.

Staples conducted thorough marketing research to find out what its online customers liked and disliked about its Web site services. The results yielded ideas for new systems that could make customer's lives easier. Staples developed a new service called "Easy Reorder" that analyzes a customer's order history, looking for patterns, and creates an inventory list that is updated with each order. Another system named "Easy Rebate" simplifies the process for claiming product rebates.

The investment in new e-commerce systems has provided Staples with significant returns. An important statistic in e-commerce is the conversion rate—the share of online shoppers that start by browsing and end by buying. Since the system upgrade, the Staples.com conversion rate has improved by 60 percent. Staples.com is also much more stable than it was earlier. When it experienced a surge of 9,000 orders in one hour on the day after Thanksgiving—the so-called Black Friday—it suffered no degradation in performance.

Staples realizes that online competition poses a serious threat to its market dominance. Selling online is no longer considered an accessory to a brick-and-mortar business, but has become a major sales channel that can make or break a business. E-commerce tactics and strategies have become an important consideration in meeting a company's primary goals and objectives. Staples and most other large corporations are engaged in serious e-commerce battles online to gain or maintain rank in their respective markets.

As you read this chapter, consider the following:

- What advantages do e-commerce and m-commerce offer sellers and vendors over traditional shopping venues?
- What are the limitations of m-commerce and e-commerce? What doesn't sell well online, and why are some shoppers uncomfortable shopping online?

Why Learn About Electronic and Mobile Commerce?

Electronic and mobile commerce have transformed many areas of our lives and careers. One fundamental change has been the manner in which companies interact with their suppliers, customers, government agencies, and other business partners. As a result, most organizations today have or are considering setting up business on the Internet. To be successful, all members of the organization need to participate in that effort. As a sales or marketing manager, you will be expected to help define your firm's e-commerce business model. Customer service employees can expect to participate in the development and operation of their firm's Web site. As a human resource or public relations manager, you will likely be asked to provide Web site content for use by potential employees and investors. Analysts in finance need to know how to measure the business impact of their firm's Web operations and how to compare that to competitors' efforts. Clearly, as an employee in today's organization, you must understand what the potential role of e-commerce is, how to capitalize on its many opportunities, and how to avoid its pitfalls. The emergence of m-commerce adds an exciting new dimension to these opportunities and challenges. This chapter begins by providing a brief overview of the dynamic world of e-commerce and defining its various components.

AN INTRODUCTION TO ELECTRONIC COMMERCE

electronic commerce
Conducting business activities (e.g., distribution, buying, selling, marketing, and servicing of products or services) electronically over computer networks such as the Internet, extranets, and corporate networks.

Electronic commerce is the conducting of business activities (e.g., distribution, buying, selling, marketing, and servicing of products or services) electronically over computer networks such as the Internet, extranets, and corporate networks. Business activities that are strong candidates for conversion to e-commerce are paper-based, time-consuming, and inconvenient activities for customers. Thus, some of the first business processes that companies converted to an e-commerce model were those related to buying and selling. For example, after Cisco Systems, the maker of Internet routers and other telecommunications equipment, put its procurement operation online, the company reported that it halved cycle times and saved an additional $170 million in material and labor costs. Similarly, Charles Schwab & Co. slashed transaction costs by more than half by shifting brokerage transactions from traditional channels such as retail and phone centers to the Internet.

Business-to-Business (B2B) E-Commerce

business-to-business (B2B) e-commerce
A subset of e-commerce where all the participants are organizations.

Business-to-business (B2B) e-commerce is a subset of e-commerce where all the participants are organizations. B2B e-commerce is a useful tool for connecting business partners in a virtual supply chain to cut resupply times and reduce costs. Although the business-to-consumer market grabs more of the news headlines, the B2B market is considerably larger and is growing more rapidly. As early as 2003, over 80 percent of U.S. companies had already experimented with some form of B2B online procurement.[1]

Covisint operates a Web portal that supports B2B by performing data translations and code conversions to enable auto makers and parts suppliers to collaborate on orders, scheduling, shipping, and other manufacturing-related tasks. Covisint is expanding its data translation and collaboration services into the healthcare industry to enable sharing of patient care data among healthcare providers and insurance companies.[2]

Business-to-Consumer (B2C) E-Commerce

Early **business-to-consumer (B2C) e-commerce** pioneers competed with the traditional "brick-and-mortar" retailers in an industry selling their products directly to consumers. For example, in 1995, upstart Amazon.com challenged well-established booksellers Waldenbooks and Barnes and Noble. Although Amazon did not become profitable until 2003, the firm has grown from selling only books on a U.S.-based Web site to selling a wide variety of products (including apparel, CDs, DVDs, home and garden supplies, and consumer electronic devices) from international Web sites in Canada, China, France, Germany, Japan, and the United Kingdom. Although it is estimated that B2C e-commerce represents only about 3.4 percent of retail sales in the U.S., the rate of growth of online purchases is three times faster than the growth in total retail sales.[3] One reason for the rapid growth is that shoppers find that many goods and services are cheaper when purchased via the Web, including stocks, books, newspapers, airline tickets, and hotel rooms. They can also compare information about automobiles, cruises, loans, insurance, and home prices to find better values.

business-to-consumer (B2C) e-commerce
A form of e-commerce in which customers deal directly with an organization and avoid intermediaries.

More than just a tool for placing orders, the Internet is an extremely useful way to compare prices, features, and value. Internet shoppers can, for example, unleash shopping bots or access sites such as eBay Shopping.com, Google Froogle, Shopzilla, PriceGrabber, Yahoo! Shopping, or Excite to browse the Internet and obtain lists of items, prices, and merchants. Yahoo! is adding what it calls "social commerce" to its Web site by creating a new section of Yahoo! where users can go to see only those products that have been reviewed and listed by other shoppers. As mentioned in Chapter 7, bots are software programs that can follow a user's instructions; they can also be used for search and identification.

By using B2C e-commerce to sell directly to consumers, producers or providers of consumer products can eliminate the middlemen, or intermediaries, between them and the consumer. In many cases, this squeezes costs and inefficiencies out of the supply chain and can lead to higher profits and lower prices for consumers. The elimination of intermediate organizations between the producer and the consumer is called *disintermediation.*

Dell is an example of a manufacturer that has successfully embraced this model to achieve a strong competitive advantage. People can specify a unique computer online, and Dell assembles the components and ships the computer directly to the consumer within five days.

Many retailers have elected to increase their sales by adding a Web site component to their operations. For example, American Eagle Outfitters launched a B2C Web site for Martin + OSA, its brand targeting 28- to 40-year old men and women. Says Laura Dubin-Wander, president of Martin + Osa: "We're excited to introduce Martin + Osa as a global brand through our e-commerce Web site. Free shipping and returns, along with unique shopping tools, give customers a world-class online shopping experience that's both frictionless and fun."[4]

Consumer-to-Consumer (C2C) E-Commerce

Consumer-to-consumer (C2C) e-commerce is a subset of e-commerce that involves consumers selling directly to other consumers. eBay is an example of a C2C e-commerce site; customers buy and sell items directly to each other through the site. Founded in 1995, eBay has become one of the most popular Web sites in the world; in 2007, 2.3 billion items were listed for sale and 276 million registered users bought and sold items valued at more than $57 billion.[5]

consumer-to-consumer (C2C) e-commerce
A subset of e-commerce that involves consumers selling directly to other consumers.

Many C2C sites are on the Web, with some of the more popular being Bidzcom, Craigslist, eBid, ePier, Ibidfree, Ubid, and Tradus. The growth of C2C is responsible for reducing the use of the classified pages of a newspaper to advertise and sell personal items. Many people make a living out of selling items on auction Web sites.

C2C is highly popular among college students because they represent a large community of low-income people in the same geographical region who watch for values. Universities often set up Web sites for students to sell textbooks and other items to other students. EachNet.com trains students on how to open online stores in monthly promotions in universities across China. Students are the most active traders, though they have low average buying power. Still, "it indicates the huge market potential out there when the young users grow up and are able to pay more," according to Song Xing, an analyst with Analysys, a global telecommunications consultancy and research firm.[6]

e-Government

e-Government
The use of information and communications technology to simplify the sharing of information, speed formerly paper-based processes, and improve the relationship between citizens and government.

e-Government is the use of information and communications technology to simplify the sharing of information, speed formerly paper-based processes, and improve the relationship between citizens and government. Government-to-consumer (G2C), government-to-business (G2B), and government-to-government (G2G) are all forms of e-Government, each with different applications.

Citizens can use G2C applications to submit their state and federal tax returns online, renew auto licenses, apply for student loans, and make campaign contributions. Information about the 2008 economic stimulus payments that were sent to over 130 million taxpayers was available on the IRS Web site for months before the rebates were mailed out.

G2B applications support the purchase of materials and services from private industry by government procurement offices, enable firms to bid on government contracts, and help businesses receive current government regulations related to their operations. Business.gov allows businesses to access information about laws and regulations and relevant forms needed to comply with federal requirements for their business.

G2G applications are designed to improve communications among the various levels of government. For example, the E-Vital initiative establishes common electronic processes for federal and state agencies to collect, process, analyze, verify, and share death record information. Geospatial One-Stop's Web portal, GeoData.gov, makes it easier, faster, and less expensive to find, share, and access geospatial information for all levels of government.

The next section describes a basic model that supports products for purchase via e-commerce methods.

Multistage Model for E-Commerce

A successful e-commerce system must address the many stages that consumers experience in the sales life cycle. At the heart of any e-commerce system is the user's ability to search for and identify items for sale; select those items and negotiate prices, terms of payment, and delivery date; send an order to the vendor to purchase the items; pay for the product or service; obtain product delivery; and receive after-sales support. Figure 8.1 shows how e-commerce can support each of these stages. Product delivery can involve tangible goods delivered in a traditional form (e.g., clothing delivered via a package service) or goods and services delivered electronically (e.g., software downloaded over the Internet).

Search and Identification

An employee ordering parts for a storeroom at a manufacturing plant would follow the steps shown in Figure 8.1. Such a storeroom stocks a wide range of office supplies, spare parts, and maintenance supplies. The employee prepares a list of needed items—for example, fasteners, piping, and plastic tubing. Typically, for each item carried in the storeroom, a corporate buyer has already identified a preferred supplier based on the vendor's price competitiveness, level of service, quality of products, and speed of delivery. The employee then logs on to the Internet and goes to the Web site of the preferred supplier.

From the supplier's home page, the employee can access a product catalog and browse until finding the items that meet the storeroom's specifications. The employee fills out a request-for-quotation form by entering the item codes and quantities needed. When the

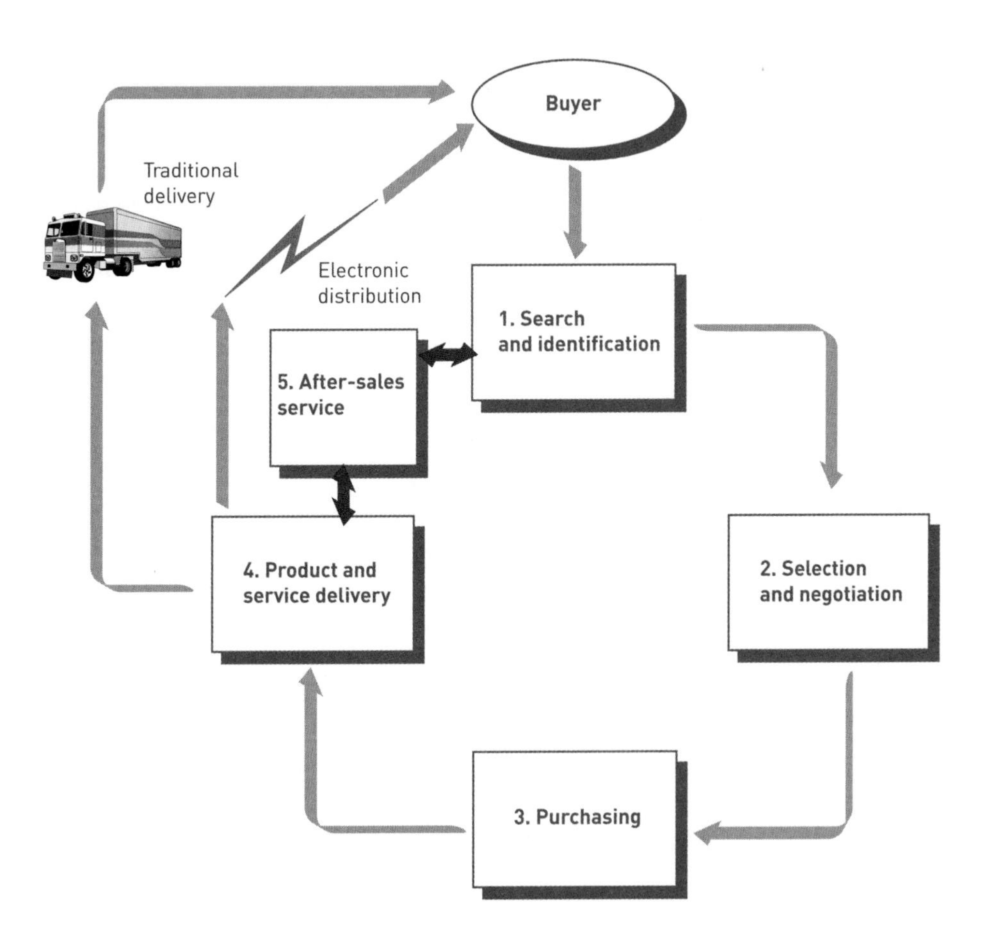

Figure 8.1

Multistage Model for E-Commerce (B2B and B2C)

employee completes the quotation form, the supplier's Web application prices the order with the most current prices and shows the additional cost for various forms of delivery—overnight, within two working days, or the next week. The employee might elect to visit other suppliers' Web home pages and repeat this process to search for additional items or obtain competing prices for the same items.

Selection and Negotiation

After the price quotations have been received from each supplier, the employee examines them and indicates, by clicking the request-for-quotation form, which items to order from a given supplier. The employee also specifies the desired delivery date. This data is used as input into the supplier's order-processing TPS. In addition to price, an item's quality and the supplier's service and speed of delivery can be important in the selection and negotiation process.

B2B e-commerce systems need to support negotiation between a buyer and the selected seller over the final price, delivery date, delivery costs, and any extra charges. However, this is not a fundamental requirement of most B2C systems, which offer their products for sale on a "take-it-or-leave-it basis."

Purchasing Products and Services Electronically

The employee completes the purchase order specifying the final agreed-to terms and prices by sending a completed electronic form to the supplier. Complications can arise in paying for the products. Typically, a corporate buyer who makes several purchases from the supplier each year has established credit with the supplier in advance, and all purchases are billed to a corporate account. But when individual consumers make their first, and perhaps only, purchase from the supplier, additional safeguards and measures are required. Part of the purchase transaction can involve the customer providing a credit card number. Another approach to paying for goods and services purchased over the Internet is using electronic money, which can be exchanged for hard cash, as discussed later in the chapter.

The Department of Education and Training for Victoria, Australia, chose three primary suppliers for desktop computers to be used in Victorian government schools. Staff can purchase products directly from these preferred suppliers with no need for requesting separate price quotes because the terms of purchase have already been negotiated. Staff need only to download negotiated price lists from a Web site and complete online purchase orders to order equipment. This process ensures competitive pricing from financially viable providers who have agreed to provide three-year, on-site warranty of equipment evaluated to be technically cost effective. It also eliminates days or weeks of delay in completing necessary paperwork and obtaining approvals.[7]

Product and Service Delivery

Electronic distribution can be used to download software, music, pictures, video, and written material through the Internet faster and for less expense than shipping the items via a package delivery service. Most products cannot be delivered over the Internet, so they are delivered in a variety of other ways: overnight carrier, regular mail service, truck, or rail. In some cases, the customer might elect to drive to the supplier and pick up the product.

Many manufacturers and retailers have outsourced the physical logistics of delivering merchandise to cybershoppers—the storing, packing, shipping, and tracking of products. To provide this service, DHL, Federal Express, United Parcel Service, and other delivery firms have developed software tools and interfaces that directly link customer ordering, manufacturing, and inventory systems with their own system of highly automated warehouses, call centers, and worldwide shipping networks. The goal is to make the transfer of all information and inventory—from the manufacturer to the delivery firm to the consumer—fast and simple.

For example, when a customer orders a printer at the Hewlett-Packard (HP) Web site, that order actually goes to FedEx, which stocks all the products that HP sells online at a dedicated e-distribution facility in Memphis, Tennessee, a major FedEx shipping hub. FedEx ships the order, which triggers an e-mail notification to the customer that the printer is on its way and an inventory notice to HP that the FedEx warehouse now has one less printer in stock (see Figure 8.2). For product returns, HP enters return information into its own system, which is linked to FedEx. This signals a FedEx courier to pick up the unwanted item at the customer's house or business. Customers don't need to fill out shipping labels or package the item. Instead, the FedEx courier uses information transmitted over the Internet to a computer in his truck to print a label from a portable printer attached to his belt. FedEx has control of the return, and HP can monitor its progress from start to finish.

Figure 8.2

Product and Information Flow for HP Printers Ordered over the Web

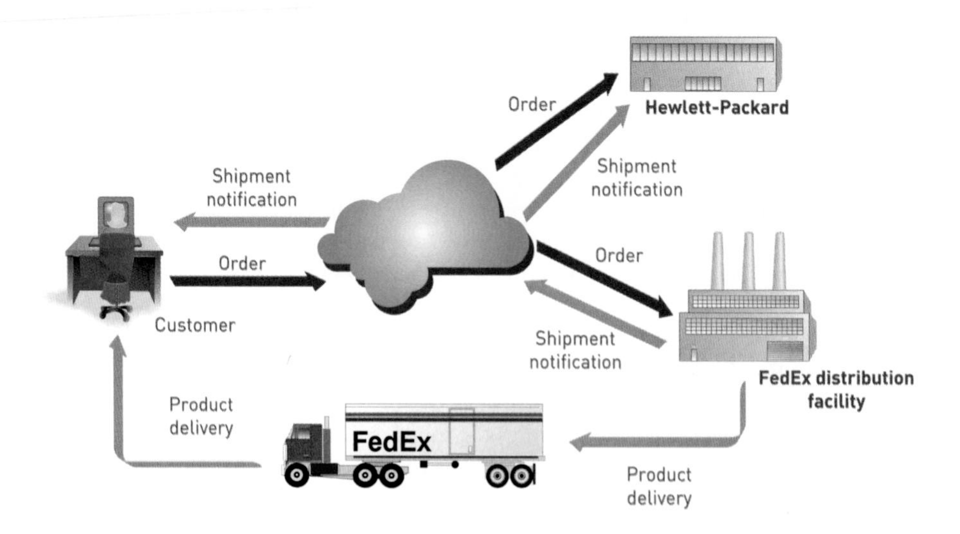

After-Sales Service

In addition to capturing the information to complete the order, comprehensive customer information is captured from the order and stored in the supplier's customer database. This information can include customer name, address, telephone numbers, contact person, credit history, and some order details. For example, if the customer later contacts the supplier to complain that not all items were received, that some arrived damaged, or even that the product provides unclear instructions, all customer service representatives can retrieve the order information from the database via a computing/communications device. Companies are adding the capability to answer many after-sales questions to their Web sites, such as how to maintain a piece of equipment, how to effectively use the product, and how to receive repairs under warranty.

The preceding sections discuss how a successful e-commerce system must address the many stages that consumers experience in the sales life cycle. In addition, looking at an e-commerce system from the perspective of the provider of goods or services, the system must support the activities associated with supply chain management and customer relationship management. These aspects of the e-commerce system are discussed next.

Supply Chain Management

As mentioned in Chapter 2, supply chain management (SCM) is increasingly accomplished using the Internet exchanges. An organization with many suppliers can use Internet exchanges to negotiate competitive prices and service. SCM is becoming a global issue, as companies have parts and products made around the world.[8] One example of an electronic marketplace is Aviall, a wholly owned subsidiary of the Boeing Company that provides after-market supply-chain management services for the aerospace, defense, and marine industries. Aviall's mission is to be the global leader in aircraft parts sales through world-class customer service to every customer, every time. The firm markets and distributes products for more than 225 manufacturers and offers approximately 1 million catalog items from 39 customer service centers located in North America, Europe, and the Asia-Pacific region. Its Inventory Locator Service (ILS) unit provides buyers and sellers immediate access via its Web site to aircraft and marine inventory 24 hours a day, seven days a week. Some 20,000 ILS subscribers around the globe access the ILS databases 60,000 times per day to complete transactions, from purchase initiation and order tracking to fulfillment. Subscribers can negotiate online, place orders, send and receive purchase orders and invoices, and track their negotiation history. Over 3,500 customer shipments are created daily.[9]

Customer Relationship Management

As discussed in Chapter 2, customer relationship management (CRM) involves managing every aspect of an organization's interactions with its customers or clients including marketing and advertising, sales, customer service after the sale, and programs to retain loyal customers. CRM systems enable a company to collect customer data, contact customers, educate them about new products, and actively sell products to existing and new customers. CRM systems can also obtain and analyze customer feedback to help design new or improved products and services.

Superior Industries manufactures a complete line of portable and stationary conveying equipment used in ship, barge, and rail loading and unloading applications for sugar, rock, coal, and wood. The firm markets through a dealer network serving the United States and Canada, and recently established an international presence by installing equipment at mines in Chile, Russia, Israel, Aruba, and Mexico. Superior employs 300 people at its Morris, Minnesota, headquarters with additional manufacturing operations in Prescott Valley, Arizona. Superior uses a CRM system to keep information about its distributor and dealer networks, generate quotes for customers, store customer lead and contact data, and save every quote and document associated with the sales process.[10]

E-Commerce Challenges

A company must overcome many challenges to convert its business processes from the traditional form to e-commerce processes, especially for B2C e-commerce. This section summarizes three key challenges: 1) defining an effective e-commerce model and strategy, 2) dealing with consumer privacy concerns, and 3) overcoming consumers' lack of trust.

The first major challenge is for the company to define an effective e-commerce model and strategy. Although companies can select from a number of approaches, the most successful e-commerce models include three basic components: community, content, and commerce, as shown in Figure 8.3. Message boards and chat rooms can build a loyal *community* of people who are interested in and enthusiastic about the company and its products and services. Providing useful, accurate, and timely *content*—such as industry and economic news and stock quotes—is a sound approach to encourage people to return to your Web site time and again. *Commerce* involves consumers and businesses paying to purchase physical goods, information, or services that are posted or advertised online.

Figure 8.3

Three Basic Components of a Successful E-Commerce Model

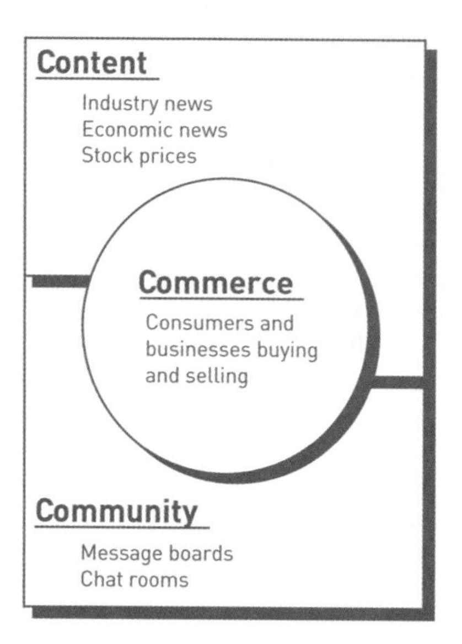

While the number of people shopping online and the dollar volume of online shopping continue to increase, about one-third of all adult Internet users will not buy anything online because they have privacy concerns or lack trust in online merchants.[11] In addition to having an effective e-commerce model and strategy, companies must carefully address consumer privacy concerns and overcome their lack of trust.

According to the Privacy Rights Clearinghouse, the approximate number of computer records containing sensitive personal information involved in security breaches in the United States from January 2005 to March 2008 is nearly 224 million![12] This represents the approximate number of records, not people affected. Some people have been the victim of more than one breach. Following are a few examples of security beaches in which personal data was compromised.

- One of TD Ameritrade's databases was hacked, and the e-mail addresses, phone numbers, and home addresses for more than 6.3 million customers were stolen.
- Customer names, addresses, telephone numbers, and credit card numbers were compromised by an intrusion into the Web site of online retailer Geeks.com.
- An international gang of cybercriminals hacked into the computer records of the OmniAmerican Bank of Fort Worth, Texas. They stole account numbers, created new PINs, fabricated debit cards, and withdrew cash from ATMs around the world.
- Attacks on Web servers hosted by a third-party service provider compromised the names, addresses, credit card data, debit card data, and passwords of people who shopped on Major League Soccer's MLSgear.com Web site.

In some cases, the compromise of personal data can lead to identity theft. According to the Federal Trade Commission (FTC), "**Identity theft** occurs when someone uses your personally identifying information, like your name, Social Security number, or credit card number, without your permission, to commit fraud or other crimes."[13] Thieves may use consumers' credit card numbers to charge items to their account, use identification information to apply for a new credit card or a loan in their name, or use their name and Social Security number to receive government benefits.

identify theft
Someone using your personally identifying information without your permission to commit fraud or other crimes.

The CardersMarket was a Web site where people's stolen credit card information was bought and sold like a commodity. Purchasers either sold the information to others or used it to make fraudulent cards for in-store purchases that were sold on auction sites to generate cash. After stealing tens of thousands of credit card numbers, the person who ran CardersMarket was indicted on wire fraud and identity theft charges carrying a maximum of 40 years in prison and a fine of $1.5 million.[14]

Companies must be prepared to make a substantial investment to safeguard their customers' data privacy or run the risk of losing customers and generating potential class action law suits should the data be compromised. Most Web sites invest in the latest security technology and employ highly trained security experts to protect their consumers' data.

Lack of trust in online sellers is one of the most frequently cited reasons for consumers not willing to purchase online. Can they be sure that the company or person with which they are dealing is legitimate and will send the item(s) they purchase from them? What if there is a problem with the product or service when it is received—for example, if it does not match the description on the Web site, is the wrong size or wrong color, is damaged during the delivery process, or does not work as advertised?

Online marketers must create specific trust-building strategies for their Web sites by analyzing their customers, products, and services. A perception of trustworthiness can be created by implementing one or more of the following strategies:

- Demonstrate a strong desire to build an ongoing relationship with customers by giving first-time price incentives, offering loyalty programs, or eliciting and sharing customer feedback.
- Demonstrate that the company has been in business for a long time.
- Make it clear that considerable investment has been made in the Web site.
- Provide brand endorsements from well-known experts or well-respected individuals.
- Demonstrate participation in appropriate regulatory programs or industry associations.
- Display Web site accreditation by the Better Business Bureau Online or TRUSTe programs.

Here are some tips to help online shoppers to avoid problems:

- Only buy from a well-known Web site you can trust—one that advertises on national media, is recommended by a friend, or receives strong ratings in the media.
- Look for a seal of approval from organizations such as the Better Business Bureau Online or TRUSTe (see Figure 8.4).
- Review the Web site's privacy policy to be sure that you are comfortable with its conditions before you provide personal information.
- Determine what the Web site policy is for return of products purchased.
- Be wary if you must enter any personal information other than what's required to complete the purchase (credit card number, address, and telephone number).
- Do not, under any conditions, ever provide information such as your Social Security number, bank account numbers, or your mother's maiden name.
- When you open the Web page where you enter credit card information or other personal data, make sure that the Web address begins with https and check to see if a locked padlock icon appears in the Address bar or status bar, as shown in Figure 8.5.
- Consider using virtual credit cards, which expire after one use when doing business.
- Before downloading music, change your browser's advanced settings to disable access to all computer areas that contain personal information.

Figure 8.4

Better Business Bureau Online and TRUSTe Seals of Approval

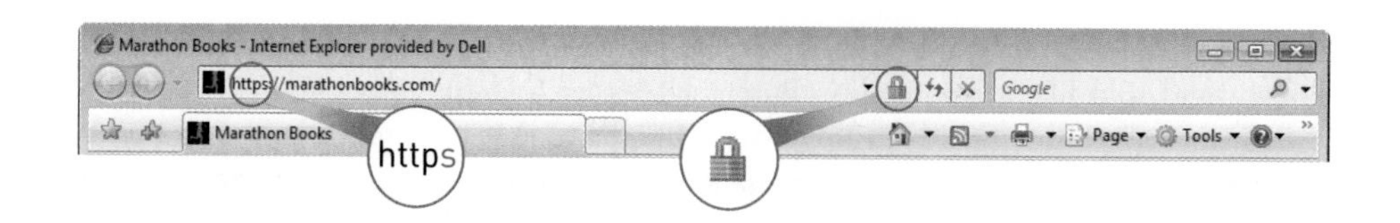

Figure 8.5

Web site that uses *https* in the address and a secure site lock icon

AN INTRODUCTION TO MOBILE COMMERCE

As discussed briefly in Chapter 1, mobile commerce (m-commerce) relies on the use of mobile, wireless devices, such as personal digital assistants, cell phones, and smartphones, to place orders and conduct business. Handset manufacturers such as Ericsson, Motorola, Nokia, and Qualcomm are working with communications carriers such as AT&T, Cingular, Sprint/Nextel, and Verizon to develop such wireless devices, related technology, and services. The Internet Corporation for Assigned Names and Numbers (ICANN) created a .mobi domain to help attract mobile users to the Web. mTLD Top Level Domain Ltd of Dublin, Ireland, administers this domain and helps to ensure that the .mobi destinations work fast, efficiently, and effectively with user handsets.

Mobile Commerce in Perspective

The market for m-commerce in North America is maturing much later than in Western Europe and Japan for several reasons. In North America, responsibility for network infrastructure is fragmented among many providers, consumer payments are usually made by credit card, and many Americans are unfamiliar with mobile data services. In most Western European countries, communicating via wireless devices is common, and consumers are much more willing to use m-commerce. Japanese consumers are generally enthusiastic about new technology and are much more likely to use mobile technologies for making purchases.

M-commerce spending in the United States is expected to exceed $500 million in 2008 and grow to almost $2 billion by 2010 according to Juniper Research. For perspective, U.S. e-commerce exceeded $100 billion in 2006.[15]

It is estimated that 40 percent of U.S. companies with annual revenue exceeding $50 million have established mobile Web sites.[16] The number of mobile Web sites is expected to grow because of advances in wireless broadband technologies, the development of new and useful applications, and the availability of less costly but more powerful handsets. For example, Yahoo's oneSearch 2.0 mobile search service includes a predictive text-search completion capability as well as voice recognition technology that adapts to a user's vocal patterns.[17] However, the relative clumsiness of mobile browsers and security concerns must be overcome to ensure rapid m-commerce growth.[18]

When it comes to mobile Web sites and mobile Web browsing capabilities, "just because you build it, doesn't mean they'll come," says Nikki Baird, managing partner at Retail Systems Research LLC. "You have to make consumers aware. It's all about getting people to try something new in the hope they'll come back for more."[19]

M-Commerce Web Sites

A number of retailers have established special Web sites for users of mobile devices. FlowerShop.com launched its m-commerce site, FlowerShopMobile.com, just in time to take advantage of one of its biggest shopping days of the year, Valentine's Day. Mobile device users can browse and buy floral gifts, plants, gift baskets, and gourmet foods. "The decision to go mobile was a natural one for FlowerShop.com," says Eric Luoma, the firm's president. "Flowers tend to be an on-the-fly purchase. If you're in an airport and it's your anniversary, it makes sense to pull out your phone and order flowers for your wife."[20]

mdog.com is a portal for your mobile device's Web browser. You direct your browser to mdog.com and many of your favorite Web sites (e.g., eBay, Craigslist, Wikipedia, Citysearch, and MySpace) and blogs are displayed in a format convenient for your mobile device.

ELECTRONIC AND MOBILE COMMERCE APPLICATIONS

E-commerce and m-commerce are being used in innovative and exciting ways. This section examines a few of the many B2B, B2C, C2C, and m-commerce applications in the retail and wholesale, manufacturing, marketing, investment and finance, online real estate services, and auction arenas.

Retail and Wholesale

E-commerce is being used extensively in retailing and wholesaling. **Electronic retailing**, sometimes called *e-tailing*, is the direct sale of products or services by businesses to consumers through electronic storefronts, which are typically designed around the familiar electronic catalog and shopping cart model. Companies such as Office Depot, Wal-Mart, and many others have used the same model to sell wholesale goods to employees of corporations. Tens of thousands of electronic retail Web sites sell everything from soup to nuts.

electronic retailing (e-tailing)
The direct sale from business to consumer through electronic storefronts, typically designed around an electronic catalog and shopping cart model.

Cybermalls are another means to support retail shopping. A **cybermall** is a single Web site that offers many products and services at one Internet location—similar to a regular shopping mall. An Internet cybermall pulls multiple buyers and sellers into one virtual place, easily reachable through a Web browser.

cybermall
A single Web site that offers many products and services at one Internet location.

Sears, the company that pioneered the use of the mail-order catalog back in the 1890s, is making a major investment in B2C e-commerce, employing more than 100 technology workers to improve its online sales. It ranks as the second largest mass merchant retailer online with recent sales of $2.6 billion (Amazon.com is ranked number one). With the number of unique visitors per month growing at over 20 percent, Sears is the second fastest growing site among mass retailers (Costco is ranked number one). Some industry experts believe that Sears.com may turn into a cybermall that sells all kinds of products and competes with companies such as Amazon.com.[21]

A key sector of wholesale e-commerce is spending on manufacturing, repair, and operations (MRO) goods and services—from simple office supplies to mission-critical equipment, such as the motors, pumps, compressors, and instruments that keep manufacturing facilities running smoothly. MRO purchases often approach 40 percent of a manufacturing company's total revenues, but the purchasing system can be haphazard, without automated controls. In addition to these external purchase costs, companies face significant internal costs resulting from outdated and cumbersome MRO management processes. For example, studies show that a high percentage of manufacturing downtime is often caused by not having the right part at the right time in the right place. The result is lost productivity and capacity. E-commerce software for plant operations provides powerful comparative searching capabilities to enable managers to identify functionally equivalent items, helping them spot opportunities to combine purchases for cost savings. Comparing various suppliers, coupled with consolidating more spending with fewer suppliers, leads to decreased costs. In addition, automated workflows are typically based on industry best practices, which can streamline processes.

Manufacturing

One approach taken by many manufacturers to raise profitability and improve customer service is to move their supply chain operations onto the Internet. Here they can form an **electronic exchange** to join with competitors and suppliers alike, using computers and Web sites to buy and sell goods, trade market information, and run back-office operations, such as inventory control, as shown in Figure 8.6. With such an exchange, the business center is not a physical building but a network-based location where business interactions occur. This approach has greatly speeded up the movement of raw materials and finished products among all members of the business community, thus reducing the amount of inventory that must be maintained. It has also led to a much more competitive marketplace and lower prices. Private exchanges are owned and operated by a single company. The owner uses the exchange

electronic exchange
An electronic forum where manufacturers, suppliers, and competitors buy and sell goods, trade market information, and run back-office operations.

Dell sells its products through the Dell.com Web site.

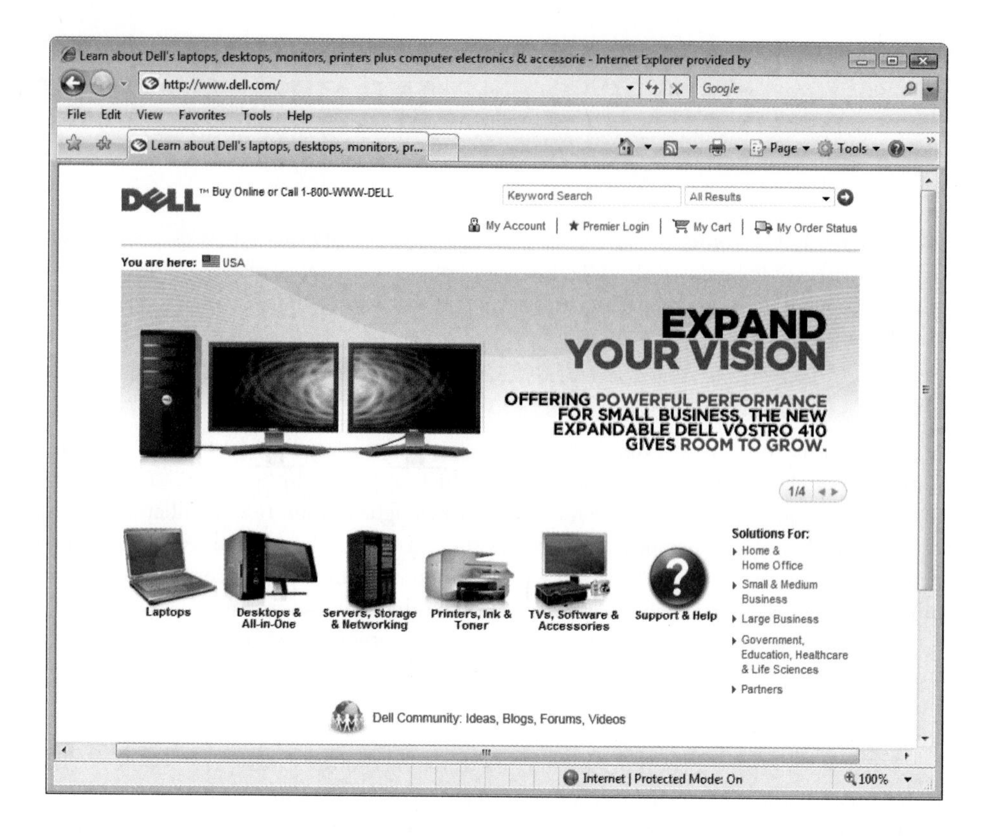

to trade exclusively with established business partners. Public exchanges are owned and operated by industry groups. They provide services and a common technology platform to their members and are open, usually for a fee, to any company that wants to use them.

Figure 8.6

Model of an Electronic Exchange

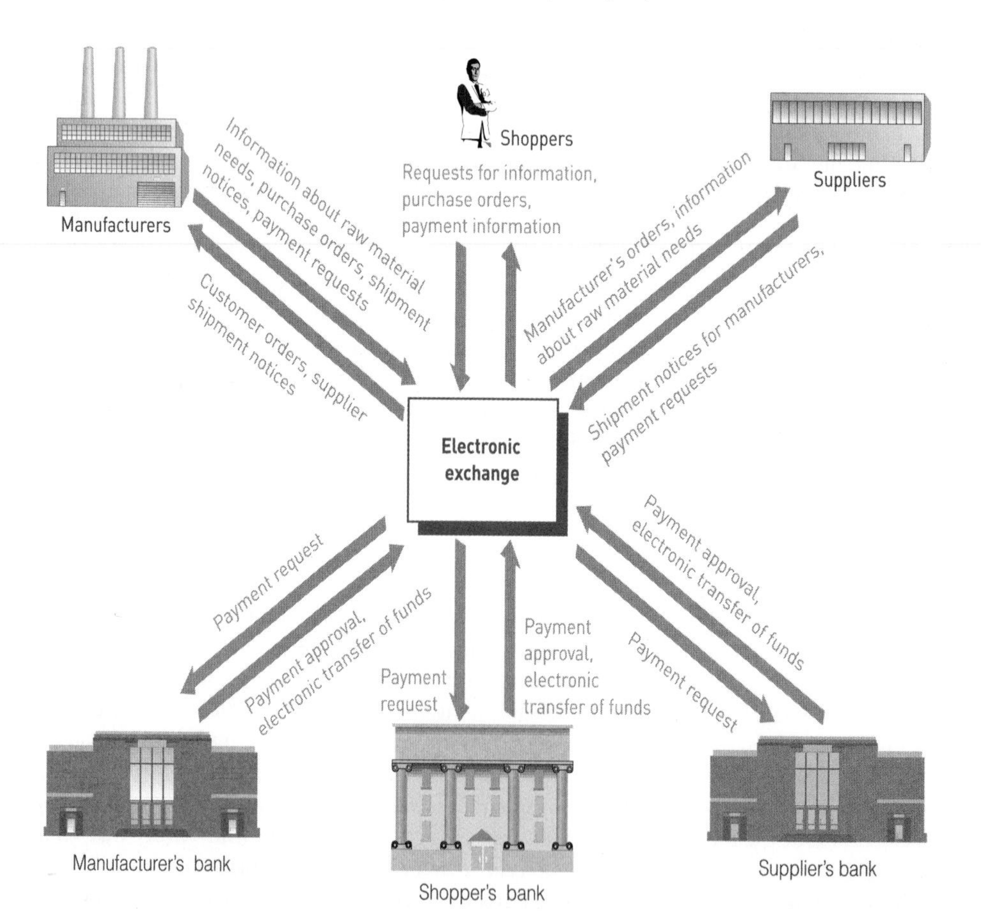

The Detroit Trading Exchange lets auto dealers and others bid to buy more than 300,000 sales leads generated from consumers who visit a host of auto-related Web sites. The sales leads can be sorted by zip code, financial factors, and other parameters so buyers tailor the sales leads they receive.[22]

Several strategic and competitive issues are associated with the use of exchanges. Many companies distrust their corporate rivals and fear they might lose trade secrets through participation in such exchanges. Suppliers worry that the online marketplaces and their auctions will drive down the prices of goods and favor buyers. Suppliers also can spend a great deal of money in the setup to participate in multiple exchanges. For example, more than a dozen new exchanges have appeared in the oil industry, and the printing industry is up to more than 20 online marketplaces. Until a clear winner emerges in particular industries, suppliers are more or less forced to sign on to several or all of them. Yet another issue is potential government scrutiny of exchange participants—when competitors get together to share information, it raises questions of collusion or antitrust behavior.

Many companies that already use the Internet for their private exchanges have no desire to share their expertise with competitors. At Wal-Mart, the world's number-one retail chain, executives turned down several invitations to join exchanges in the retail and consumer goods industries. Wal-Mart is pleased with its in-house exchange, Retail Link, which connects the company to 7,000 worldwide suppliers that sell everything from toothpaste to furniture.

Marketing

The nature of the Web allows firms to gather much more information about customer behavior and preferences than they could using other marketing approaches. Marketing organizations can measure many online activities as customers and potential customers gather information and make their purchase decisions. Analysis of this data is complicated because of the Web's interactivity and because each visitor voluntarily provides or refuses to provide personal data such as name, address, e-mail address, telephone number, and demographic data. Internet advertisers use the data they gather to identify specific portions of their markets and target them with tailored advertising messages. This practice, called **market segmentation**, divides the pool of potential customers into subgroups, which are usually defined in terms of demographic characteristics, such as age, gender, marital status, income level, and geographic location.

market segmentation
The identification of specific markets to target them with advertising messages.

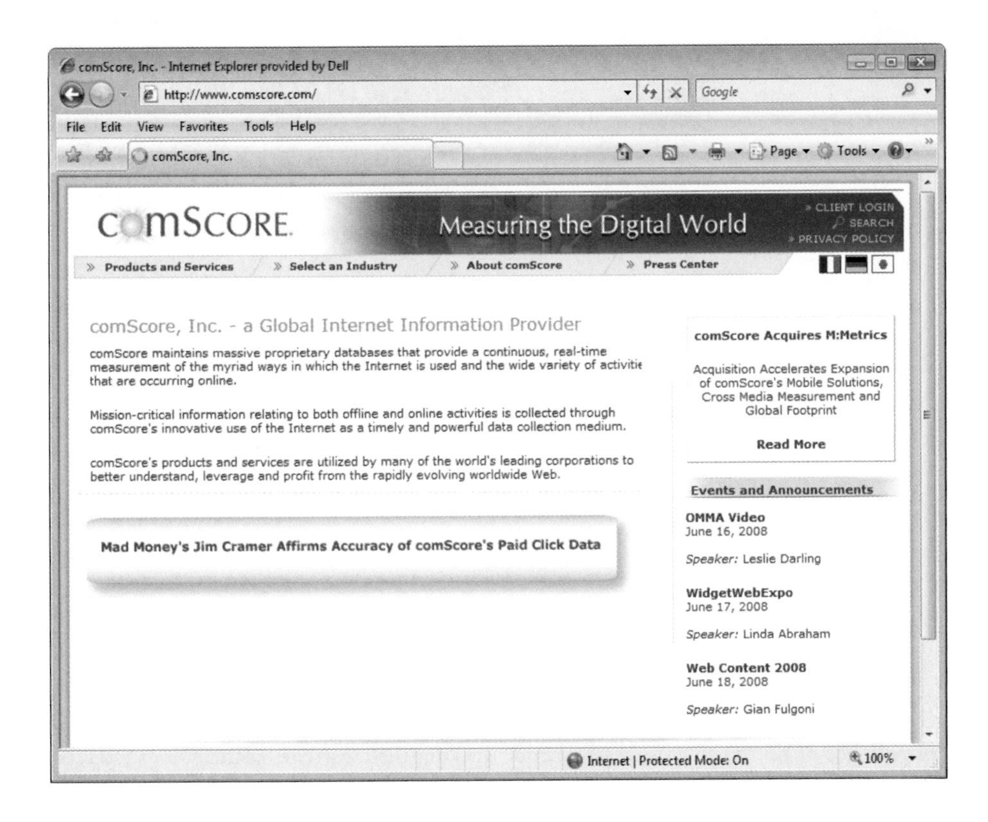

comScore Networks is a global information provider to large companies seeking information on consumer behavior to boost their marketing, sales, and trading strategies.

technology-enabled relationship management
Occurs when a firm obtains detailed information about a customer's behavior, preferences, needs, and buying patterns and uses that information to set prices, negotiate terms, tailor promotions, add product features, and otherwise customize its entire relationship with that customer.

Technology-enabled relationship management is a new twist on establishing direct customer relationships made possible when firms promote and sell on the Web. **Technology-enabled relationship management** occurs when a firm obtains detailed information about a customer's behavior, preferences, needs, and buying patterns and uses that information to set prices, negotiate terms, tailor promotions, add product features, and otherwise customize its entire relationship with that customer.

Cliff Conneighton, senior vice president of e-commerce platform provider Art Technology Group (ATG), says: "The secret to improved sales on the Web is to deliver the right offer to someone at the right time. [You have] to know something about who you're selling to, and try to show them the goods and the offer that's more relevant." American Eagle Outfitters, an ATG client, followed this advice and doubled the revenue generated at its Web site in only one year.[23]

Investment and Finance

The Internet has revolutionized the world of investment and finance. Perhaps the changes have been so significant because this industry had so many built-in inefficiencies and so much opportunity for improvement.

The brokerage business adapted to the Internet faster than any other arm of finance. The allure of online trading that enables investors to do quick, thorough research and then buy shares in any company in a few seconds and at a fraction of the cost of a full-commission firm has brought many investors to the Web. Online brokerage firms have consolidated, with Ameritrade acquiring TD Waterhouse, and E-Trade acquiring Harrisdirect and the online brokerage services of JP Morgan. In spite of the wealth of information available online, the average consumer buys stocks based on a tip or a recommendation rather than as the result of research and analysis. It is the more sophisticated investor that really takes advantage of the data and tools available on the Internet.

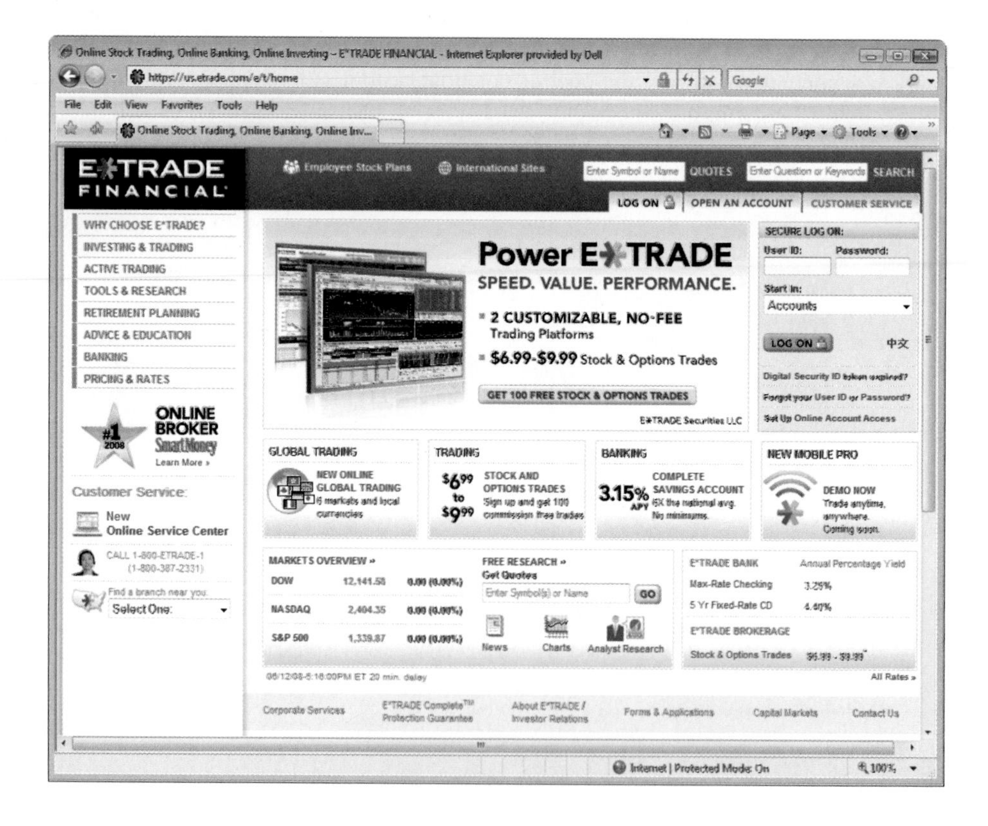

E-Trade is an online brokerage site that offers information, tools, and account-management services for investors.

Online banking customers can check balances of their savings, checking, and loan accounts; transfer money among accounts; and pay their bills. These customers enjoy the convenience of not writing checks by hand, tracking their current balances, and reducing expenditures on envelopes and stamps. In addition, the PayItGreen Alliance reports that paying bills online is good for the environment. By its estimates, the average household makes seven paper bill payments per month. If just 10 percent of the U.S. population converted to online bill payment, the environmental savings would total more than 75 million pounds of paper, nearly 1 million trees and 2 million pounds of greenhouse gases.[24]

Internet banking in Asia, Europe, and Japan is considerably advanced compared to the United States. For example, the Industrial and Commercial Bank of China (Asia), or ICBC Asia, offers secure personal Internet banking services that allow customers to manage their personal finances safely and reliably anywhere and anytime. Customers can view their account summary and detailed transactions; transfer funds; pay bills; submit applications for loans, insurance, and ATM card services; inquire about interest rates and exchange rates; view their checking account; stop check payments; request new checks; buy and sell securities; and even more functions.[25]

All of the major banks and many of the smaller banks in the U.S. enable their customers to pay bills online; many support bill payment via cell phone or other wireless device. Banks are eager to gain more customers who pay bills online because such customers tend to stay with the bank longer, have higher cash balances, and use more of the bank's products. To encourage the use of this service, many banks have eliminated all fees associated with online bill payment.

The next advance in online bill paying is **electronic bill presentment**, which eliminates all paper, down to the bill itself. With this process, the vendor posts an image of your statement on the Internet and alerts you by e-mail that your bill has arrived. You then direct your bank to pay it. ePower is an electronic bill presentment and payment solution provider that enables utility firms to provide interactive financial statements to their customers via e-mail and on the Internet at *www.payabill.com*.[26]

electronic bill presentment
A method of billing whereby a vendor posts an image of your statement on the Internet and alerts you by e-mail that your bill has arrived.

INFORMATION SYSTEMS @ WORK

MoneyAisle.com Puts Customers in Charge

E-commerce has shaken up traditional forms of commerce, and in some cases turned them completely upside down. Consider the consumer's ability to comparison shop online, sampling prices from businesses of all sizes located around the world. Consider how easy it is to find rare items, out-of-print books, and collectibles. Online auction houses such as eBay have created an entirely new kind of marketplace.

While e-commerce has dramatically affected retail sales, other types of transactions have remained relatively stable. For example, consider choosing a bank for financial services. In the days before the popularity of the Internet, if you wanted to invest in a high-yield savings account or a certificate of deposit (CD), you would visit a number of local banks and find the best interest rate for the amount of money you planned to invest. The banks simply advertised their rates, and the customer did the work of collecting the data and making the decision based on value and bank reputation.

The Web has simplified this task, making thousands of bank quotes available online, though the process is still time consuming, and when you're done, it's hard to know if you've really found the best deal. Web sites such as LendingTree.com and Bankrate.com aggregate quotes from numerous banks, reducing the customer's research time, but the rates are still inflexible and the banks are in control.

MoneyAisle.com is working to change these factors by providing a service that puts the consumer in charge. At MoneyAisle.com, more than 100 reputable banks compete for your business. Unlike other services that merely give the impression of competition between banks, at MoneyAisle.com the banks actually work to outbid each other for your business in a live, real-time auction. Customers use the form at MoneyAisle.com to provide the amount they want to invest and their state of residence, and then click the Start Auction button. After a few minutes, the customer watches banks bid in real time, round after round, until all but one bank has dropped out, offering the best interest rate.

MoneyAisle.com chief executive, Mukesh Chatter, thought of the idea for the business after noticing that prices for high-definition TVs varied significantly depending on the vendor. He saw similar variations in pricing elsewhere as well, including in banks. Chatter worked with partners to develop the algorithms to allow banks to place their bids for investor dollars, which is how MoneyAisle works. The site now earns revenue from charging participating banks a small fee. This provides a benefit to investors by finding the best return on investment with the lowest amount of effort. It also provides benefits to smaller banks with less advertising capital. It is ordinarily difficult for smaller banks to compete with big banks with big advertising budgets. MoneyAisle.com levels the playing field giving banks of all sizes an equal opportunity.

The service offered by MoneyAisle.com meets the needs of smaller banks looking to increase business through online tools. The challenge for MoneyAisle.com will be to generate enough traffic to let banks know that using the service is worth the effort. So far, the idea seems to be catching on. In its first week of business, MoneyAisle.com was used as a tool for investing over $1 million at small and mid-size banks.

Discussion Questions

1. How does MoneyAisle.com turn the process of investing upside down?
2. Who benefits from the service offered by MoneyAisle.com? Who is negatively affected?

Critical Thinking Questions

1. For what other types of products might reverse bidding be useful? What makes a product a good subject for reverse bidding?
2. How does reverse bidding impact the way that a bank operates and its budget and profit margins?

SOURCES: Rosencrance, Linda, "MoneyAisle launches 'reverse' consumer auction Web site for banks," *Computerworld*, June 9, 2008, *www.computerworld.com/action/article.do?command=viewArticleBasic&taxonomyName=internet_business&articleId=9094758&taxonomyId=71&intsrc=kc_top;* MoneyAisle.com Web site, *https://www.moneyaisle.com*, accessed June 21, 2008; Rosencrance, Linda, "$1M deposited in banks via MoneyAisle in first week," *Computerworld*, June 17, 2008, *www.computerworld.com/action/article.do?command=viewArticleBasic&articleId=9099458*; Johnson, Carolyn, "Website lets banks bid for customers," *Boston Globe*, June 9, 2008, *www.boston.com/business/technology/articles/2008/06/09/website_lets_banks_bid_for_customers.*

Online Real Estate Services

Cyberhomes, FHA Anonymous, Loanexa, Realtor.com, Redfin, Terabitz, Trulia, and Zillow are just a few of the hundreds of Web sites that offer interesting services for those looking to buy a home. Many of the sites offer the capability to search the U.S. for homes based on geographic location, price range, number of bedrooms or bathrooms, and special features such as a pool or hot tub.

Online real estate service Zillow has set up a large number of Web sites based on specific communities and enables users to exchange data such as demographics and the crime rate in neighborhoods. This is data that for various legal and ethical reasons, real estate agents can't freely discuss.[27]

Redfin is an online real estate company that provides both online real estate search capabilities and access to live agents. The firm employs its agents so it can better manage customer service—unlike traditional real estate firms that license their names to independent agents. Redfin pays bonuses to agents when they receive high customer satisfaction ratings. It claims to reimburse home buyers roughly two-thirds of their real estate fees immediately upon closing, thus reducing the purchase price by many thousands of dollars.[28]

From the customer's viewpoint, an important service is the ability to receive competitive quotes from lenders without giving out personally identifying information that makes them a target of aggressive loan officers. Consumers can anonymously request loan quotes through several Web sites including FHA Anonymous, Loanexa, and Zillow.

E-Boutiques

An increasing number of Web sites offer personalized shopping consultations for shoppers interested in upscale, contemporary clothing—dresses, sportswear, denim, handbags, jewelry, shoes, and gifts. Key to the success of Web sites such as ShopLaTiDa is a philosophy of high customer service and strong, personal client relationships. Online boutique shoppers complete a personal shopping profile by answering questions about body measurements, profession, interests, preferred designers, and areas of shopping where they would welcome assistance.[29] Shoppers are then given suggestions on what styles and designers might work best and where they can be found—online or in brick-and-mortar shops.

Auctions

eBay has become synonymous with online auctions for both private sellers and small companies. Other popular online auction Web sites include Craigslist, uBid, Auctions, Onsale, WeBidz, and many others. The most frequent complaints about online auctions are increases in fees and problems with unscrupulous buyers. As mentioned in Chapter 7, auction sites are used by criminals to unload stolen, diverted, and counterfeit products. Law enforcement organizations regularly monitor such Web sites to capture criminals and recover stolen goods. Another frequent problem with online auctions is inaccurate or incomplete representation of the item for sale. Descriptions may omit important aspects or photos may not be clear enough to show the item's features.

There are two common types of online auctions. In an English auction, the initial price starts low and is bid up by successive bidders. In a reverse auction, sellers compete to obtain business by submitting successively lower prices for their goods or services. Reverse auctions are frequently used in B2B procurement.

Blair Corporation is a multichannel direct marketer of fashions for men, women, and homes. The firm worked with eDynaQuote to conduct its first reverse auction and ensure broad supplier participation. Blair achieved significant cost savings on its first reverse auction for $1 million in packaging supplies.[30]

Anywhere, Anytime Applications of Mobile Commerce

Because m-commerce devices usually have a single user, they are ideal for accessing personal information and receiving targeted messages for a particular consumer. Through m-commerce, companies can reach individual consumers to establish one-to-one marketing relationships and communicate whenever it is convenient—in short, anytime and anywhere. Following are just a few examples of potential m-commerce applications:

Mobile Banking

With mobile banking, consumers can manage their finances from anywhere without driving to their bank or credit union or booting their computer. Consumers can use mobile banking to access multiple banks, accounts, and financial services to:

- View account balances (checking, savings, Money Market, and credit cards)
- Transfer funds between accounts
- View and pay bills
- Review a history of account transactions

Such capability allows consumers to check their credit card balance before making a major purchase and avoid having the credit provider rejecting the purchase. They can also transfer funds from savings to checking accounts to avoid an overdraft.

To begin using mobile banking with their wireless phones, consumers must visit their bank's online banking site and enroll in mobile banking. They then download the mobile application to their phone. As a security measure, mobile banking users must enter their personal PIN to unlock the application each time they use it.

Mobile banking from AT&T is available to AT&T wireless users who bank with Bancorp South, Wachovia, Sun Trust, Synovus, Arvest, and First Bank. BlackBerry, LG, Motorola, Nokia, Samsung, and Sony Ericsson manufacture several models of phones that support AT&T's mobile banking.[31]

Mobile Price Comparison

A growing number of companies are employing a strategy that encourages shoppers to do Web-based price comparisons while they are in the stores. The idea is to drive the shopper who is ready to make a purchase from one retailer to another based on price and product comparisons. Web sites, like Google Maps, can be used to locate stores, restaurants, gas stations, and other retailers while you are on the move.

AbeBooks.com is a Web-only retailer and will accept text messages from college students containing the International Standard Book Number (ISBN) of a textbook. AbeBooks replies with a text message containing its lowest price for a new copy of the textbook. If the students decide to buy from AbeBooks after reviewing the price, they reply by texting "fwd" and their e-mail address. AbeBooks sends an e-mail to the address containing a link to the AbeBooks.com page where the book is listed. The students can then log on to a personal computer, receive the e-mail, link to the AbeBooks page, and buy the text.[32]

BikeSomeWhere.com offers an m-commerce Web site that enables shoppers to do product and price comparisons as well as buy bikes and biking gear via their cell phones. BikeSomeWhere wants bikers to use the Web site as a tool to make an informed purchasing decision. The firm offers free shipping on orders over $75 and consumers do not have to pay sales tax, which usually makes BikeSomeWhere very price competitive.[33]

Barcle allows shoppers using any mobile device with a Web browser to enter the 12-digit bar code of a product and receive search results showing prices for the same product at Web-only and brick-and-mortar retailers.[34]

ShopLocal offers product location and comparison on mobile devices via a service called Where from mobile technology vendor uLocate. Shoppers can download the Where application using a text message from uLocate. The application works with GPS-enabled phones and provides comparison shoppers with product, price, and retailer information including step-by-step directions to the selected retailer's store. The Where service is available for $2.99 per month with users who have wireless phone plans with Alltel, Boost, or Sprint Nextel.[35]

Mobile Advertising

While some 58 million U.S. wireless subscribers viewed an ad on their cell phones in February 2008, many advertisers are not yet convinced that mobile advertising is effective and are taking a wait-and-see approach.[36]

Traditional Web sites designed for access by users with personal computers place cookies on your computer to track your browsing behavior and pass the data on to advertisers and ad-placement networks. However, the wireless industry service providers block cookies before they get to the cell phone out of concern that the cookies could provide access to their networks for computer viruses. They also fear that the cookies might cause a dramatic increase in the volume of data traffic as the cookies report back to the advertisers and ad-placement networks. The increase in volume could be enough to choke the network and seriously degrade performance. Thus advertisers are frustrated in their attempt to gather data to measure the number of views or effectiveness of mobile ads.

Mobile Coupons

About 2 percent of advertisers surveyed by Jupiter Research are using mobile coupons.[37] The Clorox Company, Del Monte Corporation, General Mills, Kimberly-Clark, and Procter & Gamble are collaborating with grocery retailer Kroger to test how consumers will react to using mobile coupons. Users in the test must first download a mobile marketing application to their cell phones so that coupons from the companies can be loaded onto their cell phones. While in a Kroger store, a shopper can choose an item, select the appropriate coupon from the cell phone, and have the coupon information sent to Kroger's in-store computer, which identifies the shopper by her loyalty card. At checkout, the coupon discount is applied when the loyalty card is scanned.[38]

As with any new technology, m-commerce will succeed only if it provides users with real benefits. Companies involved in m-commerce must think through their strategies carefully and ensure that they provide services that truly meet customers' needs.

Advantages of Electronic and Mobile Commerce

Conversion to an e-commerce or m-commerce system enables organizations to reduce the cost of doing business, speed the flow of goods and information, increase the accuracy of order processing and order fulfillment, and improve the level of customer service.

Reduce Costs

By eliminating or reducing time-consuming and labor-intensive steps throughout the order and delivery process, more sales can be completed in the same period and with increased accuracy. With increased speed and accuracy of customer order information, companies can reduce the need for inventory—from raw materials to safety stocks and finished goods—at all the intermediate manufacturing, storage, and transportation points.

Speed the Flow of Goods and Information

When organizations are connected via e-commerce, the flow of information is accelerated because electronic connections and communications are already established. As a result, information can flow easily, directly, and rapidly from buyer to seller.

Increase Accuracy

By enabling buyers to enter their own product specifications and order information directly, human data-entry error on the part of the supplier is eliminated.

Improve Customer Service

Increased and more detailed information about delivery dates and current status can increase customer loyalty. In addition, the ability to consistently meet customers' desired delivery dates with high-quality goods and services eliminates any incentive for customers to seek other sources of supply.

Global Challenges for E-Commerce and M-Commerce

E-commerce and m-commerce offer enormous opportunities by allowing manufacturers to buy at a low cost worldwide. They also offer enterprises the chance to sell to a global market right from the start of their business. Moreover, they offer great promise for developing countries, helping them to enter the prosperous global marketplace, and hence helping reduce the gap between rich and poor countries. People and companies can get products and services from around the world, instead of around the corner or across town. These opportunities, however, come with numerous obstacles and issues, first identified in Chapter 1 as challenges associated with all global systems.

Cultural Challenges

Countries and regional areas have their own cultures and customs that can significantly affect people and organizations involved in global trade. A Web site must be designed carefully if it will be viewed by different cultural groups outside or within a country. Great care must be taken to ensure that the site is appealing, easy to use, and inoffensive to people around the world.

Language Challenges

Obviously, language differences can make it difficult for visitors to understand the information and directions posted on a Web site. Thus, many Web sites add an entrance page that lets visitors select a language for viewing the Web site. Sometimes, it is not enough to have multilingual versions of the text; a complete redesign may be called for. For example, if your Web site design includes a vertical menu bar, you may place it in the left margin of your pages for English visitors but on the right for Arabic visitors who start reading pages from right to left. In addition, measurement conversions for quantities used in recipes, distances, and temperatures are necessary. U.S. measurements such as cups, miles, and degrees Fahrenheit must be converted to liters, kilometers, and degrees Celsius.

Time and Distance Challenges

Time and distance issues can be barriers for people and organizations involved with global trade in remote locations. Significant time differences make it difficult for customers to speak directly with salespeople or customer service representatives in other locations unless your business schedules staff to work around the clock. Because of the great distances involved, it can take days for customers to receive a product, a critical part, or a piece of equipment. For this reason, many Web sites provide customers with a means to track the shipment progress of their order via a connection to the shipper's order tracking system.

Infrastructure Challenges

The Web site must be displayed correctly in all the major Web browsers including Internet Explorer, Firefox, Safari, Opera, Netscape, Mozilla, and others. If it does not, visitors will quickly switch to competitors' Web sites. The Web site must also support access from laptops, PDAs, cell phones, and other devices.

Currency Challenges

Prices for all items offered for sale on the Web site must clearly indicate the currency. If the Web site is to support sales to multiple countries, it must indicate whether other currencies are acceptable and provide an easy means for customers to convert from their currency to the currency in which the price is quoted.

Product and Service Challenges

E-products such as software, music, and books and e-services such as customer support and advice can be delivered to customers electronically over the Internet. The Web site must operate reliably to allow fast, consistent delivery of such products and services.

State, Regional, and National Laws

Every state, region, and country has a set of laws that governs commercial transactions. These laws cover a variety of issues, including the protection of trademarks and patents, the sale of copyrighted material, the collection and safeguarding of personal or financial data, the payment of sales taxes and fees, and much more. Keeping track of these laws and incorporating them into the operation of a global Web site is extremely complex and time consuming, requiring expert legal advice.

THREATS TO ELECTRONIC AND MOBILE COMMERCE

Businesses must deal with a host of issues to ensure that e-commerce and m-commerce transactions are safe and consumers are protected. The following sections summarize a number of threats to the continued growth and success of e-commerce and m-commerce and present practical ideas on how to minimize their impact.

Security

Many organizations that accept credit cards to pay for items purchased via e-commerce have adopted the Payment Card Industry security standard. This standard spells out measures and security procedures to safeguard the card issuer, the cardholder, and the merchant. Some of the measures include installing and maintaining a firewall configuration to control access to computers and data; never using software/hardware vendor-supplier defaults for system passwords; and requiring merchants to protect stored data, encrypt transmission of cardholder information across public networks, use and regularly update antivirus software, and restrict access to sensitive data on a need-to-know basis.

Various measures are being implemented to increase the security associated with the use of credit cards at the time of purchase. Address Verification System is a check built into the payment authorization request that compares the address on file with the card issuer to the billing address provided by the cardholder. The Card Verification Number technique is a check of the additional digits printed on the back of the card. Visa has Advanced Authorization, a Visa-patented process that provides an instantaneous rating of that transaction's potential for fraud to the financial institution that issued the card. The card issuer can then send an immediate response to the merchant whether to accept or decline the transaction. The technology is now being applied to every Visa credit and check card purchase today. Visa estimates that this technique will reduce fraudulent credit card charges by 40 percent

The Federal Financial Institutions Examination Council has developed a new set of guidelines called "Authentication in an Internet Banking Environment," recommending two-factor authorization. This approach adds another identity check along with the password system. A number of multifactor authentication schemes can be used, such as biometrics, one-time passwords, or hardware tokens that plug into a USB port on the computer and generate a password that matches the ones used by a bank's security system. Currently, the use of biometric technology to secure online transactions is rare for both cost and privacy reasons. It can be expensive to outfit every merchant with a biometric scanner, and it is difficult to convince consumers to supply something as personal and distinguishing as a fingerprint. In spite of this, a growing number of financial service firms from large (e.g., Citibank) to small (e.g., Perdue Employees Federal Credit Union) are considering biometric systems.

Theft of Intellectual Property

Intellectual property includes works of the mind such as books, films, music, processes, and software, which are distinct somehow and are owned or created by a single entity. The owner of the intellectual property is entitled to certain rights in relation to the subject matter of the intellectual property. Thus, copyright law protects authored works such as books, film, images, music, and software from unauthorized copying. Patents can also protect software as well as business processes, formulae, compounds, and inventions. Information that has significant value for a firm and for which strong measures are taken to protect it are trade secrets. They too are protected under various laws. Although concerns about intellectual property and digital rights management (discussed next) apply to creative works distributed traditionally through brick-and-mortar retailers and libraries, these issues are more urgent for e-commerce because computers and the Internet make it easy to access, copy, and distribute digital content.

intellectual property
Includes works of the mind such as books, films, music, processes, and software, which are distinct somehow and are owned and/or created by a single entity.

Digital rights management (DRM) refers to the use of any of several technologies to enforce policies for controlling access to digital media such as movies, music, and software. Many digital content publishers state that DRM technologies are needed to prevent revenue

digital rights management (DRM)
Refers to the use of any of several technologies to enforce policies for controlling access to digital media such as movies, music, and software.

loss due to illegal duplication of their copyrighted works. For example, the Motion Picture Association of America (MPAA) estimates that the film industry lost approximately $7 billion in movie piracy in 2005.[39] On the other hand, many digital content users argue that DRM and associated technologies lead to a loss of user rights. For example, users can purchase a music track online for under a dollar through Apple's iTunes music store. They can then burn that song to a CD and transfer it to an iPod. However, the purchased music files are encoded in the AAC format supported by iPods and protected by FairPlay, a DRM technology developed by Apple. To the consternation of music lovers, many music devices are not compatible with the AAC format and cannot play iTunes' protected files.

Fraud

The first wave of Internet crime consisted mostly of online versions of offline hoaxes, the usual get-rich-quick schemes. For example, many people received pleas from desperate Nigerians trying to enlist their help in transferring funds out of their country. More recently, however, fraud artists have begun to exploit the Internet to execute more sophisticated ploys, using fake Web sites and spam.

phishing
A practice that entails sending bogus messages purportedly from a legitimate institution to pry personal information from customers by convincing them to go to a "spoofed" Web site.

Phishing entails sending bogus messages purportedly from a legitimate institution to pry personal information from customers by convincing them to go to a "spoofed" Web site. The spoofed Web site appears to be a legitimate site but actually collects personal information from unsuspecting victims. Phishing scams are frequently disguised as requests for donations from a charitable organization. Sadly, criminals take advantage of the generosity of others following every natural disaster by sending out tens of thousands of bogus requests for donations from charitable organizations. Unfortunately, many generous but naive people provide personal information or bank account data.[40] Another frequent phishing ploy involves the use of phony e-mail requests from the U.S. Internal Revenue Service requesting personal information to help speed the processing of tax refund checks. In the spring of 2008, tens of thousands of phishing messages were sent stating the fastest way to receive the economic stimulus tax rebate was through direct deposit. The e-mail included a Web link to an online submission form designed to steal submitted information from those fooled by the phishing scam. The IRS never initiates taxpayer communications via e-mail.[41]

click fraud
A problem arising in a pay-per-click online advertising environment where additional clicks are generated beyond those that come from actual, legitimate users.

Click fraud can arise in a pay-per-click online advertising environment when additional clicks are generated beyond those that come from actual, legitimate users. In pay-per-click advertising, the advertiser pays when a user clicks its ad to visit its Web site. The additional clicks may be generated by an illegitimate user, automated script, or some other means. These bogus clicks generate revenue for the advertising network such as Google or Yahoo!. Bigreds.com, an online seller of collectibles, employed a pay-per-click advertising service run by Yahoo and sued the firm for more than $1 million in damages and penalties. Bigreds.com alleged that it paid more than $900,000 for clicks that its ads received on sites affiliated with Yahoo, but that many of those clicks were fraudulent because they were generated by software programs and people other than actual customers.[42]

Online auction fraud represents a major source of complaints both in the United States and abroad. In 2007, the Internet Crime Complaint Center at the U.S. Federal Trade Commission received 124,130 complaints related to Internet auction fraud and nondelivery of merchandise.[43] The majority of the problems come from so-called person-to-person auctions, which account for roughly half the auction sites. On these sites, it is up to the buyer and seller to resolve details of payment and delivery; the auction sites offer no guarantees. Sticking with auction sites like eBay (*www.ebay.com*) that ensure the delivery and quality of all the items up for bidding can help buyers avoid trouble.

Another Internet auction-related problem is fake goods that find their way onto virtually all of the online auctions. eBay, as the world's largest online auction site, is constantly battling counterfeiters. For example, eBay and Montres Rolex S.A. have been engaged in court battles for more than six years over the sale of counterfeit Rolex watches at the eBay site.[44] In another example, seven people were charged with selling counterfeit limited edition prints of works by Pablo Picasso, Andy Warhol, Marc Chagall, and others for over $5 million on eBay. eBay says it is simply not possible for them to distinguish between a legitimate item and a fake among the millions of items sold on its site each year.[45]

Invasion of Consumer Privacy

Online consumers are more at risk today than ever before. One of the primary factors causing higher risk is *online profiling*—the practice of Web advertisers recording online behavior for the purpose of producing targeted advertising. **Clickstream data** is the data gathered based on the Web sites you visit and the items you click. From the marketers' perspective, the use of online profiling allows businesses to market to customers electronically one to one. The benefit to customers is personalized, more effective service; the benefit to providers is the increased business that comes from building relationships and encouraging customers to return for subsequent purchases. However, what may be considered as one person's relevant ad can be viewed by others as a manipulative and potentially harmful marketing technique. For example, the Center for Digital Democracy and the U.S. Public Interest Research Group accused General Mills, MasterFoods USA, and Pepsi of targeting youths with online ads that contributed to their obesity.[46]

clickstream data
The data gathered based on the Web sites you visit and the items you click.

Lack of Internet Access

The *digital divide* is a term that describes the difference between people who do and do not have the access or the capability to use high-quality, modern information and communications technology such as computers, the Internet, telephone, and television to improve their standard of living. For example, it is estimated that of the roughly 1 billion Internet users worldwide, only 20 million (2 percent) are in the less-developed nations. The lack of universal Internet access makes it impossible to conduct e-commerce with many of the world's people. The digital divide exists not only between more and less developed countries but within countries between economic classes, the educated and uneducated, and those who live in cities and those who live in rural areas. Obviously, those who lack Internet access form a barrier to further e-commerce expansion.

Return on Investment

Often the investment required for a large firm to establish and operate a B2B or B2C Web site can be in the millions of dollars. For example, Starwood Hotels and Resorts Worldwide plans to move its 700 hotels to a new Web-based reservation system at an estimated cost of between $10 million and $60 million with annual cost savings of $15 million. Using the low-cost estimate, the payback period is $10 million/$15 million or .67 years, while the high-cost estimate yields a payback period of four years, not nearly as economically attractive. The example illustrates a common problem with determining return on investment—it is difficult to forecast project costs and benefits.

Legal Jurisdiction

Companies engaging in e-commerce must be careful that their sales do not violate the rules of various county, state, or country legal jurisdictions. For example, New York and six other states forbid the possession of stun guns and similar devices. The New York state attorney general received a tip that such guns were being sold to New Yorkers through eBay. A subsequent investigation led to the arrest of 16 sellers allegedly responsible for the sale of more than 1,100 stun guns and Tasers. Other examples of illegal sales are sales to those who would not be able to obtain cigarettes or wine because of their age.

Taxation

United States businesses and consumers must be aware of taxation issues when conducting e-commerce. Based on U.S. Supreme Court rulings (Quill Corp. vs. North Dakota), Internet-based merchants must apply sales tax only when buyers live in a state where the company has physical facilities, or "nexus." Most businesses want to avoid the complexity of dealing with the nonstandard rules of the more than 7,500 taxing districts nationwide. To avoid this complexity of paying sales taxes, businesses set up their Internet sales operations as legally separate companies with no physical presence outside of where their computers and warehouses are located. This leaves it up to the consumers to voluntarily remit the sales taxes; but because it is almost impossible to enforce this practice, few people pay them. Thus, despite having a legal basis to do so, the states find it very difficult to collect sales taxes on Internet purchases. Total e-commerce B2C sales were estimated to be about $136 billion in 2007 according to the U.S. Census statistics.[47] An average sales tax rate of 6% yields an estimate of $8 billion in lost state and local sales tax revenue.

STRATEGIES FOR SUCCESSFUL E-COMMERCE AND M-COMMERCE

With all the constraints to e-commerce just covered, a company must develop an effective Web site—one that is easy to use and accomplishes the goals of the company, yet is safe, secure, and affordable to set up and maintain. The next sections examine several issues for a successful e-commerce site.

Defining the Web Site Functions

When building a Web site, you should first decide which tasks the site must accomplish. Most people agree that an effective Web site is one that creates an attractive presence and that meets the needs of its visitors, including the following:

- Obtaining general information about the organization
- Obtaining financial information for making an investment decision in the organization
- Learning the organization's position on social issues
- Learning about the products or services that the organization sells
- Buying the products or services that the company offers
- Checking the status of an order
- Getting advice or help on effective use of the products
- Registering a complaint about the organization's products
- Registering a complaint concerning the organization's position on social issues
- Providing a product testimonial or idea for a product improvement or new product
- Obtaining information about warranties or service and repair policies for products
- Obtaining contact information for a person or department in the organization

After a company determines which objectives its site should accomplish, it can proceed to the details of actually developing a site.

As the number of e-commerce shoppers increases and they become more comfortable—and more selective—making online purchases, you might need to redefine your site's basic business model to capture new business opportunities. For example, consider the major travel sites such as Expedia, Travelocity, CheapTickets, Orbitz, and Priceline. These sites used to specialize in one area of travel—inexpensive airline tickets. Now they offer a full range of travel products, including airline tickets, auto rentals, hotel rooms, tours, and last-minute trip packages. Expedia provides in-depth hotel details to help comparison shoppers and even offers 360-degree visual tours and expanded photo displays. It also entices flexible travelers to search for rates, compare airfares, and configure hotel and air prices at the same time. Expedia has developed numerous hotel partnerships to reduce costs and help secure great values for consumers. Meanwhile, Orbitz has launched a special full-service program for corporate business travelers.

Establishing a Web Site

Companies large and small can establish Web sites. Some companies elect to develop their sites in house, but this requires learning the intricacies of HTML, Java, and Web design software. Many firms, especially those with few or no experienced Web developers, have decided that to outsource the building of their Web site and get the Web site up and running is faster and cheaper than doing it themselves.

Web site hosting companies such as HostWay and BroadSpire make it possible to set up a Web page and conduct e-commerce within a matter of days and with little up-front cost.

These companies can also provide free hosting for your store, but to allow visitors to pay for merchandise with credit cards, you need a merchant account with a bank. If your company doesn't already have one, it must establish one. Table 8.1 lists some corporate customers for HostWay and BroadSpire.

Table 8.1

Customers of Web Site Hosting Companies

HostWay Customers	BroadSpire Customers
Sony BMG Music	British Petroleum
Coca-Cola Company	CNBC
McGraw-Hill	Sheraton
Bank of Montreal	Kmart
Hershey's Food	Symantec
Campbell Soup Company	Pardee Homes
Walt Disney Company	BrightHand
Infinity Broadcasting	CD Warehouse
FOX News	GWI Electric

Macronimous.com is a venture of AES Technologies (India) with 110 employees who design and implement Web sites for some 170 clients around the world. Their design process includes independent market research to understand exactly what sort of online offering their clients' customers desire. Based on this research, they build a Web site with their clients' continual input and feedback.[48]

Web site designer Corporate Communications built a Global Diversity and Inclusion Web site for Eastman Kodak to address the company's policy on embracing diversity to an external audience including potential employees, customers, and suppliers.[49]

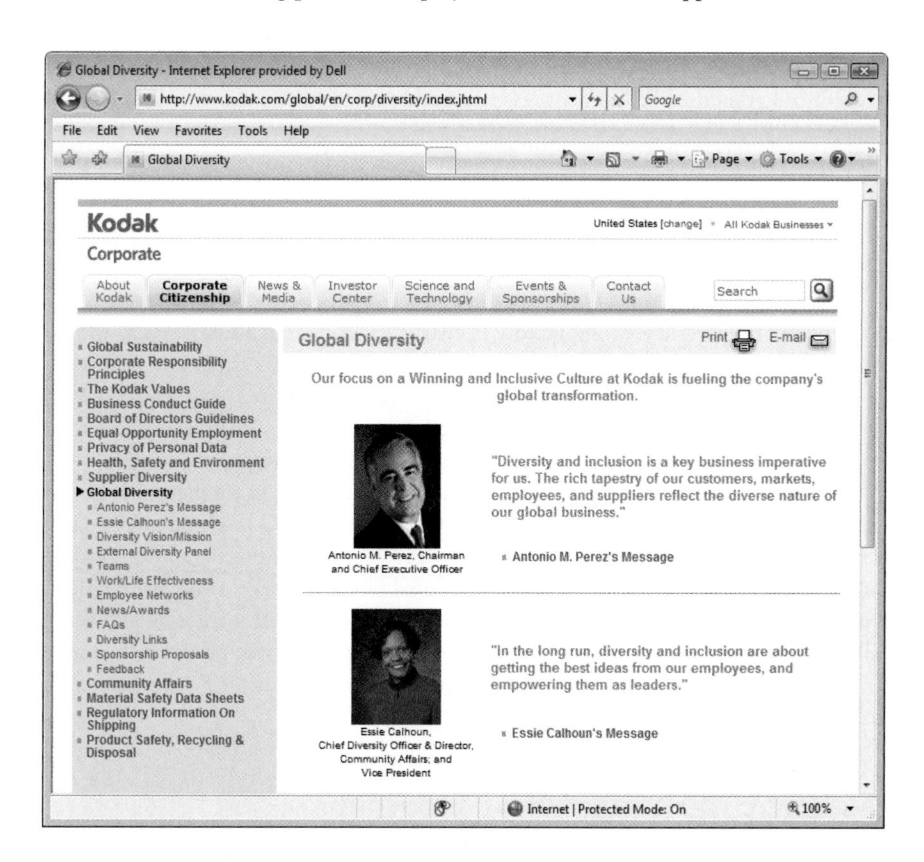

Eastman Kodak's Global Diversity and Inclusion Web site addresses the company's policy on embracing diversity.

storefront broker
A company that acts as an intermediary between your Web site and online merchants who have the products and retail expertise.

Another model for setting up a Web site is the use of a **storefront broker**, which serves as an intermediary between your Web site and online merchants who have the actual products and retail expertise. The storefront broker deals with the details of the transactions, including who gets paid for what, and is responsible for bringing together merchants and reseller sites. The storefront broker is similar to a distributor in standard retail operations, but in this case no product moves—only electronic data flows back and forth. Products are ordered by a customer at your site, orders are processed through a user interface provided by the storefront broker, and the product is shipped by the merchant.

Building Traffic to Your Web Site

The Internet includes hundreds of thousands of e-commerce Web sites. With all those potential competitors, a company must take strong measures to ensure that the customers it wants to attract can find its Web site. The first step is to obtain and register a domain name, and your domain name should say something about your business. For instance, stuff4u might seem to be a good catchall, but it doesn't describe the nature of the business—it could be anything. If you want to sell soccer uniforms and equipment, then you'd try to get a domain name such as *www.soccerstuff4u.com*, *www.soccerequipment.com*, or *www.stuff4soccercoaches.com*. The more specific the Web address, the better.

The next step to attracting customers is to make your site search-engine friendly by improving its rankings. Following are several ideas on how to do this.

meta tag
A special HTML tag, not visible on the displayed Web page, that contains keywords representing your site's content, which search engines to use to build indexes pointing to your Web site.

- Include a meta tag in your store's home page. A **meta tag** is a special HTML tag, not visible on the displayed Web page, that contains keywords representing your site's content, which search engines to use to build indexes pointing to your Web site. Again, selecting keywords is critical to attracting customers, so they should be chosen carefully.
- Use Web site traffic data analysis software to turn the data captured in the Web log file into useful information. This data can tell you the URLs from which your site is being accessed, the search engines and keywords that find your site, and other useful information. Using this data can help you identify search engines to which you need to market your Web site, allowing you to submit your Web pages to them for inclusion in the search engine's index.
- Provide quality, keyword-rich content. Be careful not to use too many keywords as this can get you banned from the search engines. Judiciously place keywords throughout your site ensuring that the Web content is sensible and easy to read by humans as well as search engines.
- Add new content to the Web site on a regular basis. Again, this makes the site attractive to humans as well as search engines.
- Acquire links to your site from other reputable Web sites that are popular and actually related to your Web site. Avoid the use of low-quality links as they can actually hurt your Web site's rating.

The use of the Internet is growing rapidly in markets throughout Europe, Asia, and Latin America. Obviously, companies that want to succeed on the Web cannot ignore this global shift. Companies must be aware that consumers outside the United States will access sites with different devices and modify their site design accordingly. In Europe, for example, closed-system iDTVs (integrated digital televisions) are becoming popular for accessing online content, with some 80 million European households now using them. Because such devices have better resolution and more screen space than the PC monitors that U.S. consumers use to access the Internet, iDTV users expect more ambitious graphics. Successful global firms operate with a portfolio of sites designed for each market, with shared sourcing and infrastructure to support the network of stores, and with local marketing and business development teams to take advantage of local opportunities. Service providers continue to emerge to solve the cross-border logistics, payments, and customer service needs of these global retailers.

Maintaining and Improving Your Web Site

Web site operators must constantly monitor the traffic to their site and the response times experienced by visitors. Internet shoppers expect service to be better than or equal to their in-store experience, says AMR Research (a Boston-based, independent research analysis firm).

Nothing will drive potential customers away faster than if they experience unbearable delays while trying to view or order your products or services. To keep pace with technology and increasing traffic, it might be necessary over time to modify the software, databases, or hardware on which the Web site runs to ensure acceptable response times.

Web site operators must also continually be alert to new trends and developments in the area of e-commerce and be prepared to take advantage of new opportunities. For example, recent studies show that customers more frequently visit Web sites they can customize. **Personalization** is the process of tailoring Web pages to specifically target individual consumers. The goal is to meet the customer's needs more effectively, make interactions faster and easier, and, consequently, increase customer satisfaction and the likelihood of repeat visits. Building a better understanding of customer preferences also can aid in cross-selling related products and more expensive products. The most basic form of personalization involves using the consumer's name in an e-mail campaign or in a greeting on the Web page. Amazon uses a more advanced form of personalization, in which each repeat customer is greeted by name, and a list of new products is recommended based on the customer's previous purchases.

personalization
The process of tailoring Web pages to specifically target individual consumers.

Businesses use two types of personalization techniques to capture data and build customer profiles. *Implicit personalization* techniques capture data from actual customer Web sessions—primarily based on which pages were viewed and which weren't. *Explicit personalization* techniques capture user-provided information, such as information from warranties, surveys, user registrations, and contest entry forms completed online. Data can also be gathered through access to other data sources such as the Bureau of Motor Vehicles, Bureau of Vital Statistics, and marketing affiliates (firms that share marketing data). Marketing firms aggregate this information to build databases containing a huge amount of consumer behavioral data. During each customer interaction, powerful algorithms analyze both types of data in real time to predict the consumer's needs and interests. This analysis makes it possible to deliver new, targeted information before the customer leaves the site. Because personalization depends on gathering and using personal user information, privacy issues are a major concern.

These tips and suggestions are only a few ideas that can help a company set up and maintain an effective e-commerce site. With technology and competition changing constantly, managers should read articles in print and on the Web to keep up to date on ever-evolving issues.

Now that we've examined how to establish e-commerce effectively, let's look at some of the technical issues related to e-commerce systems and technology that make it possible.

TECHNOLOGY INFRASTRUCTURE REQUIRED TO SUPPORT E-COMMERCE AND M-COMMERCE

Successful implementation of e-business requires significant changes to existing business processes and substantial investment in IS technology. These technology components must be chosen carefully and integrated to support a large volume of transactions with customers, suppliers, and other business partners worldwide. Online consumers complain that poor Web site performance (e.g., slow response time, inadequate customer support, and lost orders) drives them to abandon some e-commerce sites in favor of those with better, more reliable performance. This section provides a brief overview of the key technology infrastructure components (see Figure 8.7).

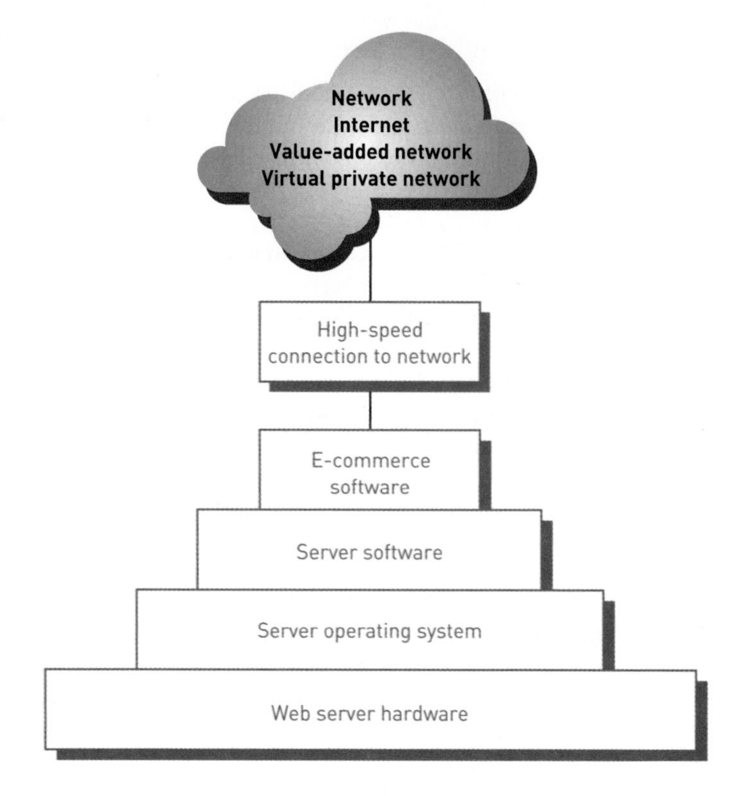

Figure 8.7

Key Technology Infrastructure Components

Hardware

A Web server hardware platform complete with the appropriate software is a key e-commerce infrastructure ingredient. The amount of storage capacity and computing power required of the Web server depends primarily on two things: the software that must run on the server and the volume of e-commerce transactions that must be processed. Although IS staff can sometimes define the software to be used, they can only estimate how much traffic the site will generate. As a result, the most successful e-commerce solutions are designed to be highly scalable so that they can be upgraded to meet unexpected user traffic.

A key decision facing new e-commerce companies is whether to host their own Web site or to let someone else do it. Many companies decide that using a third-party Web service provider is the best way to meet initial e-commerce needs. The third-party company rents space on its computer system and provides a high-speed connection to the Internet, which minimizes the initial out-of-pocket costs for e-commerce start-up. The third party can also provide personnel trained to operate, troubleshoot, and manage the Web server. Of course, many companies decide to take full responsibility for acquiring, operating, and supporting the Web server hardware and software themselves, but this approach requires considerable up-front capital and a set of skilled and trained workers. No matter which approach a company takes, it must have adequate hardware backup to avoid a major business disruption in case of a failure of the primary Web server.

Web Server Software

In addition to the Web server operating system, each e-commerce Web site must have Web server software to perform fundamental services, including security and identification, retrieval and sending of Web pages, Web site tracking, Web site development, and Web page development. The two most popular Web server software packages are Apache HTTP Server and Microsoft Internet Information Services.

Security and Identification

Security and identification services are essential for intranet Web servers to identify and verify employees accessing the system from the Internet. Access controls provide or deny access to files based on the username or URL. Web servers support encryption processes for transmitting private information securely over the public Internet.

Retrieving and Sending Web Pages

The fundamental purpose of a Web server is to process and respond to client requests that are sent using HTTP. In response to such a request, the Web server program locates and fetches the appropriate Web page, creates an HTTP header, and appends the HTML document to it. For dynamic pages, the server involves other programs, retrieves the results from the back-end process, formats the response, and sends the pages and other objects to the requesting client program.

Web Site Tracking

Web servers capture visitors' information, including who is visiting the Web site (the visitor's IP address), what search engines and keywords they used to find the site, how long their Web browser viewed the site, the date and time of each visit, and which pages were displayed. This data is placed into a Web log file for future analysis.

Web Site Development

Web site development tools include features such as an HTML/visual Web page editor (e.g., Microsoft Expression Web, Adobe Dreamweaver, NetStudio Easy Web Graphics, and SoftQuad HoTMetaL Pro), software development kits that include sample code and code development instructions for languages such as Java or Visual Basic, and Web page upload support to move Web pages from a development PC to the Web site. The tools bundled with the Web server software depend on which Web server software you select.

Web site development tools
Tools used to develop a Web site, including HTML or visual Web page editor, software development kits, and Web page upload support.

Web Page Construction

Web page construction software uses HTML editors and extensions to produce Web pages—either static or dynamic. **Static Web pages** always contain the same information—for example, a page that provides text about the history of the company or a photo of corporate headquarters. **Dynamic Web pages** contain variable information and are built to respond to a specific Web site visitor's request. For example, if a Web site visitor inquires about the availability of a certain product by entering a product identification number, the Web server searches the product inventory database and generates a dynamic Web page based on the current product information it found, thus fulfilling the visitor's request. This same request by another visitor later in the day might yield different results due to ongoing changes in product inventory. A server that handles dynamic content must be able to access information from a variety of databases. The use of open database connectivity enables the Web server to assemble information from different database management systems, such as SQL Server, Oracle, and Informix.

Web page construction software
Software that uses Web editors and extensions to produce both static and dynamic Web pages.

static Web pages
Web pages that always contain the same information.

dynamic Web pages
Web pages containing variable information that are built to respond to a specific Web visitor's request.

E-Commerce Software

After you have located or built a host server, including the hardware, operating system, and Web server software, you can begin to investigate and install e-commerce software. E-commerce software must support five core tasks: catalog management, product configuration, shopping cart facilities, e-commerce transaction processing, and Web traffic data analysis.

The specific e-commerce software you choose to purchase or install depends on whether you are setting up for B2B or B2C transactions. For example, B2B transactions do not include sales tax calculations if they involve items purchased for resale, and software to support B2B must incorporate electronic data transfers between business partners, such as purchase orders, shipping notices, and invoices. B2C software, on the other hand, must handle the complication of accounting for sales tax based on the current state laws and rules. However, it does not need to support negotiation between buyer and seller.

Catalog Management

Any company that offers a wide range of products requires a real-time interactive catalog to deliver customized content to a user's screen. *Catalog management software* combines different product data formats into a standard format for uniform viewing, aggregating, and integrating catalog data. It also provides a central repository for easy access, retrieval, and updating of pricing and availability changes. The data required to support large catalogs is almost always stored in a database on a computer that is separate from, but accessible to, the e-commerce server machine.

Product Configuration

Customers need help when an item they are purchasing has many components and options. *Product configuration software* tools were originally developed in the 1980s to assist B2B salespeople to match their company's products to customer needs. Buyers use the new Web-based product configuration software to build the product they need online with little or no help from salespeople. For example, Dell customers use product configuration software to build the computer that meets their needs. Such software is also used in the service arena to help people decide what sort of consumer loan or insurance is best for them.

Shopping Cart

Today many e-commerce sites use an *electronic shopping cart* to track the items selected for purchase, allowing shoppers to view what is in their cart, add new items to it, or remove items from it, as shown in Figure 8.8. To order an item, shoppers simply click an item. All the details about it—including its price, product number, and other identifying information—are stored automatically. If shoppers later decide to remove one or more items from the cart, they can view the cart's contents and remove any unwanted items. When shoppers are ready to pay for the items, they click a button (usually labeled "proceed to checkout") and begin a purchase transaction. Clicking the Checkout button opens another window that usually asks shoppers to fill out billing, shipping, and payment method information and to confirm the order.

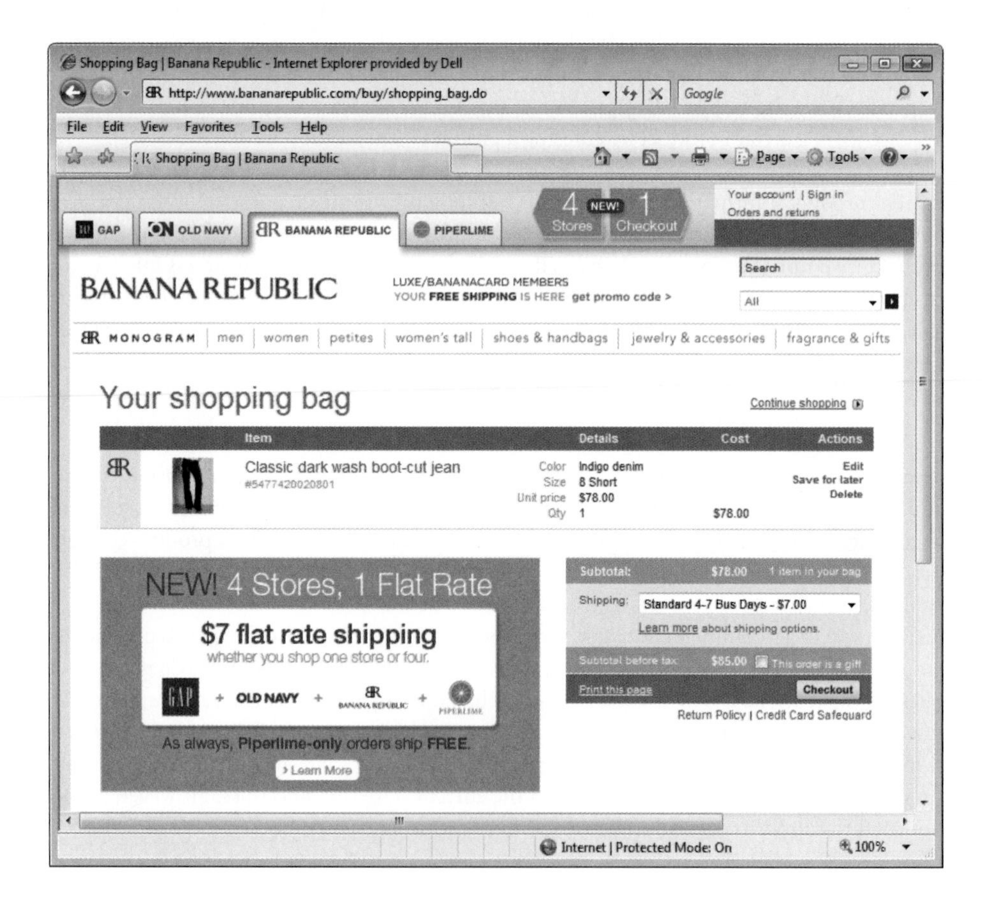

Figure 8.8

Electronic Shopping Cart

An electronic shopping cart (or bag) allows online shoppers to view their selections and add or remove items.

Web Services

Web services are software modules supporting specific business processes that users can interact with over a network (such as the Internet) as necessary. Web services can combine software and services from different companies to provide an integrated way to communicate. For example, an organization could use a supplier-provided Web service to streamline the payment of vendor invoices. The Web service could be developed so that when the user moves the mouse over a purchase order number in an e-mail from the supplier, the amount of funds remaining in the purchase order are displayed. The user can then approve payment by clicking a button or link.

Web services
Software modules supporting specific business processes that users can interact with over a network (such as the Internet) on an as-needed basis.

Software manufacturers are scrambling to meet customer demands by offering software applications for use over the Web as services supported by advertising or subscription fees. SAP, for example, offers more than 500 components that run as Web services to support business functions such as finance, human resources, logistics, manufacturing, procurement, and product development. Dun & Bradstreet provides an address verification service called GlobalAccess that checks and completes the addresses of prospects or customers to ensure the accuracy and completeness of this key information. Oanda.com offers a currency exchange rate service that downloads the most current set of rates to support the running of accounting processes such as consolidation reporting that require the translation of multicurrency transactions into a single corporate currency. UPS provides a shipment tracking service for determining the cost, current location, and the receiving party of a specific package to enhance the order fulfillment process and provide shippers with greater visibility into the shipping process.[50] In addition to these strategies for increasing positive PR for an organization's Web site, it is also necessary to minimize negative PR, as discussed in the Ethical and Societal Issues special feature.

ETHICAL AND SOCIETAL ISSUES

Manipulating Cyberstatus

The Web provides an uncensored platform for public opinion. Blogs, social networks, and other Web tools allow you to speak your mind on any issue. Web sites such as complaints.com and Ripoff Report provide an easy means to express dissatisfaction with products and services. While this is generally viewed as a positive aspect of the Internet, it has proved a challenge for businesses looking to control their image. A disgruntled employee, customer, or even a business competitor can plant seeds of discontent on the Web that can quickly grow into a serious problem for a business's reputation. Cyberstatus is an important factor in the success of a product or business.

Consider lock manufacturer Kryptonite, a company that has a good reputation for manufacturing high-end bicycle locks. That is, until their cyberstatus was knocked down several notches by a blogger who revealed a secret for cracking an expensive Kryptonite lock in seconds using a ballpoint pen. Within days the word had spread to thousands of cyclists through online news services and Kryptonite's entire business was in jeopardy. The company quickly created a lock exchange program that replaced over 400,000 locks in 21 countries for free. Kryptonite ultimately was forced to redesign nine years worth of locks in ten months to save its reputation.

Over the past few years, businesses have started using strategies to control their cyberstatus. Controlling cyberstatus, however, is like controlling radio signals—you can't prevent them from being broadcast, but you can tune out the signals you don't want, and tune in the signal you do want to hear. Tuning out harmful online publicity for a business and focusing on good publicity is a practice referred to as Online Reputation Management. Online Reputation Management focuses on Search Engine Optimization (SEO)—controlling the Web sites listed on the first page of Web search engine results and using Web monitoring to track what is being said about a company or product on blogs, forums, podcasts, and comments on Web sites.

SEO is used to bump negative publicity such as customer complaints out of top search results and load positive publicity in its place, such as positive customer testimonials. Many consider SEO to be "gaming the system." SEO service providers learn how the systems used by search giants such as Google, Yahoo!, and MSN work and alter Web content to control search results. SEO companies conduct metadata analysis of a corporate Web site and make suggestions about content and keyword tags to help a product or company rise to the top of the search engine results. Without the technology, the product might have been listed hundreds of entries down in search results, where the public would never see it.

In a more ethically questionable technique called "Google Bombing," a number of bogus links to a particular Web site or sites are created to make it appear more popular than it really is, assuring it a higher ranking in Google. Google Bombing is often used to attack a product, company, or person. For example, in 2008 a Google Bomb was launched against then presidential candidate John McCain. A political blogger encouraged fellow partisan bloggers to link to nine negative newspaper articles about McCain. The thousands of resulting links caused the negative articles to rise to the top of search results when someone searched for the name John McCain using the Google search engine.

Another technique in Online Reputation Management is to monitor the Web for negative commentary and work to block it. Online reputation management firms invest significant resources in continuously monitoring the Web through a number of techniques. Employees might be hired to monitor popular social networks such as MySpace, Facebook, YouTube, and Twitter, watching for positive and negative commentary on a client. Automated natural language processing software may be employed to watch for product and company names that come up through search engine results. London-based Reputica Ltd provides its clients with an online dashboard that displays the range of positive and negative content about the company on the Web in the form of graphs and statistics.

The faster a company can react to negative publicity, the more it can control the damage. Often, an Online Reputation Management company directly contacts the person posting the negative comments and works to convince that person to voluntarily remove the content. In other cases, the company might be able to threaten a law suit. If the company has no leverage with which to convince the person to remove the content, SEO is used to keep the negative comments from the public.

The act of manipulating search engine results and exercising control over Web content is not illegal. However, many question the ethics of the practice. Internet purists feel that the Internet, being an uncensored public network, should provide an accurate account of true public sentiment. Businesses looking to control their message to the public to maximize profits and competitive advantage feel that they should be able to work within the legal system to benefit their shareholders.

Discussion Questions

1. Should businesses be allowed to manipulate search engine results?
2. Is the practice of eradicating negative publicity on the Web ethically sound? What methods do you endorse?

Critical Thinking Questions

1. What might the search engine companies do to minimize SEO techniques?
2. What forms of Web 2.0 technologies (blogs, social networks, wikis) are less vulnerable to corporate manipulation? Why?

SOURCES: Hoffman, Thomas, "Online reputation management is hot—but is it ethical?" *Computerworld*, February 12, 2008, *www.computerworld.com/action/article.do?command=viewArticleBasic&taxonomyName=Internet_Business&articleId=9060960&taxonomyId=71&intsrc=kc_li_story*; Havenstein, Heather, "Blogger launches 'Google bomb' at McCain," *Computerworld*, June 19, 2008, *www.computerworld.com/action/article.do?command=viewArticleBasic&articleId=91012 18&intsrc=news_ts_head*; Warner, Bernhard, "How to be unGoogleable," TimesOnline, May 28, 2008, *http://technology.timesonline.co.uk/tol/news/tech_and_web/the_web/article4022374.ece* .

Technology Needed for Mobile Commerce

For m-commerce to work effectively, the interface between the wireless device and its user must improve to the point that it is nearly as easy to purchase an item on a wireless device as it is to purchase it on a PC. In addition, network speed must improve so that users do not become frustrated. Security is also a major concern, particularly in two areas: the security of the transmission itself and the trust that the transaction is being made with the intended party. Encryption can provide secure transmission. Digital certificates, discussed later in this chapter, can ensure that transactions are made between the intended parties.

The handheld devices used for m-commerce have several limitations that complicate their use. Their screens are small, perhaps no more than a few square inches, and might be able to display only a few lines of text. Their input capabilities are limited to a few buttons, so entering data can be tedious and error prone. They also have less processing power and less bandwidth than desktop computers, which are usually hardwired to a high-speed LAN. They also operate on limited-life batteries. For these reasons, it is currently impossible to directly access many Web sites with a handheld device. Web developers must rewrite Web applications so that users with handheld devices can access them.

To address the limitations of wireless devices, the industry has undertaken a standardization effort for their Internet communications. The Wireless Application Protocol (WAP) is a standard set of specifications for Internet applications that run on handheld, wireless devices. It effectively serves as a Web browser for such devices. WAP is a key underlying technology of m-commerce that is supported by an entire industry association of over 200 vendors of wireless devices, services, and tools. In the future, devices and service systems based on WAP and its derivatives (including WAP 2.0 and Wireless Internet Protocol) will be able to interoperate. Japan's largest wireless network provider, DoCoMo, developed a competing standard called the i-mode system.

For equipment and service providers, the existence of competing standards makes it much more difficult to meet the needs of their customers. In many cases, the providers must develop their services or products based on one standard and forfeit the market for customers who elect to adopt the competing standard. Of course, multiple standards also create problems for customers who must make a decision on which set of services and equipment to adopt. Early adopters may find to their dismay that they have chosen a standard that falls out of favor.

WAP uses the Wireless Markup Language (WML), which is designed for effectively displaying information on small devices. A user with a WAP-compliant device uses the built-in microbrowser to make a WML request. The request is forwarded to a special WAP gateway to fetch the information from the appropriate Internet server. If the information is already in WML format, it can be passed from the Internet server through the gateway directly to the user's device. If the information is in HTML format, the gateway translates the HTML content into WML so it can be displayed on the user's device.

Electronic Payment Systems

Electronic payment systems are a key component of the e-commerce infrastructure. Current e-commerce technology relies on user identification and encryption to safeguard business transactions. Actual payments are made in a variety of ways, including electronic cash, electronic wallets, and smart, credit, charge, and debit cards. Web sites that accept multiple payment types convert more visitors to purchasing customers than merchants who offer only a single payment method.

Authentication technologies are used by many organizations to confirm the identity of a user requesting access to information or assets. A **digital certificate** is an attachment to an e-mail message or data embedded in a Web site that verifies the identity of a sender or Web site. A **certificate authority** (CA) is a trusted third-party organization or company that issues digital certificates. The CA is responsible for guaranteeing that the people or organizations granted these unique certificates are, in fact, who they claim to be. Digital certificates thus create a trust chain throughout the transaction, verifying both purchaser and supplier identities.

digital certificate
An attachment to an e-mail message or data embedded in a Web site that verifies the identity of a sender or Web site.

certificate authority (CA)
A trusted third-party organization or company that issues digital certificates.

Secure Sockets Layer

Secure Sockets Layer (SSL)
A communications protocol is used to secure sensitive data during e-commerce.

All online shoppers fear the theft of credit card numbers and banking information. To help prevent this type of identity theft, the **Secure Sockets Layer (SSL)** communications protocol is used to secure sensitive data. The SSL communications protocol includes a handshake stage, which authenticates the server (and the client, if needed), determines the encryption and hashing algorithms to be used, and exchanges encryption keys. Following the handshake stage, data might be transferred. The data is always encrypted, ensuring that your transactions are not subject to interception or "sniffing" by a third party. Although SSL handles the encryption part of a secure e-commerce transaction, a digital certificate is necessary to provide server identification.

Electronic Cash

electronic cash
An amount of money that is computerized, stored, and used as cash for e-commerce transactions.

Electronic cash is an amount of money that is computerized, stored, and used as cash for e-commerce transactions. Typically, consumers must open an account with an electronic cash service provider by providing identification information. When the consumers want to withdraw electronic cash to make a purchase, they access the service provider via the Internet and present proof of identity—a digital certificate issued by a certification authority or a username and password. After verifying a consumer's identity, the system debits the consumer's account and credits the seller's account with the amount of the purchase. PayPal, BillMeLater, MoneyZap, and TeleCheck are four popular forms of electronic cash.

The SSL communications protocol assures customers that information they provide to retailers, such as credit card numbers, cannot be viewed by anyone else on the Web.

The PayPal service of eBay enables any person or business with an e-mail address to securely, easily, and quickly send and receive payments online. To send money, you enter the recipient's e-mail address and the amount you want to send. You can pay with a credit card, debit card, or funds from a checking account. The recipient gets an e-mail that says, "You've Got Cash!" Recipients can then collect their money by clicking a link in the e-mail that takes them to *www.paypal.com*. To receive the money, the user also must have a credit card or checking account to accept fund transfers. To request money for an auction, invoice a customer, or send a personal bill, you enter the recipient's e-mail address and the amount you are requesting. The recipient gets an e-mail and instructions on how to pay you using PayPal. PayPal serves more than 60 million active accounts worldwide. It is available in 190 markets and processes payments in 17 currencies around the world. [51]

Bill Me Later by I4 Commerce is for customers who do not have a credit card or prefer not to use a credit card online. To make a purchase, an existing account owner provides basic information, such as the last four digits of a Social Security number and date of birth. Within seconds, Bill Me Later qualifies the customer, completes the purchase, and sends a bill. The customer can pay the cost in full or finance the purchase over time. Bill Me Later is currently available at over 750 leading stores, including Apple, Champs Sports, FTD.com, JetBlue, Overstock, Reebok, ToysRUs, and Walmart.com, with more stores expected to participate.[52]

MoneyZap is a service offered by Western Union that enables consumers and businesses to pay retailers from an existing checking account. They must complete a one-time registration and provide their name, address, checking account information, Social Security number, e-mail address, and home phone number. After successfully registering, they can authorize payments using their username and password. An electronic funds transfer is initiated to debit money from their account and transfer money to the merchant.[53]

The TeleCheck Electronic Check Acceptance Verification service from First Data offers merchants a safe option for accepting and processing checks at the point-of-sale that avoids the high bank service fees associated with credit cards. When a customer presents the merchant with a paper check, the merchant uses the service to perform a risk assessment using check writer negative and activity databases to assess the risk of accepting the check. If the check passes the verification criteria, the TeleCheck ECA Verification service converts the paper check into an electronic transaction at the point-of-sale. The Automated Clearing House (ACH) network is used to process the transaction, and funds are deposited directly into the merchant's bank account within two business days.[54]

According to the International Air Transport Association (IATA), the airline industry earned a slim profit margin of $5.6 billion (1.1 percent) on sales of $490 billion in 2007. Over 80 percent of passengers pay for their tickets using credit cards, whose fees cost the airlines $1.5 billion. In an attempt to improve their profit margin, the airline Web sites now offer a variety of lower-fee payment options including PayPal, Bill Me Later, MoneyZap, and TeleCheck.[55]

Credit, Charge, Debit, and Smart Cards

Many online shoppers use credit and charge cards for most of their Internet purchases. A credit card, such as Visa or MasterCard, has a preset spending limit based on the user's credit history, and each month the user can pay part or all of the amount owed. Interest is charged on the unpaid amount. A charge card, such as American Express, carries no preset spending limit, and the entire amount charged to the card is due at the end of the billing period. Charge cards do not involve lines of credit and do not accumulate interest charges. American Express became the first company to offer disposable credit card numbers in 2000. Other banks, such as Citibank, protect the consumer by providing a unique number for each transaction. Debit cards look like credit cards or automated teller machine (ATM) cards, but they operate like cash or a personal check. Credit, charge, and debit cards currently store limited information about you on a magnetic strip. This information is read each time the card is swiped to make a purchase. All credit card customers are protected by law from paying more than $50 for fraudulent transactions.

The **smart card** is a credit card–sized device with an embedded microchip to provide electronic memory and processing capability. Smart cards can be used for a variety of purposes, including storing a user's financial facts, health insurance data, credit card numbers, and network identification codes and passwords. They can also store monetary values for spending.

smart card
A credit card–sized device with an embedded microchip to provide electronic memory and processing capability.

Smart cards are better protected from misuse than conventional credit, charge, and debit cards because the smart-card information is encrypted. Conventional credit, charge, and debit cards clearly show your account number on the face of the card. The card number, along with a forged signature, is all that a thief needs to purchase items and charge them against your card. A smart card makes credit theft practically impossible because a key to unlock the encrypted information is required, and there is no external number that a thief can identify and no physical signature a thief can forge.

The smart card connects to a reader with direct physical contact or via remote contactless radio frequency interface. Smart cards have been around for over a decade and are widely used in Europe, Australia, and Japan. UK credit card giant Barclaycard is conducting a pilot test of contactless retail and transit payment using mobile phones that support near field communications (NFC).[56] Smart card use has not caught on in the United States because there are few smart-card readers to record payments and U.S. banking regulations have slowed smart-card marketing and acceptance as well. Table 8.2 compares various types of payment systems.

Table 8.2

Comparison of Payment Systems

Payment System	Description	Advantages	Disadvantages
Credit card	Carries preset spending limit based on the user's credit history.	Each month the user can pay part or all of the amount owed.	Unpaid balance accumulates interest charges—often at a high rate of interest.
Charge card	Looks like a credit card but carries no preset spending limit.	Charge cards do not involve lines of credit and do not accumulate interest charges.	The entire amount charged to the card is due at the end of the billing period.
Debit card	Look like a credit card or automated teller machine (ATM) card.	Operates like cash or a personal check	Money is immediately deducted from user's account balance.
Smart card	Credit card device with embedded microchip capable of storing facts about card holder	Better protected from misuse than conventional credit, charge, and debit cards because the smart-card information is encrypted	Not widely used in the U.S.

Payments Using Cell Phones

The retail and banking industries are keenly interested in using a cell phone like a credit card by waving the end of the phone near a scanner device to pay for purchases. Some people believe that mobile device-based transactions will exceed card-based transactions.

U.S. Bank began testing the concept of a credit card "buried" inside a cell phone and the use of no contact scanners. When two NFC devices (the scanner and cell phone) come within about three inches of each other, they can exchange data using radio signals including encrypted credit card account numbers. The U.S. Bank pilot supports the use of only one credit card; however, if successful, banks and wireless service providers may allow customers to load their "tap and go" phone with multiple credit cards or merchant reward cards. Says Dominic Venturo, the bank vice president helping to manage the U.S. Bank pilot test: "Anytime you can combine the phone, which most of us have in our wallet, with the bank payment card many of us carry in our wallets, into a single system, you've created a simpler and easier way for your customers to manage their lives."[57] Japan, Australia, and Korea are also experimenting with "tap and go" phones.

SUMMARY

Principle

Electronic commerce and mobile commerce are evolving, providing new ways of conducting business that present both opportunities for improvement and potential problems.

E-commerce is the conducting of business activities electronically over networks. Business-to-business (B2B) e-commerce allows manufacturers to buy at a low cost worldwide, and it offers enterprises the chance to sell to a global market. B2B e-commerce is currently the largest type of e-commerce. Business-to-consumer (B2C) e-commerce enables organizations to sell directly to consumers, eliminating intermediaries. In many cases, this squeezes costs and inefficiencies out of the supply chain and can lead to higher profits and lower prices for consumers. Consumer-to-consumer (C2C) e-commerce involves consumers selling directly to other consumers. Online auctions are the chief method by which C2C e-commerce is currently conducted. e-Government involves the use of information and communications technology to simplify the sharing of information, speed formerly paper-based processes, and improve the relationship between citizens and government.

A successful e-commerce system must address the many stages consumers experience in the sales life cycle. At the heart of any e-commerce system is the ability of the user to search for and identify items for sale; select those items; negotiate prices, terms of payment, and delivery date; send an order to the vendor to purchase the items; pay for the product or service; obtain product delivery; and receive after-sales support.

Looking at things from the perspective of the provider of goods and/or services, an effective e-commerce system must be able to support the activities associated with supply chain management and customer relationship management.

A firm must overcome three key challenges to convert its business processes from the traditional form to e-commerce processes: 1) it must define an effective e-commerce model and strategy, 2) it must deal effectively with consumer privacy concerns, and 3) it must successfully overcome consumers' lack of trust.

Mobile commerce is the use of wireless devices such as PDAs, cell phones, and smartphones to facilitate the sale of goods or services—anytime, anywhere. The market for m-commerce in North America is expected to mature much later than in Western Europe and Japan. Numerous retailers have established special Web sites for users of mobile devices.

Principle

E-commerce and m-commerce can be used in many innovative ways to improve the operations of an organization.

Electronic retailing (e-tailing) is the direct sale from a business to consumers through electronic storefronts designed around an electronic catalog and shopping cart model.

A cybermall is a single Web site that offers many products and services at one Internet location.

Manufacturers are joining electronic exchanges, where they can work with competitors and suppliers to use computers and Web sites to buy and sell goods, trade market information, and run back-office operations such as inventory control. They are also using e-commerce to improve the efficiency of the selling process by moving customer queries about product availability and prices online.

The Web allows firms to gather much more information about customer behavior and preferences than they could using other marketing approaches. This new technology has greatly enhanced the practice of market segmentation and enabled companies to establish closer relationships with their customers. Technology relationship management enables an organization to gain detailed information about a customer's behavior, preferences, needs, and buying patterns to allow companies to set prices, negotiate terms, tailor promotions, add product features, and otherwise customize a relationship with a customer.

The Internet has revolutionized the world of investment and finance, especially online stock trading and online banking. The Internet has also created many options for electronic auctions, where geographically dispersed buyers and sellers can come together.

Online real estate services and e-boutiques are readily available.

The numerous m-commerce applications include mobile banking, mobile price comparison, mobile advertising and mobile coupons.

Principle

Although e-commerce and m-commerce offer many advantages, users must be aware of and protect themselves from many threats associated with use of this technology.

Businesses and people use e-commerce and m-commerce to reduce transaction costs, speed the flow of goods and information, improve the level of customer service, and enable the close coordination of actions among manufacturers, suppliers, and customers.

E-commerce and m-commerce also enable consumers and companies to gain access to worldwide markets. They offer great promise for developing countries, enabling them to enter the prosperous global marketplace, and hence helping to reduce the gap between rich and poor countries.

Since e-commerce and m-commerce are global systems, they face cultural; language; time and distance; infrastructure; currency; product and service; and state, regional, and national law challenges.

Revolutionary change always raises new issues, and e-commerce is no exception. Among the issues that must be addressed are security, theft of intellectual property, fraud, invasion of consumer privacy, lack of Internet access, return on investment, legal jurisdiction, and taxation.

Principle

Organizations must define and execute a strategy to be successful in e-commerce and m-commerce.

Most people agree that an effective Web site is one that creates an attractive presence and meets the needs of its visitors. E-commerce start-ups must decide whether they will build and operate the Web site themselves or outsource this function. Web site hosting services and storefront brokers provide alternatives to building your own Web site.

To build traffic to your Web site, you should register a domain name that is relevant to your business, make your site search-engine friendly by including a meta tag in your home page, use Web site traffic data analysis software to attract additional customers, and modify your Web site so that it supports global commerce. Web site operators must constantly monitor the traffic and response times associated with their site and adjust software, databases, and hardware to ensure that visitors have a good experience when they visit the site.

Principle

E-commerce and m-commerce require the careful planning and integration of a number of technology infrastructure components.

A number of infrastructure components must be chosen and integrated to support a large volume of transactions with customers, suppliers, and other business partners worldwide. These components include hardware, Web server software, and e-commerce software.

M-commerce presents additional infrastructure challenges including improving the ease of use of wireless devices, addressing the security of wireless transactions, and improving network speed. The Wireless Application Protocol (WAP) is a standard set of specifications to enable development of m-commerce software for wireless devices. WAP uses the Wireless Markup Language, which is designed for effectively displaying information on small devices. The development of WAP and its derivatives addresses many m-commerce issues.

Electronic payment systems are a key component of the e-commerce infrastructure. A digital certificate is an attachment to an e-mail message or data embedded in a Web page that verifies the identity of a sender or a Web site. To help prevent the theft of credit card numbers and banking information, the Secure Sockets Layer (SSL) communications protocol is used to secure all sensitive data. Several electronic cash alternatives require the purchaser to open an account with an electronic cash service provider and to present proof of identity whenever payments are to be made. Payments can also be made by credit, charge, debit, and smart cards. Retail and banking industries are developing means to enable payments by using the cell phone like a credit card by waving the end of the phone near a scanner device to pay for purchases.

CHAPTER 8: SELF-ASSESSMENT TEST

Electronic commerce and mobile commerce are evolving, providing new ways of conducting business that present both opportunities for improvement and potential problems.

1. Successful implementation of e-business requires ________ and ________.
   a. changes to existing business processes; substantial investment in IS technology
   b. conversion to XML software standards; Java programming scripts
   c. implementation of tight security standards; Web site personalization
   d. market segmentation; Web site globalization
2. Covisint is an example of which of the following forms of e-commerce?
   a. B2B
   b. B2C
   c. C2C
   d. G2C
3. Which form of e-commerce is the largest?
4. What is the elimination of intermediate organizations between the producer and the consumer called?
5. The sole objective of e-Government is to improve communications between citizens and the federal government. True or False?
6. The market for m-commerce in North America is far advanced relative to Western Europe and Japan. True or False?

E-commerce and m-commerce can be used in many innovative ways to improve the operations of an organization.

7. A(n) _________ is a single Web site that offers many products and services at one Internet location.
   a. e-tailer
   b. Web service
   c. cybermall
   d. none of the above
8. _________________ occurs when a firm obtains detailed information about a customer's behavior, preferences, needs, and buying patterns and uses that information to set prices, negotiate terms, tailor promotions, add product features, and otherwise customize its entire relationship with that customer.
9. The practice of _________ divides the pool of potential customers into subgroups, which are usually defined in terms of demographic characteristics.
10. An advancement in online bill payment that uses e-mail for the biller to post an image of your statement on the Internet so you can direct your bank to pay it is called _____________.

Although e-commerce and m-commerce offer many advantages, users must be aware of and protect themselves from many threats associated with use of this technology.

11. Which of the following is a frequent advantage of converting to an e-commerce supply chain?
   a. a decrease in transportation costs
   b. an increase in available product inventory
   c. acquisition of expensive information systems technology
   d. an improved level of customer service
12. The use of any of several technologies to enforce policies for controlling access to digital media is called _________________.

Organizations must define and execute a strategy to be successful in e-commerce and m-commerce.

13. After your Web site is established and successful, there is no need to redefine your site's basic business model. True or False?
14. Web site operators can take several actions to improve their rankings by search engines. True or False?

E-commerce and m-commerce require the careful planning and integration of a number of technology infrastructure components.

15. Poor Web site performance can drive consumers to abandon your Web site in favor of those with better, more reliable performance. True or False?
16. _________ contain variable information and are built to respond to a specific Web site visitor's request for information.

CHAPTER 8: SELF-ASSESSMENT TEST ANSWERS

(1) a (2) a (3) B2B (4) disintermediation (5) false (6) false (7) c (8) technology-enabled relationship management (9) market segmentation (10) electronic bill presentment (11) d (12) digital rights management (13) false (14) true (15) true (16) dynamic Web pages

REVIEW QUESTIONS

1. Identify three primary challenges that a company must overcome to convert its business processes from the traditional form to e-commerce processes, especially for B2C e-commerce.
2. Identify and briefly describe three forms of e-Government.
3. What is identity theft? Provide several tips for online shoppers to avoid identity theft.
4. Identify the six stages consumers experience in the sales life cycle that must be supported by a successful e-commerce system.
5. Outline some specific trust-building strategies for an organization to create an image of trustworthiness for its Web site.
6. What are Web services? Provide a brief example of how you might use a Web service in conducting e-commerce.
7. What benefits can a firm achieve by converting to an e-commerce supply chain system?
8. What is the Wireless Application Protocol? Is it universally accepted? Why or why not?
9. Why is it necessary to continue to maintain and improve an existing Web site?
10. What role do digital certificates and certificate authorities play in e-commerce?
11. What is the Secure Sockets Layer and how does it support e-commerce?
12. Briefly explain the differences among smart, credit, charge, and debit cards.
13. Since e-commerce and m-commerce systems are global systems, what are some of the global challenges that they face?

14. What is technology-enabled relationship management?
15. Identify the key elements of the technology infrastructure required to successfully implement e-commerce within an organization.

DISCUSSION QUESTIONS

1. Describe the process of electronic bill presentment. Outline some potential problems in using this form of billing customers.
2. Why are many manufacturers and retailers outsourcing the physical logistics of delivering merchandise to shoppers? What advantages does such a strategy offer? Are there any potential issues or disadvantages?
3. What does it mean to define the functions of a Web site? What are some of the possible functions?
4. What do you think are the biggest barriers to wide-scale adoption of m-commerce by consumers? Who do you think is working on solutions to these problems and what might the solutions entail?
5. Wal-Mart, the world's largest retail chain, has turned down several invitations to join exchanges in the retail and consumer goods industries. Is this good or bad for the overall U.S. economy? Why?
6. Identify and briefly describe three m-commerce applications you or a friend have used.
7. Discuss the use of e-commerce to improve spending on manufacturing, repair, and operations (MRO) of goods and services.
8. Discuss the pros and cons of e-commerce companies capturing data about you as you visit their Web sites.
9. Outline the key steps in developing a corporate global e-commerce strategy.
10. Identify three kinds of business organizations that would have difficulty in becoming a successful e-commerce organization.
11. Discuss how you might gather Web traffic data for analysis of your firm's Web site. For what decisions might this data be useful?

PROBLEM-SOLVING EXERCISES

1. Develop a set of criteria you would use to evaluate various business-to-consumer Web sites based on factors such as ease of use, protection of consumer data, and security of payment process. Develop a simple spreadsheet containing these criteria. Evaluate five popular Web sites using the criteria you developed. What changes would you recommend to the Web developer of the site that scored lowest?
2. Research the growth of B2B and B2C e-commerce and retail sales for the period 2000 to present. Use the graphics capability of your spreadsheet software to plot the growth of all three. Using current growth rates, predict the year that B2C e-commerce will exceed 10 percent of retail sales.
3. Do research to learn more about the use of WAP and other specifications being developed to support m-commerce. Briefly describe the specifications you uncover. Who is behind the development of these standards? Which standards seem to be gaining the broadest acceptance? Prepare a one- to two-page report for your instructor.

TEAM ACTIVITIES

1. Imagine that your team has been hired as consultants to provide recommendations to boost the traffic to a Web site that sells environmentally friendly household cleaning products. Identify as many ideas as possible for how you can increase traffic to this Web site. Next, rank your ideas from best to worst.
2. As a team, develop a set of criteria that you would use in selecting a firm to design and implement your organization's Web site. Identify and evaluate three different providers of Web services to support business processes associated with the Accounting and Finance business function. Prepare a table outlining the pros and cons of each service provider.

WEB EXERCISES

1. Do research on the Web to find three recent examples where the consumer records of a major e-tailer were compromised. How did each e-tailer handle its situation with its customers? Which e-tailer do you think reacted best? Why did you choose this e-tailer?
2. Research and document the current status of the Motion Picture Association of America (MPAA) efforts to prevent the illegal copying and/or sharing of movies. Write a brief report for your instructor. Include a discussion of your opinion on the legality of sharing movies.

CAREER EXERCISES

1. Do research to identify several new interesting m-commerce applications.
2. For your chosen career field, describe how you might use or be involved with e-commerce. If you have not chosen a career yet, answer this question for someone in marketing, finance, or human resources.

CASE STUDIES

Case One

The NFL and B2B

The National Football League (NFL) is the biggest business in sports. Established in 1920, the NFL sets the standard for a successful profit-making sports league. It has been called one of America's best-run businesses by *BusinessWeek* magazine.

The NFL knows that its success depends strongly on its positive relationship with the media—news organizations that cover the sport on television, radio, the Web, and in magazines and newspapers. The NFL has worked hard to provide information to the media efficiently to increase coverage of the league, games, teams, and players.

Recently the NFL set out to improve its business-to-business transactions with its media customers. The products that the NFL provides to media include game video, highlights, team and player stats, and other league information—all digital products.

The NFL was one of the first sports organizations to implement a media-only Web site. Media companies originally paid for a subscription to the information provided on the password-protected site. In 1997, the site was state-of-the-art and unique, but ten years later the site is out-of-date and inefficient compared to what is being provided by new technologies. The NFL decided to invest in major improvements to its B2B system.

The system that the NFL worked to design uses an extranet to provide secure access for media customers to massive amounts of NFL information custom-designed to suit each need. The NFL required the system to be secure, robust, scalable, and flexible to allow for continuous growth of users and content.

The NFL calls the resulting system the media portal at NFLMedia.com. The portal provides a "one-stop-shop for concise, intuitive, searchable, and immediate consumable league, game, and team information that helps reporters write better stories." The new portal is also easy to administer and maintain. As with many development efforts, the NFL solicited suggestions from its customers, the media that it serves. Personalization is an important new capability. Each user has a unique profile that keeps track of the user's history and interests and automatically provides the information that is of most interest to that user. For example, the Chicago Tribune is offered news that focuses on the Chicago Bears.

The NFLMedia.com site provides a wealth of information, including current news and press releases, a link to the archives, league standings, and links to each of the 32 team Web sites. The site includes information on prominent events, history, policies of the NFL, press credentials, calendars and schedules, community relation activities, league statements and transcripts, and much more, according to the cited article. All of this information is organized so that it can be custom sorted. Much of the information is labeled for "Media Access" only, allowing reporters to get scoops that are otherwise unavailable to the public.

The NFL employs a number of people in its PR department to filter information that continuously pours in from league departments and transform it into content for the media site. A Web content management system helps to simplify and streamline this task. Departments posting stories to the site can specify run dates determining when the story is available to the media and for how long. The PR staff adds metadata content tags to each article so that the information can appear on several related sites in different contexts. On game day, the site displays a special page that draws information from many locations within the NFL systems, including game schedules, injury reports, and team standings.

The new NFL media portal offers improvements for both reporters and those that manage the resource. Reporters find it much easier to find timely information to fuel their stories. About 3,500 media staff signed up to use the new service within the first month it was offered. This fosters goodwill between the media and NFL. The NFL PR staff appreciates the automation that allows them to easily manage the flow of information to the portal even while traveling. As increasing amounts of products are digital in nature, streamlining the process of acquiring those goods and automating the task of assembling them is key to success for the vendor.

Discussion Questions

1. What makes NFLMedia.com a B2B e-commerce system?
2. Why has the NFL invested so much in NFLMedia.com? What benefits does the investment provide the NFL?

Critical Thinking Questions

1. What improvements were made to NFLMedia.com that the press appreciates?
2. What other portals might the NFL consider for providing information and services to other groups?

SOURCES: IBM Staff, "The NFL scores a win with extranet media portal," IBM Case Studies, February 15, 2008, *www-01.ibm.com/software/success/cssdb.nsf/CS/CCLE-7BUR5T?OpenDocument&Site=wssoftware&cty=en_us*; WebHost TALK Staff, "IBM Powers NFL Media Portal," WebHosting TALK, February 1, 2008, *www.webhostingtalk.com/news/ibm-powers-nfl-media-portal*; NFL Media Portal, *www.nflmedia.com*, accessed June 21, 2008.

Case Two

Paying with Cell Phones in Canada

Companies in Canada are racing to see who can motivate Canadians to start paying at the register with their cell phones. You are probably familiar with MasterCard's PayPass technology that allows people in many countries to make payments at checkouts by touching their credit card to a pad. The cards are equipped with a chip that supports near-field communication (NFC), passing credit card information to the receiver in the payment pad. Now vendors want to do away with plastic credit cards altogether by embedding NFC chips in cell phones and allowing customers to make payments by touching their cell phones to pads rather than credit cards.

In Canada, MasterCard and Visa are rolling out mobile wireless payment pilot programs this year. However, credit card companies need the support of cell phone handset manufacturers and carriers to successfully launch the programs. MasterCard is working with carrier Bell Canada and using handsets from an unnamed vendor. Visa is working with the Royal Bank of Canada (RBC) and is determining who the carrier and handset manufacturer will be.

MasterCard might have an advantage over Visa in that it already has significant penetration in this market with its PayPass technology. More than 28 million MasterCard PayPass cards are being used at more than 109,000 merchants worldwide. MasterCard is starting a trial with Bell Canada employees who will be paying with their phones and testing the benefits, which include faster checkouts and additional services. For example, financial services are provided that allow you to check your transaction history and bank balance and conduct online banking transactions. Another benefit is avoiding having to carry a wallet packed with plastic.

The big question in this race is whether consumers are interested in paying with their phones. A 2007 survey of 15–29 year-old Canadians showed that only 8.8 percent of those surveyed were interested in contactless payments via cell phone. Some people have voiced concern over security and privacy issues surrounding wireless payments. The technology might also decrease battery life in cell phones.

Providers are concerned about the legal risks of offering m-commerce services over near-field communication technologies. "Who's responsible for liability issues?" asked Anne Koski, head of payment innovations at the Royal Bank of Canada at a recent conference on the topic. If money is lost due to inefficiencies in the technology, who foots the bill—the handset manufacturer, the carrier, or the bank? Legal teams are devising the answer to this important question. Those in the industry know that convincing customers of the security of the new system is important in winning them over to the technology.

Data reliability, authentication, fraud, theft, and privacy protection are all issues that these companies are confronting as they begin planning their marketing campaigns. Some believe that creating a standard for mobile payments is the most important factor in launching mobile payments. Being able to advertise the stability and security features of an agreed upon standard would help in winning over consumers. However, a standard might mean that companies such as MasterCard have to overhaul their entire PayPass network.

Many believe that consumer education is also important. In the early days of Internet-based e-commerce, many people were afraid to purchase products and services on the Web due to concerns over privacy and security. Most of those people have overcome those fears in exchange for the convenience and opportunities that online shopping offers. Proponents of m-commerce and contactless payment systems are hoping that once educated, consumers will choose the convenience of paying by swiping your cell phone over any perceived risk.

Discussion Questions

1. What are the benefits and concerns associated with cell phone payments?
2. Why does MasterCard have a head start in the cell phone payment race taking place in Canada?

Critical Thinking Questions

1. What assurances would you need from the service provider prior to buying into a cell phone wireless payment system?
2. Do you think that a cell phone payment system is inevitable in Canada and North America? If so, how long do you think it will take before it becomes the norm? If not, why?

SOURCES: Smith, Briony, "MasterCard Gets moving on mobile payments," *IT World Canada*, May 28, 2008, *www.itworldcanada.com/a/Enterprise-Business-Applications/36ee2d57-2fad-4a2a-8d41-03f6f6bb6a34.html*; Smith, Briony, "RBC, Visa prepare to jump mobile payment hurdles," *IT World Canada*, June 11, 2008, *www.itworldcanada.com/a/Enterprise-Business-Applications/d628b121-3309-47fd-a927-c8f9408c87e5.html*; Smith, Briony, "The biggest legal risks around mobile payments," *IT World Canada*, June 11, 2008, *www.itworldcanada.com/a/Enterprise-Business-Applications/e3425f71-90f6-416a-9121-edd7a4ccf154.html*.

Whitmann Price Consulting: E-Commerce Considerations

Discussion Questions

1. How will the AMCI system contribute to the e-commerce components of Whitmann Price?
2. How can the new system save Whitmann Price employees time and improve the reliability of billing data?

Critical Thinking Questions

1. How might the new system be extended to partners who contract with Whitmann Price to serve its clients?
2. What types of security and privacy concerns may arise over the ability to send billing information directly over the wireless network?

NOTES

Sources for the opening vignette: IBM Staff, "Staples makes it easy for online customers and becomes a more flexible and successful business," IBM Case Studies, August 30, 2007, *www-01.ibm.com/software/success/cssdb.nsf/CS/JSTS-765Q96?OpenDocument&Site=wssoftware&cty=en_us*; Staples Web site, *www.staples.com*, accessed June 21, 2008; IBM Websphere Commerce Web site, *www-306.ibm.com/software/genservers/commerceproduct-line*, accessed June 21, 2008.

1 Chabrow, Eric, "E-Commerce Continues to Grow Very Nicely," *Web Design & Technology News*, May 1, 2003.
2 Messmer, Ellen, "Dot-Com Survivor Covisint Finds B2B Niche, *Network World*, November 1, 2007.
3 Broache, Anne, "Tax-Free Internet Shopping Days Could Be Numbered," *CNET News.com*, April 15, 2008.
4 "American Eagle Outfitters Launches E-Commerce Web site for Martin + OSA," *DataMonitor*, April 4, 2008.
5 "eBay Reports Fourth Quarter and Full Year 2007 Results," at *http://files.shareholder.com/downloads/ebay*, accessed April 7, 2008.
6 "Auction Site to Hammer on University Students," *China Economic Net*, December 21, 2007.
7 "Victorian Government Purchasing Board—Desktop Products & Service," *www.vgpb.vic.gov.au*, accessed May 26, 2008.
8 Abboud, Leila, "Global Suppliers Play Catch-Up in Information Age," *The Wall Street Journal*, January 4, 2007, p. B3.
9 Aviall Web site, *www.aviall.com*, accessed May 25, 2008.
10 Sarrel, Matthew D., "Frugal CRM," *PC Magazine*, March 6, 2007.
11 Hulme, George, "Identity Theft Is a Drag for Everyone," *InformationWeek*, January 17, 2008.
12 "A Chronology of Data Breaches," The Privacy Rights Clearing House *www.privacyrights.org/ar/ChrondataBreaches.htm*, accessed April 13, 2008.
13 The Federal Trade Commission's Identity Theft Web site, *www.ftc.gov/bcp/edu/microsites/idtheft/consumers/about-identity-theft.html*, accessed April 8, 2008.
14 Gaudin, Sharon, "'Iceman' Hacker Indicted for Running Identity Theft Scheme," *Information Week*, September 11, 2007.
15 Regan, Keith, "Sprint Cuts Ribbon on Mobile Shopping Service," *E-Commerce Times*, September 13, 2007.
16 "40% of Big U.S. Companies Boast Mobile Sites," *Internet Retailer*, January 18, 2008.
17 Offner, Jim, "Yahoo to Mobile Searchers: Talk to the Handset," *E-Commerce Times*, April 3, 2008.
18 Regan, Keith, "Amazon Aims to Light M-Commerce Fire with TextBuyIt," *E-Commerce Times*, April 2, 2008.
19 Siwicki, Bill, "Merchant Beware," *Internet Retailer*, January 2008.
20 "FlowerShop.com Goes Mobile in Time for Valentine's Day," *Internet Retailer*, February 8, 2008.
21 Jones, Sandra M., "Sears' Quiet E-Commerce Revolution," *E-commerce Times*, April 13, 2008.
22 Regan, Keith, "Detroit Trading Exchange: Swinging the Door Wide Open for Auto Leads," *E-Commerce Times*, April 4, 2008.
23 Crum, Rachelle, "ATG's Cliff Conneighton: E-Tailers 'Can't Wing It Anymore'," *E-Commerce Times*, April 6, 2008.
24 Offner, Jim, "Report: Paying Bills Online Saves Gas, Trees, Water," *E-Commerce Times*, March 27, 2008.
25 "Personal Internet Banking," ICBC Asia Web site at *www.icbcasia.com* accessed on June 21, 2008.
26 "ePower Systems Executive Summary," Center for Business Planning Web site, *www.businessplans.org/ePower/ePower00.html*, accessed May 3, 2008.
27 "Zillow Goes After an Area That Real Estate Agents Can't Touch," *TechDirt*, July 11, 2007.
28 "Redfin," Crunchbase, *www.crunchbase.com/company/redfin*, accessed May 5, 2008.
29 Noyes, Katherine, "ShopLaTiDa.com: A Boutique with a Personalized Focus," *E-Commerce Times*, May 2, 2008.
30 "Blair Uses e-Sourcing to Contain Costs," eDyanQuote Case Study, *www.edynaquote.com*, accessed May 24, 2008.

31 "AT&T Mobile Banking," AT&T Web site, *www.wireless.att.com*, accessed April 5, 2008.
32 Siwicki, Bill, "Merchant Beware," *Internet Retailer*, January 2008.
33 Siwicki, Bill, "Merchant Beware," *Internet Retailer*, January 2008.
34 Siwicki, Bill, "Merchant Beware," *Internet Retailer*, January 2008.
35 "Walk 100 Yards North, Turn Right, Enter Store," *Internet Retailer*, December 31, 2007.
36 Kharif, Olga, "Building a Case for Mobile Advertising," *E-Commerce Times*, March 16, 2008.
37 "40% of Big U.S. Companies Boast Mobile Sites," *Internet Retailer*, January 18, 2008.
38 Rosencrance, Linda, "5 Companies to Test Mobile Coupons at Grocery Chain," *Computerworld*, April 18, 2008.
39 "MPAA Accuses Pullmylink.com of Aiding Movie Piracy," WinXpFix.com Web site, *www.winxpfix.com /NEWS/04-17-2008*, accessed May 31, 2008.
40 Claburn, Thomas, "US-CERT Warns About Phishers Scamming Disaster Donors," *InformationWeek*, May 20, 2008.
41 Claburn, Thomas, "Phishing Campaign Targets Rebate Checks," *InformationWeek*, May 9, 2008.
42 McDougall, Paul, "Yahoo Hit with $1 Million Click Fraud Lawsuit," *InformationWeek*, April 16, 2008.
43 Claburn, Thomas, "Internet Fraud Loss for 2007 Tops $239 Million," *InformationWeek*, April 4, 2008.
44 Gardner, David, "German Court Says eBay Must Work to Halt Sales of Fake Rolexes," *InformationWeek*, July 27, 2007.
45 Jones, K.C., "Seven Charged in International Counterfeit Art Scheme on eBay," *InformationWeek*, March 20, 2008.
46 Claburn, Thomas, "Privacy Vs. Personalization: Can Advertisers Ward Off Looming Threat of Do Not Track List?," *InformationWeek*, November 10, 2007.
47 Broache, Anne, "Tax-Free Internet Shopping Days Could be Numbered," *CNET News.com*, April 15, 2008.
48 "Who We Are—About Macronimous", Macronimous Web site, *www.macronimous.com/aboutus.asp*, accessed May 31, 2008.
49 "Case Studies: Eastman Kodak Company," Corporate Communications Web site, *www.corporatecomm.com/caseStudies/CaseStudies09.php*, accessed May 31, 2008.
50 UPS: Trade Tools and Terminology, UPS Web site, *www.ups.com/content/us/en/resources/advisor/tools_terms/index.html*, accessed June 2, 2008.
51 About Us page, PayPal Web site, *www.paypal.com*, accessed May 31, 2008.
52 About Us page, BillMeLater Web site, *www.billmelater.com*, accessed May 31, 2008.
53 About Us page, MoneyZap Web site, *www.moneyzap.com*, accessed May 31, 2008.
54 TeleCheck Electronic Check Verification at First Data Web site, *www.firstdata.com/product_solutions/payment_solutions/telecheck/eca_verification.htm*, accessed May 31, 2008.
55 "Rising Fuel Costs Pit Airlines Against Credit Card Companies," *Travel Daily News*, April 14, 2008.
56 "Barclaycard to Test Cobranded Transit Card on NFC Phones," *Card Technology*, November 2007.
57 Sowa, Tom, "Wading into the M-Commerce Waters," *E-Commerce Times*, April 14, 2008.

CHAPTER • 9 •

Enterprise Systems

PRINCIPLES	LEARNING OBJECTIVES
▪ An organization must have information systems that support the routine, day-to-day activities that occur in the normal course of business and help a company add value to its products and services.	▪ Identify the basic activities and business objectives common to all transaction processing systems. ▪ Describe the transaction processing systems associated with the order processing, purchasing, and accounting business functions. ▪ Identify key control and management issues associated with transaction processing systems.
▪ A company that implements an enterprise resource planning system is creating a highly integrated set of systems, which can lead to many business benefits.	▪ Discuss the advantages and disadvantages associated with the implementation of an enterprise resource planning system. ▪ Identify the challenges multinational corporations must face in planning, building, and operating their TPSs.

Information Systems in the Global Economy Maporama, France

Maporama Gains Ground Through New Enterprise System

Based in Paris, France, Maporama is a world leader in developing location-based services for businesses. Maporama develops GPS-based systems that allow businesses to track mobile work forces, sales outlets, and competition. Maporama's Web site states that its custom-designed products "empower existing mission-critical applications or processes within an organization and display results enterprise-wide, as needed, via connected devices (computers, mobile phones, PDAs, etc.)" The company claims 500 customers supporting 26 languages on five continents, and boasts the most complete pan-European map coverage.

Recently Maporama had a pressing need to gain tighter control over its global mobile workforce and associated customer information. The company's rapid growth to 500 customers—with 10,000 contacts to track—left its European and North American sales force with more information than it could manage. The company decided to invest in a customer relationship management system that could help manage its customer relations and synchronize sales information across the entire enterprise.

When the company was smaller, Maporama used Microsoft Outlook and Exchange to manage customer information. Over time, Maporama outgrew that system and could no longer easily manage customer accounts. It became difficult to synchronize the information between departments, which resulted in disconnected islands of information. The company needed a system that would collect all information into a central database, which could be leveraged to improve sales and customer service.

Initially, Maporama investigated systems that they could implement themselves through an internal server. The company decided that such a system would be too costly and a burden to maintain. When Maporama discovered a hosted system accessed over the Internet, the company knew that it was the perfect solution. Software as a Service (SaaS) allowed Maporama to use a full-service CRM system for a monthly fee delivered to any Internet-connected device around the world.

Maporama's new CRM system was set up and ready to go in 15 days. (It would have taken over a year for Maporama to set up such a system in-house.) Maporama's sales, marketing, and support teams use the system at headquarters and worldwide offices. According to Dominique Grillet, CEO of Maporama, the new system allows staff and management to find answers to questions such as: "Who should a salesperson call on today?", "Which products have customers bought in the past, and when?", "What was the last contact with a customer and by whom?", "Has a sister company bought our products in another country?", "How effective are our telesales and marketing campaign?", "What are our sales forecasts?", "In what industry sectors are we winning the most business?"

With the new enterprise system, Maporama can quickly respond to customers' needs and boost revenue opportunities. The company can also maximize the productivity of its sales force as individuals and a team, and implement more effective marketing campaigns for less money by improving customer targeting.

As you read this chapter, consider the following:

- How can an effective enterprise system affect the overall performance of a business?
- What types of information systems are critical to the success of a business and how are the systems related to one another?

Why Learn About Enterprise Systems?

Organizations today are moving from a collection of nonintegrated transaction processing systems to highly integrated enterprise systems to perform routine business processes and maintain records about them. These systems support a wide range of business activities associated with supply chain management and customer relationship management, as mentioned in Chapter 1. Although they were initially thought to be cost-effective only for very large companies, even small and midsized companies are now implementing these systems to reduce costs and improve service.

In our increasingly service-oriented economy, outstanding customer service has become a goal of virtually all companies. Employees who work directly with customers—whether in sales, customer service, or marketing—require high-quality transaction processing systems and their associated information to provide good customer service. Such workers might use an enterprise system to check the inventory status of ordered items, view the production planning schedule to tell the customer when the item will be in stock, or enter data to schedule delivery to the customer.

No matter what your role, it is very likely that you will provide input to or use the output from your organization's enterprise systems. Your effective use of these systems will be essential to raise the productivity of your firm, improve customer service, and enable better decision making. Thus, it is important that you understand how these systems work and what their capabilities and limitations are.

enterprise system
A system central to the organization that ensures information can be shared across all business functions and all levels of management to support the running and managing of a business.

An **enterprise system** is central to an organization and ensures that information can be shared across all business functions and all levels of management to support the running and managing of a business. Enterprise systems employ a database of key operational and planning data that can be shared by all. This eliminates the problems of lack of information and inconsistent information caused by multiple transaction processing systems that support only one business function or one department in an organization. Examples of enterprise systems include enterprise resource planning systems that support supply-chain processes, such as order processing, inventory management, and purchasing and customer relationship management systems that support sales, marketing, and customer service–related processes.

As demonstrated in the opening vignette, businesses rely on enterprise systems to perform many of their daily activities in areas such as product supply, distribution, sales, marketing, human resources, manufacturing, accounting, and taxation so that work is performed quickly, while avoiding waste and mistakes. Without such systems, recording and processing business transactions would consume huge amounts of an organization's resources. This collection of processed transactions also forms a storehouse of data invaluable to decision making. The ultimate goal is to satisfy customers and provide a competitive advantage by reducing costs and improving service.

This chapter begins by presenting an overview of enterprise systems.

AN OVERVIEW OF ENTERPRISE SYSTEMS: TRANSACTION PROCESSING SYSTEMS AND ENTERPRISE RESOURCE PLANNING

Every organization has many *transaction processing systems (TPS)*, which capture and process the detailed data necessary to update records about the fundamental business operations of the organization. These systems include order entry, inventory control, payroll, accounts payable, accounts receivable, and the general ledger, to name just a few. The input to these systems includes basic business transactions, such as customer orders, purchase orders, receipts, time cards, invoices, and customer payments. The processing activities include data collection, data editing, data correction, data manipulation, data storage, and document production. The result of processing business transactions is that the organization's records are updated to reflect the status of the operation at the time of the last processed transaction.

A TPS also provides employees involved in other business processes—via management information systems/decision support systems (MIS/DSS), the special-purpose information systems, and knowledge management systems—with data to help them achieve their goals. (MIS/DSS systems are discussed in Chapter 10.) A transaction processing system serves as the foundation for these other systems (see Figure 9.1).

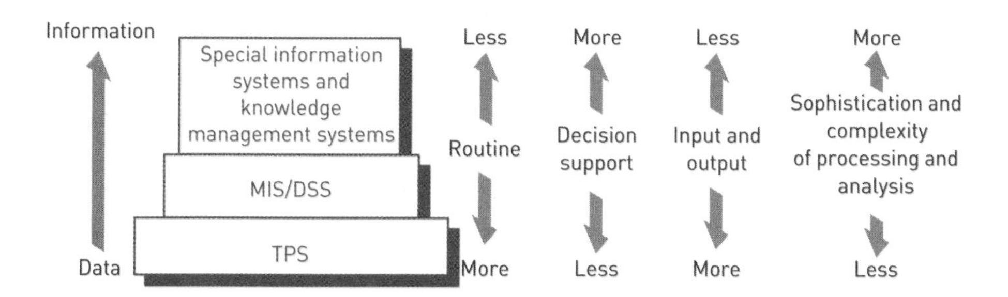

Figure 9.1

TPS, MIS/DSS, and Special Information Systems in Perspective

Transaction processing systems support routine operations associated with customer ordering and billing, employee payroll, purchasing, and accounts payable. The amount of support for decision making that a TPS directly provides managers and workers is low.

TPSs work with a large amount of input and output data and use this data to update the official records of the company about such things as orders, sales, customers, and so on. As systems move from transaction processing to management information/decision support and special-purpose information systems, they involve less routine, more decision support, less input and output, and more sophisticated and complex processing and analysis. These higher-level systems require the basic business transaction data captured by the TPS.

Because TPSs often perform activities related to customer contacts—such as order processing and invoicing—these information systems play a critical role in providing value to the customer. For example, by capturing and tracking the movement of each package, shippers such as Federal Express and United Parcel Service (UPS) can provide timely and accurate data on the exact location of a package. Shippers and receivers can access an online database and, by providing the airbill number of a package, find the package's current location. If the package has been delivered, they can see who signed for it (a service that is especially useful in large companies where packages can become "lost" in internal distribution systems and mailrooms). Such a system provides the basis for added value through improved customer service.

Traditional Transaction Processing Methods and Objectives

With **batch processing systems**, business transactions are accumulated over a period of time and prepared for processing as a single unit or batch (see Figure 9.2a). Transactions are accumulated for the length of time needed to meet the needs of the users of that system. For example, it might be important to process invoices and customer payments for the accounts receivable system daily. On the other hand, the payroll system might receive time cards and process them biweekly to create checks, update employee earnings records, and distribute labor costs. The essential characteristic of a batch processing system is that there is some delay between an event and the eventual processing of the related transaction to update the organization's records.

batch processing system
A form of data processing where business transactions are accumulated over a period of time and prepared for processing as a single unit or batch.

With **online transaction processing (OLTP)**, each transaction is processed immediately, without the delay of accumulating transactions into a batch (see Figure 9.2b). Consequently, at any time, the data in an online system reflects the current status. This type of processing is essential for businesses that require access to current data such as airlines, ticket agencies, and stock investment firms. Many companies find that OLTP helps them provide faster, more efficient service—one way to add value to their activities in the eyes of the customer. Increasingly, companies are using the Internet to capture and process transaction data such as customer orders and shipping information from e-commerce applications.

online transaction processing (OLTP)
A form of data processing where each transaction is processed immediately, without the delay of accumulating transactions into a batch.

FedEx adds value to its service by providing timely and accurate data online about the exact location of a package.

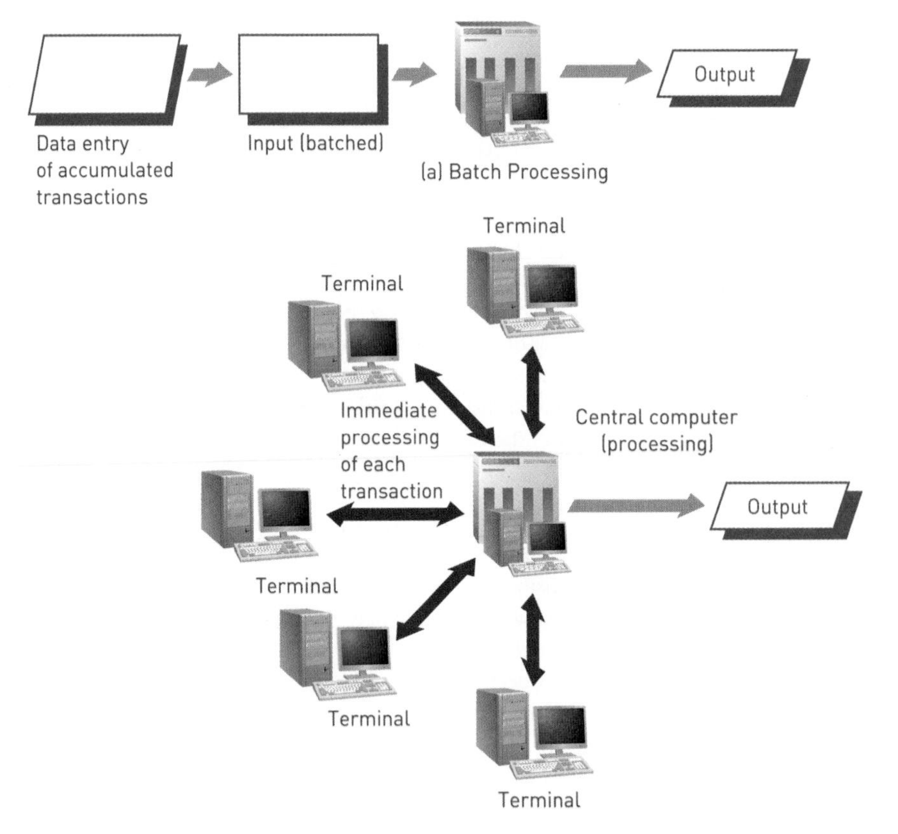

Figure 9.2

Batch Versus Online Transaction Processing

(a) Batch processing inputs and processes data in groups. (b) In online processing, transactions are completed as they occur.

TPS applications do not always run using online processing. For many applications, batch processing is more appropriate and cost-effective. Payroll transactions and billing are typically done via batch processing. Specific goals of the organization define the method of transaction processing best suited for the various applications of the company.

Figure 9.3 shows the traditional flow of key pieces of information from one TPS to another for a typical manufacturing organization.

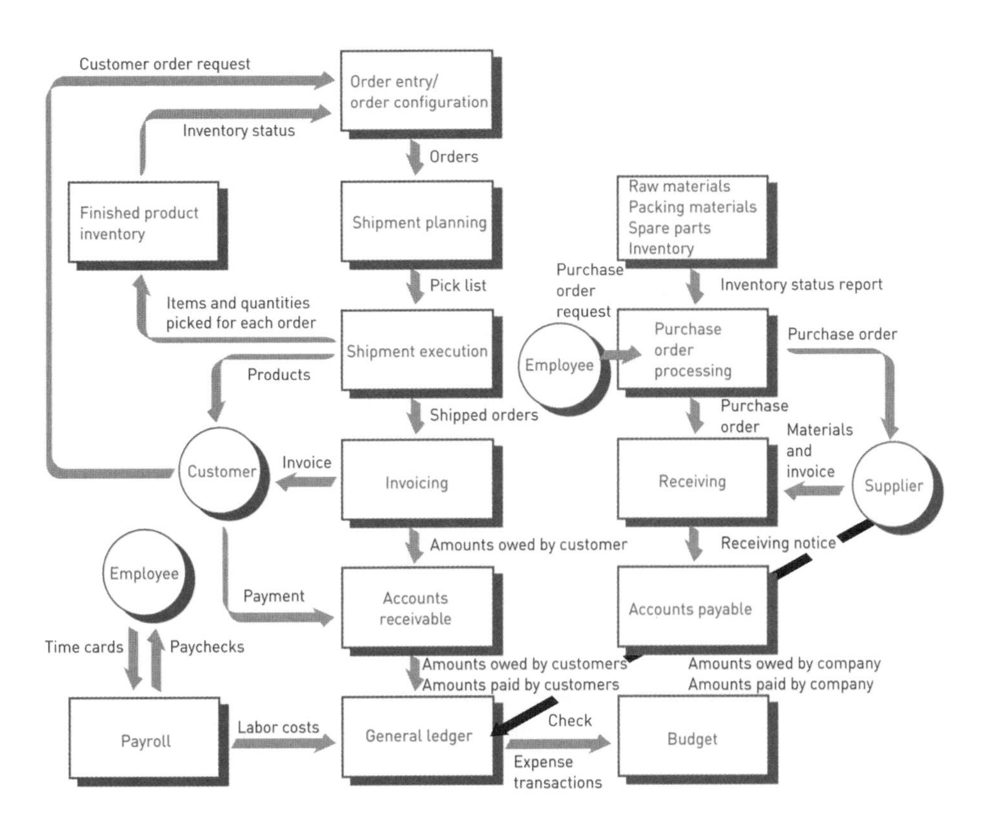

Figure 9.3

Integration of a Firm's TPSs

Because of the importance of transaction processing, organizations expect their TPSs to accomplish a number of specific objectives including:

- Capture, process, and update databases of business data required to support routine business activities. Wal-Mart probably holds the world record, outside of the federal government, processing some 800 million business transactions per day.[1]
- Ensure that the data is processed accurately and completely.
- Avoid processing fraudulent transactions.
- Produce timely user responses and reports.
- Reduce clerical and other labor requirements.
- Help improve customer service.
- Achieve competitive advantage.

A TPS typically includes the following types of systems:

- **Order processing systems.** Running these systems efficiently and reliably is so critical that the order processing systems are sometimes referred to as the "lifeblood of the organization." The processing flow begins with the receipt of a customer order. The finished product inventory is checked to see if sufficient inventory is on hand to fill the order. If sufficient inventory is available, the customer shipment is planned to meet the customer's desired receipt date. A product pick list is printed at the warehouse from which the order is to be filled on the day the order is planned to be shipped. At the warehouse, workers gather the items needed to fill the order, and enter the item identifier and quantity for each item to update the finished product inventory. When the order is complete and sent on its way, a customer invoice is created with a copy included in the customer shipment.
- **Accounting systems.** The accounting systems must track the flow of data related to all the cash flows that affect the organization. As mentioned earlier, the order processing system generates an invoice for customer orders to include with the shipment. This information is also sent to the accounts receivable system to update the customer's

account. When the customer pays the invoice, the payment information is also used to update the customer's account. The necessary accounting transactions are sent to the general ledger system to keep track of amounts owed and amounts paid. Similarly, as the purchasing systems generate purchase orders and those items are received, information is sent to the accounts payable system to manage the amounts owed by the company. Data about amounts owed and paid by customers to the company and from the company to vendors and others are sent to the general ledger system that records and reports all financial transactions for the company.

- **Purchasing systems.** The traditional transaction processing systems that support the purchasing business function include inventory control, purchase order processing, receiving, and accounts payable. Employees place purchase order requests in response to shortages identified in inventory control reports. Purchase order information flows to the receiving system and accounts payable systems. A record of receipt is created upon receipt of the items ordered. When the invoice arrives from the supplier, it is matched to the original order and the receiving report, and a check is generated if all data is complete and consistent.

In the past, organizations knitted together a hodgepodge of systems to accomplish the transaction processing activities shown in Figure 9.3. Some of the systems might have been applications developed using in-house resources, some may have been developed by outside contractors, and others may have been off-the-shelf software packages. Much customization and modification of this diversity of software was necessary for all the applications to work together efficiently. In some cases, it was necessary to print data from one system and manually reenter it into other systems. Of course, this increased the amount of effort required and increased the likelihood of processing delays and errors.

The approach taken today by many organizations is to implement an integrated set of transaction processing systems from a single or limited number of software vendors that handles most or all of the transaction processing activities shown in Figure 9.3. The data flows automatically from one application to another with no delay or need to reenter data.

Table 9.1 summarizes some of the ways that companies can use transaction processing systems to achieve competitive advantage.

Table 9.1

Examples of Transaction Processing Systems for Competitive Advantage

Competitive Advantage	Example
Customer loyalty increased	Customer interaction system to monitor and track each customer interaction with the company
Superior service provided to customers	Tracking systems that customers can access to determine shipping status
Better relationship with suppliers	Internet marketplace to allow the company to purchase products from suppliers at discounted prices
Superior information gathering	Order configuration system to ensure that products ordered will meet customer's objectives
Costs dramatically reduced	Warehouse management system employing RFID technology to reduce labor hours and improve inventory accuracy
Inventory levels reduced	Collaborative planning, forecasting, and replenishment to ensure the right amount of inventory is in stores

Depending on the specific nature and goals of the organization, any of these objectives might be more important than others. By meeting these objectives, TPSs can support corporate goals such as reducing costs; increasing productivity, quality, and customer satisfaction; and running more efficient and effective operations. For example, overnight delivery companies such as FedEx expect their TPSs to increase customer service. These systems can locate a client's package at any time—from initial pickup to final delivery. This improved customer information allows companies to produce timely information and be more responsive to customer needs and queries.

Transaction Processing Systems for Small and Medium-Size Enterprises (SMEs)

Many software packages provide integrated transaction processing system solutions for small and medium-size enterprises (SMEs), where small is an enterprise with less than 50 employees and medium is one with fewer than 250 employees. These systems are typically easy to install, easy to operate, and have a low total cost of ownership with an initial cost of a few hundred to a few thousand dollars. Such solutions are highly attractive to firms that have outgrown their current software but cannot afford a complex, high-end integrated system solution. Table 9.2 presents some of the dozens of such software solutions available.

Table 9.2

Sample of Integrated TPS Solutions for SMEs

Vendor	Software	Type of TPS Offered	Target Customers
AccuFund	AccuFund	Financial reporting and accounting	Nonprofit, municipal, and government organizations
OpenPro	OpenPro	Complete ERP solution including financials, supply chain management, e-commerce, customer relationship management, and retail POS system	Manufacturers, distributors, and retailers
Intuit	QuickBooks	Financial reporting and accounting	Manufacturers, professional services, contractors, nonprofits, and retailers
Sage	Timberline	Financial reporting, accounting, and operations	Contractors, real estate developers, and residential builders
Redwing	TurningPoint	Financial reporting and accounting	Professional services, banks, and retailers

The city of Lexington, Kentucky (2006 population 275,000), implemented the AccuFund software and decreased the time to close the books at the end of each month by as much as 20 percent and reduced the number of corrections needed to the general ledger.[2]

TRANSACTION PROCESSING ACTIVITIES

Along with having common characteristics, all TPSs perform a common set of basic data-processing activities. TPSs capture and process data that describes fundamental business transactions. This data is used to update databases and to produce a variety of reports used by people both within and outside the enterprise. The business data goes through a **transaction processing cycle** that includes data collection, data editing, data correction, data manipulation, data storage, and document production (see Figure 9.4).

transaction processing cycle
The process of data collection, data editing, data correction, data manipulation, data storage, and document production.

Data Collection

Capturing and gathering all data necessary to complete the processing of transactions is called **data collection**. In some cases, it can be done manually, such as by collecting handwritten sales orders or changes to inventory. In other cases, data collection is automated via special input devices such as scanners, point-of-sale devices, and terminals.

data collection
Capturing and gathering all data necessary to complete the processing of transactions.

Data collection begins with a transaction (e.g., taking a customer order) and results in data that serves as input to the TPS. Data should be captured at its source and recorded accurately in a timely fashion, with minimal manual effort, and in an electronic or digital form that can be directly entered into the computer. This approach is called *source data automation*. An example of source data automation is an automated device at a retail store that speeds the checkout process—either UPC codes read by a scanner or RFID signals picked

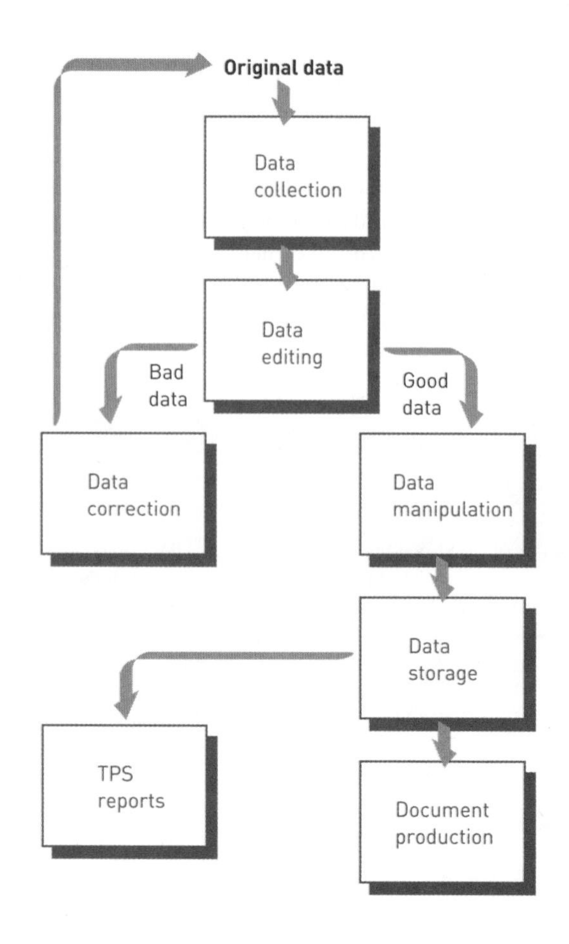

Figure 9.4

Data-Processing Activities Common to Transaction Processing Systems

up when the items approach the checkout stand. Using both UPC bar codes and RFID tags is quicker and more accurate than having a clerk enter codes manually at the cash register. The product ID for each item is determined automatically, and its price retrieved from the item database. The point-of-sale TPS uses the price data to determine the customer's bill. The store's inventory and purchase databases record the number of units of an item purchased, the date, the time, and the price. The inventory database generates a management report notifying the store manager to reorder items that have fallen below the reorder quantity. The detailed purchases database can be used by the store or sold to marketing research firms or manufacturers for detailed sales analysis (see Figure 9.5).

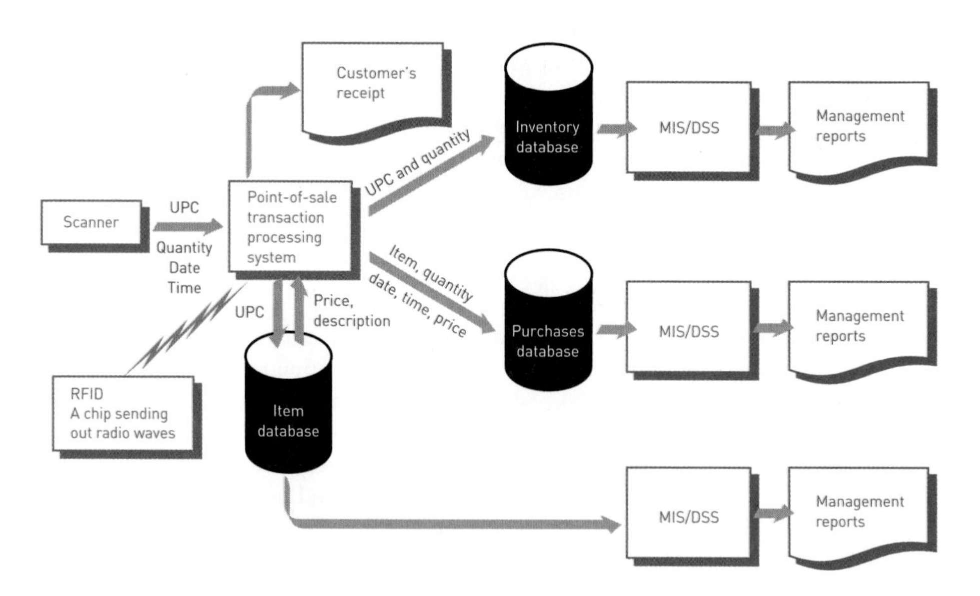

Figure 9.5

Point-of-Sale Transaction Processing System

The purchase of items at the checkout stand updates a store's inventory database and its database of purchases.

INFORMATION SYSTEMS @ WORK

Georgia Aquarium Controls Crowds with Online TPS

Museums, zoos, and aquaria are wonderful places to visit on the weekend except for one historic nuisance: the crowds! A visit to some public places on a Saturday morning often leaves you waiting in line for an hour or more to reach the ticket counter. One of America's most popular and prestigious aquaria, Georgia Aquarium, has beaten this long-standing problem with an online transaction processing system, or TPS.

Billed as the world's largest aquarium, Georgia Aquarium was a gift to the Atlanta community, the state of Georgia, and the world from Bernie Marcus, cofounder of The Home Depot, and his wife Billi. The $320 million facility opened in 2005, offering visitors a look at the world's largest collection of aquatic animals. The facility houses the world's largest single aquarium habitat: a whale shark tank and environment that holds 8 million gallons of water.

When designing information systems for the facility, Beach Clark, vice president of information technology, knew that the systems needed to accommodate record numbers of visitors and transactions. Above all, the aquarium administration wanted to avoid long lines at ticket counters and exhibits that are all too common for such businesses. The administration decided to make strong use of Web-delivered services to shoulder the demands of transactions and other customer needs.

Using off-the-shelf software customized to meet the aquarium's needs, Beach Clark and his team designed a transaction processing system and a ticketing system unlike any previously used for this venue. The team modified the shopping cart software to manage ticket sale transactions and to control the number of visitors at the aquarium during each hour of operation. Now, when you purchase a general admissions ticket at georgiaaquarium.org, you select the date and time you plan to arrive. When a time fills to capacity, such as 10:00 on a Saturday morning, that time becomes unavailable online. After purchasing tickets, visitors can print the tickets at home so upon arrival, they simply present their tickets at the door and begin the tour.

Aquarium administrators unexpectedly found that, in the first months of operation, 90 percent of tickets were purchased online, an unprecedented percentage. Visitors, anticipating large crowds at the new facility, used the online service to guarantee admission. While the percentage of tickets purchased online has decreased over the years, most visitors still purchase and print tickets prior to arrival.

Advance ticket sales provide the aquarium with a number of benefits. The biggest benefit is being the first aquarium to claim "no lines, no waiting." Also, the aquarium can keep crowd congestion inside the exhibits to a reasonable limit. Advance ticket sales also mean management can plan ahead and determine how many staff and other resources to have on hand at any time.

In addition to ticket sales, Georgia Aquarium uses its Web site for scheduling volunteer work hours. The personnel department manages nearly 1,000 volunteers online. The Web site has also processed more than $2 million from over 40,000 donors. Recently, Georgia Aquarium has started providing audio tours of the facility for download to iPods from the Web site. Visitors arrive with their iPods loaded with audio files designed to lead them through the exhibits.

The automation of transaction processing provides buyers and sellers with many benefits. The smart implementation of transaction processing systems exhibited by Georgia Aquarium provides its customers and management with more services than traditional forms of transaction processing could ever offer.

Discussion Questions

1. How has Georgia Aquarium's Web site provided easier transactions for visitors?
2. How does Georgia Aquarium's management benefit from its unique approach to online ticket sales?

Critical Thinking Questions

1. How might an aquarium, zoo, or museum use technology to speed transaction processing for patrons who visit but have not purchased tickets in advance?
2. How do you think Georgia Aquarium deals with customers that arrive at an hour when advance tickets have sold to capacity?

Sources: *Computerworld* staff, "No lines, no waiting with the Georgia Aquarium's Web-based reservation and ticketing system," *Computerworld*, August 14, 2007, *www.computerworld.com/action/article.do?command=viewArticleBasic&articleId=9030 680&intsrc=cs_li_latest*; Georgia Aquarium Web site, *www.georgiaaquarium.org*, accessed June 29, 2008.

Many grocery stores combine point-of-sale scanners and coupon printers. The systems are programmed so that each time a specific product—for example, a box of cereal—crosses a checkout scanner, an appropriate coupon—perhaps a milk coupon—is printed. Companies can pay to be promoted through the system, which is then reprogrammed to print those companies' coupons if the customer buys a competitive brand. These TPSs help grocery stores increase profits by improving their repeat sales and bringing in revenue from other businesses.

Data Editing

data editing
The process of checking data for validity and completeness.

An important step in processing transaction data is to check data for validity and completeness to detect any problems, a task called **data editing.** For example, quantity and cost data must be numeric, and names must be alphabetic; otherwise, the data is not valid. Often, the codes associated with an individual transaction are edited against a database containing valid codes. If any code entered (or scanned) is not present in the database, the transaction is rejected.

Data Correction

data correction
The process of reentering data that was not typed or scanned properly.

It is not enough simply to reject invalid data. The system should also provide error messages that alert those responsible for editing the data. Error messages must specify the problem so proper corrections can be made. A **data correction** involves reentering data that was not typed or scanned properly. For example, a scanned UPC code must match a code in a master table of valid UPCs. If the code is misread or does not exist in the table, the checkout clerk is given an instruction to rescan the item or type the information manually.

Data Manipulation

data manipulation
The process of performing calculations and other data transformations related to business transactions.

Another major activity of a TPS is **data manipulation**, the process of performing calculations and other data transformations related to business transactions. Data manipulation can include classifying data, sorting data into categories, performing calculations, summarizing results, and storing data in the organization's database for further processing. In a payroll TPS, for example, data manipulation includes multiplying an employee's hours worked by the hourly pay rate. Overtime pay, federal and state tax withholdings, and deductions are also calculated.

Data Storage

data storage
The process of updating one or more databases with new transactions.

Data storage involves updating one or more databases with new transactions. After being updated, this data can be further processed and manipulated by other systems so that it is available for management reporting and decision making. Thus, although transaction databases can be considered a by-product of transaction processing, they have a pronounced effect on nearly all other information systems and decision-making processes in an organization.

Document Production and Reports

document production
The process of generating output records and reports.

Document production involves generating output records, documents, and reports. These can be hard-copy paper reports or displays on computer screens (sometimes referred to as *soft copy*). Printed paychecks, for example, are hard-copy documents produced by a payroll TPS, whereas an outstanding balance report for invoices might be a soft-copy report displayed by an accounts receivable TPS. Often, results from one TPS flow downstream to become input to other systems (as shown in Figure 9.5), which might use the results of updating the inventory database to create the stock exception report (a type of management report) of items whose inventory level is below the reorder point.

In addition to major documents such as checks and invoices, most TPSs provide other useful management information and decision support, such as printed or on-screen reports that help managers and employees perform various activities. A report showing current inventory is one example; another might be a document listing items ordered from a supplier to help a receiving clerk check the order for completeness when it arrives. A TPS can also produce reports required by local, state, and federal agencies, such as statements of tax withholding and quarterly income statements.

CONTROL AND MANAGEMENT ISSUES

Transaction processing systems process the fundamental business transactions that are the lifeblood of the firm's operation. They capture facts about basic business operations of the organization—facts without which orders cannot be shipped, customers cannot be invoiced, and employees and suppliers cannot be paid. In addition, the data captured by TPSs flows downstream to other systems in the organization where it is used to support analysis and decision making. TPSs are so critical to the operation of most firms that many business activities would come to a halt if the supporting TPSs failed. Because firms must ensure the reliable operation of their TPSs, they must also engage in disaster recovery planning and TPS audits.

Disaster Recovery Plan

Unfortunately, recent history reminds us of the need to be prepared in the event of a natural or man-made accident or disaster. The **disaster recovery plan (DRP)** is a firm's strategy to recover data, technology, and tools that support critical information systems and necessary information systems components such as the network, databases, hardware, software, and operating systems.

disaster recovery plan (DRP)
A formal plan describing the actions that must be taken to restore computer operations and services in the event of a disaster.

Those TPSs that directly affect the cash flow of the firm (such as order processing, accounts receivable, accounts payable, and payroll) are typically identified as critical business information systems. A lengthy disruption in the operation of any of those systems could create a serious cash flow problem for the firm and potentially put it out of business. Companies vary widely in the thoroughness and effectiveness of their disaster recovery planning, and, as a result, some have a harder time resuming business than others.

Fires, hurricanes, floods, earthquakes, and tornados are the most dramatic causes of business disasters. TiVo operates with 600 employees and 700 servers that store more than 100 TB of data. Its headquarters is in a part of the country where there are occasional earthquakes and other natural disasters. As a result, the firm set up its disaster recovery site in Las Vegas, a location relatively free of natural disasters and with an infrastructure proven reliable for operating the city's many casinos.[3] If a disaster strikes its headquarters, computer operations will be relocated to this site until normal operations can be restored.

However, according to Bob Vieraitis, vice president of marketing for change control software vendor Solidcore Systems, "Up to 80 percent of IT outages are caused by improper changes to the IT environment."[4] Such changes would include ill-planned upgrades to operating systems and applications or hardware that cause a system failure instead of improving matters. For example, a well-meaning system administrator at Web conferencing company WebEx Communications made a minor change to a file on one of the company's more than 2,000 servers spread across seven data centers. As soon as the change was made, the server went offline, making the service unavailable to a number of WebEx customers.[5]

ETHICAL AND SOCIETAL ISSUES

JetBlue—Trial by Fire and Ice

Today's global society depends on air travel for business and pleasure more than ever before. Services offered by airlines have come under scrutiny due to incidents that point to unreliability. Some of these incidents are caused by the inefficient use of information systems.

On the list of the major corporate disaster recovery challenges of recent times, JetBlue and the St. Valentine's Day ice storm appears in the top ten. Jet Blue has built a reputation as an airline that caters to the needs of its customers. Plush leather seats, expanded leg room, complimentary beverages and snacks, snooze kits, seat-back displays providing 36 channels of entertainment, satellite radio, first-run movies, wireless Internet, and smiling crew members, all at affordable rates, are amenities rare at traditional airlines. On February 14, 2007, the honeymoon seemed to be over for JetBlue and its customers.

Weather forecasters predicted an ice storm would hit the east coast on Valentine's Day. While it was unclear how much it might affect air traffic, most airlines took precautionary measures, canceling dozens of flights. In its efforts to please passengers, JetBlue gambled and waited it out until it was too late. Rather than improving, conditions only worsened over the course of the day, leaving hundreds of JetBlue passengers stranded in planes on tarmacs at JFK International airport in New York and other major airports including Washington, DC, and Newark, New Jersey—some for as long as 11 hours. Around 3:00 pm, JetBlue gave up hope and called in buses to rescue the passengers from the planes. By then the damage was done.

Thousands of passengers waiting in airline terminals were hoping to complete their trips despite the storms. More passengers were arriving at airports unaware of delays and cancellations. Still other passengers were returning to the terminals by bus from stranded aircraft. JetBlue wound up with thousands of irate passengers at their counters and no flights departing or arriving on the east coast. Chief executive officer David Neeleman admitted to doing a horrible job. "We got ourselves into a situation where we were doing rolling cancellations instead of a massive cancellation. Communications broke down, we weren't able to reach out to passengers, and they continued to arrive at the airports... it had a cascading effect."

Charles "Duffy" Mees won't ever forget that day. Duffy Mees was vice president and CIO of JetBlue Airways at the time. He came to JetBlue a few months prior to the disaster with years of experience in the airline industry. During his first few months, he oversaw the completion of an enterprise resource planning (ERP) installation at JetBlue. However, his experience did not prepare him for handling the Valentine's Day crisis.

The impact of the storm on JetBlue's information systems lasted a week. During the days that followed, many systems were pushed beyond their limits. Massive flight cancellations and rescheduling placed an unprecedented amount of traffic on JetBlue's reservation systems. Since JetBlue did not support rebooking flights online or at airport kiosks, customers had only one option for rebooking: call the JetBlue reservation office. JetBlue's Salt Lake City reservation agents were flooded with calls from angry passengers. Limitations in the system allowed only up to 650 agents to work at a time, plenty for normal days, but not for an exceptional demand. Many customers were stranded on hold waiting to rebook flights. Mees worked with their software provider to boost the limit to 950, which helped to open the bottleneck. Still, it was days before many passengers could get through to an agent.

Meanwhile, customer luggage was piling up in huge mounds at airports. JetBlue had no computerized system in place for tracking bags. The company had placed that system on the back burner while concentrating on its new ERP system. JetBlue had to haul mountains of luggage to off-site locations, where extra workers were employed to sort and identify bags. An information system was developed on the fly to scan bag tags and identify the owners from passenger records.

Besides reservations and baggage problems, managers faced outages and failure from important systems that control core operations. SkySolver software, which operations planners use to redeploy planes and crews, could not transfer new schedules to the primary flight scheduling systems. Programmers from the vendors attacked the problem and solved it within hours, but the delay caused more havoc. JetBlue was caught in a tailspin of system failures triggered by too much information all at once.

Mees and his crew spent three days and nights working to bring JetBlue systems back online. They pushed systems to their limits and created databases, tools, and applications on the fly in their efforts to find solutions.

During the crisis, JetBlue management learned many lessons and discovered many solutions, including preventative measures. A new system has now been implemented that allows passengers to rebook canceled flights online. Computer terminals have been installed at airports to allow passengers to rebook onsite. Software allows double the number of booking agents to respond in emergencies. A lost-bag system has been installed to track luggage—which is particularly valuable when flights are cancelled. A new system has been implemented that notifies passengers by e-mail, phone, or Web when flights are cancelled or changed.

Most significantly, the crisis motivated JetBlue to create a customer bill of rights offering compensation to customers whose flights have been cancelled or are left sitting too long on planes.

The cost to JetBlue for the Valentine's Day disaster has been estimated at around $30 million. What about the cost to JetBlue's reputation? After JetBlue offered many apologies and fired several top-level executives, it appears that JetBlue is still loved by its customers. J.D. Power and Associates 2007 airline satisfaction survey ranked JetBlue Number 1 by far for the third year in a row. In this case, good intentions seem to have won out over poor management.

Discussion Questions

1. What could JetBlue have done to prevent the Valentine's Day disaster in terms of information systems and management decisions?
2. What information systems does JetBlue use to manage air travel?

Critical Thinking Questions

1. How do you think JetBlue's disaster affected the airline industry from the airlines' perspective and the traveler's perspective?
2. JetBlue offers many amenities to its customers that other airlines have discontinued in order to cut costs. What are the benefits and dangers of JetBlue's approach, and how does this incident illustrate the dangers?

Sources: Duvall, Mel, "What Really Happened At JetBlue," *CIO Insight*, April 5, 2007, *www.cioinsight.com/c/a/Past-News/What-Really-Happened-At-JetBlue/1*; Ho, David, "Fans stand behind JetBlue," *Atlanta Journal-Constitution*, June 10, 2008, *www.ajc.com/business/content/business/stories/2008/06/10/jet_blue.html*; JetBlue Web site, *www.jetblue.com*, accessed June 29, 2008.

A key requirement in the event of a disaster is the ability to contact employees and others to inform them of the disaster and what action they should take. The MessageOne service from Dell offers e-mail continuity and storage services. It also offers an emergency warning and crisis communication service. In the event of a disaster, the system will send e-mail, fax, pager, and recorded SMS (Short Message Service, the communications protocol used to exchange short text messages via mobile phone) messages to specified users telling them what action to take.[6]

Companies such as Iron Mountain provide a secure, off-site environment for records storage. In the event of a disaster, vital data can be recovered.

(Source: Geostock/Getty Images.)

Transaction Processing System Audit

The Sarbanes-Oxley Act, enacted as a result of several major accounting scandals, requires public companies to implement procedures to ensure their audit committees can document financial data, validate earnings reports, and verify the accuracy of information. The Financial Services Modernization Act (Gramm-Leach-Bliley) requires systems security for financial service providers, including specific standards to protect customer privacy. The Health Insurance Portability and Accountability Act (HIPAA) defines regulations covering healthcare providers to ensure that their patient data is adequately protected. Many organizations conduct ongoing **transaction processing system audits** to prevent the kind of accounting irregularities or loss of data privacy that can put their firm in violation of these acts and erase investor confidence. The audit can be performed by the firm's own internal audit group, or an outside auditor might be hired to provide a higher degree of objectivity. A transaction processing system audit attempts to answer four basic questions:

transaction processing system audit
A check of a firm's TPS systems to prevent accounting irregularities and/or loss of data privacy.

- Does the system meet the business need for which it was implemented?
- What procedures and controls have been established?
- Are these procedures and controls being used properly?
- Are the information systems and procedures producing accurate and honest reports?

A typical audit also examines the distribution of output documents and reports, determines if only appropriate people can execute key system functions (e.g., approve the payment of an invoice), assesses the training and education associated with existing and new systems, and determines the effort required to perform various tasks and to resolve problems in the system. General areas of improvement are also identified and reported during the audit.

ENTERPRISE RESOURCE PLANNING, SUPPLY CHAIN MANAGEMENT, AND CUSTOMER RELATIONSHIP MANAGEMENT

As defined in Chapter 4, enterprise resource planning (ERP) is a set of integrated programs that manage a company's vital business operations for an entire multisite, global organization. Recall that a business process is a set of coordinated and related activities that takes one or more types of input and creates an output of value to the customer of that process. The customer might be a traditional external business customer who buys goods or services from the firm. An example of such a process is capturing a sales order, which takes customer input and generates an order. The customer of a business process might also be an internal customer such as a worker in another department of the firm. For example, the shipment process generates the internal documents workers need in the warehouse and shipping departments to pick, pack, and ship orders. At the core of the ERP system is a database that is shared by all users so that all business functions have access to current and consistent data for operational decision making and planning, as shown in Figure 9.6.

Figure 9.6

Enterprise Resource Planning System

An ERP integrates business processes and the ERP database.

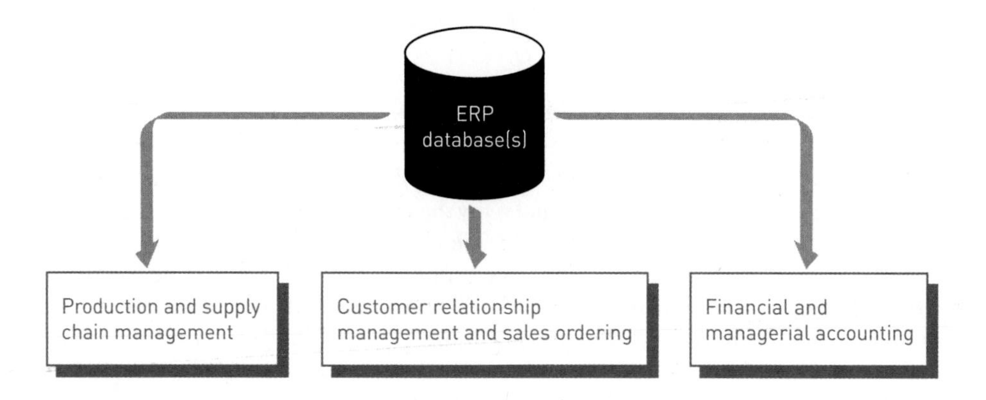

An Overview of Enterprise Resource Planning

ERP systems evolved from materials requirement planning systems (MRP) developed in the 1970s. These systems tied together the production planning, inventory control, and purchasing business functions for manufacturing organizations. During the late 1980s and early 1990s, many organizations recognized that their legacy transaction processing systems lacked the integration needed to coordinate activities and share valuable information across all the business functions of the firm. As a result, costs were higher and customer service poorer than desired. Large organizations, members of the *Fortune* 1000, were the first to take on the challenge of implementing ERP. As they did, they uncovered many advantages as well as some disadvantages summarized in the following sections.

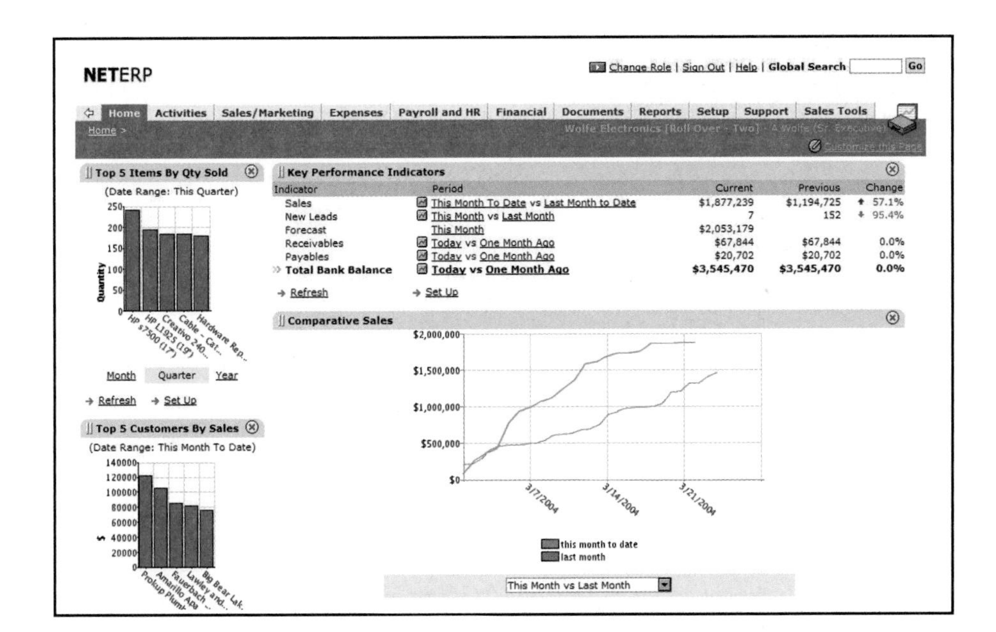

NetERP software from NetSuite provides tightly integrated, comprehensive ERP solutions for businesses, giving them access to real-time business intelligence and thus enabling better decision making.

(Source: Courtesy of NetSuite Inc.)

Advantages of ERP

Increased global competition, new needs of executives for control over the total cost and product flow through their enterprises, and ever-more-numerous customer interactions drive the demand for enterprise-wide access to real-time information. ERP offers integrated software from a single vendor to help meet those needs. The primary benefits of implementing ERP include improved access to data for operational decision making, elimination of inefficient or outdated systems, improvement of work processes, and technology standardization. ERP vendors have also developed specialized systems that provide effective solutions for specific industries and market segments.

Improved Access to Data for Operational Decision Making

ERP systems operate via an integrated database, using one set of data to support all business functions. The systems can support decisions on optimal sourcing or cost accounting, for instance, for the entire enterprise or business units from the start, rather than gathering data from multiple business functions and then trying to coordinate that information manually or reconciling data with another application. The result is an organization that looks seamless, not only to the outside world but also to the decision makers who are deploying resources within the organization. The data is integrated to facilitate operational decision making and allows companies to provide greater customer service and support, strengthen customer and supplier relationships, and generate new business opportunities.

For example, success in the retail industry is determined by a retailer's ability to have the right products on store shelves and priced correctly when customers come to shop. As a result, retailers need accurate, current, point-of-sale data and inventory data to match the merchandise assortment in their stores to the needs of their local markets. Tumi is a manufacturer and retailer of luggage, business cases, handbags, wallets, writing instruments, and watches. Its products are available at top department and specialty stores as well as over 50 Tumi stores around the world. The firm replaced its legacy systems with an integrated ERP system and achieved the following benefits from improved operational decision making: reduced days of sales outstanding in accounts receivable by 44 percent, increased sales by 100 percent with no increase in the number of employees, cut inventory levels by 30 percent leading to a reduction of warehouse space requirements of 38 percent, and sliced its month-end close process by five days.[7]

Elimination of Costly, Inflexible Legacy Systems

Adoption of an ERP system enables an organization to eliminate dozens or even hundreds of separate systems and replace them with a single, integrated set of applications for the entire enterprise. In many cases, these systems are decades old, the original developers are long gone, and the systems are poorly documented. As a result, the systems are extremely difficult to fix when they break, and adapting them to meet new business needs takes too long. They become an anchor around the organization that keeps it from moving ahead and remaining competitive. An ERP system helps match the capabilities of an organization's information systems to its business needs—even as these needs evolve.

Gujarat Reclaim and Rubber Products Ltd (GRRP) is an India-based company that produces reclaimed rubber from the scrap of whole tires, tread peelings, natural rubber tubes, butyl tubes, and other rubber products. The reclaimed rubber is then used to make other tire and non-tire products. The company implemented an ERP system that improved its work processes and eliminated dozens of separate systems. As a result, GRRP can now deliver products to customers on time. The ERP system also provided support for the currencies and financial laws of multiple countries, thus simplifying the firm's financial accounting processes. Perhaps most importantly, the ERP system made it possible for GRRP to increase the number of suppliers and customers as well as reduce its inventory of waste rubber.[8]

Improvement of Work Processes

Competition requires companies to structure their business processes to be as effective and customer oriented as possible. ERP vendors do considerable research to define the best business processes. They gather requirements of leading companies within the same industry and combine them with research findings from research institutions and consultants. The individual application modules included in the ERP system are then designed to support these **best practices**, the most efficient and effective ways to complete a business process. Thus, implementation of an ERP system ensures good work processes based on best practices. For example, for managing customer payments, the ERP system's finance module can be configured to reflect the most efficient practices of leading companies in an industry. This increased efficiency ensures that everyday business operations follow the optimal chain of activities, with all users supplied the information and tools they need to complete each step.

best practices
The most efficient and effective ways to complete a business process.

Amgen is a human therapeutics company in the biotechnology industry. It pioneered the development of innovative products based on advances in recombinant DNA and molecular biology, and launched some of the biotechnology industry's first major medicines. The firm is implementing a worldwide ERP program to manage business processes that are global, standardized, cross-functional, and scalable. With success, Amgen will increase its productivity and improve decision making. It expects to modify some 800 business processes in the areas of customer order to cash, finance, human resources, supply chain, and procurement, affecting more than 18,000 employees working in 45 countries.[9]

Upgrade of Technology Infrastructure

When implementing an ERP system, an organization has an opportunity to upgrade the information technology (hardware, operating systems, databases, etc.) that it uses. While centralizing and formalizing these decisions, the organization can eliminate the hodgepodge of multiple hardware platforms, operating systems, and databases it is currently using—most likely from a variety of vendors. Standardizing on fewer technologies and vendors reduces ongoing maintenance and support costs as well as the training load for those who must support the infrastructure.

BNSF Railway Company is a subsidiary of Burlington Northern Santa Fe Corporation and is in an industry that is anticipating an increase in demand for freight transportation of some 67 percent over the next 20 years. According to Jeff Campbell, vice president of technology services and CIO at BNSF, "It was a business imperative that we transform our entire back office to prepare for this kind of growth and to establish a fresh technology platform that would serve this company for the next 15 years. So we decided to replace our legacy systems with an ERP solution." BNSF is replacing its core financial systems, human resources, and payroll with an ERP system. When the implementation is complete, customers will be able to view and pay their freight invoices online.[10]

Disadvantages of ERP Systems

Unfortunately, implementing ERP systems can be difficult and error-prone. Some of the major disadvantages of ERP systems are the expense and time required for implementation, the difficulty in implementing the many business process changes that accompany the ERP system, the problems with integrating the ERP system with other systems, the risks associated with making a major commitment to a single vendor, and the risk of implementation failure.

Expense and Time in Implementation

Getting the full benefits of ERP takes time and money. Although ERP offers many strategic advantages by streamlining a company's TPSs, large firms typically need three to five years and spend tens of millions of dollars to implement a successful ERP system. Waste Management Inc. sued its ERP vendor to recover more than $100 million in project-related expenses plus unrealized savings and benefits from a failed ERP software implementation.[11]

Difficulty Implementing Change

In some cases, a company has to radically change how it operates to conform to the ERP's work processes—its best practices. These changes can be so drastic to long-time employees that they retire or quit rather than go through the change. This exodus can leave a firm short of experienced workers. Sometimes, the best practices simply are not appropriate for the firm and cause great work disruptions. American LaFrance, a manufacturer of emergency vehicles and equipment, filed bankruptcy in part due to operational disruptions caused by the installation of a new ERP system.[12]

Difficulty Integrating with Other Systems

Most companies have other systems that must be integrated with the ERP system, such as financial analysis programs, e-commerce operations, and other applications. Many companies have experienced difficulties making these other systems operate with their ERP system. Other companies need additional software to create these links.

Risks in Using One Vendor

The high cost to switch to another vendor's ERP system makes it extremely unlikely that a firm will do so. After a company has adopted an ERP system, the vendor has less incentive to listen and respond to customer concerns. The high cost to switch also increases risk—in the event the ERP vendor allows its product to become outdated or goes out of business. Selecting an ERP system involves not only choosing the best software product but also the right long-term business partner. It was unsettling for many companies that had implemented PeopleSoft, J.D. Edwards, or Siebel Systems enterprise software when these firms were acquired by Oracle.

Risk of Implementation Failure

Implementing an ERP system for a large organization is extremely challenging and requires tremendous amounts of resources, the best IS and businesspeople, and plenty of management support. Unfortunately, large ERP installations occasionally fail, and problems with an ERP implementation can require expensive solutions.

The following list provides tips for avoiding many common causes for failed ERP implementations:

- Assign a full-time executive to manage the project.
- Appoint an experienced, independent resource to provide project oversight and to verify and validate system performance.
- Allow sufficient time for transition from the old way of doing things to the new system and new processes.
- Plan to spend considerable time and money training people; many project managers recommend that $10,000–$20,000 per employee be budgeted for training of personnel.
- Define metrics to assess project progress and to identify project-related risks.
- Keep the scope of the project well defined and contained to essential business processes.
- Be wary of modifying the ERP software to conform to your firm's business practices.

ERP for Small and Medium-Size Enterprises (SMEs)

Organizations that are successful in implementing ERP are not limited to large *Fortune* 1000 companies. SMEs (both for-profit and not-for-profit) can achieve real business benefits from their ERP efforts. Many SMEs elect to implement open source ERP systems. With open source software, anyone can see and modify the source code to customize it to meet their needs. Such systems are much less costly to acquire and are relatively easy to modify to meet business needs. A wide range of organizations can perform the system development and maintenance. Table 9.3 lists some of the open source ERP systems geared for SMEs.

Table 9.3

Open Source ERP Systems

Vendor	ERP Solutions
Apache	Open For Business ERP
Compiere	Compiere Open Source ERP
Openbravo	Openbravo Open Source ERP
WebERP	WebERP

The lower cost of open source ERP systems is a powerful advantage for SMEs like frozen-food maker Cedarlane. IT director Daniel Baroco says the firm saved "a couple of hundred thousand dollars" by choosing an open source ERP system. Such cost savings were critical for the then-$40-million business.[13]

Vertex Distribution is a medium-sized manufacturer and distributor of rivets, screws, and other fasteners. Mark Alperin, COO with CIO responsibilities for the firm, chose the Compiere open source ERP suite due to its low cost and the flexibility it offered to modify the software to meet the needs of his business. Alperin says: "We have our own programming staff, and the ability because of that to customize services on our own and respond to customer needs is an advantage." Prevention Partners, Inc., a maker of health program posters, buttons, and related signage, also chose an open source ERP program, WebERP, for the same reasons. Galenicum, a three-year-old supplier of raw material for the pharmaceutical industry, chose the open source Openbravo ERP system.[14]

Recognizing that cost plays such a heavy factor in choosing an ERP vendor for SMEs, SAP offers an appliance-like system that comes preconfigured with its ERP software, a database and Linux operating system running on hardware from IBM or Hewlett-Packard. The offering lowers the total cost of ownership for customers.[15]

The following sections outline how an ERP system can support the various major business processes.

Business Intelligence

As discussed in Chapter 5, business intelligence (BI) involves gathering enough of the right information in a timely manner and usable form and analyzing it to shine a spotlight on the organization's performance. BI has become recognized as an essential component of an organization's ERP system. BI tools are used to access all the operational data captured in the ERP database and analyze performance on a daily basis, highlight areas for improvement, and monitor the results of business strategies. The most widely used BI software comes from SAP, IBM, Oracle, and Microsoft, with JasperSoft and Pentaho offering open source solutions.

In the retail industry, BI can enable retailers to gain customer knowledge and improve sales visibility across the enterprise so the firm can react to and better predict customer demand and maximize sales. Each week more than 13 million customers visit one of Lowe's 14,000 stores, which are stocked with more than 40,000 items. All this business generates billions of customer sales transaction records each year.[16] Lowe's uses BI to track sales for each item it carries at each of its stores to help plan the appropriate level of inventory to meet customer demand. Lowe's also uses BI to analyze customer product returns in real time to identify potential fraudulent returns. The amount of data and transaction processing power required is so great that Lowe's BI operation runs on some 3,000 servers.[17]

Production and Supply Chain Management

ERP systems follow a systematic process for developing a production plan that draws on the information available in the ERP system database.

The process starts with *sales forecasting* to develop an estimate of future customer demand. This initial forecast is at a fairly high level with estimates made by product group rather than by each individual product item. The sales forecast extends for months into the future; it might be developed using an ERP software module or produced by other means using specialized software and techniques. Many organizations are moving to a collaborative process with major customers to plan future inventory levels and production rather than relying on an internally generated sales forecast.

Oberto Sausage Company is a manufacturer of meat snacks and sausage products. Its products are sold directly to mass merchandisers and major supermarket chains in the United States. Oberto Sausage products are also distributed globally by Frito-Lay. Eric Kapinos, Director of Forecasting and Planning at Oberto, leads a forecasting process that is executed by a team of individuals using sales forecasting tools. "We have a full complement of Forecast Pro products at Oberto," explains Kapinos, "We use Forecast Pro Unlimited as the main foundation for our demand forecasting process—it's where the forecast is generated and maintained. After we establish the forecast, it is fed into our ERP system where it drives procurement, planning, scheduling, and plant execution."[18]

The *sales and operations plan* (S&OP) takes demand and current inventory levels into account and determines the specific product items that need to be produced and when to meet the forecast future demand. Production capacity and any seasonal variability in demand must also be considered. The result is a high-level production plan that balances market demand to production capacity. Air Products and Chemicals, Inc. serves technology, energy, industrial, and healthcare customers globally with a portfolio of products, services, and solutions that include atmospheric gases, process and specialty gases, performance materials, equipment, and services. The firm operates in 30 countries with about 18,500 employees. It uses software to support its sales and operations planning process. During S&OP meetings, various sales scenarios and options for meeting those scenarios are evaluated based on revenue, profit, and inventory impact, and then compared to operating plans. The advantages of this rigorous approach are "aligned strategy and operations, improved visibility and control over the global supply chain, fact-based planning, concentrating on exceptions and consistent performance management."[19]

Demand management refines the production plan by determining the amount of weekly or daily production needed to meet the demand for individual products. The output of the demand management process is the master production schedule, which is a production plan for all finished goods.

Detailed scheduling uses the production plan defined by the demand management process to develop a detailed production schedule specifying production scheduling details, such as which item to produce first and when production should be switched from one item to another. A key decision is how long to make the production runs for each product. Longer production runs reduce the number of machine setups required, thus reducing production costs. Shorter production runs generate less finished product inventory and reduce inventory holding costs.

Materials requirement planning determines the amount and timing for placing raw material orders with suppliers. The types and amounts of raw materials required to support the planned production schedule are determined based on the existing raw material inventory and the bill of materials or BOM, a sort of "recipe" of ingredients needed to make each product item. The quantity of raw materials to order also depends on the lead time and lot sizing. Lead time is the amount of time it takes from the placement of a purchase order until the raw materials arrive at the production facility. Lot size has to do with discrete quantities that the supplier will ship and the amount that is economical for the producer to receive and/or store. For example, a supplier might ship a certain raw material in units of 80,000 pound rail cars. The producer might need 95,000 pounds of the raw material. A decision must be made to order one or two rail cars of the raw material.

Walters Metal Fabrication is a structural steel fabrication and erection company. As the firm experienced a recent growth spurt, it became clear that it needed software tools to support an increase in production volume. Walters used to depend on suppliers to tell it how much material was required for its parts. The firm implemented MRP software from FabTrol to help with estimating, material management, production control, and shipping management. According to Don Porter, project manager: "The seamless import of the bill of materials from the [software] saves us a lot of time as well. In fact, importing the bill of materials, determining the changes, and producing an estimate now takes minutes instead of what used to take us days. Plus, the estimates are more accurate."[20]

Purchasing uses the information from materials requirement planning to place purchase orders for raw materials and transmit them to qualified suppliers. Typically, the release of these purchase orders is timed so that raw materials arrive just in time to be used in production and minimize warehouse and storage costs. Often, producers will allow suppliers to tap into data via an extranet that enables them to determine what raw materials the supplier needs, thus minimizing the effort and lead time to place and fill purchase orders.

Production uses the detailed schedule to plan the details of running and staffing the production operation.

ERP systems do not work directly with production machines, so they need a way to capture information about what was produced. This data must be passed to the ERP accounting modules to keep an accurate count of finished product inventory. Many companies have personal computers on the production floor that count the number of cases of each product item by scanning a UPC code on the packing material. Other approaches for capturing production quantities include the use of RFID chips and manually entering the data via a PDA.

Separately, production-quality data can be added based on the results of quality tests run on a sample of the product for each batch of product produced. Typically, this data includes the batch identification number, which identifies the production run and the results of various product quality tests.

Customer Relationship Management and Sales Ordering

Customer Relationship Management

As discussed in Chapter 2, a **customer relationship management (CRM) system** helps a company manage all aspects of customer encounters, including marketing and advertising, sales, customer service after the sale, and programs to keep and retain loyal customers (see Figure 9.7). The goal of CRM is to understand and anticipate the needs of current and potential customers to increase customer retention and loyalty while optimizing the way that products and services are sold. CRM is used primarily by people in the sales, marketing, and service organizations to capture and view data about customers and improve communications. Businesses implementing CRM systems report benefits such as improved customer satisfaction, increased customer retention, reduced operating costs, and the ability to meet customer demand.

customer relationship management (CRM) system
A system that helps a company manage all aspects of customer encounters, including marketing and advertising, sales, customer service after the sale, and programs to retain loyal customers.

CRM software automates and integrates the functions of sales, marketing, and service in an organization. The objective is to capture data about every contact a company has with a customer through every channel and store it in the CRM system so the company can truly understand customer actions. CRM software helps an organization build a database about its customers that describes relationships in sufficient detail so that management, salespeople, customer service providers—and even customers—can access information to match customer needs with product plans and offerings, remind them of service requirements, and know what other products they have purchased. Figure 9.8 shows contact manager software from SAP that fills this CRM role.

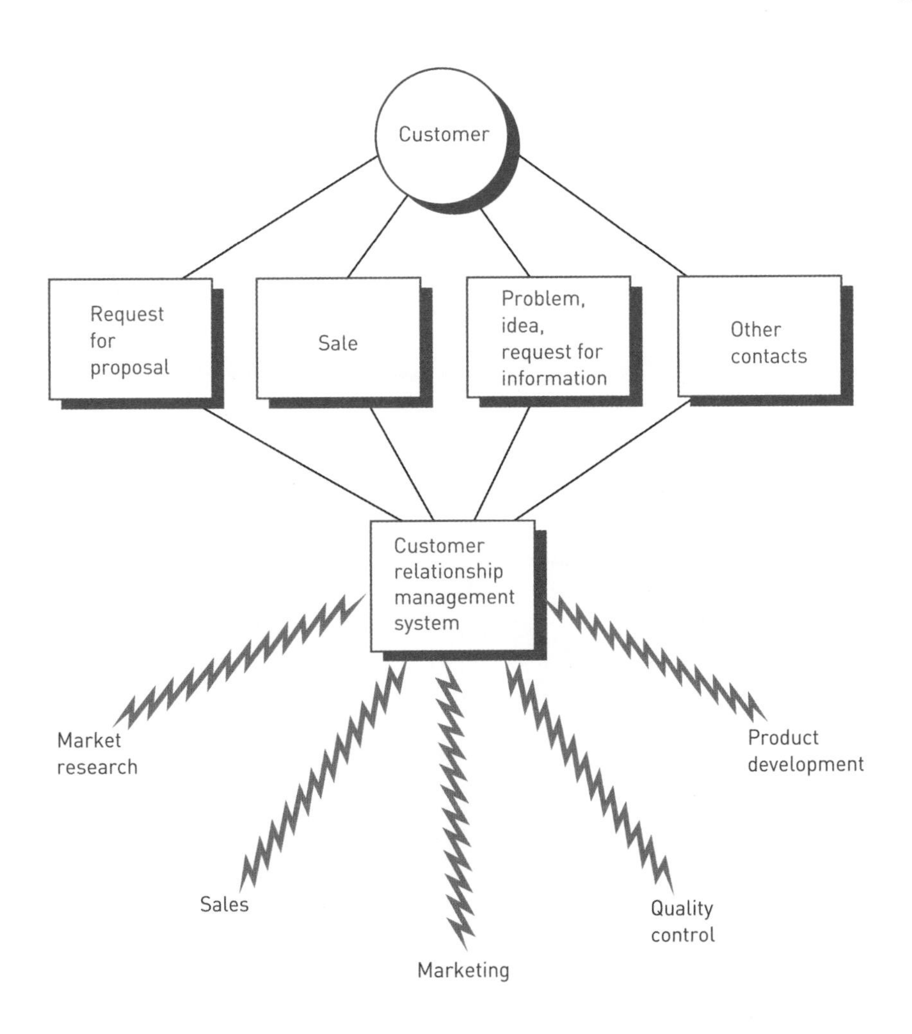

Figure 9.7

Customer Relationship Management System

The key features of a CRM system include the following:

- *Contact management:* The ability to track data on individual customers and sales leads and access that data from any part of the organization
- *Sales management:* The ability to organize data about customers and sales leads and then to prioritize the potential sales opportunities and identify appropriate next steps
- *Customer support:* The ability to support customer service reps so that they can quickly, thoroughly, and appropriately address customer requests and resolve customers' issues while at the same time collecting and storing data about those interactions
- *Marketing automation:* The ability to capture and analyze all customer interactions, generate appropriate responses, and gather data to create and build effective and efficient marketing campaigns
- *Analysis:* The ability to analyze customer data to identify ways to increase revenue and decrease costs, identify the source of the firm's "best customers," and determine how to retain them and find even more of them

CRM software vendors are racing to add new features and capabilities, as outlined in the following examples:

- Siebel CRM On Demand from Oracle provides social networking features to assist salespeople in reaching their sales goals. One feature identifies sales opportunities by combining sales data from internal systems with external information. Another feature assists salespeople in creating and joining groups like Facebook where they can make contacts with potential customers.[21]
- Sugar CRM has a version of its Web-based customer relationship management software that can be accessed by either the BlackBerry or Apple iPhone devices. Not to be outdone, SAP and Research in Motion have been working on a joint development that will enable customers to access SAP applications from BlackBerry mobile devices.[22]

- SugarCRM and other CRM vendors have also provided analytical reporting capabilities through a reporting wizard that can be used to create complex reports on contacts, accounts, sales opportunities, and cases.[23]

Most CRM software enables users to import contact data from various data service providers such as Jigsaw, which offers company-level contact data that can be downloaded for free directly into the CRM application. It also sells access to its more than 8 million contact records.[24] See Figure 9.8 for a view of the SAP Contact Manager.

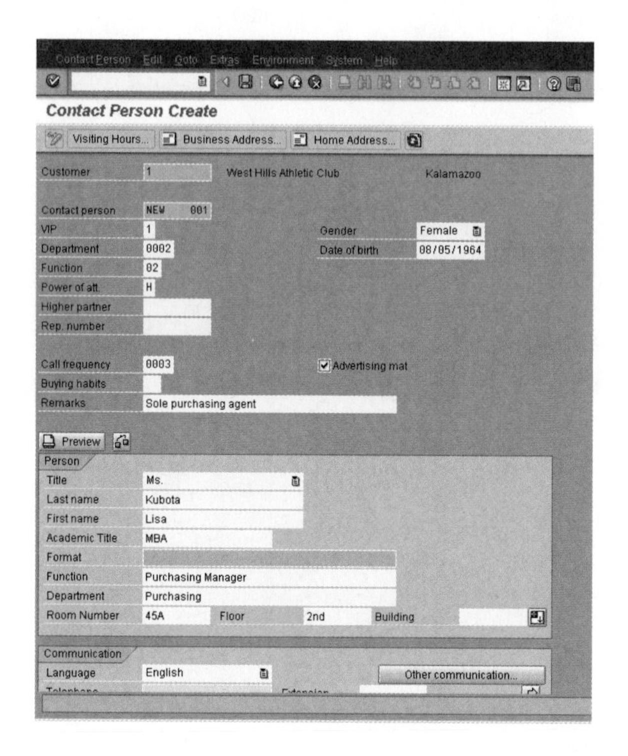

Figure 9.8

SAP Contact Manager

(Source: Copyright © by SAP AG.)

The focus of CRM involves much more than installing new software. Moving from a culture of simply selling products to placing the customer first is essential to a successful CRM deployment. Before any software is loaded onto a computer, a company must retrain employees. Who handles customer issues and when must be clearly defined, and computer systems need to be integrated so that all pertinent information is available immediately, whether a customer calls a sales representative or customer service representative. In addition to using stationary computers, most CRM systems can now be accessed via wireless devices.

ISM, Inc., is a CRM strategic advisor that rigorously tests the available CRM packages each year. Table 9.4 lists ISM's top-rated packages in alphabetical order by vendor for both large enterprises and SMEs. These packages scored the highest according to 217 selection criteria: 103 business functions, 52 technical features, 36 implementation capabilities, 9 real-time criteria, and 17 user-support features.

Organizations choose to implement CRM for a variety of reasons depending on their needs. American Eagle implemented a CRM system to improve marketers' ability to interact with customers via multiple channels including stores, the Web, mobile devices, and other means.[25] American of Martinsville, a contract furniture manufacturer, implemented CRM to automate its process for developing customer quotes and to improve customer communications.[26] Central Michigan University implemented a CRM system to improve its operational efficiency by reducing the elapsed time it takes to convert a request for information into an applicant and to retain those students through matriculation.[27] Kabel, a provider of cable TV, Internet, and telephone services, implemented CRM to improve its communications with customers and to provide support for an anticipated increase in customers.[28]

Even charitable organizations are employing CRM to track information about their supporters. The Salvation Army raises around $1.5 billion annually. It is implementing a CRM

ISM Top CRM for Large Enterprise	ISM Top CRM for Small and Medium Enterprise
Amdocs CRM CES v. 7.5; Amdocs Limited	Ardexus MODE v. 6.0; Ardexus, Inc.
Pivotal CRM v. 6.0; CDC Software	Powertrak v. 8.04; Axonom, Inc.
Saratoga CRM 6.6; CDC Software	C2 CRM v. 8.6; Clear C2, Inc.
C2 CRM v. 8.6; Clear C2, Inc.	Goldmine Enterprise Edition; FrontRange Solutions, Inc.
Consona CRM; Consona Corporation	Salesplace 2008; Interchange Solutions
Firstwave CRM v. 3.1; Firstwave Technologies, Inc.	Maximizer CRM 10; Maximizer Software Inc.
Infor CRM Epiphany; Infor	Microsoft CRM 4.0; Microsoft Corporation
CMS v. 9.0/OnContact CRM V v. 6.1; Oncontact Software Corporation	NetSuite CRM 2007.1 & NetSuite 2007.1; NetSuite, Inc.
ExSellence 5.5; Optima Technologies, Inc.	CMS v. 9.0/OnContact CRM V v. 6.1; Oncontact Software Corporation
PeopleSoft CRM; Oracle Corporation	Siebel CRM OnDemand; Oracle Corporation
Siebel 8.0; Oracle Corporation	Relavis CRM; Relavis Corporation
RightNow CRM v.8.2; RightNow Technologies, Inc.	Sage CRM 6.1; Sage Software
Salesforce.com; Salesforce.com	Sage SalesLogix v. 7.2; Sage Software
SAP CRM 2007; SAP AG	Salesforce.com; Salesforce.com
growBusiness Solutions; Software Innovation ASA update 7.0 STRIKE!; update software AG	Salespage CRM; Salespage Technologies, LLC
	StayinFront CRM v. 10; StayinFront, Inc.

Table 9.4

Top-Rated CRM Systems

Source: "ISM Announces Winners of 2008 Top 15 CRM Software Awards for Enterprise and Small & Medium Business Solutions," ISM Web site, *www.crm2day.com/news/crm/124848.php*, accessed June 10, 2008.

to aid in its fundraising activities to clothe, feed, and provide temporary living facilities for thousands of dispossessed people. John Herring, the Salvation Army's supporter relationship program director says: "We have a proliferation of databases and information silos that we need to bring together into an integrated CRM solution that will ensure we employ consistent methods of managing our valuable supporter data and better understand our relationships with our supporters."[29]

Sales Ordering

Sales ordering is the set of activities that must be performed to capture a customer sales order. A few of the essential steps include recording the items to be purchased, setting the sales price, recording the order quantity, determining the total cost of the order including delivery costs, and confirming the customer's available credit. The determination of the sales prices can become quite complicated and include quantity discounts, promotions, and incentives. After the total cost of the order is determined, it is necessary to check the customer's available credit to see if this order puts the customer over his credit limit. Figure 9.9 shows a sales order entry window in SAP business software.

Many small-to-midsize businesses are turning to ERP software to make it easier for their large customers to place orders with them. Car distributor smart USA is the only authorized distributor in the U.S. for a car called the smart fortwo. This rear-engine, two-passenger auto can achieve more than 50 miles per gallon. The firm implemented an ERP system with features to manage vehicle and parts ordering plus manage warranty claims from the dealer network to the manufacturer in Germany.[30]

Financial and Managerial Accounting

The general ledger is the main accounting record of a business. It is often divided into different categories, including assets, liabilities, revenue, expenses, and equity. These categories, in turn, are subdivided into subledgers to capture details such as cash, accounts payable, accounts receivable, and so on. In an ERP system, input to the general ledger occurs simultaneously with the input of a business transaction to a specific module. Here are several examples of how this occurs:

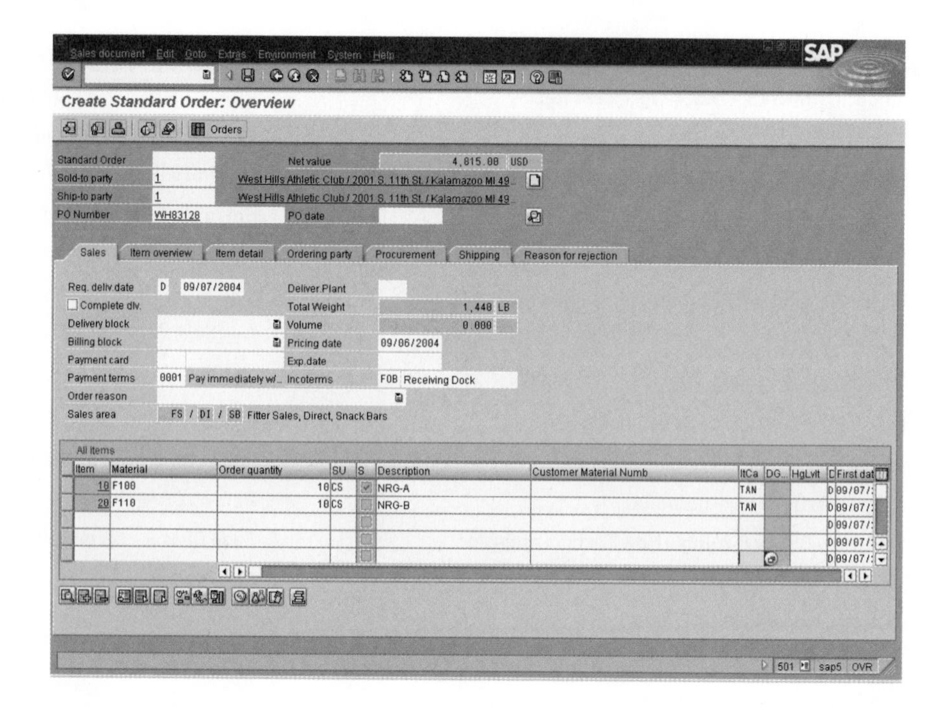

Figure 9.9

Sales Order Entry Window

(Source: Copyright © by SAP AG.)

- An order clerk records a sale and the ERP system automatically creates an accounts receivable entry indicating that a customer owes money for goods received.
- A buyer enters a purchase order and the ERP system automatically creates an accounts payable entry in the general ledger registering that the company has an obligation to pay for goods that will be received at some time in the future.
- A dock worker enters a receipt of purchased materials from a supplier, and the ERP system automatically creates a general ledger entry to increase the value of inventory on hand.
- A production worker withdraws raw materials from inventory to support production, and the ERP system generates a record to reduce the value of inventory on hand.

Thus, the ERP system captures transactions entered by workers in all functional areas of the business. The ERP system then creates the associated general ledger record to track the financial impact of the transaction. This set of records is an extremely valuable resource that companies can use to support financial accounting and managerial accounting.

Financial accounting consists of capturing and recording all the transactions that affect a company's financial state and then using these documented transactions to prepare financial statements to external decision makers, such as stockholders, suppliers, banks, and government agencies. These financial statements include the profit and loss statement, balance sheet, and cash flow statement. They must be prepared in strict accordance to rules and guidelines of agencies such as the Securities and Exchange Commission, the Internal Revenue Service, and the Financial Accounting Standards Board. Data gathered for financial accounting can also form the basis for tax accounting because this involves external reporting of a firm's activities to the local, state, and federal tax agencies.

Managerial accounting involves using "both historical and estimated data in providing information that management uses in conducting daily operations, in planning future operations, and in developing overall business strategies."[31] Managerial accounting provides data to enable the firm's managers to assess the profitability of a given product line or specific product, identify underperforming sales regions, establish budgets, make profit forecasts, and measure the effectiveness of marketing campaigns.

All transactions that affect the financial state of the firm are captured and recorded in the database of the ERP system. This data is used in the financial accounting module of the ERP system to prepare the statements required by various constituencies. The data can also be used in the managerial accounting module of the ERP system along with various assumptions and forecasts to perform various analyses such as generating a forecasted profit and loss statement to assess the firm's future profitability.

Hosted Software Model for Enterprise Software

Many business application software vendors are pushing the use of the hosted software model for SMEs. The goal is to help customers acquire, use, and benefit from the new technology while avoiding much of the associated complexity and high start-up costs. SAP, Microsoft, NetSuite, Intacct, Oracle, BizAutomation.com, Salesforce.com, NetBooks, and Workday are among the software vendors who offer hosted versions of their ERP or CRM software at a cost of $50–$200 per month per user.[32]

This pay-as-you-go approach is appealing to SMEs because they can experiment with powerful software capabilities without making a major financial investment. Organizations can then dispose of the software without large investments if the software fails to provide value or otherwise misses expectations. Also, using the hosted software model means the small business firm does not need to employ a full-time IT person to maintain key business applications. The small business firm can expect additional savings from reduced hardware costs and costs associated with maintaining an appropriate computer environment (such as air conditioning, power, and an uninterruptible power supply).

Potential problems can occur if the hosted software vendor cannot provide a reliable operation environment that ensures both that the software is available when needed and that company-sensitive data is safe from compromise. Car Toys, Inc., experienced an outage of its hosted BI software that disrupted month-end reporting when the hosted software provider decided to move the firm's data to new hardware with no prior warning.[33] Table 9.5 lists the advantages and disadvantages of hosted software.

Table 9.5

Advantages and Disadvantages of Hosted Software Model

Advantages	Disadvantages
Decreased total cost of ownership	Potential availability and reliability issues
Faster system startup	Potential data security issues
Lower implementation risk	Potential problems integrating the hosted products of different vendors
Management of systems outsourced to experts	Savings anticipated from outsourcing may be offset by increased effort to manage vendor

Not only is the hosted software model attractive to small and medium-sized firms, even some large companies are experimenting with it. Flextronics, a large contracted manufacturer, selected Workday to provide human capital management software as a service for 200,000 employees worldwide. Chiquita Brands, with 26,000 employees, also selected Workday to handle its HR system needs. Japan Post and Citibank adopted Salesforce.com's CRM software with 40,000 and 30,000 users, respectively.

INTERNATIONAL ISSUES ASSOCIATED WITH ENTERPRISE SYSTEMS

Enterprise systems must support businesses that interoperate with customers, suppliers, business partners, shareholders, and government agencies in multiple countries. Different languages and cultures, disparities in IS infrastructure, varying laws and customs rules, and multiple currencies are among the challenges that must be met by an enterprise system of a multinational company. The following sections highlight these issues.

Different Languages and Cultures

Teams composed of people from several countries speaking different languages and familiar with different cultures might not agree on a single work process. In some cultures, people do not routinely work in teams in a networked environment. Despite these complications, many multinational companies can establish close connections with their business partners and roll out standard IS applications for all to use. However, those standard applications often don't account for all the differences among business partners and employees operating in other parts of the world. So, sometimes they require extensive and costly customization. For example, even though English has become a standard business language among executives and senior managers, many people within organizations do not speak English. As a result, software might need to be designed with local language interfaces to ensure the successful implementation of a new system. Customization might also be needed for date fields: The U.S. date format is month/day/year, the European format is day/month/year, and Japan uses year/month/day. Sometimes, users might also have to implement manual processes to override established formatting to enable systems to function correctly.

Disparities in Information System Infrastructure

The lack of a robust or a common information infrastructure can also create problems. The U.S. telecommunications industry is highly competitive, with many options for high-quality service at relatively low rates. Many other countries' telecommunications services are controlled by a central government or operated as a monopoly, with no incentives to provide fast and inexpensive customer service. For example, much of Latin America lags the rest of the world in Internet usage, and online marketplaces are rare. This gap makes it difficult for multinational companies to get online with their Latin American business partners. Even something as mundane as the power plug on a piece of equipment built in one country might not fit into the power socket of another country.

Varying Laws and Customs Rules

Numerous laws can affect the collection and dissemination of data. For example, labor laws in some countries prohibit the recording of worker performance data. Also, some countries have passed laws limiting the transborder flow of data linked to individuals. Specifically, European Community Directive 95/96/EC of 1998 requires that any company doing business within the borders of the 25 European Union member nations protect the privacy of customers and employees. It bars the export of data to countries that do not have data-protection standards comparable to the European Union's.

Trade custom rules between nations are international laws that set practices for two or more nations' commercial transactions. They cover imports and exports and the systems and procedures dealing with quotas, visas, entry documents, commercial invoices, foreign trade zones, payment of duty and taxes, and many other related issues. For example, the North American Free Trade Agreement (NAFTA) of 1994 created trade custom rules to address the flow of goods throughout the North American continent. Most of these custom rules and their changes over time create significant complications for people who must keep enterprise systems consistent with the rules.

Multiple Currencies

The enterprise system of multinational companies must conduct transactions in multiple currencies. To do so, a set of exchange rates is defined, and the information systems apply these rates to translate from one currency to another. The systems must be current with foreign currency exchange rates, handle reporting and other transactions such as cash receipts, issue vendor payments and customer statements, record retail store payments, and generate financial reports in the currency of choice.

ERP software vendors are working to help meet these challenges. For example, Brazil has one of the strongest and fastest growing economies in Latin America. Many Brazilian organizations are implementing ERP systems to support and manage their operations. Brazil is a very large country and has many unique tax requirements that must be met. The major ERP vendors (SAP, Oracle, and Microsoft) are compliant with Brazilian tax laws and offer software translated into Portuguese.[34] As another example, Ufida, China's largest supplier of ERP software, has hired Lionbridge Technologies, a provider of translation and localization services, to translate its ERP software to English. It is also adding functionality to deal with U.S. taxes, regulations, and business practices.[35]

Leading ERP Systems

ERP systems are commonly used in manufacturing companies, colleges and universities, professional service organizations, retailers, and healthcare organizations. SMEs represent the greatest growth opportunity for ERP companies. Table 9.6 identifies the current leading vendors of ERP systems for both large organizations and SMEs.

Table 9.6

Leading ERP Software Vendors

Vendor	ERP Solutions	Customer Focus
Consoria	Intuitive ERP Made2Manage ERP	Small to mid-size manufacturers
Epicor	Epicor Vantage Epicor Enterprise Epicor iScala	Mid-size organizations and divisions and subsidiaries of Global 1000 firms
Exact	Macola ES eSyngery MAX	Small to mid-size organizations
Infor	Infor ERP Solutions Suite	Customers of all sizes with tailored solutions for aerospace companies, apparel and footwear companies, automotive suppliers, electrical distributors
Microsoft	Microsoft Dynamics GP Microsoft Dynamics NAV Microsoft Dynamics AX	Small, medium, and large organizations; government and educational institutions
NetSuite	NetSuite Accounting/EEP NetSuite Small Business NetSuite Wholesale/Distribution NetSuite Services NetSuite Software Company Edition	Hosted system aimed at growing and medium-size businesses in e-commerce, wholesale and distribution, software and retail sectors
Oracle	Oracle E-business Suite PeopleSoft Enterprise JD Edwards Enterprise One Oracle Transportation Management	Customers of all sizes
Ross Enterprise	Ross ERP Suite	Small, medium, and large organizations in the food and beverage, life sciences, consumer packaged goods, chemicals, and natural products industries
Sage	Sage MAS 500 ERP Sage MAS 90 and 200 Sage PFW ERP Manufacturing Sage Pro ERP	Small and medium-sized organizations
SAP	SAP 6.0 mySAP All-in-One	Customers of all sizes
Syspro	SYSPRO ERP SYSPRO Analytics SYSPRO e.net Solution S?YSPRO Planning and Scheduling	Mid-size organizations

Interestingly, although Microsoft and SAP market competing enterprise software, they also partnered with one another in the development of Duet, a set of technologies that enables SAP users to access and interact with their back-end ERP system via a familiar Microsoft Outlook interface. The goal is to boost worker productivity without additional training. Easy access to key back-end data helps SAP and Microsoft users make faster and more informed business decisions.

SUMMARY

Principle

An organization must have information systems that support the routine, day-to-day activities that occur in the normal course of business and help a company add value to its products and services.

Transaction processing systems (TPSs) are at the heart of most information systems in businesses today. A TPS is an organized collection of people, procedures, software, databases, and devices used to capture fundamental data about events that affect the organization (transactions) and use that data to update the official records of the organization.

The methods of transaction processing systems include batch and online. Batch processing involves the collection of transactions into batches, which are entered into the system at regular intervals as a group. Online transaction processing (OLTP) allows transactions to be entered as they occur.

Order processing systems capture and process customer order data from receipt of order through creation of a customer invoice.

Accounting systems track the flow of data related to all the cash flows that affect the organization.

Purchasing systems support the inventory control, purchase order processing, receiving, and accounts payable business functions.

Organizations today, including SMEs, typically implement an integrated set of TPSs from a single or limited number of software vendors to meet their transaction processing needs.

Organizations expect TPSs to accomplish a number of specific objectives, including processing data generated by and about transactions, maintaining a high degree of accuracy and information integrity, compiling accurate and timely reports and documents, increasing labor efficiency, helping provide increased and enhanced service, and building and maintaining customer loyalty. In some situations, an effective TPS can help an organization gain a competitive advantage.

All TPSs perform the following basic activities: data collection, which involves the capture of source data to complete a set of transactions; data editing, which checks for data validity and completeness; data correction, which involves providing feedback of a potential problem and enabling users to change the data; data manipulation, which is the performance of calculations, sorting, categorizing, summarizing, and storing data for further processing; data storage, which involves placing transaction data into one or more databases; and document production, which involves outputting records and reports.

Because of the importance of TPSs to ongoing operations, organizations must develop a disaster recovery plan that focuses on the actions that must be taken to restore computer operations and services in the event of a disaster.

Organizations conduct ongoing transaction processing system audits to avoid accounting irregularities and loss of data privacy that could put their firm into legal difficulty or destroy investor confidence.

The TPS audit attempts to answer four basic questions: (1) Does the system meet the business need for which it was implemented? (2) What procedures and controls have been established? (3) Are these procedures and controls being used properly? and (4) Are the information systems and procedures producing accurate and honest reports?

Principle

A company that implements an enterprise resource planning system is creating a highly integrated set of systems, which can lead to many business benefits.

Enterprise resource planning (ERP) is software that supports the efficient operation of business processes by integrating activities throughout a business, including sales, marketing, manufacturing, logistics, accounting, and staffing.

Implementation of an ERP system can provide many advantages, including providing access to data for operational decision making; elimination of costly, inflexible legacy systems; providing improved work processes; and creating the opportunity to upgrade technology infrastructure.

Some of the disadvantages associated with an ERP system are that they are time consuming, difficult, and expensive to implement.

Many SMEs are implementing ERP systems to achieve organizational benefits. In many cases, they choose to implement open source systems because of the lower total cost of ownership and their ability to be easily modified.

Although the scope of ERP implementation can vary from firm to firm, most firms use ERP systems to support business intelligence, production and supply chain management, customer relationship management and sales ordering, and financial and managerial accounting.

The production and supply chain management process starts with sales forecasting to develop an estimate of future customer demand. This initial forecast is at a fairly high level, with estimates made by product group rather than by each individual product item. The sales and operations plan takes demand and current inventory levels into account and determines the specific product items that need to be produced and when to meet the forecast future demand. Demand management refines the production plan by determining the amount of weekly or daily production needed to meet the demand for individual products. Detailed scheduling uses the production plan defined by the demand management process to develop a detailed production schedule specifying details such as which item to produce first and when production should be switched from one item to another. Materials requirement planning determines the amount and timing for placing raw material orders with suppliers. Purchasing uses the information from materials requirement planning to place purchase orders for raw materials and transmit them to qualified suppliers. Production uses the detailed schedule to plan the logistics of running and staffing the production operation.

The individual application modules included in the ERP system are designed to support best practices, the most efficient and effective ways to complete a business process.

Business application software vendors are experimenting with the hosted software model to see if the approach meets customer needs and is likely to generate significant revenue. This approach is especially appealing to SMEs due to the low initial cost, which makes it possible to experiment with powerful software capabilities.

Numerous complications arise that multinational corporations must address in planning, building, and operating their TPSs. These challenges include dealing with different languages and cultures, disparities in IS infrastructure, varying laws and customs rules, and multiple currencies.

CHAPTER 9: SELF-ASSESSMENT TEST

An organization must have information systems that support the routine, day-to-day activities that occur in the normal course of business and help a company add value to its products and services.

1. Identify the missing TPS basic activity: data collection, data editing, data __________, data manipulation, data storage, and document production.
2. The amount of support for decision making that a TPS directly provides managers and workers is low. True or False?
3. Which of the following is *not* one of the basic components of a TPS?
   a. databases
   b. networks
   c. procedures
   d. analytical models
4. A form of TPS where business transactions are accumulated over a period of time and prepared for processing as a single unit is called ______________.
5. Capturing data at its source and recording it accurately in a timely fashion, with minimal manual effort, and in an electronic or digital form that can be directly entered into the computer are the principles behind __________.
6. Which of the following is a set of transaction processing systems sometimes referred to as the "lifeblood of the organization?"
   a. purchasing systems
   b. accounting systems
   c. order processing systems
   d. none of the above
7. Many organizations conduct ongoing transaction processing system ______ to prevent accounting irregularities or loss of data privacy that might violate federal acts.
8. Inventory control, purchase order processing, receiving, and accounts payable systems make up a set of systems that support the ____________________ business function.
9. The ____________ transaction processing system manages the cash flow of the company by keeping track of the money owed the company.

A company that implements an enterprise resource planning system is creating a highly integrated set of systems, which can lead to many business benefits.

10. Many multinational companies roll out standard IS applications for all to use. However, standard applications often don't account for all the differences among business partners and employees operating in other parts of the world. Which of the following is a frequent modification that is needed for standard software?
    a. Software might need to be designed with local language interfaces to ensure the successful implementation of a new IS.
    b. Customization might be needed to handle date fields correctly.
    c. Users might also have to implement manual processes and overrides to enable systems to function correctly.
    d. all of the above
11. Which of the following is a primary benefit of implementing an ERP system?
    a. elimination of inefficient systems
    b. easing adoption of improved work processes
    c. improving access to data for operational decision making
    d. all of the above
12. The individual application modules included in an ERP system are designed to support the ______ ___________, the most efficient and effective ways to complete a business process.

13. Most companies can implement an ERP system without major difficulty. True or False?
14. Only large, multinational companies can justify the implementation of ERP systems. True or False?

CHAPTER 9: SELF-ASSESSMENT TEST ANSWERS

(1) correction (2) True (3) d (4) batch processing system (5) source data automation (6) c (7) audits (8) purchasing (9) accounts receivable (10) d (11) d (12) best practices (13) False (14) False

REVIEW QUESTIONS

1. Identify six specific objectives organizations expect their TPSs to accomplish.
2. What basic transaction processing activities are performed by all transaction processing systems?
3. Distinguish between a batch processing system and an online processing system.
4. Identify four significant international issues associated with the use of enterprise systems.
5. What special needs does an SME have in selecting an ERP system that is different from a large organization?
6. Identify four complications that multinational corporations must address in planning, building, and operating their ERP systems.
7. How does materials requirement planning support the purchasing process? What are some of the issues and complications that arise in materials requirement planning?
8. A disaster recovery plan should place emphasis on recovery of what sort of transaction processing systems?
9. What is the role of a CRM system? What sort of business benefits can such a system produce?
10. What systems are included in the traditional TPS systems that support the accounting business function?
11. What is the purpose of the transaction processing system audit? Who typically performs the audit?
12. Why is the general ledger application key to the generation of accounting information and reports?
13. What is the difference between managerial and financial accounting?
14. What is the chart of accounts? How is it used?
15. List and briefly describe the set of activities that must be performed by the sales ordering module of an ERP system to capture a customer sales order.

DISCUSSION QUESTIONS

1. Identify at least three ways a TPS can provide a firm with a competitive advantage. Develop an example of one way a firm could gain a competitive advantage from a TPS.
2. Assume that you are the owner of a large landscaping firm serving hundreds of customers in your area. Identify the kinds of customer information you would like to have captured by your firm's CRM system. How might this information be used to provide better service or increase revenue?
3. Imagine that you are the new IS manager for a *Fortune* 1000 company. Surprisingly, the firm still operates with a hodgepodge of transaction processing systems—some are software packages from various vendors and some are in-house developed systems. Prepare a brief outline of a talk you will make to senior company managers to convince them that it is time to implement a comprehensive ERP system. What sort of resistance and objections do you expect to encounter? How would you overcome these?
4. In what ways is the implementation of an ERP system simpler and less risky for an SME than for a large, multinational corporation?
5. The text mentioned that Lowe's uses business intelligence to analyze customer product returns in real time to identify potential fraudulent returns. Develop a description of how this process might work and what data is required.
6. What are some of the challenges and potential problems of implementing a CRM system and CRM mindset in a firm's employees? How might you overcome these?
7. You are the key user of the firm's accounts receivable system and have been asked to lead an internal audit of this system. Outline the steps you would take to complete the audit. Identify specific problems you would look for.
8. Your friend has been appointed the project manager of your firm's ERP implementation plan. What advice would you offer to help ensure the success of the project?
9. What sort of benefits should the suppliers and customers of a firm that has successfully implemented an ERP system

see? What sort of issues might arise for suppliers and customers during an ERP implementation?

10. Many organizations are moving to a collaborative process with their major customers to get their input on planning future inventory levels and production rather than relying on an internally generated demand forecast. Explain how such a process might work. What issues and concerns might a customer have in entering into an agreement to do this?

PROBLEM-SOLVING EXERCISES

1. Assume that you are forming a consulting firm to perform an external audit of firms' accounting transaction processing systems. Use a word processing software package to develop a survey questionnaire. Develop a list of at least five questions you would ask as part of your audit that cover the firm's overall approach to control and security. Develop another set of at least five questions specific to each of the accounts payable, accounts receivable, and general ledger systems. Now use a spreadsheet program to devise an audit scorecard that rates how well a firm performs on the audit. Consider assigning weights to the various questions so that 100 points is a "perfect score." Develop some sort of scale for each firm's score (e.g., 95–100 is excellent, 90–95 is very good, etc).
2. Use a spreadsheet program to develop a sales forecasting system for a new car dealership that can estimate monthly sales for each make and model based on historical sales data and various parameters. *Suggestion*: Assume that this month's sales will be the same as the sales for this month last year except for adjustments due to the cost of gas and each make of car's miles per gallon. You can further refine the model to take into account change in interest rates for new cars or other parameters you wish to include. Document the assumptions you make in building your model.

TEAM ACTIVITIES

1. Your team members should interview several business managers at a firm that has implemented a CRM system. Try to define the scope and schedule for the overall project. Make a list of what they see as the primary benefits of the implementation. What were the biggest hurdles they had to overcome? Did the firm need to retrain its employees to place greater emphasis on putting the customer first?
2. As a team, develop a list of at least seven key criteria that a small to medium-sized manufacturing firm should consider in selecting an ERP system. Discuss each criterion and assign a weight representing the relative importance of that criterion. Develop a simple spreadsheet to use in scoring various ERP alternatives.

WEB EXERCISES

1. Do research on the Web and find a Web site that offers a demo of an ERP or CRM system. View the demo, perhaps more than once. Write a review of the software based on the demo. What are its strengths and weaknesses? What additional questions about the software do you have? E-mail your questions to the vendor and document their response to your questions.
2. Using the Web, identify several companies that have implemented an ERP system in the last two years. Classify the implementations as a success, partial success, or failure. What is your basis for making this classification? Do you see any common reasons for success? For failure?

CAREER EXERCISES

1. Initially thought to be cost-effective for only very large companies, CRM systems are now being implemented in small and mid-sized companies to reduce costs and improve service. A firm's operations and accounting personnel play a dual role in the implementation of such a system: (1) They must ensure a good payback on the investment in CRM, and (2) they must also ensure that the system meets the needs of the operations and accounting organizations. Identify three or four tasks that the operations and accounting personnel need to perform to ensure that both these goals are met.
2. ERP software vendors need business systems analysts that understand both information systems and business processes. Make a list of six or more specific qualifications needed to be a strong business systems analyst supporting an ERP implementation within a medium-sized, but global organization.

CASE STUDIES

Case One

Aselsan Overhauls Core Systems

Based in Ankara, Turkey, Aselsan is the largest military electronics manufacturer in Turkey. The company is divided into four divisions: Communications (HC), Radar, Electronic Warfare and Intelligence Systems (REHIS), Defense Systems Technologies (SST), and Microelectronics, Guidance & Electro-Optics Division (MGEO).

Historically, Aselsan has been a production-focused business. It received orders from customers for particular electronics components and filled them. Recently, the company has taken on more responsibility. Aselsan customers are increasingly requiring Aselsan to provide project management services. The company now needs to extend its expertise into areas of design, system engineering, and the coordination of exchanges among third-party subcontractors.

Aselsan's information systems were not originally designed to support project management activities. The company needed an ERP that could provide integrated views of finance and budgeting, supply planning and scheduling, and coordinate production centered on specific projects. "The information needed for effective project management is getting more and more complex, and we needed an integrated solution to support that," explained Fatih Bilgi, IT director at Aselsan.

Aselsan evaluated ERP products from a number of vendors and identified the product and company that presented the best fit for its needs. The systems engineers had an additional challenge: The project needed to be completed within 18 months, before Turkey adopted a new currency. Aselsan's old systems could not accommodate the new currency, so if the new system wasn't in place, Aselsan would be out of business.

As in many ERP installations, Aselsan had to overcome many obstacles along the way. Systems engineers had to adapt to changes within the business while designing the core business systems. For example, over the 18-month installment, Aselsan's revenue doubled—so the problem changed while the solution was being developed. Aselsan also formed a new subsidiary, Aselsan Net, that the new system needed to accommodate.

The new financial systems were installed and running in enough time for the currency switch. In the months that followed, the company implemented a new planning and optimization component and a data warehousing component.

Aselsan introduced parts of the new system over time. During the year after the initial installation, the user base grew from 1,300 to 2,150 across seven locations and continues to grow. Dramatic improvements in the way Aselsan does business are occurring gradually and steadily.

The new integrated system has provided many benefits to Aselsan. The system has eliminated data redundancy in the organization, lowering administrative overhead. Many time-consuming operations such as preparation of annual plans and month-end closure operations have been cut in half. Product design change approval time has been reduced by 40 percent. Data warehousing systems allow employees across the organization to access information and reports with a mouse click. Managers can compare and balance project budgets for optimum return on investment.

In Aselsan's so-called Advanced Planning & Optimization system, suppliers can collaborate with Aselsan engineers on projects. The system provides a Web interface that allows all involved in a project to communicate and share documents and information.

Aselsan is working hard to become a leading global player in the defense industry. Expanding into project management backed by a strong ERP system will give the company the organizational fuel to achieve its goals.

Discussion Questions

1. Why did Aselsan need to make the huge investment in an ERP system? Why was its old system insufficient?
2. What challenges did Aselsan face while developing its ERP system?

Critical Thinking Questions

1. How does the quality of an ERP system contribute to a business's ability to compete in the global market?
2. How do information systems support collaboration among companies involved in a project?

Sources: SAP Staff, "Aselsan, Defense Manufacturer Supports New Project-Based Business Model with SAP Software", SAP Customer Success Story, September 2007, *http://download.sap.com/solutions/business-suite/erp/customersuccess/download.epd?context=96175C87E66495B39AE005-DAEE62545D775C2CE0BC2F4F8760AAF7F5EC70-CA2A44171385155615C2F30DD3179A2A13EA136F5F1219B19DE5*; SAP Web site, *www.sap.com*, accessed June 30, 2008; Aselsan Web site, *www.aselsan.com.tr*, accessed June 30, 2008.

Case Two

Delhi Government Embraces Enterprise Systems

Delhi is the second largest city in India, with a population of over 17 million. Located on the banks of the Yamuna River, Delhi was established in 1000 BC, making it one of the oldest cities in the world. Delhi is governed in part by the Municipal Corporation of Delhi (MCD), which has a staff of over 100,000 working in 107 offices across 12 geographic zones. Recently, the MCD has been working to move its staff online, migrating from a paper form-driven system to online automation. Those in charge of the project are learning that changing the work processes of an organization of this size is no small undertaking.

One of the MCD's most challenging tasks has been procuring contractors, services, and products for use on city projects. The traditional procedure for procurement involved getting the word out to let the community know that a project was planned and contractors were needed. Businesses that wanted to bid on the contract would then travel to the MCD, sometimes over hundreds of miles, to bid on the project. The process of bidding and negotiation might take months, until one contractor finally won the contract.

The time-honored bidding system, referred to as tendering, was far from convenient or fair. In many cases, eligible companies lost contracts because they missed the deadline by minutes due to travel delays. Corruption in the system also treated participants unfairly, with some companies using intimidation techniques to keep competitors from bidding on contracts. It was time to bring the MCD's tendering system into the digital age.

Arun Kumar, an executive engineer at the MCD, took on the responsibility of making that transition. Arun has completed the installation of an e-tendering system that automates much of the bidding process and is working to go farther and establish a complete e-procurement system.

The new e-tendering system allows contractors to download and upload tender documents online, track the status of tenders, and receive e-mail alerts. The system was developed and deployed in stages. First, the MCD required companies to submit tenders and bids online. Secondly, the MCD set up an online and offline backup system to safeguard the tender information against equipment failure. The MCD also provided a telephone help desk available to contractors 24 hours a day.

The MCD leases data center space from the Center for Development of Advanced Computing (C-DAC). The Web-based e-tendering system was developed by information systems company Wipro, which has since installed the system in several other government agencies in India.

The new system has eliminated the need for companies to physically send representatives to the MCD headquarters. It has also provided bidders with privacy and the government with transparency. Bidders need no longer fear intimidation from competitors since no one knows who is bidding. Bids are placed in an anonymous fashion, freeing contracting decision makers from outside influences. Now contracts are awarded on a bidder's merit and bid, and not according to who knows whom.

The MCD went to great lengths to sell contractors and government staff on the new way of doing business. It contracted Wipro to train hundreds of users and thousands of contractors. At first, only 70 percent of the engineers used the system, with the remaining engineers not willing to touch a computer. After about six months, the advantages of the system won over the holdouts.

The new e-tendering system has been a huge success, with over 30,000 tenders placed over the system—the world's highest volume in numbers by any government organization. The MCD increased the number of transactions it handles each week and reduced the time it takes to award a contract from 90 days to 30 days.

Still, Arun Kumar sees other areas that need improvement. He and his team are testing an e-procurement system that they hope will streamline approvals. The contract approval cycle currently takes two to three months because it's a manual task that involves a certifying authority auditing the process and paperwork. If Arun can reduce this time by two-thirds, that would really make a big difference.

Discussion Questions

1. What were some of the biggest challenges in implementing the new e-tendering system at the MCD?
2. What benefits does the new system provide for the MCD and its contractors?

Critical Thinking Questions

1. Compared to a business, what considerations might be different for a government agency designing an enterprise system?
2. Why do you think some engineers were hesitant to cooperate with the MCD in using the new system?

Sources: Talgeri, Kunal, "Delhi, India to set up 'comprehensive' e-procurement system," *itWorld Canada*, September 3, 2007, *www.itworldcanada.com/Pages/Docbase/ViewArticle.aspx?ID=idgml-a9370738-e371-4cee-8dcb-d52207a095b1&Portal=2e6e7040-2373-432d-b393-91e487ee7d70&ParaStart=0&ParaEnd=15&direction=next&Next=Next*; MCD Online, *www.mcdonline.gov.in*, accessed June 30, 2008; Google Maps, *maps.google.com/maps?q=Delhi,+India&ie=UTF-8&oe=utf-8&rls=org.mozilla:en-US:official&client=firefox-a&um=1&sa=X&oi=geocode_result&resnum=1&ct=title*, accessed June 30, 2008; Delhi India City Guide Web site, *delhi.clickindia.com*, accessed June 30, 2008.

Questions for Web Case

See the Web site for this book to read about the Whitmann Price Consulting case for this chapter. Following are questions concerning this Web case.

Whitmann Price Consulting: Enterprise Systems

Discussion Questions

1. What would be the danger of Josh and Sandra developing the Advanced Mobile Communications and Information System without considering other systems within Whitmann Price?
2. What are the advantages of ERP systems that provide an integrated one-vendor approach over multiple systems from multiple vendors?

Critical Thinking Questions

1. What are the pros and cons of buying predesigned software from a vendor such as SAP instead of a company developing software itself to exactly meet its own needs?
2. How might a CRM system assist Whitmann Price consultants in the field?

NOTES

Sources for the opening vignette: Salesforce.com staff, "Maporama International Identifies New Revenue Opportunities After Implementing Salesforce in 15 Days," Hi-Tech Software & Services Case Study, *www.salesforce.com/customers/hi-tech-software/case-studies/maporama.jsp*, accessed June 29, 2008; Maporama Web site, *www.maporama.com/home/En/default.asp*, accessed June 29, 2008.

1 Havenstein, Heather, "Wal-Mart CTO Details HP Data Warehouse Move," *Computerworld*, August 2, 2007.
2 AccuFund News, *www.accufund.com/news.htm*, accessed June 17, 2008.
3 Cummings, Joanne, "TiVo's Disaster Recovery Plan," *NetworkWorld*, August 21, 2007.
4 Grememeier, Larry, "Business Continuity: To Err is Human, To Plan is Divine," *InformationWeek*, August 9, 2007.
5 Grememeier, Larry, "Business Continuity: To Err is Human, To Plan is Divine," *InformationWeek*, August 9, 2007.
6 Marks, Howard, "MessageOne's Emergency Alert System Goes Global," *InformationWeek*, June 13, 2008.
7 "Midmarket ERP Solutions Buyer's Guide," Inside ERP @2008, Tippit, Inc.
8 "The Computerworld Honors Program: Gujarat Reclaim and Rubber Products Limited," *www.cwhonors.org/viewCaseStudy.asp?NominationID=106*, accessed June 2, 2008.
9 "The Computerworld Honors Program: Amgen," *Computerworld*, *www.cwhonors.org/viewCaseStudy2008.asp?NominationID=2031* accessed June 9, 2008.
10 "BSNF Railway to Upgrade its Back Office with SAP Software," *Computerworld*, September 17, 2007.
11 Kanaracus, Chris, "Waste Management Sues SAP over ERP Implementation, *Computerworld*, March 27, 2008.
12 Reed, Brad, "IBM Blamed for Fire-Truck Maker's Bankruptcy," *NetworkWorld*, January 30, 2008.
13 Lemos, Robert, "Open Sources ERP Grows Up," *CIO*, April 22, 2008.
14 Gruman, Galen, "More Midmarket Firms Choose Open-Source ERP," *Computerworld*, February 22, 2007.
15 Weier, Mary Hayes, "SAP Teams with HP, IBM for Business Software Appliances," *InformationWeek*, May 5, 2008.
16 Caulfield, John, "Lowe's Looks to Exploit Opportunities in Downturn," *Builder*, September 26, 2007.
17 Havenstein, Heather, "Lowe's Builds Up Infrastructure to Support BI," *NetworkWorld*, January 24, 2007.
18 "Oberto Sausage Finds the Right Recipe for Forecasting," ForecastPro Web site, *www.forecastpro.com/customers/success/oberto.html*, accessed June 10, 2008.
19 Reekie, Stuart, "How Air Products Used Steelwedge to Extend SAP for Integrated S&OP," Steelwedge Software Web page at *www.steelwedge.com/customers/index.php?z=air_products*, accessed June 10, 2008.
20 "Case Study Maximizing Equipment Investments," FabTol MRP Web site, *www.fabtol.com/cs/cs_cnc.asp*, accessed June 10, 2008.
21 Lai, Eric, "OpenWorld: Siebel On Demand to Become 'Social CRM'," *Computerworld*, November 13, 2007.
22 Fonseca, Brian, "SAP, RIM to Let BlackBerry Users Access ERP Apps," *Computerworld*, May 2, 2008.
23 Morphy, Erika, "SugarCRM Adds a New Layer of Functionality," *E-Commerce Times*, May 2, 2008.
24 Kanaracus, Chris, "CRM Systems to Get Free Company Contact Data," *NetworkWorld*, June 4, 2006.
25 Staff, "American Eagle Outfitters Selects Terradata's CRM Site," *CBR*, May 21, 2008.
26 Staff, "American of Martinsville Implements Microsoft Dynamics AX and CRM Solutions," *CBR*, May 23, 2008.
27 Staff, "Central Michigan University Improves Recruitment and Retention with Talisma," *CBR*, May 2, 2008.
28 Staff, "Kabel Selects SAP's CRM Application," *CBR*, May 20, 2008.
29 Rossi, Sandra, "Salvation Army Implements CRM for Fundraising Operations," *Computerworld*, December 3, 2007.
30 Staff, "Smart USA Selects IBS Automotive ERP System, *CBR*, May 15, 200.
31 Glossary of terms at *www.crfonline.org/oc/glossary/m.html*, accessed June 11, 2008.
32 "Top 5 SaaS ERP Software Applications," ERPSoftware 360 Web site at *www.erpsoftware360.com*, accessed June 11, 2008.
33 Fonseca, Brian, "SaaS Benefits Starting to Outweigh Risks for Some," *Computerworld*, May 5, 2008.
34 Karasev, Andres, "Oracle Financials Implementation: Brazil – Notes for International ERP Consultant," Ezine Articles at *http://ezinearticles.com/?Oracle-Financials-Implementation:-Brazil*, accessed June 18, 2008.
35 Murphy, Chris, "Could Your Next ERP System Come from China?" *InformationWeek*, October 31, 2007.

CHAPTER • 10 •

Information and Decision Support Systems

PRINCIPLES	LEARNING OBJECTIVES
■ Good decision-making and problem-solving skills are the key to developing effective information and decision support systems.	■ Define the stages of decision making. ■ Discuss the importance of implementation and monitoring in problem solving.
■ The management information system (MIS) must provide the right information to the right person in the right format at the right time.	■ Explain the uses of MISs and describe their inputs and outputs. ■ Discuss information systems in the functional areas of business organizations.
■ Decision support systems (DSSs) are used when the problems are unstructured.	■ List and discuss important characteristics of DSSs that give them the potential to be effective management support tools. ■ Identify and describe the basic components of a DSS.
■ Specialized support systems, such as group support systems (GSSs) and executive support systems (ESSs), use the overall approach of a DSS in situations such as group and executive decision making.	■ State the goals of a GSS and identify the characteristics that distinguish it from a DSS. ■ Identify the fundamental uses of an ESS and list the characteristics of such a system. ■ List and discuss the use of other special-purpose systems.

Information Systems in the Global Economy General Mills, United States

Food Giant General Mills Relies on Management Information Systems to Maintain Product Specifications

General Mills markets many well-known food brands including Betty Crocker, Pillsbury, Green Giant, Yoplait, Häagen-Dazs, and Cheerios. One of the company's greatest challenges is consistently providing high quality in every product it produces, packages, and delivers around the world, while adhering to strict regulations imposed by various governments. Management information systems (MISs) allow General Mills to do just that.

General Mills stores the specifications for each of its products in a huge database. The data includes specifications for ingredients, formulas, processing, and packaging. In total, the company stores over 22,000 product specifications, which are released to all of its manufacturing sites on a regular schedule. Over the course of a year, the company makes over 10,000 modifications to product specifications.

The process of modifying specifications is much more complicated than it sounds. Many specifications are shared across multiple products. For example, if General Mills changes a design feature of its cereal box, it affects the specifications for dozens of products. If Germany changes its regulations regarding the use of partially hydrogenated oils, the change could affect the ingredients, formulas, and processing specifications of dozens of products.

Because of the interrelated nature of General Mills' product specifications and the size of its database, one change to specifications might require several hours of design work followed by a lengthy review and approval process. What the company needed was an MIS that could automate the process of updating specifications and undo changes as necessary to return to the previously approved specifications.

General Mills worked with IS professionals from an outside MIS development company to design "mass-change and undo" functionality for their product specifications system. The MIS professionals worked with General Mills employees in the United States and abroad to fully understand the nature of the challenge. After rigorous testing and user training, the new features were integrated into the existing system with impressive results. The first modification to a product affected 332 related specifications in the database. Prior to the system improvement, the change would have required over 5,000 keystrokes and more than one day's work to implement. With the new system, the change took six keystrokes and five minutes.

Not only does the new system save a significant amount of time, but it reduces errors. Reducing keystrokes from 5,000 to five obviously reduces data entry errors. The system also reduces logical errors; by requiring changes to be made in a specific order following instructions provided in the software, operators are much less likely to overlook important considerations when changing specifications.

The development of General Mills' new mass-change and undo system gives the company an advantage in the highly competitive and closely regulated food industry. When market conditions change due to new regulations or public opinion, the first company to adjust and bring the changes to market achieves a competitive advantage. General Mills now has a much greater chance of being the first to introduce products that customers want. Management information systems and decision support systems give businesses in all industries a path to success through more efficient and effective operations.

As you read this chapter, consider the following:

- How is an MIS used in the various functional areas of a business?
- How do an MIS and DSS affect a company's business practices and its ability to compete in a market?

Why Learn About Information and Decision Support Systems?

You have seen throughout this book how information systems can make you more efficient and effective through the use of database systems, the Internet, e-commerce, transaction processing systems, and many other technologies. The true potential of information systems, however, is in helping you and your coworkers make more informed decisions. This chapter shows you how to slash costs, increase profits, and uncover new opportunities for your company using management information and decision support systems. Transportation coordinators can use management information reports to find the least expensive way to ship products to market and to solve bottlenecks. A loan committee at a bank or credit union can use a group support system to help them determine who should receive loans. Store managers can use decision support systems to help them decide what and how much inventory to order to meet customer needs and increase profits. An entrepreneur who owns and operates a temporary storage company can use vacancy reports to help determine what price to charge for new storage units. Everyone wants to be a better problem solver and decision maker. This chapter shows you how information systems can help. It begins with an overview of decision making and problem solving.

As shown in the opening vignette, information and decision support are the lifeblood of today's organizations. Thanks to information and decision support systems, managers and employees can obtain useful information in real time. As discussed in Chapter 9, TPS and ERP systems capture a wealth of data. When this data is filtered and manipulated, it can provide powerful support for managers and employees. The ultimate goal of management information and decision support systems is to help managers and executives at all levels make better decisions and solve important problems. The result can be increased revenues, reduced costs, and the realization of corporate goals. Many of today's information and decision support systems are built into the organization's TPS or ERP systems. In other cases, they are developed separately. No matter what type of information and decision support system you use, a system's primary goal is to help you and others become better decision makers and problem solvers.

DECISION MAKING AND PROBLEM SOLVING

Every organization needs effective decision making. The U.S. Coast Guard, for example, uses a formal decision process called the Ports and Waterways Safety Assessment (PAWSA) model to determine what resources it needs to secure the nation's coastlines and waterways.[1] As a result of its formal decision-making process, the Coast Guard demonstrated it needed four additional vessel-traffic centers.

In most cases, strategic planning and the overall goals of the organization set the course for decision making, helping employees and business units achieve their objectives and goals. Often, information systems also assist with problem solving, helping people make better decisions and save lives. For example, an information system at Hackensack University Medical Center (*www.humed.com*) in New Jersey analyzes possible drug interactions. In one case, an AIDS patient taking drugs for depression avoided therapeutic medication that could have dangerously interacted with the depression medication. The hospital has invested millions of dollars into its information system.

Decision Making as a Component of Problem Solving

In business, one of the highest compliments you can receive is to be recognized by your colleagues and peers as a "real problem solver." Problem solving is a critical activity for any business organization. After identifying a problem, the process of solving the problem begins with decision making. A well-known model developed by Herbert Simon divides the **decision-making phase** of the problem-solving process into three stages: intelligence, design, and choice. This model was later incorporated by George Huber into an expanded model of the entire problem-solving process (see Figure 10.1).

decision-making phase
The first part of problem solving, including three stages: intelligence, design, and choice.

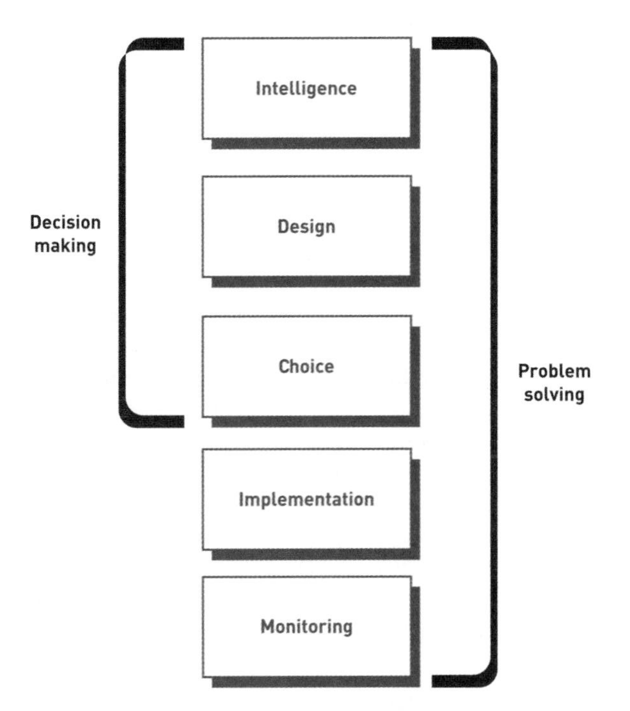

Figure 10.1

How Decision Making Relates to Problem Solving

The three stages of decision making—intelligence, design, and choice—are augmented by implementation and monitoring to result in problem solving.

The first stage in the problem-solving process is the **intelligence stage**. During this stage, you identify and define potential problems or opportunities. You also investigate resource and environmental constraints. For example, if you were a Hawaiian farmer, during the intelligence stage you would explore the possibilities of shipping tropical fruit from your farm in Hawaii to stores in Michigan. The perishability of the fruit and the maximum price that consumers in Michigan are willing to pay for the fruit are problem constraints.

intelligence stage
The first stage of decision making, in which potential problems or opportunities are identified and defined.

In the **design stage**, you develop alternative solutions to the problem and evaluate their feasibility. In the tropical fruit example, you would consider the alternative methods of shipment, including the transportation times and costs associated with each. During this stage, you might determine that shipment by freighter to California and then by truck to Michigan is not feasible because the fruit would spoil.

design stage
The second stage of decision making, in which alternative solutions to the problem are developed.

The last stage of the decision-making phase, the **choice stage**, requires selecting a course of action. In the tropical fruit example, you might select the method of shipping fruit by air from your Hawaiian farm to Michigan as the solution. The choice stage would then conclude with selection of an air carrier. As you will see later, various factors influence choice; the act of choosing is not as simple as it might first appear.

choice stage
The third stage of decision making, which requires selecting a course of action.

Problem solving includes and goes beyond decision making. It also includes the **implementation stage**, when the solution is put into effect. For example, if your decision is to ship tropical fruit to Michigan as air freight using a specific carrier, implementation involves informing your field staff of the new activity, getting the fruit to the airport, and actually shipping the product to Michigan. As another example, the Operations Research Group at British Airways used quantitative problem-solving techniques to help the airline achieve better departure statistics for its 750 flights per day to about 130 destinations.[2] On-time departure is a complex task, involving the coordination of cabin crews, airline

problem solving
A process that goes beyond decision making to include the implementation stage.

implementation stage
A stage of problem solving in which a solution is put into effect.

cleaners, maintenance, catering, cargo, baggage, passengers, pilots, fuel, and air traffic clearance. British Airways' quantitative analysis helped them achieve a better departure record.

monitoring stage
The final stage of the problem-solving process, in which decision makers evaluate the implementation.

The final stage of the problem-solving process is the **monitoring stage**. In this stage, decision makers evaluate the implementation to determine whether the anticipated results were achieved and to modify the process in light of new information. Monitoring can involve feedback and adjustment. For example, after the first shipment of fruit from Hawaii to Michigan, you might learn that the flight of your chosen air freight firm routinely stops in Phoenix, Arizona, where the plane sits on the runway for a number of hours while loading additional cargo. If this unforeseen fluctuation in temperature and humidity adversely affects the fruit, you might have to readjust your solution to include a new carrier that does not make such a stop, or perhaps you would consider a change in fruit packaging.

Good decision makers monitor their decisions and make changes if necessary. After monitoring a decision to place its video programming on its Internet site called Innertube, CBS decided to change course and place its sports, news, and entertainment content on a wide range of video Web sites to get wider coverage.[3]

Programmed versus Nonprogrammed Decisions

programmed decision
A decision made using a rule, procedure, or quantitative method.

In the choice stage, various factors influence the decision maker's selection of a solution. One such factor is whether the decision can be programmed. **Programmed decisions** are made using a rule, procedure, or quantitative method. For example, to say that inventory should be ordered when inventory levels drop to 100 units is a programmed decision because it adheres to a rule. Programmed decisions are easy to computerize using traditional information systems. For example, you can easily program a computer to order more inventory when levels for a certain item reach 100 units or less. Most of the processes automated through enterprise resource planning or transaction processing systems share this characteristic: The relationships between system elements are fixed by rules, procedures, or numerical relationships. Management information systems can also reach programmed decisions by providing reports on problems that are routine and in which the relationships are well defined. (In other words, they are structured problems.)

Ordering more products when inventory levels drop to specified levels is an example of a programmed decision.

(Source: © Andersen Ross/Getty Images.)

Nonprogrammed decisions deal with unusual or exceptional situations. In many cases, these decisions are difficult to quantify. Determining the appropriate training program for a new employee, deciding whether to develop a new type of product line, and weighing the benefits and drawbacks of installing an upgraded pollution control system are examples. Each of these decisions contains unique characteristics, and standard rules or procedures might not apply to them. Today, decision support systems help solve many nonprogrammed decisions, in which the problem is not routine and rules and relationships are not well defined (unstructured or ill-structured problems). These problems can include deciding the best location for a manufacturing plant or whether to rebuild a hospital that was severely damaged from a hurricane or tornado.

nonprogrammed decision
A decision that deals with unusual or exceptional situations.

Optimization, Satisficing, and Heuristic Approaches

In general, computerized decision support systems can either optimize or satisfice. An **optimization model** finds the best solution, usually the one that will best help the organization meet its goals. For example, an optimization model can find the appropriate number of products that an organization should produce to meet a profit goal, given certain conditions and assumptions. Optimization models use problem constraints. A limit on the number of available work hours in a manufacturing facility is an example of a problem constraint. Some spreadsheet programs, such as Excel, have optimizing features (see Figure 10.2). A business such as an appliance manufacturer can use an optimization program to reduce the time and cost of manufacturing appliances and increase profits by millions of dollars. The Scheduling Appointments at Trade Events (SATE) software package is an optimization program that schedules appointments between buyers and sellers at trade shows and meetings. Optimization software also allows decision makers to explore various alternatives.[4]

optimization model
A process to find the best solution, usually the one that will best help the organization meet its goals.

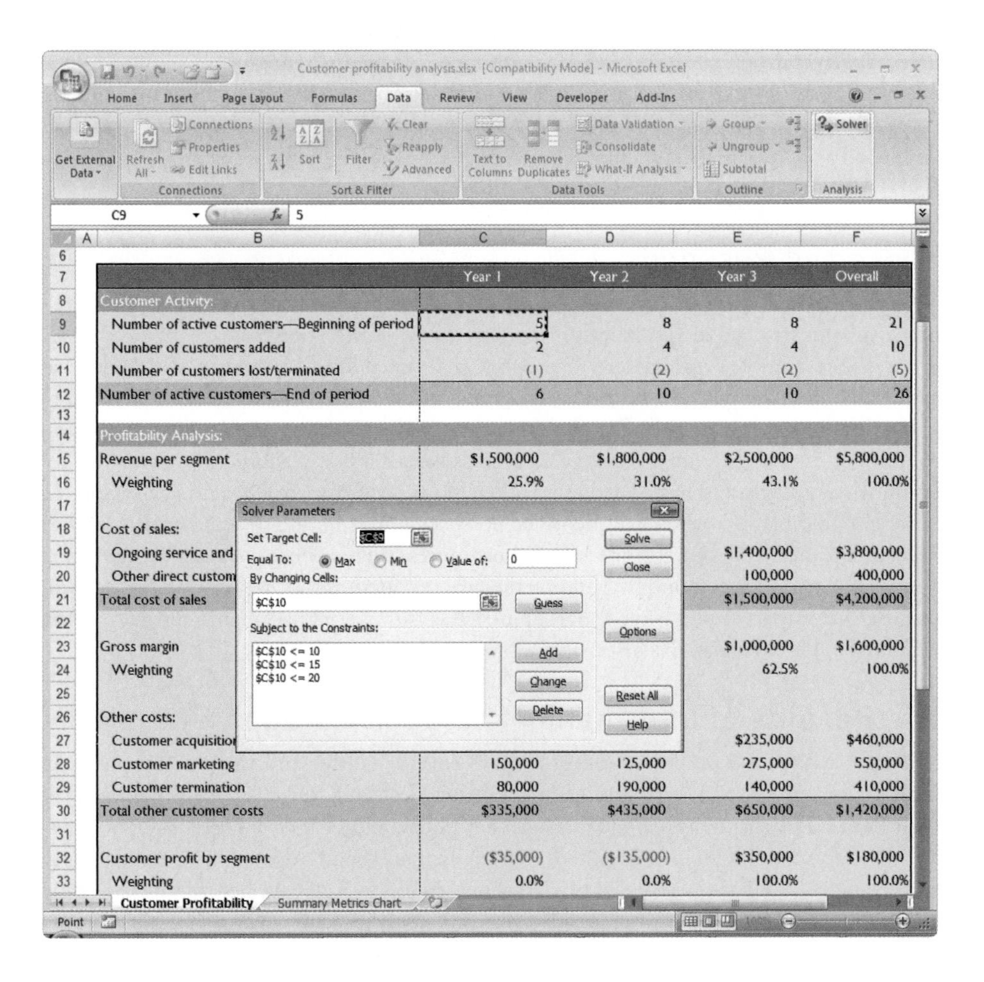

Figure 10.2

Optimization Software

Some spreadsheet programs, such as Microsoft Excel, have optimizing routines. This figure shows Solver, which can find an optimal solution given certain constraints.

Consider a few examples of how you can use optimization to achieve huge savings. Coca-Cola, for example, used optimization to schedule and route about 10,000 trucks used to deliver its soft drinks and products to save about $45 million annually.[5] Bombardier Flexjet (*www.flexjet.com*), a company that sells fractional ownership of jets, used an optimization program to better schedule its aircraft and crews, saving almost $30 million annually. Hutchison Port Holdings (*www.hph.com.hk*), the world's largest container terminal, saved even more—over $50 million annually. The company processes 10,000 trucks and 15 ships every day, and used optimization to maximize the use of its fleet. Deere & Company, a manufacturer of commercial vehicles and equipment, increased shareholder value by over $100 million annually by using optimization to minimize inventory levels and enhance customer satisfaction.

satisficing model
A model that will find a good—but not necessarily the best—problem solution.

A **satisficing model** is one that finds a good—but not necessarily the best—problem solution. Satisficing is usually used because modeling the problem properly to get an optimal decision would be too difficult, complex, or costly. Satisficing normally does not look at all possible solutions but only at those likely to give good results. Consider a decision to select a location for a new manufacturing plant. To find the optimal (best) location, you must consider all cities in the United States or the world. A satisficing approach is to consider only five or ten cities that might satisfy the company's requirements. Limiting the options might not result in the best decision, but it will likely result in a good decision, without spending the time and effort to investigate all cities. Satisficing is a good alternative modeling method because it is sometimes too expensive to analyze every alternative to find the best solution.

heuristics
Commonly accepted guidelines or procedures that usually find a good solution.

Heuristics, often referred to as "rules of thumb"—commonly accepted guidelines or procedures that usually find a good solution—are often used in decision making. A heuristic that baseball team managers use is to place batters most likely to get on base at the top of the lineup, followed by the power hitters who can drive them in to score. An example of a heuristic used in business is to order four months' supply of inventory for a particular item when the inventory level drops to 20 units or less; although this heuristic might not minimize total inventory costs, it can serve as a good rule of thumb to avoid stockouts without maintaining excess inventory. Trend Micro (*www.trendmicro.com*), a provider of antivirus software, has developed an antispam product that is based on heuristics. The software examines e-mails to find those most likely to be spam. It doesn't examine all e-mails.

Sense and Respond

Sense and Respond (SaR) involves determining problems or opportunities (sense) and developing systems to solve the problems or take advantage of the opportunities (respond). SaR often requires nimble organizations that replace traditional lines of authority with those that are flexible and dynamic. IBM, for example, used SaR with its Microelectronics Division to help with inventory control. The division used mathematical models and optimization routines to control inventory levels. The models sensed when a shortage of inventory for customers was likely and responded by backlogging and storing extra inventory to avoid the shortages. In this application, SaR identified potential problems and solved them before they became a reality. SaR can also identify opportunities, such as new products or marketing approaches, and then respond by building the new products or starting new marketing campaigns. One way to implement the SaR approach is through management information and decision support systems, discussed next.

The Benefits of Information and Decision Support Systems

The information and decision support systems covered in this and the next chapter help individuals, groups, and organizations make better decisions, solve problems, and achieve their goals.[6] These systems include management information systems, decision support systems, group support systems, executive support systems, knowledge management systems, and a variety of special-purpose systems. As shown in Figure 10.3, the benefits are a measure of increased performance of these systems versus the cost to deliver them. The plus sign (+) by the arrow from *performance* to *benefits* indicates that increased performance has a positive impact on benefits. The minus sign (-) from *cost* to *benefits* indicates that increased cost has a negative impact on benefits.

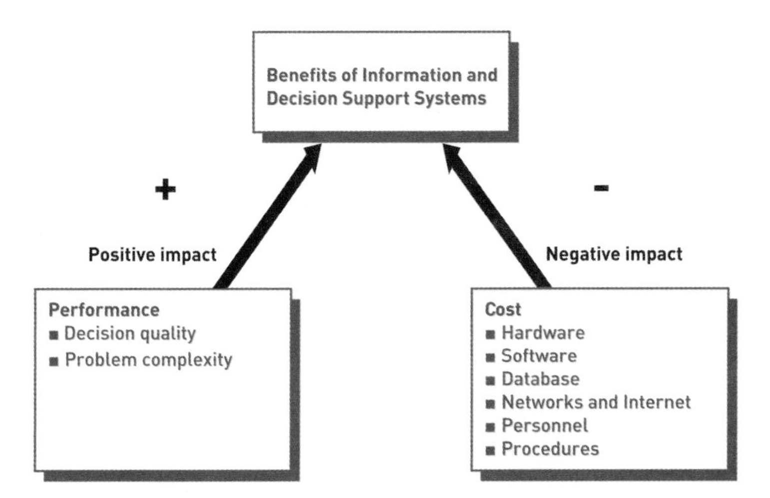

Figure 10.3

The Benefits of Information and Decision Support Systems

The performance of these systems is typically a function of decision quality and problem complexity. Decision quality can result in increased effectiveness, increased efficiency, higher productivity, and many other measures first introduced in Chapter 2. Problem complexity depends on how hard the problem is to solve and implement. The cost of delivering these systems are the expenditures of the information technology components covered in Part II of this book, including hardware, software, databases, networks and the Internet, people, and procedures. But how do these systems actually deliver benefits to the individuals, groups, and organizations that use them? It depends on the type of information system. We begin our discussion with traditional management information systems.

AN OVERVIEW OF MANAGEMENT INFORMATION SYSTEMS

A management information system (MIS) is an integrated collection of people, procedures, databases, and devices that provides managers and decision makers with information to help achieve organizational goals. MISs can often give companies and other organizations a competitive advantage by providing the right information to the right people in the right format and at the right time. For example, a shipping department could develop a spreadsheet to generate a report on possible delays that must be addressed to increase the number of on-time deliveries for the day. A music store might use a database system to develop a report that summarizes profits and losses for the month to make sure that the store is on track to make a 10 percent profit for the year.

Management Information Systems in Perspective

The primary purpose of an MIS is to help an organization achieve its goals by providing managers with insight into the regular operations of the organization so that they can control, organize, and plan more effectively. One important role of the MIS is to provide the right information to the right person in the right format at the right time. In short, an MIS provides managers with information, typically in reports, that supports effective decision making and provides feedback on daily operations. Figure 10.4 shows the role of MISs within the flow of an organization's information. Note that business transactions can enter the organization through traditional methods or via the Internet or an extranet connecting customers and suppliers to the firm's ERP or transaction processing systems. The use of MISs spans all levels of management. That is, they provide support to and are used by employees throughout the organization.

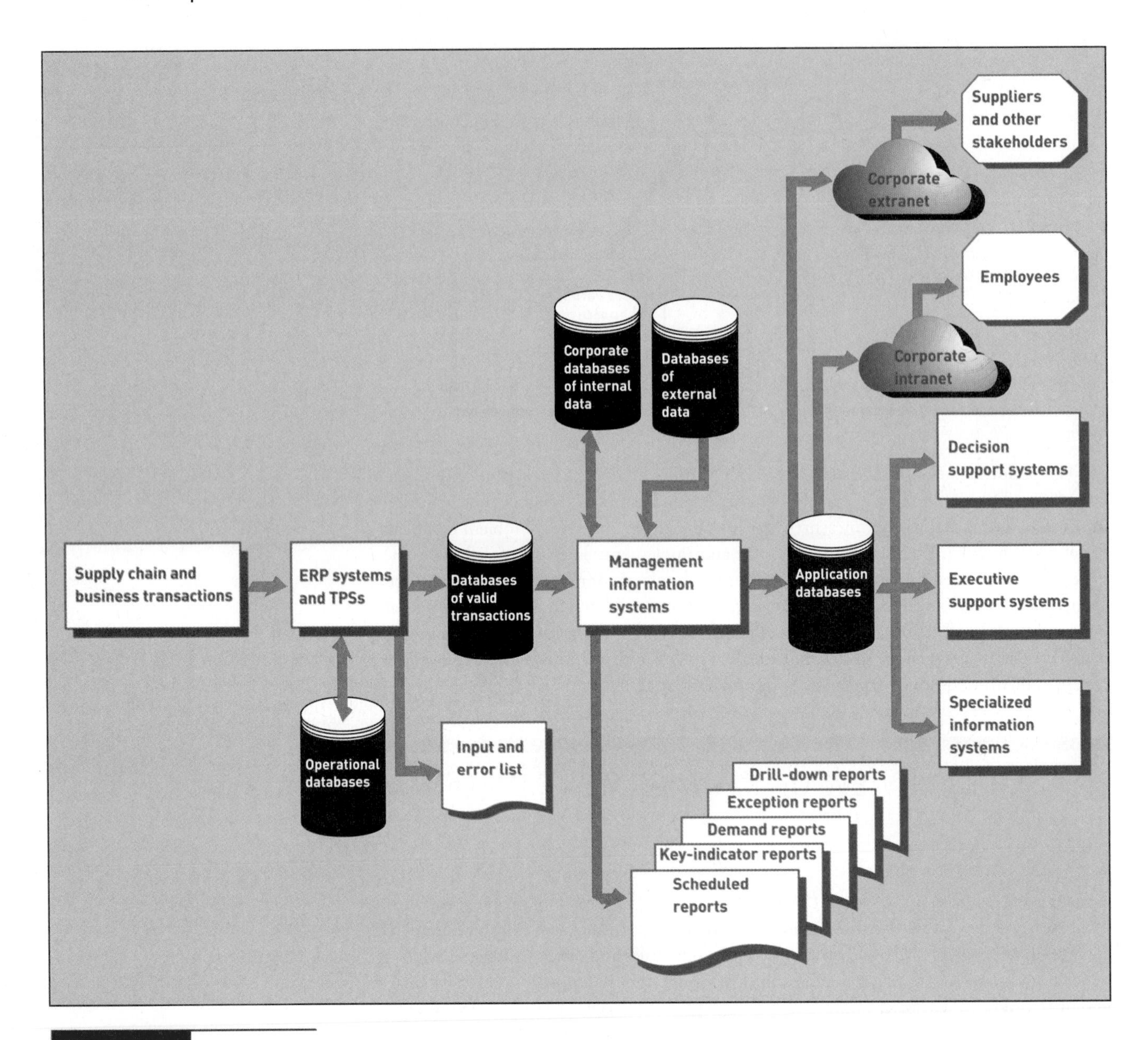

Figure 10.4

Sources of Managerial Information

The MIS is just one of many sources of managerial information. Decision support systems, executive support systems, and expert systems also assist in decision making.

Inputs to a Management Information System

As shown in Figure 10.4, data that enters an MIS originates from both internal and external sources, including a company's supply chain, first discussed in Chapter 2. The most significant internal data sources for an MIS are the organization's various TPS and ERP systems and related databases. As discussed in Chapter 5, companies also use data warehouses and data marts to store valuable business information. Business intelligence, also discussed in Chapter 5, can be used to turn a database into useful information throughout the organization. Other internal data comes from specific functional areas throughout the firm.

External sources of data can include customers, suppliers, competitors, and stockholders, whose data is not already captured by the TPS, as well as other sources, such as the Internet. In addition, many companies have implemented extranets to link with selected suppliers and other business partners to exchange data and information.

The MIS uses the data obtained from these sources and processes it into information that is more readily usable by managers, primarily in the form of predetermined reports. For example, rather than simply obtaining a chronological list of sales activity over the past week, a national sales manager might obtain her organization's weekly sales data in a format that allows her to see sales activity by region, by local sales representative, by product, and even in comparison with last year's sales.

ETHICAL AND SOCIETAL ISSUES

Web 2.0 MIS Finds Compromise Between Service and Privacy

WhitePages.com maintains an MIS fed by a huge database that provides information on 180 million adults in the United States—80 percent of the U.S. population. Look up an old friend on WhitePages.com and you may find his new address, the name of his wife, his age, e-mail address, phone number, even a map to his house. Using associated services, you could find police records for your old friend and find out how much his house is worth. You can guess the obvious concern that many people have about WhitePages.com—privacy!

Information systems that post personal information on the Web without a person's consent run the risk of negative press and lawsuits. Consider the Beacon program provided by social-networking giant, Facebook. Beacon was designed to collect information about a Facebook user's activities on partner Web sites and use those members to endorse products. For example, you might see an announcement in your Facebook news feed informing you that your friend Shannon just rented the movie "Iron Man" from Blockbuster Online.

When Facebook users learned about how the Beacon system worked, many were enraged. Facebook quickly responded by giving users the ability to opt out of the Beacon system. The difference between Facebook and WhitePages.com is that Facebook is set up to serve registered users, and they have certain expectations of the Facebook service. WhitePages.com, however, was not designed to service members, but rather the entire Internet population. WhitePages.com does not obtain information about Internet users through an online profile that the user submits. It collects information from freely available public records on the Internet. This makes WhitePages.com less liable to privacy violations than Facebook. However, that level of liability might be changing.

WhitePages.com sees many similarities between today's popular social networks and its own directory service. It also sees the potential for substantial profits. In moving to a design that reflects a social network, WhitePages.com now includes a way for those listed in its directory to correct and add information about themselves. It is also setting up a service that allows members to fill out a profile and find others in the directory with similar interests. Users can send an anonymous message to others listed in the directory to find out if they are interested in striking up a friendship or renewing an old one. Now WhitePages.com finds itself in the same quandary as the big social networks: what information to share and what to keep private.

WhitePages.com decided to allow those listed in its directory to select the information they want to make public—if any. The company has decided to give its users control over their information. In doing so, WhitePages hopes to transform its directory service into an Internet-wide social network. While some users are likely to omit information that WhitePages previously made available, others might add to their information. By being conscious of privacy concerns, WhitePages hopes to improve its reputation and draw more visitors. The company has also released a software development platform that will allow developers to publish useful applications based on the WhitePages directory.

Businesses such as WhitePages.com need to control access to the private information in their information systems. Whether it's an information system available only to employees of the company or one that's accessible on the public Web, the reputation of a business depends on the trust of its clientele. If customers don't trust one company due to mismanagement of private information, they are likely to select a competitor that promises better security and privacy.

Discussion Questions

1. Why is WhitePages.com more concerned about customer privacy today than it has been in the past?
2. What are the differences between the services offered by WhitePages.com and Facebook.com?

Critical Thinking Questions

1. What benefits does WhitePages.com have compared to Facebook.com in regards to its customer base?
2. What types of applications might be developed using the WhitePages.com software development platform?

SOURCES: Vaughn-Nichols, Steven, "WhitePages.com grapples with privacy in a Web 2.0 world," *Computerworld*, May 16, 2008, *http://computerworld.com/action/article.do?command=viewArticleBasic&taxonomyName=security&articleId=9085718&taxonomyId=17&intsrc=kc_feat*; Cheng, Jacqui, "Facebook reevaluating Beacon after privacy outcry, possible FTC complaint," *Ars Technica*, November 29, 2007, *http://arstechnica.com/news.ars/post/20071129-facebook-reevaluating-beacon-after-privacy-outcry-possible-ftc-complaint.html*; WhitePages.com Web site, *www.whitepages.com*, accessed May 19, 2008.

Outputs of a Management Information System

The output of most management information systems is a collection of reports that are distributed to managers. These reports can be tailored for each user and delivered in a timely fashion. Providence Washington Insurance Company, for example, uses ReportNet from Cognos (*www.cognos.com*), an IBM company, to reduce the number of paper reports they produce and the associated costs.[7] The new reporting system creates an "executive dashboard" that shows current data, graphs, and tables to help managers make better real-time decisions.[8] Executives from Dunkin' Donuts use a dashboard to see the status of new stores.[9] The dashboard displays geographic areas and the new stores that are being developed. By clicking on a store, executives can see the details of how new stores are being constructed and if any stores are being delayed. The company hopes to grow to 15,000 franchises around the globe in the next several years.[10] The city of Atlanta, Georgia also uses Cognos to measure the performance of its various departments and to keep track of its expenditures and budgets.[11] See Figure 10.5 for an example of an executive dashboard. In 2007, IBM announced that it would acquire Cognos.[12] Microsoft makes a reporting system called Business Scorecard Manager to give decision makers timely information about sales and customer information.[13] The software, which competes with Business Objects and Cognos, can integrate with other Microsoft software products, including Microsoft Office Excel. Hewlett-Packard's OpenView Dashboard is another MIS package that can quickly and efficiently render pictures, graphs, and tables that show how a business is functioning. In addition, some software packages and the Internet can be used to produce, gather, and distribute reports from different computer systems. Ace Hardware, for example, decided to use a more flexible report system called WebFocus from Information Builders (*www.informationbuilders.com*).[14] Referring to the old reporting system, one executive said, "People were getting tied in knots trying to develop reports in that tool. The tool was very rigid and had a lot of requirements as far as the way you did reporting." The new reporting system helped overcome some of these problems.

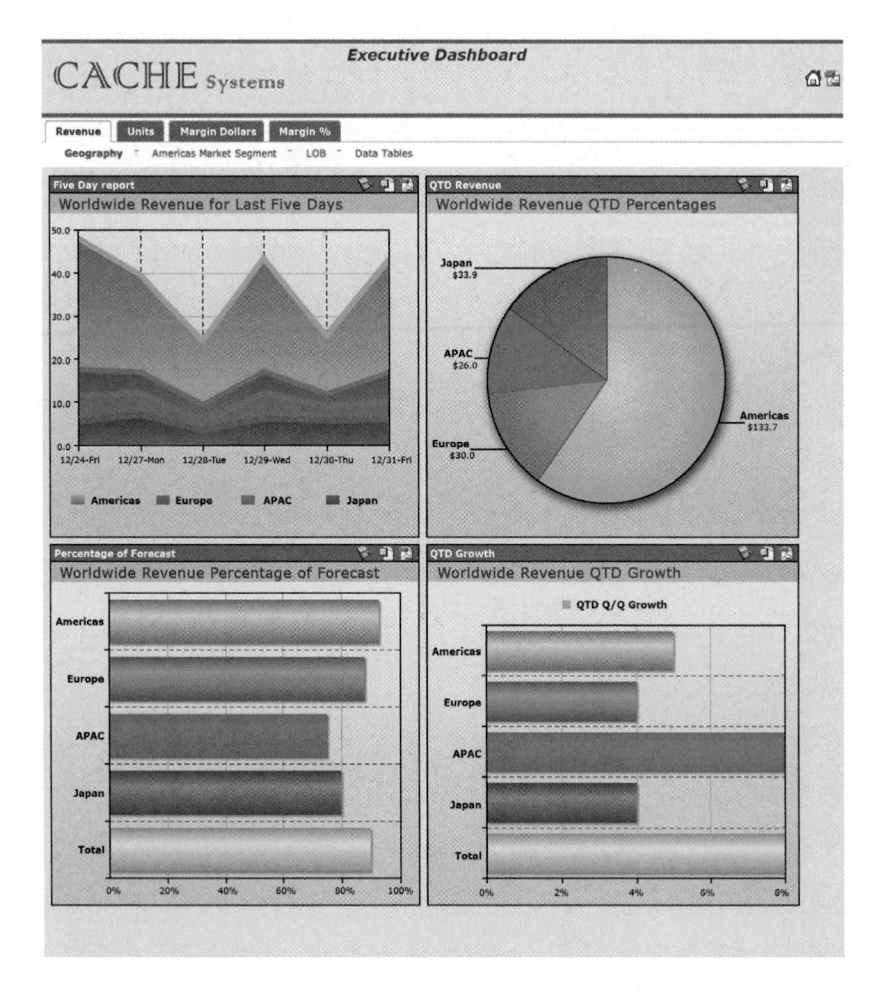

Figure 10.5

An Executive Dashboard

This MIS reporting system puts many kinds of real-time information at managers' fingertips to aid in decision making.

(Source: Courtesy of CORDA Technologies, Inc.)

Management reports can come from various company databases, data warehouses, and other sources. These reports include scheduled reports, key-indicator reports, demand reports, exception reports, and drill-down reports (see Figure 10.6).

(a) Scheduled Report

Daily Sales Detail Report — Prepared: 08/10/08

Order #	Customer ID	Salesperson ID	Planned Ship Date	Quantity	Item #	Amount
P12453	C89321	CAR	08/12/08	144	P1234	$3,214
P12453	C89321	CAR	08/12/08	288	P3214	$5,660
P12454	C03214	GWA	08/13/08	12	P4902	$1,224
P12455	C52313	SAK	08/12/08	24	P4012	$2,448
P12456	C34123	JMW	08/13/08	144	P3214	$720
....	....	....	....	......	......	......

(b) Key-Indicator Report

Daily Sales Key-Indicator Report

	This Month	Last Month	Last Year
Total Orders Month to Date	$1,808	$1,694	$1,914
Forecasted Sales for the Month	$2,406	$2,224	$2,608

(c) Demand Report

Daily Sales by Salesperson Summary Report — Prepared: 08/10/08

Salesperson ID	Amount
CAR	$42,345
GWA	$38,950
SAK	$22,100
JWN	$12,350
....	....
....	....

(d) Exception Report

Daily Sales Exception Report—Orders Over $10,000 — Prepared: 08/10/08

Order #	Customer ID	Salesperson ID	Planned Ship Date	Quantity	Item #	Amount
P12345	C89321	GWA	08/12/08	576	P1234	$12,856
P22153	C00453	CAR	08/12/08	288	P2314	$28,800
P23023	C32832	JMN	08/11/08	144	P2323	$14,400
....	....	....	....	......	......	....
....	....	....	....	......	......	....

(e) First-Level Drill-Down Report

Earnings by Quarter (Millions)

	Actual	Forecast	Variance
2nd Qtr. 2008	$12.6	$11.8	6.8%
1st Qtr. 2008	$10.8	$10.7	0.9%
4th Qtr. 2008	$14.3	$14.5	-1.4%
3rd Qtr. 2008	$12.8	$13.3	-3.8%

(f) Second-Level Drill-Down Report

Sales and Expenses (Millions)

Qtr: 2nd Qtr. 2008	Actual	Forecast	Variance
Gross Sales	$110.9	$108.3	2.4%
Expenses	$ 98.3	$ 96.5	1.9%
Profit	$ 12.6	$ 11.8	6.8%

(g) Third-Level Drill-Down Report

Sales by Division (Millions)

Qtr: 2nd Qtr. 2008	Actual	Forecast	Variance
Beauty Care	$ 34.5	$ 33.9	1.8%
Health Care	$ 30.0	$ 28.0	7.1%
Soap	$ 22.8	$ 23.0	-0.9%
Snacks	$ 12.1	$ 12.5	-3.2%
Electronics	$ 11.5	$ 10.9	5.5%
Total	$110.9	$108.3	2.4%

(h) Fourth-Level Drill-Down Report

Sales by Product Category (Millions)

Qtr: 2nd Qtr. 2008 Division: Health Care	Actual	Forecast	Variance
Toothpaste	$12.4	$10.5	18.1%
Mouthwash	$ 8.6	$ 8.8	-2.3%
Over-the-Counter Drugs	$ 5.8	$ 5.3	9.4%
Skin Care Products	$ 3.2	$ 3.4	-5.9%
Total	$30.0	$28.0	7.1%

Figure 10.6

Reports Generated by an MIS

The types of reports are (a) scheduled, (b) key indicator, (c) demand, (d) exception, and (e–h) drill down.

(Source: George W. Reynolds, *Information Systems for Managers*, Third Edition. St. Paul, MN: West Publishing Co., 1995.)

Scheduled Reports

Scheduled reports are produced periodically, or on a schedule, such as daily, weekly, or monthly. For example, a production manager could use a weekly summary report that lists total payroll costs to monitor and control labor and job costs. A manufacturing report generated once per day to monitor the production of a new item is another example of a scheduled report. Other scheduled reports can help managers control customer credit, performance of sales representatives, inventory levels, and more.

A **key-indicator report** summarizes the previous day's critical activities and is typically available at the beginning of each workday. These reports can summarize inventory levels, production activity, sales volume, and the like. Key-indicator reports are used by managers and executives to take quick, corrective action on significant aspects of the business.

scheduled report
A report produced periodically, or on a schedule, such as daily, weekly, or monthly.

key-indicator report
A summary of the previous day's critical activities; typically available at the beginning of each workday.

Demand Reports

Demand reports are developed to provide certain information upon request. In other words, these reports are produced on demand. Like other reports discussed in this section, they often come from an organization's database system. For example, an executive might want to know the production status of a particular item—a demand report can be generated to provide the requested information by querying the company's database. Suppliers and customers can also use demand reports. FedEx, for example, provides demand reports on its Web site to allow customers to track packages from their source to their final destination. Other examples of demand reports include reports requested by executives to show the hours worked by a particular employee, total sales to date for a product, and so on. Many companies are putting some medical records on the Internet to make them available on demand.[15] Software and

demand report
A report developed to give certain information at someone's request.

Internet companies like Microsoft are developing systems that allow people to search the Internet to get medical information, including lab tests, drug records, and X-rays, from different sources. Doctors use these systems to make a diagnosis and prescribe treatment from remote locations.

Exception Reports

exception report
A report automatically produced when a situation is unusual or requires management action.

Exception reports are reports that are automatically produced when a situation is unusual or requires management action. For example, a manager might set a parameter that generates a report of all inventory items with fewer than the equivalent of five days of sales on hand. This unusual situation requires prompt action to avoid running out of stock on the item. The exception report generated by this parameter would contain only items with fewer than five days of sales in inventory. Exception reports are used by businesses and nonprofit organizations. British Petroleum (BP), for example, uses a variety of data sources to get exception reports about potential damage from hurricanes to their facilities in the Gulf of Mexico.[16] The company collects data from geospatial map Internet sites and its own company resources. It then mashes up, or integrates, the different data sources into exception reports that show the location of its oil facilities that may have been damaged from a hurricane or severe storm. According to one company executive, "If there is any kind of damage after a storm, we want to know about it very quickly." Some companies, universities, and law enforcement agencies use cell phones and text messages to send exception reports to employees, students, or neighborhood residents. If a campus threat is detected or a crime committed, a text message can be sent to students and staff with the details. The University of Texas, for example, sent a text message to faculty, staff, and students about a potential ice storm, warning them that the university would be closed the next day and to stay home.[17] The next day, there were almost no students at the university, and no storm-related injuries were reported on campus. After police in the Netherlands sent a text message to neighborhood residents about a stolen boat, a resident called to say that she found a boat that met the description of the text message.[18] The police found the boat and arrested the criminal.

As with key-indicator reports, exception reports are most often used to monitor aspects important to an organization's success. In general, when an exception report is produced, a manager or executive takes action. Parameters, or *trigger points*, for an exception report should be set carefully. Trigger points that are set too low might result in too many exception reports; trigger points that are too high could mean that problems requiring action are overlooked. For example, if a manager wants a report that contains all projects over budget by $100 or more, the system might retrieve almost every company project. The $100 trigger point is probably too low. A trigger point of $10,000 might be more appropriate.

Drill-Down Reports

drill-down report
A report providing increasingly detailed data about a situation.

Drill-down reports provide increasingly detailed data about a situation. Using these reports, analysts can see data at a high level first (such as sales for the entire company), then at a more detailed level (such as the sales for one department of the company), and then at a very detailed level (such as sales for one sales representative). Boehringer Ingelheim (*www.boehringer-ingelheim.com/corporate/home/home.asp*), a large German drug company with over $7 billion in revenues and thousands of employees in 60 countries, uses a variety of drill-down reports so it can respond rapidly to changing market conditions. Managers can drill down into more levels of detail to individual transactions if they want. Companies and organizations of all sizes and types use drill-down reports.[19] The military, for example, is using software from Business Objects to determine if a defective battery could explode and damage or destroy military vehicles. According to an Army spokesman, "We can drill down, go into every contract we ordered that battery on … and get them away from units so no one will get hurt."

Characteristics of a Management Information System

Scheduled, key-indicator, demand, exception, and drill-down reports have all helped managers and executives make better, more timely decisions. In general, MISs perform the following functions:

- Provide reports with fixed and standard formats
- Produce hard-copy and soft-copy reports

- Use internal data stored in the computer system
- Allow users to develop their own custom reports
- Require user requests for reports developed by systems personnel

FUNCTIONAL ASPECTS OF THE MIS

Most organizations are structured along functional lines or areas. This functional structure is usually apparent from an organization chart, which typically shows a hierarchy in roles or positions. Some traditional functional areas include finance, manufacturing, marketing, human resources, and other specialized information systems. The MIS can also be divided along those functional lines to produce reports tailored to individual functions (see Figure 10.7).

Figure 10.7

An Organization's MIS

The MIS is an integrated collection of functional information systems, each supporting particular functional areas.

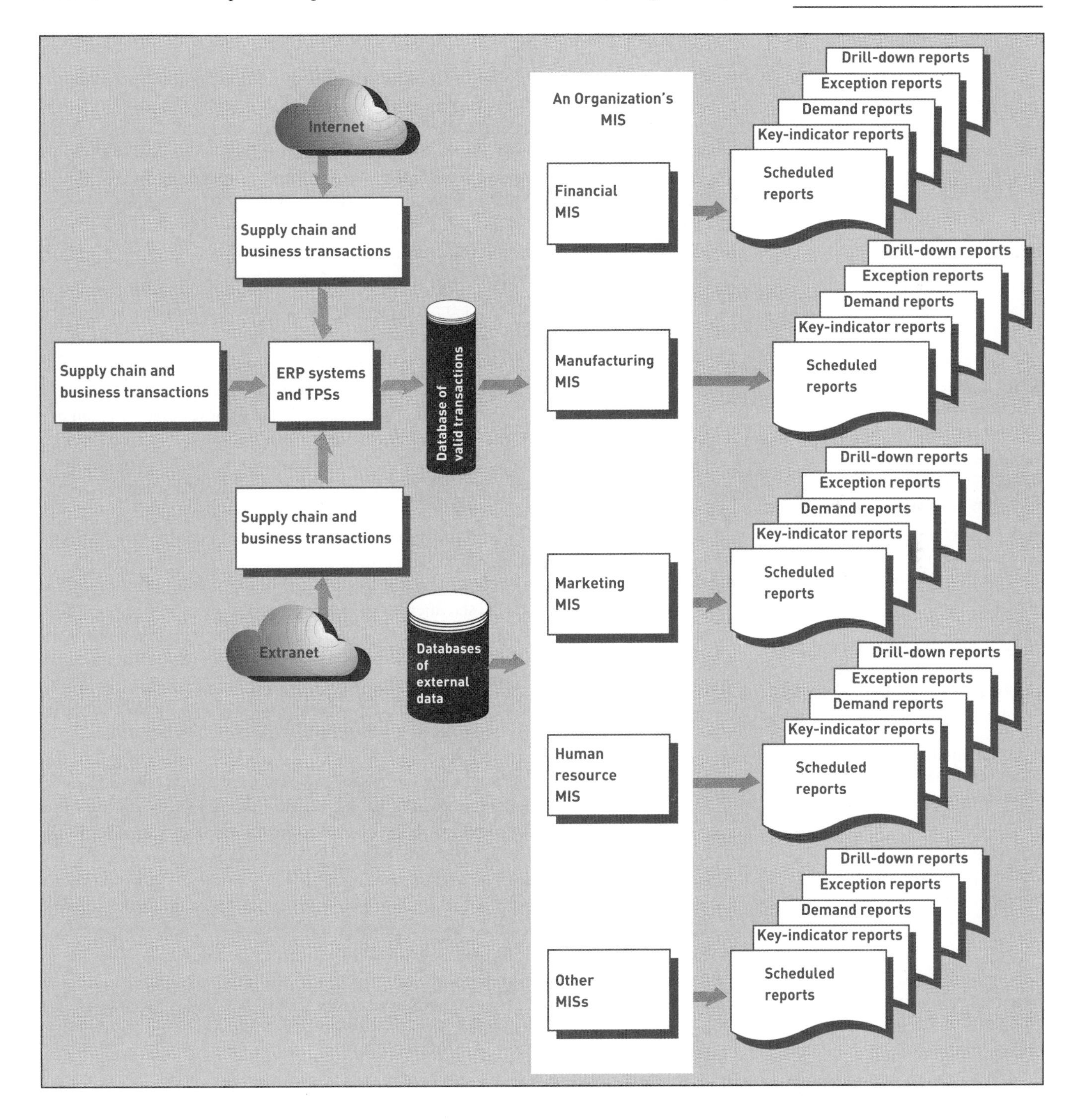

Financial Management Information Systems

financial MIS
An information system that provides financial information not only for executives but also for a broader set of people who need to make better decisions on a daily basis.

A **financial MIS** provides financial information not only for executives but also for a broader set of people who need to make better decisions on a daily basis. Reuters, for example, has developed an automated reporting system that scans articles about companies for its stock traders to determine if the news is favorable or unfavorable.[20] The reports can result in buy orders if the news is positive or sell orders if the news is negative. Eventually, the system will be tied into machine trading that doesn't require trade orders generated by people. Web sites can also provide financial information. For example, Web sites like *www.kiva.com* can provide information for people seeking small loans, called microloans.[21] Many microloans are for $100 or less and are made for a period of several months. The Internet is also used for larger loans. When Jeff Walsh wanted to find funds for his small business, he went to *www.prosper.com*.[22] According to Mr. Walsh, "I just bought a house in 2007 and was a little nervous about what the bank would say about my debt-to-income ratio." Financial MISs are often used to streamline reports of transactions. Most financial MISs perform the following functions:

- Integrate financial and operational information from multiple sources, including the Internet, into a single system
- Provide easy access to data for both financial and nonfinancial users, often through the use of a corporate intranet to access corporate Web pages of financial data and information
- Make financial data immediately available to shorten analysis turnaround time
- Enable analysis of financial data along multiple dimensions—time, geography, product, plant, and customer
- Analyze historical and current financial activity
- Monitor and control the use of funds over time

Figure 10.8 shows typical inputs, function-specific subsystems, and outputs of a financial MIS, including profit and loss, auditing, and uses and management of funds. Some of the financial MIS subsystems and outputs are outlined below.

profit center
A department within an organization that focuses on generating profits.

revenue center
A division within a company that generates sales or revenues.

cost center
A division within a company that does not directly generate revenue.

auditing
Analyzing the financial condition of an organization and determining whether financial statements and reports produced by the financial MIS are accurate.

internal auditing
Auditing performed by individuals within the organization.

external auditing
Auditing performed by an outside group.

- **Profit/loss and cost systems.** Many departments within an organization are **profit centers**, which means that they focus on generating profits. An investment division of a large insurance or credit card company is an example of a profit center. Other departments can be **revenue centers**, which are divisions within the company that focus primarily on sales or revenues, such as a marketing or sales department. Still other departments can be **cost centers**, which are divisions within a company that do not directly generate revenue, such as manufacturing or research and development. In most cases, information systems are used to compute revenues, costs, and profits.
- **Auditing. Auditing** involves analyzing the financial condition of an organization and determining whether financial statements and reports produced by the financial MIS are accurate. **Internal auditing** is performed by individuals within the organization. For example, the finance department of a corporation might use a team of employees to perform an audit. **External auditing** is performed by an outside group, usually an accounting or consulting firm such as PricewaterhouseCoopers, Deloitte & Touche, or one of the other major, international accounting firms. Computer systems are used in all aspects of internal and external auditing.
- **Uses and management of funds.** Internal uses of funds include purchasing additional inventory, updating plants and equipment, hiring new employees, acquiring other companies, buying new computer systems, increasing marketing and advertising, purchasing raw materials or land, investing in new products, and increasing research and development. External uses of funds are typically investment related. Companies often invest excess funds in such external revenue generators as bank accounts, stocks, bonds, bills, notes, futures, options, and foreign currency using financial MISs. Some individuals and companies are exploring making loans over the Internet. Lending Club, for example, facilitates loans made between people using Facebook, the social-networking site.[23] The company has facilitated about $1 million in loans that average about $5,000 per loan with interest rates that vary from about 7 to 17 percent. The loan default rate has been less than 1 percent.

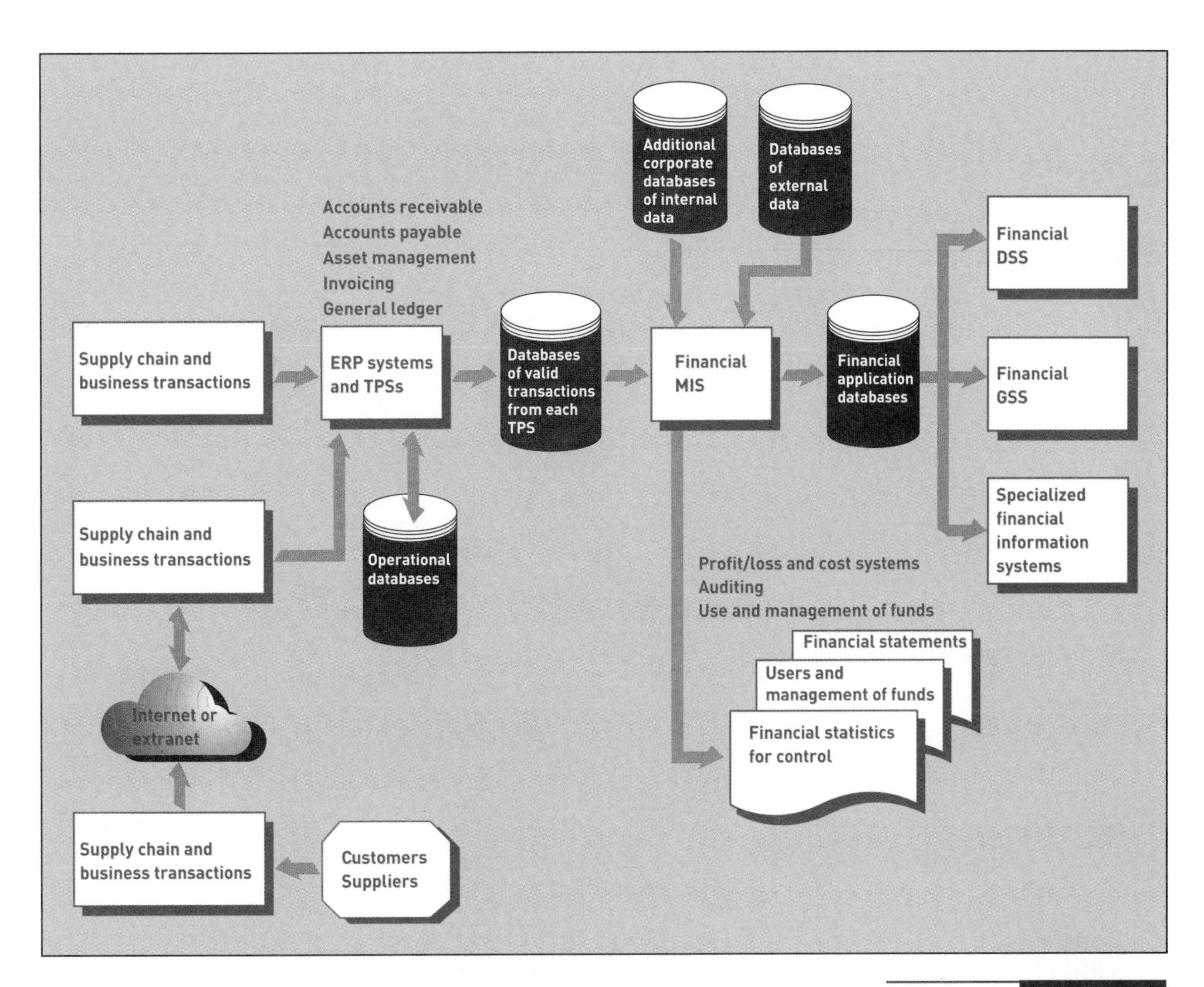

Figure 10.8

Overview of a Financial MIS

Financial institutions use information systems to shorten turnaround time for loan approvals.

(Source: © Royalty-Free/Corbis.)

Manufacturing Management Information Systems

More than any other functional area, advances in information systems have revolutionized manufacturing.[24] As a result, many manufacturing operations have been dramatically improved over the last decade. Also, with the emphasis on greater quality and productivity, having an effective manufacturing process is becoming even more critical. The use of computerized systems is emphasized at all levels of manufacturing—from the shop floor to the executive suite. Increasingly, companies are outsourcing the manufacturing process. With almost 300,000 employees, the Hon Hai company in China is one of the world's largest manufacturers of electronic products, including music players, cell phones, game consoles, and many other electronic products.[25] Some believe the company is China's largest exporter. Dell Computer has used both optimization and heuristic software to help it manufacture a larger variety of products.[26] Dell was able to double its product variety, while saving about $1 million annually in manufacturing costs. Figure 10.9 gives an overview of some of the manufacturing MIS inputs, subsystems, and outputs.

Figure 10.9

Overview of a Manufacturing MIS

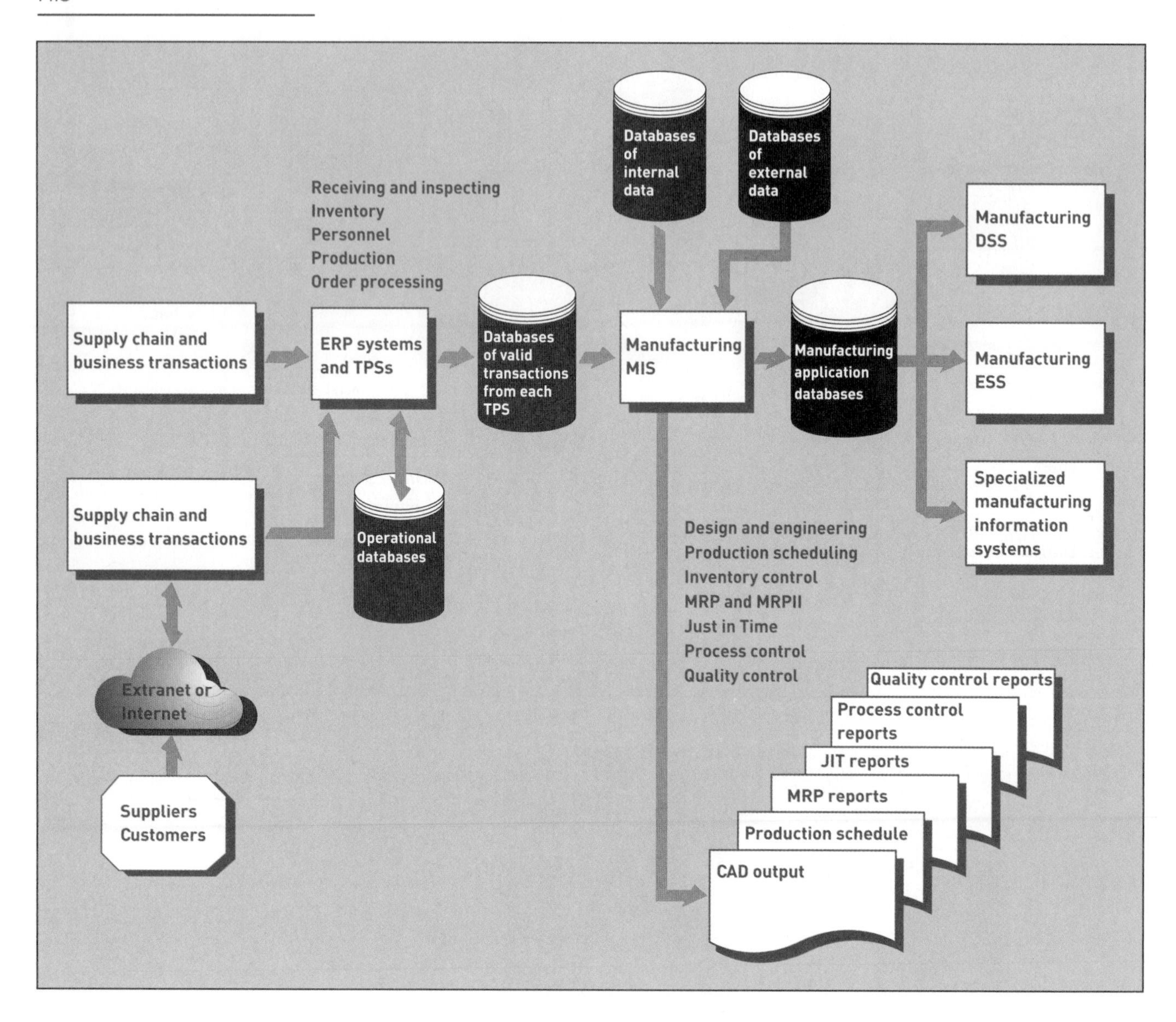

The manufacturing MIS subsystems and outputs are used to monitor and control the flow of materials, products, and services through the organization. As raw materials are converted to finished goods, the manufacturing MIS monitors the process at almost every stage. New technology could make this process easier. Using specialized computer chips and tiny radio transmitters, companies can monitor materials and products through the entire manufacturing process. Procter & Gamble, Wal-Mart, and Target have funded research into this manufacturing MIS. Car manufacturers, which convert raw steel, plastic, and other materials into a finished automobile, also monitor their manufacturing processes. Auto manufacturers add thousands of dollars of value to the raw materials they use in assembling a car. If the manufacturing MIS also lets them provide additional service, such as customized paint colors, it has added further value for customers. In doing so, the MIS helps provide the company the edge that can differentiate it from competitors. The success of an organization can depend on the manufacturing function. Some common information subsystems and outputs used in manufacturing are discussed next.

- **Design and engineering.** Manufacturing companies often use computer-aided design (CAD) with new or existing products. For example, Boeing (*www.boeing.com*) uses a CAD system to develop a complete digital blueprint of an aircraft before it begins the manufacturing process. As mock-ups are built and tested, the digital blueprint is constantly revised to reflect the most current design. Using such technology helps Boeing reduce manufacturing costs and the time to design a new aircraft.

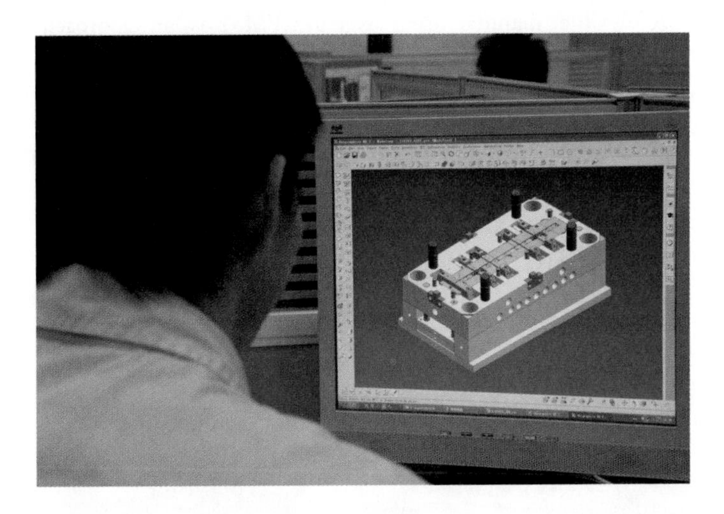

Computer-aided design (CAD) is used in the development and design of complex products or structures.

(Source: © Kim Steele/Getty Images.)

- **Master production scheduling and inventory control.** Scheduling production and controlling inventory are critical for any manufacturing company.[27] The overall objective of master production scheduling is to provide detailed plans for both short-term and long-range scheduling of manufacturing facilities. Some companies hire outside companies to help them with inventory control. Delta Airlines, for example, has a long-term, $1 billion agreement with Chromalloy Gas Turbine to help in providing inventory parts and maintenance of its jet engines.[28] Apparel company Tween Brands, Inc. uses a number of software packages to help it control inventory and reduce costs.[29]
  Most techniques are used to minimize inventory costs. They determine when and how much inventory to order. One method of determining the amount of inventory to order is called the **economic order quantity (EOQ)**. This quantity is calculated to minimize the total inventory costs. The "When to order?" question is based on inventory usage over time. Typically, the question is answered in terms of a **reorder point (ROP)**, which is a critical inventory quantity level. When the inventory level for a particular item falls to the reorder point, or critical level, the system generates a report so that an order is immediately placed for the EOQ of the product. Another inventory technique used when demand for one item depends on the demand for another is called **material requirements planning (MRP)**. The basic goal of MRP is to determine when finished products, such as automobiles or airplanes, are needed and then to work backward to determine deadlines and resources needed, such as engines and tires, to complete the final product on schedule. **Just-in-time (JIT) inventory** and manufacturing is an approach that maintains inventory at the lowest levels without sacrificing the availability of finished products. With this approach, inventory and materials are delivered just before they are used in a product. A JIT inventory system would arrange for a car windshield to be delivered to the assembly line only a few moments before it is secured to the automobile, rather than storing it in the manufacturing facility while the car's other components are being assembled. JIT, however, can result in some organizations running out of inventory when demand exceeds expectations. Even so, companies like Toyota continue to embrace JIT.[30] According to the president of the company, "We've been implementing this strategy for decades, and

economic order quantity (EOQ)
The quantity that should be reordered to minimize total inventory costs.

reorder point (ROP)
A critical inventory quantity that determines when to order more inventory.

material requirements planning (MRP)
A set of inventory-control techniques that help coordinate thousands of inventory items when the demand of one item is dependent on the demand for another.

just-in-time (JIT) inventory
A philosophy of inventory management in which inventory and materials are delivered just before they are used in manufacturing a product.

we will keep on with it." The car manufacturer had to momentarily close more than ten plants when an earthquake prevented a supplier from producing $1.50 piston ring parts.

- **Process control.** Managers can use a number of technologies to control and streamline the manufacturing process. For example, computers can directly control manufacturing equipment, using systems called **computer-assisted manufacturing** (CAM). CAM systems can control drilling machines, assembly lines, and more. **Computer-integrated manufacturing** (CIM) uses computers to link the components of the production process into an effective system. CIM's goal is to tie together all aspects of production, including order processing, product design, manufacturing, inspection and quality control, and shipping. A **flexible manufacturing system** (FMS) is an approach that allows manufacturing facilities to rapidly and efficiently change from making one product to another. In the middle of a production run, for example, the production process can be changed to make a different product or to switch manufacturing materials. By using an FMS, the time and cost to change manufacturing jobs can be substantially reduced, and companies can react quickly to market needs and competition.

computer-assisted manufacturing (CAM)
A system that directly controls manufacturing equipment.

computer-integrated manufacturing (CIM)
Using computers to link the components of the production process into an effective system.

flexible manufacturing system (FMS)
An approach that allows manufacturing facilities to rapidly and efficiently change from making one product to making another.

Computer-assisted manufacturing systems control complex processes on the assembly line and provide users with instant access to information.

(Source: © Phototake/Alamy.)

- **Quality control and testing.** With increased pressure from consumers and a general concern for productivity and high quality, today's manufacturing organizations are placing more emphasis on **quality control**, a process that ensures that the finished product meets the customers' needs. Information systems are used to monitor quality and take corrective steps to eliminate possible quality problems.

quality control
A process that ensures that the finished product meets the customers' needs.

INFORMATION SYSTEMS @ WORK

Pharmaceutical Company Reduces Time-to-Market

AstraZeneca is one of the world's leading pharmaceutical companies and truly a global corporation. With a presence in over 100 countries, AstraZeneca is based in London, England, with research and development sites in Sweden, the United States, and the United Kingdom. The company has more than 67,000 employees, who mostly work in Europe. AstraZeneca totaled $29.6 billion in sales in 2007.

The pharmaceutical industry is highly competitive, with many companies racing to be the first to market with drug remedies for common ailments and diseases. It typically takes 8–12 years to bring a new drug to market. Furthermore, a drug patent lasts 20–25 years, much of which is taken up in development time. The less time a company spends developing a drug, the more years it can reap in profits before generic alternatives are made available. Every day saved in development can mean millions of dollars in profits. The key to shortening the development time of drugs lies in efficient project management.

AstraZeneca developed a project management system to allow its research facilities around the world to share research and development (R&D) information. The system is called Matrix, and it gathers and analyzes research and development information stored in a large corporate data warehouse. Approximately 5,000 researchers working at the six AstraZeneca R&D sites have access to the Matrix system. The ability to access research information is transforming how research is conducted. Researchers can collaborate on projects from different sites, tracking each other's progress. Matrix has also eliminated wasteful duplication of effort caused by miscommunication.

AstraZeneca researchers create a high volume of project data every day. The Matrix system allows researchers and managers to track, understand, and manage that data. Top-level executives use an executive dashboard manager to view key performance indicators and keep their finger on the pulse of the business to make quick, confident decisions. The system helps the finance department get a clear picture of project costs. Managers can easily manage project scheduling, budgeting, and resource allocation. Matrix makes AstraZeneca more nimble, which is a significant achievement considering the size of the company. AstraZeneca can launch products quickly, often faster than the competition, which lets them establish leadership in a variety of specialty areas.

AstraZeneca business manager David Scanion believes that the new MIS has improved the company's project management, cost control, and resource usage, improving the company's ability to compete with world-class research done in record-breaking timescales.

Discussion Questions

1. What time factors affect the product life of new drugs?
2. What benefits does the Matrix system provide to researchers at AstraZeneca?

Critical Thinking Questions

1. How does the Matrix system improve project management at AstraZeneca?
2. What financial savings and benefits are provided to the company from the Matrix system?

SOURCES: Business Objects Staff, "AstraZeneca", Business Objects Customers in the Spotlight, 2008, *www.businessobjects.com/company/customers/spotlight/astrazeneca.asp*; AstraZeneca Web Site, accessed May 20, 2008, www.astrazeneca.com.

Marketing Management Information Systems

marketing MIS
An information system that supports managerial activities in product development, distribution, pricing decisions, and promotional effectiveness.

A **marketing MIS** supports managerial activities in product development, distribution, pricing decisions, promotional effectiveness, and sales forecasting. Marketing functions are increasingly being performed on the Internet.[31] When an English teacher in Japan put a number of fun family projects for children and their families on *www.Digg.com*, a popular Web site, it had an unexpected marketing consequence.[32] In only a few days, it dramatically increased sales of rivets used to complete the projects. According to Andy McGrew, owner of a company that sells blueprints for many home projects, "It would have taken me a year to sell that many rivets." Many companies are developing Internet marketplaces to advertise and sell products. The amount spent on online advertising is worth billions of dollars annually. Newer marketing companies, such as AdMob, Inc. (*www.admob.com*), are placing ads on cell phones and mobile devices with Internet access.[33] According to an executive of AdMob, "Everybody's just trying to dip their toes in the water and figure out what's going to work." Software can measure how many customers see the advertising. Some companies use a software product called SmartLoyalty to analyze customer loyalty.

Some marketing departments are actively using the Internet to advertise their products and services and keep customers happy. Companies, for example, are starting to advertise their products and services on Facebook (*www.facebook.com*), a social-networking site.[34] YouTube, the video-sharing Internet site, sells video ads on its site to companies, including Ford, BMW, Time Warner, and others.[35] After about ten seconds, the video ad disappears unless the user clicks it.[36] Corporate marketing departments also use social-networking sites, such as Second Life (*www.secondlife.com*), to advertise their products and perform marketing research.[37]

Customer relationship management (CRM) programs, available from some ERP vendors, help a company manage all aspects of customer encounters. CRM software can help a company collect customer data, contact customers, educate customers on new products, and sell products to customers through a Web site. An airline, for example, can use a CRM system to notify customers about flight changes. New Zealand's Jade Stadium uses CRM software from GlobalTech Solutions to give a single entry point to its marketing efforts and customer databases, instead of using about 20 spreadsheets. The CRM software will help Jade Stadium develop effective marketing campaigns, record and track client contacts, and maintain an accurate database of clients. Yet, not all CRM systems and marketing sites on the Internet are successful. Customization and ongoing maintenance of a CRM system can be expensive. Figure 10.10 shows the inputs, subsystems, and outputs of a typical marketing MIS.

Subsystems for the marketing MIS include marketing research, product development, promotion and advertising, and product pricing. These subsystems and their outputs help marketing managers and executives increase sales, reduce marketing expenses, and develop plans for future products and services to meet the changing needs of customers.

- **Marketing research.** The purpose of marketing research is to conduct a formal study of the market and customer preferences.[38] Computer systems are used to help conduct and analyze the results of surveys, questionnaires, pilot studies, and interviews. eCourier, for example, uses Crystal Reports from Business Objects to determine customer habits and preferences.[39] The company can perform marketing research using its Web site to help determine which customers are happy and still buying and which ones might switch to another company. According to an executive at eCourier, "We know of ten cases where there were problems, for whatever reason, and we got in there early. As a result, we were able to save the client." In addition to knowing what you buy, market research can determine where you buy.[40] This can help in developing new products and services and tailoring ads and promotions. With the use of GPS positioning systems, marketing firms can promote products to you over cell phones and other mobile devices by knowing your location.
- **Product development.** Product development involves the conversion of raw materials into finished goods and services and focuses primarily on the physical attributes of the product. Many factors, including plant capacity, labor skills, engineering factors, and

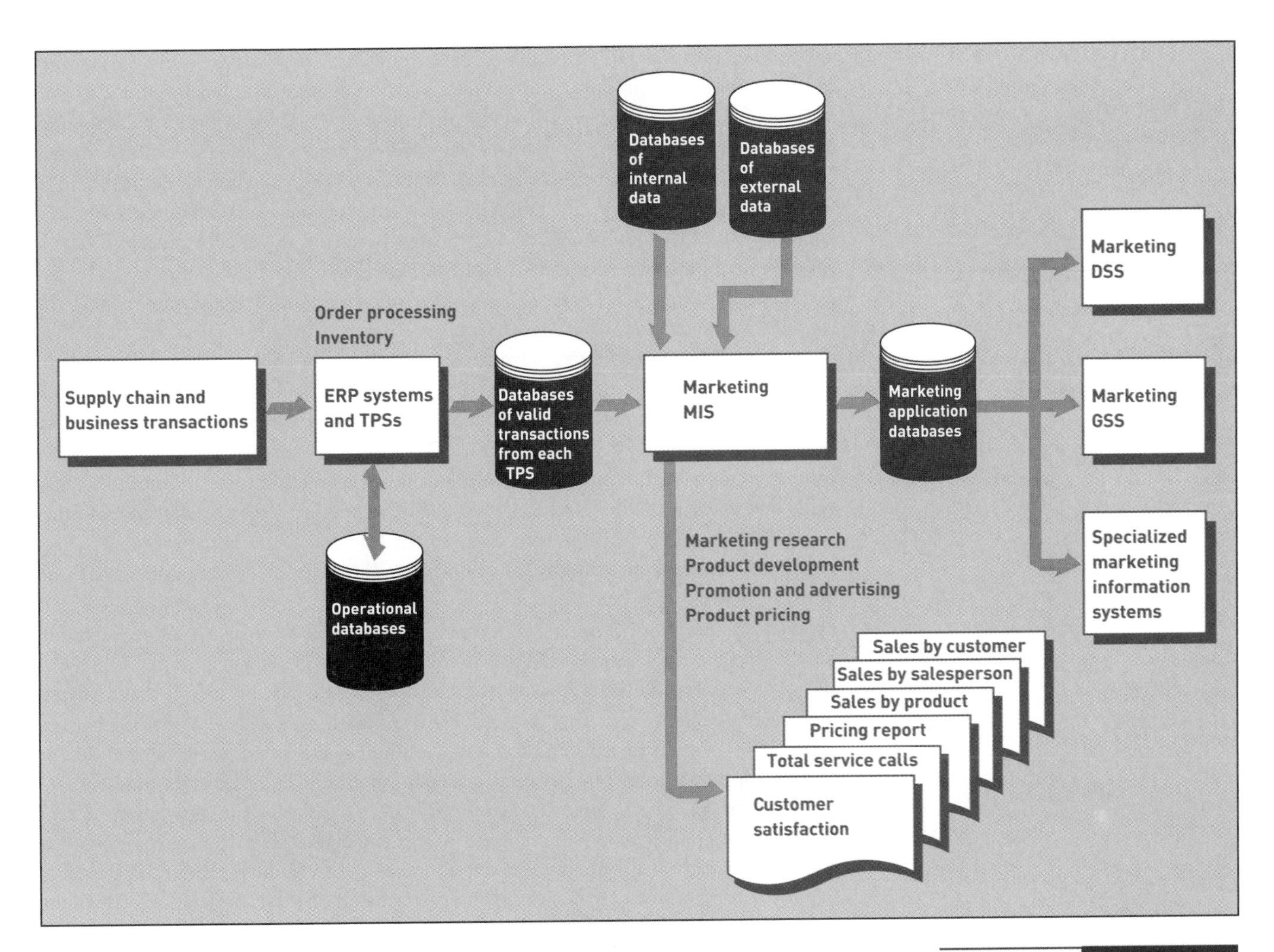

Figure 10.10

Overview of a Marketing MIS

Marketing research data yields valuable information for the development and marketing of new products.

(Source: © Michael Newman/ PhotoEdit.)

materials are important in product development decisions. In many cases, a computer program analyzes these various factors and selects the appropriate mix of labor, materials, plant and equipment, and engineering designs. Make-or-buy decisions can also be made with the assistance of computer programs. To get additional revenues, some TV programs and movies promote products and services during their programs.[41] Movies, for example, can show actors driving luxury cars and wearing expensive watches. Food companies can sponsor food and cooking TV programs that give their products more exposure. The approach is called "branded entertainment."

- **Promotion and advertising.** One of the most important functions of any marketing effort is promotion and advertising. Product success is a direct function of the types of advertising and sales promotion done. Increasingly, organizations are using the Internet to advertise and sell products and services. Johnson & Johnson used Internet cartoons instead of extensive TV advertising to promote a popular baby lotion.[42] Yahoo has launched Brand Universe and advertises specific products around various interest groups.[43] According to a Yahoo executive, "We can speak to a particular audience." The goal is to target ads to a specific group of people that are likely to purchase the advertised goods and services. Companies are also trying to measure the effectiveness of different advertising approaches, such as TV and Internet advertising.[44] According to a Toyota marketing executive, "We wanted to be able to have a tool to really start judging the Internet compared to TV." Toyota uses IAG Research to help it measure the effectiveness of TV and Internet advertising. Several companies, including ScanScout (*www.scanscout.com*) and YuMeNetworks (*www.yumenetworks.com*), are trying to match up video content on the Internet with specific ads that cater to those watching the videos.[45] Companies, such as furniture maker Ikea, are increasingly hiring digital ad companies to make sure their products get promoted on the Internet.[46] Some individuals and companies are willing to tolerate advertising to get free software or Internet service.[47] Companies are also using blogs on the Internet to advertise products.[48] eBay, the popular Internet auction site, is working with Bid4Spots to auction radio advertising spots to clients.[49]

- **Product pricing.** Product pricing is another important and complex marketing function. Retail price, wholesale price, and price discounts must be set. Chrysler, for example, was able to save about $500 million by using a sophisticated pricing model that analyzed incentives, financing, and other factors.[50] Most companies try to develop pricing policies that will maximize total sales revenues. Computers are often used to analyze the relationship between prices and total revenues. Traditionally, executives used costs to determine prices. They simply added a profit margin to total costs to guarantee a decent profit. Today, however, more executives look at the marketplace to determine product pricing. In one case, a company was able to increase its revenue by about $200 million in part by using a more aggressive pricing policy based on what the market was willing to pay.[51]
- **Sales analysis.** Computerized sales analysis is important to identify products, sales personnel, and customers that contribute to profits and those that do not. Several reports can be generated to help marketing managers make good sales decisions (see Figure 10.11). The sales-by-product report lists all major products and their sales for a specified period of time. This report shows which products are doing well and which need improvement or should be discarded altogether. The sales-by-salesperson report lists total sales for each salesperson for each week or month. This report can also be subdivided by product to show which products are being sold by each salesperson. The sales-by-customer report is a tool that can be used to identify high- and low-volume customers.

Human Resource Management Information Systems

human resource MIS
An information system that is concerned with activities related to employees and potential employees of an organization, also called a personnel MIS.

A **human resource MIS (HRMIS)**, also called the *personnel MIS*, is concerned with activities related to previous, current, and potential employees of the organization. Because the personnel function relates to all other functional areas in the business, the human resource (HR) MIS plays a valuable role in ensuring organizational success. Some of the activities performed by this important MIS include workforce analysis and planning, hiring, training, job and task assignment, and many other personnel-related issues. An effective human resource MIS allows a company to keep personnel costs at a minimum, while serving the required business processes needed to achieve corporate goals. Although human resource information systems focus on cost reduction, many of today's HR systems concentrate on hiring and managing existing employees to get the total potential of the human talent in the organization. According to the High Performance Workforce Study conducted by Accenture, the most important HR initiatives include improving worker productivity, improving adaptability to new opportunities, and facilitating organizational change. Figure 10.12 shows some of the inputs, subsystems, and outputs of the human resource MIS.

(a) Sales by Product

Product	August	September	October	November	December	Total
Product 1	34	32	32	21	33	152
Product 2	156	162	177	163	122	780
Product 3	202	145	122	98	66	633
Product 4	345	365	352	341	288	1,691

(b) Sales by Salesperson

Salesperson	August	September	October	November	December	Total
Jones	24	42	42	11	43	162
Kline	166	155	156	122	133	732
Lane	166	155	104	99	106	630
Miller	245	225	305	291	301	1,367

(c) Sales by Customer

Customer	August	September	October	November	December	Total
Ang	234	334	432	411	301	1,712
Braswell	56	62	77	61	21	277
Celec	1,202	1,445	1,322	998	667	5,634
Jung	45	65	55	34	88	287

Figure 10.11

Reports Generated to Help Marketing Managers Make Good Decisions

(a) This sales-by-product report lists all major products and their sales for the period from August to December. (b) This sales-by-salesperson report lists total sales for each salesperson for the same time period. (c) This sales-by-customer report lists sales for each customer for the period. Like all MIS reports, totals are provided automatically by the system to show managers at a glance the information they need to make good decisions.

Human resource subsystems and outputs range from the determination of human resource needs and hiring through retirement and outplacement. Most medium and large organizations have computer systems to assist with human resource planning, hiring, training and skills inventorying, and wage and salary administration. Outputs of the human resource MIS include reports, such as human resource planning reports, job application review profiles, skills inventory reports, and salary surveys, are discussed next.

- **Human resource planning.** One of the first aspects of any human resource MIS is determining personnel and human needs. The overall purpose of this MIS subsystem is to put the right number and types of employees in the right jobs when they are needed, including internal employees that work exclusively for the organization and outside workers that are hired when they are needed. Some experts believe that workers should be managed like a supply chain, using supply chain management (SCM) and just-in-time techniques, first discussed in Chapter 2.[52] Effective human resource planning often requires computer programs, such as SPSS and SAS, to forecast the future number of employees needed and anticipate the future supply of people for these jobs. IBM used an HR pilot program, called Professional Marketplace, to plan for workforce needs, including the supplies and tools the workforce needs to work efficiently. Professional Marketplace helps IBM to catalog employees into a glossary of skills and abilities. Like many other companies, HR and workforce costs are IBM's biggest expense.
- **Personnel selection and recruiting.** If the human resource plan reveals that additional personnel are required, the next logical step is recruiting and selecting personnel. Companies seeking new employees often use computers to schedule recruiting efforts and trips and to test potential employees' skills. Many companies now use the Internet to screen for job applicants. Applicants use a template to load their résumé onto the

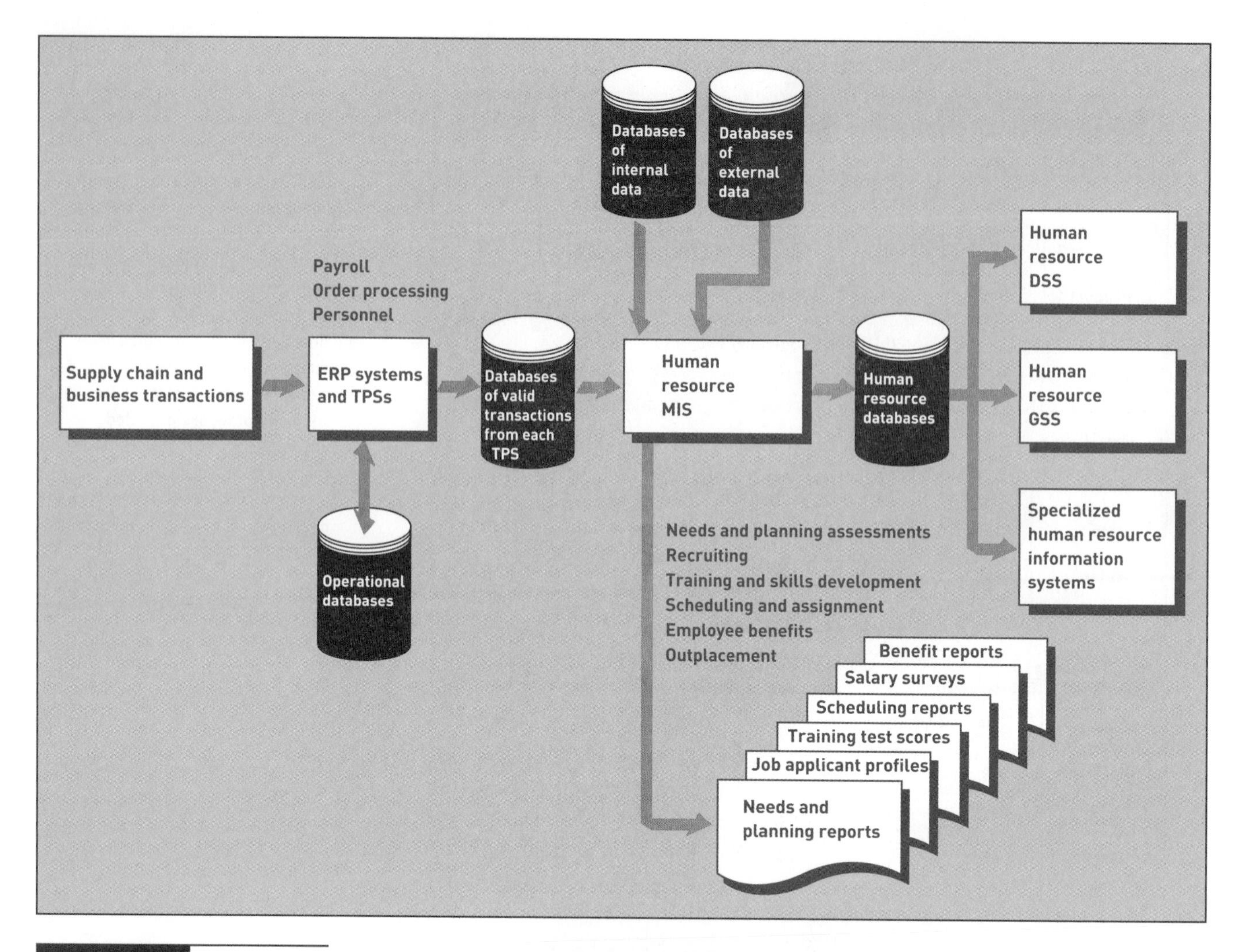

Figure 10.12

Overview of a Human Resource MIS

Human resource MIS subsystems help to determine personnel needs and match employees to jobs.

(Source: © Bambu Productions/ Getty Images.)

Internet site. HR managers can then access these résumés and identify applicants they are interested in interviewing.

- **Training and skills inventory.** Some jobs, such as programming, equipment repair, and tax preparation, require very specific training for new employees.[53] Other jobs may require general training about the organizational culture, orientation, dress standards, and expectations of the organization. When training is complete, employees often take computer-scored tests to evaluate their mastery of skills and new material.

- **Scheduling and job placement.** Employee schedules are developed for each employee, showing his job assignments over the next week or month. Job placements are often determined based on skills inventory reports showing which employee might be best suited to a particular job. Sophisticated scheduling programs are often used in the airline industry, the military, and many other areas to get the right people assigned to the right jobs at the right time.
- **Wage and salary administration.** Another human resource MIS subsystem involves determining wages, salaries, and benefits, including medical payments, savings plans, and retirement accounts. Wage data, such as industry averages for positions, can be taken from the corporate database and manipulated by the human resource MIS to provide wage information and reports to higher levels of management.
- **Outplacement.** Employees leave a company for a variety of reasons. Outplacement services are offered by many companies to help employees make the transition. *Outplacement* can include job counseling and training, job and executive search, retirement and financial planning, and a variety of severance packages and options. Many employees use the Internet to plan their future retirement or to find new jobs, using job sites such as *www.monster.com*.

Other Management Information Systems

In addition to finance, manufacturing, marketing, and human resource MISs, some companies have other functional management information systems. For example, most successful companies have well-developed accounting functions and a supporting accounting MIS. Also, many companies use geographic information systems for presenting data in a useful form.

Accounting MISs

In some cases, accounting works closely with financial management. An **accounting MIS** performs a number of important activities, providing aggregate information on accounts payable, accounts receivable, payroll, and many other applications. The organization's enterprise resource planning and transaction processing system captures accounting data, which is also used by most other functional information systems.

accounting MIS
An information system that provides aggregate information on accounts payable, accounts receivable, payroll, and many other applications.

Some smaller companies hire outside accounting firms to assist them with their accounting functions. These outside companies produce reports for the firm using raw accounting data. In addition, many excellent integrated accounting programs are available for personal computers in small companies. Depending on the needs of the small organization and its staff's computer experience, using these computerized accounting systems can be a very cost-effective approach to managing information.

Geographic Information Systems

Increasingly, managers want to see data presented in graphical form. A **geographic information system** (GIS) is a computer system capable of assembling, storing, manipulating, and displaying geographically referenced information, that is, data identified according to its location. A GIS enables users to pair maps or map outlines with tabular data to describe aspects of a particular geographic region. For example, sales managers might want to plot total sales for each county in the states they serve. Using a GIS, they can specify that each county be shaded to indicate the relative amount of sales—no shading or light shading represents no or little sales, and deeper shading represents more sales. Staples Inc., the large office supply store chain, used a geographic information system to select new store locations. Finding the best location is critical. It can cost up to $1 million for a failed store because of a poor location. Staples uses a GIS tool from Tactician Corporation (*www.tactician.com*) along with software from SAS. Although many software products have seen declining revenues, the use of GIS software is increasing.

geographic information system (GIS)
A computer system capable of assembling, storing, manipulating, and displaying geographic information, that is, data identified according to its location.

We saw earlier in this chapter that management information systems (MISs) provide useful summary reports to help solve structured and semistructured business problems. Decision support systems (DSSs) offer the potential to assist in solving both semistructured and unstructured problems. These systems are discussed next.

AN OVERVIEW OF DECISION SUPPORT SYSTEMS

A DSS is an organized collection of people, procedures, software, databases, and devices used to help make decisions that solve problems. The focus of a DSS is on decision-making effectiveness when faced with unstructured or semistructured business problems. As with a TPS and an MIS, a DSS should be designed, developed, and used to help an organization achieve its goals and objectives. Decision support systems offer the potential to generate higher profits, lower costs, and better products and services. For example, healthcare organizations use DSSs to improve patient care and reduce costs.

Decision support systems, although skewed somewhat toward the top levels of management, are used at all levels. To some extent, today's managers at all levels are faced with less structured, nonroutine problems, but the quantity and magnitude of these decisions increase as a manager rises higher in an organization. Many organizations contain a tangled web of complex rules, procedures, and decisions. DSSs are used to bring more structure to these problems to aid the decision-making process. In addition, because of the inherent flexibility of decision support systems, managers at all levels are able to use DSSs to assist in some relatively routine, programmable decisions in lieu of more formalized management information systems. DSSs are also used in government, law enforcement, and nonprofit organizations. See Figure 10.13.

Figure 10.13

Decision support systems are also used by nonprofit organizations and in government, such as in police departments.

(Source: © Spencer C. Grant/ PhotoEdit.)

Characteristics of a Decision Support System

Decision support systems have many characteristics that allow them to be effective management support tools. Of course, not all DSSs work the same. The following list shows some important characteristics of a DSS.

- Provide rapid access to information.
- Handle large amounts of data from different sources.
- Provide report and presentation flexibility.
- Offer both textual and graphical orientation.
- Support drill-down analysis.
- Perform complex, sophisticated analysis and comparisons using advanced software packages.
- Support optimization, satisficing, and heuristic approaches. (See Figure 10.14.)
- Perform simulation analysis—the ability of the DSS to duplicate the features of a real system, where probability or uncertainty is involved.

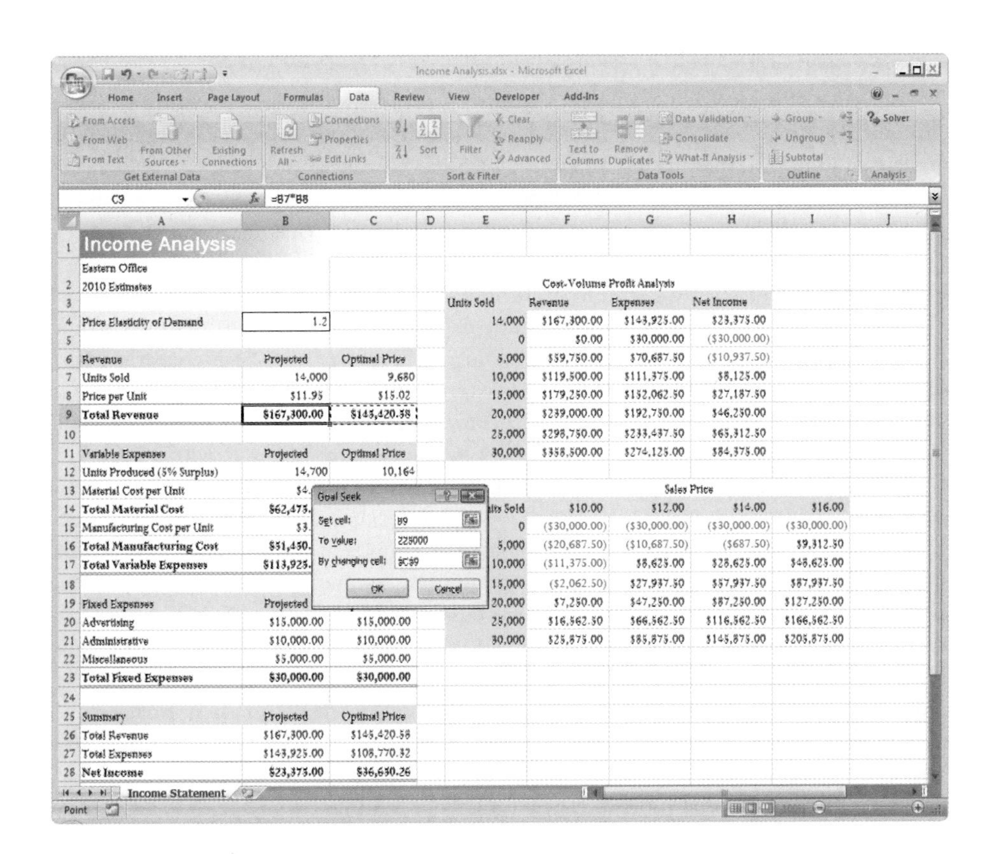

Figure 10.14

With a spreadsheet program, a manager can enter a goal and the spreadsheet will determine the input needed to achieve the goal.

Capabilities of a Decision Support System

Developers of decision support systems strive to make them more flexible than management information systems and to give them the potential to assist decision makers in a variety of situations. Table 10.1 lists a few DSS applications. DSSs can assist with all or most problem-solving phases, decision frequencies, and different degrees of problem structure. DSS approaches can also help at all levels of the decision-making process. A single DSS might provide only a few of these capabilities, depending on its uses and scope.

Table 10.1

Selected DSS Applications

Company or Application	Description
ING Direct	The financial services company uses a DSS to summarize the bank's financial performance. The bank needed a measurement and tracking mechanism to determine how successful it was and to make modifications to plans in real time.
Cinergy Corporation	The electric utility developed a DSS to reduce lead time and effort required to make decisions in purchasing coal.
U.S. Army	It developed a DSS to help recruit, train, and educate enlisted forces. The DSS uses a simulation that incorporates what-if features.
National Audubon Society	It developed a DSS called Energy Plan (EPLAN) to analyze the impact of U.S. energy policy on the environment.
Hewlett-Packard	The computer company developed a DSS called Quality Decision Management to help improve the quality of its products and services.
State of Virginia	The State of Virginia developed the Transportation Evacuation Decision Support System (TEDSS) to determine the best way to evacuate people in case of a nuclear disaster at its nuclear power plants.

Support for Problem-Solving Phases

The objective of most decision support systems is to assist decision makers with the phases of problem solving. As previously discussed, these phases include intelligence, design, choice, implementation, and monitoring. A specific DSS might support only one or a few phases. By supporting all types of decision-making approaches, a DSS gives the decision maker a great deal of flexibility in getting computer support for decision-making activities.

Support for Different Decision Frequencies

ad hoc DSS
A DSS concerned with situations or decisions that come up only a few times during the life of the organization.

institutional DSS
A DSS that handles situations or decisions that occur more than once, usually several times per year or more. An institutional DSS is used repeatedly and refined over the years.

Decisions can range on a continuum from one-of-a-kind to repetitive decisions. One-of-a-kind decisions are typically handled by an **ad hoc DSS**. An ad hoc DSS is concerned with situations or decisions that come up only a few times during the life of the organization; in small businesses, they might happen only once. For example, a company might need to decide whether to build a new manufacturing facility in another area of the country. Repetitive decisions are addressed by an institutional DSS. An **institutional DSS** handles situations or decisions that occur more than once, usually several times per year or more. An institutional DSS is used repeatedly and refined over the years. Examples of institutional DSSs include systems that support portfolio and investment decisions and production scheduling. These decisions might require decision support numerous times during the year. For example, DSSs are used to help solve computer-related problems that can occur multiple times throughout the day. With this approach, the DSS monitors computer systems second by second for problems and takes action to prevent problems, such as slowdowns and crashes, and to recover from them when they occur. One IBM engineer believes that this approach, called *autonomic computing*, is the key to the future of computing. Between these two extremes are decisions that managers make several times, but not regularly or routinely.

Support for Different Problem Structures

highly structured problems
Problems that are straightforward and require known facts and relationships.

semistructured or unstructured problems
More complex problems in which the relationships among the pieces of data are not always clear, the data might be in a variety of formats, and the data is often difficult to manipulate or obtain.

As discussed previously, decisions can range from highly structured and programmed to unstructured and nonprogrammed. **Highly structured problems** are straightforward, requiring known facts and relationships. **Semistructured** or **unstructured problems**, on the other hand, are more complex. The relationships among the pieces of data are not always clear, the data might be in a variety of formats, and it is often difficult to manipulate or obtain. In addition, the decision maker might not know the information requirements of the decision in advance. For example, a DSS has been used to support sophisticated and unstructured investment analysis and make substantial profits for traders and investors. Some DSS trading software is programmed to place buy and sell orders automatically without a trader manually entering a trade, based on parameters set by the trader.

Support for Various Decision-Making Levels

Decision support systems can provide help for managers at different levels within the organization. Operational managers can get assistance with daily and routine decision making. Tactical decision makers can use analysis tools to ensure proper planning and control. At the strategic level, DSSs can help managers by providing analysis for long-term decisions requiring both internal and external information (see Figure 10.15).

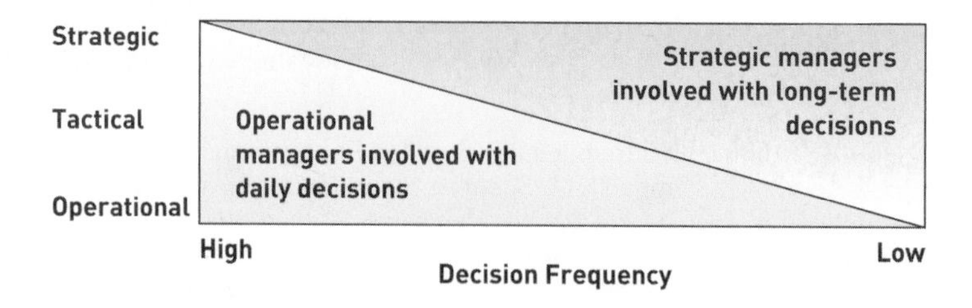

Figure 10.15

Decision-Making Level

Strategic managers are involved with long-term decisions, which are often made infrequently. Operational managers are involved with decisions that are made more frequently.

A Comparison of DSS and MIS

A DSS differs from an MIS in numerous ways, including the type of problems solved, the support given to users, the decision emphasis and approach, and the type, speed, output, and development of the system used. Table 10.2 lists brief descriptions of these differences.

Table 10.2

Comparison of DSSs and MISs

Factor	DSS	MIS
Problem Type	A DSS can handle unstructured problems that cannot be easily programmed.	An MIS is normally used only with structured problems.
Users	A DSS supports individuals, small groups, and the entire organization. In the short run, users typically have more control over a DSS.	An MIS supports primarily the organization. In the short run, users have less control over an MIS.
Support	A DSS supports all aspects and phases of decision making; it does not replace the decision maker—people still make the decisions.	Some MIS systems make automatic decisions and replace the decision maker.
Emphasis	A DSS emphasizes actual decisions and decision-making styles.	An MIS usually emphasizes information only.
Approach	A DSS is a direct support system that provides interactive reports on computer screens.	An MIS is typically an indirect support system that uses regularly produced reports.
System	The computer equipment that provides decision support is usually online (directly connected to the computer system) and related to real time (providing immediate results). Computer terminals and display screens are examples—these devices can provide immediate information and answers to questions.	An MIS, using printed reports that might be delivered to managers once per week, cannot provide immediate results.
Speed	Because a DSS is flexible and can be implemented by users, it usually takes less time to develop and is better able to respond to user requests.	An MIS's response time is usually longer.
Output	DSS reports are usually screen oriented, with the ability to generate reports on a printer.	An MIS typically is oriented toward printed reports and documents.
Development	DSS users are usually more directly involved in its development. User involvement usually means better systems that provide superior support. For all systems, user involvement is the most important factor for the development of a successful system.	An MIS is frequently several years old and often was developed for people who are no longer performing the work supported by the MIS.

COMPONENTS OF A DECISION SUPPORT SYSTEM

At the core of a DSS are a database and a model base. In addition, a typical DSS contains a user interface, also called **dialogue manager**, that allows decision makers to easily access and manipulate the DSS and to use common business terms and phrases. Finally, access to the Internet, networks, and other computer-based systems permits the DSS to tie into other powerful systems, including the TPS or function-specific subsystems. Internet software agents, for example, can be used in creating powerful decision support systems. Figure 10.16 shows a conceptual model of a DSS. Specific DSSs might not have all the components shown in Figure 10.16.

dialogue manager
A user interface that allows decision makers to easily access and manipulate the DSS and to use common business terms and phrases.

The Database

The database management system allows managers and decision makers to perform *qualitative analysis* on the company's vast stores of data in databases, data warehouses, and data marts, discussed in Chapter 5. A *data-driven DSS* primarily performs qualitative analysis based on the company's databases. Data-driven DSSs tap into vast stores of information contained in the corporate database, retrieving information on inventory, sales, personnel, production, finance, accounting, and other areas. Tween Brands, Inc. specialty retail store uses the Oracle database to provide decision support to reduce inventory costs.[54] Jo-Ann Stores uses its database to support decision making. According to a vice president and CIO

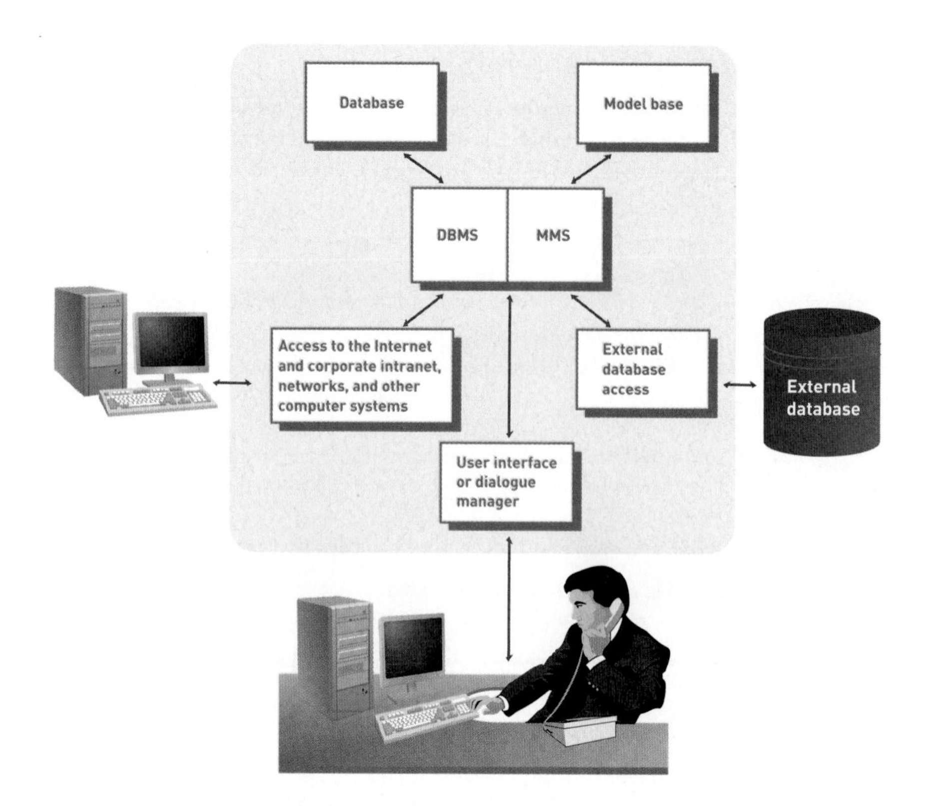

Figure 10.16

Conceptual Model of a DSS

DSS components include a model base; database; external database access; access to the Internet and corporate intranet, networks, and other computer systems; and a user interface or dialogue manager.

of the company, "The people in the business need to understand what to do with the data and how the data impacts other parts of the organization." Data mining and business intelligence, introduced in Chapter 5, are often used in a data-driven DSS.[55] A casino can use a data-driven DSS to search large databases to get detailed information on patrons. It can tell how much each patron spends per day on gambling, and more. Swiss telecommunications company Cablecom uses a data-driven DSS from SPSS to identify customers that might leave the company.[56] Data-driven DSSs can also be used in emergency medical situations to make split-second, life-or-death treatment decisions. Data-driven medical DSSs allow doctors to access the complete medical records of a patient. Some medical record systems also allow patients to enter their own health information into the database, such as medicines, allergies, and family health histories. WebMD, iHealthRecord, Walgreens, and PersonalHealthKey allow people to put their medical records online for rapid access.

Not everyone is satisfied with data-driven DSSs, however. One survey revealed that many middle-level managers spend about two hours each workday trying to find the data they need to perform their jobs.[57] The biggest problems are being deluged with too much data, other departments not sharing their data, and not knowing if the data they get is current and accurate. Some people also have privacy concerns. A few firms mine personal data on purchasing habits and then sell the information to online advertisers that want you to buy their products and services.[58] Some people believe this is an invasion of their privacy. Today, companies are spending over $500 million on online ads. This amount is expected to grow to about $4 billion over the next ten years. Data mining firms help these companies target their advertising to people likely to buy their products and services

A database management system can also connect to external databases to give managers and decision makers even more information and decision support. External databases can include the Internet, libraries, government databases, and more. The combination of internal and external database access can give key decision makers a better understanding of the company and its environment. Schumacher Group, for example, uses software to mash up, or tie together, information from TV reports, maps, computerized phone books, and other sources to analyze the impact of hurricanes on how doctors are to be scheduled at different emergency rooms in Lafayette and other cities in Louisiana.[59] Other companies, including Audi and AccuWeather, are using similar software packages to integrate data from different sources into data-driven DSSs.

The Model Base

The **model base** allows managers and decision makers to perform *quantitative analysis* on both internal and external data. A *model-driven DSS* primarily performs mathematical or quantitative analysis. The model base gives decision makers access to a variety of models so that they can explore different scenarios and see their effects. Ultimately, it assists them in the decision-making process. Procter & Gamble, maker of Pringles potato chips, Pampers diapers, and hundreds of other consumer products, uses a model-driven DSS to streamline how raw materials and products flow from suppliers to customers. The model-driven DSS has saved the company hundreds of millions of dollars in supply chain-related costs. Model-driven DSSs are excellent at predicting customer behaviors. LoanPerformance (*www.loanperformance.com*), for example, uses models to help it forecast which customers might be late with payments or might default on their loans.[60] Other financial service and insurance firms, such as health insurer HighMark, use model-driven DSSs to predict fraud. Some stock trading and investment firms use sophisticated model-driven DSSs to make trading decisions and huge profits.[61] Some experts believe that a slight time advantage in computerized trading programs can result in millions of dollars of trading profits.

model base
Part of a DSS that provides decision makers access to a variety of models and assists them in decision making.

Model management software (MMS) can coordinate the use of models in a DSS, including financial, statistical analysis, graphical, and project-management models. Depending on the needs of the decision maker, one or more of these models can be used (see Table 10.3).

model management software
Software that coordinates the use of models in a DSS.

Model Type	Description	Software
Financial	Provides cash flow, internal rate of return, and other investment analysis	Spreadsheet, such as Microsoft Excel
Statistical	Provides summary statistics, trend projections, hypothesis testing, and more	Statistical programs, such as SPSS or SAS
Graphical	Assists decision makers in designing, developing, and using graphic displays of data and information	Graphics programs, such as Microsoft PowerPoint
Project Management	Handles and coordinates large projects; also used to identify critical activities and tasks that could delay or jeopardize an entire project if they are not completed in a timely and cost-effective fashion	Project management software, such as Microsoft Project

Table 10.3

Model Management Software

DSSs often use financial, statistical, graphical, and project-management models.

The User Interface or Dialogue Manager

The user interface or dialogue manager allows users to interact with the DSS to obtain information. It assists with all aspects of communications between the user and the hardware and software that constitute the DSS. In a practical sense, to most DSS users, the user interface is the DSS. Upper-level decision makers are often less interested in where the information came from or how it was gathered than that the information is both understandable and accessible.

GROUP SUPPORT SYSTEMS

The DSS approach has resulted in better decision making for all levels of individual users. However, many DSS approaches and techniques are not suitable for a group decision-making environment. Although not all workers and managers are involved in committee meetings and group decision-making sessions, some tactical and strategic-level managers can spend more than half their decision-making time in a group setting. Such managers need assistance with group decision making. A **group support system** (GSS), also called a *group decision support system* and a *computerized collaborative work system*, consists of most of the elements in a DSS, plus software to provide effective support in group decision-making settings (see Figure 10.17).

group support system (GSS)
Software application that consists of most elements in a DSS, plus software to provide effective support in group decision making; also called *group decision support system* or *computerized collaborative work system*.

Figure 10.17

Configuration of a GSS

A GSS contains most of the elements found in a DSS, plus software to facilitate group member communications.

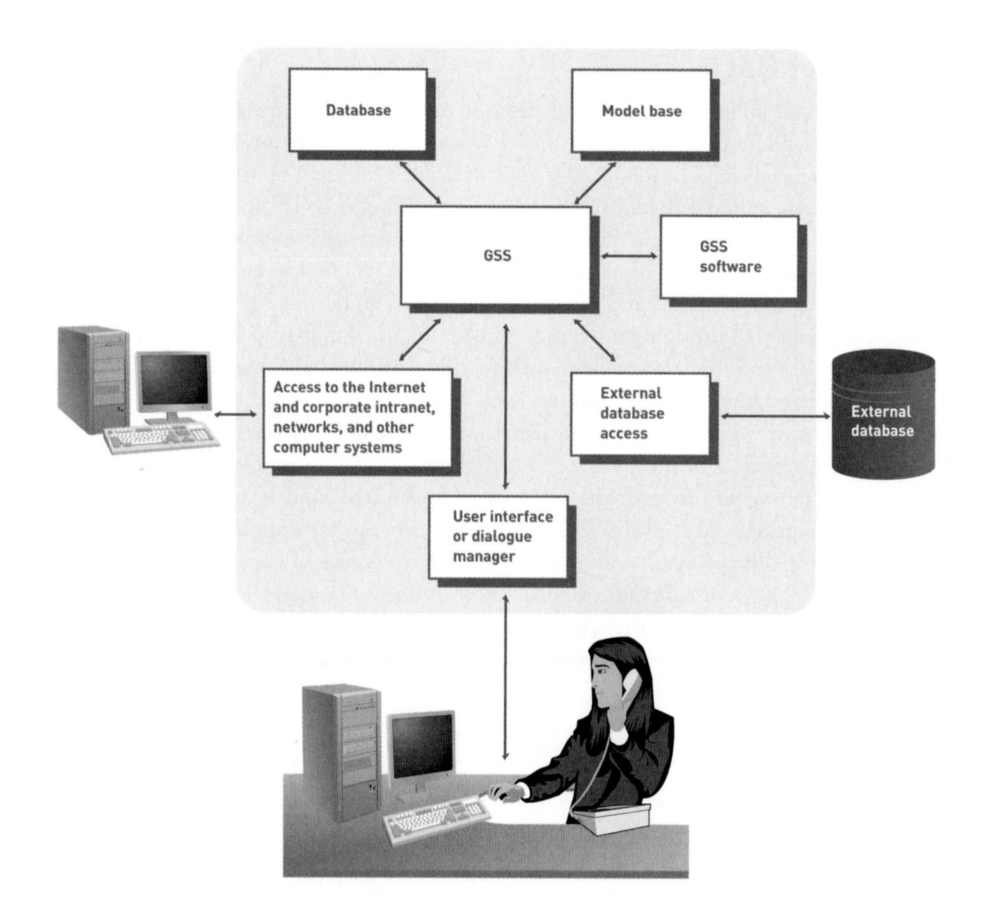

Group support systems are used in most industries, governments, and the military.[62] Architects are increasingly using GSSs to help them collaborate with other architects and builders to develop the best plans and to compete for contracts. Manufacturing companies use GSSs to link raw material suppliers to their own company systems. Engineers can use Mathcad Enterprise, another GSS. The software allows engineers to create, share, and reuse calculations.

Social-networking Internet sites can be used to support group decision making.[63] Serena, a software company in California, uses the Facebook social-networking site to collaborate on projects and exchange documents.[64] The company believes this type of collaboration is so important that it has instituted "Facebook Fridays" to encourage its employees to use the social-networking site to collaborate and make group decisions.[65] Facebook Fridays also help the company work with clients and recruit new employees. Many other organizations have used Facebook or have developed their own social-networking sites to help their employees collaborate on important projects.[66] Popular snow skier Bode Miller helped found Ski Space (*www.skispace.com*), a social-networking site for people interested in skiing and winter sports.[67] Some executives, however, believe that social-networking Internet sites waste time and corporate resources.[68]

Characteristics of a GSS That Enhance Decision Making

It is often said that two heads are better than one. When it comes to decision making, a GSS's unique characteristics have the potential to result in better decisions. Developers of these systems try to build on the advantages of individual support systems while adding new approaches unique to group decision making. For example, some GSSs can allow the exchange of information and expertise among people without direct face-to-face interaction, although some face-to-face meeting time is usually beneficial.[69] The following sections describe some characteristics that can improve and enhance decision making.

Special Design

The GSS approach acknowledges that special procedures, devices, and approaches are needed in group decision-making settings. These procedures must foster creative thinking, effective communications, and good group decision-making techniques.

Ease of Use

Like an individual DSS, a GSS must be easy to learn and use. Systems that are complex and hard to operate will seldom be used. Many groups have less tolerance than do individual decision makers for poorly developed systems.

Flexibility

Two or more decision makers working on the same problem might have different decision-making styles and preferences. Each manager makes decisions in a unique way, in part because of different experiences and cognitive styles. An effective GSS not only has to support the different approaches that managers use to make decisions, but also must find a means to integrate their different perspectives into a common view of the task at hand.

Decision-Making Support

A GSS can support different decision-making approaches, including the **delphi approach**, in which group decision makers are geographically dispersed throughout the country or the world. This approach encourages diversity among group members and fosters creativity and original thinking in decision making. Another approach, called **brainstorming**, in which members offer ideas "off the top of their heads," fosters creativity and free thinking. The **group consensus approach** forces members in the group to reach a unanimous decision. The Shuttle Project Engineering Office at the Kennedy Space Center has used the consensus-ranking organizational-support system (CROSS) to evaluate space projects in a group setting. The group consensus approach analyzes the benefits of various projects and their probabilities of success. CROSS is used to evaluate and prioritize advanced space projects. With the **nominal group technique**, each decision maker can participate; this technique encourages feedback from individual group members, and the final decision is made by voting, similar to a system for electing public officials.

delphi approach
A decision-making approach in which group decision makers are geographically dispersed; this approach encourages diversity among group members and fosters creativity and original thinking in decision making.

brainstorming
A decision-making approach that often consists of members offering ideas "off the top of their heads."

group consensus approach
A decision-making approach that forces members in the group to reach a unanimous decision.

nominal group technique
A decision-making approach that encourages feedback from individual group members, and the final decision is made by voting, similar to the way public officials are elected.

Anonymous Input

Many GSSs allow anonymous input, where the person giving the input is not known to other group members. For example, some organizations use a GSS to help rank the performance of managers. Anonymous input allows the group decision makers to concentrate on the merits of the input without considering who gave it. In other words, input given by a top-level manager is given the same consideration as input from employees or other members of the group. Some studies have shown that groups using anonymous input can make better decisions and have superior results compared with groups that do not use anonymous input. Anonymous input, however, can result in flaming, where an unknown team member posts insults or even obscenities on the GSS.

Reduction of Negative Group Behavior

One key characteristic of any GSS is the ability to suppress or eliminate group behavior that is counterproductive or harmful to effective decision making. In some group settings, dominant individuals can take over the discussion, which can prevent other members of the group from presenting creative alternatives. In other cases, one or two group members can sidetrack or subvert the group into areas that are nonproductive and do not help solve the problem at hand. Other times, members of a group might assume they have made the right decision without examining alternatives—a phenomenon called *groupthink*. If group sessions are poorly planned and executed, the result can be a tremendous waste of time. Today, many GSS designers are developing software and hardware systems to reduce these types of problems. Procedures for effectively planning and managing group meetings can be incorporated into the GSS approach. A trained meeting facilitator is often employed to help lead the group decision-making process and to avoid groupthink. See Figure 10.18.

Figure 10.18

Using the GSS Approach

A trained meeting facilitator can help lead the group decision-making process and avoid groupthink.

(Source: © Bill Bachmann/Getty Images.)

Parallel and Unified Communication

With traditional group meetings, people must take turns addressing various issues. One person normally talks at a time. With a GSS, every group member can address issues or make comments at the same time by entering them into a PC or workstation. These comments and issues are displayed on every group member's PC or workstation immediately. *Parallel communication* can speed meeting times and result in better decisions. Increasingly, organizations are using unified communications to support group decision making. *Unified communications* ties together and integrates different communication systems, including traditional phones, cell phones, e-mail, text messages, the Internet, and more.[70] With unified communications, members of a group decision-making team use a wide range of communications methods to help them collaborate and make better decisions. According to Microsoft founder Bill Gates, "Unified communications jumped out as one of those great opportunities to integrate something into the software you use for everything else you do. Even by Microsoft standards, it's a huge opportunity."

Automated Recordkeeping

Most GSSs can keep detailed records of a meeting automatically. Each comment that is entered into a group member's PC or workstation can be anonymously recorded. In some cases, literally hundreds of comments can be stored for future review and analysis. In addition, most GSS packages have automatic voting and ranking features. After group members vote, the GSS records each vote and makes the appropriate rankings.

GSS Software

GSS software, often called *groupware* or *workgroup software*, helps with joint work group scheduling, communication, and management. Software from Autodesk, for example, has GSS capabilities that allow groups to work together on the design. Designers, for example, can use Autodesk's Buzzsaw Professional Online Collaboration Service, which works with AutoCAD, a design and engineering software product from Autodesk. The U.S. Navy uses Virtual Office from Groove Networks to help it manage critical information in delivering humanitarian relief in disaster areas. The software is used for collaboration and communications in transmitting critical information between field offices. Virtual Office also has encryption capabilities to keep sensitive information safe and secure.

One popular package, Lotus Notes, can capture, store, manipulate, and distribute memos and communications that are developed during group projects. It can also incorporate knowledge management, discussed in Chapter 5, into the Lotus Notes Package. Some companies standardize on messaging and collaboration software, such as Lotus Notes. Lotus Connections is a newer feature of Lotus Notes that allows people to post documents and information on the Internet.[71] The new feature is similar to popular social-networking sites like Facebook and MySpace but is designed for business use.[72] Microsoft has invested billions of dollars in GSS software to incorporate collaborative features into its Office suite and related products. Office Communicator, for example, is a Microsoft product developed to allow better and faster collaboration. Other companies are also heavily investing in GSS software. In addition to Lotus Notes, IBM has developed Workplace to allow workers to collaborate more efficiently in doing their jobs. Microsoft's NetMeeting product supports application sharing in multiparty calls. NetDocuments Enterprise can be used for Web collaboration. The groupware is intended for legal, accounting, and real-estate businesses. A Breakout Session feature allows two people to take a copy of a document to a shared folder for joint revision and work. The software also permits digital signatures and the ability to download and work on shared documents on handheld computers. Other GSS software packages include Collabnet, Collabra Share, OpenMind, and TeamWare. All of these tools can aid in group decision making. *Shared electronic calendars* can be used to coordinate meetings and schedules for decision-making teams.[73] Using electronic calendars, team leaders can block out time for all members of the decision-making team. Some employees, however, don't like the use of shared electronic calendars. A member of one team said, "It's an intrusion. It's just a theft of your time."

A number of additional collaborative tools are available on the Internet.[74] Sharepoint (*www.microsoft.com*), WebOffice (*www.weboffice.com*), and BaseCamp (*www.basecamphq.com*) are just a few examples.[75] Twitter (*www.twitter.com*) and Jaiku (*www.jaiku.com*) are Internet sites that some organizations use to help people and groups stay connected and coordinate work schedules.[76] Sermo (*www.sermo.com*) is a social-networking site used by doctors to collaborate with other doctors, share their medical experiences, and even help make diagnoses.[77] Many of these Internet packages embrace the use of Web 2.0. Some executives, however, worry about security and corporate compliance issues in adopting Web 2.0 technologies.[78]

In addition to stand-alone products, GSS software is increasingly being incorporated into existing software packages. Today, some transaction processing and enterprise resource planning packages include collaboration software. Some ERP producers (see Chapter 9), for example, have developed groupware to facilitate collaboration and to allow users to integrate applications from other vendors into the ERP system of programs. Today, groupware can interact with wireless devices. Research In Motion, the maker of BlackBerry software, offers mobile communications, access to group information, meeting schedules, and other services that can be directly tied to groupware software and servers. In addition to groupware, GSSs use a number of tools discussed previously, including the following:

- E-mail, instant messaging (IM), and text messaging (TM)
- Videoconferencing
- Group scheduling
- Project management
- Document sharing

GSS Alternatives

Group support systems can take on a number of network configurations, depending on the needs of the group, the decision to be supported, and the geographic location of group members. GSS alternatives include a combination of decision rooms, local area networks, teleconferencing, and wide area networks.

- The **decision room** is ideal for situations in which decision makers are located in the same building or geographic area and the decision makers are occasional users of the GSS approach. In some cases, the decision room might have a few computers and a projector for presentations. In other cases, the decision room can be fully equipped with a network of computers and sophisticated GSS software. A typical decision room is shown in Figure 10.19.
- *The local area decision network* can be used when group members are located in the same building or geographic area and under conditions in which group decision making is frequent. In these cases, the technology and equipment for the GSS approach is placed directly into the offices of the group members.

decision room
A room that supports decision making, with the decision makers in the same building, combining face-to-face verbal interaction with technology to make the meeting more effective and efficient.

GSS software allows work teams to collaborate and reach better decisions—even if they work across town, in another region, or on the other side of the globe.

(Source: © Flying Colours Ltd./Getty Images.)

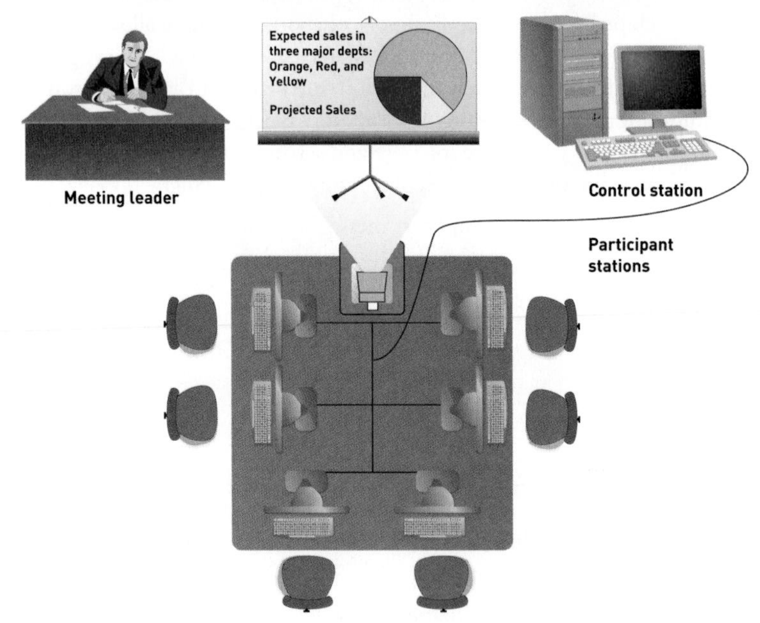

Figure 10.19

The GSS Decision Room

For group members who are in the same location, the decision room is an optimal GSS alternative. This approach can use both face-to-face and computer-mediated communication. By using networked computers and computer devices, such as project screens and printers, the meeting leader can pose questions to the group, instantly collect their feedback, and, with the help of the governing software loaded on the control station, process this feedback into meaningful information to aid in the decision-making process.

- *Teleconferencing* is used when the decision frequency is low and the location of group members is distant. These distant and occasional group meetings can tie together multiple GSS decision-making rooms across the country or around the world.
- The *wide area decision network* is used when the decision frequency is high and the location of group members is distant. In this case, the decision makers require frequent or constant use of the GSS approach. This GSS alternative allows people to work in **virtual workgroups**, where teams of people located around the world can work on common problems.

virtual workgroups
Teams of people located around the world working on common problems.

EXECUTIVE SUPPORT SYSTEMS

Because top-level executives often require specialized support when making strategic decisions, many companies have developed systems to assist executive decision making. This type of system, called an **executive support system** (ESS), is a specialized DSS that includes all hardware, software, data, procedures, and people used to assist senior-level executives within the organization. In some cases, an ESS, also called an *executive information system (EIS)*, supports decision making of members of the board of directors, who are responsible to stockholders. These top-level decision-making strata are shown in Figure 10.20.

executive support system (ESS)
Specialized DSS that includes all hardware, software, data, procedures, and people used to assist senior-level executives within the organization.

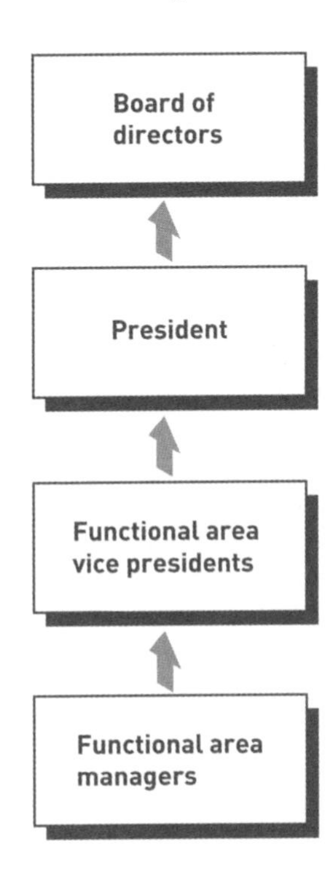

Figure 10.20

The Layers of Executive Decision Making

An ESS can also be used by individuals at middle levels in the organizational structure. Once targeted at the top-level executive decision makers, ESSs are now marketed to—and used by—employees at other levels in the organization. In the traditional view, ESSs give top executives a means of tracking critical success factors. Today, all levels of the organization share information from the same databases. However, for our discussion, assume ESSs remain in the upper-management levels, where they highlight important corporate issues, indicate new directions the company might take, and help executives monitor the company's progress.

Executive Support Systems in Perspective

An ESS is a special type of DSS and, like a DSS, is designed to support higher-level decision making in the organization. The two systems are, however, different in important ways. DSSs provide a variety of modeling and analysis tools to enable users to thoroughly analyze problems—that is, they allow users to *answer* questions. ESSs present structured information about aspects of the organization that executives consider important. The characteristics of an ESS are summarized in the following list.

- Are tailored to individual executives
- Are easy to use
- Have drill-down abilities
- Support the need for external data
- Can help with situations that have a high degree of uncertainty
- Have a future orientation
- Are linked with value-added business processes

Capabilities of Executive Support Systems

The responsibility given to top-level executives and decision makers brings unique problems and pressures to their jobs. This section discusses some of the characteristics of executive decision making that are supported through the ESS approach. ESSs take full advantage of data mining, the Internet, blogs, podcasts, executive dashboards, and many other technological innovations. As you will note, most of these decisions are related to an organization's overall profitability and direction. An effective ESS should have the capability to support executive decisions with components such as strategic planning and organizing, crisis management, and more.

Support for Defining an Overall Vision

One of the key roles of senior executives is to provide a broad vision for the entire organization. This vision includes the organization's major product lines and services, the types of businesses it supports today and in the future, and its overriding goals.

Support for Strategic Planning

strategic planning
Determining long-term objectives by analyzing the strengths and weaknesses of the organization, predicting future trends, and projecting the development of new product lines.

ESSs also support strategic planning. **Strategic planning** involves determining long-term objectives by analyzing the strengths and weaknesses of the organization, predicting future trends, and projecting the development of new product lines. It also involves planning the acquisition of new equipment, analyzing merger possibilities, and making difficult decisions concerning downsizing and the sale of assets if required by unfavorable economic conditions.

Support for Strategic Organizing and Staffing

Top-level executives are concerned with organizational structure. For example, decisions concerning the creation of new departments or downsizing the labor force are made by top-level managers. Overall direction for staffing decisions and effective communication with labor unions are also major decision areas for top-level executives. ESSs can be employed to help analyze the impact of staffing decisions, potential pay raises, changes in employee benefits, and new work rules.

Support for Strategic Control

Another type of executive decision relates to strategic control, which involves monitoring and managing the overall operation of the organization. Goal seeking can be done for each major area to determine what performance these areas need to achieve to reach corporate expectations. Effective ESS approaches can help top-level managers make the most of their existing resources and control all aspects of the organization.

Support for Crisis Management

Even with careful strategic planning, a crisis can occur. Major incidents, including natural disasters, fires, and terrorist activities, can totally shut down major parts of the organization. Handling these emergencies is another responsibility for top-level executives. In many cases, strategic emergency plans can be put into place with the help of an ESS. These contingency plans help organizations recover quickly if an emergency or crisis occurs.

Decision making is a vital part of managing businesses strategically. IS systems such as information and decision support, group support, and executive support systems help employees by tapping existing databases and providing them with current, accurate information. The increasing integration of all business information systems—from TPSs to MISs to DSSs—can help organizations monitor their competitive environment and make better-informed decisions. Organizations can also use specialized business information systems, discussed in the next chapter, to achieve their goals.

SUMMARY

Principle

Good decision-making and problem-solving skills are the key to developing effective information and decision support systems.

Every organization needs effective decision making and problem solving to reach its objectives and goals. Problem solving begins with decision making. A well-known model developed by Herbert Simon divides the decision-making phase of the problem-solving process into three stages: intelligence, design, and choice. During the intelligence stage, potential problems or opportunities are identified and defined. Information is gathered that relates to the cause and scope of the problem. Constraints on the possible solution and the problem environment are investigated. In the design stage, alternative solutions to the problem are developed and explored. In addition, the feasibility and implications of these alternatives are evaluated. Finally, the choice stage involves selecting the best course of action. In this stage, the decision makers evaluate the implementation of the solution to determine whether the anticipated results were achieved and to modify the process in light of new information learned during the implementation stage.

Decision making is a component of problem solving. In addition to the intelligence, design, and choice steps of decision making, problem solving also includes implementation and monitoring. Implementation places the solution into effect. After a decision has been implemented, it is monitored and modified if necessary.

Decisions can be programmed or nonprogrammed. Programmed decisions are made using a rule, procedure, or quantitative method. Ordering more inventory when the level drops to 100 units or fewer is an example of a programmed decision. A nonprogrammed decision deals with unusual or exceptional situations. Determining the best training program for a new employee is an example of a nonprogrammed decision.

Decisions can use optimization, satisficing, or heuristic approaches. Optimization finds the best solution. Optimization problems often have an objective such as maximizing profits given production and material constraints. When a problem is too complex for optimization, satisficing is often used. Satisficing finds a good, but not necessarily the best, decision. Finally, a heuristic is a "rule of thumb" or commonly used guideline or procedure used to find a good decision.

Principle

The management information system (MIS) must provide the right information to the right person in the right format at the right time.

A management information system is an integrated collection of people, procedures, databases, and devices that provides managers and decision makers with information to help achieve organizational goals. An MIS can help an organization achieve its goals by providing managers with insight into the regular operations of the organization so that they can control, organize, and plan more effectively and efficiently. The primary difference between the reports generated by the TPS and those generated by the MIS is that MIS reports support managerial decision making at the higher levels of management.

Data that enters the MIS originates from both internal and external sources. The most significant internal sources of data for the MIS are the organization's various TPSs and ERP systems. Data warehouses and data marts also provide important input data for the MIS. External sources of data for the MIS include extranets, customers, suppliers, competitors, and stockholders.

The output of most MISs is a collection of reports that are distributed to managers. These reports include scheduled reports, key-indicator reports, demand reports, exception reports, and drill-down reports. Scheduled reports are produced periodically, or on a schedule, such as daily, weekly, or monthly. A key-indicator report is a special type of scheduled report. Demand reports are developed to provide certain information at a manager's request. Exception reports are automatically produced when a situation is unusual or requires management action. Drill-down reports provide increasingly detailed data about situations.

Management information systems have a number of common characteristics, including producing scheduled, demand, exception, and drill-down reports; producing reports with fixed and standard formats; producing hard-copy and soft-copy reports; using internal data stored in organizational computerized databases; and having reports developed and implemented by IS personnel or end users. Increasingly, MIS reports are being delivered over the Internet and through mobile devices, such as cell phones.

Most MISs are organized along the functional lines of an organization. Typical functional management information systems include financial, manufacturing, marketing, human resources, and other specialized systems. Each system is composed of inputs, processing subsystems, and outputs. The primary sources of input to functional MISs include the corporate strategic plan, data from the ERP system and TPS, information from supply chain and business transactions, and external sources including the Internet and extranets. The primary output of these functional MISs are summary reports that assist in managerial decision making.

A financial management information system provides financial information to all financial managers within an organization, including the chief financial officer (CFO). Subsystems are profit/loss and cost systems, auditing, and use and management of funds.

A manufacturing MIS accepts inputs from the strategic plan, the ERP system and TPS, and external sources, such as supply chain and business transactions. The systems involved support the business processes associated with the receiving and inspecting of raw material and supplies; inventory tracking of raw materials, work in process, and finished goods; labor and personnel management; management of assembly lines, equipment, and machinery; inspection and maintenance; and order processing. The subsystems involved are design and engineering, master production scheduling and inventory control, process control, and quality control and testing.

A marketing MIS supports managerial activities in the areas of product development, distribution, pricing decisions, promotional effectiveness, and sales forecasting. Subsystems include marketing research, product development, promotion and advertising, and product pricing.

A human resource MIS is concerned with activities related to employees of the organization. Subsystems include human resource planning, personnel selection and recruiting, training and skills inventories, scheduling and job placement, wage and salary administration, and outplacement.

An accounting MIS performs a number of important activities, providing aggregate information on accounts payable, accounts receivable, payroll, and many other applications. The organization's ERP system or TPS captures accounting data, which is also used by most other functional information systems. Geographic information systems provide regional data in graphical form.

Principle

Decision support systems (DSSs) are used when the problems are unstructured.

A decision support system (DSS) is an organized collection of people, procedures, software, databases, and devices working to support managerial decision making. DSS characteristics include the ability to handle large amounts of data; obtain and process data from different sources; provide report and presentation flexibility; support drill-down analysis; perform complex statistical analysis; offer textual and graphical orientations; support optimization, satisficing, and heuristic approaches; and perform what-if, simulation, and goal-seeking analysis.

DSSs provide support assistance through all phases of the problem-solving process. Different decision frequencies also require DSS support. An ad hoc DSS addresses unique, infrequent decision situations; an institutional DSS handles routine decisions. Highly structured problems, semistructured problems, and unstructured problems can be supported by a DSS. A DSS can also support different managerial levels, including strategic, tactical, and operational managers. A common database is often the link that ties together a company's TPS, MIS, and DSS.

The components of a DSS are the database, model base, user interface or dialogue manager, and a link to external databases, the Internet, the corporate intranet, extranets, networks, and other systems. The database can use data warehouses and data marts. A data-driven DSS primarily performs qualitative analysis based on the company's databases. Data-driven DSSs tap into vast stores of information contained in the corporate database, retrieving information on inventory, sales, personnel, production, finance, accounting, and other areas. Data mining is often used in a data-driven DSS. The model base contains the models used by the decision maker, such as financial, statistical, graphical, and project-management models. A model-driven DSS primarily performs mathematical or quantitative analysis. Model management software (MMS) is often used to coordinate the use of models in a DSS. The user interface provides a dialogue management facility to assist in communications between the system and the user. Access to other computer-based systems permits the DSS to tie into other powerful systems, including the TPS or function-specific subsystems.

Principle

Specialized support systems, such as group support systems (GSSs) and executive support systems (ESSs), use the overall approach of a DSS in situations such as group and executive decision making.

A group support system (GSS), also called a *computerized collaborative work system*, consists of most of the elements in a DSS, plus software to provide effective support in group decision-making settings. GSSs are typically easy to learn and use and can offer specific or general decision-making support. GSS software, also called *groupware*, is specially designed to help generate lists of decision alternatives and perform data analysis. These packages let people work on joint documents and files over a network. Newer Web 2.0 technologies are being used to a greater extent in delivering group decision-making support. Text messages and the Internet are also commonly used in a GSS.

The frequency of GSS use and the location of the decision makers will influence the GSS alternative chosen. The decision room alternative supports users in a single location who meet infrequently. Local area networks can be used when group members are located in the same geographic area and users meet regularly. Teleconferencing is used when decision frequency is low and the location of group members is distant. A wide area network is used when the decision frequency is high and the location of group members is distant.

Executive support systems (ESSs) are specialized decision support systems designed to meet the needs of senior management. They serve to indicate issues of importance to the organization, indicate new directions the company might take, and help executives monitor the company's progress. ESSs are typically easy to use, offer a wide range of computer resources, and handle a variety of internal and external data. In addition, the ESS performs sophisticated data analysis, offers a high degree of specialization, and provides flexibility and comprehensive communications capabilities. An ESS also supports individual decision-making styles. Some of the major decision-making areas that can be supported through an ESS are providing an overall vision, strategic planning and organizing, strategic control, and crisis management.

CHAPTER 10: SELF-ASSESSMENT TEST

Good decision-making and problem-solving skills are the key to developing effective information and decision support systems.

1. Developing decision alternatives is done during what decision-making stage?
   a. initiation stage
   b. intelligence stage
   c. design stage
   d. choice stage
2. Problem solving is one of the stages of decision making. True or False?
3. The final stage of problem solving is ___________.
4. A decision that inventory should be ordered when inventory levels drop to 500 units is an example of a(n) _______________.
   a. synchronous decision
   b. asynchronous decision
   c. nonprogrammed decision
   d. programmed decision
5. A(n) ___________ model will find the best solution to help the organization meet its goals.
6. A satisficing model is one that will find a good problem solution, but not necessarily the best problem solution. True or False?

The management information system (MIS) must provide the right information to the right person in the right format at the right time.

7. What summarizes the previous day's critical activities and is typically available at the beginning of each workday?
   a. key-indicator report
   b. demand report
   c. exception report
   d. database report
8. MRP and JIT are a subsystem of the _______________.
   a. marketing MIS
   b. financial MIS
   c. manufacturing MIS
   d. auditing MIS
9. Another name for the _______________ MIS is the personnel MIS because it is concerned with activities related to employees and potential employees of the organization.

Decision support systems (DSSs) are used when the problems are unstructured.

10. The focus of a decision support system is on decision-making effectiveness when faced with unstructured or semistructured business problems. True or False?
11. _______________ is used to find the best solution.
12. What component of a decision support system allows decision makers to easily access and manipulate the DSS and to use common business terms and phrases?
    a. the knowledge base
    b. the model base
    c. the user interface or dialogue manager
    d. the expert system

Specialized support systems, such as group support systems (GSSs) and executive support systems (ESSs), use the overall approach of a DSS in situations such as group and executive decision making.

13. What decision-making technique allows voting group members to arrive at a final group decision?
    a. groupthink
    b. anonymous input
    c. nominal group technique
    d. delphi
14. A type of software that helps with joint work group scheduling, communication, and management is called _______________.

15. The local area decision network is the ideal GSS alternative for situations in which decision makers are located in the same building or geographic area and the decision makers are occasional users of the GSS approach. True or False?
16. A(n) ______________ supports the actions of members of the board of directors, who are responsible to stockholders.

CHAPTER 10: SELF-ASSESSMENT TEST ANSWERS

(1) c (2) False (3) monitoring (4) d (5) optimization (6) True (7) a (8) c (9) human resource (10) True (11) optimization (12) c (13) c (14) groupware or workgroup software (15) False (16) executive information system (EIS)

REVIEW QUESTIONS

1. What is a satisficing model? Describe a situation when it should be used.
2. What is the difference between intelligence and design in decision making?
3. What is the difference between a programmed decision and a nonprogrammed decision? Give several examples of each.
4. What are the basic kinds of reports produced by an MIS?
5. How can text messaging be used to develop MIS reports?
6. What are the functions performed by a financial MIS?
7. Describe the functions of a marketing MIS.
8. List and describe some other types of MISs.
9. What are the stages of problem solving?
10. What is the difference between decision making and problem solving?
11. What is a geographic information system?
12. Describe the difference between a structured and an unstructured problem and give an example of each.
13. Define *decision support system*. What are its characteristics?
14. Describe the difference between a data-driven and a model-driven DSS.
15. What is the difference between what-if analysis and goal-seeking analysis?
16. What are the components of a decision support system?
17. State the objective of a group support system (GSS) and identify three characteristics that distinguish it from a DSS.
18. How can social-networking sites be used in a GSS?
19. Identify three group decision-making approaches often supported by a GSS.
20. What is an executive support system? Identify three fundamental uses for such a system.

DISCUSSION QUESTIONS

1. Select an important problem you had to solve during the last two years. Describe how you used the decision-making and problem-solving steps discussed in this chapter to solve the problem.
2. Describe how a GSS can be used at your school or university.
3. How can management information systems be used to support the objectives of the business organization?
4. Describe a financial MIS for a *Fortune* 1000 manufacturer of food products. What are the primary inputs and outputs? What are the subsystems?
5. How can a strong financial MIS provide strategic benefits to a firm?
6. Why is auditing so important in a financial MIS? Give an example of an audit that failed to disclose the true nature of the financial position of a firm. What was the result?
7. Describe two industries where a marketing MIS is critical to sales and success.
8. You have been hired to develop a management information system and a decision support system for a manufacturing company. Describe what information you would include in printed reports and what information you would provide using a screen-based decision support system.
9. Pick a company and research its human resource management information system. Describe how the system works. What improvements could be made to the company's human resource MIS?
10. You have been hired to develop a DSS for a car company such as Ford or GM. Describe how you would use both data-driven and model-driven DSSs.
11. Describe how you would develop a social-networking Internet site to help a company collaborate on an important project or decision. What features would you include in the Internet site?
12. What functions do decision support systems support in business organizations? How does a DSS differ from a TPS and an MIS?
13. How is decision making in a group environment different from individual decision making, and why are information systems that assist in the group environment different? What are the advantages and disadvantages of making decisions as a group?

14. You have been hired to develop group support software. Describe the features you would include in your new GSS software.

15. Imagine that you are the vice president of manufacturing for a *Fortune* 1000 manufacturing company. Describe the features and capabilities of your ideal ESS.

PROBLEM-SOLVING EXERCISES

1. Use the Internet to research the use of Web 2.0 technologies in collaborative group decision making. Use a word processor to describe what you discovered. Develop a set of slides using a graphics program to deliver a presentation on the use of Web 2.0 in GSS.
2. Review the summarized consolidated statement of income for the manufacturing company whose data is shown here. Use graphics software to prepare a set of bar charts that shows the data for this year compared with the data for last year.

   a. This year, operating revenues increased by 3.5 percent, while operating expenses increased 2.5 percent.
   b. Other income and expenses decreased to $13,000.
   c. Interest and other charges increased to $265,000.

Operating Results (in millions)

Operating Revenues	$2,924,177
Operating Expenses (including taxes)	2,483,687
Operating Income	440,490
Other Income and Expenses	13,497
Income Before Interest and Other Charges	453,987
Interest and Other Charges	262,845
Net Income	191,142
Average Common Shares Outstanding	147,426
Earnings per Share	1.30

If you were a financial analyst tracking this company, what detailed data might you need to perform a more complete analysis? Write a brief memo summarizing your data needs.

3. As the head buyer for a major supermarket chain, you are constantly being asked by manufacturers and distributors to stock their new products. Over 50 new items are introduced each week. Many times, these products are launched with national advertising campaigns and special promotional allowances to retailers. To add new products, the amount of shelf space allocated to existing products must be reduced or items must be eliminated altogether. Develop a marketing MIS that you can use to estimate the change in profits from adding or deleting an item from inventory. Your analysis should include input such as estimated weekly sales in units, shelf space allocated to stock an item (measured in units), total cost per unit, and sales price per unit. Your analysis should calculate total annual profit by item and then sort the rows in descending order based on total annual profit.

TEAM ACTIVITIES

1. Use a social-networking site to help your team collaborate on a decision of your choice. Each team member should write a brief report on his or her experiences in using the social-networking site. In addition to the individual reports, your team should collaborate on a group report that describes the different perceptions of each team member and makes recommendations about the use of social-networking sites in group decision making.
2. Have your team make a group decision about how to solve the most frustrating aspect of college or university life. Appoint one or two members of the team to disrupt the meeting with negative group behavior. After the meeting, have your team describe how to prevent this negative group behavior. What GSS software features would you suggest to prevent the negative group behavior your team observed?
3. Imagine that you and your team have decided to develop an ESS software product to support senior executives in the music recording industry. What are some of the key decisions these executives must make? Make a list of the capabilities that such a system must provide to be useful. Identify at least six sources of external information that will be useful to its users.

WEB EXERCISES

1. Use a search engine, such as Yahoo! or Google, to explore two or more companies that produce and sell MIS or DSS software. Describe what you found and any problems you had in using search engines on the Internet to find information. You might be asked to develop a report or send an e-mail message to your instructor about what you found.
2. Use the Internet to explore two or more software packages described in this chapter. Summarize your findings in a report.
3. Software, such as Microsoft Excel, is often used to find an optimal solution to maximize profits or minimize costs. Search the Internet using Yahoo!, Google, or another search engine to find other software packages that offer optimization features. Write a report describing one or two of the optimization software packages. What are some of the features of the package?

CAREER EXERCISES

1. What decisions are critical for success in a career that interests you? What specific types of reports could help you make better decisions on the job? Give three specific examples.
2. Use two or more social-networking sites to explore careers that interest you. Describe what you found and the differences between the sites. What features would you like to see in a social-networking site to help you find a good job?

CASE STUDIES

Case One

Enterprise Rent-A-Car and Business Process Management

Enterprise Rent-A-Car is based in St. Louis, Missouri, and runs approximately 6,900 branch offices around the world. Enterprise routes all of its information system requests through its information systems group housed in the home office. Until recently, Enterprise used an outdated process for managing corporate requests. The Enterprise Requests Online system assisted the Requests department in handling product and service requests. The system automated everything from setting up a new laptop to opening a new Enterprise Rent-A-Car office. To make a request, Enterprise employees would navigate to a corporate Web site where a system request form was located. Fifteen categories of requests were provided for employees, who entered the details of the problem into a text box. The Web form generated an e-mail to the Requests department that managed its job queue from its e-mail inbox.

While such a system was considered state of the art ten years ago, it has inherent problems that more modern systems have addressed. For one thing, e-mail is a difficult communications medium to manage. Individual messages must be opened to examine details of the request. Messages are also easy to lose. Those seeking help cannot see how or whether their request is being handled. Maintaining a history of work requests over many years is next to impossible with such a system.

Enterprise decided to revamp and improve its work request process to streamline the process and improve service to its branch offices. The information system that Enterprise wanted is called a business process management (BPM) system—an automated method of streamlining business processes. Enterprise turned to APPIAN, a company that specializes in BPM systems. The two worked together to produce a powerful BPM system for managing incoming information system requests.

Enterprise's new Requests Online system provides users with detailed options to narrow a request to one of 200 request types. The system recognizes the user and lists only options for that particular branch office. For example, the system detects software used by that branch office and provides options only for that software. This focuses the options to those that interest the user, saving time. The software also fills in the user data, such as name, phone, and location, saving even more time. Once submitted, users of the system can view the progress of their request in a job queue page.

On the back end, Requests staff uses an executive dashboard application to keep track of their work. The system tracks all jobs in the queue and produces useful reports, which provide graphical information to the dashboard that indicates how smoothly operations are running. Using these visual cues, employees can tell if they are keeping up with the work, and managers can decide how many workers are required to meet the load. Using other reports, managers can determine which months of the year are busiest and which weeks, days, and hours require the most or least amount of staffing.

The new system is expected to save Enterprise between 15 and 20 percent in costs and time in administrative and data entry activities. Already the savings are being felt. Enterprise has redeployed its staff from maintaining the old legacy system to jobs that are more "strategically valuable."

Discussion Questions

1. What is the purpose of business process management systems? What benefits do they provide?
2. What problems did the original Enterprise Requests Online system have?

Critical Thinking Questions

1. What attributes do you think make up a system that is optimized for the greatest convenience to users?
2. What factors would lead a business to decide that it is time to improve its business processes with a new MIS?

SOURCES: Ruffolo, Rafael, "Enterprise Rent-A-Car drives service request app," *ITWorldCanada.com*, March 27, 2008, *www.itworldcanada.com//Pages/Docbase/ViewArticle.aspx?ID=idgml-924c5e06-673e-4b68-86de-a83dbfcb0f37*; Appian Corporate Web site, *www.appian.com*, accessed May 20, 2008; Enterprise Rent-A-Car Web site, *www.enterprise.com*, accessed May 20, 2008.

Case Two

Keiper Watches Production Like a Hawk

Keiper GmbH & Co. KG is a leading manufacturer of the metal components of car seats. The company runs 11 production sites scattered around the world and employs 6,000 workers. Keiper struggles with the common challenges facing all international manufacturing businesses. One of its largest problems is synchronizing production across manufacturing plants separated by many miles and time zones.

A few years ago, Keiper made the wise move to connect its production sites over the Internet using production management software. The software allowed system specialists and production managers at Keiper headquarters in Kaiserslautern, Germany, to monitor production systems at all 11 sites. This ability made all the difference in the world to Keiper's production quality. Managers can troubleshoot problems as they arise, upgrade system software, and make database entries at the same time across all locations. In short, the system made it seem as though the 11 production facilities were actually one big manufacturing plant.

More recently, Keiper upgraded the system to allow it to react to emergencies more quickly. One of the biggest problems in manufacturing occurs when flaws are introduced into the process. Flaws might be the result of a defective part received from a supplier and used on the assembly line. Sometimes these defects are not noticed until many products have been manufactured and shipped. Keiper wanted a traceability system so it could track the car seat components and assembled car seats through the production line to their destinations.

Keiper partnered with an information system company that specialized in traceability systems for the automotive industry. The company designed a system that collected and connected information from production facilities around the world. The information included specific information about the parts used in the manufacturing process. The system tracked each step in the production process at all 11 manufacturing plants in real time. If an employee on the line notices that a certain type of screw is defective, an alarm is sounded and an investigation immediately launched. The defective screw is traced back to the batch from which it came. Each screw used from that batch is traced to a specific seat in the production line, shipment center, or automotive plant where the cars are assembled. Recalls of seats containing the bad screw can occur within hours of the discovery.

Using MISs to control production lines around the world provides Keiper management with more control over its business. The ability to quickly catch defects in its products minimizes the extent of the damage they cause. Keiper customers appreciate the corporation's ability to minimize problems before they grow to an unmanageable scope.

Discussion Questions

1. What is one of the biggest production challenges facing global manufacturing corporations?
2. How did Keiper management gain more control over its 11 manufacturing facilities?

Critical Thinking Questions

1. What type of management information system does Keiper use, and what functional unit of the company is it designed for?
2. What information components do you think were used in Keiper's new traceability system?

SOURCES: Staff SAP, "Keiper GmbH & Co. KG," SAP Customer Success Story, *http://download.sap.com/solutions/manufacturing/customersuccess/download.epd?cont ext=BAC27560689AE-F8040ED89D61B269843913D4B9555561819698A5E0F5A5BF40C229A297AA3C9C20AA4384FDE88B9A968C706CA6FF70055CD*, accessed May 20, 2008.

Questions for Web Case

See the Web site for this book to read about the Whitmann Price Consulting case for this chapter. The following are questions concerning this Web case.

Whitmann Price Consulting: MIS and DSS Considerations

Discussion Questions

1. What different types of needs can MISs and DSSs fulfill in Whitmann Price's new system?
2. Why did Whitmann Price decide to design its own Calendar and Contacts MIS rather than using standard BlackBerry software?

Critical Thinking Questions

1. How does the source of input differ between the MISs being designed for all consulting areas and the DSSs uniquely designed for each consulting area?
2. How might a GSS be useful for consultants at Whitmann Price?

NOTES

Sources for the opening vignette: SAP Staff, "General Mills," SAP Customer Success Story, *http://download.sap.com/solutions/business-suite/plm/customersuccess/download.epd?context=DF359538DBDBF-B3739D0C646A8 BF36C42C2999E7D28C06CB095A5648B35773-CA7A03AFD504A5C7CA79A98EB720 AFFA28877CC33DEAF2BD52*, accessed May 18, 2008; General Mills Web site, *www.generalmills.com/corporate/company/index.aspx*, accessed May 18, 2008.

1 Merrick, Jason, and Harrald, John, "Making Decisions About Safety in US Ports and Waterways," *Interfaces*, June 2007, p. 240.
2 Ross, Alex and Swain, Alison, "Fighting Flight Delays," *OR/MS Today*, April 2007, p. 22.
3 Barnes, Brooks, "Can CBS Put the Net into the Network?" *The Wall Street Journal*, May 14, 2007, p. B1.
4 *www.cmis.csiro.au/OR/MEDIA/sate.htm*, accessed November 16, 2007.
5 Kant, Goos, "Coca-Cola Enterprises Optimizes Vehicle Routes for Efficient Product Delivery," *Interfaces*, January 2008, p. 40.
6 Clark, T., et al., "The Dynamic Structure of Management Support Systems," *MIS Quarterly*, September 2007, p. 579.
7 *www.cognos.com*, accessed November 14, 2007.
8 Biddick, Michael, "Hunting the Elusive CIO Dashboard," *InformationWeek*, March 3, 2008, p. 47.
9 Hayes Weier, Mary, "A Steaming Cup of Insight," *InformationWeek*, May 7, 2007, p. 47.
10 *www.cognos.com/news/releases/2007/1112.html?mc=-web_hp*, accessed November 15, 2007.
11 Havenstein, Heather, "Atlanta to Roll Out Cognos Analysis Software to City Agency Workers," *Computerworld*, January 15, 2007, p. 15.
12 *www.cognos.com/news/releases/2007/1112.html?mc=-web_hp*, accessed November 15, 2007.
13 *www.microsoft.com/dynamics/product/business_scorecard_manager.mspx*, accessed November 14, 2007.
14 Havenstein, Heather, "Ace Hardware Shifts BI Tools After Early Struggles," *Computerworld*, April 23, 2007, p. 11.
15 Greene, Jay, "Microsoft Wants Your Health Records," *Business Week*, October 15, 2007, p. 44.
16 Havenstein, Heather, "Military, Oil Firm Use BI to Avert Disaster," *Computerworld*, October 1, 2007, p. 8.
17 Yuan, Li and Prada, Paulo, "Texting When There's Trouble," *The Wall Street Journal*, April 18, 2007, p. B1.
18 Yuan, Li, "Murder, She Texted," *The Wall Street Journal*, July 2, 2007, p. B1.
19 Havenstein, Heather, "Military, Oil Firm Use BI to Avert Disaster," *Computerworld*, October 1, 2007, p. 8.
20 Patrick, Aaron O., "Reuters to Launch Trade Software," *The Wall Street Journal*, April 30, 2007, p. C2.
21 Ferraro, Nicole, "Lending and Philanthropy 2.0," *InformationWeek*, February 4, 2008, p. 40.
22 Kim, Jane, "Where Either a Borrower Or a Lender Can Be," *The Wall Street Journal*, March 12, 2008, p. D1.
23 Brown, Erika, "Widget Work," *Forbes*, October 29, 2007, p. 66.
24 Nagali, Jerry, et al., "Procurement Risk Management at Hewlett-Packard Company," *Interfaces*, January 2008, p. 51.
25 Dean, Jason, "The Forbidden City of Terry Gou," *The Wall Street Journal*, August 11, 2007, p. A1.
26 Loveland, Jennifer, et al., "Dell Uses a New Production Scheduling Algorithm," *Interfaces*, May 2007, p. 209.
27 Khawam, John, et al., "Warranty Inventory Optimization for Hitachi Global Storage," *Interfaces*, September 2007, p. 455.
28 Prada, Paulo, "Delta Maintains Its Maintenance," *The Wall Street Journal*, November 7, 2007, p. A16.
29 Songini, Marc, "BI Helping Retailers Control Inventory," *Computerworld*, January 22, 2007, p. 14.
30 Chozick, Amy, "Toyota Sticks by Just In Time Strategy After Quake," *The Wall Street Journal*, July 24, 2007, p. A2.
31 Vranica, Suzanne, "Internet Reshapes Role of Media Buyers," *The Wall Street Journal*, February 28, 2007, p. B3.
32 Warren, Jamin and Jurgensen, John, "A New Kind of Web Site," *The Wall Street Journal*, February 10, 2007, p. P1.
33 Sharma, Amol, "Companies Vie for Ad Dollars on Mobile Web," *The Wall Street Journal*, January 17, 2007, p. D1.
34 Vara, Vauhini, "Facebook Gets Personal with Ad Targeting Plan," *The Wall Street Journal*, August 23, 2007, p. B1.
35 Steel, Emily, "YouTube to Start Selling Ads in Video," *The Wall Street Journal*, August 22, 2007, p. B3.
36 Jesdanun, Anick, "YouTube's Overlay Ads Aim to Retain Viewership," *Rocky Mountain News*, August 8, 2007, p. 4.
37 Fass, Allison, "Sex, Pranks, and Reality," *Forbes*, July 2, 2007, p. 48.
38 McMillan, Robert, "Every Click You Make ..." *Computerworld*, January 7, 2008, p. 8.
39 Hayes Weier, Mary, "Dear Customer: Please Don't Leave," *Information Week*, June 18, 2007, p. 49.
40 Copeland, Michael, "Location, Location, Location," *Fortune*, November 12, 2007, p. 147.
41 Vranica, Suzanne, "Hellmann's Targets Yahoo for Its Spread," *The Wall Street Journal*, June 27, 2007, p. B4.

42 Steel, Emily, "J&J's Web Ads Depart from TV Formula," *The Wall Street Journal,* February 12, 2008, p. B3.
43 Gonsalves, Antone, "Yahoo's Brand Grab Bag," *Information Week,* February 5, 2007, p. 25.
44 Steel, Emily, "Web vs. TV," *The Wall Street Journal,* March 19, 2008, p. B3.
45 Delaney, Kevin, "Start-ups Seek to Cash in on Web-Video Advertising," *The Wall Street Journal,* March 2, 2007, p. B1.
46 Patrick, Aaron, "Internet Ad Shops Are Crossing the Digital Divide," *The Wall Street Journal,* February 9, 2007, p. B3.
47 Vara, Vauhini, "Companies Tolerate Ads to Get Free Software," *The Wall Street Journal,* March 27, 2007, p. B1.
48 Banjo, Shelly, "Attention, Bloggers," *The Wall Street Journal,* March 17, 2008, p. R5.
49 Vara, Vauhini, "Ebay to Broker Radio Ad Time," *The Wall Street Journal,* June 6, 2007, p. A13.
50 Silva-Risso, Jorge, et al., "Chrysler and J.D. Powers: Pioneering Scientific Price Customization in the Automotive Industry," *Interfaces,* January 2008, p. 26.
51 Aeppel, Timothy, "Seeking Perfect Prices, CEO Tears Up the Rules," *The Wall Street Journal,* March 27, 2007, p. A1.
52 Melymuka, Kathleen, "Just-In-Time Talent," *Computerworld,* March 24, 2008, p. 34.
53 Needleman, Sarah, "Demand Rises for Talent-Management Software," *The Wall Street Journal,* January 15, 2008, p. B8.
54 Songini, Marc, "BI Helping Retailers Control Inventory," *Computerworld,* January 22, 2007, p. 14.
55 Howson, Cindi, "The Road to Pervasive BI," *Information Week,* February 25, 2008, p. 39.
56 Hayes Weier, Mary, "Dear Customer: Please Don't Leave," *Information Week,* June 18, 2007, p. 49.
57 Kolbasuk McGee, Marianne, "The Useless Hunt for Data," *Information Week,* January 8, 2007, p. 19.
58 Delaney, Devin and Steel, Emily, "Firm Mines Offline Data to Target Online Ads," *The Wall Street Journal,* October 17, 2007, p. B1.
59 Worthen, Ben, "'Mashups' Sew Data Together," *The Wall Street Journal,* July 31, 2007, p. B4.
60 LoanPerformance Web site, *www.loanperformance.com*, accessed November 15, 2007.
61 Martin, Richard, "Business at Light Speed," *Information Week,* April 23, 2007, p. 42.
62 Retchless, Todd, et al., "A Group Decision-Making Application of the Analytic Hierarchy Process," *Interfaces,* April 2007, p. 163.
63 Shepherd, Lauren, "MySpace Becoming a Business Launch Pad," *Rocky Mountain News,* June 18, 2007, p. 4.
64 Green, Heather, "Now, Social Networking Fridays," *Business Week,* November 12, 2007, p. 24.
65 Gomes, Lee, "More Firms Create Own Social Networks," *The Wall Street Journal,* February 19, 2008, p. B3.
66 Green, Heather, "The Water Cooler Is Now on the Web," *Business Week,* October 1, 2007, p. 78.
67 Lehman, Paula, "Social Networks That Break a Sweat," *Business Week,* February 4, 2008, p. 66.
68 Hoover, Nicholas, "Social Experiment," *Information Week,* September 24, 2007, p. 40.
69 Cowen, Tyler, "In Favor of Face Time," *Forbes,* October 1, 2007, p. 30.
70 Schlender, Brent, "And What Does Mr. Gates Think?" *Fortune,* November 1, 2007, p. 102.
71 Bulkeley, William, "IBM Plans to Join Social-Networking Field," *The Wall Street Journal,* January 22, 2007, p. B5.
72 Hoover, Nicholas, "Lotus' Leap," *Information Week,* January 29, 2007, p. 29.
73 Sandberg, Jared, "Shared Calendars Mean Never Getting to Fib, I'm Booked," *The Wall Street Journal,* June 19, 2007, p. B1.
74 Babcock, Charles, "Microsoft Gets BI Focused," *Information Week,* September 24, 2007, p. 28.
75 Spanbauer, Scott, "Comparing Collaborative Web Services," *PC World,* January 2007, p. 61.
76 Dejean, David, "Pick Your Presence Tool," *Information Week,* May 21, 2007, p. 51.
77 Vascellaro, Jessica, "Social Networking Goes Professional," *The Wall Street Journal,* August 28, 2007, p. D1.
78 Havenstein, Heather, "Advocates Overcoming IT Resistance to Web 2.0," *Computerworld,* October 1, 2007, p. 14.

CHAPTER • 11 •

Knowledge Management and Specialized Information Systems

PRINCIPLES	LEARNING OBJECTIVES
▪ Knowledge management allows organizations to share knowledge and experience among their managers and employees.	▪ Discuss the differences among data, information, and knowledge. ▪ Describe the role of the chief knowledge officer (CKO). ▪ List some of the tools and techniques used in knowledge management.
▪ Artificial intelligence systems form a broad and diverse set of systems that can replicate human decision making for certain types of well-defined problems.	▪ Define the term *artificial intelligence* and state the objective of developing artificial intelligence systems. ▪ List the characteristics of intelligent behavior and compare the performance of natural and artificial intelligence systems for each of these characteristics. ▪ Identify the major components of the artificial intelligence field and provide one example of each type of system.
▪ Expert systems can enable a novice to perform at the level of an expert but must be developed and maintained very carefully.	▪ List the characteristics and basic components of expert systems. ▪ Identify at least three factors to consider in evaluating the development of an expert system. ▪ Outline and briefly explain the steps for developing an expert system. ▪ Identify the benefits associated with the use of expert systems.
▪ Virtual reality systems can reshape the interface between people and information technology by offering new ways to communicate information, visualize processes, and express ideas creatively.	▪ Define the term *virtual reality* and provide three examples of virtual reality applications.
▪ Specialized systems can help organizations and individuals achieve their goals.	▪ Discuss examples of specialized systems for organizational and individual use.

Information Systems in the Global Economy
Ericsson, Sweden
Telecom Giant Uses Knowledge Management and Expert Systems

A corporation's success depends on the knowledge it maintains and uses through its employees, executives, board of directors, and information systems. Building that body of corporate knowledge is a primary goal of most businesses—a goal that requires a concerted effort. If left untended, corporate knowledge pools in certain people, who gain it through time and experience, and is no longer accessible when those people leave the organization. Information systems, such as knowledge management systems and expert systems, can store knowledge gained by those within a corporation over time. For expert systems, artificial intelligence (AI) techniques can automate expert reasoning and activities. Consider how an expert system allows telecom giant Ericsson to monitor telecom networks in a manner that would overwhelm a human expert.

Established in 1876 and headquartered in Stockholm, Sweden, Ericsson is a world leader in telecommunications services, networking, multimedia solutions, and core technologies for mobile handsets. In fact, ten of the world's largest mobile phone operators use Ericsson technologies and 40 percent of all mobile traffic travels through Ericsson systems. Ericsson's 50 percent share of Sony Ericsson Mobile Communications has further expanded its power in the market.

Ericsson's busy telecom networks, as with all complex networks, encounter many obstacles over the course of the day. Lines may go down; some percentage of the thousands of routers, junction boxes, transmitters, and other telecom hardware devices used may fail; and systems software may experience bugs. Each time a problem, or fault, occurs, network administrators are alerted through dashboards that test and monitor the network.

Ericsson network administrators work in a stressful environment, where streams of alarms indicating network faults need to be evaluated and acted upon, sometimes immediately. Therein lays the challenge. It is difficult, if not impossible, for network administrators to determine which alarms are important and need immediate attention and which can wait. Given time, the administrators could read, interpret, and analyze each alarm to determine its level of importance, but with hundreds of thousands of alarms flowing in each day, there isn't even enough time to read one before ten more have arrived. This type of problem is well suited for an expert system.

Ericsson worked with an expert system company to assist with network fault management. The expert system company used real-time rule technology to automate the process of monitoring network alarms and determining which alarms needed immediate attention and which could wait. The human experts at Ericsson assisted the company in designing the software by sharing the secrets to the complicated process of alarm interpretation. With more than 50 types of equipment sending more than 500,000 alarms each day, the expert knowledge that was collected needed to be complete and executed quickly.

Today the expert system for managing network faults is deployed in 500 Ericsson systems in over 100 countries. The stream of alarms faced by human experts has been slowed to a tolerable amount of only the most important problems that require immediate intervention. The performance and quality of Ericsson networks has improved, its administrators can focus on priorities, and important corporate knowledge has been digitized to benefit the company over generations of employees.

As you read this chapter, consider the following:

- What steps can a business take to retain corporate knowledge within the business?
- How does computer intelligence compare to human intelligence?
- How can people and businesses make the best use of artificial intelligence and other specialized systems?

Why Learn About Knowledge Management and Specialized Information Systems?

Knowledge management and specialized information systems are used in almost every industry. If you are a manager, you might use a knowledge management system to support decisive action to help you correct a problem. If you are a production manager at an automotive company, you might oversee robots that attach windshields to cars or paint body panels. As a young stock trader, you might use a special system called a *neural network* to uncover patterns and make millions of dollars trading stocks and stock options. As a marketing manager for a PC manufacturer, you might use virtual reality on a Web site to show customers your latest laptop and desktop computers. If you are in the military, you might use computer simulation as a training tool to prepare you for combat. In a petroleum company, you might use an expert system to determine where to drill for oil and gas. You will see many additional examples of using these specialized information systems throughout this chapter. Learning about these systems will help you discover new ways to use information systems in your day-to-day work.

Like other aspects of an information system, the overall goal of knowledge management and the specialized systems discussed in this chapter is to help people and organizations achieve their goals. In some cases, knowledge management and these specialized systems can help an organization achieve a long-term, strategic advantage. In this chapter, we explore knowledge management, artificial intelligence, and many other specialized information systems, including expert systems, robotics, vision systems, natural language processing, learning systems, neural networks, genetic algorithms, intelligent agents, and virtual reality.

KNOWLEDGE MANAGEMENT SYSTEMS

Chapter 1 defines and discusses data, information, and knowledge. Recall that *data* consists of raw facts, such as an employee number, number of hours worked in a week, inventory part numbers, or sales orders. A list of the quantity available for all items in inventory is an example of data. When these facts are organized or arranged in a meaningful manner, they become information. *Information* is a collection of facts organized so that they have additional value beyond the value of the facts themselves. An exception report of inventory items that might be out of stock in a week because of high demand is an example of information. *Knowledge* is the awareness and understanding of a set of information and the ways that information can be made useful to support a specific task or reach a decision. Knowing the procedures for ordering more inventory to avoid running out is an example of knowledge. In a sense, information tells you what has to be done (low inventory levels for some items), while knowledge tells you how to do it (make two important phone calls to the right people to get the needed inventory shipped overnight). See Figure 11.1.

Figure 11.1

The Differences Among Data, Information, and Knowledge

Data	There are 20 PCs in stock at the retail store.
Information	The store will run out of inventory in a week unless more is ordered today.
Knowledge	Call 800-555-2222 to order more inventory.

According to Carol Csanda, director of knowledge management at State Farm Insurance, "We feel at State Farm that everybody's role is in some way about managing and transferring knowledge."[1] A *knowledge management system (KMS)* is an organized collection of people, procedures, software, databases, and devices used to create, store, share, and use the organization's knowledge and experience.[2] KMSs cover a wide range of systems, from software that contains some KMS components to dedicated systems designed specifically to capture, store, and use knowledge.[3] According to a software engineer for the Tata Group, a large Indian company, "We have learned that every time staff leaves our companies, their knowledge has gone with them and our knowledge is gradually reduced. Hence, we put knowledge management in the group's strategy to ensure our staff at all levels can maintain knowledge at all times."[4] Using KMS, the steel, software, coffee, chemicals, watches, and power company was able to win the Most Admired Knowledge Enterprise (MAKE) award in Asia. Other organizations also face the loss of knowledge when workers leave or retire. According to a spokesperson for the Aerospace and Defense (A&D) organization, "Knowledge management represents a major issue for A&D organizations, as they stand to lose decades of accumulated institutional knowledge and expertise each time an employee retires. This is particularly true for program managers and engineers, whose retirement rate nearly doubled between 2004 and 2005."[5]

Overview of Knowledge Management Systems

Like the other systems discussed throughout the book, including information and decision support systems, knowledge management systems attempt to help organizations achieve their goals.[6] For businesses, this usually means increasing profits or reducing costs. For nonprofit organizations, it can mean providing better customer service or providing special needs to people and groups. Many types of firms use KMSs to increase profits or reduce costs. Pratt & Whitney, for example, uses knowledge management systems to help it deliver information and knowledge about its jet engine parts to airlines, including Delta and United.[7] The World Bank uses knowledge management to obtain, share, and use knowledge to help fight worldwide poverty and disease.[8] According to a survey of CEOs, firms that use KMSs are more likely to innovate and perform better.

Pratt & Whitney uses knowledge management systems to help it deliver information and knowledge about its jet engine parts to airlines, including Delta and United.

(Source: Courtesy of Pratt & Whitney, a United Technologies Company.)

A KMS can involve different types of knowledge.[9] *Explicit knowledge* is objective and can be measured and documented in reports, papers, and rules. For example, knowing the best road to take to minimize drive time from home to the office when a major highway is closed due to an accident is explicit knowledge. It can be documented in a report or a rule, as in "If I-70 is closed, take Highway 6 to town and the office." *Tacit knowledge*, on the other hand, is hard to measure and document and typically is not objective or formalized. Knowing the best way to negotiate with a foreign government about nuclear disarmament or a volatile hostage situation often requires a lifetime of experience and a high level of skill. These are examples of tacit knowledge. It is difficult to write a detailed report or a set of rules that would always work in every hostage situation. Many organizations actively attempt to convert tacit knowledge to explicit knowledge to make the knowledge easier to measure, document, and share with others.

Data and Knowledge Management Workers and Communities of Practice

The personnel involved in a KMS include data workers and knowledge workers. Secretaries, administrative assistants, bookkeepers, and similar data-entry personnel are often called *data workers*. As mentioned in Chapter 1, *knowledge workers* are people who create, use, and disseminate knowledge. They are usually professionals in science, engineering, or business, and work in offices and belong to professional organizations. Other examples of knowledge workers include writers, researchers, educators, and corporate designers.[10] See Figure 11.2.

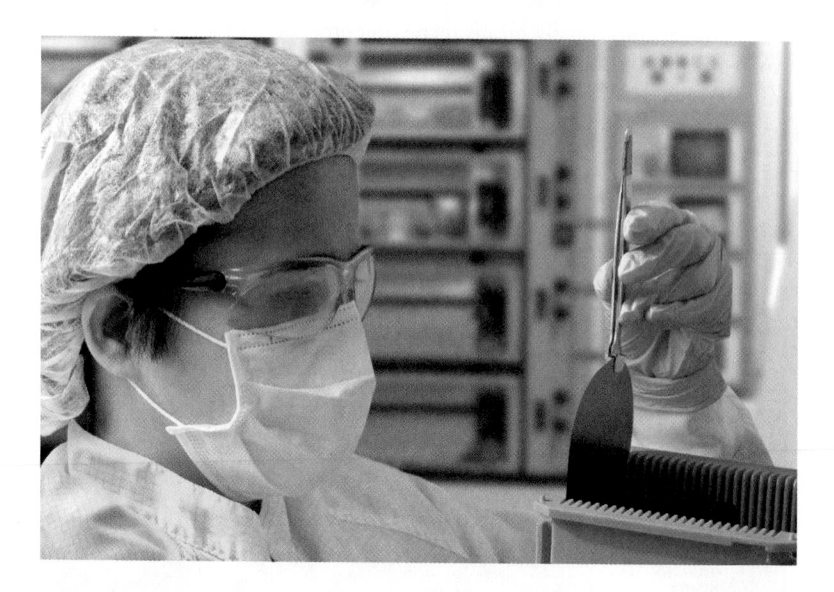

Figure 11.2

Knowledge Workers

Knowledge workers are people who create, use, and disseminate knowledge, including professionals in science, engineering, business, and other areas.

(Source: Eliza Snow/iStockphoto.)

chief knowledge officer (CKO)
A top-level executive who helps the organization use a KMS to create, store, and use knowledge to achieve organizational goals.

The **chief knowledge officer (CKO)** is a top-level executive who helps the organization work with a KMS to create, store, and use knowledge to achieve organizational goals.[11] The CKO is responsible for the organization's KMS, and typically works with other executives and vice presidents, including the chief executive officer (CEO), chief financial officer (CFO), and others. According to Jay Kostrzewa, assistant vice president of knowledge management at CNA, "My role as leader of the knowledge management area is to make certain the company has the right tools, the right information, and the right processes in place to share information."[12] CNA is a Chicago financial services company.

Some organizations and professions use *communities of practice (COP)* to create, store, and share knowledge. A COP is a group of people dedicated to a common discipline or practice, such as open-source software, auditing, medicine, or engineering. A group of oceanographers investigating climate change or a team of medical researchers looking for new ways to treat lung cancer are examples of COPs. COPs excel at obtaining, storing, sharing, and using knowledge. A COP from the International Conference on Knowledge Management in Nuclear Facilities investigates the use of knowledge management systems in the development and control of nuclear facilities.[13] According to Yuri Sokolov, IAEA deputy director general and head of the Department of Nuclear Energy, "All applications of nuclear technology are based on nuclear knowledge, so managing, preserving, and building on the knowledge we have accumulated is both wise [in the] near term and an important intergenerational responsibility."

Obtaining, Storing, Sharing, and Using Knowledge

Obtaining, storing, sharing, and using knowledge is the key to any KMS.[14] MWH Global, located in Colorado, uses a KMS to create, disseminate, and use knowledge specializing in environmental engineering, construction, and management activities worldwide.[15] The company has about 7,000 employees and 170 offices around the world. A KMS can help an organization increase profits or achieve its goals, but obtaining, storing, sharing, and using knowledge can be difficult.[16] In one survey, almost 60 percent of the respondents indicated that they couldn't find the information and knowledge that they need to do their jobs every day. Using a KMS often leads to additional knowledge creation, storage, sharing, and usage. According to Richard Cantor, knowledge management team manager for Chubb Commercial Insurance, "At Chubb, we're focusing on using our intranet as the vehicle that delivers shared knowledge. Many of our knowledge management efforts are packaged within that veil."[17] Business professors often conduct research in marketing strategies, management practices, corporate and individual investments and finance, effective accounting and auditing practices, and much more. Drug companies and medical researchers invest billions of dollars in creating knowledge on cures for diseases. Although knowledge workers can act alone, they often work in teams to create or obtain knowledge. See Figure 11.3.

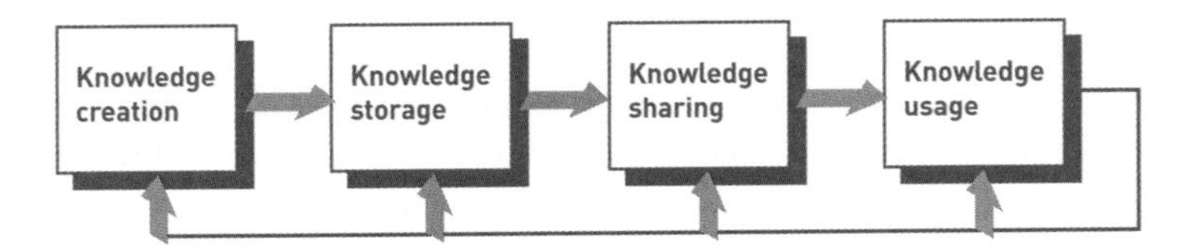

Figure 11.3

Knowledge Management System

Obtaining, storing, sharing, and using knowledge is the key to any KMS.

After knowledge is created, it is often stored in a *knowledge repository* that includes documents, reports, files, and databases. The knowledge repository can be located both inside the organization and outside. Some types of software can store and share knowledge contained in documents and reports. Adobe Acrobat PDF files, for example, allow you to store corporate reports, tax returns, and other documents and send them to others over the Internet. This publisher and the authors of this book used PDF files to store, share, and collaborate on each chapter. Traditional databases, data warehouses, and data marts, discussed in Chapter 5, often store the organization's knowledge. Specialized knowledge bases in expert systems, discussed later in this chapter, can also be used.

Because knowledge workers often work in groups or teams, they can use collaborative work software and group support systems (discussed in Chapter 10) to share knowledge, such as groupware, meeting software, and collaboration tools.[18] Intranets and password-protected Internet sites also provide ways to share knowledge. The social services department of the Surrey County Council in the United Kingdom, for example, used an intranet to help it create and manipulate knowledge. Because knowledge can be critical in maintaining a competitive advantage, businesses should be careful in how they share knowledge. Although they want important decision makers inside and outside the organization to have complete and easy access to knowledge, they also need to protect knowledge from competitors, hackers, and others who shouldn't obtain the organization's knowledge. As a result, many businesses use patents, copyrights, trade secrets, Internet firewalls, and other measures to keep prying eyes from seeing important knowledge that is often expensive and hard to create.

In addition to using information systems and collaborative software tools to share knowledge, some organizations use nontechnical approaches. These include corporate retreats and gatherings, sporting events, informal knowledge worker lounges or meeting places, kitchen facilities, daycare centers, and comfortable workout centers.

Using a knowledge management system begins with locating the organization's knowledge. This is often done using a *knowledge map* or directory that points the knowledge worker to the needed knowledge. Drug companies have sophisticated knowledge maps that include database and file systems to allow scientists and drug researchers to locate previous medical studies. The Army Defense Ammunition Center has signed an $8 million contract with SI International to provide it with knowledge management tools to help evaluate its training courses.[19] Medical researchers, university professors, and even textbook authors use Lexis-Nexis to locate important knowledge. Corporations often use the Internet or corporate Web portals to help their knowledge workers find knowledge stored in documents and reports.

Technology to Support Knowledge Management

KMSs use a number of tools discussed throughout the book. In Chapter 2, for example, we explored the importance of *organizational learning* and *organizational change*. An effective KMS is based on learning new knowledge and changing procedures and approaches as a result. A manufacturing company, for example, might learn new ways to program robots on the factory floor to improve accuracy and reduce defective parts. The new knowledge will likely cause the manufacturing company to change how it programs and uses its robots. In Chapter 5, we investigated the use of *data mining* and *business intelligence*. These powerful tools can be important in capturing and using knowledge. Enterprise resource planning tools, such as SAP, include knowledge management features. In Chapter 10, we showed how *groupware* could improve group decision making and collaboration. Groupware can also be used to help capture, store, and use knowledge. Of course, hardware, software, databases, telecommunications, and the Internet, discussed in Part 2, are important technologies used to support most knowledge management systems.

Hundreds of organizations provide specific KM products and services. See Figure 11.4. In addition, researchers at colleges and universities have developed tools and technologies to support knowledge management. The University of South Carolina, for example, has joined with Collexis to develop and deliver new knowledge management software, based on Collexis's Knowledge Discovery Platform.[20] It has been estimated that American companies would pay about $70 billion on knowledge management technology in 2007.[21] This amount is expected to increase by more than 15 percent in 2008. Companies such as IBM have many knowledge management tools in a variety of products, including Lotus Notes and Domino. Lotus Notes is a collection of software products that help people work together to create, share, and store important knowledge and business documents. Its knowledge management features include domain search, content mapping, and Lotus Sametime. Domain search allows people to perform sophisticated searches for knowledge in Domino databases using a single, simple query. Content mapping organizes knowledge by categories, like a table of contents for a book. Lotus Sametime helps people communicate, collaborate, and share ideas in real time. Lotus Domino Document Manager, formerly called Lotus Domino, helps people and organizations store, organize, and retrieve documents. The software can be used to write, review, archive, and publish documents throughout the organization.

Microsoft offers a number of knowledge management tools, including Digital Dashboard, which is based on the Microsoft Office suite. Digital Dashboard integrates information from different sources, including personal, group, enterprise, and external information and documents. Other tools from Microsoft include Web Store Technology, which uses wireless technology to deliver knowledge to any location at any time; Access Workflow Designer, which helps database developers create effective systems to process transactions and keep work flowing through the organization; and related products. Some additional knowledge management organizations and resources are summarized in Table 11.1. In addition to these tools, several artificial intelligence and special-purpose technologies and tools, discussed next, can be used in a KMS.

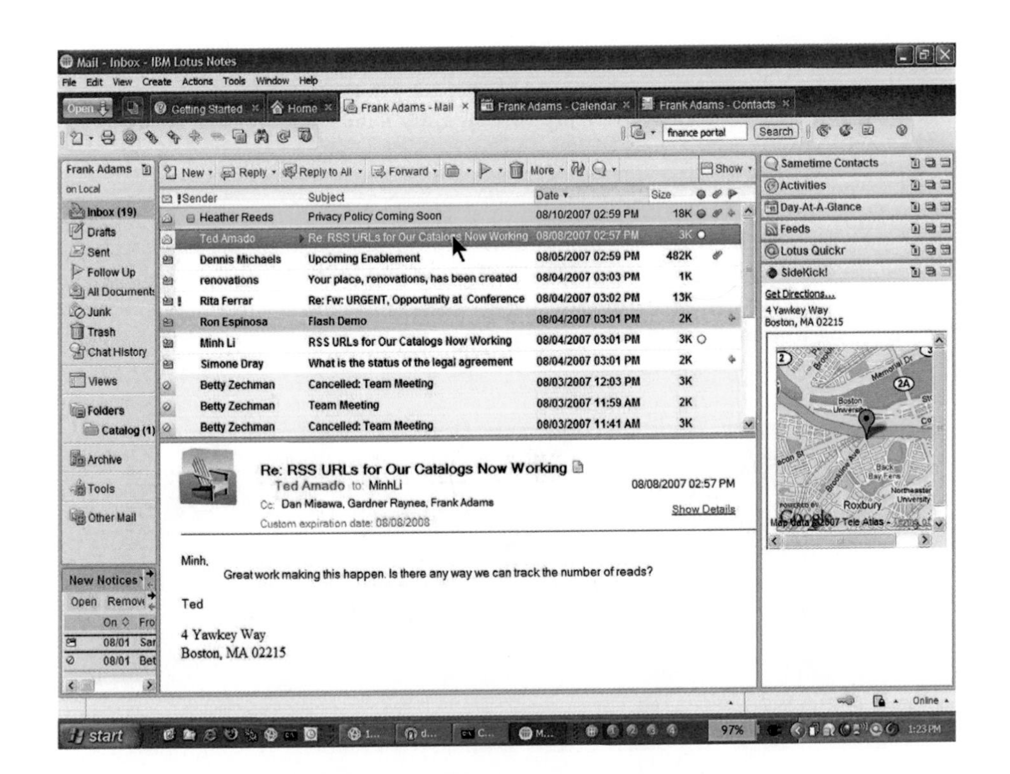

Figure 11.4

Knowledge Management Technology

Lotus Notes helps people communicate, collaborate, and streamline their work.

(Source: Courtesy of IBM Corporation)

Table 11.1

Additional Knowledge Management Organizations and Resources

Company	Description	Web Site
CortexPro	Knowledge management collaboration tools	www.cortexpro.com[22]
Delphi Group	A knowledge management consulting company	www.delphigroup.com[23]
Knowledge Management Resource Center	Knowledge management sites, products and services, magazines, and case studies	www.kmresource.com[24]
Knowledge Management Solutions, Inc.	Tools to create, capture, classify, share, and manage knowledge	www.kmsi.us[25]
Knowledge Management Web Directory	A directory of knowledge management Web sites	www.knowledge-manage.com[26]
KnowledgeBase	Content creation and management	www.knowledgebase.net
Law Clip Knowledge Manager	A service that collects and organizes text, Web links, and more from law-related Web sites	www.lawclip.com[27]
Meta KM	Knowledge management articles, resources, and opinions	www.metakm.com[28]

AN OVERVIEW OF ARTIFICIAL INTELLIGENCE

At a Dartmouth College conference in 1956, John McCarthy proposed the use of the term **artificial intelligence** (AI) to describe computers with the ability to mimic or duplicate the functions of the human brain. For example, advances in AI have led to systems that work like the human brain to recognize complex patterns.

artificial intelligence (AI)
The ability of computers to mimic or duplicate the functions of the human brain.

Many AI pioneers attended this first conference; a few predicted that computers would be as "smart" as people by the 1960s. The prediction has not yet been realized, but the benefits of artificial intelligence in business and research can be seen today, and research continues.

Science fiction novels and popular movies have featured scenarios of computer systems and intelligent machines taking over the world. Stephen Hawking, who is the Lucasian professor of mathematics at Cambridge University (a position once held by Isaac Newton) and author of *A Brief History of Time*, said, "In contrast with our intellect, computers double their performance every 18 months. So the danger is real that they could develop intelligence and take over the world." Computer systems such as Hal in the classic movie *2001: A Space Odyssey* and those in the movie *A.I.* are futuristic glimpses of what might be. These accounts are fictional, but they show the real application of many computer systems that use the notion of AI. These systems help to make medical diagnoses, explore for natural resources, determine what is wrong with mechanical devices, and assist in designing and developing other computer systems.

Science fiction movies give us a glimpse of the future, but many practical applications of artificial intelligence exist today, among them medical diagnostics and development of computer systems.

(Source: WALL-E, 2008. © Walt Disney Studios Motion Pictures/ Courtesy Everett Collection.)

Artificial Intelligence in Perspective

artificial intelligence systems
People, procedures, hardware, software, data, and knowledge needed to develop computer systems and machines that demonstrate the characteristics of intelligence.

Artificial intelligence systems include the people, procedures, hardware, software, data, and knowledge needed to develop computer systems and machines that demonstrate characteristics of intelligence.[29] Artificial intelligence can be used by most industries and applications. According to University of California-Santa Cruz professor Michael Mateas, "As graphics improvements top out, artificial intelligence will [drive] game innovation." Researchers, scientists, and experts on how human beings think are often involved in developing these systems.

The Nature of Intelligence

From the early AI pioneering stage, the research emphasis has been on developing machines with intelligent behavior.[30] In a book called *The Singularity Is Near* and articles by and about him, Ray Kurzweil predicts computers will have humanlike intelligence in 20 years.[31] The author also foresees that, by 2045, human and machine intelligence might merge. According to Kurzweil, "The Singularity Institute for Artificial Intelligence (SIAI) is playing a critical role in advancing humanity's understanding of the profound promise and peril of strong AI."[32] Machine intelligence, however, is hard to achieve.

intelligent behavior
The ability to learn from experiences and apply knowledge acquired from experience, handle complex situations, solve problems when important information is missing, determine what is important, react quickly and correctly to a new situation, understand visual images, process and manipulate symbols, be creative and imaginative, and use heuristics.

The *Turing Test* attempts to determine whether the responses from a computer with intelligent behavior are indistinguishable from responses from a human being. No computer has passed the Turing Test, developed by Alan Turing, a British mathematician. The Loebner Prize offers money and a gold medal for anyone developing a computer that can pass the Turing Test (see *www.loebner.net*). Some of the specific characteristics of **intelligent behavior** include the ability to do the following:

- **Learn from experience and apply the knowledge acquired from experience.** Learning from past situations and events is a key component of intelligent behavior and is a natural ability of humans, who learn by trial and error. This ability, however, must be carefully programmed into a computer system. Today, researchers are developing systems that can learn from experience. For instance, computerized AI chess software can learn to improve while playing human competitors. In one match, Garry Kasparov competed against a personal computer with AI software developed in Israel, called Deep Junior. This match was a 3–3 tie, but Kasparov picked up something the machine would have no interest in—$700,000. The 20 questions (20q) Web site, *www.20q.net*, is another example of a system that learns.[33] The Web site is an artificial intelligence game that learns as people play.

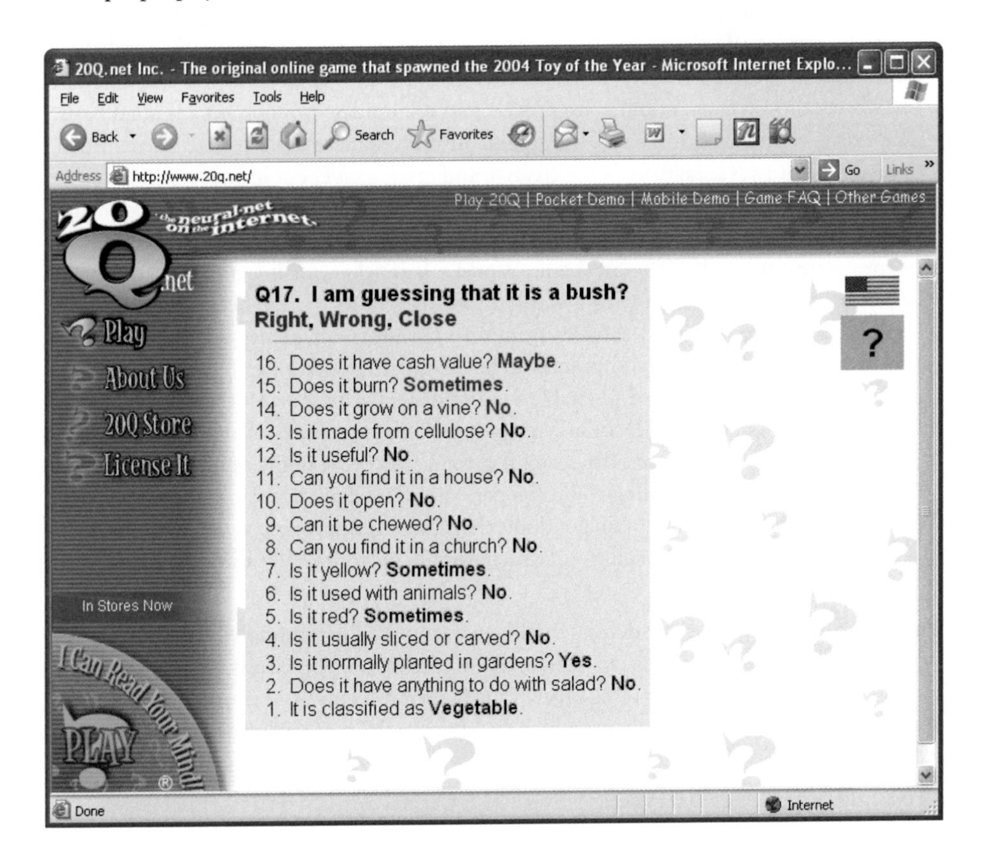

20Q is an online game where users play the popular game *Twenty Questions* against an artificial intelligence foe.

(Source: *www.20q.net.*)

- **Handle complex situations.** People are often involved in complex situations. World leaders face difficult political decisions regarding terrorism, conflict, global economic conditions, hunger, and poverty. In a business setting, top-level managers and executives must handle a complex market, challenging competitors, intricate government regulations, and a demanding workforce. Even human experts make mistakes in dealing with these situations. Developing computer systems that can handle perplexing situations requires careful planning and elaborate computer programming.
- **Solve problems when important information is missing.** The essence of decision making is dealing with uncertainty. Often, decisions must be made with little information or inaccurate information because obtaining complete information is too costly or impossible. Today, AI systems can make important calculations, comparisons, and decisions even when information is missing.
- **Determine what is important.** Knowing what is truly important is the mark of a good decision maker. Developing programs and approaches to allow computer systems and machines to identify important information is not a simple task.
- **React quickly and correctly to a new situation.** A small child, for example, can look over a ledge or a drop-off and know not to venture too close. The child reacts quickly and correctly to a new situation. Computers, on the other hand, do not have this ability without complex programming.
- **Understand visual images.** Interpreting visual images can be extremely difficult, even for sophisticated computers. Moving through a room of chairs, tables, and other objects can be trivial for people but extremely complex for machines, robots, and computers. Such machines require an extension of understanding visual images, called a **perceptive system.** Having a perceptive system allows a machine to approximate the way a person sees, hears, and feels objects. Military robots, for example, use cameras and perceptive systems to conduct reconnaissance missions to detect enemy weapons and soldiers. Detecting and destroying them can save lives.

perceptive system
A system that approximates the way a person sees, hears, and feels objects.

- **Process and manipulate symbols.** People see, manipulate, and process symbols every day. Visual images provide a constant stream of information to our brains. By contrast, computers have difficulty handling symbolic processing and reasoning. Although computers excel at numerical calculations, they aren't as good at dealing with symbols and three-dimensional objects. Recent developments in machine-vision hardware and software, however, allow some computers to process and manipulate symbols on a limited basis.
- **Be creative and imaginative.** Throughout history, some people have turned difficult situations into advantages by being creative and imaginative. For instance, when shipped defective mints with holes in the middle, an enterprising entrepreneur decided to market these new mints as LifeSavers instead of returning them to the manufacturer. Ice cream cones were invented at the St. Louis World's Fair when an imaginative store owner decided to wrap ice cream with a waffle from his grill for portability. Developing new and exciting products and services from an existing (perhaps negative) situation is a human characteristic. Few computers can be imaginative or creative in this way, although software has been developed to enable a computer to write short stories.
- **Use heuristics.** For some decisions, people use heuristics (rules of thumb arising from experience) or even guesses. In searching for a job, you might rank the companies you are considering according to profits per employee. Today, some computer systems, given the right programs, obtain good solutions that use approximations instead of trying to search for an optimal solution, which would be technically difficult or too time consuming.

This list of traits only partially defines intelligence. Unlike the terminology used in virtually every other field of IS research, in which the objectives can be clearly defined, the term *intelligence* is a formidable stumbling block. One of the problems in AI is arriving at a working definition of real intelligence against which to compare the performance of an AI system.

The Difference Between Natural and Artificial Intelligence

Since the term *artificial intelligence* was defined in the 1950s, experts have disagreed about the difference between natural and artificial intelligence. Can computers be programmed to have common sense? Profound differences separate natural from artificial intelligence, but they are declining in number (see Table 11.2). One of the driving forces behind AI research is an attempt to understand how people actually reason and think. Creating machines that can reason is possible only when we truly understand our own processes for doing so.

The Major Branches of Artificial Intelligence

AI is a broad field that includes several specialty areas, such as expert systems, robotics, vision systems, natural language processing, learning systems, and neural networks (see Figure 11.5). Many of these areas are related; advances in one can occur simultaneously with or result in advances in others.

Expert Systems

expert system
Hardware and software that stores knowledge and makes inferences, similar to a human expert.

An **expert system** consists of hardware and software that stores knowledge and makes inferences, similar to those of a human expert.[34] Because of their many business applications, expert systems are discussed in more detail in the next several sections of the chapter.

Robotics

robotics
Mechanical or computer devices that perform tasks requiring a high degree of precision or that are tedious or hazardous for humans.

Robotics involves developing mechanical or computer devices that can paint cars, make precision welds, and perform other tasks that require a high degree of precision or are tedious or hazardous for human beings.[35] The word "robot" comes from a play by Karel Capek in the 1920s, when he used the word "robota" to describe factory machines that do drudgery work and revolt.[36] The use of robots has expanded and is likely to increase in the future. According to Takeo Kande, "Someday, robots will do more than vacuum your floors. They'll train you and advise you—and maybe even help out with the cooking."[37] Some robots are

Table 11.2

A Comparison of Natural and Artificial Intelligence

Ability to	Natural Intelligence (Human)		Artificial Intelligence (Machine)	
	Low	High	Low	High
Use sensors (eyes, ears, touch, smell)		√	√	
Be creative and imaginative		√	√	
Learn from experience		√	√	
Adapt to new situations		√	√	
Afford the cost of acquiring intelligence		√	√	
Acquire a large amount of external information		√		√
Use a variety of information sources		√		√
Make complex calculations	√			√
Transfer information	√			√
Make a series of calculations rapidly and accurately	√			√

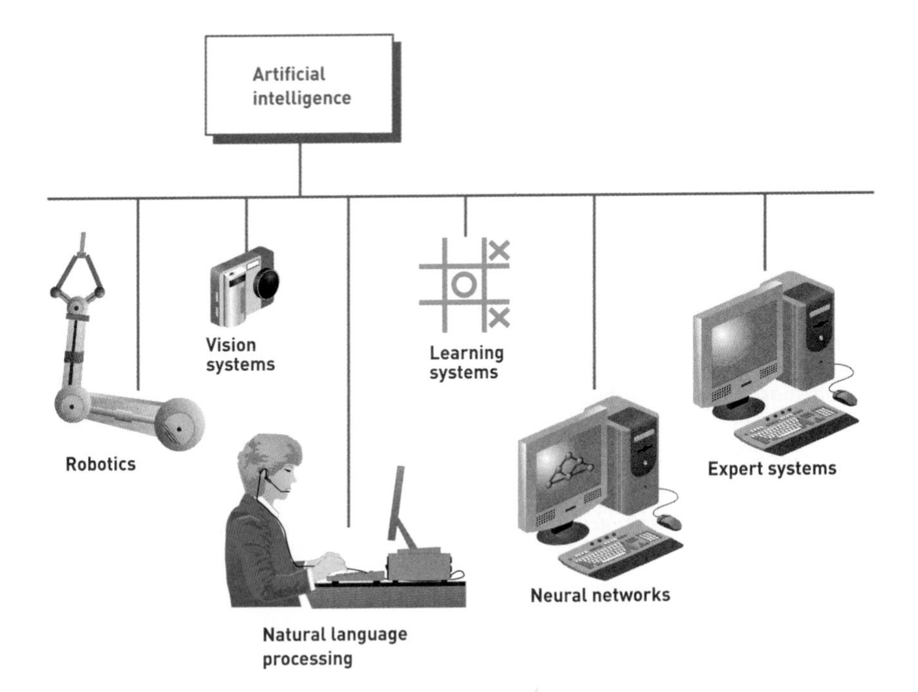

Figure 11.5

A Conceptual Model of Artificial Intelligence

mechanical devices that don't use the AI features discussed in this chapter. Others are sophisticated systems that use one or more AI features or characteristics, such as vision systems, learning systems, or neural networks discussed later in the chapter. For many businesses, robots are used to do the three Ds—dull, dirty, and dangerous jobs. Manufacturers use robots to assemble and paint products.[38] The NASA shuttle crash of 2003 has led some people to recommend using robots instead of people to explore space and perform scientific research. Some robots, such as the ER series by Intelitek (*www.intelitek.com*), can be used for training or entertainment. Contemporary robotics combine both high-precision machine capabilities and sophisticated controlling software. The controlling software in robots is what is most important in terms of AI.

The field of robotics has many applications, and research into these unique devices continues. The following are a few examples:

- The Robot Learning Laboratory, part of the computer science department and the Robotics Institute at Carnegie Mellon University (*www.ri.cmu.edu*), conducts research into the development and use of robotics.[39]
- IRobot (*www.irobot.com)* is a company that builds a number of robots, including the Roomba Floorvac for cleaning floors and the PackBot, an unmanned vehicle used to assist and protect soldiers.
- Robots are used in a variety of ways in medicine. The Porter Adventist Hospital (*www.porterhospital.org*) in Denver, Colorado, uses a $1.2 million Da Vinci Surgical System to perform surgery on prostate cancer patients.[40] The robot has multiple arms that hold surgical tools. According to one doctor at Porter, "The biggest advantage is it improves recovery time. Instead of having an eight-inch incision, the patient has a 'band-aid' incision. It's much quicker." The Heart-Lander is a very small robot that is inserted below the rib cage and used to perform delicate heart surgery.[41] Cameron Riviere at the Carnegie Mellon Robotics Institute (*www.ri.cmu.edu*) developed the robot along with help from John Hopkins University.
- DARPA (The Defense Advanced Research Project Agency) sponsors the DARPA Grand Challenge (*www.darpagrandchallenge.com*), a 132-mile race over rugged terrain for computer-controlled cars. The agency also sponsors other races and challenges.[42]
- The Hybrid Assisted Limb (HAL) lab is developing a robotic suit to help paraplegics and stroke victims move and perform basic functions.[43] The suit helps with lifting heavy objects, walking long distances, or performing other basic movements that can't be done otherwise. HAL was also the name of an artificial-intelligence computer in the classic movie *2001: A Space Odyssey.* The letters in HAL are one letter up from the letters in IBM.
- In the military, robots are moving beyond movie plots to become real weapons.[44] The Air Force is developing a smart robotic jet fighter. Often called *unmanned combat air vehicles (UCAVs)*, these robotic war machines, such as the X-45A, will be able to identify and destroy targets without human pilots. UCAVs send pictures and information to a central command center and can be directed to strike military targets. These machines extend the current Predator and Global Hawk technologies the military used in Afghanistan after the September 11 terrorist attacks and Iraq. Big Dog, made by Boston Dynamics (*www.bostondynamics.com*), is a robot that can carry up to 200 pounds of military gear in field conditions.

Although most of today's robots are limited in their capabilities, future robots will find wider applications in banks, restaurants, homes, doctors' offices, and hazardous working environments such as nuclear stations. The Repliee Q1 and Q2 robots from Japan are ultra-humanlike robots or androids that can blink, gesture, speak, and even appear to breathe (*www.ed.ams.eng.osakau.ac.jp/development/Android_RplieeQ2_e.html*). See Figure 11.6. Microrobotics, also called *micro-electro-mechanical systems (MEMS)*, are also being developed (*www.memsnet.org/mems/what-is.html*). MEMS can be used in a person's blood to monitor the body, and for other purposes in air bags, cell phones, refrigerators, and more.

Vision Systems

vision systems
The hardware and software that permit computers to capture, store, and manipulate visual images.

Another area of AI involves vision systems. **Vision systems** include hardware and software that permit computers to capture, store, and manipulate visual images. The U.S. Justice Department uses vision systems to perform fingerprint analysis with almost the same level of precision as human experts. The speed with which the system can search a huge database of fingerprints has brought quick resolution to many long-standing mysteries. Vision systems are also effective at identifying people based on facial features. In another application, a California wine bottle manufacturer uses a computerized vision system to inspect wine bottles for flaws. The company produces about 2 million wine bottles per day, and the vision system

Big Dog, manufactured by Boston Dynamics, is a robot that can carry up to 200 pounds of military gear in field conditions.

(Source: Courtesy of Boston Dynamics.)

Figure 11.6

The Repliee Q2 Robot from Japan

(Source: AP Photo/Katsumi Kasahara.)

saves the bottle producer both time and money. According to Takeo Kanade, Professor of Computer Science and Robotics at Carnegie Mellon University, "The trend toward computer vision is clear, and it will accelerate. In ten years, I wouldn't be surprised to see computers recognizing certain levels of emotions, expressions, gestures, and behaviors, all through vision."[45]

Natural Language Processing and Voice Recognition

natural language processing
Processing that allows the computer to understand and react to statements and commands made in a "natural" language, such as English.

As discussed in Chapter 4, **natural language processing** allows a computer to understand and react to statements and commands made in a "natural" language, such as English. Google, for example, has a service called Google Voice Local Search that allows you to dial a toll-free number and search for local businesses using voice commands and statements.[46] Restoration Hardware (*www.restorationhardware.com*) has developed a Web site that uses natural language processing to allow its customers to quickly find what they want. The natural language processing system corrects spelling mistakes, converts abbreviations into words and commands, and allows people to ask questions in English.

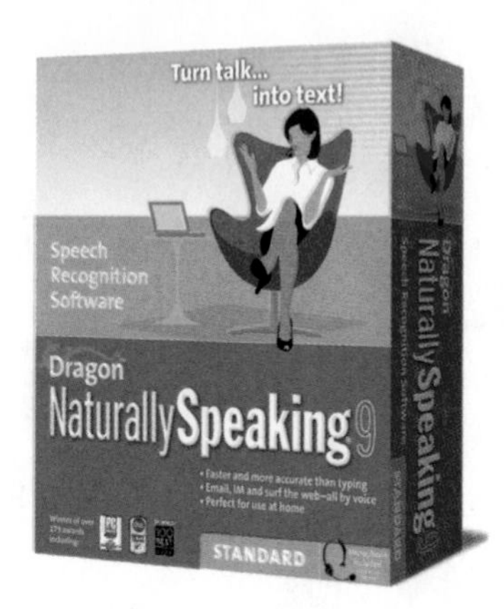

Dragon Systems' Naturally Speaking 9 Essentials uses continuous voice recognition, or natural speech, allowing the user to speak to the computer at a normal pace without pausing between words. The spoken words are transcribed immediately onto the computer screen.

(Source: Courtesy of Nuance Communications, Inc.)

In some cases, voice recognition is used with natural language processing. *Voice recognition* involves converting sound waves into words.[47] After converting sounds into words, natural language processing systems react to the words or commands by performing a variety of tasks. Brokerage services are a perfect fit for voice-recognition and natural language processing technology to replace the existing "press 1 to buy or sell a stock" touchpad telephone menu system. People buying and selling stock use a vocabulary too varied for easy access through menus and touchpads, but still small enough for software to process in real time. Several brokerages—including Charles Schwab & Company, Fidelity Investments, DLJdirect, and TD Waterhouse Group—offer these services. These systems use voice recognition and natural language processing to let customers access retirement accounts, check balances, and find stock quotes. Eventually, the technology will allow people to make transactions using voice commands over the phone and to use search engines to have their questions answered through the brokerage firm's call center. Using voice recognition to convert recordings into text is also possible.[48] Some companies claim that voice-recognition and natural language processing software is so good that customers forget they are talking to a computer and start discussing the weather or sports scores.

Learning Systems

learning systems
A combination of software and hardware that allows the computer to change how it functions or reacts to situations based on feedback it receives.

Another part of AI deals with **learning systems**, a combination of software and hardware that allows a computer to change how it functions or reacts to situations based on feedback it receives.[49] For example, some computerized games have learning abilities. If the computer does not win a game, it remembers not to make the same moves under the same conditions again. DARPA is investing about $10 million into a learning system called Bootstrapped Learning that will help military computers learn from human instructors.[50] If successful, the project could help the development and control of unmanned aircraft. The Center for Automated Learning and Discovery at Carnegie Mellon University (*www.cald.cs.cmu.edu*) is experimenting with two learning software packages that help each other learn. The hope is that two learning software packages that cooperate are better than separate learning packages.

Learning systems software requires feedback on the results of actions or decisions. At a minimum, the feedback needs to indicate whether the results are desirable (winning a game) or undesirable (losing a game). The feedback is then used to alter what the system will do in the future.

Neural Networks

An increasingly important aspect of AI involves neural networks, also called neural nets. A **neural network** is a computer system that can act like or simulate the functioning of a human brain.[51] The systems use massively parallel processors in an architecture that is based on the human brain's own mesh-like structure. In addition, neural network software simulates a neural network using standard computers. Neural networks can process many pieces of data at the same time and learn to recognize patterns. A chemical company, for example, can use neural network software to analyze a large amount of data to control chemical reactors. Neural network analysis has also helped some medical clinics diagnose cardiovascular disease.[52] Some oil and gas exploration companies use a program called the Rate of Penetration based on neural networks to monitor and control drilling operations.[53] The neural network program helps engineers slow or speed drilling operations to help increase drilling accuracy and reduce costs. Some of the specific abilities of neural networks include the following:

neural network
A computer system that can simulate the functioning of a human brain.

- Retrieving information even if some of the neural nodes fail
- Quickly modifying stored data as a result of new information
- Discovering relationships and trends in large databases
- Solving complex problems for which all the information is not present

A particular skill of neural nets is analyzing detailed trends.[54] Large amusement parks and banks use neural networks to determine staffing needs based on customer traffic—a task that requires precise analysis, down to the half-hour. Increasingly, businesses are firing up neural nets to help them navigate ever-thicker forests of data and make sense of a myriad of customer traits and buying habits. Computer Associates has developed Neugents (*www.neugents.com*), neural intelligence agents that "learn" patterns and behaviors and predict what will happen next. For example, Neugents can track the habits of insurance customers and predict which ones will not renew an automobile policy, for example. They can then suggest to an insurance agent what changes to make in the policy to persuade the consumer to renew it. The technology also can track individual users at e-commerce sites and their online preferences so that they don't have to enter the same information each time they log on—their purchasing history and other data is recalled each time they access a Web site.

AI Trilogy, available from the Ward Systems Group (*www.wardsystems.com*), is a neural network software program that can run on a standard PC. The software can make predictions with NeuroShell Predictor and classify information with NeuroShell Classifier. See Figure 11.7. The software package also contains GeneHunter, which uses a special type of algorithm called a genetic algorithm to get the best result from the neural network system. (Genetic algorithms are discussed later in this chapter.) Some pattern-recognition software uses neural networks to analyze hundreds of millions of bank, brokerage, and insurance accounts involving a trillion dollars to uncover money laundering and other suspicious money transfers.

Other Artificial Intelligence Applications

A few other artificial intelligence applications exist in addition to those just discussed. A **genetic algorithm**, also called a genetic program, is an approach to solving large, complex problems in which many repeated operations or models change and evolve until the best one emerges. The approach is based on the theory of evolution that requires (1) variation and (2) natural selection. The first step is to change or vary competing solutions to the problem. This can be done by changing the parts of a program or by combining different program segments into a new program, mimicking the evolution of species, in which the genetic makeup of a plant or animal mutates or changes over time. The second step is to select only the best models or algorithms, which continue to evolve. Programs or program segments that are not as good as others are discarded, similar to natural selection or "survival of the fittest," in which only the best species survive and continue to evolve. This process of variation and natural selection continues until the genetic algorithm yields the best possible solution to the original problem. For example, some investment firms use genetic algorithms to help select the best stocks or bonds. Genetic algorithms can help companies control inventory levels and get the best usage of warehouse space.[55] Genetic algorithms are also being used to monitor patient health.[56]

genetic algorithm
An approach to solving large, complex problems in which a number of related operations or models change and evolve until the best one emerges.

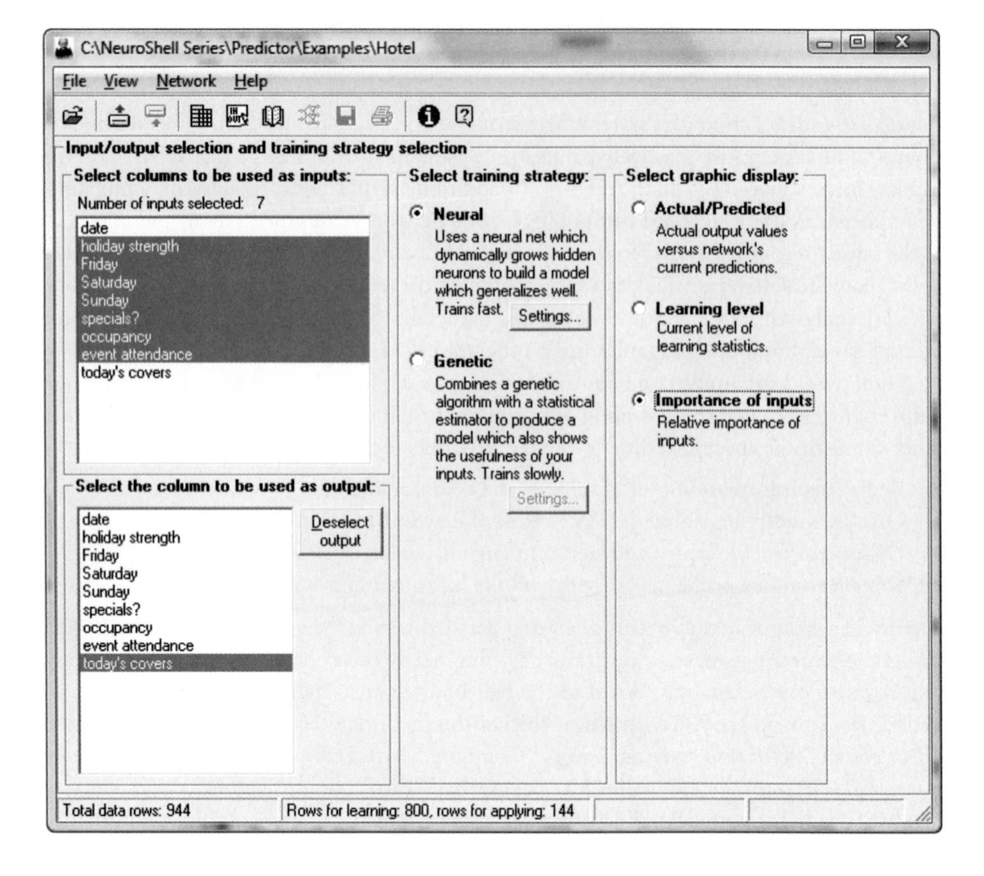

Figure 11.7

Neural Network Software

NeuroShell Predictor uses recognized forecasting methods to look for future trends in data.

(Source: Courtesy of Ward Systems Group, Inc.)

intelligent agent
Programs and a knowledge base used to perform a specific task for a person, a process, or another program; also called *intelligent robot* or *bot*.

An **intelligent agent** (also called an *intelligent robot* or *bot*) consists of programs and a knowledge base used to perform a specific task for a person, a process, or another program.[57] Like a sports agent who searches for the best endorsement deals for a top athlete, an intelligent agent often searches to find the best price, schedule, or solution to a problem. The programs used by an intelligent agent can search large amounts of data as the knowledge base refines the search or accommodates user preferences. Often used to search the vast resources of the Internet, intelligent agents can help people find information on an important topic or the best price for a new digital camera. Intelligent agents can also be used to make travel arrangements, monitor incoming e-mail for viruses or junk mail, and coordinate meetings and schedules of busy executives. The U.S. Army uses intelligent agents to help in its recruiting efforts.[58] Called Sgt. Star, the intelligent agent personalizes responses to visitors and potential recruits to its Web site, *www.goarmy.com*.

Some prosthetic limbs use AI to improve "virtual touch," which improves sensation and mobility. The Power Knee, for example, receives information from a sensor on the shoe of the sound leg to accurately mimic movement.

(Source: AP Photo/Dima Gavrysh.)

Providing Knowledge to Physicians Just in Time

Few professions are more complex and continuously changing and expanding than medicine. The quality of healthcare provided to a community depends on physicians being equipped with the latest medical knowledge on wide-ranging ailments and treatments. Information systems that feed this knowledge to physicians are the foundation on which life-and-death decisions are based. The responsibility of acquiring and managing the knowledge of the most up-to-date discoveries by the greatest minds in medicine is a daunting task, one that Partners Healthcare takes very seriously.

Partners HealthCare is an integrated healthcare system founded by Brigham and Women's Hospital and Massachusetts General Hospital. The system includes primary care and specialty physicians, community hospitals, academic medical centers, and other health-related entities. For years, Partners has invested heavily in medical knowledge management systems that provide physicians with information about the latest drugs and treatments for illnesses and diseases. In recent years, the amount of knowledge needed to make healthcare decisions has become so immense and changes so frequently that it has become unmanageable through traditional systems that rely on committee meetings and e-mail. In the near future, as physicians begin to practice personalized gene-based medicine, the amount of information to manage will explode in size.

Partners HealthCare's main objective is to maintain the quality of knowledge and information in medical systems. Partners involves hundreds of physicians in the process of storing and checking information in its medical knowledge management system. Gathering physicians together to build a knowledge base is difficult enough. Establishing a way to keep the content updated is even more challenging. Rather than focusing on the knowledge, Partners Healthcare began focusing on improving the efficiency of acquiring and maintaining the knowledge.

Starting with one person, Partners built a knowledge management (KM) team that has grown to more than 50 people in the last five years. Those involved include analysts, project managers, knowledge engineers, and software developers. The goal was to develop policies and processes for maintaining clinical knowledge content. The team focused on building a collaborative system that allowed domain knowledge experts to communicate without attending meetings or conference calls.

The team created a central repository for knowledge based on a product from EMC called Documentum. Documentum is a content management platform accessed through eRoom collaboration software. Together these products provide physicians with a robust, Web-based content management infrastructure that is flexible and scalable.

Prior to the Documentum system, physicians organized medical documents in folders on file systems. The location of the file and its name provided all the hierarchical and organizational information for storing and retrieving the file. Files were often updated and over time the organization of the system degraded. Files were also lost and mismanaged. With the Documentum system, knowledge is stored in a database. All interaction with the data is tracked and archived. For example, if a pharmacist reads about new findings related to dosages of ibuprofen for geriatric patients, she can share that article with colleagues, making it available through the database management system. Colleagues can then comment on the article and work to a consensus decision on what dosage is best for patients. The article, the discussion, and the vote are all catalogued in the database and can be referenced in a few years if someone wants to re-evaluate the dosage.

The Documentum and eRoom system has substantially reduced the cost of maintaining the knowledge management system and increased the speed at which Partners Healthcare can acquire knowledge. Physicians have more confidence in the information provided by the system. Rather than attending monthly meetings, clinicians are spending time poring over the information provided by the knowledge management system. To maintain the quality of data, participants log on to the system at the end of the day to comment on or approve new guidelines. Allowing physicians to work at their convenience saves everyone time and makes an unmanageable amount of information manageable.

Discussion Questions

1. What was the main challenge facing Partners Healthcare for managing clinical knowledge and information?
2. What functionality does Documentum and eRoom provide that was missing in Partners Healthcare's previous system?

Critical Thinking Questions

1. How does the quality of a medical knowledge management system affect a community?
2. How might the Partners Healthcare knowledge management system be expanded to benefit medical organizations nationwide or even worldwide in developing countries?

SOURCES: *Computerworld* Staff, "Managing clinical evidence at the speed of change," *Computerworld*—Honors Program, 2008, *www.cwhonors.org/viewCaseStudy2008.asp?NominationID=365*; Partners Healthcare Web site, *www.partners.org* accessed July 4, 2008; EMC Web site, *www.emc.com*, accessed July 4, 2008.

AN OVERVIEW OF EXPERT SYSTEMS

As mentioned earlier, an expert system behaves similarly to a human expert in a particular field. Computerized expert systems have been developed to diagnose problems, predict future events, and solve energy problems. Like human experts, computerized expert systems use heuristics, or rules of thumb, to arrive at conclusions or make suggestions. The research conducted in AI during the past two decades is resulting in expert systems that explore new business possibilities, increase overall profitability, reduce costs, and provide superior service to customers and clients. Blagg & Johnson uses the Lantek expert system to cut and fabricate metal into finished products for the automotive, construction, and mining industries.[59] The expert system helps reduce raw material waste and increase profits. The U.S. Army uses the Knowledge and Information Fusion Exchange (KnIFE) expert system to help soldiers in the field make better military decisions based on successful decisions made in previous military engagements.[60]

Expert systems are used in metal fabrication plants to aid in decision making.

(Source: © H. Mark Weidman Photography/Alamy.)

When to Use Expert Systems

Sophisticated expert systems can be difficult, expensive, and time consuming to develop. This is especially true for large expert systems implemented on mainframes. The following is a list of factors that normally make expert systems worth the expenditure of time and money. People and organizations should develop an expert system if it can do any of the following:

- Provide a high potential payoff or significantly reduce downside risk
- Capture and preserve irreplaceable human expertise
- Solve a problem that is not easily solved using traditional programming techniques
- Develop a system more consistent than human experts
- Provide expertise needed at a number of locations at the same time or in a hostile environment that is dangerous to human health
- Provide expertise that is expensive or rare

- Develop a solution faster than human experts can
- Provide expertise needed for training and development to share the wisdom and experience of human experts with many people

Components of Expert Systems

An expert system consists of a collection of integrated and related components, including a knowledge base, an inference engine, an explanation facility, a knowledge base acquisition facility, and a user interface. A diagram of a typical expert system is shown in Figure 11.8. In this figure, the user interacts with the interface, which interacts with the inference engine. The inference engine interacts with the other expert system components. These components must work together to provide expertise. This figure shows the inference engine coordinating the flow of knowledge to other components of the expert system. Note that different knowledge flows can exist, depending on what the expert system is doing and the specific expert system involved.

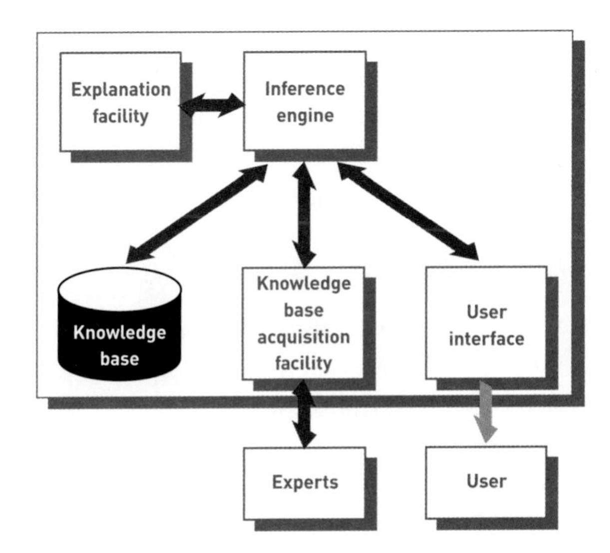

Figure 11.8

Components of an Expert System

The Knowledge Base

The **knowledge base** stores all relevant information, data, rules, cases, and relationships that the expert system uses. As shown in Figure 11.9, a knowledge base is a natural extension of a database (presented in Chapter 5) and an information and decision support system (presented in Chapter 10). A knowledge base must be developed for each unique application. For example, a medical expert system contains facts about diseases and symptoms. The following are some tools and techniques that can be used to create a knowledge base.

knowledge base
A component of an expert system that stores all relevant information, data, rules, cases, and relationships used by the expert system.

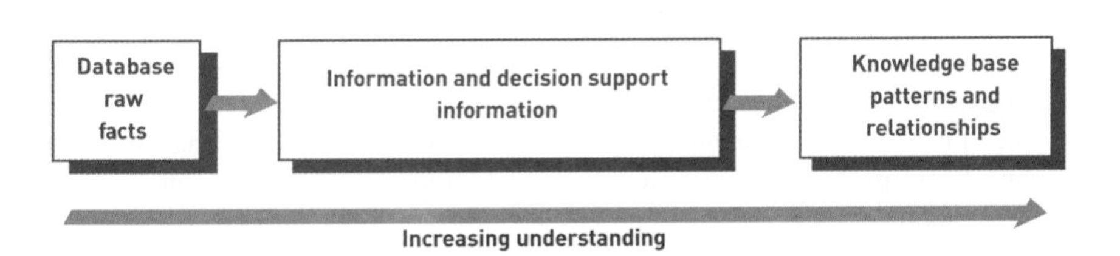

Figure 11.9

The Relationships Among Data, Information, and Knowledge

- **Assembling human experts.** One challenge in developing a knowledge base is to assemble the knowledge of multiple human experts. Typically, the objective in building a knowledge base is to integrate the knowledge of people with similar expertise (for example, many doctors might contribute to a medical diagnostics knowledge base).
- **Using fuzzy logic.** Another challenge for designers and developers of expert systems is capturing knowledge and relationships that are not precise or exact. Instead of the black-and-white, yes/no, or true/false conditions of typical computer decisions, fuzzy logic

rule
A conditional statement that links conditions to actions or outcomes.

IF-THEN statements
Rules that suggest certain conclusions.

allows shades of gray, or what is known as "fuzzy sets." Fuzzy logic rules help computers evaluate the imperfect or imprecise conditions they encounter and make educated guesses based on the probability of correctness of the decision.

- **Using rules.** A **rule** is a conditional statement that links conditions to actions or outcomes. In many instances, these rules are stored as **IF-THEN statements**, such as "If a certain set of network conditions exists, then a certain network problem diagnosis is appropriate." In an expert system for a weather forecasting operation, for example, the rules could state that, if certain temperature patterns exist with a given barometric pressure and certain previous weather patterns over the last 24 hours, then a specific forecast will be made, including temperatures, cloud coverage, and wind-chill factor. Figure 11.10 shows how to use expert system rules in determining whether a person should receive a mortgage loan from a bank. These rules can be placed in almost any standard program language discussed in Chapter 4 using "IF-THEN" statements or into special expert systems shells and products, discussed later in the chapter. In general, as the number of rules that an expert system knows increases, the precision of the expert system also increases.

Figure 11.10
Rules for a Credit Application

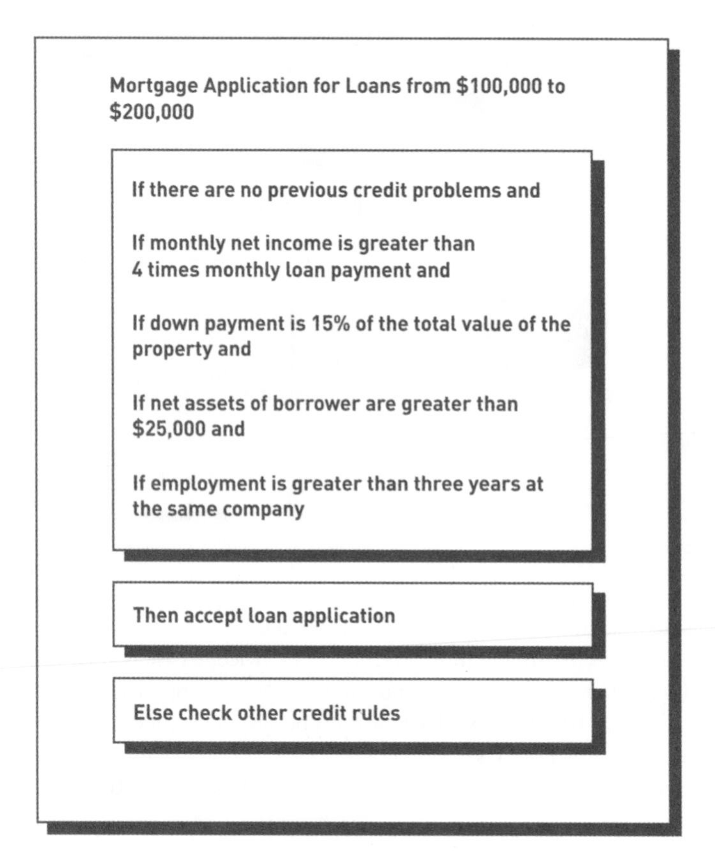

Mortgage Application for Loans from $100,000 to $200,000
If there are no previous credit problems and
If monthly net income is greater than 4 times monthly loan payment and
If down payment is 15% of the total value of the property and
If net assets of borrower are greater than $25,000 and
If employment is greater than three years at the same company
Then accept loan application
Else check other credit rules

- **Using cases.** An expert system can use cases in developing a solution to a current problem or situation. This process involves (1) finding cases stored in the knowledge base that are similar to the problem or situation at hand and (2) modifying the solutions to the cases to fit or accommodate the current problem or situation. For example, a company might use an expert system to determine the best location for a new service facility in the state of New Mexico. The expert system might identify two previous cases involving the location of a service facility where labor and transportation costs were also important—one in the state of Colorado and the other in the state of Nevada. The expert system can modify the solution to these two cases to determine the best location for a new facility in New Mexico.

The Inference Engine

The overall purpose of an **inference engine** is to seek information and relationships from the knowledge base and to provide answers, predictions, and suggestions similar to the way a human expert would. In other words, the inference engine is the component that delivers the expert advice. To provide answers and give advice, expert systems can use backward and forward chaining. **Backward chaining** is the process of starting with conclusions and working backward to the supporting facts. If the facts do not support the conclusion, another conclusion is selected and tested. This process is continued until the correct conclusion is identified. **Forward chaining** starts with the facts and works forward to the conclusions. Consider the expert system that forecasts future sales for a product. Forward chaining starts with a fact such as "The demand for the product last month was 20,000 units." With the forward-chaining approach, the expert system searches for rules that contain a reference to product demand. For example, "IF product demand is over 15,000 units, THEN check the demand for competing products." As a result of this process, the expert system might use information on the demand for competitive products. Next, after searching additional rules, the expert system might use information on personal income or national inflation rates. This process continues until the expert system can reach a conclusion using the data supplied by the user and the rules that apply in the knowledge base.

inference engine
Part of the expert system that seeks information and relationships from the knowledge base and provides answers, predictions, and suggestions similar to the way a human expert would.

backward chaining
The process of starting with conclusions and working backward to the supporting facts.

forward chaining
The process of starting with the facts and working forward to the conclusions.

The Explanation Facility

An important part of an expert system is the **explanation facility**, which allows a user or decision maker to understand how the expert system arrived at certain conclusions or results. A medical expert system, for example, might reach the conclusion that a patient has a defective heart valve given certain symptoms and the results of tests on the patient. The explanation facility allows a doctor to find out the logic or rationale of the diagnosis made by the expert system. The expert system, using the explanation facility, can indicate all the facts and rules that were used in reaching the conclusion. This facility allows doctors to determine whether the expert system is processing the data and information correctly and logically.

explanation facility
Component of an expert system that allows a user or decision maker to understand how the expert system arrived at certain conclusions or results.

The Knowledge Acquisition Facility

A difficult task in developing an expert system is the process of creating and updating the knowledge base. In the past, when more traditional programming languages were used, developing a knowledge base was tedious and time consuming. Each fact, relationship, and rule had to be programmed into the knowledge base. In most cases, an experienced programmer had to create and update the knowledge base.

Today, specialized software allows users and decision makers to create and modify their own knowledge bases through the knowledge acquisition facility (see Figure 11.11). The purpose of the **knowledge acquisition facility** is to provide a convenient and efficient means for capturing and storing all components of the knowledge base. Knowledge acquisition software can present users and decision makers with easy-to-use menus. After entering the appropriate attributes, the knowledge acquisition facility correctly stores information and relationships in the knowledge base, making the knowledge base easier and less expensive to set up and maintain. Knowledge acquisition can be a manual process or a mixture of manual and automated procedures. Regardless of how the knowledge is acquired, the knowledge base must be validated and updated frequently to make sure that it is still accurate.

knowledge acquisition facility
Part of the expert system that provides convenient and efficient means of capturing and storing all the components of the knowledge base.

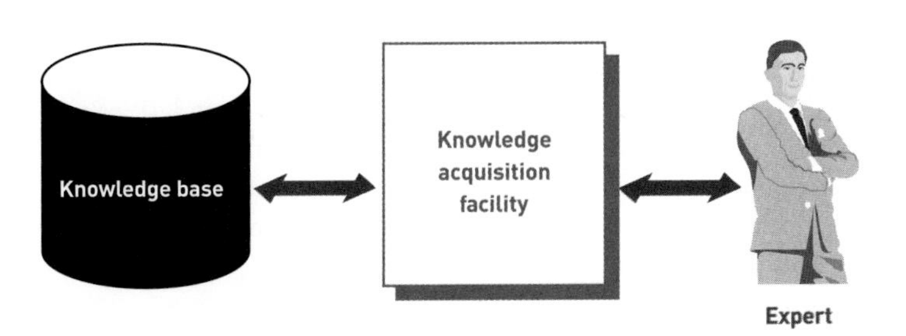

Figure 11.11

Knowledge Acquisition Facility

The knowledge acquisition facility acts as an interface between experts and the knowledge base.

The User Interface

Specialized user interface software is employed for designing, creating, updating, and using expert systems. The main purpose of the user interface is to make an expert system easier for users and decision makers to develop and use. At one time, skilled computer personnel created and operated most expert systems; today, the user interface permits decision makers to develop and use their own expert systems. Because expert systems place more emphasis on directing user activities than do other types of systems, text-oriented user interfaces (using menus, forms, and scripts) might be more common in expert systems than the graphical interfaces often used with DSSs. A typical user interface between a user and the expert system for the credit application shown in Figure 11.10 follows.

EXPERT SYSTEM: Are there any previous credit problems?

USER: No.

EXPERT SYSTEM: Is the applicant's net income greater than four times the monthly loan payment?

USER: Yes.

EXPERT SYSTEM: Is the down payment greater than 15% of the property value?

USER: Yes.

EXPERT SYSTEM: Are the net assets of the borrower greater than $25,000?

USER: Yes.

EXPERT SYSTEM: Has the borrower worked more than three years at the same company?

USER: Yes.

EXPERT SYSTEM: **Accept the loan application.**

Participants in Developing and Using Expert Systems

Typically, several people are involved in developing and using an expert system (see Figure 11.12).

Figure 11.12

Participants in Expert Systems Development and Use

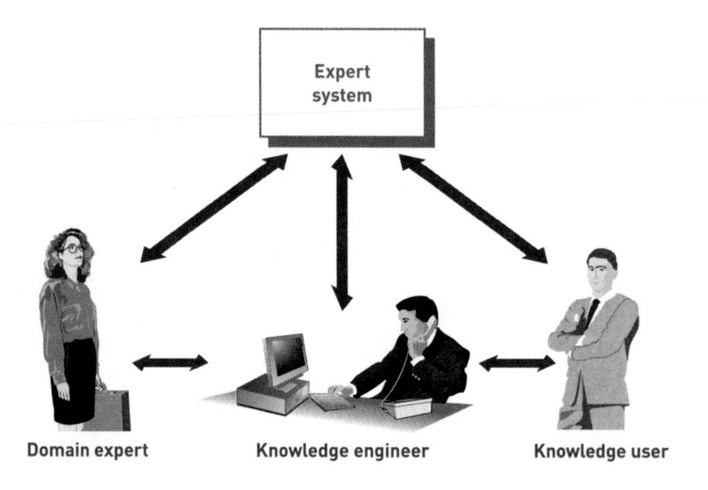

The Domain Expert

Because of the time and effort involved in the task, an expert system is developed to address only a specific area of knowledge. This area of knowledge is called the **domain.** The **domain expert** is the person or group with the expertise or knowledge the expert system is trying to capture. In most cases, the domain expert is a group of human experts. The domain expert (individual or group) usually can do the following:

- Recognize the real problem
- Develop a general framework for problem solving

domain
The area of knowledge addressed by the expert system.

domain expert
The person or group who has the expertise or knowledge the expert system is trying to capture.

- Formulate theories about the situation
- Develop and use general rules to solve a problem
- Know when to break the rules or general principles
- Solve problems quickly and efficiently
- Learn from experience
- Know what is and is not important in solving a problem
- Explain the situation and solutions of problems to others

The Knowledge Engineer and Knowledge Users

A **knowledge engineer** is a person who has training or experience in the design, development, implementation, and maintenance of an expert system, including training or experience with expert system shells. The **knowledge user** is the person or group who uses and benefits from the expert system. Knowledge users do not need any previous training in computers or expert systems.

knowledge engineer
A person who has training or experience in the design, development, implementation, and maintenance of an expert system.

knowledge user
The person or group who uses and benefits from the expert system.

Expert Systems Development Tools and Techniques

Theoretically, expert systems can be developed from any programming language. Since the introduction of computer systems, programming languages have become easier to use, more powerful, and increasingly able to handle specialized requirements. In the early days of expert systems development, traditional high-level languages, including Pascal, FORTRAN, and COBOL, were used (see Figure 11.13). LISP was one of the first special languages developed and used for artificial intelligence applications. PROLOG was also developed for AI applications. Since the 1990s, however, other expert system products (such as shells) have become available that remove the burden of programming, allowing nonprogrammers to develop and benefit from the use of expert systems.

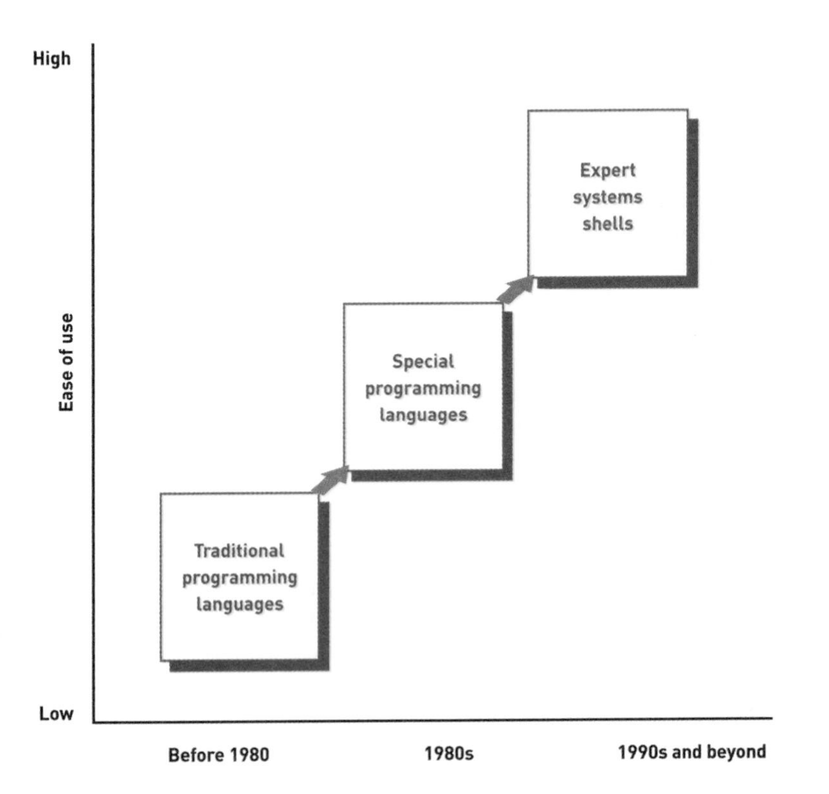

Figure 11.13

Expert Systems Development

Software for expert systems development has evolved greatly since 1980, from traditional programming languages to expert system shells.

Expert System Shells and Products

An *expert system shell* is a collection of software packages and tools used to design, develop, implement, and maintain expert systems. Expert system shells are available for both personal computers and mainframe systems. Some shells are inexpensive, costing less than $500. In addition, off-the-shelf expert system shells are complete and ready to run. The user enters the appropriate data or parameters, and the expert system provides output to the problem or situation. Table 11.3 lists a few expert system products.

Table 11.3

Popular Expert System Products

Name of Product	Application and Capabilities
G2	Assists in oil and gas operations. Transco, a British company, uses it to help in the transport of gas to more than 20 million commercial and domestic customers.
HazMat Loader	Analyzes hazardous materials in truck shipments (http://hazmat.dot.gov).
Imprint Business Systems	This company has an expert system that helps printing and packaging companies manage their businesses (www.imprint-mis.co.uk).
Lantek Expert System	Helps metal fabricators reduce waste and increase profits (www.lantek.es).
RAMPART	Developed by Sandia National Laboratories, the U.S. General Services Administration (GSA) uses it to analyze risk to the approximately 8,000 federal buildings it manages (www.sandia.gov).

Applications of Expert Systems and Artificial Intelligence

Expert systems and artificial intelligence have wide applications in business and government. A few additional applications of expert systems that are being used today or have been used in the past are summarized next.

- **Credit granting and loan analysis.** KPMG Peat Marwick uses an expert system called Loan Probe to review its reserves to determine whether sufficient funds have been set aside to cover the risk of some uncollectible loans.
- **Catching cheats and terrorists.** Some gambling casinos use expert system software to catch gambling cheats.
- **Plant layout and manufacturing.** FLEXPERT was an expert system that uses fuzzy logic to perform plant layout. The software helped companies determine the best placement for equipment and manufacturing facilities.
- **Hospitals and medical facilities.** Hospitals, pharmacies, and other healthcare providers can use Alineo by MEDecision to determine possible high-risk or high-cost patients. MYCIN is an expert system developed at Stanford University to analyze blood infections. UpToDate is another expert system used to diagnose patients. To help doctors in the diagnosis of thoracic pain, MatheMEDics has developed THORASK, a straightforward, easy-to-use program, requiring only the input of carefully obtained clinical information. The program helps the less experienced to distinguish the three principal categories of chest pain from each other.
- **Employee performance evaluation.** An expert system developed by Austin-Hayne, called Employee Appraiser, provides managers with expert advice for use in employee performance reviews and career development.
- **Repair and maintenance.** ACE is an expert system used by AT&T to analyze the maintenance of telephone networks. IET-Intelligent Electronics uses an expert system to diagnose maintenance problems related to aerospace equipment. General Electric Aircraft Engine Group uses an expert system to enhance maintenance performance levels at all sites and improve diagnostic accuracy.
- **Shipping.** CARGEX cargo expert system is used by Lufthansa, a German airline, to help determine the best shipping routes.
- **Marketing.** CoverStory is an expert system that extracts marketing information from a database and automatically writes marketing reports.

Computer manufacturers often use expert systems to monitor manufacturing processes and improve yields.

(Source: © Colorblind/Getty Images.)

VIRTUAL REALITY

The term *virtual reality* was initially coined by Jaron Lanier, founder of VPL Research, in 1989. Originally, the term referred to *immersive virtual reality* in which the user becomes fully immersed in an artificial, three-dimensional world that is completely generated by a computer. Immersive virtual reality can represent any three-dimensional setting, real or abstract, such as a building, an archaeological excavation site, the human anatomy, a sculpture, or a crime scene reconstruction. Through immersion, the user can gain a deeper understanding of the virtual world's behavior and functionality. The Media Grid at Boston College has a number of initiatives in the use of immersive virtual reality in education.[61]

A **virtual reality system** enables one or more users to move and react in a computer-simulated environment. Virtual reality simulations require special interface devices that transmit the sights, sounds, and sensations of the simulated world to the user. These devices can also record and send the speech and movements of the participants to the simulation program, enabling users to sense and manipulate virtual objects much as they would real objects. This natural style of interaction gives the participants the feeling that they are immersed in the simulated world. For example, an auto manufacturer can use virtual reality to help it simulate and design factories.

virtual reality system
A system that enables one or more users to move and react in a computer-simulated environment.

Interface Devices

To see in a virtual world, often the user wears a head-mounted display (HMD) with screens directed at each eye. The HMD also contains a position tracker to monitor the location of the user's head and the direction in which the user is looking. Using this information, a computer generates images of the virtual world—a slightly different view for each eye—to match the direction that the user is looking, and displays these images on the HMD. Many companies sell or rent virtual-reality interface devices, including Virtual Realities (*www.vrealities.com*), Amusitronix (*www.amusitronix.com*), I-O Display Systems (*www.i-glassesstore.com*), and others.

The PowerWall is a virtual reality system that displays large models in accurate dimensions.

(Source: Courtesy of Fakespace Systems, Inc.)

The Electronic Visualization Laboratory at the University of Illinois at Chicago introduced a room constructed of large screens on three walls and the floor on which the graphics are projected. The CAVE, as this room is called, provides the illusion of immersion by projecting stereo images on the walls and floor of a room-sized cube (*http://cave.ncsa.uiuc.edu*). Several persons wearing lightweight stereo glasses can enter and walk freely inside the CAVE. A head-tracking system continuously adjusts the stereo projection to the current position of the leading viewer.

Military personnel train in an immersive CAVE system.

(Source: Courtesy of Fakespace Systems, Inc.)

Users hear sounds in the virtual world through earphones. The information reported by the position tracker is also used to update audio signals. When a sound source in virtual space is not directly in front of or behind the user, the computer transmits sounds to arrive at one ear a little earlier or later than at the other and to be a little louder or softer and slightly different in pitch.

The *haptic* interface, which relays the sense of touch and other physical sensations in the virtual world, is the least developed and perhaps the most challenging to create.[62] Currently, with the use of a glove and position tracker, the computer locates the user's hand and measures finger movements. The user can reach into the virtual world and handle objects; however, it is difficult to generate the sensations of a person tapping a hard surface, picking up an object, or running a finger across a textured surface. Touch sensations also have to be synchronized with the sights and sounds users experience.

Forms of Virtual Reality

Aside from immersive virtual reality, which we just discussed, virtual reality can also refer to applications that are not fully immersive, such as mouse-controlled navigation through a three-dimensional environment on a graphics monitor, stereo viewing from the monitor via stereo glasses, stereo projection systems, and others.

Some virtual reality applications allow views of real environments with superimposed virtual objects. Motion trackers monitor the movements of dancers or athletes for subsequent studies in immersive virtual reality. Telepresence systems (such as telemedicine and telerobotics) immerse a viewer in a real world that is captured by video cameras at a distant location and allow for the remote manipulation of real objects via robot arms and manipulators. Many believe that virtual reality will reshape the interface between people and information technology by offering new ways to communicate information, visualize processes, and express ideas creatively.

Computer-generated image technology and simulation are used by companies to determine plant capacity, manage bottlenecks, and optimize production rates.

(Source: © Lester Lefkowitz/Getty Images.)

Virtual Reality Applications

You can find thousands of applications of virtual reality, with more being developed as the cost of hardware and software declines and people's imaginations are opened to the potential of virtual reality. The following are a few virtual reality applications in medicine, education and training, business, and entertainment. See Figure 11.14.

Medicine

Barbara Rothbaum, the director of the Trauma and Recovery Program at Emory University School of Medicine and cofounder of Virtually Better, uses an immersive virtual reality system to help in the treatment of anxiety disorders.[63] "For most of our applications, we use a head-mounted display that's kind of like a helmet with a television screen in front of each eye and has position trackers and sensors," says Rothbaum. One VR program, called SnowWorld, helps treat burn patients.[64] Using VR, the patients can navigate through icy terrain and frigid waterfalls. VR helps because it gets a patient's mind off the pain.

Figure 11.14

Virtual Reality Applications

Virtual reality has been used to increase real estate sales in several powerful ways. RealSpace Vision Communication, for example, helps real estate developers showcase their properties with virtual reality tours.

(Source: Courtesy of RealSpace Vision Communication Inc.)

Virtual reality technology can also link stroke patients to their physical therapists. Patients put on special gloves and other virtual reality devices at home that are linked to the physical therapist's office. The physical therapist can then see whether the patient is performing the correct exercises without having to travel to the patient's home or hospital room. In this way, using virtual reality can cut travel time and costs.

Education and Training

Virtual environments are used in education to bring exciting new resources into the classroom. Students can stroll among digital bookshelves, develop communication skills in front of a virtual audience, learn anatomy on a simulated cadaver, or participate in historical events—all virtually. Virtual reality, for example, has been used to help those with disabilities.[65] The Archaeology Technologies Laboratory at North Dakota State University has developed a three-dimensional virtual reality system that displays an eighteenth-century American Indian village.[66] Third-grade students at John Cotton Tayloe School in Washington, North Carolina, can take a virtual trip down the Nile for a cross-disciplinary lesson on ancient Egypt. This interactive virtual reality computer lesson integrates social studies, geography, music, art, science, math, and language arts.

Some virtual reality systems help train people to overcome their fear of public speaking.

(Source: Copyright © Virtually Better, Inc.)

Virtual technology has also been applied by the military. To help with aircraft maintenance, a virtual reality system has been developed to simulate an aircraft and give a user a sense of touch, while computer graphics provide a sense of sight and sound. The user sees, touches, and manipulates the various parts of the virtual aircraft during training. The Virtual Aircraft Maintenance System simulates real-world maintenance tasks that are routinely performed on the AV-8B vertical takeoff and landing aircraft used by the U.S. Marines. Also, the Pentagon is using a virtual reality training lab to prepare for a military crisis. The virtual reality system simulates various war scenarios.

Business and Commerce

Virtual reality has been used in all areas of business. Kimberly-Clark Corporation, for example, has developed a virtual reality system to view store aisles carrying its products.[67] This allows executives to see how Kimberly-Clark products look in stores aside competing products. The virtual reality view of store aisles should help executives monitor customer behavior and determine the best packaging and placement of their products on store aisles. Boeing uses virtual reality to help it design and manufacture airplane parts and new planes, including the 787 Dreamliner. Boeing uses 3D PLM from Dassault Systems.[68] Clothing and fashion companies, such as Neiman Marcus and Saks Fifth Avenue, are using virtual reality on the Internet to display and promote new products and fashions.[69] In another Web application, virtual reality was used to design a $90 million addition to the Denver Art Museum. The software can also show the picture, length, and diameter of the 50,000 bolts that are being used. Palomar Pomerado Health used Second Life to create a virtual hospital when it started construction of a real $700 million hospital in California in 2007.[70] The purpose of the Second Life virtual hospital was to show clients and staff the layout and capabilities of the new hospital. Second Life has also been used in business and recruiting.

Boeing uses virtual reality to collaborate with customers during aircraft design.

(Source: AP Photo/Ted S. Warren.)

Entertainment

Computer-generated image technology, or CGI, has been around since the 1970s. Many movies use this technology to bring realism to the silver screen, including *Finding Nemo*, *Spider-Man II*, and *Star Wars Episode II—Attack of the Clones*. A team of artists rendered the roiling seas and crashing waves of *Perfect Storm* almost entirely on computers using weather reports, scientific formulas, and their imagination. Other films include *Dinosaur* with its realistic talking reptiles, *Titan A.E.*'s beautiful 3-D space-scapes, and the casts of computer-generated crowds and battles in *Gladiator* and *The Patriot*. CGI can also be used for sports simulation to enhance the viewers' knowledge and enjoyment of a game. SimCity (*http://simcity.ea.com/*), a virtual reality game, allows people to experiment with decisions related to urban planning. Natural and man-made disasters test decisions on designing buildings and the surrounding area. Other games can display a 3-D view of the world and allow people to interact with simulated people or avatars in the game. Second Life (*www.secondlife.com*) allows people to play games, interact with avatars, and build structures, such as homes.[71]

INFORMATION SYSTEMS @ WORK

Realtors Rely on Virtual Reality

Virtual reality allows you to experience places, to some degree, without the inconvenience of travel. Although a trip to a virtual location is not as rich of an experience as actually being there, virtual reality sometimes provides valuable information all the same. For example, consider the information provided by virtual earth software from Microsoft and Google.

In its initial release, virtual earth software stitched together high-resolution satellite imagery available from commercial providers to let users scroll and pan around the Earth from a satellite view. The effect was breathtaking, though flat. Virtual earth developers began building photorealistic, geospecific, 3-D landscapes that allowed users to zoom in on satellite images and fly horizontally through virtual landscapes that replicated the landscape of the Earth and its cities. The effort to virtualize the world is ongoing, with thousands of people providing assistance in adding cities and buildings to the virtual landscape.

It didn't take long for virtual earth applications to move from novelty to serious business tool. Professionals in the real-estate industry were quick to acknowledge the value of visiting neighborhoods virtually. Seain Conover points out that while a real-estate agent's photo may show a quaint house in the country, virtual earth would let you see that it's actually in the shadow of a five-story apartment building.

Conover works for Terasoft Corporation, a Canadian company based in British Columbia, that specializes in Multiple Listing System (MLS) systems. The MLS allows realtors to list houses for other Realtors and house-hunters to find. Through the MLS, realtors can communicate their needs and recommend properties to others in the business. Until now, the MLS has provided home photos and specifications, but virtual earth is changing all of that.

Terasoft and other software companies around the world are working with virtual earth providers such as Microsoft and Google to build commercial applications using virtual earth as a foundation. Using an application developed by Terasoft on Microsoft Virtual Earth, a realtor can take a client to a prospective property virtually, zoom down to the rooftop, and then turn up 45 degrees to view the building, property, and neighborhood from all sides. What used to take days of driving around is now condensed into an hour. Clients and realtors can quickly narrow the market to a few houses that match the client's interests.

Terasoft has built an overlay for Microsoft Virtual Earth that specifies items of interest such as school districts, demographics, and crime rate. Using color-coding, home shoppers can find neighborhoods they desire. Marking a geographic area on the map with the mouse quickly displays the available houses in that area in the shopper's price range. Rather than scrolling through thousands of listings in the MLS, realtors can search the virtual landscape for specific needs, see them as pinpoints on a map, zoom in, and view the house in 3-D while analyzing information about the house and its neighborhood.

Terasoft chose Microsoft for this project over Google due to Microsoft's long history of developing software for the real-estate industry. Another contributing factor is that Microsoft's mapping and virtual earth software has clearly been defined as an enterprise platform.

The use of virtual reality in real estate doesn't end at a house's front door. Now through 360-degree photography, customers can inspect a home's interior as well. A virtual walkthrough combined with a floor plan reveals the layout and condition of the interior. As virtual reality technologies mature, finding the home of your dreams from your computer screen or even a VR headset is quickly becoming possible. You only need to visit your dream house to make sure that the virtual-reality experience is a true one.

Real-estate professionals all over the world are turning to virtual earth software to revolutionize their business. Professionals in the public sector, hospitality and travel, retail, financial services, manufacturing, utilities, oil and gas, and media and entertainment are also applying virtual earth technology to their industries. The ability to view remote locations through virtual reality is proving to help boost productivity and build customer satisfaction.

Discussion Questions

1. What conveniences does a virtual reality system such as the one developed by Terasoft provide for realtors and home shoppers?
2. Describe the work required to develop and maintain a virtual model of the Earth and its towns and cities.

Critical Thinking Questions

1. If home shoppers can access tools such as the one provided by Terasoft, why are realtors necessary? How might realtors change their job description to maintain their value to customers?
2. Many industries use Virtual Earth. Provide a few examples of how you think they might apply its technology. How might they customize Virtual Earth to their needs?

SOURCES: Lau, Kathleen, "ISV aims Virtual Earth at realtors," *IT World Canada*, December 12, 2007, *www.itworldcanada.com/a/Enterprise-Business-Applications/cf429ae4-e618-4281-90b5-0cc5596cd234.html*; Terasoft Web site, *www.terasoft.com/company*, accessed July 4, 2008; Microsoft Virtual Earth Web site, *www.microsoft.com/virtualearth/industry/realestate.aspx*, accessed July 4, 2008.

OTHER SPECIALIZED SYSTEMS

In addition to artificial intelligence, expert systems, and virtual reality, other interesting specialized systems have appeared. Segway, for example, is an electric scooter that uses sophisticated software, sensors, and gyro motors to transport people through warehouses, offices, downtown sidewalks, and other spaces (*www.segway.com*). Originally designed to transport people around a factory or around town, more recent versions are being tested by the military for gathering intelligence and transporting wounded soldiers to safety. The military and DARPA are developing energy-efficient, mechanical computers that have the ability to operate in environments that are too harsh for traditional chip-based computers.[72] A Japanese company is experimenting with using specially designed floor mats that contain wires and other electronic components to generate electricity when people step on them.[73] The 3VR Security (*www.3vr.com*) system makes a video face-recognition system to identify people from pictures or images.[74] According to a security officer for the Bank of Hawaii, "It seemed too good to be true, but since being installed at bank branches in December, it has reduced our surveillance time, and it's especially useful in tracking multiple transactions by an ID thief."

A number of special-purpose systems are now available in vehicles. Ford Motor Company and Microsoft have developed a voice-activated system called *Sync* that can play music, make phones calls, and more.[75] The Advanced Warning System by Mobileye warns drivers to keep a safe distance from other vehicles and drivers.[76] Automotive software allows cars and trucks to connect to the Internet. The software can track a driver's speed and location, allow gas stations to remotely charge for fuel and related services, and more.

Many new computing devices, such as Microsoft's Surface, are also becoming available.[77] The Surface is a touch-screen computer that uses a glass-top display. It looks like a coffee table or dining room table with a built-in computer. Microsoft's Smart Personal Objects Technology (SPOT) allows small devices to transmit data and messages over the air. SPOT is being used in wrist watches to transmit data and messages over FM radio broadcast bands. The new technology, however, requires a subscription to the Microsoft MSN Direct information service. Some manufacturing is also being done with inkjet printers to allow them to "print" 3-D parts. For example, the printer sprays layers of polymers onto circuit boards to form transistors and other electronic components. Some new computers can even be worn on your body. Smith Drug, for example, uses a wearable computer by Vocollect, Inc. (*www.vocollect.com*) to help its employees monitor inventory levels.[78] The waist-worn computer that includes a headset with a microphone and speaker dramatically increases productivity and helps eliminate errors. According to a corporate executive, "Previously, they had a clipboard with 25 items per sheet. Now, they don't have to look at the paper. Their hands are free, and all they have to do is listen and think."

Increasingly, companies are using special-purpose tracking devices, chips, and bar codes.[79] As mentioned previously, *Radio Frequency Identification (RFID)* tags that contain small chips with information about products or packages can be quickly scanned to perform inventory control or trace a package as it moves from a supplier to a company to its customers.[80] Many companies have used RFID tags to reduce costs, improve customer service, and achieve a competitive advantage.[81] When attached to clothing and worn close to a mirror, some RFID tags will display sizes, styles, color, suggested accessories, and images of models wearing the clothing on the mirror or a display screen.[82] RFID tags are even used to help track lost airline luggage.[83] The state of Colorado uses RFID to track elk herds. Farmers are looking into using these tags to track cattle to help identify and control mad cow disease. An Italian cheese consortium uses RFID tags in the crust of cheese wheels. The RFID tags contain information about when and where the cheese was made to ensure freshness and avoid spoilage. Two German students have developed a smart beer mat, which uses sensor chips to help determine the weight or amount of beer in a glass or beer mug. When the chips sense that the beer mug is nearly empty, the sensor chip sends an alert to a computer monitor telling the bartender that a customer needs more beer. The endorsement of an electronic product code standard will likely make RFID even more popular.[84]

Special-purpose bar codes are also being introduced in a variety of settings. For example, to manage office space efficiently, a company gives each employee and office a bar code. Instead of having permanent offices, the employees are assigned offices and supplies as needed, and the bar codes help to make sure that an employee's work, mail, and other materials are routed to the right place. Companies can save millions of dollars by reducing office space and supplies. Another technology is being used to create "smart containers" for ships, railroads, and trucks. NaviTag (*http://navitag.com/*) and other companies are developing communications systems that allow containers to broadcast the contents, location, and condition of shipments to shipping and cargo managers. A railroad company can use standard radio messages to generate shipment and tracking data for customers and managers.

Navitag, an electronic security device, is attached to a cargo container door and monitors whether the door opens and whether light, radiation, or carbon monoxide enters the container.

(Source: AP Images.)

game theory
The use of information systems to develop competitive strategies for people, organizations, or even countries.

One special application of computer technology is derived from a branch of mathematics called game theory. **Game theory** involves the use of information systems to develop competitive strategies for people, organizations, or even countries. Two competing businesses in the same market can use game theory to determine the best strategy to achieve their goals. The military could also use game theory to determine the best military strategy to win a conflict against another country, and individual investors could use game theory to determine the best strategies when competing against other investors in a government auction of bonds. Groundbreaking work on game theory was pioneered by John Nash, the mathematician whose life was profiled in the book and film *A Beautiful Mind.* Game theory has also been used to develop approaches to deal with terrorism. The Los Angeles airport is experimenting with the use of game theory to help security guards do a better job patrolling sensitive areas.[85]

informatics
A specialized system that combines traditional disciplines, such as science and medicine, with computer systems and technology.

Informatics, another specialized system, combines traditional disciplines, such as science and medicine, with information systems and technology. *Bioinformatics*, for example, combines biology and computer science. Also called *computational biology*, bioinformatics has been used to help map the human genome and conduct research on biological organisms. Using sophisticated databases and artificial intelligence, bioinformatics helps unlock the secrets of the human genome, which could eventually prevent diseases and save lives. Stanford University has a course on bioinformatics and offers a bioinformatics certification. Medical informatics combines traditional medical research with computer science. Journals, such as *Healthcare Informatics*, report current research on applying computer systems and technology to reduce medical errors and improve healthcare. The University of Edinburgh even has a School of Informatics (*www.ed.ac.uk/about/structure/informatics.html*). The school has courses on the structure, behavior, and interactions of natural and artificial computational systems. The program combines artificial intelligence, computer science, engineering, and science.

SUMMARY

Principle

Knowledge management allows organizations to share knowledge and experience among their managers and employees.

Knowledge is an awareness and understanding of a set of information and the ways that information can be made useful to support a specific task or reach a decision. A knowledge management system (KMS) is an organized collection of people, procedures, software, databases, and devices used to create, store, share, and use the organization's knowledge and experience. Explicit knowledge is objective and can be measured and documented in reports, papers, and rules. Tacit knowledge is hard to measure and document and is typically not objective or formalized.

Knowledge workers are people who create, use, and disseminate knowledge. They are usually professionals in science, engineering, business, and other areas. The chief knowledge officer (CKO) is a top-level executive who helps the organization use a KMS to create, store, and use knowledge to achieve organizational goals. Some organizations and professions use communities of practice (COP) to create, store, and share knowledge. A COP is a group of people or a community dedicated to a common discipline or practice, such as open-source software, auditing, medicine, engineering, and other areas.

Obtaining, storing, sharing, and using knowledge is the key to any KMS. The use of a KMS often leads to additional knowledge creation, storage, sharing, and usage. Many tools and techniques can be used to create, store, and use knowledge. These tools and techniques are available from IBM, Microsoft, and other companies and organizations.

Principle

Artificial intelligence systems form a broad and diverse set of systems that can replicate human decision making for certain types of well-defined problems.

The term *artificial intelligence* is used to describe computers with the ability to mimic or duplicate the functions of the human brain. The objective of building AI systems is not to replace human decision making completely but to replicate it for certain types of well-defined problems.

Intelligent behavior encompasses several characteristics, including the abilities to learn from experience and apply this knowledge to new experiences; handle complex situations and solve problems for which pieces of information might be missing; determine relevant information in a given situation, think in a logical and rational manner, and give a quick and correct response; and understand visual images and process symbols. Computers are better than people at transferring information, making a series of calculations rapidly and accurately, and making complex calculations, but human beings are better than computers at all other attributes of intelligence.

Artificial intelligence is a broad field that includes several key components, such as expert systems, robotics, vision systems, natural language processing, learning systems, and neural networks. An expert system consists of the hardware and software used to produce systems that behave as a human expert would in a specialized field or area (e.g., credit analysis). Robotics uses mechanical or computer devices to perform tasks that require a high degree of precision or are tedious or hazardous for humans (e.g., stacking cartons on a pallet). Vision systems include hardware and software that permit computers to capture, store, and manipulate images and pictures (e.g., face-recognition software). Natural language processing allows the computer to understand and react to statements and commands made in a "natural" language, such as English. Learning systems use a combination of software and hardware to allow a computer to change how it functions or reacts to situations based on feedback it receives (e.g., a computerized chess game). A neural network is a computer system that can simulate the functioning of a human brain (e.g., disease diagnostics system). A genetic algorithm is an approach to solving large, complex problems in which a number of related operations or models change and evolve until the best one emerges. The approach is based on the theory of evolution, which requires variation and natural selection. Intelligent agents consist of programs and a knowledge base used to perform a specific task for a person, a process, or another program.

Principle

Expert systems can enable a novice to perform at the level of an expert but must be developed and maintained very carefully.

An expert system consists of a collection of integrated and related components, including a knowledge base, an inference engine, an explanation facility, a knowledge acquisition facility, and a user interface. The knowledge base is an extension of a database, discussed in Chapter 5, and an information and decision support system, discussed in Chapter 10. It contains all the relevant data, rules, and relationships used in the expert system. The rules are often composed of if-then statements, which are used for drawing conclusions. Fuzzy logic allows expert systems to incorporate facts and relationships into expert system knowledge bases that might be imprecise or unknown.

The inference engine processes the rules, data, and relationships stored in the knowledge base to provide answers, predictions, and suggestions the way a human expert would. Two common methods for processing include backward and forward chaining. Backward chaining starts with a conclusion, then searches for facts to support it; forward chaining starts with a fact, then searches for a conclusion to support it.

The explanation facility of an expert system allows the user to understand what rules were used in arriving at a decision. The knowledge acquisition facility helps the user add or update knowledge in the knowledge base. The user interface makes it easier to develop and use the expert system.

The people involved in the development of an expert system include the domain expert, the knowledge engineer, and the knowledge users. The domain expert is the person or group who has the expertise or knowledge being captured for the system. The knowledge engineer is the developer whose job is to extract the expertise from the domain expert. The knowledge user is the person who benefits from the use of the developed system.

The steps involved in the development of an expert system include determining requirements, identifying experts, constructing expert system components, implementing results, and maintaining and reviewing the system.

Expert systems can be implemented in several ways. Previously, traditional high-level languages, including Pascal, FORTRAN, and COBOL, were used. LISP and PROLOG are two languages specifically developed for creating expert systems from scratch. A faster and less-expensive way to acquire an expert system is to purchase an expert system shell or existing package. The shell program is a collection of software packages and tools used to design, develop, implement, and maintain expert systems.

The benefits of using an expert system go beyond the typical reasons for using a computerized processing solution. Expert systems display "intelligent" behavior, manipulate symbolic information and draw conclusions, provide portable knowledge, and can deal with uncertainty. Expert systems can be used to solve problems in many fields or disciplines and can assist in all stages of the problem-solving process. Past successes have shown that expert systems are good at strategic goal setting, planning, design, decision making, quality control and monitoring, and diagnosis.

Applications of expert systems and artificial intelligence include credit granting and loan analysis, catching cheats and terrorists, budgeting, games, information management and retrieval, AI and expert systems embedded in products, plant layout, hospitals and medical facilities, help desks and assistance, employee performance evaluation, virus detection, repair and maintenance, shipping, and warehouse optimization.

Principle

Virtual reality systems can reshape the interface between people and information technology by offering new ways to communicate information, visualize processes, and express ideas creatively.

A virtual reality system enables one or more users to move and react in a computer-simulated environment. Virtual reality simulations require special interface devices that transmit the sights, sounds, and sensations of the simulated world to the user. These devices can also record and send the speech and movements of the participants to the simulation program. Thus, users can sense and manipulate virtual objects much as they would real objects. This natural style of interaction gives the participants the feeling that they are immersed in the simulated world.

Virtual reality can also refer to applications that are not fully immersive, such as mouse-controlled navigation through a three-dimensional environment on a graphics monitor, stereo viewing from the monitor via stereo glasses, stereo projection systems, and others. Some virtual reality applications allow views of real environments with superimposed virtual objects. Virtual reality applications are found in medicine, education and training, real estate and tourism, and entertainment.

Principle

Specialized systems can help organizations and individuals achieve their goals.

A number of specialized systems have recently appeared to assist organizations and individuals in new and exciting ways. Segway, for example, is an electric scooter that uses sophisticated software, sensors, and gyro motors to transport people through warehouses, offices, downtown sidewalks, and other spaces. Originally designed to transport people around a factory or around town, more recent versions are being tested by the military for gathering intelligence and transporting wounded soldiers to safety. Radio Frequency Identification (RFID) tags are used in a variety of settings. Game theory involves the use of information systems to develop competitive strategies for people, organizations, and even countries. Informatics combines traditional disciplines, such as science and medicine, with computer science. Bioinformatics and medical informatics are examples. A number of special-purpose telecommunications systems can be placed in products for varied uses.

CHAPTER 11: SELF-ASSESSMENT TEST

Knowledge management allows organizations to share knowledge and experience among their managers and employees.

1. _______________ are people who create, use, and disseminate knowledge and are typically professionals in business, science, engineering, or another area.
2. What type of knowledge is objective and can be measured and documented in reports, papers, and rules?
   a. tacit
   b. descriptive
   c. prescriptive
   d. explicit
3. A community of practice (COP) is a group of people or a community dedicated to a common discipline or practice, such as open-source software, auditing, medicine, engineering, and other areas. True or False?

Artificial intelligence systems form a broad and diverse set of systems that can replicate human decision making for certain types of well-defined problems.

4. The Turing Test attempts to determine whether the responses from a computer with intelligent behavior are indistinguishable from responses from a human. True or False?
5. _______________ are rules of thumb arising from experience or even guesses.
6. What is *not* an important attribute for artificial intelligence?
   a. the ability to understand visual images
   b. the ability to learn from experience
   c. the ability to be creative
   d. the ability to make complex calculations
7. _______________ involves mechanical or computer devices that can paint cars, make precision welds, and perform other tasks that require a high degree of precision or are tedious or hazardous for human beings.
8. What branch of artificial intelligence involves a computer understanding and reacting to statements in English or another language?
   a. expert systems
   b. neural networks
   c. natural language processing
   d. vision systems
9. A(n) _______________ is a combination of software and hardware that allows the computer to change how it functions or reacts to situations based on feedback it receives.

Expert systems can enable a novice to perform at the level of an expert but must be developed and maintained very carefully.

10. What is a disadvantage of an expert system?
    a. the inability to solve complex problems
    b. the inability to deal with uncertainty
    c. limitations to relatively narrow problems
    d. the inability to draw conclusions from complex relationships
11. A(n) _______________ is a collection of software packages and tools used to develop expert systems that can be implemented on most popular PC platforms to reduce development time and costs.
12. A heuristic consists of a collection of software and tools used to develop an expert system to reduce development time and costs. True or False?
13. What stores all relevant information, data, rules, cases, and relationships used by the expert system?
    a. the knowledge base
    b. the data interface
    c. the database
    d. the acquisition facility
14. A disadvantage of an expert system is the inability to provide expertise needed at a number of locations at the same time or in a hostile environment that is dangerous to human health. True or False?
15. What allows a user or decision maker to understand how the expert system arrived at a certain conclusion or result?
    a. domain expert
    b. inference engine
    c. knowledge base
    d. explanation facility
16. An important part of an expert system is the _______________, which allows a user or decision maker to understand how the expert system arrived at certain conclusions or results.
17. In an expert system, the domain expert is the individual or group who has the expertise or knowledge one is trying to capture in the expert system. True or False?

Virtual reality systems can reshape the interface between people and information technology by offering new ways to communicate information, visualize processes, and express ideas creatively.

18. A(n) ____________________ enables one or more users to move and react in a computer-simulated environment.
19. What type of virtual reality is used to make human beings feel as though they are in a three-dimensional setting, such as a building, an archaeological excavation site, the human anatomy, a sculpture, or a crime scene reconstruction?
    a. chaining
    b. relative
    c. immersive
    d. visual

Specialized systems can help organizations and individuals achieve their goals.

20. ____________________ involves the use of information systems to develop competitive strategies for people, organizations, or even countries.

CHAPTER 11: SELF-ASSESSMENT TEST ANSWERS

(1) knowledge workers (2) d (3) True (4) True (5) Heuristics (6) d (7) Robotics (8) c (9) learning system (10) c (11) expert system shell (12) False (13) a (14) False (15) d (16) explanation facility (17) True (18) virtual reality system (19) c (20) game theory

REVIEW QUESTIONS

1. What is a knowledge management system?
2. What is a community of practice?
3. What is the difference between knowledge and information?
4. What is a vision system? Discuss two applications of such a system.
5. What is natural language processing? What are the three levels of voice recognition?
6. Describe three examples of the use of robotics. How can a microrobot be used?
7. What is a learning system? Give a practical example of such a system.
8. What is a neural network? Describe two applications of neural networks.
9. Under what conditions is the development of an expert system likely to be worth the effort?
10. Identify the basic components of an expert system and describe the role of each.
11. What is fuzzy logic?
12. What is virtual reality? Give several examples of its use.
13. Expert systems can be built based on rules or cases. What is the difference between the two?
14. Describe the roles of the domain expert, the knowledge engineer, and the knowledge user in expert systems.
15. What is informatics? Give a few examples.
16. Describe three applications of expert systems or artificial intelligence.
17. Identify three special interface devices developed for use with virtual reality systems.
18. Identify and briefly describe three specific virtual reality applications.
19. What is informatics? How is it used?
20. Give three examples of other specialized systems.

DISCUSSION QUESTIONS

1. What are the requirements for a computer to exhibit human-level intelligence? How long will it be before we have the technology to design such computers? Do you think we should push to try to accelerate such a development? Why or why not?
2. You work for an insurance company as an entry-level manager. The company contains both explicit and tacit knowledge. Describe the types of explicit and tacit knowledge that might exist in your insurance company. How would you capture each type of knowledge?
3. Describe a knowledge management system for a college or university.
4. What are some of the tasks at which robots excel? Which human tasks are difficult for them to master? What fields of AI are required to develop a truly perceptive robot?
5. Describe how natural language processing could be used in a university setting.
6. Discuss how learning systems can be used in a military war simulation to train future officers and field commanders.
7. You have been hired to develop an expert system for a university career placement center. Develop five rules a student could use in selecting a career.
8. What is the relationship between a database and a knowledge base?

9. Imagine that you are developing the rules for an expert system to select the strongest candidates for a medical school. What rules or heuristics would you include?
10. Describe how informatics can be used in a business setting.
11. Which interface is the least developed and most challenging to create in a virtual reality system? Why do you think this is so?
12. What application of virtual reality has the most potential to generate increased profits in the future?
13. Describe a situation where game theory would be appropriate and could be used.

PROBLEM-SOLVING EXERCISES

1. You are a senior vice president of a company that manufactures kitchen appliances. You are considering using robots to replace up to ten of your skilled workers on the factory floor. Using a spreadsheet, analyze the costs of acquiring several robots to paint and assemble some of your products versus the cost savings in labor. How many years would it take to pay for the robots from the savings in fewer employees? Assume that the skilled workers make $20 per hour, including benefits.
2. Assume that you have just won a lottery worth $100,000. You have decided to invest half the amount in the stock market. Develop a simple expert system to pick ten stocks to consider. Using your word processing program, create seven or more rules that could be used in such an expert system. Create five cases and use the rules you developed to determine the best stocks to pick.
3. Using a graphics program, develop a diagram that shows a KMS for a college or university.

TEAM ACTIVITIES

1. Do research with your team to identify KMSs in three different businesses or nonprofit organizations. Describe the types of tacit and explicit knowledge that would be needed by each organization or business.
2. Form a team and debate other teams from your class on the following topic: "Are expert systems superior to human beings when it comes to making objective decisions?" Develop several points supporting either side of the debate.
3. Have your team members explore the use of a special-purpose system in an industry of your choice. Describe the advantages and disadvantages of this special-purpose system.

WEB EXERCISES

1. Use the Internet to find information about the use of robotics. Describe three examples of how this technology is used.
2. This chapter discussed several examples of expert systems. Search the Internet for two examples of the use of expert systems. Which one has the greatest potential to increase profits for the firm? Explain your choice.
3. Use the Internet to get information about the application of game theory in business or the military. Write a report about what you found.

CAREER EXERCISES

1. Describe how a COP can be used to help advance your career.
2. Describe the future of artificial intelligence in a career area of your choice.

CASE STUDIES

Case One

Bird & Bird Have Knowledge in Hand

Bird & Bird (B&B) is an international commercial law firm that focuses on industries including aviation and aerospace, financial services, communications, e-commerce, IT, life sciences, media, and sport. The firm has offices in Beijing, Brussels, Dusseldorf, Frankfurt, The Hague, Helsinki, Hong Kong, London, Lyon, Madrid, Milan, Paris, Rome, and Stockholm.

B&B lawyers wanted to be able to access case histories and other legal reference materials online from any of the firm's 14 offices. The firm had a system called Solutions Lab designed to perform this task, but the system had failed to keep up with improvements in search technologies. B&B lawyers wanted a more powerful system.

The knowledge management (KM) team at B&B collaborated with the information systems specialists to incorporate cutting-edge technologies into a new knowledge management system for the firm. They began by conducting focus groups to learn about the needs of the lawyers. All 14 offices were involved, and the results provided innovative ideas for serving the law firm's needs.

Next, the team evaluated off-the-shelf KM products to see if any would suit their needs—unfortunately, none did. The KM team decided to custom design the KM system in-house and commission an external company to build it. They selected UCLogic, a document and knowledge management systems company.

B&B lawyers wanted more powerful searching capabilities for finding topic-related content within documents stored in the KM system. The KM team found a search technology called conceptSearching that provides more flexibility than traditional keyword search. ConceptSearching allows the user to enter natural sentences that might include several key terms or topics and then applies artificial intelligence for impressive search results.

After a few months of testing and refining the conceptSearching technology to the favor of the firm, the team designed the user interface for the new KM system. The resulting system was designed by the KM team, implemented by UCLogic, and incorporated by conceptSearching. The KM team saved time and effort by using the internal and external repositories designed for use in the previous document management system. The new system was implemented gradually to ensure a smooth transition.

Using the new "know-how" KM system, lawyers can now search in two areas: the firm's own internal document repositories, holding the relevant experience of those working in the firm, and external sources to which the firm subscribes such as LexisNexis. Using powerful AI technologies applied through conceptSearching, the system yields tabulated results ranked by relevance. A second list of related topics is presented in a sidebar. Articles are ranked by "know-who"—the amount of hours invested in the work. The system allows the users to adjust search relevancy rankings to further refine the quality of results.

Lawyers in the firm receive a one-page guide on using the system along with one-on-one training sessions as needed. Besides basic guides to the steps, the training has allowed the development team to meet with the lawyers and solicit new suggestions for the system. It has also increased overall usage by encouraging some lawyers who do not use online systems to try it at least once. Due to the increased contact between developers and users, the number of documents being submitted to the system is reaching record highs. Shortly after its successful roll out, the KM team is already hard at work on improvements to add in the next round of development.

Discussion Questions

1. How do knowledge management systems assist B&B lawyers in their research for cases?
2. What additional power does B&B's new "know-how" system provide for its lawyers?

Critical Thinking Questions

1. Why do you think the lawyers at B&B found traditional keyword searching insufficient for their needs?
2. What further benefits might the KM team at B&B design into its next system?

SOURCES: McQuay, Martha, "Know-how: Ushering in the next generation," *PLC Magazine*, July 24, 2007, *http://plc.practicallaw.com/0-374-0976* and *www.uclogic.com/Articles/TwoBirds.pdf*; UCLogic Web site, http://www.uclogic.com, accessed July 5, 2008; Bird & Bird Web site, *www.twobirds.com*, accessed July 5, 2008; ConceptSearching Web site, *www.conceptsearching.com/web*, accessed July 5, 2008.

Case Two

Where Virtual Worlds and AI Collide

The use of information systems saves workers and businesses countless hours of tedious labor. Still, users of information systems often become frustrated with the system's inability to grasp simple common sense knowledge. "Why can't you understand what I need?" is a typical response to a computer incapable of working outside the framework for which it was designed.

The solution is to create computer systems that use AI and guess what the user needs provided with little input or prompting. However, true artificial intelligence has challenged computer scientists for decades. Most researchers believe that to create a thinking machine, the machine must have a physical presence with which to experience its environment. Researchers have used robotics to try this approach, but it is costly, especially since robotics technology is still in its infancy.

Many AI researchers are turning to virtual worlds to provide their systems with an environment from which to experience life. Virtual worlds such as Second Life provide a virtual landscape for people to explore through the use of avatars, characters within the environment that the user controls. Second Life allows users to build houses and businesses and even sell products to other avatars. It strives to mirror the physical, social, and economic sense of the real world. Because Second Life is not bound by the laws of physics, users perform actions that aren't possible in real life, such as flying.

Second Life presents the perfect environment for AI systems to experience the world and then learn from those experiences. Novamente LLC is one AI company that has created AI-driven avatars in a virtual world. These avatars appear as animals that are eager to learn. For example, a Novamente dog avatar can be taught to play soccer. Through the use of praise and correction, the AI system will learn how to play the game, including its rules and strategies. Novamente also has a Parrot avatar that is learning language skills by talking with people.

While in Second Life, you might run into Edd, an AI avatar created by researchers at Rensselaer Polytechnic Institute. Edd can converse and reason, although he has the intelligence of a four year old. Even that much intelligence takes an immense amount of complex calculus to accomplish. Still, Edd can communicate and influence a real user's actions.

Many AI avatars are moving to Second Life and other virtual worlds. AI researchers find it to be an ideal environment for training AI systems and allowing them to interact with real people through virtual avatars. Michael Mateas, a computer science professor at the University of California, Santa Cruz, says, "It's a fantastic sweet spot—not too simple, not too complicated, high cultural value." But how will these AI systems service people and businesses?

Selmer Bringsjord, head of Rensselaer's Cognitive Science Department and leader of the research project, sees the research applying to practical needs in other virtual environments such as entertainment and gaming, as well as immersive training and education. "The apps, frankly, are endless," Bringsjord said. "Imagine being able to step into a simulation environment in which you interact with synthetic characters as sophisticated as those seen in Star Trek's holodeck."

Consider other uses of AI avatars in virtual worlds. AI systems might be used to work on the behalf of businesses. For example, a salesperson could create a thousand avatars of himself and send the team out to sell products. Other AI systems could be used to collect information and survey the population for marketing or other uses. At the same time, rules must be developed to govern the use of virtual reality in business. For example, an avatar should identify itself as a virtual being so that people do not assume it is human.

Discussion Questions

1. Why are AI features and avatars proliferating in Second Life and other virtual worlds?
2. What types of research are being conducted using AI avatars?

Critical Thinking Questions

1. Why might AI avatars eventually make Second Life users uncomfortable? What might be done to calm the fears?
2. What types of business applications might be provided by AI avatars in Second Life?

SOURCES: Tay, Liz, "Child-like intelligence created in Second Life," *ITNews*, March 14, 2008, *www.itnews.com.au/News/72057,childlike-intelligence-created-in-second- life.aspx*; Hill, Michael, "'Second Life' is frontier for AI research," MSNBC, May 18, 2008, *www.msnbc.msn.com/id/24668099*; Havenstein, Heather, "Virtual worlds making artificial intelligence apps 'smarter'," *Computerworld*, September 13, 2007, *www.computerworld.com/action/article.do?command=viewArticleBasic&taxonomyId=11&articleId=9036438&intsrc=hm_topic.*

Questions for Web Case

See the Web site for this book to read about the Whitmann Price Consulting case for this chapter. The following are questions concerning this Web case.

Whitmann Price Consulting: Knowledge Management and Specialized Information Systems

Discussion Questions

1. List three forms of AI that are being considered for the AMCI system and how they will be used.
2. List the advantages and disadvantages of implementing the AI systems in the AMCI system.

Critical Thinking Questions

1. What types of considerations might Josh and Sandra take into account when deciding which AI system to include?
2. How might Whitmann Price consultants react when they learn about the Presence system that will track their location? Why?

NOTES

Sources for the opening vignette: Gensym Staff, "Ericsson", Gensym Success Stories, 2008, *www.gensym.com/?p=success_stories&id=13*; Gensym Web site, *www.gensym.com*, accessed July 2, 2008; Ericsson Web site, *www.wricsson.com* accessed July 2, 2008.

1 Vorro, Alex, "Knowledge Management: Searching For Knowledge," *Insurance Networking News*, October 1, 2007, p. 1.
2 Chua, Alton, "The Curse of Success," *The Wall Street Journal*, April 28, 2007, p. R8.
3 King, William, "Knowledge Management and Organizational Learning," *Omega*, April 2007, p. 167.
4 Asawanipont, Nitida, "Timely Tips from Tata," *The Nation*, January 22, 2007, p. 1B.
5 Short, Tim and Schwendinger, Jim, "How To Win the War For Talent," *Aviation Week*, February 5, 2007, p. 54.
6 Seneviratne, Vienna, "IAEA Team Finds Canada Plants Using Comprehensive KM Practices," *Inside NRC*, May 28, 2007, p. 16.
7 McCormick, John, "5 Big Companies That Got Knowledge Management Right," *CIO Magazine*, October 5, 2007.
8 McCormick, John, "5 Big Companies That Got Knowledge Management Right," *CIO Magazine*, October 5, 2007.
9 Nguyen, Le, et al., "Acquiring Tacit and Explicit Marketing Knowledge from Foreign Partners in IJVs," *Journal of Business Research*, November 2007, p. 1152.
10 McGregor, Jena and Hamm, Steve, "Managing the Global Workforce," *Business Week*, January 28, 2008, p. 34.
11 Staff, "Remedy Interactive Adds CFO and CKO to Management Team," *Market Wire*, March 12, 2007.
12 Vorro, Alex, "Knowledge Management: Searching for Knowledge," *Insurance Networking News*, October 1, 2007, p. 1.
13 Seneviratne, Gamini, "Knowledge Management Now Seen as a Priority," *Nuclear News*, September 2007, p. 59.
14 King, William, "Motivating Knowledge Sharing Through a Knowledge Management System," *Omega*, February 2007, p. 131.
15 Davenport, T., et al., "Knowledge Management Can Make a Difference," *The Wall Street Journal*, March 10, 2008, p. R11.
16 Blackman, Andrew, "Dated and Confused," *The Wall Street Journal*, May 14, 2007, p. R5.
17 Vorro, Alex, "Knowledge Management: Searching for Knowledge," *Insurance Networking News*, October 1, 2007, p. 1.
18 Martin, Richard, "Collaboration Cisco Style," *Information Week*, January 28, 2008, p. 30.
19 Beizer, Doug, "SI Gets Army Knowledge Management Deal," *Washington Technology*, September 26, 2007.
20 Staff, "University of South Carolina, Collexis Announce Collaboration on Knowledge Management Applications," *Biotech Business Week*, February 19, 2007, p. 168.
21 McCormick, John, "5 Big Companies That Got Knowledge Management Right," *CIO Magazine*, October 5, 2007.
22 *www.cortexpro.com*, accessed November 21, 2007.
23 *www.delphigroup.com*, accessed November 21, 2007.
24 *www.kmresource.com*, accessed November 21, 2007
25 *www.kmsi.us*, accessed November 25, 2007.
26 *www.knowledge-manage.com*, accessed November 25, 2007.
27 *www.lawclip.com*, accessed November 25, 2007.
28 *www.metakm.com*, accessed November 25, 2007.
29 Snider, Mike, "AI is OK in New Games," *USA Today*, September 25, 2007, p. 3D.
30 Miller, Stephen, "MIT's Professor's Work Led Him to Preach the Evils of Computers," *The Wall Street Journal*, March 15, 2008, p. A6.
31 O'Keefe, Brian, "The Smartest, the Nuttiest Futurist on Earth," *Fortune*, May 14, 2007, p. 60.
32 Staff, "Ray Kurzweil Joins the Singularity Institute's Board of Directors," *Business Wire*, May 20, 2007.
33 *www.20q.net*, accessed November 26, 2007.
34 Staff, "VIASPACE Leveraging SHINE Technology," *PR Newswire US*, February 14, 2007.
35 Anders, George, "The Winding Road to the Robotic Future," *The Wall Street Journal*, March 16, 2007, p. W6.
36 Abate, Tom, "Future Moving From I, Robot to My Robot," *Rocky Mountain News*, February 26, 2007, p. 8.
37 Anthes, Gary "I, Coach," *Computerworld*, May 21, 2007, p. 38.
38 *www.irobot.com*, accessed November 23, 2007.
39 *www.cs.cmu.edu/~rll*, accessed July 18, 2006.
40 *www.porterhospital.org*, accessed November 23, 2007.
41 Staff, "Worming Its Way into Our Hearts," *Business Week*, May 7, 2007, p. 79.
42 *www.darpagrandchallenge.com*, accessed November 23, 2007.
43 Kelly, Tim, "Robot Race," *Forbes*, September 3, 2007, p. 39.
44 Staff, "The Military Machine," *Rocky Mountain News*, July 14, 2007, p. 27.
45 Anthes, Gary "I, Coach," *Computerworld*, May 21, 2007, p. 38.
46 Clark, Don, "Google Targets Voice Searches," *The Wall Street Journal*, April 12, 2007, p. B3.
47 Gomes, Lee, "Voice Recognition Is Starting to Work," *The Wall Street Journal*, January 10, 2007, p. B1.
48 Ali, Sarmad, "New Services Turn Recordings into Text," *The Wall Street Journal*, May 24, 2007, p. B1.
49 Staff, "Designed to Meet the Special Needs of Cooperative Learning," *School Planning and Management*, September 1, 2007, p. 170.
50 Johnson, Colin, "DARPA Contract Tries Improved Take on AI," *Electronic Engineering Times*, September 10, 2007, p. 18.
51 Srinivasan, D., "Freeway Incident Detection Using Hybrid Fuzzy Neural Network," *IET Intelligent Transportation Systems*, "December 2007, p. 249.
52 Ninad, Patil and Smith, Timothy, "Neural Network Analysis Speeds Disease Risk Predictions," *Scientific Computing*, July 1, 2007, p. 36.
53 Mendes, J., et al., "Applying a Genetic Neuro-Model Reference Adaptive Controller in Drilling Optimization," *World Oil*, October 2007.
54 Preminger, Arie, "Forecasting Exchange Rates," *International Journal of Forecasting*, "January 2007, p. 71.
55 Ming-Jong, Yao, "A Genetic Algorithm for Determining Optimal Replenishment Cycles," *Omega*, August 2008, p. 619.
56 Kuang, K., "An Application of a Plastic Optical Fiber Sensor and Genetic Algorithm for Structural Health Monitoring," *Fiber Optic Sensors and Systems*, June 1, 2007, p. 5.
57 Staff, "Intelligent Agent Technology; Proceedings," *SciTech Book News*, March 1, 2007.
58 Staff, "Meet Sgt Star, The U.S. Army's Software Recruiting Agent," *COMMWEB*, January 3, 2007.
59 Staff, "Less Waste, More Profits," *Metalworking Production*, July 31, 2007, p. 72.
60 Rendleman, John, "Joint Command, Microsoft Team on R&D," *Government Computer News*, May 11, 2007.
61 *www.mediagrid.org*, accessed November 24, 2007.
62 *http://osl-www.colorado.edu/Research/haptic/hapticInterface.shtml*, accessed November 24, 2007.
63 *www.emory.edu/EMORY_MAGAZINE/winter96/rothbaum.html*, accessed November 24, 2007.
64 *www.temple.edu/ispr/examples/ex03_07_23.html*, accessed November 24, 2007.
65 Walsh, Aaron, "Advocate of Virtual Reality," *Computerworld*, July 9, 2007, p. 52.

66 Staff, "History Is Virtually Alive," *Rocky Mountain News*, May 30, 2007, p. D7.
67 Byron, Ellen, "A Virtual View of Store Aisle," *The Wall Street Journal*, October 3, 2007, B1.
68 *www.3ds.com/home*, accessed November 24, 2007.
69 Peng, Tina, "Why Trunk Shows Are Going Virtual," *The Wall Street Journal*, August 11, 2007, p. P1.
70 Gaudin, Sharon, "Real-World Hospital Makes Virtual Debut in Second Life," *Computerworld*, March 3, 2008, p.12.
71 Anthes, Gary, "Second Life: Is There Any There?" *Computerworld*, December 3, 2007, p. 30.
72 Anthes, Gary, "Different Engines," *Computerworld*, March 10, 2008, p. 34.
73 Bergstein, Brian, "New Wrinkle," *Rocky Mountain News*, February 18, 2008, .3.
74 Bulkeley, William, "Recognizing the Face of Fraud," *The Wall Street Journal*, March 22, 2007, p. B3.
75 White, Joseph, "Digital Driving a Reality," *The Wall Street Journal*, August 28, 2007, p. D5.
76 Ramirez, Elva, "A Warning System As Your Co-Pilot," *The Wall Street Journal*, June 21, 2007, p. D3.
77 Guth, Robert and Clark, Don, "The Shape of Computers to Come," *The Wall Street Journal*, May 30, 2007, p. B1.
78 Wood, Lamont, "Wearable Computers," *Computerworld*, December 17, 2007, p. 32.
79 Weier-Hayes, Mary, "Wal-Mart Gets Tough on RFID," *Information Week*, January 21, 2008, p. 26.
80 Weier-Hayes, Mary, "RFID Tags Are on the Menu," *Information Week*, February 5, 2007, p. 49.
81 Mitchell, Robert, "No Contact," *Computerworld*, June 11, 2007, p. 20.
82 Woyke, Elizabeth, "Mirror, Mirror, Talk to Me," *Business Week*, July 9, 2007, p. 12.
83 McCartney, Scott, "A New Way to Prevent Lost Luggage," *The Wall Street Journal*, February 22, 2007, p. D1.
84 Weier-Hayes, Mary, "Will This Rev Up RFID," *Information Week*, April 23, 2007, p. 30.
85 Staff, "Airport Uses Game Theory to Boost Security," *New Scientist*, October 13, 2007, p. 27.

PART 4

Systems Development

CHAPTER • 12 •

Systems Development: Investigation and Analysis

PRINCIPLES	LEARNING OBJECTIVES
▪ Effective systems development requires a team effort from stakeholders, users, managers, systems development specialists, and various support personnel, and it starts with careful planning.	▪ Identify the key participants in the systems development process and discuss their roles. ▪ Define the term *information systems planning* and list several reasons for initiating a systems project.
▪ Systems development often uses tools to select, implement, and monitor projects, including net present value (NPV), prototyping, rapid application development, CASE tools, and object-oriented development.	▪ Discuss the key features, advantages, and disadvantages of the traditional, prototyping, rapid application development, and end-user systems development life cycles. ▪ Identify several factors that influence the success or failure of a systems development project. ▪ Discuss the use of CASE tools and the object-oriented approach to systems development.
▪ Systems development starts with investigation and analysis of existing systems.	▪ State the purpose of systems investigation. ▪ Discuss the importance of performance and cost objectives. ▪ State the purpose of systems analysis and discuss some of the tools and techniques used in this phase of systems development.

Information Systems in the Global Economy GRUMA, Mexico

Systems Development South of the Border

GRUMA is a global business based in Monterrey, Nuevo Leòn, Mexico. GRUMA is the world leader in corn flour and tortilla production. It runs operations in the United States, Mexico, Central America, Venezuela, Europe, and China. About 19,000 employees work for GRUMA, and the company earns $3.2 billion in annual revenue.

GRUMA has only recently expanded its operation to Europe and China. One factor that allowed the expansion is a complete redesign of GRUMA's core information systems. The systems development process contributes to the success of the new systems.

GRUMA began operations in Mexico in 1949 based on progressive ideas. The company set goals to revolutionize the corn flour and tortilla industry through modern industrialization that was ecologically sound and efficient. Before long, the company grew to be the largest in the country and planned to expand beyond its borders. After establishing branches in the United States and South America, GRUMA faced challenges to further growth.

The company's information systems were designed to handle only one country's currency, taxes, and regulations. Different systems were designed for different countries—GRUMA in Mexico used one system, another system in the United States, and others in South American countries. To expand further, GRUMA needed a flexible centralized system that adapted to the economic requirements of many countries. Information systems are designed and implemented to meet the primary goals and strategic plans of a business. During the systems investigation phase, GRUMA discovered that its current system could not support its goal of becoming a global business. The company decided to conduct research into developing a new system that could help to achieve its goals.

Through a process called systems analysis, GRUMA studied their existing systems to discover what changes they needed. GRUMA's information systems team interviewed internal stakeholders—employees who interacted with the system and others who were otherwise impacted by the system. Through the interviews, the information systems team learned how the current system was used, what operations were effective, and what operations needed an overhaul. Realizing that significant changes were necessary, GRUMA decided to contract with an information systems firm to design the new system. It developed a request for proposals (RFP) to find a company to provide assistance at a reasonable cost.

After negotiations with several companies, GRUMA selected the information systems company SAP to develop an ERP system for the company. SAP dedicated systems analysts to work with GRUMA's information systems team to design a new system on which to base its operations. At this point, the systems development process progressed from investigation to analysis and then to design. Together, the two companies designed a system that accommodated "country-specific variances for taxes, product requirements, and currencies, languages, and cultural differences," according to SAP.

The team designed a template that GRUMA could distribute to international companies it acquired to standardize operations in all corporate facilities. The template would minimize work to be done setting up new facilities. The system also supported a variety of character sets so it could work with international languages, including Chinese.

During the systems implementation stage, members of the project team at GRUMA tested several prototypes before rolling out the new system for use. The team decided to use a phase-in approach to implement the system gradually. If problems occurred, business would not be interrupted. The new system was deployed at GRUMA's headquarters in Mexico for testing. After a successful launch, the system was deployed to GRUMA operations worldwide.

GRUMA management is thrilled with the new system, saying it allows them to better control growth and more quickly merge new acquisitions into in-house resources. The company can manage global operations in real time, reacting to market changes and new customer requirements as they arise.

Now in the operations and maintenance phase, GRUMA's new system is a huge success. GRUMA's information systems team is reviewing the system to measure its success and identify areas that can be improved. The team is looking to fine-tune features to better coordinate with systems managed by its business partners. It also plans to expand customer relationship management (CRM) tools. The work by GRUMA's systems team and SAP systems analysts has so impressed the executives at GRUMA that they now view the team as a value-creation unit with many more projects to investigate.

As you read this chapter, consider the following:

- What situations can arise in a business to trigger new systems development initiatives?
- What are the best methods for a business to use in approaching new systems development projects?

Why Learn About Systems Development?

Throughout this book, you have seen many examples of the use of information systems in a variety of careers. But where do you start to acquire these systems or have them developed? How can you work with IS personnel, such as systems analysts and computer programmers, to get what you need to succeed on the job or in your own business? This chapter, the first of two chapters on systems development, provides the answers to these questions. You will see how you can initiate the systems development process and analyze your needs with the help of IS personnel. Systems investigation and systems analysis are the first two steps of the systems development process. This chapter provides specific examples of how new or modified systems are initiated and analyzed in a number of industries. In this chapter, you will learn how your project can be planned, aligned with corporate goals, rapidly developed, and much more. We start with an overview of the systems development process.

When an organization needs to accomplish a new task or change a work process, how does it do so? It develops a new system or modifies an existing one. Systems development is the activity of creating new systems or modifying existing systems. It refers to all aspects of the process—from identifying problems to solve or opportunities to exploit to implementing and refining the chosen solution.

AN OVERVIEW OF SYSTEMS DEVELOPMENT

In today's businesses, managers and employees in all functional areas work together and use business information systems. As a result, they are helping with development and, in many cases, leading the way. Users might request that a systems development team determine whether they should purchase a few PCs or create an attractive Web site, using the tools discussed in Chapter 7. In another case, an entrepreneur might use systems development to build an Internet site to compete with large corporations.

This chapter and the next provide you with a deeper appreciation of the systems development process. Individuals can also use systems development to their advantage. Systems development skills and techniques discussed in this chapter and the next can help people launch their own businesses.[1] When Marc Mallow couldn't find off-the-shelf software to schedule workers, he took a few years to develop his own program. The software he created became the core of a New York-based company he founded. Corporations and nonprofit organizations use systems development to achieve their goals. First Health of the Carolinas, for example, upgraded its old imaging system to slash costs and provide better healthcare for patients. The nonprofit health organization reduced costs by more than 30 percent and offered doctors better radiological images to improve patient care.[2]

This chapter will also help you avoid systems development failures or projects that go over budget. In one example, a large $4 billion systems development effort to convert older, paper-based medical records to electronic records for a large healthcare company ran into trouble when it exceeded its budget.[3] In some cases, poorly executed systems development efforts can be costly. A tax system developed for the District of Columbia at a cost of $100 million didn't prevent tax fraud of about $20 million.[4] The fraud involved cashing refund checks sent to fictitious corporate accounts. In other cases, systems development failures can be deadly.[5] According to the CIO of Duke University's Health System, "Issues arising from badly designed and poorly integrated healthcare IT systems harm and kill more patients every year than do medications and medical devices."

To stay competitive in today's global economy, some cities and counties, including Chattanooga, Tennessee, are investing in high-speed fiber-optic cables that have the potential to deliver greater speed compared to existing cable and phone company offerings.[6] In the United States, less than 60 percent of the population has broadband Internet access, while some countries like Denmark and the Netherlands have more than 75 percent of their population with broadband Internet access. South Korea has over 90 percent of its citizens on broadband Internet.

Participants in Systems Development

Effective systems development requires a team effort. The team usually consists of stakeholders, users, managers, systems development specialists, and various support personnel. This team, called the *development team*, is responsible for determining the objectives of the information system and delivering a system that meets these objectives. Many development teams use a project manager to head the systems development effort combined with the project management approach to help coordinate the systems development process. A *project* is a planned collection of activities that achieves a goal, such as constructing a new manufacturing plant or developing a new decision support system. All projects have a defined starting point and ending point, normally expressed as dates such as August 4 and December 11. Most have a budget, such as $150,000. A *project manager* is responsible for coordinating all people and resources needed to complete a project on time. The project manager can make the difference between project success and failure. According to Tyrone Howard, founder of BizNova Consulting, "A project management system is just a tool. It is like this: A carpenter can buy a hammer, but the hammer won't build a house.... In IT, it's the people who do the building, not the technology."[7] In systems development, the project manager can be an IS person inside the organization or an external consultant hired to complete the project. Project managers need technical, business, and people skills. In addition to completing the project on time and within the specified budget, the project manager is usually responsible for controlling project quality, training personnel, facilitating communications, managing risks, and acquiring any necessary equipment, including office supplies and sophisticated computer systems. Research studies have shown that project management success factors include good leadership from executives and project managers, a high level of trust in the project and its potential benefits, and the commitment of the project team and organization to successfully complete the project and implement its results. Project escalation, where the size and scope of a new systems development effort greatly expands over time, is a major problem for project managers.[8] Project escalation often causes projects to go over budget and behind schedule.

In the context of systems development, **stakeholders** are people who, either themselves or through the area of the organization they represent, ultimately benefit from the systems development project. **Users** are people who will interact with the system regularly. They can be employees, managers, or suppliers. For large-scale systems development projects, where the investment in and value of a system can be high, it is common for senior-level managers,

stakeholders
People who, either themselves or through the organization they represent, ultimately benefit from the systems development project.

users
People who will interact with the system regularly.

including the functional vice presidents (of finance, marketing, and so on), to be part of the development team.

Because stakeholders ultimately benefit from the systems development project, they often work with others in developing a computer application.

(Source: © Reza Estakhrian/Getty Images.)

systems analyst
A professional who specializes in analyzing and designing business systems.

programmer
A specialist responsible for modifying or developing programs to satisfy user requirements.

Depending on the nature of the systems project, the development team might include systems analysts and programmers, among others. A **systems analyst** is a professional who specializes in analyzing and designing business systems. Systems analysts play various roles while interacting with the stakeholders and users, management, vendors and suppliers, external companies, programmers, and other IS support personnel (see Figure 12.1). Like an architect developing blueprints for a new building, a systems analyst develops detailed plans for the new or modified system. The **programmer** is responsible for modifying or developing programs to satisfy user requirements. Like a contractor constructing a new building or renovating an existing one, the programmer takes the plans from the systems analyst and builds or modifies the necessary software. The demand for systems analysts and computer programmers is expected to increase.[9] In Canada, the unemployment rate for IS professionals is about one-third the national average. According to the chairman of the Computer Science Department at the University of Toronto, "The numbers are quite stark. It's clear the demand in the workforce is there."

The other support personnel on the development team are mostly technical specialists, including database and telecommunications experts, hardware engineers, and supplier representatives. One or more of these roles might be outsourced to outside experts or consultants. Depending on the magnitude of the systems development project and the number of IS systems development specialists on the team, one or more IS managers might also belong to the team. The composition of a development team can vary over time and from project to project. For small businesses, the development team might consist of a systems analyst and the business owner as the primary stakeholder. For larger organizations, formal IS staff can include hundreds of people involved in a variety of activities, including systems development. Every development team should have a team leader. This person can be from the IS department, a manager from the company, or a consultant from outside the company. The team leader needs both technical and people skills.

Today, companies are using innovative ways to build new systems or modify existing ones without using in-house programmers. Outsourcing, which is discussed later in the chapter, is one approach.[10] Constellation Energy, a $19 billion utility company, is using another approach that asks programmers from around the world to get involved. The approach, called *crowd sourcing*, asks programmers to contribute code to the project.[11] Winning programmers that submit excellent code can be given from $500 to more than $2,000. Constellation is hoping to save time and money by using crowd sourcing, but neither result is guaranteed. In addition, the resulting programming code may not be consistent with what the company is expecting.

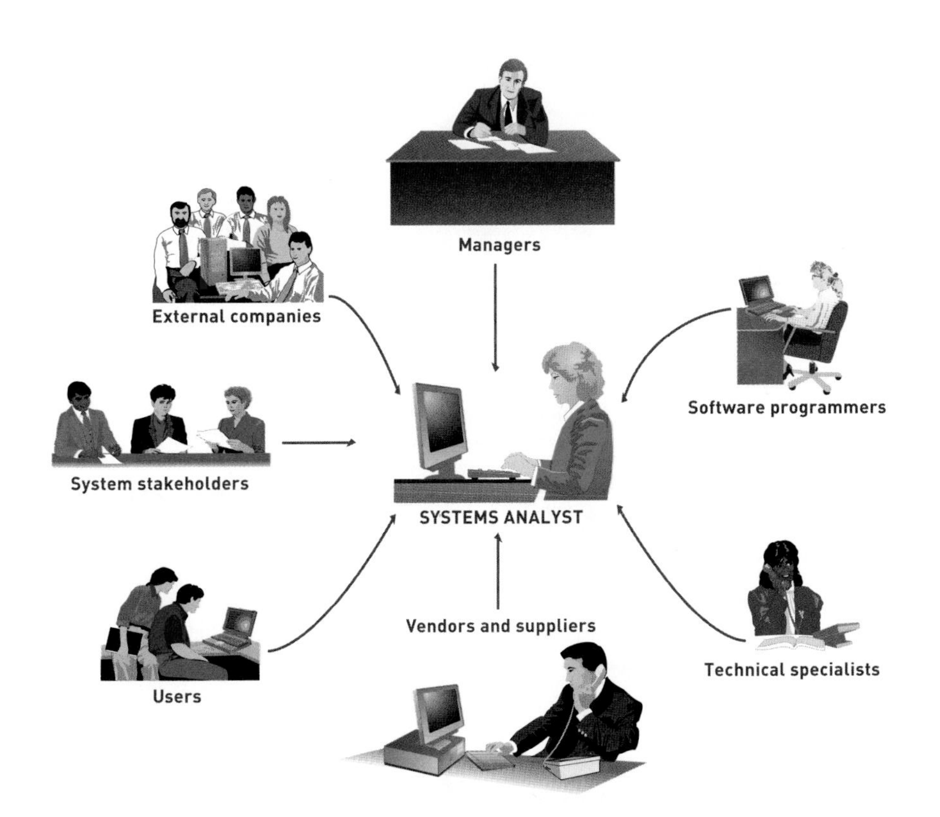

Figure 12.1

Role of the Systems Analyst

The systems analyst plays an important role in the development team and is often the only person who sees the system in its totality. The one-way arrows in this figure do not mean that there is no direct communication between other team members. These arrows just indicate the pivotal role of the systems analyst—a person who is often called on to be a facilitator, moderator, negotiator, and interpreter for development activities.

Regardless of the specific nature of a project, systems development creates or modifies systems, which ultimately means change. Managing this change effectively requires development team members to communicate well. Because you probably will participate in systems development during your career, you must learn communication skills. You might even be the individual who initiates systems development.

Initiating Systems Development

Systems development initiatives arise from all levels of an organization and are both planned and unplanned. Systems development projects are initiated for many reasons, as shown in Figure 12.2.

As shown in Figure 12.2, problems with the existing system can initiate systems development activity. Hannaford Brothers, a large grocer located in Maine, decided to upgrade its security system after millions of credit and debit card records were stolen from its computer system. The security upgrade is expected to cost millions of dollars.[12] The desire to exploit new opportunities is another cause of systems initiation. The increased use of the *cloud computing* approach, discussed in Chapter 7, has many IS professionals looking into using the Internet for applications, such as word processing and spreadsheet analysis, instead of putting these applications on desktop or laptop computers. According to Internet pioneer Marc Andreessen, "The cloud is a smart, complex, powerful computing system in the sky that people can just plug into."[13] Mergers and acquisitions can trigger many systems development projects.[14] Because information systems often vary within a company, a large systems development effort is typically required to unify systems. Even with similar information systems, the procedures, culture, training, and management of the information systems are often different, requiring a realignment of the IS departments. In another case, Six Flags, one of the largest amusement park companies in the world with about $1 billion in annual sales, initiated a systems development project to build a sophisticated inventory control system to increase revenues.[15] According to CIO Michael Israel, "If a food stand is running low, we know at mid-day instead of the end of the day."

A company's customers or suppliers can trigger systems development. Daisy Brand, a dairy products company, was asked by one of its major customers, Wal-Mart, to start using special RFID tags.[16] Although the Wal-Mart RFID initiative was never fully implemented

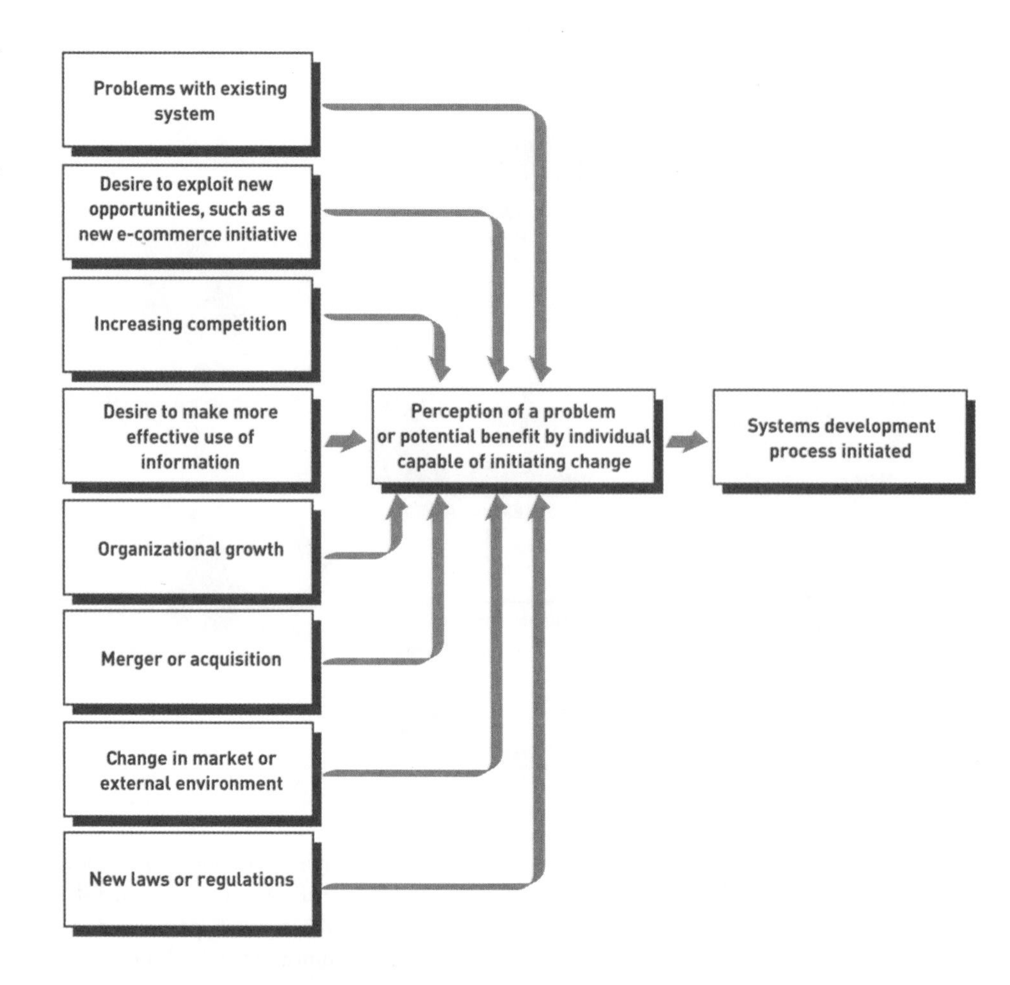

Figure 12.2

Typical Reasons to Initiate a Systems Development Project

by all of its customers, Dairy Products benefitted from the technology by streamlining its inventory processing. By putting RFID tags on every pallet of dairy products that it ships to customers, the company cut in half the time it used to take to load pallets onto delivery trucks.

Systems development can also be initiated when a vendor no longer supports an older system or older software. When this support is no longer available, companies are often forced to upgrade to new software and systems, which can be expensive and require additional training.[17] Major systems and application software companies, for example, often stop supporting their older software a few years after new software has been introduced. Some printer and computer vendors do the same. They stop providing support for their older systems after newer ones are introduced and sold in the market. This lack of support is a dilemma for many companies trying to keep older systems operational.

The federal government can foster new systems development projects in the private sector. As a result of some financial scandals, the government has instituted new corporate financial reporting rules under the Sarbanes-Oxley Act. These regulations have caused many U.S. companies to initiate systems development efforts. To comply with this law, companies can spend hundreds of thousands or millions of dollars in new systems development efforts.

Information Systems Planning and Aligning Corporate and IS Goals

Information systems planning and aligning corporate and IS goals are important aspects of any systems development project.[18] Achieving a competitive advantage is often the overall objective of systems development.

Information Systems Planning

The term **information systems planning** refers to translating strategic and organizational goals into systems development initiatives (see Figure 12.3).[19] Proper IS planning ensures that specific systems development objectives support organizational goals. Long-range planning can also be important and result in getting the most from a systems development effort. It can also align IS goals with corporate goals and culture, which is discussed next.[20] Hess Corporation, a large energy company with over 1,000 retail gasoline stations, uses long-range planning to determine what computer equipment they need and the IS personnel needed to run it.[21] According to Hess's CIO, "It became pretty clear that we needed to lay out a long-term strategy that would allow us to figure out how IT could support our business strategy over the next five years."

information systems planning
Translating strategic and organizational goals into systems development initiatives.

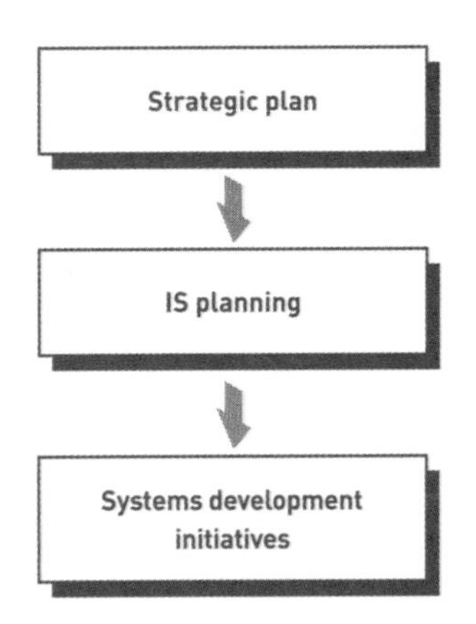

Figure 12.3

Information Systems Planning

Information systems planning transforms organizational goals outlined in the strategic plan into specific systems development activities.

Aligning Corporate and IS Goals

Aligning organizational goals and IS goals is critical for any successful systems development effort. Because information systems support other business activities, IS staff and people in other departments need to understand each other's responsibilities and tasks. Determining whether organizational and IS goals are aligned can be difficult, so researchers have increasingly tackled the problem. Most corporations, for example, have profits and return on investment (ROI), first introduced in Chapter 2, as primary goals. Procter & Gamble (P&G) uses ROI to measure the success of its projects and systems development efforts.[22] P&G produces Tide, Pringles, Pampers, and many other consumer products. The huge consumer-products company has a $76 billion annual supply chain. ROI calculations help companies like P&G prioritize systems development projects and align them with corporate goals. Providing outstanding service is another important corporate goal.[23] Coca-Cola Enterprises, which is Coca-Cola's largest bottler and distributor, decided to use online services from Microsoft and SharePoint to speed its systems development process.[24] According to the company CIO, "This is not a head-count reduction for us. Services are complementary to our IT strategy."

Specific systems development initiatives can spring from the IS plan, but the IS plan must also provide a broad framework for future success. The IS plan should guide development of the IS infrastructure over time. Another benefit of IS planning is that it ensures better use of IS resources—including funds, personnel, and time for scheduling specific projects. The steps of IS planning are shown in Figure 12.4.

Developing a Competitive Advantage

In today's business environment, many companies seek systems development projects that will provide them with a competitive advantage. Thinking competitively usually requires creative and critical analysis. By looking at problems in new or different ways and by introducing innovative methods to solve them, many organizations have gained significant competitive advantage.

Creative analysis involves investigating new approaches to existing problems. By looking at problems in new or different ways and by introducing innovative methods to solve them, many firms have gained a competitive advantage. Typically, these new solutions are inspired

creative analysis
The investigation of new approaches to existing problems.

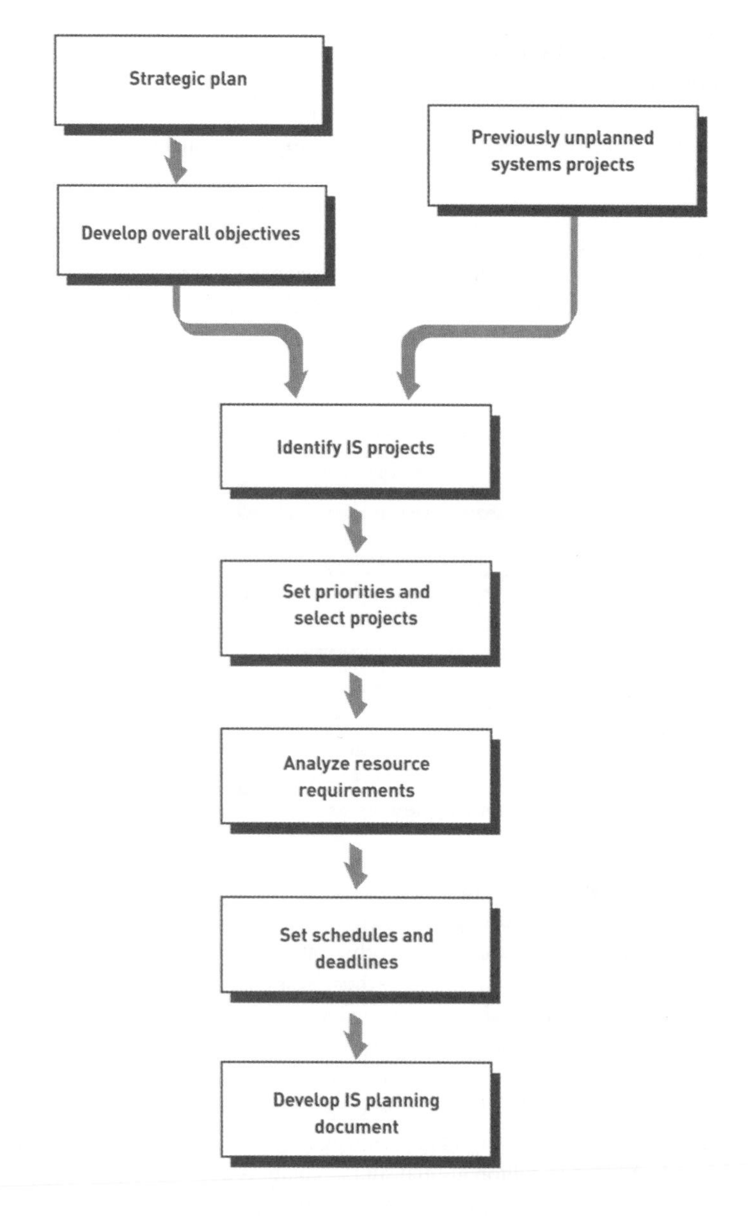

Figure 12.4

The Steps of IS Planning

Some projects are identified through overall IS objectives, whereas additional projects, called *unplanned projects*, are identified from other sources. All identified projects are then evaluated in terms of their organizational priority.

by people and events not directly related to the problem. Creative analysis can help organizations achieve their performance goals. According to Michael Hugos, principle at the Center for Systems Innovation and one of *Computerworld*'s 2006 Premier 100 IT Leaders, "Creativity is where we come up with ideas for combining available resources to create systems that could meet performance requirements."[25]

critical analysis
The unbiased and careful questioning of whether system elements are related in the most effective ways.

Critical analysis requires unbiased and careful questioning of whether system elements are related in the most effective ways. It involves considering the establishment of new or different relationships among system elements and perhaps introducing new elements into the system. Critical analysis in systems development involves the following actions:

- **Questioning statements and assumptions.** Questioning users about their needs and clarifying their initial responses can result in better systems and more accurate predictions. Too often, stakeholders and users specify certain system requirements because they assume that their needs can only be met that way.
- **Identifying and resolving objectives and orientations that conflict.** Each department in an organization can have different objectives and orientations. The buying department might want to minimize the cost of spare parts by always buying from the lowest-cost supplier, but engineering might want to buy more expensive, higher-quality spare parts to reduce the frequency of replacement. These differences must be identified and resolved before a new purchasing system is developed or an existing one modified.

INFORMATION SYSTEMS @ WORK

Investigating Conversion at Art.com

Art.com was an early Web pioneer, launching in 1995 with the purpose of selling all kinds of visual art online. Since then, Art.com has assisted over 4 million customers in decorating their walls by providing a virtual gallery of approximately 400,000 images. The company operates in both the United States and Europe and employs more than 500 worldwide.

Art.com has over 12 million visitors to its Web sites per month, most of whom visit the site without making a purchase. Art.com wanted to increase the percentage of visitors that make purchases—known as the conversion rate—by improving its Web site. With 12 million visitors, even a small improvement could mean a major increase in profits. The challenge was that the Web site had been continuously revised during the company's many years in business, and Art.com's management did not know what changes would improve the visitor's experience. They certainly didn't want to risk changes that might inadvertently turn visitors away.

With the goal clearly articulated to "increase the conversion rate by offering the best customer experience," Art.com systems analysts began to investigate what portions of the current system worked well and what portions could be improved.

The systems investigation proved to be no small task. Art.com draws thousands of images from product lines offered by many online properties. The company had been using a traditional Web analytics information system that recorded information such as number of visitors and which products were most popular; however, the system did not evaluate information on site obstacles that might be discouraging sales. Art.com needed a system that could provide more telling information such as key performance indicators (KPI) that suggested what customers did not like about its site.

The systems team found an off-the-shelf solution that performed more detailed Web analysis. The "online customer experience management solution" allowed systems investigators to view key performance indicators and then review the qualitative details of individual customer sessions on the site. Viewing the basic analytics allowed the investigators to quickly find trends in customer activity. Drilling down into those trends allowed the investigators to "play back" a customer's activities on the site to determine where the customer experienced problems or decided to leave the site. Rather than having to guess what was happening on the site, investigators could track the action in real time.

Using the new online customer experience management solution allowed Art.com to make several improvements to its Web site that contributed to a significant increase in conversion rate and prevented possible disasters.

One example of disaster recovery took place when Art.com sent sale coupons to many of its customers. Unfortunately, the coupon numbers were not entered into the back-end system, so when customers with coupons checked out, they received an "invalid coupon code" error message. Most abandoned their purchase at that point. Art.com's new Web analytics tool alerted management to the problem within hours. The coupon codes were added to the back-end system, and because user data was collected by the system, Art.com contacted those who were frustrated by the error and enticed them back. The quick correction of the problem probably saved Art.com $25,000 of revenue per day.

In another example, the checkout process at Art.com's French site was displaying error messages to customers using outdated browsers. Art.com's new system caught the problem when the alarm was raised and management corrected the problem within days. Again, customers who experienced difficulties were contacted and enticed to return.

In a third example, Art.com's new system showed investigators that up to 20,000 visitors referred by Web search engines were greeted with a page that informed them that "sorry, this product is no longer available." Web developers at Art.com changed the message to be less negative and more inviting by providing alternative products that might interest those visitors.

Systems analysts depend on tools to provide them with information on which portions of systems are working and which are not. Analyzing a Web site such as Art.com is like analyzing pedestrian traffic in a major city—it's impossible without appropriate tools. Using powerful Web analytics and an online customer experience management system, systems analysts can continuously review and investigate the effect that the system is having on Web site visitors, launching systems development projects as needed.

Discussion Questions

1. What was Art.com's biggest challenge in improving their customers' online experience?
2. How does the new online system allow Art.com to launch systems development projects that can improve sales?

Critical Thinking Questions

1. What are some useful functions of a good Web analytics and online customer experience management system?
2. During which stages of the systems development life cycle can Web analytics be useful, and why?

SOURCES: Tealeaf staff, "Art.com: Purveyor of the World's Largest Selection of Wall Décor," Computerworld/TeaLeaf, 2007, *http://zones.computerworld.com/tealeaf_customer_exp/registration.php?item=13&from= cw&src=cwlp*; Art.com Web site, *www.art.com*, accessed July 12, 2008; Tealeaf Web site, *www.tealeaf.com*, accessed July 12, 2008.

Establishing Objectives for Systems Development

The overall objective of systems development is to achieve business goals, not technical goals, by delivering the right information to the right person at the right time. The impact a particular system has on an organization's ability to meet its goals determines the true value of that system to the organization. Southern States, which sells farm equipment in over 20 states and is owned by about 300,000 farmers, decided to use Skyway Software, Inc.'s Visual Workplace to develop a new pricing application to help increase revenue.[26] The use of this service-oriented architecture (SOA) tool allowed Southern States to generate $1.4 million more in revenue the year after it was placed into operation.

Although all systems should support business goals, some systems are more pivotal in continued operations and goal attainment than others. These systems are called **mission-critical systems**. An order processing system, for example, is usually considered mission-critical. Without it, few organizations could continue daily activities, and they clearly would not meet set goals.

mission-critical systems
Systems that play a pivotal role in an organization's continued operations and goal attainment.

The goals defined for an organization also define the objectives that are set for a system. A manufacturing plant, for example, might determine that minimizing the total cost of owning and operating its equipment is critical to meet production and profit goals. **Critical success factors** (**CSFs**) are factors that are essential to the success of certain functional areas of an organization. The CSF for manufacturing—minimizing equipment maintenance and operating costs—would be converted into specific objectives for a proposed system. One specific objective might be to alert maintenance planners when a piece of equipment is due for routine preventative maintenance (e.g., cleaning and lubrication). Another objective might be to alert the maintenance planners when the necessary cleaning materials, lubrication oils, or spare parts inventory levels are below specified limits. These objectives could be accomplished either through automatic stock replenishment via electronic data interchange or through the use of exception reports.

critical success factors (CSFs)
Factors that are essential to the success of a functional area of an organization.

Regardless of the particular systems development effort, the development process should define a system with specific performance and cost objectives. The success or failure of the systems development effort will be measured against these objectives.

Performance Objectives

The extent to which a system performs as desired can be measured through its performance objectives. System performance is usually determined by factors such as the following:

- **The quality or usefulness of the output.** Is the system generating the right information for a value-added business process or by a goal-oriented decision maker?
- **The accuracy of the output.** Is the output accurate and does it reflect the true situation? As a result of the accounting scandals of the early 2000s, when some companies overstated revenues or understated expenses, accuracy is becoming more important, and top corporate officers are being held responsible for the accuracy of all corporate reports.
- **The speed at which output is generated.** Is the system generating output in time to meet organizational goals and operational objectives? Objectives such as customer response time, the time to determine product availability, and throughput time are examples. For Six Flags, speed is critical.[27] According to the CIO of Six Flags, "Speed per attendee is everything."
- **The scalability of the resulting system.** As mentioned in Chapter 4, *scalability* allows an information system to handle business growth and increased business volume. For example, if a midsized business realizes an annual 10 percent growth in sales for several years, an information system that is scalable will be able to efficiently handle the increase by adding processing, storage, software, database, telecommunications, and other information systems resources to handle the growth.
- **The risk of the system.** One important objective of many systems development projects is to reduce risk.[28] The BRE Bank in Poland (*www.brebank.pl/en*), for example, used systems development to create a model-based DSS to analyze and reduce loan risk and a variety of related risks associated with bank transactions. The project uses a mathematical algorithm, called FIRST (Financial Institutions Risk Scenario Trends), to reduce risk.

In some cases, the achievement of performance objectives can be easily measured (e.g., by tracking the time it takes to determine product availability). In other cases, it is sometimes more difficult to ascertain in the short term. For example, it might be difficult to determine how many customers are lost because of slow responses to customer inquiries regarding product availability. These outcomes, however, are often closely associated with corporate goals and are vital to the long-term success of the organization. Senior management usually dictates their attainment.

Cost Objectives

Organizations can spend more than is necessary during a systems development project. The benefits of achieving performance goals should be balanced with all costs associated with the system, including the following:

- **Development costs.** All costs required to get the system up and running should be included. Some computer vendors give cash rewards to companies using their systems to reduce costs and act as an incentive.
- **Costs related to the uniqueness of the system application.** A system's uniqueness has a profound effect on its cost. An expensive but reusable system might be preferable to a less costly system with limited use.
- **Fixed investments in hardware and related equipment.** Developers should consider costs of such items as computers, network-related equipment, and environmentally controlled data centers in which to operate the equipment.
- **Ongoing operating costs of the system.** Operating costs include costs for personnel, software, supplies, and resources such as the electricity required to run the system. Tridel Corporation (*www.tridel.com*) used systems development to build a new invoicing application, called Invoice Zero, to save over $20,000 in operating costs.[29] The new invoicing application, which consolidated invoices and sent them out once a month, cut the number of monthly invoices from 2,400 to just 17. Reducing costs was also an important factor for Cincinnati Bell. By switching from dedicated PCs to thin client computers and virtualization software, Cincinnati Bell expects to see a large reduction in help desk costs.[30] Some experts predict that help desk costs could be reduced by 70 percent or more. For many IS operations, ongoing operating costs are much higher than development or acquisition costs. According to a Gartner study, acquisition or development cost is only 20 percent of the total cost of a new information system.[31]

Balancing performance and cost objectives within the overall framework of organizational goals can be challenging. Setting objectives is important, however, because they allow an organization to allocate resources effectively and measure the success of a systems development effort. For PC manufacturers, for example, parts and components of a typical PC can cost under $500, which includes about $130 for the processor, $100 for a CD or DVD, $100 for memory, $45 for the Windows operating system, and the rest for other hardware parts and components. Some believe these low costs will eventually lead to lower costs for PCs.

Cincinnati Bell reduced help desk costs by switching from dedicated PCs to thin client computers and virtualization software.

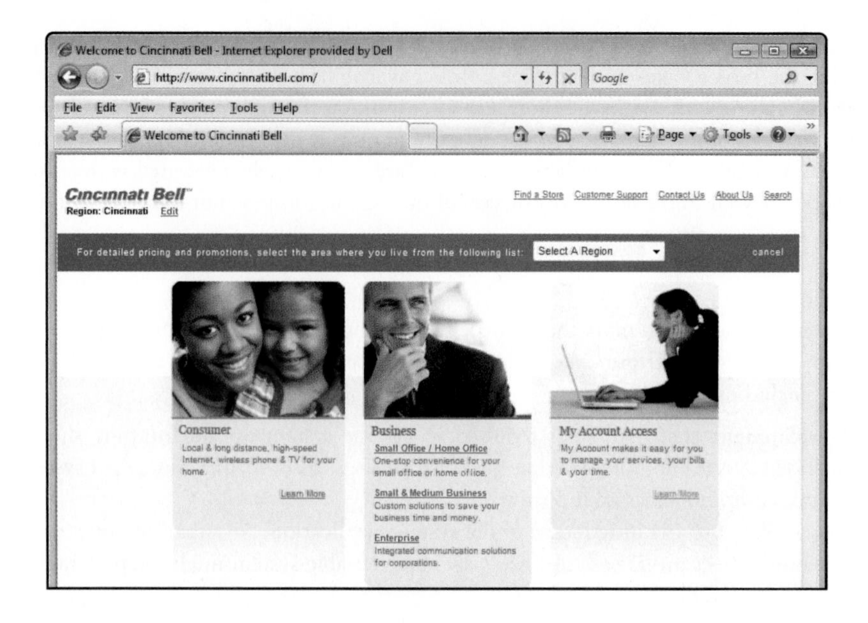

SYSTEMS DEVELOPMENT LIFE CYCLES

The systems development process is also called a *systems development life cycle (SDLC)* because the activities associated with it are ongoing. As each system is built, the project has timelines and deadlines, until at last the system is installed and accepted. The life of the system continues as it is maintained and reviewed. If the system needs significant improvement beyond the scope of maintenance, if it needs to be replaced because of a new generation of technology, or if the IS needs of the organization change significantly, a new project will be initiated and the cycle will start over.

A key fact of systems development is that the later in the SDLC an error is detected, the more expensive it is to correct (see Figure 12.5). One reason for the mounting costs is that, if an error is found in a later phase of the SDLC, the previous phases must be reworked to some extent. Another reason is that the errors found late in the SDLC affect more people. For example, an error found after a system is installed might require retraining users when a "work-around" to the problem has been found. Thus, experienced systems developers prefer an approach that will catch errors early in the project life cycle.

Figure 12.5

Relationship Between Timing of Errors and Costs

The later that system changes are made in the SDLC, the more expensive these changes become.

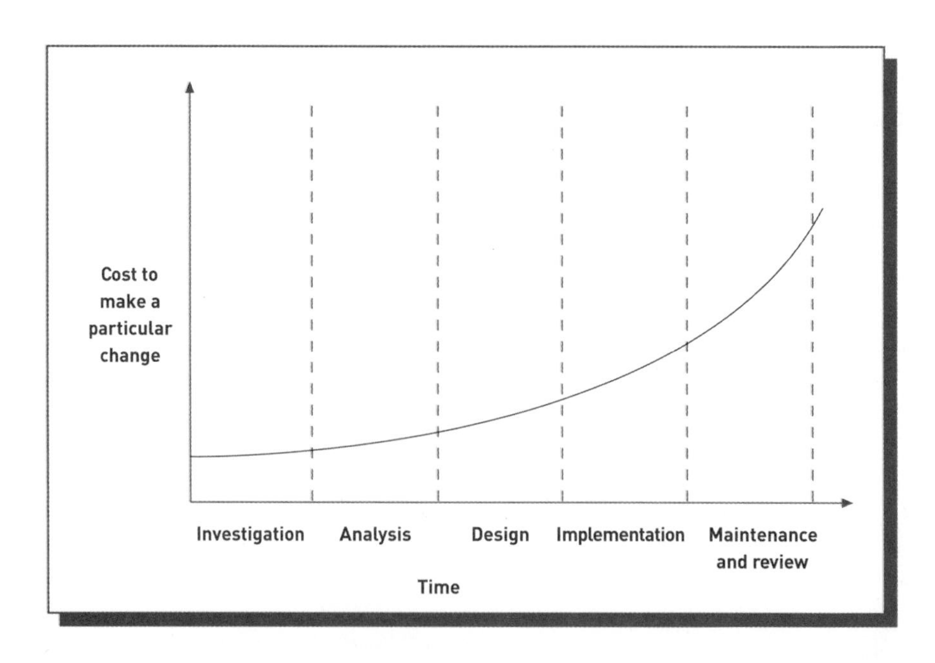

Several common systems development life cycles exist: traditional, prototyping, rapid application development (RAD), and end-user development. In addition, companies can outsource the systems development process. With some companies, these approaches are formalized and documented so that systems developers have a well-defined process to follow; other companies use less formalized approaches. Keep Figure 12.5 in mind as you are introduced to alternative SDLCs in the next section.

The Traditional Systems Development Life Cycle

Traditional systems development efforts can range from a small project, such as purchasing an inexpensive computer program, to a major undertaking. The steps of traditional systems development might vary from one company to the next, but most approaches have five common phases: investigation, analysis, design, implementation, and maintenance and review (see Figure 12.6).

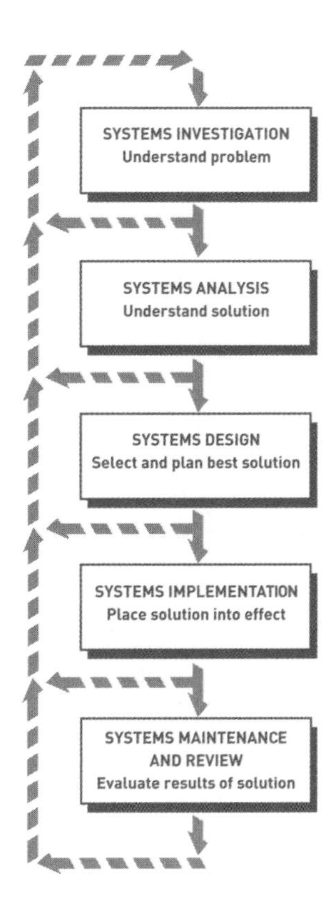

Figure 12.6

The Traditional Systems Development Life Cycle

Sometimes, information learned in a particular phase requires cycling back to a previous phase.

In the **systems investigation** phase, potential problems and opportunities are identified and considered in light of the goals of the business. Systems investigation attempts to answer the questions "What is the problem, and is it worth solving?" The primary result of this phase is a defined development project for which business problems or opportunity statements have been created, to which some organizational resources have been committed, and for which systems analysis is recommended. **Systems analysis** attempts to answer the question "What must the information system do to solve the problem?" This phase involves studying existing systems and work processes to identify strengths, weaknesses, and opportunities for improvement. The major outcome of systems analysis is a list of requirements and priorities. **Systems design** seeks to answer the question "How will the information system do what it must do to obtain the problem solution?" The primary result of this phase is a technical design that either describes the new system or describes how existing systems will be modified. The system design details system outputs, inputs, and user interfaces; specifies hardware,

systems investigation
The systems development phase during which problems and opportunities are identified and considered in light of the goals of the business.

systems analysis
The systems development phase that determines what the information system must do to solve the problem by studying existing systems and work processes to identify strengths, weaknesses, and opportunities for improvement.

systems design
The systems development phase that defines how the information system will do what it must do to obtain the problem solution.

systems implementation
The systems development phase involving the creation or acquisition of various system components detailed in the systems design, assembling them, and placing the new or modified system into operation.

systems maintenance and review
The systems development phase that ensures the system operates and modifies the system so that it continues to meet changing business needs.

software, database, telecommunications, personnel, and procedure components; and shows how these components are related. **Systems implementation** involves creating or acquiring the various system components detailed in the systems design, assembling them, and placing the new or modified system into operation.[32] An important task during this phase is to train the users. Systems implementation results in an installed, operational information system that meets the business needs for which it was developed. It can also involve phasing out or removing old systems, which can be difficult for existing users, especially when the systems are free. In 2005, Walt Disney developed the *Virtual Magic Kingdom (VMK)* game to celebrate the fiftieth anniversary of Disneyland.[33] The VMK game used Disney avatars and offered virtual rewards to game players. When Disney decided to remove or terminate the game, some players were outraged and protested outside Disney offices in California.

The purpose of **systems maintenance and review** is to ensure that the system operates and to modify the system so that it continues to meet changing business needs. As shown in Figure 12.6, a system under development moves from one phase of the traditional SDLC to the next.

The traditional SDLC allows for a large degree of management control. However, a major problem is that the user does not use the solution until the system is nearly complete. Table 12.1 lists advantages and disadvantages of the traditional SDLC.

Table 12.1

Advantages and Disadvantages of Traditional SDLC

Advantages	Disadvantages
Formal review at the end of each phase allows maximum management control.	Users get a system that meets the needs as understood by the developers; this might not be what is really needed.
This approach creates considerable system documentation.	Documentation is expensive and time consuming to create. It is also difficult to keep current.
Formal documentation ensures that system requirements can be traced back to stated business needs.	Often, user needs go unstated or are misunderstood.
It produces many intermediate products that can be reviewed to see whether they meet the users' needs and conform to standards.	Users cannot easily review intermediate products and evaluate whether a particular product (e.g., data flow diagram) meets their business requirements.

Prototyping

Prototyping takes an iterative approach to the systems development process. During each iteration, requirements and alternative solutions to the problem are identified and analyzed, new solutions are designed, and a portion of the system is implemented.[34] Users are then encouraged to try the prototype and provide feedback (see Figure 12.7). Prototyping begins with creating a preliminary model of a major subsystem or a scaled-down version of the entire system. For example, a prototype might show sample report formats and input screens. After they are developed and refined, the prototypical reports and input screens are used as models for the actual system, which can be developed using an end-user programming language such as Visual Basic. The first preliminary model is refined to form the second- and third-generation models, and so on until the complete system is developed (see Figure 12.8).

Prototypes can be classified as operational or nonoperational. An *operational prototype* is a prototype that works—accesses real data files, edits input data, makes necessary computations and comparisons, and produces real output. A *nonoperational prototype* is a mock-up, or model, that includes output and input specifications and formats. The advantages and disadvantages of prototyping are summarized in Table 12.2.

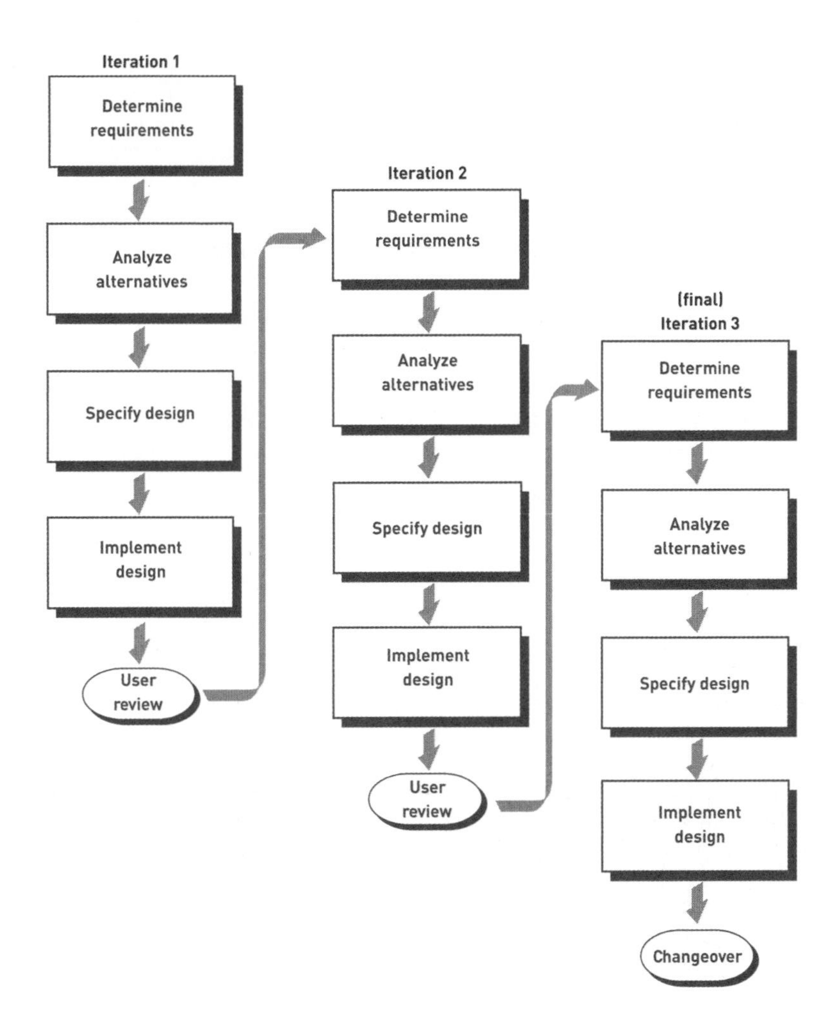

Figure 12.7

Prototyping

Prototyping is an iterative approach to systems development.

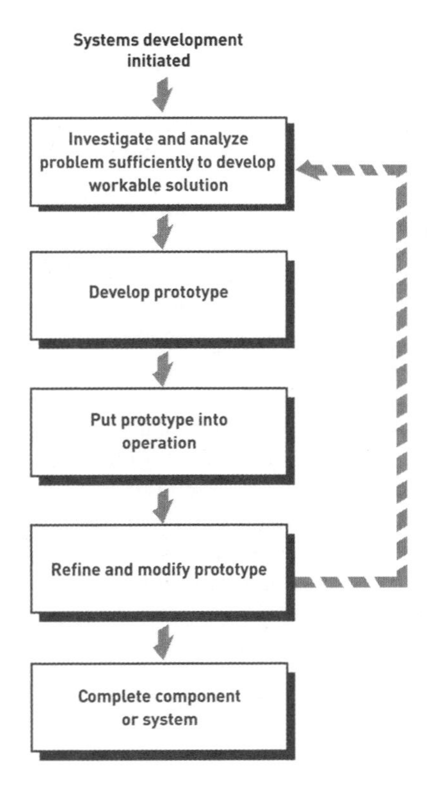

Figure 12.8

Refining During Prototyping

Each generation of prototype is a refinement of the previous generation based on user feedback.

Table 12.2

Advantages and Disadvantages of Prototyping

Advantages	Disadvantages
Users can try the system and provide constructive feedback during development.	Each iteration builds on the previous one. The final solution might be only incrementally better than the initial solution.
An operational prototype can be produced in weeks.	Formal end-of-phase reviews might not occur. Thus, it is very difficult to contain the scope of the prototype, and the project never seems to end.
As solutions emerge, users become more positive about the process and the results.	System documentation is often absent or incomplete because the primary focus is on development of the prototype.
Prototyping enables early detection of errors and omissions.	System backup and recovery, performance, and security issues can be overlooked in the haste to develop a prototype.

Rapid Application Development, Agile Development, Joint Application Development, and Other Systems Development Approaches

rapid application development (RAD)
A systems development approach that employs tools, techniques, and methodologies designed to speed application development.

Rapid application development (**RAD**) employs tools, techniques, and methodologies designed to speed application development. Vendors, such as Computer Associates International, IBM, and Oracle, market products targeting the RAD market. Rational Software, a division of IBM, has a RAD tool, called Rational Rapid Developer, to make developing large Java programs and applications easier and faster. Locus Systems, a program developer, used a RAD tool called OptimalJ to generate more than 60 percent of the computer code for three applications it developed. Royal Bank of Canada used OptimalJ to develop some customer-based applications. According to David Hewick, group manager of application architecture for the bank, "It was an opportunity to improve the development life cycle, reduce costs, and bring consistency." Advantage Gen, formerly known as COOL:Gen, is a RAD tool from Computer Associates International. It can be used to rapidly generate computer code from business models and specifications.[35]

Other approaches to rapid development, such as *agile development* or *extreme programming (XP)*, allow the systems to change as they are being developed. Agile development requires frequent face-to-face meetings with the systems developers and users as they modify, refine, and test how the system meets users' needs and what its capabilities are. Microsoft, for example, has adopted a more agile development process in its server development division.[36] According to a Microsoft senior vice president, "We just realized that we're building products for customers, not just for technology's sake. So the sooner we could engage with our customers, the better we could make it from an architecture, feature, quality, and scalability perspective." BT Group, a large British telecommunications company, uses agile systems development to substantially reduce development time and increase customer satisfaction.[37] According to BT's managing director of service design, "BT's shift from traditional waterfall development techniques to an agile approach has led to significant productivity and business benefits, but it didn't happen overnight, nor was it easy for a company as massive and widespread as BT." Extreme programming (XP) uses pairs of programmers who work together to design, test, and code parts of the systems they develop.[38] The iterative nature of XP helps companies develop robust systems with fewer errors. Sabre Airline Solutions, a $2 billion computer company serving the airline travel industry, used XP to eliminate programming errors and shorten program development times.

joint application development (JAD)
A process for data collection and requirements analysis in which users, stakeholders, and IS professionals work together to analyze existing systems, propose possible solutions, and define the requirements of a new or modified system.

RAD makes extensive use of the **joint application development** (**JAD**) process for data collection and requirements analysis. Originally developed by IBM Canada in the 1970s, JAD involves group meetings in which users, stakeholders, and IS professionals work together to analyze existing systems, propose possible solutions, and define the requirements of a new or modified system. Today, JAD often uses *group support systems (GSS)* software to foster positive group interactions, while suppressing negative group behavior. Boeing, for example, used RAD and JAD to help develop software for its airplanes.[39] Group support systems were introduced in Chapter 10.

RAD should not be used on every software development project. In general, it is best suited for DSSs and MISs and less well suited for TPSs. During a RAD project, the level of participation of stakeholders and users is much higher than in other approaches. Table 12.3 lists advantages and disadvantages of RAD.

Table 12.3

Advantages and Disadvantages of RAD

Advantages	Disadvantages
For appropriate projects, this approach puts an application into production sooner than any other approach.	This intense SDLC can burn out systems developers and other project participants.
Documentation is produced as a by-product of completing project tasks.	This approach requires systems analysts and users to be skilled in RAD systems development tools and RAD techniques.
RAD forces teamwork and lots of interaction between users and stakeholders.	RAD requires a larger percentage of stakeholders' and users' time than other approaches.

In addition to the systems development approaches discussed previously, a number of other systems development approaches are available, including adaptive software development, lean software development, Rational Unified Process (RUP), Feature-Driven Development (FDD), and dynamic systems development methods. Often created by computer vendors and authors of systems development books, these approaches all attempt to deliver better systems. The Ohio Casualty Corporation, for example, uses RUP from IBM and Rational Software. RUP uses an iterative approach to software development that concentrates on software quality as it is changed and updated over time.[40] Many other companies have also used RUP to their advantage.[41]

The End-User Systems Development

The term **end-user systems development** describes any systems development project in which business managers and users assume the primary effort. User-developed systems range from the very small (such as a software routine to merge form letters) to those of significant organizational value (such as customer contact databases for the Web). With end-user systems development, managers and other users can get the systems they want without having to wait for IS professionals to develop and deliver them.[42] End-user systems development, however, does have some disadvantages. Some end users don't have the training to effectively develop and test a system. Multimillion-dollar mistakes, for example, can be made using faulty spreadsheets that were never tested. Some end-user systems are also poorly documented. When these systems are updated, problems can be introduced that make the systems error-prone. In addition, some end users spend time and corporate resources developing systems that were already available.

end-user systems development
Any systems development project in which business managers and users assume the primary effort.

Many end users today are demonstrating their systems development capability by designing and implementing their own PC-based systems.

(Source: © Daniel Allan/Getty Images.)

Outsourcing and On-Demand Computing

Many companies hire an outside consulting firm or computer company that specializes in systems development to take over some or all of its development and operations activities.[43] Some companies, such as General Electric, have their own outsourcing subunits or have spun off their outsourcing subunits as separate companies. As mentioned in Chapter 2, *outsourcing* and *on-demand computing* are often used.[44] Table 12.4 describes the circumstances in which outsourcing is a good idea.

Reason	Example
When a company believes it can cut costs	PacifiCare outsourced its IS operations to IBM and Keane, Inc. PacifiCare hopes the outsourcing will save it about $400 million over ten years.
When a firm has limited opportunity to distinguish itself competitively through a particular IS operation or application	Kodak outsourced its IS operations, including mainframe processing, telecommunications, and personal computer support, because it had limited opportunity to distinguish itself through these IS operations. Kodak kept application development and support in-house because it thought that these activities had competitive value.
When outsourcing does not strip the company of technical know-how required for future IS innovation	Firms must ensure that their IS staffs remain technically up-to-date and have the expertise to develop future applications.
When the firm's existing IS capabilities are limited, ineffective, or technically inferior	A company might use outsourcing to help it make the transition from a centralized mainframe environment to a distributed client/server environment.
When a firm is downsizing	First Fidelity, a major bank, used outsourcing as part of a program to reduce the number of employees by 1,600 and slash expenses by $85 million.

Table 12.4

When to Use Outsourcing for Systems Development

Increasingly, small and medium-sized firms are using outsourcing to cut costs and acquire needed technical expertise that would be difficult to afford with in-house personnel. Millennium Partners Sports Club Management, for example, used Center Beam to outsource many of its IS functions, including its help desk operations. The Boston-based company plans to spend about $30,000 a month on outsourcing services, which it estimates to be less than it would have to pay in salaries for additional employees.[45] According to a company vice president, "If we hadn't outsourced, I couldn't focus 100 percent on things that can drive the company forward." The market for outsourcing services for small and medium-sized firms is expected to increase by 15 percent annually through 2010 and beyond.

Reducing costs, obtaining state-of-the-art technology, eliminating staffing and personnel problems, and increasing technological flexibility are reasons that companies have used the outsourcing and on-demand computing approaches.[46] Reducing costs is a primary reason for outsourcing. One American computer company, for example, estimated that a programmer with three to five years of experience in China would cost about $13 per hour, while a programmer with similar experience in the United States would cost about $56 per hour. U.S. companies also provide outsourcing services. Aelera Corporation spent about six months looking for the best outsourcing deal and determined that a company in Savannah, Georgia was the best. McKesson Corporation saved about $10 million by outsourcing jobs from San Francisco to Dubuque, Iowa. Mattel outsourced to rural Jonesboro, Arkansas. Increasingly, companies are looking to American outsourcing companies to reduce costs and increase services. Individuals, including students, are also outsourcing tasks they have to perform.

Companies often use several outsourcing services. GM, the large automotive company, has used six outsourcing companies since its outsourcing agreement with EDS expired.[47] Using more than one outsourcing company can increase competition and reduce outsourcing costs. According to one GM executive, "That's really just Economics 101."

A number of companies and nonprofit organizations offer outsourcing and on-demand computing services—from general systems development to specialized services.[48] IBM's Global Services, for example, is one of the largest full-service outsourcing and consulting services.[49] IBM has consultants located in offices around the world. In India, IBM has increased its employees from less than 10,000 people to more than 30,000.[50] The company looks for skilled and talented workers. According to Amitabh Ray, a vice president of consulting and application services at IBM in India, "We're not a body shop. We need the right kind of people." Electronic Data Systems (EDS) is another large company that specializes in consulting and outsourcing.[51] EDS has approximately 140,000 employees in almost 60 countries and more than 9,000 clients worldwide. Accenture is another company that specializes in consulting and outsourcing.[52] The company has more than 75,000 employees in 47 countries. Amazon, the large online retailer of books and other products, will offer on-demand computing to individuals and other companies of all sizes, allowing them to use Amazon's computer expertise and database capacity.[53] Individuals and companies will only pay for the computer services they use. See Figure 12.9.

Figure 12.9

With more than 75,000 employees in 47 countries, Accenture specializes in consulting and outsourcing.

(Source: © Namas Bhojani/ Bloomberg News/Landov.)

Outsourcing has some disadvantages, however. Some companies, such as J. Crew, are starting to reduce their use of outsourcing and bring systems development back in-house.[54] Internal expertise can be lost and loyalty can suffer under an outsourcing arrangement. When a company outsources, key IS personnel with expertise in technical and business functions are no longer needed. When these IS employees leave, their experience with the organization and expertise in information systems is lost. For some companies, it can be difficult to achieve a competitive advantage when competitors are using the same computer or consulting company. When the outsourcing or on-demand computing is done offshore or in a foreign country, some people have raised security concerns. How will important data and trade secrets be guarded? U.S. federal authorities often investigate defense contractors for improper outsourcing.[55] In one case, a company was fined about $100 million for violating a federal arms export control law.

FACTORS AFFECTING SYSTEMS DEVELOPMENT SUCCESS

Successful systems development means delivering a system that meets user and organizational needs—on time and within budget. Getting users and stakeholders involved in systems development is critical for most systems development projects. Having the support of top-level managers is also important. In addition to user involvement and top management support, other factors can contribute to successful systems development efforts—at a reasonable cost. These factors are discussed next.

Degree of Change

A major factor that affects the quality of systems development is the degree of change associated with the project. The scope can vary from enhancing an existing system to major reengineering. The project team needs to recognize where they are on this spectrum of change.

Continuous Improvement versus Reengineering

As discussed in Chapter 2, continuous improvement projects do not require significant business process or IS changes or retraining of people; thus, they have a high degree of success. Typically, because continuous improvements involve minor improvements, these projects also have relatively modest benefits. On the other hand, reengineering involves fundamental changes in how the organization conducts business and completes tasks. The factors associated with successful reengineering are similar to those of any development effort, including top management support, clearly defined corporate goals and systems development objectives, and careful management of change. Major reengineering projects tend to have a high degree of risk but also a high potential for major business benefits (see Figure 12.10).

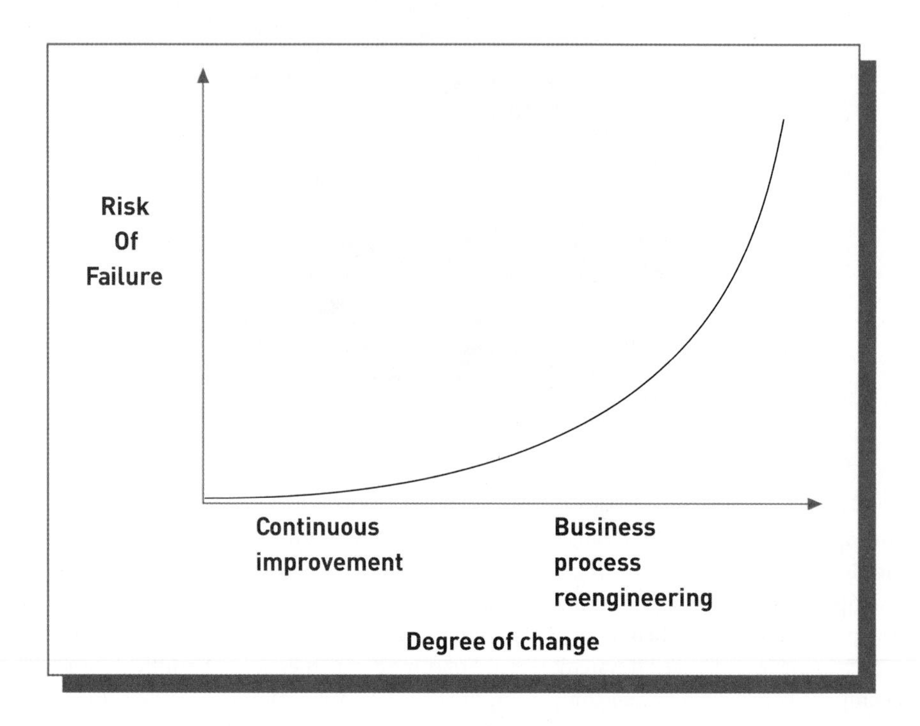

Figure 12.10

The degree of change can greatly affect the probability of a project's success.

Managing Change

The ability to manage change is critical to the success of systems development. New systems inevitably cause change. For example, the work environment and habits of users are invariably affected by the development of a new information system. Unfortunately, not everyone adapts easily, and the increasing complexity of systems can multiply the problems. Managing change requires the ability to recognize existing or potential problems (particularly the concerns of users) and deal with them before they become a serious threat to the success of the new or modified system. Here are several of the most common problems that often need to be addressed as a result of new or modified systems:

- Fear that the employee will lose his job, power, or influence within the organization
- Belief that the proposed system will create more work than it eliminates
- Reluctance to work with "computer people"
- Anxiety that the proposed system will negatively alter the structure of the organization
- Belief that other problems are more pressing than those solved by the proposed system or that the system is being developed by people unfamiliar with "the way things need to get done"
- Unwillingness to learn new procedures or approaches

ETHICAL AND SOCIETAL ISSUES

When Systems Development Fails

Systems analysts and developers carry a significant amount of responsibility on their shoulders. Information systems play an important role in the success of today's businesses, and a faulty system can mean the end of a business or worse. When information systems mean life and death to clients, much more is at stake than a business's reputation.

Such was the case with a major healthcare organization, which for the sake of anonymity is called HCO in this article. In 2004, HCO decided it would be more economical to handle all of its kidney transplants itself rather than using a nearby university medical center. HCO built a new kidney transplant center to handle their kidney transplant patients and named a director. The director began to transfer all the patient records from the university medical center to the new center—over 1,500 patients in all.

However, rather than coordinating with the university medical center to transfer patients and their data from one information system to the other, HCO decided to forgo the usual systems development process and rush the transition.

The staff at the previous medical facility found themselves ill-equipped to process and transfer the large number of patient records to the new center in the necessary amount of time. They discovered that the data in many patient records was incorrect, and until they corrected it, the center's staff could not process the patient records. Managing kidney transplants is complex and time sensitive. Kidneys are in rare supply and those eligible for transplants spend time on a waiting list, hoping they will be called before their own kidneys give out. Due to the glitch in data transfer, hundreds of patient records were lost.

To make matters worse, the new transplant center was understaffed and underfunded. Because it did not have proper information systems, the staff at the new center maintained medical records primarily on paper. They did not have a system to determine if any patient records were lost in the transfer, nor could patients use a system to voice concern or lodge a complaint.

Over two years, patients whose records had failed to transfer to the new facility were still waiting for the call for a new kidney that would never come. Finally, based on a whistleblower's story, a local TV station and newspaper began pressing the new center to reveal why patients were waiting longer than usual for transplants. The investigation quickly led to formal litigation against HCO on a number of counts, not the least of which was HCO's failure to adhere to five state and 15 federal regulations mostly dealing with the management of patient records. The state Department of Managed Health Care (DMHC) has concluded that the problems experienced by HCO and its patients are due to "lack of effective planning" and that the absence of proper information management posed "potentially life-threatening delays in care."

In fact, "potential" appears to be "actual" as further investigation shows that in the first year of operations, twice the typical number of patient deaths were caused by an extended wait for kidneys. Professionals in the transplant business say that this is the worst problem the industry has ever seen.

Eventually HCO abandoned its plans for a new center and returned all of its kidney patients to their previous care. The organization has paid $2 million in fines to the state Department of Managed Health Care (DMHC) and volunteered another $3 million in contributions to a transplant education group. Meanwhile, over 50 patients and families of people who died waiting for kidneys are suing the organization in separate cases, mostly for negligence or wrongful death.

As investigators sort through this case seeking an explanation of exactly what went wrong, those involved are accepting some blame, but also pointing fingers at each other. One thing is clear: Had proper systems development practices been put into place, the new kidney transplant center would be operational, patients lives would have been saved, and the reputation of the previously well-respected HCO would still be sparkling.

Discussion Questions

1. What went wrong at HCO? Who paid the price?
2. How is HCO responding to its mistakes, and how might it further regain its good reputation?

Critical Thinking Questions

1. What legitimate reasons might HCO's director provide for the failure of the new center? Is there any acceptable excuse? Who within HCO is ultimately to blame?
2. What other life-threatening or life-saving information systems are at risk of similar catastrophes?

SOURCES: Gage, Deborah, "We Really Did Screw Up," *CIO Insight*, May 14, 2008, *www.cioinsight.com/c/a/Past-News/QTEWe-Really-Did-Screw-UpQTE*; Kaiser Permanente Web site, *www.kaiserpermanente.org*, accessed July 12, 2008.

Quality and Standards

Quality and standards are other key success factors for systems development. Increasingly, corporations are expanding their standards to include many different computer platforms. While many companies try to standardize their operations on one operating system, others have multiple systems and platforms to take advantage of the strengths of each.[56] In these cases, many IS managers seek one tool to manage everything. According to a Clear Channel IS executive, "I don't care what enterprise you walk into, they're not going to be single-platform. I want to manage it all from one spot." Today, many companies, including Microsoft, are developing software and systems that can be used to manage different operating systems and software products. In addition, organizations that do business around the globe may be required to meet certain international standards, such as ISO 9000, a set of international quality standards originally developed in Europe in 1987.

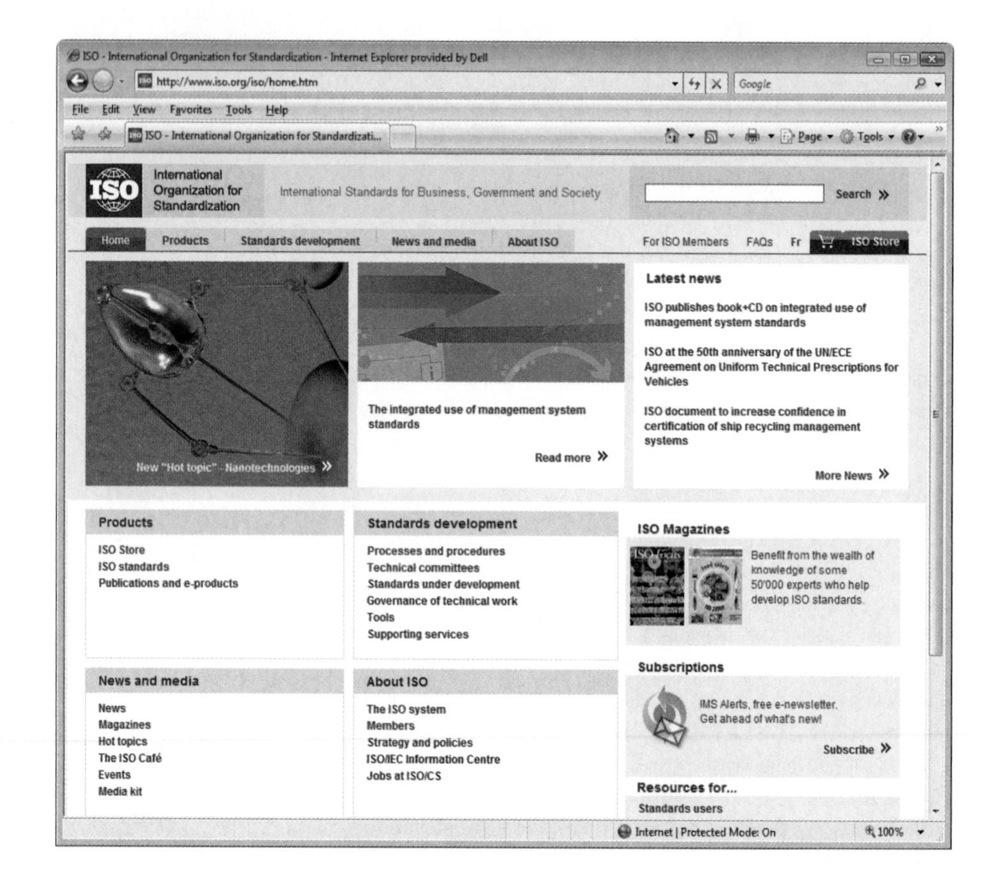

ISO 9000 is a set of international quality standards used by IS and other organizations to ensure the quality of products and services.

(Source: *www.iso.org*.)

The bigger the project, the more likely that poor planning will lead to significant problems. Many companies find that large systems projects fall behind schedule, go over budget, and do not meet expectations. Although proper planning cannot guarantee that these types of problems will be avoided, it can minimize the likelihood of their occurrence. Good systems development is not automatic. Certain factors contribute to the failure of systems development projects. These factors and the countermeasures to eliminate or alleviate the problem are summarized in Table 12.5.

Organizational experience with the systems development process is also an important factor for systems development success.[57] The *Capability Maturity Model (CMM)* is one way to measure this experience. It is based on research done at Carnegie Mellon University and work by the Software Engineering Institute (SEI). CMM is a measure of the maturity of the software development process in an organization. CMM grades an organization's systems development maturity using five levels: initial, repeatable, defined, managed, and optimized.[58]

Table 12.5

Project Planning Issues Frequently Contributing to Project Failure

Factor	Countermeasure
Solving the wrong problem	Establish a clear connection between the project and organizational goals.
Poor problem definition and analysis	Follow a standard systems development approach.
Poor communication	Set up communications procedures and protocols.
Project is too ambitious	Narrow the project focus to address only the most important business opportunities.
Lack of top management support	Identify the senior manager who has the most to gain from the success of the project, and recruit this person to champion the project.
Lack of management and user involvement	Identify and recruit key stakeholders to be active participants in the project.
Inadequate or improper system design	Follow a standard systems development approach.
Lack of standards	Implement a standards system, such as ISO 9001.

Use of Project Management Tools

Project management involves planning, scheduling, directing, and controlling human, financial, and technological resources for a defined task whose result is achievement of specific goals and objectives.[59] Corporations and nonprofit organizations use these important tools and techniques. As an academic exercise, for example, Purdue University undertook a project to build a supercomputer using off-the-shelf PCs. The project was completed in a day and required more than 800 PCs.[60]

A **project schedule** is a detailed description of what is to be done. Each project activity, the use of personnel and other resources, and expected completion dates are described. A **project milestone** is a critical date for the completion of a major part of the project. The completion of program design, coding, testing, and release are examples of milestones for a programming project. The **project deadline** is the date the entire project is to be completed and operational—when the organization can expect to begin to reap the benefits of the project. One company offers a 20 percent refund if it doesn't meet a client's project deadline. In addition, any additional work done after the project deadline is performed free of charge.

project schedule
A detailed description of what is to be done.

project milestone
A critical date for the completion of a major part of the project.

project deadline
The date the entire project is to be completed and operational.

In systems development, each activity has an earliest start time, earliest finish time, and slack time, which is the amount of time an activity can be delayed without delaying the entire project. The **critical path** consists of all activities that, if delayed, would delay the entire project. These activities have zero slack time. Any problems with critical-path activities will cause problems for the entire project. To ensure that critical-path activities are completed in a timely fashion, formalized project management approaches have been developed. Tools such as Microsoft Project are available to help compute these critical project attributes.

critical path
Activities that, if delayed, would delay the entire project.

Although the steps of systems development seem straightforward, larger projects can become complex, requiring hundreds or thousands of separate activities. For these systems development efforts, formal project management methods and tools become essential. A formalized approach called **Program Evaluation and Review Technique** (**PERT**) creates three time estimates for an activity: shortest possible time, most likely time, and longest possible time. A formula is then applied to determine a single PERT time estimate. A **Gantt chart** is a graphical tool used for planning, monitoring, and coordinating projects; it is essentially a grid that lists activities and deadlines. Each time a task is completed, a marker such as a darkened line is placed in the proper grid cell to indicate the completion of a task (see Figure 12.11).

Program Evaluation and Review Technique (PERT)
A formalized approach for developing a project schedule that creates three time estimates for an activity.

Gantt chart
A graphical tool used for planning, monitoring, and coordinating projects.

Both PERT and Gantt techniques can be automated using project management software. Several project management software packages are identified in Table 12.6. This software monitors all project activities and determines whether activities and the entire project are on time and within budget. Project management software also has workgroup capabilities to handle multiple projects and to allow a team to interact with the same software. Project management software helps managers determine the best way to reduce project completion

Figure 12.11

Sample Gantt Chart

A Gantt chart shows progress through systems development activities by putting a bar through appropriate cells.

PROJECT PLANNING DOCUMENTATION — Page 1 of 1

System: Warehouse Inventory System (Modification) — Date: 12/10

System: — Scheduled activity; ▬ Completed activity — Analyst: Cecil Truman — Signature:

Activity*	Individual assigned	Week 1	2	3	4	5	6	7	8	9	10	11	12	13	14
R — Requirements definition															
R.1 Form project team	VP, Cecil, Bev														
R.2 Define obj. and constraints	Cecil														
R.3 Interview warehouse staff for requirements report	Bev														
R.4 Organize requirements	Team														
R.5 VP review	VP, Team														
D — Design															
D.1 Revise program specs.	Bev														
D. 2. 1 Specify screens	Bev														
D. 2. 2 Specify reports	Bev														
D. 2. 3 Specify doc. changes	Cecil														
D. 4 Management review	Team														
I — Implementation															
I. 1 Code program changes	Bev														
I. 2. 1 Build test file	Team														
I. 2. 2 Build production file	Bev														
I. 3 Revise production file	Cecil														
I. 4. 1 Test short file	Bev														
I. 4. 2 Test production file	Cecil														
I. 5 Management review	Team														
I. 6 Install warehouse**															
I. 6. 1 Train new procedures	Bev														
I. 6. 2 Install	Bev														
I. 6. 3 Management review	Team														

*Weekly team reviews not shown here
**Report for warehouses 2 through 5

Table 12.6

Selected Project Management Software

Software	Vendor
AboutTime	NetSQL Partners *(www.netsql.com)*
OpenPlan	Welcom *(www.welcom.com)*
Microsoft Project	Microsoft *(www.microsoft.com)*
Unifier	Skire *(www.skire.com)*
Project Scheduler	Scitor *(www.scitor.com)*

time at the least cost. In what some people believe is the largest private construction project in the United States, MGM Mirage and others used project management software to help them embark on an ambitious $8 billion construction project on 76 acres with over 4,000 hotel rooms, retail spaces, and other developments.[61] To complete the project, managers selected Skire's Unifier (*www.skire.com*), a powerful and flexible project management software package. The project management software should save the developer a substantial amount of money. According to a company spokesperson, "Any incremental improvement can go right to the bottom line when you're building on 76 acres of prime Las Vegas Strip real estate valued at $30 million an acre."

Use of Computer-Aided Software Engineering (CASE) Tools

Computer-aided software engineering (CASE) tools automate many of the tasks required in a systems development effort and encourage adherence to the SDLC, thus instilling a high degree of rigor and standardization to the entire systems development process. Prover Technology has developed a CASE tool that searches for programming bugs. The CASE tool searches for all possible design scenarios to make sure that the program is error free. Other CASE tools include Visible Systems (*www.visible.com*), Popkin Software (*www.popkin.com*), Rational Rose (part of IBM), and Visio, a charting and graphics program from Microsoft. Companies that produce CASE tools include Accenture, Microsoft, and Oracle. Oracle Designer and Developer CASE tools, for example, can help systems analysts automate and simplify the development process for database systems. See Table 12.7 for a list of CASE tools and their providers. The advantages and disadvantages of CASE tools are listed in Table 12.8. CASE tools that focus on activities associated with the early stages of systems development are often called *upper-CASE* tools. These packages provide automated tools to assist with systems investigation, analysis, and design activities. Other CASE packages, called *lower-CASE* tools, focus on the later implementation stage of systems development, and can automatically generate structured program code.

computer-aided software engineering (CASE)
Tools that automate many of the tasks required in a systems development effort and encourage adherence to the SDLC.

Table 12.7

Typical CASE Tools

CASE Tool	Vendor
Oracle Designer	Oracle Corporation *www.oracle.com*
Visible Analyst	Visible Systems Corporation *www.visible.com*
Rational Rose	Rational Software *www.ibm.com*
Embarcadero Describe	Embarcadero Describe *www.embarcadero.com*

Table 12.8

Advantages and Disadvantages of CASE Tools

Advantages	Disadvantages
Produce systems with a longer effective operational life	Increase the initial costs of building and maintaining systems
Produce systems that more closely meet user needs and requirements	Require more extensive and accurate definition of user needs and requirements
Produce systems with excellent documentation	Can be difficult to customize
Produce systems that need less systems support	Require more training of maintenance staff
Produce more flexible systems	Can be difficult to use with existing systems

Object-Oriented Systems Development

The success of a systems development effort can depend on the specific programming tools and approaches used. As mentioned in Chapter 4, object-oriented (OO) programming languages allow the interaction of programming objects—that is, an object consists of both data and the actions that can be performed on the data. So, an object could be data about an employee and all the operations (such as payroll, benefits, and tax calculations) that might be performed on the data.

Developing programs and applications using OO programming languages involves constructing modules and parts that can be reused in other programming projects. DTE Energy, a $7 billion Detroit-based energy company, has set up a library of software components that can be reused by its programmers. Systems developers from the company reuse and contribute to software components in the library. DTE's developers meet frequently to discuss ideas, problems, and opportunities of using the library of reusable software components.

Chapter 4 discussed a number of programming languages that use the object-oriented approach, including Visual Basic, C++, and Java. These languages allow systems developers to take the OO approach, making program development faster and more efficient, resulting in lower costs. Modules can be developed internally or obtained from an external source. After a company has the programming modules, programmers and systems analysts can modify them and integrate them with other modules to form new programs.

object-oriented systems development (OOSD)
An approach to systems development that combines the logic of the systems development life cycle with the power of object-oriented modeling and programming.

Object-oriented systems development (OOSD) combines the logic of the systems development life cycle with the power of object-oriented modeling and programming. OOSD follows a defined systems development life cycle, much like the SDLC. The life cycle phases are usually completed with many iterations. Object-oriented systems development typically involves the following tasks:

- **Identifying potential problems and opportunities within the organization that would be appropriate for the OO approach.** This process is similar to traditional systems investigation. Ideally, these problems or opportunities should lend themselves to the development of programs that can be built by modifying existing programming modules.
- **Defining what kind of system users require.** This analysis means defining all the objects that are part of the user's work environment (object-oriented analysis). The OO team must study the business and build a model of the objects that are part of the business (such as a customer, an order, or a payment). Many of the CASE tools discussed in the previous section can be used, starting with this step of OOSD.
- **Designing the system.** This process defines all the objects in the system and the ways they interact (object-oriented design). Design involves developing logical and physical models of the new system by adding details to the object model started in analysis.
- **Programming or modifying modules.** This implementation step takes the object model begun during analysis and completed during design and turns it into a set of interacting objects in a system. Object-oriented programming languages are designed to allow the programmer to create classes of objects in the computer system that correspond to the objects in the actual business process. Objects such as customer, order, and payment are redefined as computer system objects—a customer screen, an order entry menu, or a dollar sign icon. Programmers then write new modules or modify existing ones to produce the desired programs.
- **Evaluation by users.** The initial implementation is evaluated by users and improved. Additional scenarios and objects are added, and the cycle repeats. Finally, a complete, tested, and approved system is available for use.
- **Periodic review and modification.** The completed and operational system is reviewed at regular intervals and modified as necessary.

SYSTEMS INVESTIGATION

As discussed earlier in the chapter, systems investigation is the first phase in the traditional SDLC of a new or modified business information system. The purpose is to identify potential problems and opportunities and consider them in light of the goals of the company. In general, systems investigation attempts to uncover answers to the following questions:

- What primary problems might a new or enhanced system solve?
- What opportunities might a new or enhanced system provide?
- What new hardware, software, databases, telecommunications, personnel, or procedures will improve an existing system or are required in a new system?
- What are the potential costs (variable and fixed)?
- What are the associated risks?

Initiating Systems Investigation

Because systems development requests can require considerable time and effort to implement, many organizations have adopted a formal procedure for initiating systems development, beginning with systems investigation. The **systems request form** is a document that is filled out by someone who wants the IS department to initiate systems investigation. This form typically includes the following information:

systems request form
A document filled out by someone who wants the IS department to initiate systems investigation.

- Problems in or opportunities for the system
- Objectives of systems investigation
- Overview of the proposed system
- Expected costs and benefits of the proposed system

The information in the systems request form helps to rationalize and prioritize the activities of the IS department. Based on the overall IS plan, the organization's needs and goals, and the estimated value and priority of the proposed projects, managers make decisions regarding the initiation of each systems investigation for such projects.

Participants in Systems Investigation

After a decision has been made to initiate systems investigation, the first step is to determine what members of the development team should participate in the investigation phase of the project. Members of the development team change from phase to phase (see Figure 12.12). The systems investigation team can be diverse, with members located around the world. When Nokia decided to develop a new cell phone, its investigation team members were from England, Finland, and the United States.[62] Cooperation and collaboration are keys to successful investigation teams.

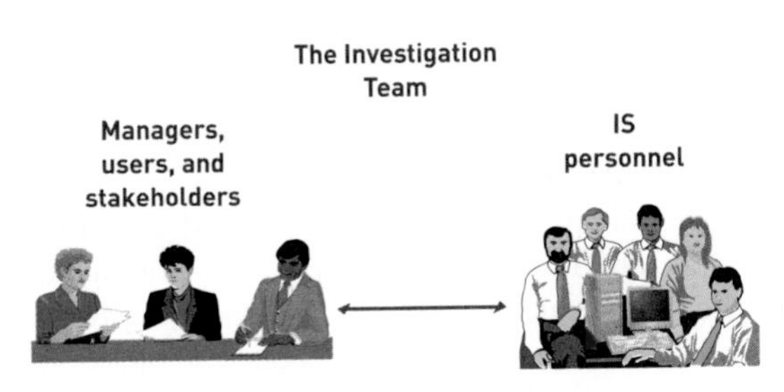
The Investigation Team
Managers, users, and stakeholders
IS personnel
• Undertakes feasibility analysis
• Establishes systems development goals
• Selects systems development methodology
• Prepares systems investigation report

Figure 12.12

The Systems Investigation Team

The team consists of upper- and middle-level managers, a project manager, IS personnel, users, and stakeholders.

In some cases, the participants in systems investigation are asked to step into big problems and fix them. When a new CIO at a major university was asked to investigate a security breach, he was a little apprehensive.[63] "I was literally walking into a river of alligators, but that's not always a bad thing. It can be a character building thing," he said.

Ideally, functional managers are heavily involved during the investigation phase. Other members could include users or stakeholders outside management, such as an employee who helped initiate systems development. The technical and financial expertise of others participating in investigation would help the team determine whether the problem is worth solving. The members of the development team who participate in investigation are then responsible for gathering and analyzing data, preparing a report justifying systems development, and presenting the results to top-level managers.

feasibility analysis
Assessment of the technical, economic, legal, operational, and schedule feasibility of a project.

Feasibility Analysis

A key step of the systems investigation phase is **feasibility analysis**, which assesses technical, economic, legal, operational, and schedule feasibility (see Figure 12.13). **Technical feasibility** is concerned with whether the hardware, software, and other system components can be acquired or developed to solve the problem.

technical feasibility
Assessment of whether the hardware, software, and other system components can be acquired or developed to solve the problem.

Figure 12.13

Technical, Economic, Legal, Operational, and Schedule Feasibility

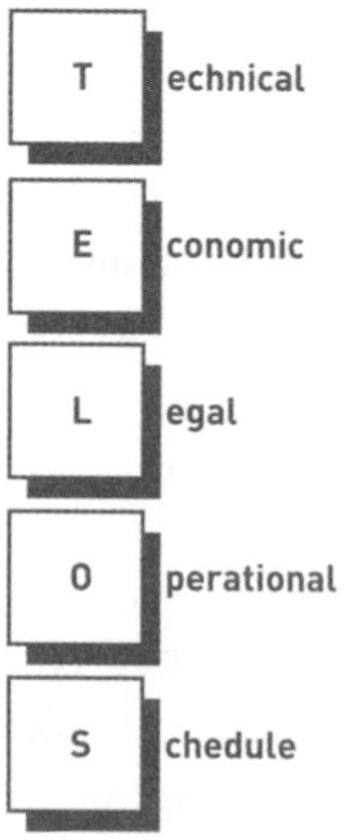

T echnical
E conomic
L egal
O perational
S chedule

economic feasibility
The determination of whether the project makes financial sense and whether predicted benefits offset the cost and time needed to obtain them.

net present value
The net amount by which project savings exceed project expenses after allowing for the cost of capital and the passage of time.

legal feasibility
The determination of whether laws or regulations may prevent or limit a systems development project.

operational feasibility
The measure of whether the project can be put into action or operation.

schedule feasibility
The determination of whether the project can be completed in a reasonable amount of time.

Economic feasibility determines whether the project makes financial sense and whether predicted benefits offset the cost and time needed to obtain them. A securities company, for example, investigated the economic feasibility of sending research reports electronically instead of through the mail. Economic analysis revealed that the new approach could save the company up to $500,000 per year. Economic feasibility can involve cash flow analysis such as that done in net present value or internal rate of return (IRR) calculations.

Net present value is an often-used approach for ranking competing projects and for determining economic feasibility. The net present value represents the net amount by which project savings exceed project expenses, after allowing for the cost of capital and the passage of time. The cost of capital is the average cost of funds used to finance the operations of the business. Net present value takes into account that a dollar returned at a later date is not worth as much as one received today, because the dollar in hand can be invested to earn profits or interest in the interim. Spreadsheet programs, such as Lotus and Microsoft Excel, have built-in functions to compute the net present value and internal rate of return.

Legal feasibility determines whether laws or regulations can prevent or limit a systems development project. For example, a Web site that allowed users to share music without paying musicians or music producers was sued. Legal feasibility should have identified this vulnerability during the Web site development project. Legal feasibility involves an analysis of existing and future laws to determine the likelihood of legal action against the systems development project and the possible consequences.

Operational feasibility is a measure of whether the project can be put into action or operation. It can include logistical and motivational (acceptance of change) considerations. Motivational considerations are important because new systems affect people and data flows and can have unintended consequences. As a result, power and politics might come into play, and some people might resist the new system. On the other hand, recall that a new system can help avoid major problems. For example, because of deadly hospital errors, a healthcare consortium looked into the operational feasibility of developing a new computerized physician order-entry system to require that all prescriptions and every order a doctor gives to staff be entered into the computer. The computer then checks for drug allergies and interactions between drugs. If operationally feasible, the new system could save lives and help avoid lawsuits.

Schedule feasibility determines whether the project can be completed in a reasonable amount of time—a process that involves balancing the time and resource requirements of the project with other projects.

Object-Oriented Systems Investigation

The object-oriented approach can be used during all phases of systems development, from investigation to maintenance and review. Consider a kayak rental business in Maui, Hawaii, where the owner wants to computerize its operations, including renting kayaks to customers and adding new kayaks into the rental program (see Figure 12.14). As you can see, the kayak rental clerk rents kayaks to customers and adds new kayaks to the current inventory available for rent. The stick figure is an example of an *actor*, and the ovals each represent an event, called a *use case*. In our example, the actor (the kayak rental clerk) interacts with two use cases (rent kayaks to customers and add new kayaks to inventory). The use case diagram is part of the Unified Modeling Language (UML) that is used in object-oriented systems development.

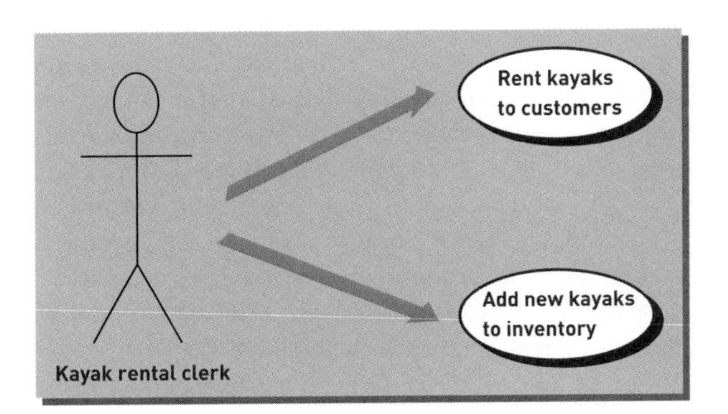

Figure 12.14

Use Case Diagram for a Kayak Rental Application

The Systems Investigation Report

The primary outcome of systems investigation is a **systems investigation report**, also called a *feasibility study*. This report summarizes the results of systems investigation and the process of feasibility analysis and recommends a course of action: continue on into systems analysis, modify the project in some manner, or drop it. A typical table of contents for the systems investigation report is shown in Figure 12.15.

systems investigation report
A summary of the results of the systems investigation and the process of feasibility analysis and recommendation of a course of action.

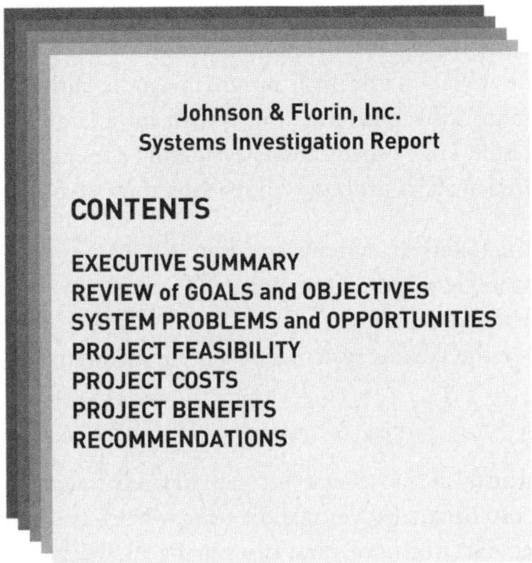

Johnson & Florin, Inc.
Systems Investigation Report

CONTENTS

EXECUTIVE SUMMARY
REVIEW of GOALS and OBJECTIVES
SYSTEM PROBLEMS and OPPORTUNITIES
PROJECT FEASIBILITY
PROJECT COSTS
PROJECT BENEFITS
RECOMMENDATIONS

Figure 12.15

A Typical Table of Contents for a Systems Investigation Report

The systems investigation report is reviewed by senior management, often organized as an advisory committee, or **steering committee**, consisting of senior management and users from the IS department and other functional areas. These people help IS personnel with their decisions about the use of information systems in the business and give authorization to pursue further systems development activities. After review, the steering committee might agree with the recommendation of the systems development team or suggest a change in project focus to concentrate more directly on meeting a specific company objective. Another alternative is that everyone might decide that the project is not feasible and cancel the project.

steering committee
An advisory group consisting of senior management and users from the IS department and other functional areas.

SYSTEMS ANALYSIS

After a project has been approved for further study, the next step is to answer the question "What must the information system do to solve the problem?" The process needs to go beyond mere computerization of existing systems. The entire system, and the business process with which it is associated, should be evaluated. Often, a firm can make great gains if it restructures both business activities and the related information system simultaneously. The overall emphasis of analysis is gathering data on the existing system, determining the requirements for the new system, considering alternatives within these constraints, and investigating the feasibility of the solutions. The primary outcome of systems analysis is a prioritized list of systems requirements. During its systems analysis phase, Mobius Management Systems (*www.mobius.com*), a company that manages databases and data resources for other companies, determined that the physical size of its data centers was an important systems requirement. Its current data centers were simply too large.[64] According to one IS administrator, "We were taking over what formerly were people's offices and making them data centers." The company analyzed the impact of replacing more than 100 of its hardware servers for software virtualization that allowed multiple applications to run on a single server, saving a tremendous amount of space.

General Considerations

Systems analysis starts by clarifying the overall goals of the organization and determining how the existing or proposed information system helps meet them. A manufacturing company, for example, might want to reduce the number of equipment breakdowns. This goal can be translated into one or more informational needs. One need might be to create and maintain an accurate list of each piece of equipment and a schedule for preventative maintenance. Another need might be a list of equipment failures and their causes.

Analysis of a small company's information system can be fairly straightforward. On the other hand, evaluating an existing information system for a large company can be a long, tedious process. As a result, large organizations evaluating a major information system normally follow a formalized analysis procedure, involving these steps:

1. Assembling the participants for systems analysis
2. Collecting appropriate data and requirements
3. Analyzing the data and requirements
4. Preparing a report on the existing system, new system requirements, and project priorities

Participants in Systems Analysis

The first step in formal analysis is to assemble a team to study the existing system. This group includes members of the original investigation team—from users and stakeholders to IS personnel and management. Most organizations usually allow key members of the development team not only to analyze the condition of the existing system but also to perform other aspects of systems development, such as design and implementation.

After the participants in systems analysis are assembled, this group develops a list of specific objectives and activities. A schedule for meeting the objectives and completing the specific activities is also devised, along with deadlines for each stage and a statement of the resources required at each stage, such as clerical personnel, supplies, and so forth. Major milestones are normally established to help the team monitor progress and determine whether problems or delays occur in performing systems analysis.

Data Collection

The purpose of data collection is to seek additional information about the problems or needs identified in the systems investigation report. During this process, the strengths and weaknesses of the existing system are emphasized.

Identifying Sources of Data

Data collection begins by identifying and locating the various sources of data, including both internal and external sources (see Figure 12.16).

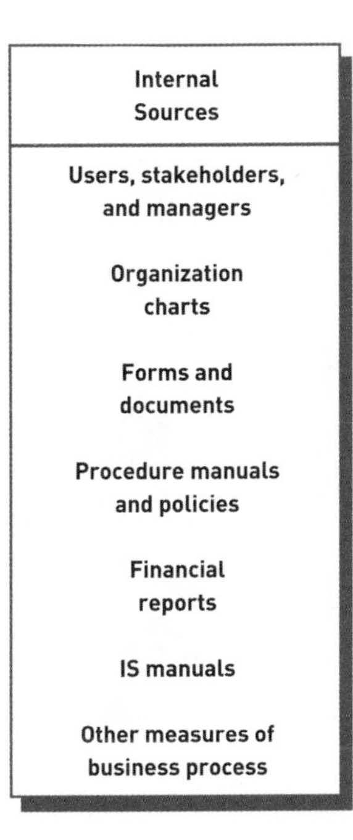

Figure 12.16

Internal and External Sources of Data for Systems Analysis

Collecting Data

After data sources have been identified, data collection begins. Figure 12.17 shows the steps involved. Data collection might require a number of tools and techniques, such as interviews, direct observation, and questionnaires.

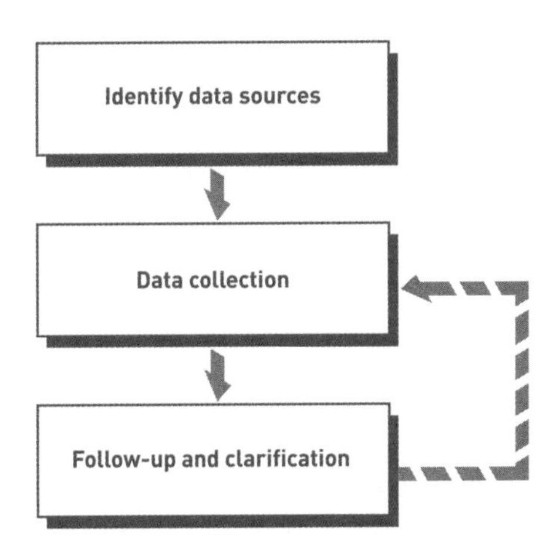

Figure 12.17

The Steps in Data Collection

Interviews can either be structured or unstructured. In a **structured interview**, the questions are written in advance. In an **unstructured interview**, the questions are not written in advance; the interviewer relies on experience in asking the best questions to uncover the inherent problems of the existing system. An advantage of the unstructured interview is that it allows the interviewer to ask follow-up or clarifying questions immediately.

With **direct observation**, one or more members of the analysis team directly observe the existing system in action. One of the best ways to understand how the existing system functions is to work with the users to discover how data flows in certain business tasks. Determining the data flow entails direct observation of users' work procedures, their reports, current screens (if automated already), and so on. From this observation, members of the analysis team determine which forms and procedures are adequate and which are inadequate

structured interview
An interview where the questions are written in advance.

unstructured interview
An interview where the questions are not written in advance.

direct observation
Watching the existing system in action by one or more members of the analysis team.

and need improvement. Direct observation requires a certain amount of skill. The observer must be able to see what is really happening and not be influenced by attitudes or feelings. This approach can reveal important problems and opportunities that would be difficult to obtain using other data collection methods. An example would be observing the work procedures, reports, and computer screens associated with an accounts payable system being considered for replacement.

Direct observation is a method of data collection. One or more members of the analysis team directly observe the existing system in action.

(Source: © Kriss Russell / iStockphoto.)

questionnaires
A method of gathering data when the data sources are spread over a wide geographic area.

When many data sources are spread over a wide geographic area, **questionnaires** might be the best method. Like interviews, questionnaires can be either structured or unstructured. In most cases, a pilot study is conducted to fine-tune the questionnaire. A follow-up questionnaire can also capture the opinions of those who do not respond to the original questionnaire.

Other data collection techniques can also be employed. In some cases, telephone calls are an excellent method. Activities can also be simulated to see how the existing system reacts. Thus, fake sales orders, stockouts, customer complaints, and data-flow bottlenecks can be created to see how the existing system responds to these situations. **Statistical sampling**, which involves taking a random sample of data, is another technique. For example, suppose that you want to collect data that describes 10,000 sales orders received over the last few years. Because it is too time consuming to analyze each of the sales orders, you can collect a random sample of 100 to 200 sales orders from the entire batch. You can assume that the characteristics of this sample apply to all 10,000 orders.

statistical sampling
Selecting a random sample of data and applying the characteristics of the sample to the whole group.

Data Analysis

The data collected in its raw form is usually not adequate to determine the effectiveness of the existing system or the requirements for the new system. The next step is to manipulate the collected data so that the development team members who are participating in systems analysis can use the data. This manipulation is called **data analysis**. Data and activity modeling and using data-flow diagrams and entity-relationship diagrams are useful during data analysis to show data flows and the relationships among various objects, associations, and activities. Other common tools and techniques for data analysis include application flowcharts, grid charts, CASE tools, and the object-oriented approach.

data analysis
The manipulation of collected data so that the development team members who are participating in systems analysis can use the data.

Data Modeling

Data modeling, first introduced in Chapter 5, is a commonly accepted approach to modeling organizational objects and associations that employ both text and graphics. How data modeling is employed, however, is governed by the specific systems development methodology.

Data modeling is most often accomplished through the use of entity-relationship (ER) diagrams. Recall from Chapter 5 that an entity is a generalized representation of an object type—such as a class of people (employee), events (sales), things (desks), or places (city)—and that entities possess certain attributes. Objects can be related to other objects in many ways. An entity-relationship diagram, such as the one shown in Figure 12.18a, describes a number of objects and the ways they are associated. An ER diagram (or any other modeling tool) cannot by itself fully describe a business problem or solution because it lacks descriptions of the related activities. It is, however, a good place to start because it describes object types and attributes about which data might need to be collected for processing.

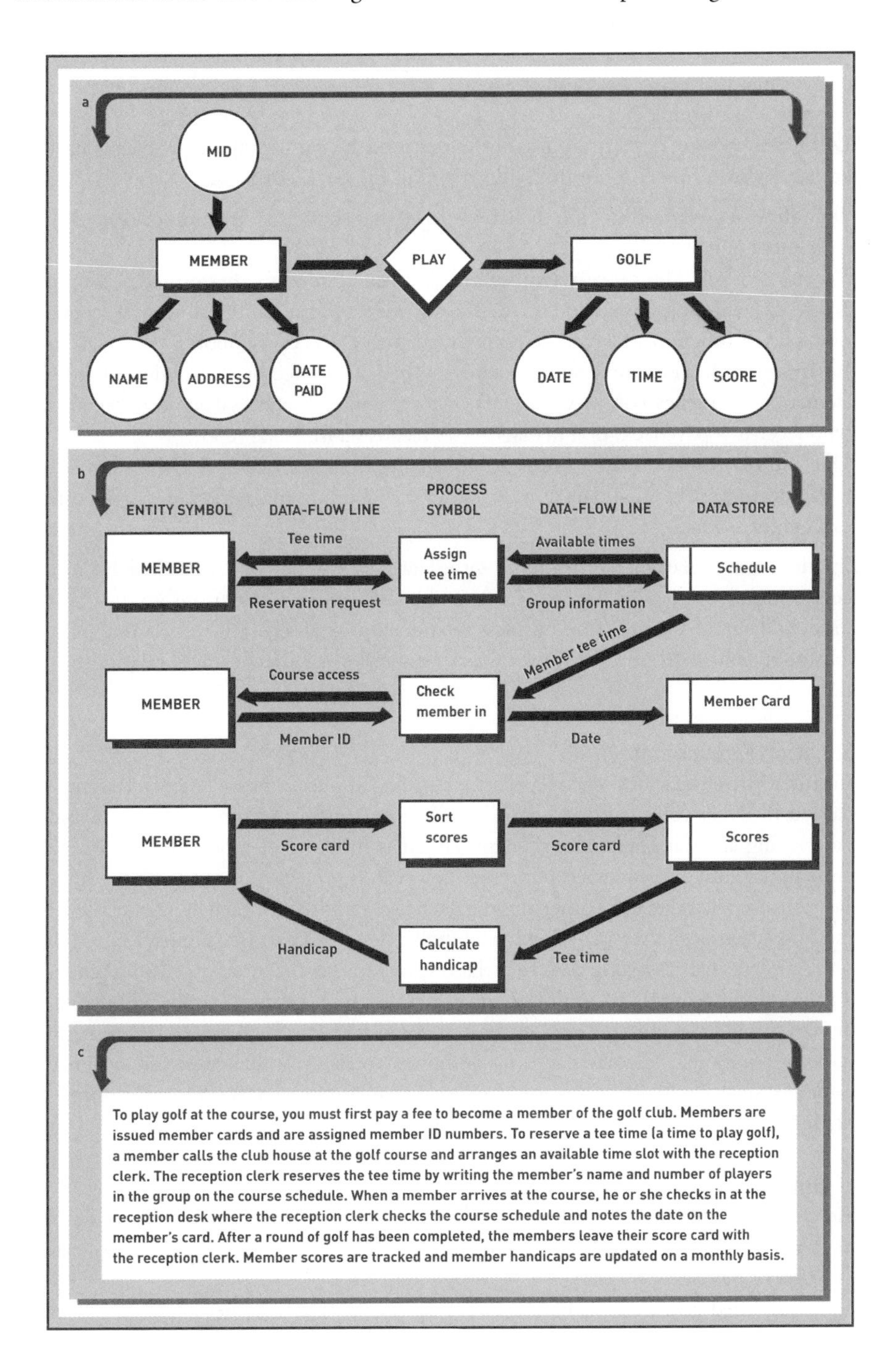

Figure 12.18

Data and Activity Modeling

(a) An entity-relationship diagram. (b) A data-flow diagram. (c) A semantic description of the business process.

(Source: G. Lawrence Sanders, *Data Modeling*, Boyd & Fraser Publishing, Danvers, MA: 1995.)

Activity Modeling

To fully describe a business problem or solution, the related objects, associations, and activities must be described. Activities in this sense are events or items that are necessary to fulfill the business relationship or that can be associated with the business relationship in a meaningful way.

data-flow diagram (DFD)
A model of objects, associations, and activities that describes how data can flow between and around various objects.

Activity modeling is often accomplished through the use of data-flow diagrams. A **data-flow diagram** (DFD) models objects, associations, and activities by describing how data can flow between and around various objects. DFDs work on the premise that every activity involves some communication, transference, or flow that can be described as a data element. DFDs describe the activities that fulfill a business relationship or accomplish a business task, not how these activities are to be performed. That is, DFDs show the logical sequence of associations and activities, not the physical processes. A system modeled with a DFD could operate manually or could be computer based; if computer based, the system could operate with a variety of technologies.

DFDs are easy to develop and easily understood by nontechnical people. Data-flow diagrams use four primary symbols, as illustrated in Figure 12.18b.

data-flow line
Arrows that show the direction of data element movement.

process symbol
Representation of a function that is performed.

entity symbol
Representation of either a source or destination of a data element.

data store
Representation of a storage location for data.

- **Data flow.** The **data-flow line** includes arrows that show the direction of data element movement.
- **Process symbol.** The **process symbol** reveals a function that is performed. Computing gross pay, entering a sales order, delivering merchandise, and printing a report are examples of functions that can be represented with a process symbol.
- **Entity symbol.** The **entity symbol** shows either the source or destination of the data element. An entity can be, for example, a customer who initiates a sales order, an employee who receives a paycheck, or a manager who receives a financial report.
- **Data store.** A **data store** reveals a storage location for data. A data store is any computerized or manual data storage location, including magnetic tape, disks, a filing cabinet, or a desk.

Comparing entity-relationship diagrams with data-flow diagrams provides insight into the concept of top-down design. Figure 12.18a and b show an entity-relationship diagram and a data-flow diagram for the same business relationship—namely, a member of a golf club playing golf. Figure 12.18c provides a brief description of the business relationship for clarification.

Application Flowcharts

application flowcharts
Diagrams that show relationships among applications or systems.

Application flowcharts show the relationships among applications or systems. Assume that a small business has collected data about its order processing, inventory control, invoicing, and marketing analysis applications. Management is thinking of modifying the inventory control application. The raw facts collected, however, do not help in determining how the applications are related to each other and the databases required for each. These relationships are established through data analysis with an application flowchart (see Figure 12.19). Using this tool for data analysis makes clear the relationships among the order processing functions.

In the simplified application flowchart in Figure 12.19, you can see that the telephone order clerk provides important data to the system about items such as versions, quantities, and prices. The system calculates sales tax and order totals. Any changes made to this order processing system could affect the company's other systems, such as inventory control and marketing.

Grid Charts

grid chart
A table that shows relationships among the various aspects of a systems development effort.

A **grid chart** is a table that shows relationships among various aspects of a systems development effort. For example, a grid chart can reveal the databases used by the various applications (see Figure 12.20).

The simplified grid chart in Figure 12.20 shows that the customer database is used by the order processing, marketing analysis, and invoicing applications. The inventory database is used by the order processing, inventory control, and marketing analysis applications. The supplier database is used by the inventory control application, and the accounts receivable database is used by the invoicing application. This grid chart shows which applications use common databases and reveals that, for example, any changes to the inventory control application must investigate the inventory and supplier databases.

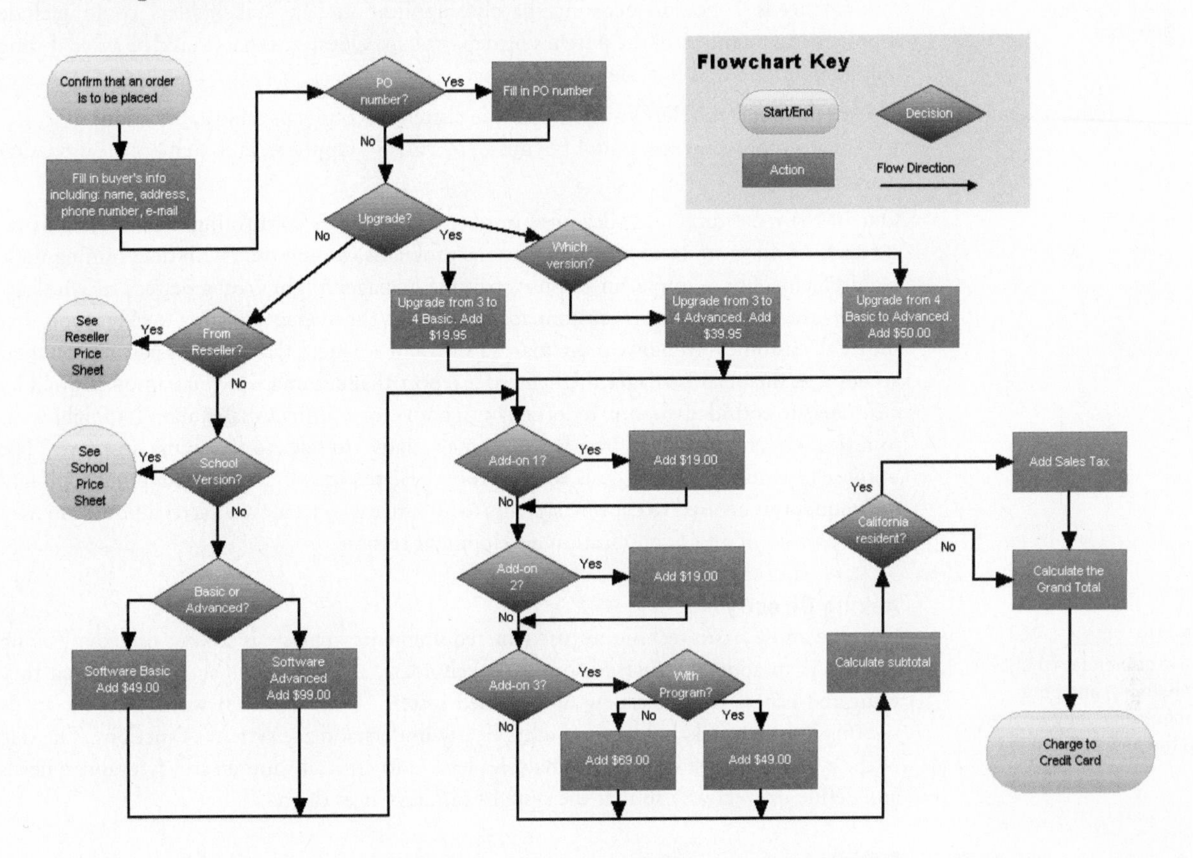

Figure 12.19

A Telephone Order Process Application Flowchart

The flowchart shows the relationships among various processes.

(Source: Courtesy of SmartDraw.com.)

Databases → / Applications ↓	Customer database	Inventory database	Supplier database	Accounts receivable database
Order processing application	X	X		
Inventory control application		X	X	
Marketing analysis application	X	X		
Invoicing application	X			X

Figure 12.20

A Grid Chart

The chart shows the relationships among applications and databases.

CASE Tools

As discussed earlier, many systems development projects use CASE tools to complete analysis tasks. Most computer-aided software engineering tools have generalized graphics programs that can generate a variety of diagrams and figures. Entity-relationship diagrams, data-flow diagrams, application flowcharts, and other diagrams can be developed using CASE graphics programs to help describe the existing system. During the analysis phase, a **CASE repository**—a database of system descriptions, parameters, and objectives—will be developed.

CASE repository
A database of system descriptions, parameters, and objectives.

Requirements Analysis

requirements analysis
The determination of user, stakeholder, and organizational needs.

The overall purpose of **requirements analysis** is to determine user, stakeholder, and organizational needs.[65] For an accounts payable application, the stakeholders could include suppliers and members of the purchasing department. Questions that should be asked during requirements analysis include the following:

- Are these stakeholders satisfied with the current accounts payable application?
- What improvements could be made to satisfy suppliers and help the purchasing department?

One of the most difficult procedures in systems analysis is confirming user or systems requirements. In some cases, communications problems can interfere with determining these requirements. For example, an accounts payable manager might want a better procedure for tracking the amount owed by customers. Specifically, the manager wants a weekly report that shows all customers who owe more than $1,000 and are more than 90 days past due on their account. A financial manager might need a report that summarizes total amount owed by customers to consider whether to loosen or tighten credit limits. A sales manager might want to review the amount owed by a key customer relative to sales to that same customer. The purpose of requirements analysis is to capture these requests in detail. Numerous tools and techniques can be used to capture systems requirements. Often, various techniques are used in the context of a joint application development session.

Asking Directly

asking directly
An approach to gather data that asks users, stakeholders, and other managers about what they want and expect from the new or modified system.

One the most basic techniques used in requirements analysis is asking directly. **Asking directly** is an approach that asks users, stakeholders, and other managers about what they want and expect from the new or modified system. This approach works best for stable systems in which stakeholders and users clearly understand the system's functions. The role of the systems analyst during the analysis phase is to critically and creatively evaluate needs and define them clearly so that the systems can best meet them.

Critical Success Factors

Another approach uses critical success factors (CSFs). As discussed earlier, managers and decision makers are asked to list only the factors that are critical to the success of their area of the organization. A CSF for a production manager might be adequate raw materials from suppliers; a CSF for a sales representative could be a list of customers currently buying a certain type of product. Starting from these CSFs, the system inputs, outputs, performance, and other specific requirements can be determined.

The IS Plan

As we have seen, the IS plan translates strategic and organizational goals into systems development initiatives. The IS planning process often generates strategic planning documents that can be used to define system requirements. Working from these documents ensures that requirements analysis will address the goals set by top-level managers and decision makers (see Figure 12.21). There are unique benefits to applying the IS plan to define systems requirements. Because the IS plan takes a long-range approach to using information technology within the organization, the requirements for a system analyzed in terms of the IS plan are more likely to be compatible with future systems development initiatives.

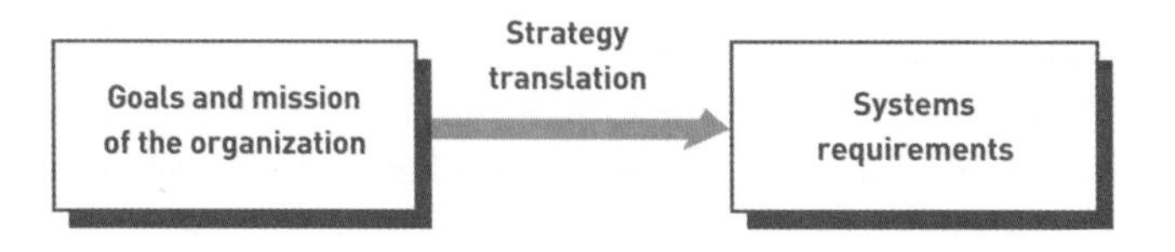

Figure 12.21

Converting Organizational Goals into Systems Requirements

Screen and Report Layout

Developing formats for printed reports and screens to capture data and display information are some of the common tasks associated with developing systems. Screens and reports relating to systems output are specified first to verify that the desired solution is being delivered. Manual or computerized screen and report layout facilities are used to capture both output and input requirements.

Using a **screen layout**, a designer can quickly and efficiently design the features, layout, and format of a display screen. In general, users who interact with the screen frequently can be presented with more data and less descriptive information; infrequent users should have more descriptive information presented to explain the data that they are viewing (see Figure 12.22).

screen layout
A technique that allows a designer to quickly and efficiently design the features, layout, and format of a display screen.

a

ORDER ENTRY

ORDER NO.	CUSTOMER NO.	SALES PERSON	REGION	COMMISSION	NET DOLLARS
XXXXX	XXXXX	XXXXX	XXX	XXX	XXXXX

ITEM NO	QTY	UNIT	PRICE	DOLLARS	DISCOUNTS
XXXXXXXXX	XXXX	XX	XXXXX	XXXXXXX	XX XX XX
XXXXXXXXX	XXXX	XX	XXXXX	XXXXXXX	XX XX XX
XXXXXXXXX	XXXX	XX	XXXXX	XXXXXXX	XX XX XX
XXXXXXXXX	XXXX	XX	XXXXX	XXXXXXX	XX XX XX
XXXXXXXXX	XXXX	XX	XXXXX	XXXXXXX	XX XX XX
XXXXXXXXX	XXXX	XX	XXXXX	XXXXXXX	XX XX XX
XXXXXXXXX	XXXX	XX	XXXXX	XXXXXXX	XX XX XX
XXXXXXXXX	XXXX	XX	XXXXX	XXXXXXX	XX XX XX
XXXXXXXXX	XXXX	XX	XXXXX	XXXXXXX	XX XX XX
XXXXXXXXX	XXXX	XX	XXXXX	XXXXXXX	XX XX XX
XXXXXXXXX	XXXX	XX	XXXXX	XXXXXXX	XX XX XX
XXXXXXXXX	XXXX	XX	XXXXX	XXXXXXX	XX XX XX

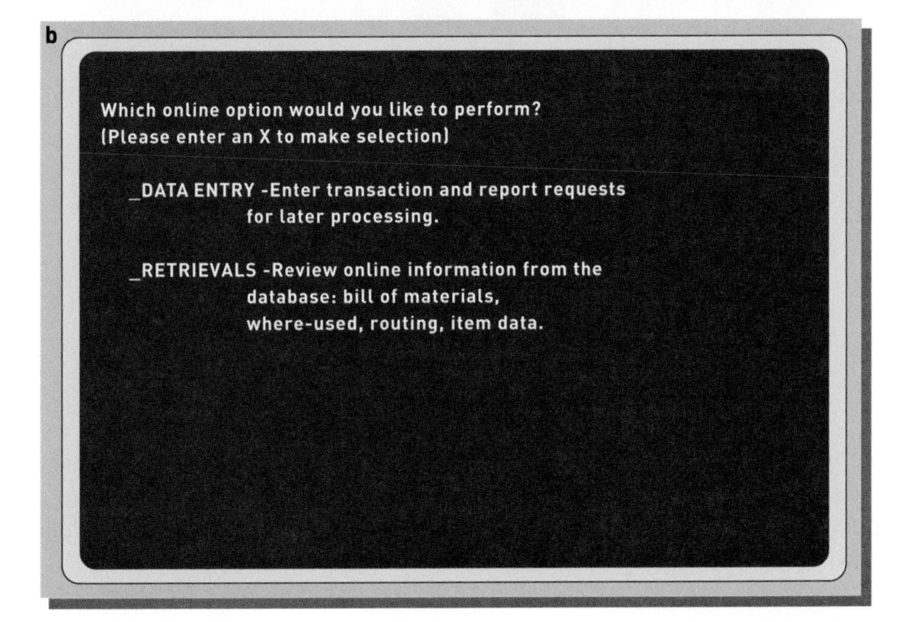

Figure 12.22

Screen Layouts

(a) A screen layout chart for frequent users who require little descriptive information.

(b) A screen layout chart for infrequent users who require more descriptive information.

Report layout allows designers to diagram and format printed reports. Reports can contain data, graphs, or both. Graphic presentations allow managers and executives to quickly view trends and take appropriate action, if necessary.

report layout
A technique that allows designers to diagram and format printed reports.

Screen layout diagrams can document the screens users desire for the new or modified application. Report layout charts reveal the format and content of various reports that the application will prepare. Other diagrams and charts can be developed to reveal the relationship between the application and outputs from the application.

Requirements Analysis Tools

A number of tools can be used to document requirements analysis, including CASE tools. As requirements are developed and agreed on, entity-relationship diagrams, data-flow diagrams, screen and report layout forms, and other types of documentation are stored in the CASE repository. These requirements might also be used later as a reference during the rest of systems development or for a different systems development project.

Object-Oriented Systems Analysis

The object-oriented approach can also be used during systems analysis. Like traditional analysis, problems or potential opportunities are identified during object-oriented analysis. Identifying key participants and collecting data is still performed. But instead of analyzing the existing system using data-flow diagrams and flowcharts, an object-oriented approach is used.

The section "Object-Oriented Systems Investigation" introduced a kayak rental example. A more detailed analysis of that business reveals that there are two classes of kayaks: single kayaks for one person and tandem kayaks that can accommodate two people. With the OO approach, a class is used to describe different types of objects, such as single and tandem kayaks. The classes of kayaks can be shown in a generalization/specialization hierarchy diagram (see Figure 12.23). KayakItem is an object that will store the kayak identification number (ID) and the date the kayak was purchased (datePurchased).

Figure 12.23

Generalization/Specialization Hierarchy Diagram for Single and Tandem Kayak Classes

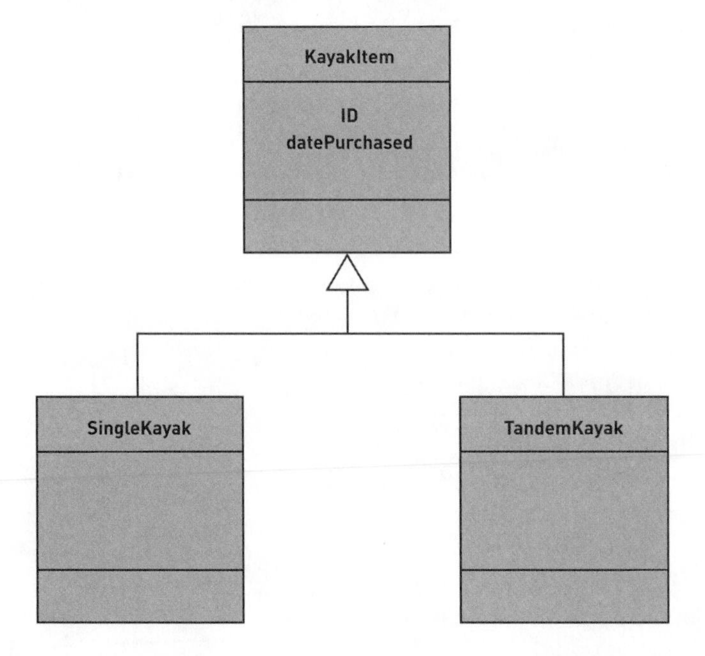

Of course, there could be subclasses of customers, life vests, paddles, and other items in the system. For example, price discounts for kayak rentals could be given to seniors (people over 65 years) and students. Thus, the Customer class could be divided into regular, senior, and student customer subclasses.

The Systems Analysis Report

Systems analysis concludes with a formal systems analysis report. It should cover the following elements:

- The strengths and weaknesses of the existing system from a stakeholder's perspective
- The user/stakeholder requirements for the new system (also called the *functional requirements*)
- The organizational requirements for the new system
- A description of what the new information system should do to solve the problem

Suppose analysis reveals that a marketing manager thinks a weakness of the existing system is its inability to provide accurate reports on product availability. These requirements and a preliminary list of the corporate objectives for the new system will be in the systems analysis report. Particular attention is placed on areas of the existing system that could be improved to meet user requirements. The table of contents for a typical report is shown in Figure 12.24.

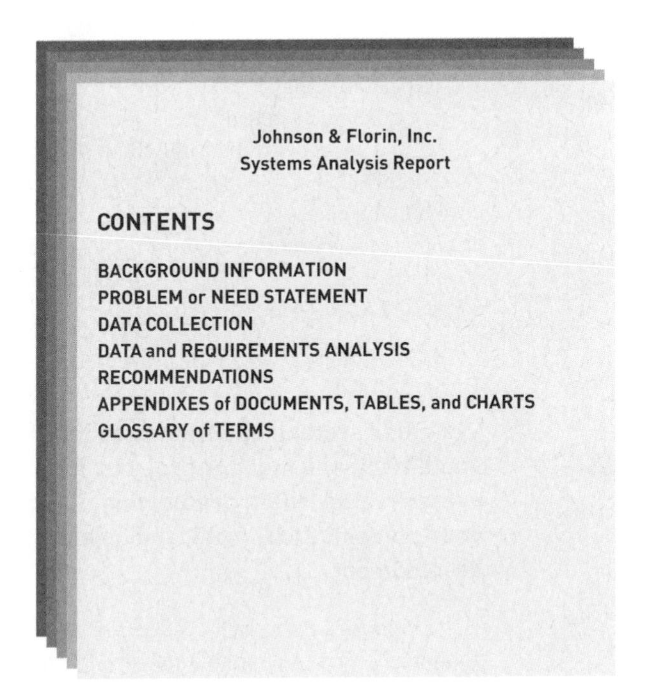

Johnson & Florin, Inc.
Systems Analysis Report
CONTENTS
BACKGROUND INFORMATION
PROBLEM or NEED STATEMENT
DATA COLLECTION
DATA and REQUIREMENTS ANALYSIS
RECOMMENDATIONS
APPENDIXES of DOCUMENTS, TABLES, and CHARTS
GLOSSARY of TERMS

Figure 12.24

A Typical Table of Contents for a Report on an Existing System

The systems analysis report gives managers a good understanding of the problems and strengths of the existing system. If the existing system is operating better than expected or the necessary changes are too expensive relative to the benefits of a new or modified system, the systems development process can be stopped at this stage. If the report shows that changes to another part of the system might be the best solution, the development process might start over, beginning again with systems investigation. Or, if the systems analysis report shows that it will be beneficial to develop one or more new systems or to make changes to existing ones, systems design, which is discussed in Chapter 13, begins.

SUMMARY

Principle

Effective systems development requires a team effort from stakeholders, users, managers, systems development specialists, and various support personnel, and it starts with careful planning.

The systems development team consists of stakeholders, users, managers, systems development specialists, and various support personnel. The development team determines the objectives of the information system and delivers to the organization a system that meets its objectives.

Stakeholders are people who, either themselves or through the area of the organization they represent, ultimately benefit from the systems development project. Users are people who will interact with the system regularly. They can be employees, managers, customers, or suppliers. Managers on development teams are typically representative of stakeholders or can be stakeholders themselves. In addition, managers are most capable of initiating and maintaining change. For large-scale systems development projects, where the investment in and value of a system can be quite high, it is common to have senior-level managers be part of the development team.

A systems analyst is a professional who specializes in analyzing and designing business systems. The programmer is responsible for modifying or developing programs to satisfy user requirements. Other support personnel on the development team include technical specialists, either IS department employees or outside consultants. Depending on the magnitude of the systems development project and the number of IS development specialists on the team, the team might also include one or more IS managers. At some point in your career, you will likely be a participant in systems development. You could be involved in a systems development team—as a user, as a manager of a business area or project team, as a member of the IS department, or maybe even as a CIO.

Systems development projects are initiated for many reasons, including the need to solve problems with an existing system, to exploit opportunities to gain competitive advantage, to increase competition, to make use of effective information, to spur organizational growth, to settle a merger or corporate acquisition, and to address a change in the market or external environment. External pressures, such as potential lawsuits or terrorist attacks, can also prompt an organization to initiate systems development.

Information systems planning refers to the translation of strategic and organizational goals into systems development initiatives. Benefits of IS planning include a long-range view of information technology use and better use of IS resources. Planning requires developing overall IS objectives; identifying IS projects; setting priorities and selecting projects; analyzing resource requirements; setting schedules, milestones, and deadlines; and developing the IS planning document. IS planning can result in a competitive advantage through creative and critical analysis.

Establishing objectives for systems development is a key aspect of any successful development project. Critical success factors (CSFs) can identify important objectives. Systems development objectives can include performance goals (quality and usefulness of the output and the speed at which output is generated) and cost objectives (development costs, fixed costs, and ongoing investment costs).

Principle

Systems development often uses tools to select, implement, and monitor projects, including net present value (NPV), prototyping, rapid application development, CASE tools, and object-oriented development.

The five phases of the traditional SDLC are investigation, analysis, design, implementation, and maintenance and review. Systems investigation identifies potential problems and opportunities and considers them in light of organizational goals. Systems analysis seeks a general understanding of the solution required to solve the problem; the existing system is studied in detail and weaknesses are identified. Systems design creates new or modifies existing system requirements. Systems implementation encompasses programming, testing, training, conversion, and operation of the system. Systems maintenance and review entails monitoring the system and performing enhancements or repairs.

Advantages of the traditional SDLC include the following: It provides for maximum management control, creates considerable system documentation, ensures that system requirements can be traced back to stated business needs, and produces many intermediate products for review. Its disadvantages include the following: Users may get a system that meets the needs as understood by the developers, the documentation is expensive and difficult to maintain, users' needs go unstated or might not be met, and users cannot easily review the many intermediate products produced.

Prototyping is an iterative approach that involves defining the problem, building the initial version, having users work with and evaluate the initial version, providing feedback, and incorporating suggestions into the second version. Prototypes can be fully operational or nonoperational, depending on how critical the system under development is and how much time and money the organization has to spend on prototyping.

Rapid application development (RAD) uses tools and techniques designed to speed application development. Its use reduces paper-based documentation, automates program source code generation, and facilitates user participation in development activities. RAD can use newer programming techniques, such as agile development or extreme programming. RAD makes extensive use of the joint application development (JAD) process to gather data and perform requirements analysis. JAD involves group meetings in which users, stakeholders, and IS professionals work together to analyze existing systems, propose possible solutions, and define the requirements for a new or modified system.

The term *end-user systems development* describes any systems development project in which the primary effort is undertaken by a combination of business managers and users.

Many companies hire an outside consulting firm that specializes in systems development to take over some or all of its systems development activities. This approach is called *outsourcing*. Reasons for outsourcing include companies' belief that they can cut costs, achieve a competitive advantage without having the necessary IS personnel in-house, obtain state-of-the-art technology, increase their technological flexibility, and proceed with development despite downsizing. Many companies offer outsourcing services, including computer vendors and specialized consulting companies.

A number of factors affect systems development success. The degree of change introduced by the project, continuous improvement and reengineering, the use of quality programs and standards, organizational experience with systems development, the use of project management tools, and the use of CASE tools and the objected-oriented approach are all factors that affect the success of a project. The greater the amount of change a system will endure, the greater the degree of risk and often the amount of reward. Continuous improvement projects do not require significant business process or IS changes, while reengineering involves fundamental changes in how the organization conducts business and completes tasks. Successful systems development projects often involve such factors as support from top management, strong user involvement, use of a proven methodology, clear project goals and objectives, concentration on key problems and straightforward designs, staying on schedule and within budget, good user training, and solid review and maintenance programs. Quality standards, such as ISO 9001, can also be used during the systems development process.

The use of automated project management tools enables detailed development, tracking, and control of the project schedule. Effective use of a quality assurance process enables the project manager to deliver a high-quality system and to make intelligent trade-offs among cost, schedule, and quality. CASE tools automate many of the systems development tasks, thus reducing an analyst's time and effort while ensuring good documentation. Object-oriented systems development can also be an important success factor. With the object-oriented systems development (OOSD) approach, a project can be broken down into a group of objects that interact. Instead of requiring thousands or millions of lines of detailed computer instructions or code, the systems development project might require a few dozen or maybe a hundred objects.

Principle

Systems development starts with investigation and analysis of existing systems.

In most organizations, a systems request form initiates the investigation process. Participants in systems investigation can include stakeholders, users, managers, employees, analysts, and programmers. The systems investigation is designed to assess the feasibility of implementing solutions for business problems, including technical, economic, legal, operational, and schedule feasibility. Net present value analysis is often used to help determine a project's economic feasibility. An investigation team follows up on the request and performs a feasibility analysis that addresses technical, economic, legal, operational, and schedule feasibility.

If the project under investigation is feasible, major goals are set for the system's development, including performance, cost, managerial goals, and procedural goals. Many companies choose a popular methodology so that new IS employees, outside specialists, and vendors will be familiar with the systems development tasks set forth in the approach. A systems development methodology must be selected. Object-oriented systems investigation is being used to a greater extent today. The use case diagram is part of the Unified Modeling Language that is used to document object-oriented systems development. As a final step in the investigation process, a systems investigation report should be prepared to document relevant findings.

Systems analysis is the examination of existing systems, which begins after a team receives approval for further study from management. Additional study of a selected system allows those involved to further understand the system's weaknesses and potential areas for improvement. An analysis team is assembled to collect and analyze data on the existing system.

Data collection methods include observation, interviews, questionnaires, and statistical sampling. Data analysis manipulates the collected data to provide information. The analysis includes grid charts, application flowcharts, and CASE tools. The overall purpose of requirements analysis is to determine user and organizational needs.

Data analysis and modeling is used to model organizational objects and associations using text and graphical diagrams. It is most often accomplished through the use of entity-relationship (ER) diagrams. Activity modeling often employs data-flow diagrams (DFDs), which model objects, associations, and activities by describing how data can flow between and around various objects. DFDs use symbols for data flows, processing, entities, and data stores. Application flowcharts, grid charts, and CASE tools are also used during systems analysis.

Requirements analysis determines the needs of users, stakeholders, and the organization in general. Asking directly, using critical success factors, and determining requirements from the IS plan can be used. Often, screen and report layout charts are used to document requirements during systems analysis.

Like traditional analysis, problems or potential opportunities are identified during object-oriented analysis. Object-oriented systems analysis can involve using diagramming techniques, such as a generalization/specialization hierarchy diagram.

CHAPTER 12: SELF-ASSESSMENT TEST

Effective systems development requires a team effort from stakeholders, users, managers, systems development specialists, and various support personnel, and it starts with careful planning.

1. ________________ is the activity of creating or modifying existing business systems. It refers to all aspects of the process—from identifying problems to be solved or opportunities to be exploited to the implementation and refinement of the chosen solution.
2. Which of the following people ultimately benefit from a systems development project?
   a. computer programmers
   b. systems analysts
   c. stakeholders
   d. senior-level manager
3. ________________ requires unbiased and careful questioning of whether systems elements are related in the most effective or efficient ways.
4. Like a contractor constructing a new building or renovating an existing one, the programmer takes the plans from the systems analyst and builds or modifies the necessary software. True or False?
5. The term ________________ refers to the translation of strategic and organizational goals into systems development initiatives.
6. What involves investigating new approaches to existing problems?
   a. critical success factors
   b. systems analysis factors
   c. creative analysis
   d. critical analysis

Systems development often uses tools to select, implement, and monitor projects, including net present value (NPV), prototyping, rapid application development, CASE tools, and object-oriented development.

7. What employs tools, techniques, and methodologies designed to speed application development?
   a. rapid application development
   b. joint optimization
   c. prototyping
   d. extended application development
8. System performance is usually determined by factors such as fixed investments in hardware and related equipment. True or False?
9. ________________ takes an iterative approach to the systems development process. During each iteration, requirements and alternative solutions to the problem are identified and analyzed, new solutions are designed, and a portion of the system is implemented.
10. Joint application development (JAD) employs tools, techniques, and methodologies designed to speed application development. True or False?
11. What consists of all activities that, if delayed, would delay the entire project?
   a. deadline activities
   b. slack activities
   c. RAD tasks
   d. the critical path

Systems development starts with investigation and analysis of existing systems.

12. The systems request form is a document that is filled out during systems analysis. True or False?
13. Feasibility analysis is typically done during which systems development stage?
   a. investigation
   b. analysis
   c. design
   d. implementation
14. Data modeling is most often accomplished through the use of ________________, whereas activity modeling is often accomplished through the use of ________________.
15. The overall purpose of requirements analysis is to determine user, stakeholder, and organizational needs. True or False?

CHAPTER 12: SELF-ASSESSMENT TEST ANSWERS

(1) Systems development (2) c (3) Critical analysis (4) True (5) information systems planning (6) c (7) a (8) False (9) Prototyping (10) False (11) d (12) False (13) a (14) entity-relationship (ER) diagrams, data-flow diagrams (15) True

REVIEW QUESTIONS

1. What is an IS stakeholder?
2. What is the goal of IS planning? What steps are involved in IS planning?
3. What are the typical reasons to initiate systems development?
4. What actions can be taken during creative analysis?
5. What is the difference between a programmer and a systems analyst?
6. What is the difference between a Gantt chart and PERT?
7. What is the difference between systems investigation and systems analysis? Why is it important to identify and remove errors early in the systems development life cycle?
8. Identify four reasons that a systems development project might be initiated.
9. List factors that have a strong influence on project success.
10. What is the purpose of systems analysis?
11. What are the steps of object-oriented systems development?
12. Define the different types of feasibility that systems development must consider.
13. What is the difference between systems investigation and systems analysis?
14. How does the JAD technique support the RAD systems development life cycle?

DISCUSSION QUESTIONS

1. Why is it important for business managers to have a basic understanding of the systems development process?
2. Briefly describe the role of a system user in the systems investigation and systems analysis stages of a project.
3. How could you use creative analysis to help you develop a better information system for your college or university?
4. Briefly describe when you would use the object-oriented approach to systems development instead of the traditional systems development life cycle.
5. Your company wants to develop or acquire a new sales program to help sales representatives identify new customers. Describe what factors you would consider in deciding whether to develop the application in-house or outsource the application to an outside company.
6. You have been hired by your university to find an outsourcing company to perform the payroll function. What are your recommendations? Describe the advantages and disadvantages of the outsourcing approach for this application.

7a. You have been hired as a project manager to develop a new Web site in the next six months for a store that sells music and books online. Describe how project management tools, such as a Gantt chart and PERT, might be used.

7b. For what types of systems development projects might prototyping be especially useful? What are the characteristics of a system developed with a prototyping technique?

8. Assume that you work for an insurance company. Describe three applications that are critical to your business. What tools would you use to develop applications?
9. How important are communications skills to IS personnel? Consider this statement: "IS personnel need a combination of skills—one-third technical skills, one-third business skills, and one-third communications skills." Do you think this is true? How would this affect the training of IS personnel?
10. You have been hired to perform systems investigation for a French restaurant owner in a large metropolitan area. She is thinking of opening a new restaurant with a state-of-the-art computer system that would allow customers to place orders on the Internet or at kiosks at restaurant tables. Describe how you would determine the technical, economic, legal, operational, and schedule feasibility for the restaurant and its new computer system.
11. Discuss three reasons why aligning overall business goals with IS goals is important.
12. You are a senior manager of a functional area in which a mission-critical system is being developed. How can you safeguard this project from mushrooming out of control?

PROBLEM-SOLVING EXERCISES

1. You are developing a new information system for The Fitness Center, a company that has five fitness centers in your metropolitan area, with about 650 members and 30 employees in each location. This system will be used by both members and fitness consultants to track participation in various fitness activities, such as free weights, volleyball, swimming, stair climbers, and aerobic and yoga classes. One of the performance objectives of the system is that it must help members plan a fitness program to meet their particular needs. The primary purpose of this system, as envisioned by the director of marketing, is to assist The Fitness Center in obtaining a competitive advantage over other fitness clubs. Use a graphics program to develop a flowchart or a grid chart to show the major components of your information system and how the components are tied together.
2. You have been hired to develop a payroll program for a medium-sized company. At a minimum, the application should have an hours-worked table that contains how many hours each employee worked and an employee table that contains information about each employee, including hourly pay rate. Design and develop these tables that could be used in a database for the payroll program.

TEAM ACTIVITIES

1. Your team should interview people involved in systems development in a local business or at your college or university. Describe the process used. Identify the users, analysts, and stakeholders for a systems development project that has been completed or is currently under development.
2. Your team has been hired to determine the requirements and layout of the Web pages for a company that sells fishing equipment over the Internet. Using the approaches discussed in this chapter, develop a rough sketch of at least five Web pages that you would recommend. Make sure to show the important features and the hyperlinks for each page.
3. Your team has been hired to determine the requirements of a new medium-cost coffee bar to compete with higher-priced coffee shops such as Starbucks. The new coffee bar will offer computer kiosks for customers to surf the Internet or order coffee and other products from the coffee bar. Using RAD and JAD techniques with your team, develop requirements for this new coffee bar and its computer system.

WEB EXERCISES

1. Cloud computing, where applications such as word processing and spreadsheet analysis are delivered over the Internet, is becoming more popular. You have been hired to analyze the potential of a cloud computing application that performs payroll and invoicing over the Internet from a large Internet company. Describe the systems development steps and procedures you would use to analyze the feasibility of this approach.
2. Using the Internet, locate an organization that is currently involved in a systems development project. Describe how they are using project management tools. Is project management software being used?

CAREER EXERCISES

1. Pick a career that you are considering. What type of information system would help you on the job? Perform technical, economic, legal, operational, and schedule feasibility for an information system you would like developed.
2. For a career you are considering, describe how you would interact with various IS personnel to acquire computer applications or a new information system that would be the most useful to you in your new career. Describe the IS personnel that would be needed to build the application and the best way to interact with them to get what you want.

CASE STUDIES

Case One

Ontario and London Hydro Move to Smart Metering

System development projects get started for a variety of reasons. Often, they are intended to support a company's strategic plans. Other times they are launched out of necessity, such as to comply with government mandates. London Hydro (LH), the electricity provider of London, Ontario, and surrounding areas, is wrapping up a lengthy and costly system upgrade that was sparked by government regulations.

Ontario prides itself on using the latest technologies to conserve electricity. In 1998 it passed two regulations that paved the way for smart-metering: the Electricity Act, 1998 and the Ontario Energy Board Act, 1998. Smart-metering uses computerized electric meters on homes and small businesses that can record electricity use on an hourly basis. So rather than totaling up kilowatt-hours on a monthly basis, as traditional meters do, smart meters provide a record of electricity use every hour. Smart meters are able to report usage directly to the utilities companies over phone lines or Internet.

The benefit of smart meters, in addition to saving the electric company the cost of sending an employee to read meters, is setting time-of-use pricing. Time-of-use pricing charges customers more for electricity during peak hours (11 a.m.–5 p.m.), less during mid-peak hours (7 a.m.–11 a.m. and 5 p.m.–10 p.m.), and even less during off-peak hours (10 p.m.–7 a.m.). Time-of-use pricing should encourage consumers to consume less during peak hours, adding up to big savings for Ontario, its citizens, and the environment.

Toronto has pledged to install smart meters in every home and small business in Ontario by 2010 (which covers 13 million citizens spread out over a million square kilometers). The province is requiring all power companies to support smart meters and has provided standards and specifications so the power companies can prepare.

Ontario's smart metering initiatives have power companies across the province scrambling to meet specifications and deadlines. Software and hardware must be purchased and installed to prepare for the arrival of a tidal wave of customer consumption data. London Hydro started to prepare early in hopes of getting a jump on the competition. Rather than adding a new system to accommodate smart metering, London Hydro decided it was time to upgrade all of its systems. London Hydro's old custom-built system could barely keep up with current usage. The company decided to shop around for a new system that could not only accommodate smart metering but could tie that data in with core business systems. Mridula Sharma, London Hydro's director of information services, stated that LH was in need of "a more integratable solution that was scalable and flexible." The company needed to "prepare for future growth as well as enhance business process workflow," Sharma said.

Sharma and her team set to work outlining the details of the new system based on government mandates and internal needs. With a systems analysis report in hand, Sharma began searching for a company that could design and implement the system. Soon she narrowed the field to three candidates: SPL Solutions (Oracle), another customer-built solution, and SAP for utilities. Sharma chose SAP primarily because the system was designed for use by a utility company and required little customization.

London Hydro selected another outside firm, Wipro Technologies, to implement the system because Wipro had extensive experience implementing utility software. The resulting system provides powerful management of smart metering data flowing from the government's central smart metering data repository. The task of assigning time-of-use prices based on customer consumption is fully automated and will cause London Hydro no additional overhead.

Now that it is ready for smart-metering in 2010 and beyond, London Hydro is investigating the integration of smart metering with geographic information systems and outage management systems. Ultimately all systems will be integrated into one centralized ERP platform.

Discussion Questions

1. Why did London Hydro initiate its smart-metering information system development project?
2. Who provided information for the systems analysis report for the new system?

Critical Thinking Questions

1. What benefits did London Hydro enjoy by purchasing an off-the-shelf system and outsourcing the implementation?
2. Who benefits from Ontario's mandate and how? What is the cost of those benefits?

SOURCES: Smith, Briony, "London Hydro prepares for smart meters," *IT World Canada*, August 21, 2007, *www.itworldcanada.com/a/Enterprise-Business-Applications/94463e97-bb12-4f5d-b9dd-1d7b2415641f.html*; Ontario Ministry of Energy and Infrastructure Web site, *www.energy.gov.on.ca/index.cfm?fuseaction=electricity.smartmeters*, accessed July 13, 2008; London Hydro's Web site, *www.londonhydro.com/lh_website/index.jsp*; accessed July 13, 2008.

Case Two

Information and Security Systems at the All England Lawn Tennis and Croquet Club

For 351 days a year, the All England Lawn Tennis and Croquet Club is a quiet private tennis club set in a sleepy suburb of London. For 14 days each year, half a million tennis enthusiasts arrive at the club from around the world to witness the

Wimbledon Championships. Needless to say, much preparation is required as the club accommodates world tennis stars, their fans, and the press and television crews. Many information systems of all types are required to support the global sporting event. Information systems specialists at the club work to support the event; maintain the Club's culture, brand, and values; and support the Club's primary mission: "to blend tradition with innovation to substantially improve the quality of the Wimbledon experience for all the key stakeholders."

A primary goal for the annual two-week event is the security and safety of all in attendance. To accomplish this goal, the Club has invested in a new electronic security and surveillance system. With so many people to watch on a vast property, Club management knew that they would need state-of-the-art automated surveillance software. The Club management wanted a system that could integrate images from video cameras, intruder alarms, and trip wires with identifying information such as license plates to provide real-time reports of suspicious activities. With these specifications in mind, the Club found a solution in a Digital and Video Security (DVS) solution designed by IBM. The system provides "real-time intelligence to automatically monitor trends and analyze events captured by security devices."

The system was tested at the 2007 Wimbledon Championships with success. At the 2008 event, the system was extended to hundreds of cameras. Because the system is scalable, it can grow each year as attendance grows at the event without any degradation in performance.

In addition to the half-million attendees, millions of fans can now view the event and related information online thanks to another new information system development project. Knowing that interest in the event was continuously growing, Wimbledon event coordinators looked for ways to provide remote coverage to more tennis fans. A system was developed called SlamTracker that provides live online scoring for matches in progress. Additional SlamTracker tools allow fans to track player progression and other important player and match statistics.

Those watching the event in person and on TV have noticed other high-tech improvements. Large LED screens have been installed on the main courts, providing the audience with statistics such as the speed of the serve. Other systems are provided to the athletes to display important statistics and trends captured during the last match so that the players can see what improvements are required and create strategies for the next match.

Systems analysts work year round to improve information systems for the next Wimbledon Championships. Security systems, media systems, Web systems, and a host of other types of systems are examined for strengths and weaknesses, searching for ways to improve the experience. In 1990, Wimbledon decided to hire one company to manage all of its systems: IBM. While outsourcing its systems to IBM may be costly, other savings make it well worth the investment. With all of its systems managed by one company, the All England Lawn Tennis and Croquet Club can integrate its systems more easily, saving money by removing redundancy that often exists with multivendor systems. Furthermore, All England Lawn Tennis and Croquet Club is not in the business of information systems. For 351 days a year, the club would rather focus on its membership than on the two-week event next summer.

Discussion Questions

1. How do information systems being implemented at the All England Lawn Tennis and Croquet Club support its primary goals?
2. How do information systems add to the enjoyment of viewing the Wimbledon Championships through multiple channels—in person, on TV, and online?

Critical Thinking Questions

1. What other sporting events might benefit from the technologies used at Wimbledon?
2. How might systems failure cause catastrophe for Wimbledon Championships organizers?

SOURCES: IBM Staff, "For two weeks a year, Wimbledon stops being a private members' club and starts welcoming the world," IBM Case Studies, June 19, 2008, *www-01.ibm.com/software/success/cssdb.nsf/cs/JGIL-7FRDYD?OpenDocument&Site=gicss67mdia&cty=en_us*; Wimbledon Web site, *http://aeltc.wimbledon.org/en_GB/about/guide/club.html*, accessed July 13, 2008.

Questions for Web Case

See the Web site for this book to read about the Whitmann Price Consulting case for this chapter. Following are questions concerning this Web case.

Whitmann Price Consulting: Systems Investigation and Analysis Considerations

Discussion Questions

1. How will the proposed AMCI system help to meet corporate goals and provide Whitmann Price with a competitive advantage?
2. Who are the stakeholders in this systems development project? Who are the primary systems analysts?

Critical Thinking Questions

1. Why did the systems investigation for the proposed AMCI system proceed so quickly and smoothly? What type of proposal might require a more time-consuming, formal investigation?
2. What reasons do you think Josh and Sandra might have for interviewing division managers and not each individual consultant in their system review? What are the pros and cons of both approaches?

NOTES

Sources for the opening vignette: SAP staff, "GRUMA, Standardizing Business Processes to Support Growth," SAP Business Transformation Study, 2008, *http://download.sap.com/solutions/business-suite/erp/customersuccess/download.epd?context=41D7EC5F1B536250F4ECA519A1F6F2E3CA2CF88579D5D63A6E8D998182804C9A0A1304D4584991E5C21230651CD5EBA4D13732442AF0C1AB*; GRUMA Web site, *www.gruma.com*, accessed July 12, 2008.

1 Pratt, Mary, "Rolling the Dice in Your Career," *Computerworld*, February 25, 2008, p. 34.
2 Hoffman, Thomas, "Beyond Film," *Computerworld*, May 5, 2008, p. 32.
3 Rundle, Rhonda, "Critical Case," *The Wall Street Journal*, April 24, 2007, p. D1.
4 Thibodeau, Patrick, "D.C.'s Tax System Won Plaudits—But Didn't Stop Alleged Fraud Scheme," *Computerworld*, March 10, 2008, p. 18.
5 Mitchell, Robert, "The Killer in the ER," *Computerworld*, April 28, 2008, p. 33.
6 Rhoads, Christopher, "Cities Start Own Efforts to Speed Up Broadband," *The Wall Street Journal*, May 19, 2008, pg. A1.
7 Demaitre, Eugene, "Project Manager Extraordinaire," *Computerworld*, July 9, 2007, p. 72.
8 Mahring, Magnus, et al., "Information Technology Project Escalation," *Decision Sciences*, May 2008, p. 239.
9 Chilton, David, "Demand for IT Workers Won't Meet Supply," *The Toronto Sun*, April 23, 2008, p. J2.
10 Hui, Pamsy, et al., "Managing Interdependence: The Effects of Outsourcing Structure on the Performance of Complex Projects," *Decision Sciences*, February, 2008, p. 5.
11 Brandel, Mary, "Crowdsourcing," *Computerworld*, March 3, 2008, p. 26.
12 Vijayan, Jaikumar, "Paying Breach Bill May Not Buy Hannaford Full Data Protection," *Computerworld*, April 28, 2008, p. 14.
13 Hamm, Steve, "Cloud Computing Made Clear," *Business Week*, May 5, 2008, p. 59.
14 Mehta, M. et al., "Strategic Alignment in Mergers and Acquisitions," *The Journal of the Association of Information Systems*, March, 2007, p. 143.
15 Mitchell, Robert, "The Grill: Michael Israel," *Computerworld*, May 5, 2008, p. 19.
16 Gaudin, Sharon, "Some Suppliers Gain from Failed Wal-Mart RFID Edict," *Computerworld*, April 28, 2008, p. 12.
17 Soat, John, "We Don't Dig Software Biz," *Information Week*, April 28, 2008, p.16.
18 Smaltz, Detlev, "Planning, Early Support Key to E-Health Success," *Computerworld*, May 21, 2007, p. 10.
19 Oh, W., et al., "On the Assessment of Strategic Value of Information Technologies," *MIS Quarterly*, June, 2007, p. 239.
20 Iivari, J., et al., "The Relationship Between Organizational Culture and the Deployment of Systems Development Methodologies," *MIS Quarterly*, March 2007, p. 35.
21 Hoffman, Thomas, "Building an IT Project Pipeline," *Computerworld*, April 7, 2008, p. 34.
22 Anthes, Gary, "What's Your Project Worth?" *Computerworld*, March 10, 2008, p. 29.
23 Jia, Ronni, "IT Service Climate," *Journal of the Association of Information Systems*, May 2008, p. 294.
24 Weier, Mary Hayes, "SaaS at Scale," *Information Week*, May 19, 2008, p. 19.
25 Hugos, Michael, "Say Goodbye to Business Analysts," *Computerworld*, May 12, 2008, p. 22.
26 Havenstein, Heather, "Farming Co-op Extends Rollout of SOA Tool," *Computerworld*, April 30, 2007, p. 13.
27 Mitchell, Robert, "The Grill: Michael Israel," *Computerworld*, May 5, 2008, pg. 19.
28 Sutton, Steve, et al., "Risk Analysis in Extended Enterprise Environments," *The Journal of the Association of Information Systems*, Special Issue, 2008, p. 151.
29 Anthes, Gary, "Pay As You Go," *Computerworld*, May 12, 2008, p. 32.
30 Babcock, Charles, "Virtualization Tipping Point," *Information Week*, May 18, 2008, p. 14.
31 Toingo, Jon, "A Costly Storage Hangover," *Information Week*, May 19, 2008, p. 37.
32 Perkins, Bart, "Post-project Review," *Computerworld*, January 15, 2007, p. 30.
33 Sanders, Peter, "Fans Resist End of Virtual Disneyland," *The Wall Street Journal*, May 20, 2008, p. B1.
34 Konke, Robin, "Debugging Embedded C," *Embedded Systems Design*, January 1, 2008, p. 34.
35 Babcock, Charles, "Speedy Web Development," *Information Week*, March 17, 2008, p. 22.
36 Lai, Eric, "Microsoft Tries to Steer a More Agile Development Course," *Computerworld*, March 3, 2008.
37 Rangaswaimi, J. P., "The Roots of Agile Development," *Computerworld*, April 28, 2008, p. 31.
38 Bradbury, Danny, "Business and Technology," *Computer Weekly*, April 1, 2008.
39 Biesecker, Calvin, "Boeing Officials Describe Project 28 Shortcomings," *Defense Daily*, February 28, 2008.
40 The Rational Unified Process, *www-306.ibm.com/software/awdtools/rup/support...*, accessed June 2, 2008.
41 Rational Case Studies, *www-01.ibm.com/software/success/cssdb.nsf/softwareL2VW?OpenView&Start=1&Count=30&RestrictToCategory=rational_RationalUnifiedProcess*, accessed June 2, 2008.
42 Au, N., et al., "Extending the Understanding of End User Information Systems Satisfaction Formulation," *MIS Quarterly*, March 2008, p. 43.
43 Xiao, T., et al., "Strategic Outsourcing Decisions for Manufacturers That Produce Partially Substitutable Products," *Decision Sciences*, February 2007, p. 81.
44 Mithas, Sunil and Whitaker, Jonathan, "Is the World Flat or Spiky?" *Information Systems Research*, September 2007, p. 237.
45 Tam, Pui-Wing, "Outsourcing Finds New Niche," *The Wall Street Journal*, April 17, 2007, p. B5.
46 Hamm, Steve, "How Accenture One-Upped Bangalore," *Business Week*, April 23, 2007, p. 98.
47 Worthen, Ben, "Outsourced Tech Work Gets Spread Around," *The Wall Street Journal*, May 20, 2008, pg. B6.
48 Hamm, Steve, "Outsourcing Heads to the Outskirts, *Business Week*, January 22, 2007, p. 56.
49 IBM Web site, *www.ibm.com*, accessed June 2, 2008.
50 Weier, Mary Hayes, "As Hiring Soars in India, Good Managers Are Hard to Find," *Information Week*, February 5, 2007, p. 33.
51 EDS Web site, *www.eds.com*, accessed June 2, 2008.
52 Accenture Web site, *www.accenture.com*, accessed June 2, 2008.
53 Menchin, Scott, "Amazon Takes on IBM, Oracle, and HP," *Business Week*, April 21, 2008, p. 25.
54 McDougal, Paul, "J. Crew Yanks EDS Work," *Information Week*, January 22, 2007, p. 18.
55 Karp, Jonathan, "U.S. to Probe Outsourcing After ITT Case," *The Wall Street Journal*, March 28, 2007, p. A3.
56 Hoover, Nicholas, "Microsoft System Management Hinges on Open Source," *Information Week*, May 5, 2008, p. 22.

57 Capability Maturity Model for Software home page, *www.sei.cmu.edu*, accessed June 2, 2008.
58 Staff, "Carnegie Mellon's Capability Maturity Model," *Oil and Gas Journal*, October 8, 2007, p. 46.
59 Feldman, Jonathan, "Victim of Success," *Information Week*, April 7, 2008, p. 43.
60 Hayes, Frank, "Stunt It," *Computerworld*, May 12, 2008, p. 40.
61 Hoffman, Thomas, "Vegas!" *Computerworld*, May 12, 2008, p. 25.
62 Mero, Jenny, "An International Affair," *Fortune*, May 12, 2008, p. 36.
63 Bible, Brice, "The Grill," *Computerworld*, September 17, 2007, p. 28.
64 Lawton, Christopher and Clark, Don, "Virtualization is Pumping Up Servers," *The Wall Street Journal*, March 6, 2007, p. B4.
65 Mathiassen, Lars, et al., "A Contingency Model of Requirements Development," *Journal of the Association of Information Systems*, November, 2007, p. 569.

CHAPTER • 13 •

Systems Development: Design, Implementation, Maintenance, and Review

PRINCIPLES	LEARNING OBJECTIVES
▪ Designing new systems or modifying existing ones should always help an organization achieve its goals.	▪ State the purpose of systems design and discuss the differences between logical and physical systems design. ▪ Describe some considerations in design modeling and the diagrams used during object-oriented design. ▪ Discuss the issues involved in environmental design. ▪ Define the term *RFP* and discuss how this document is used to drive the acquisition of hardware and software. ▪ Describe the techniques used to make systems selection evaluations.
▪ The primary emphasis of systems implementation is to make sure that the right information is delivered to the right person in the right format at the right time.	▪ State the purpose of systems implementation and discuss the activities associated with this phase of systems development. ▪ List the advantages and disadvantages of purchasing versus developing software. ▪ Discuss the software development process and some of the tools used in this process, including object-oriented program development tools.
▪ Maintenance and review add to the useful life of a system but can consume large amounts of resources. These activities can benefit from the same rigorous methods and project management techniques applied to systems development.	▪ State the importance of systems and software maintenance and discuss the activities involved. ▪ Describe the systems review process.

Information Systems in the Global Economy Carlsberg Polska, Poland

Brewery Applies SDLC to Improve Help Desk Operations

Carlsberg Polska is the Polish subsidiary of Carlsberg Breweries, a Danish global brewery that produces and sells dozens of beer brands around the world. Carlsberg Polska produces and manages seven of those brands.

Looking to support its growth while maintaining high levels of service to its workforce, Carlsberg Polska investigated how it could improve its information systems and systems support processes. The company was using an internally developed system to report trouble with technology and information systems. Recent increases in system demand were beginning to cause problems both for information system staff and other employees. The system was also outdated and lacked modern functionality. Carlsberg knew it was time to launch a new systems improvement project and enter the first of the five stages of the systems development life cycle (SDLC): investigation, analysis, design, implementation, and maintenance/review.

IT Shared Services Center (SSC) manages Carlsberg Polska information system support. SSC provides support to Carlsberg Polska's 600 information system users at six facilities in four countries and two time zones using four languages. After the initial investigation and analysis, Magdalena Cioch, the information systems director of customer service, and her team had a good idea of what changes were needed and were ready to begin designing the system.

They decided that the help desk and change management program should be merged into one transparent system. The help desk would handle employee problems with information system infrastructure: hardware, software, databases, and telecommunications. Change management refers to processes for requesting, planning, and implementing change in information systems. These two services are often provided independent of one another. Typically, an IT group fields help desk requests that employees submit, and an information systems group fields change requests—formal documents requesting a systems change. Magdalena Cioch thought that combining these activities into one system would be useful for all involved. The company also wanted the new system to be transparent, allowing everyone to track service requests from initiation to completion.

Rather than reinventing the wheel, Cioch knew that IS vendors could provide the systems Carlsberg Polska needed. She distributed a request for proposal (RFP) document to several IS companies, outlining the goals of the system. The company that provided the best proposal was SAP, with a product called Solution Manager.

SAP systems analysts worked with systems analysts from Carlsberg Polska and SSC to refine the specifications for the new system. The team decided to implement three scenarios in the new system: 1) improve help desk efficiency, 2) provide better control of change management processes, and 3) allow information systems specialists to monitor systems and solutions.

Before installing the new system, the team tested it in four countries to ensure that it functioned correctly. After successful testing, the new system was gradually installed. The team used a pilot start-up to safely and gradually introduce the system to the company. First, they installed the solutions monitoring component in Poland. When that was deemed a success, they installed the service desk and change management systems in Poland. As problems materialized, SAP developers quickly addressed and resolved them. Next, they installed all three systems in Carlsberg's facilities in the other three countries.

Now one central location receives help requests and change management requests from across the international enterprise. Magdalena Cioch says that the company has "all the information we need for managing incidents and change requests in all countries at the time they occur. This enables us to forecast our business needs immediately." As systems analysts continue to review the system, looking for ways to further improve it, they are beginning the SDLC again.

As you read this chapter, consider the following:

- After a company develops specifications for a new system, what steps should the company take to implement the system successfully?
- What important factors should a company consider when implementing a new system?

Why Learn About Systems Development?

Information systems are designed and implemented for employees and managers every day. A manager at a hotel chain can use an information system to look up client preferences. An accountant at a manufacturing company can use an information system to analyze the costs of a new plant. A sales representative for a music store can use an information system to determine which CDs to order and which to discount because they are not selling. A computer engineer can use an information system to help determine why a computer system is running slowly. Information systems have been designed and implemented for almost every career and industry. This chapter shows how you can be involved in designing and implementing an information system that will directly benefit you on the job. It also shows how to avoid errors or recover from disasters. This chapter starts with describing how systems are designed.

The way an information system is designed, implemented, and maintained profoundly affects the daily functioning of an organization. Like investigation and analysis covered in Chapter 12, design, implementation, maintenance, and review covered in this chapter strive to achieve organizational goals, such as reducing costs, increasing profits, or improving customer service. The New York Stock Exchange, for example, decided to use the Linux operating system to lower total IS costs.[1] With the high cost of many commodities today, some systems development efforts are saving money by avoiding copper wires and installing wireless telecommunications systems.[2] A single wired connection to an office can cost as much as $250. We begin this chapter with a discussion of systems design.

SYSTEMS DESIGN

systems design
The stage of systems development that answers the question "How will the information system solve a problem?"

The purpose of **systems design** is to answer the question "How will the information system solve a problem?" The primary result of the systems design phase is a technical design that details system outputs, inputs, and user interfaces; specifies hardware, software, databases, telecommunications, personnel, and procedures; and shows how these components are related. The new or modified system should take advantage of the latest developments in technology.[3] Many companies, for example, are looking into cloud computing, where applications are run on the Internet instead of being developed and run within the company or organization. Cloud computing is allowing individuals, such as racecar driving instructor Tom Dyer, to work while traveling.[4] According to Dyer, "Anywhere I go, I can hook up to the Net for whatever reason. It makes life a lot easier." Increasingly, companies and individuals are developing or purchasing systems that allow them to take advantage of the Internet.[5] Microsoft's Live Mesh, for example, allows systems developers to seamlessly coordinate data among different devices and provide data backup on the Internet.

Systems design is typically accomplished using the tools and techniques discussed in Chapter 12. Depending on the specific application, these methods can be used to support and document all aspects of systems design. Two key aspects of systems design are logical and physical design.

Logical and Physical Design

As discussed in Chapter 5, design has two dimensions: logical and physical. The **logical design** refers to what the system will do. It describes the functional requirements of a system.[6] Without logical design, the technical details of the system (such as which hardware devices should be acquired) often obscure the best solution. Logical design involves planning the purpose of each system element, independent of hardware and software considerations. The logical design specifications that are determined and documented include output, input, process, file and database, telecommunications, procedures, controls and security, and personnel and job requirements.

logical design
A description of the functional requirements of a system.

Security is always an important logical design issue for corporations and governments.[7] Rules published in September 2005, for example, require that federal agencies incorporate security procedures in the design of new or modified systems. In addition, the Federal Information Security Management Act, enacted in 2002, requires federal agencies to make sure that security protection measures are incorporated into systems provided by outside vendors and contractors. The Federal Rules of Civil Procedure requires that companies make e-mails, text messages, and other electronic communications available in some court hearings.[8] Failure to meet these electronic disclosure requirements in a timely fashion can result in executives and managers facing fines and jail time. This requirement has spurred new systems development projects that can search and find electronic communications to meet federal requirements.[9]

The **physical design** refers to how the tasks are accomplished, including how the components work together and what each component does. Physical design specifies the characteristics of the system components necessary to put the logical design into action. In this phase, the characteristics of the hardware, software, database, telecommunications, personnel, and procedure and control specifications must be described in detail. These physical design components were discussed in Part 2 on technology.

physical design
The specification of the characteristics of the system components necessary to put the logical design into action.

Object-Oriented Design

Logical and physical design can be accomplished using either the traditional approach or the object-oriented approach to systems development. Both approaches use a variety of design models to document the new system's features and the development team's understandings and agreements. Many organizations today are turning to OO development because of its increased flexibility. This section outlines a few OO design considerations and diagrams.[10]

Using the OO approach, you can design key objects and classes of objects in the new or updated system. This process includes considering the problem domain, the operating environment, and the user interface. The problem domain involves the classes of objects related to solving a problem or realizing an opportunity. In our Maui, Hawaii, kayak rental shop example first introduced in Chapter 12 and referring back to the generalization/specialization hierarchy showing classes we presented there, KayakItem in Figure 12.23 is an example of a problem domain object that will store information on kayaks in the rental program. The operating environment for the rental shop's system includes objects that interact with printers, system software, and other software and hardware devices. The user interface for the system includes objects that users interact with, such as buttons and scroll bars in a Windows program.

During the design phase, you also need to consider the sequence of events that must happen for the system to function correctly. For example, you might want to design the sequence of events for adding a new kayak to the rental program. The event sequence is often called a *scenario*, and it can be diagrammed in a sequence diagram (see Figure 13.1).

Figure 13.1

A Sequence Diagram to Add a New KayakItem Scenario

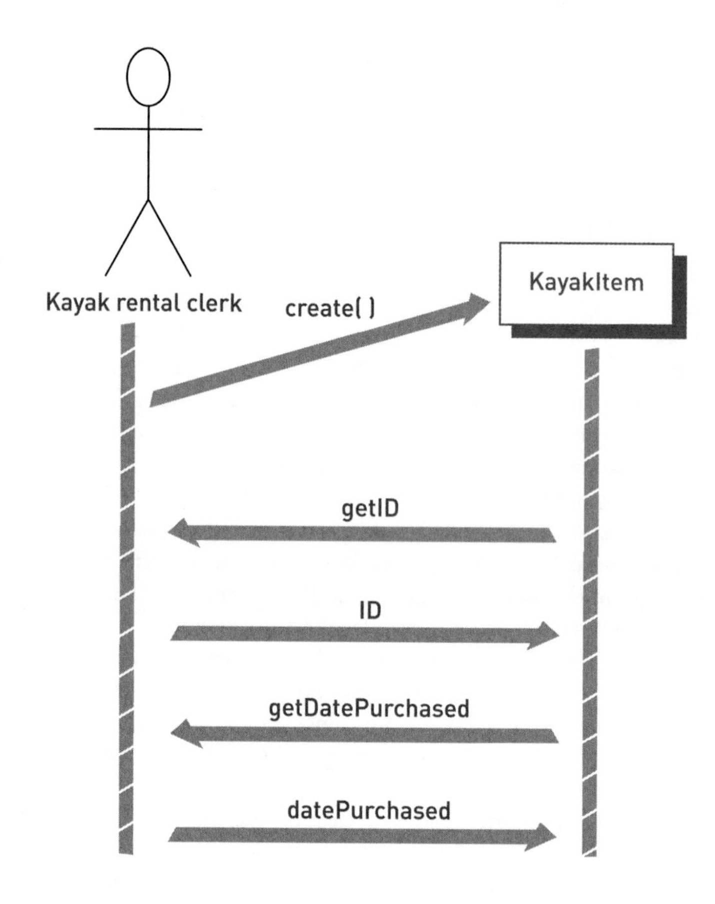

You read a sequence diagram starting at the top and moving down.

1. The Create arrow at the top is a message from the kayak rental clerk to the KayakItem object to create information on a new kayak to be placed into the rental program.
2. The KayakItem object knows that it needs the ID for the kayak and sends a message to the clerk requesting the information. See the getID arrow.
3. The clerk then types the ID into the computer. This is shown with the ID arrow. The data is stored in the KayakItem object.
4. Next, KayakItem requests the purchase date. This is shown in the getDatePurchased arrow.
5. Finally, the clerk types the purchase date into the computer. The data is also transferred to KayakItem object. This is shown in the datePurchased arrow at the bottom of Figure 13.1.

This scenario is only one example of a sequence of events. Other scenarios might include entering information about life jackets, paddles, suntan lotion, and other accessories. The same types of use case and generalization/specialization hierarchy diagrams discussed in Chapter 12 can be created for each event, and additional sequence diagrams will also be needed.

Interface Design and Controls

Some special system characteristics should be considered during both logical and physical design. These characteristics relate to how users access and interact with the system, including sign-on procedures, interactive processing, and interactive dialogue.

sign-on procedure
Identification numbers, passwords, and other safeguards needed for someone to gain access to computer resources.

- A **sign-on procedure** consists of identification numbers, passwords, and other safeguards needed for someone to gain access to computer resources. The new system or modified one should require that identification numbers and passwords be changed regularly.[11] An IS worker for a large U.S. company operating in India was caught stealing about 4,000 sensitive corporate documents using the identification number and password of another employee. See Figure 13.2.

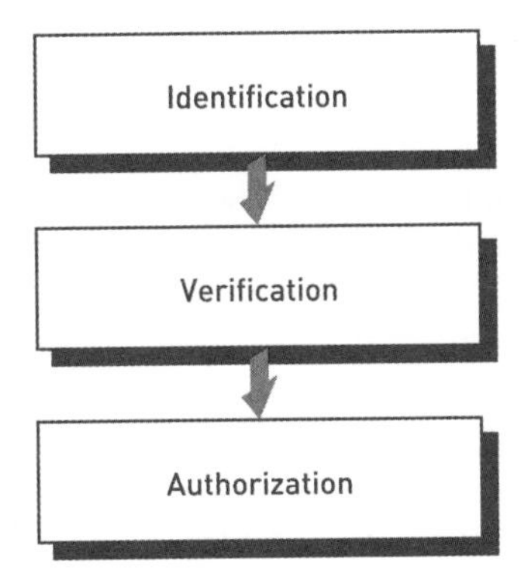

Figure 13.2

The Levels of the Sign-On Procedure

- With *interactive processing,* people directly interact with the processing component of the system through terminals or networked PCs. With a **menu-driven system** (see Figure 13.3), users simply pick what they want to do from a list of alternatives. Most people can easily operate these types of systems. They select an option or respond to questions (or prompts) from the system, and the system does the rest.

menu-driven system
A system in which users simply pick what they want to do from a list of alternatives.

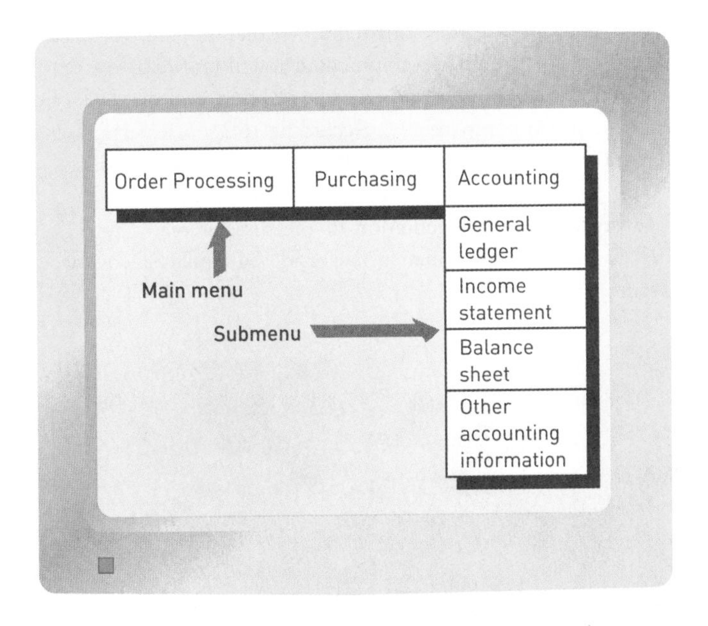

Figure 13.3

Menu-Driven System

A menu-driven system allows you to choose what you want from a list of alternatives.

- Many designers incorporate a **help facility** into the system or applications program. When users want to know more about a program or feature or what type of response is expected, they can activate the help facility.
- Computer programs can develop and use **lookup tables** to simplify and shorten data entry. For example, if you are entering a sales order for a company, you can type its abbreviation, such as ABCO. The program will then go to the customer table, normally stored on a disk, and look up all the information pertaining to the company abbreviated ABCO that you need to complete the sales order.
- With a **restart procedure,** users can restart an application where it stopped in case the application crashed or had problems.

help facility
A program that provides assistance when users want to know more about a program or feature or what type of response is expected.

lookup tables
Tables containing data that computer programs can develop and use to simplify and shorten data entry.

restart procedures
Simplified processes to access an application from where it stopped.

Design of System Security and Controls

In addition to considering the system's interface and user interactions, designers must also develop system security and controls for all aspects of the system, including hardware, software, database systems, telecommunications, and Internet operations. These key considerations involve error prevention, detection, and correction; disaster planning and recovery; and systems controls. Some small and medium-sized corporations, for example, are buying *unified threat management (UTM)* products to protect their networks from security threats and breaches.[12]

Preventing, Detecting, and Correcting Errors

The most cost-effective time to deal with potential errors is early in the design phase. Every possibility should be considered, even minor problems.[13] Hanford Brothers Company, for example, had installed backup electrical generators in case of a power failure. When a fuel truck crashed near its facility and spilled its flammable cargo, the city shut down all power to the area and didn't let Hanford Brothers use its electrical generators, fearing it could cause an explosion or severe fire. This minor incident completely shut down the IS center for the company until the spill could be cleaned up. In addition to minor problems, other important security and control measures, including disaster planning and recovery and adequate backup procedures, must be considered.

Disaster Planning and Recovery

Disaster planning is the process of anticipating and providing for disasters. A disaster can be an act of nature (a flood, fire, or earthquake) or a human act (terrorism, error, labor unrest, or erasure of an important file). Disaster planning often focuses primarily on two issues: maintaining the integrity of corporate information and keeping the information system running until normal operations can be resumed. Disaster planning, however, can be expensive. According to a director of project development for a filtration company, "The business side just isn't aware of the costs of disaster recovery projects, and that lack of understanding can pose enormous risks to companies."[14] According to a Forrester Research study, only 34 percent of IS data center managers believed they were prepared for a disaster or data center failure.[15]

disaster recovery
The implementation of the disaster plan.

Disaster recovery is the implementation of the disaster plan. According to a Harris Interactive survey, 71 percent of IS managers considered disaster recovery as important or critical.[16] See Figure 13.4.

Figure 13.4

Disaster Recovery Efforts

(Source: UPI Photo/Earl Cryer/Landov.)

The primary tools used in disaster planning and recovery are hardware; software; and database, telecommunications, and personnel backups. Most of these systems were discussed in Part 2 on information technology concepts. For some companies, personnel backup can be critical.[17] According to the IS program manager for Northrop Grumman concerning a flooding disaster in Mississippi, "From the onset, finding our employees was our No. 1 priority." Without IS employees, the IS department can't function. Hot and cold sites can be used to back up hardware. A duplicate, operational hardware system that is ready for use (or immediate access to one through a specialized vendor) is an example of a **hot site**. If the primary computer has problems, the hot site can be used immediately as a backup. It is important, however, that the hot site can't be impacted by the same disaster. The hot site for Northrop Grumman's Mississippi facility was a large ship-building company located near New Orleans. When a disaster destroyed the Mississippi IS facility, it also took out electrical power near New Orleans, making Northrop Grumman's hot site useless. Another approach is to use a **cold site**, also called a *shell*, which is a computer environment that includes rooms, electrical service, telecommunications links, data storage devices, and similar equipment. If a primary computer has a problem, backup computer hardware is brought into the cold site, and the complete system is made operational. Files and databases can be backed up by making a copy of all files and databases changed during the last few days or the last week, a technique called **incremental backup**. This approach to backup uses an **image log**, which is a separate file that contains only changes to applications. Whenever an application is run, an image log is created that contains all changes made to all files. If a problem occurs with a database, an old database with the last full backup of the data, along with the image log, can be used to re-create the current database. Organizations can also hire outside companies to help them perform disaster planning and recovery.[18] EMC, for example, offers data backup in its RecoverPoint product. For individuals and some applications, backup copies of important files can be placed on the Internet.

hot site
A duplicate, operational hardware system or immediate access to one through a specialized vendor.

cold site
A computer environment that includes rooms, electrical service, telecommunications links, data storage devices, and the like; also called a *shell*.

incremental backup
Making a backup copy of all files changed during the last few days or the last week.

image log
A separate file that contains only changes to applications.

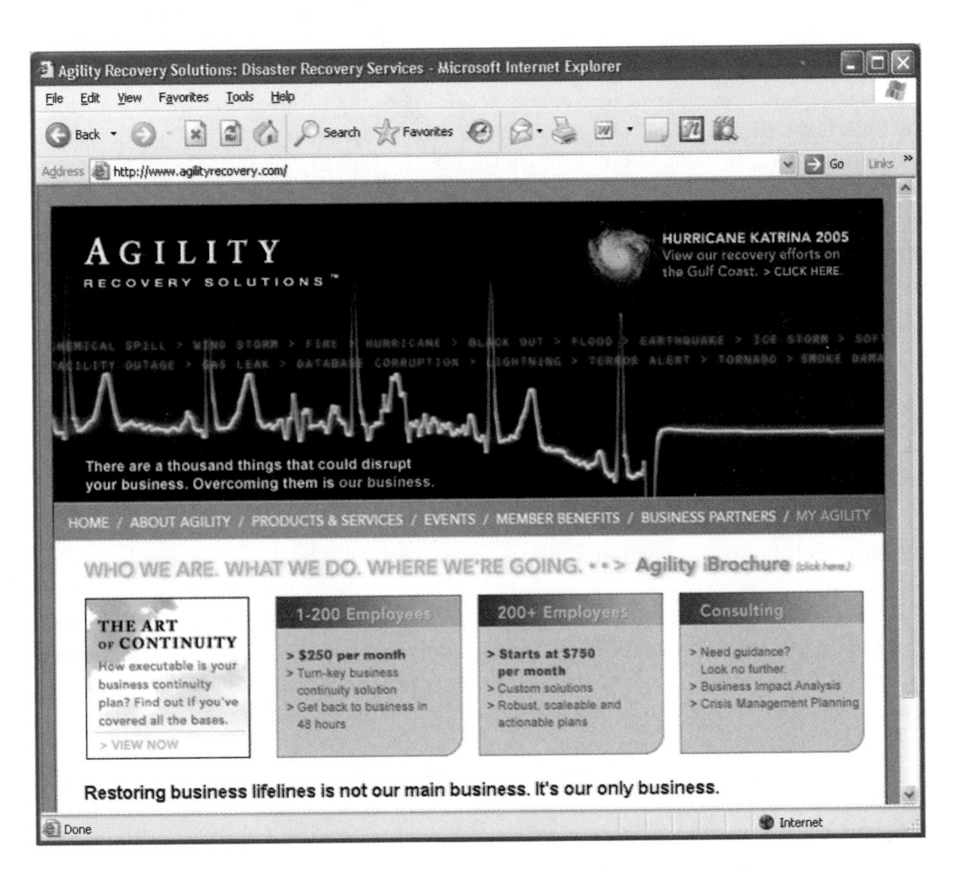

Companies that suffer a disaster can employ a disaster recovery service, which can secure critical data backup information. These service companies can also provide a facility from which to operate and communications equipment to stay in touch with customers.

(Source: *www.agilityrecovery.com*.)

The SunGard Data Systems high-availability command center oversees services that let companies continuously access and process information—with minimal downtime—even during a disaster.

(Source: © Mike Mergen/ Bloomberg News/Landov.)

Systems Controls

Security lapses, fraud, and the invasion of privacy can present difficult challenges.[19] Health care providers, for example, are now developing controls to combat medical identity theft.[20] California law enforcement officials busted a criminal ring that billed almost a million dollars in tests that were never performed. According to the national director for anti-fraud for Blue Cross Blue Shield, "Our software has become more sophisticated, particularly in identifying spikes in usage—someone who normally goes to the doctor once a year and suddenly goes 25 times in a 12-month period." In another case, a futures and options trader for a British bank lost about $1 billion. A simple systems control might have prevented a problem that caused the 200-year-old bank to collapse. Preventing and detecting these problems is an important part of systems design. Prevention includes the following:

- Determining potential problems
- Ranking the importance of these problems
- Planning the best place and approach to prevent problems
- Deciding the best way to handle problems if they occur

Every effort should be made to prevent problems, but companies must establish procedures to handle problems if they occur, including systems controls.

Most IS departments establish tight **systems controls** to maintain data security. Systems controls can help prevent computer misuse, crime, and fraud by managers, employees, and others.

systems controls
Rules and procedures to maintain data security.

Most IS departments have a set of general operating rules that helps protect the system. Some IS departments are **closed shops**, in which only authorized operators can run the computers. Other IS departments are **open shops**, in which other people, such as programmers and systems analysts, are also authorized to run the computers. Other rules specify the conduct of the IS department.

closed shops
IS departments in which only authorized operators can run the computers.

open shops
IS departments in which people, such as programmers and systems analysts, are allowed to run the computers, in addition to authorized operators.

These rules are examples of **deterrence controls**, which involve preventing problems before they occur. Good control techniques should help an organization contain and recover from problems. The objective of containment control is to minimize the impact of a problem while it is occurring, and recovery control involves responding to a problem that has already occurred.

deterrence controls
Rules and procedures to prevent problems before they occur.

Many types of systems controls can be developed, documented, implemented, and reviewed. These controls touch all aspects of the organization (see Table 13.1).

Table 13.1

Using Systems Controls to Enhance Security

Controls	Description
Input controls	Maintain input integrity and security. Their purpose is to reduce errors while protecting the computer system against improper or fraudulent input. Input controls range from using standardized input forms to eliminating data-entry errors and using tight password and identification controls.
Processing controls	Deal with all aspects of processing and storage. The use of passwords and identification numbers, backup copies of data, and storage rooms that have tight security systems are examples of processing and storage controls.
Output controls	Ensure that output is handled correctly. In many cases, output generated from the computer system is recorded in a file that indicates the reports and documents that were generated, the time they were generated, and their final destinations.
Database controls	Deal with ensuring an efficient and effective database system. These controls include the use of identification numbers and passwords, without which a user is denied access to certain data and information. Many of these controls are provided by database management systems.
Telecommunications controls	Provide accurate and reliable data and information transfer among systems. Telecommunications controls include firewalls and encryption to ensure correct communication while eliminating the potential for fraud and crime.
Personnel controls	Make sure that only authorized personnel have access to certain systems to help prevent computer-related mistakes and crime. Personnel controls can involve the use of identification numbers and passwords that allow only certain people access to particular data and information. ID badges and other security devices (such as smart cards) can prevent unauthorized people from entering strategic areas in the information systems facility.

After controls are developed, they should be documented in standards manuals that indicate how the controls are to be implemented. They should then be implemented and frequently reviewed. It is common practice to measure the extent to which control techniques are used and to take action if the controls have not been implemented.

Many companies use ID badges to prevent unauthorized access to sensitive areas in the information systems facility.

(Source: Michael Newman/Photo Edit.)

ETHICAL AND SOCIETAL ISSUES

Going Green Saves Millions of Dollars for Nationwide

Nationwide is one of the largest insurance and financial services in the world. The company offers a full range of insurance products and financial services for home, car, and family. It has garnered over $161 billion in statutory assets.

With 36,000 employees managing 16 million policies, Nationwide requires a large data center to store and manipulate policy data. Actually, Nationwide has 20 data centers and a $250 million budget for its information system infrastructure. Nationwide's primary data center in Columbus, Ohio, supports roughly 400 million transactions per month for activities such as calculating policy quotes; making policy additions, changes, and deletions; and processing claims.

Scott Miggo, vice president of Technology Solutions at Nationwide, manages the company's data centers. Scott continuously monitors demand on the data center's servers, power, and cooling. Scott has tracked a consistent 5 percent growth in data center processing from year to year. At this rate, he estimates that the data center demand will outstrip the power capacity of the company's primary data center by 2013.

Scott had a number of options for developing systems to meet future demand. He might expand by building a new data center to add to the processing power of the current center. Many companies would choose to begin construction on a new data center, retiring the old equipment in favor of using the latest energy-efficient technologies. If Nationwide began construction immediately, they could have the new data center online by 2013. However, a new data center would cost Nationwide hundreds of millions of dollars. Scott tried to find a solution that might forestall the inevitable.

Scott and his team found several solutions that would buy them two or more years beyond 2013 without having to invest in major construction. First, they began using virtualization with VMware. VMware allowed one large mainframe server to act as 20 virtual servers. By implementing virtual servers, Scott reduced the numbers of servers in the data center from 5,000 to 3,500. The VM servers were running at 65 percent usage, up from 10 percent. In essence, virtualization allowed Nationwide to get more work out of each server—freeing up space and lowering power and air conditioning demands.

Secondly, Scott and his team began replacing the oldest, energy-intensive servers with green servers. The new energy-efficient servers saved the data center $40,000 a year in electricity and cooling.

In another cost-saving effort, Scott and his team replaced tape silos with more modern, denser tapes and faster tape robots. The result was more efficient data storage and retrieval in a smaller amount of space. Although the savings from this upgrade were negligible compared to virtualization and server upgrades, every little bit helped. Scott says, "You've got to look at it holistically. We are looking at going to a massive array of idle disks that shut down and are brought up only when you need the data on a particular disk."

The total upgrade of the main data center set Nationwide back $30 million. This is a small amount compared to the hundreds of millions they would have spent constructing a new data center. Space is no longer an issue at the data center. However, Nationwide eventually needs to build an additional data center. They don't need the server space anymore, but the infrastructure of the building will no longer support the power and cooling needs of the growing number of servers—even the greenest, most energy-efficient models. Nationwide plans to continue virtualizing servers and upgrading to more energy-efficient models in its 20 data centers based on the model that Scott Miggo created.

Discussion Questions

1. What issues did Nationwide face with their data centers? What considerations determined the data center's processing capacity?
2. What three techniques did Scott Miggo implement to save Nationwide hundreds of millions of dollars?

Critical Thinking Questions

1. Why was it good for Scott Miggo to anticipate the needs of data processing ten years in advance? What luxuries did it afford him, and how did it pay off for Nationwide?
2. What other green technologies might be used to further extend the usefulness of Nationwide's primary data center and reduce the cost of operations?

SOURCES: Bartholomew, Doug, "Refurbishing Old Data Center Provides Big Savings," *CIO Insight*, September 11, 2007, *www.cioinsight.com/c/a/Case-Studies/Refurbishing-Old-Data-Center-Provides-Big-Savings*; Nationwide Web site, *www.nationwide.com*, accessed July 19, 2008.

ENVIRONMENTAL DESIGN CONSIDERATIONS

Developing new systems and modifying existing ones in an environmentally sensitive way is becoming increasingly important for many IS departments.[21] **Environmental design**, also called *green design,* involves systems development efforts that slash power consumption, require less physical space, and result in systems that can be disposed in a way that doesn't negatively affect the environment. Today, companies are using innovative ways to design efficient systems and operations, including using virtual servers to save energy and space, pushing cold air under data centers to cool equipment, using software to efficiently control cooling fans, building facilities with more insulation, and even collecting rain water from roofs to cool equipment.[22] VistaPrint (*www.vistaprint.com*), a graphics design and printing company, switched from traditional servers to virtual servers and saved about $500,000 in electricity costs over a three-year period, representing a 75 percent reduction in energy usage.[23] A *Computerworld* survey revealed that over 80 percent of IS managers considered energy efficiency when selecting new computer equipment.[24] The Environmental Protection Agency (EPA) estimates that a 10 percent cut in data center electricity usage would be enough to power about a million U.S. homes every year.[25] According to a McKinsey & Co. study, the amount of greenhouse gases generated from data centers will increase by 400 percent by 2020 and become more than the greenhouse gases emitted by U.S. airlines.[26]

environmental design
Also called *green design*, it involves systems development efforts that slash power consumption, require less physical space, and result in systems that can be disposed in a way that doesn't negatively affect the environment.

Many companies are developing products and services to help save energy. EMC, for example, has developed new disk drives that use substantially less energy.[27] Environmental design also involves developing software and systems that help organizations reduce power consumption for other aspects of their operations. Carbonetworks and Optimum Energy, for example, have developed software products to help companies reduce energy costs by helping them determine when and how to use electricity.[28] UPS developed its own software to reduce the miles its 90,000 trucks and other vehicles drive by routing them more efficiently. The new software helped UPS cut 30 million miles per year, slash fuel costs, and reduce carbon emissions by over 30,000 metric tons. Hewlett-Packard and Dell Computer have developed procedures and machines to dispose of old computers and computer equipment in environmentally friendly ways.[29] Old computers and computer equipment are fed into machines that shred them into small pieces and sort them into materials that can be reused. The process is often called *green death.*[30] One study estimates that more than 130,000 PCs in the United States are thrown out every day. The U.S. government is also involved in environmental design. It has a plan to require federal agencies to purchase energy-efficient computer systems and equipment.[31] The plan would require federal agencies to use the *Electronic Product Environmental Assessment Tool (EPEAT)* to analyze the energy usage of new systems. The U.S. Department of Energy rates products with the *Energy Star* designation to help people select products that save energy.[32]

Companies such as Hewlett-Packard and Dell Computer dispose of old computers and computer equipment in environmentally friendly ways.

(Source: © Robyn Beck/AFP/Getty Images.)

Generating Systems Design Alternatives

When people or organizations require a system to perform additional functions that an existing system cannot support, they often turn to outside vendors to design and supply their new systems. Whether an individual is purchasing a personal computer or a company is acquiring an expensive mainframe computer, the system can be obtained from a single vendor or multiple vendors. If the new system is complex, the original development team might want to involve other personnel in generating alternative designs. In addition, if new hardware and software are to be acquired from an outside vendor, a formal request for proposal (RFP) can be made.

Request for Proposals

request for proposal (RFP)
A document that specifies in detail required resources such as hardware and software.

The **request for proposal (RFP)** is an important document for many organizations involved with large, complex systems development efforts. Smaller, less-complex systems often do not require an RFP. A company that is purchasing an inexpensive piece of software that will run on existing hardware, for example, might not need to go through a formal RFP process.

In some cases, separate RFPs are developed for different needs. For example, a company might develop separate RFPs for hardware, software, and database systems. The RFP also communicates these needs to one or more vendors, and it provides a way to evaluate whether the vendor has delivered what was expected. In some cases, the RFP is part of the vendor contract. The Table of Contents for a typical RFP is shown in Figure 13.5.

Figure 13.5

A Typical Table of Contents for a Request for Proposal

Johnson & Florin, Inc.
Systems Investigation Report

Contents

COVER PAGE (with company name and contact person)
BRIEF DESCRIPTION of the COMPANY
OVERVIEW of the EXISTING COMPUTER SYSTEM
SUMMARY of COMPUTER-RELATED NEEDS and/or PROBLEMS
OBJECTIVES of the PROJECT
DESCRIPTION of WHAT IS NEEDED
HARDWARE REQUIREMENTS
PERSONNEL REQUIREMENTS
COMMUNICATIONS REQUIREMENTS
PROCEDURES to BE DEVELOPED
TRAINING REQUIREMENTS
MAINTENANCE REQUIREMENTS
EVALUATION PROCEDURES (how vendors will be judged)
PROPOSAL FORMAT (how vendors should respond)
IMPORTANT DATES (when tasks are to be completed)
SUMMARY

Financial Options

When acquiring computer systems, several choices are available, including purchase, lease, or rent. Cost objectives and constraints set for the system play a significant role in the choice, as do the advantages and disadvantages of each. In addition, traditional financial tools, including net present value and internal rate of return, can be used. Table 13.2 summarizes the advantages and disadvantages of these financial options.

Determining which option is best for a particular company in a given situation can be difficult. Financial considerations, tax laws, the organization's policies, its sales and transaction growth, marketplace dynamics, and the organization's financial resources are all important factors. In some cases, lease or rental fees can amount to more than the original purchase price after a few years.

Table 13.2

Advantages and Disadvantages of Acquisition Options

Renting (Short-Term Option)	
Advantages	**Disadvantages**
No risk of obsolescence	No ownership of equipment
No long-term financial investment	High monthly costs
No initial investment of funds	Restrictive rental agreements
Maintenance usually included	
Leasing (Intermediate to Long-Term Option)	
Advantages	**Disadvantages**
No risk of obsolescence	High cost of canceling lease
No long-term financial investment	Longer time commitment than renting
No initial investment of funds	No ownership of equipment
Less expensive than renting	
Purchasing (Long-Term Option)	
Advantages	**Disadvantages**
Total control over equipment	High initial investment
Can sell equipment at any time	Additional cost of maintenance
Can depreciate equipment	Possibility of obsolescence
Low cost if owned for a number of years	Other expenses, including taxes and insurance

Evaluating and Selecting a Systems Design

The final step in systems design is to evaluate the various alternatives and select the one that will offer the best solution for organizational goals. For example, financial concerns might make a company choose rental over equipment purchase. Specific performance objectives—for example, that the new system must perform online data processing—might result in a complex network design for which control procedures must be established. Evaluating and selecting the best design involves achieving a balance of system objectives that will best support organizational goals. Normally, evaluation and selection involves both a preliminary and a final evaluation before a design is selected.

A **preliminary evaluation** begins after all proposals have been submitted. The purpose of this evaluation is to dismiss unwanted proposals. Several vendors can usually be eliminated by investigating their proposals and comparing them with the original criteria. The **final evaluation** begins with a detailed investigation of the proposals offered by the remaining vendors. The vendors should be asked to make a final presentation and to fully demonstrate the system. The demonstration should be as close to actual operating conditions as possible. Figure 13.6 illustrates the evaluation process.

preliminary evaluation
An initial assessment whose purpose is to dismiss the unwanted proposals; begins after all proposals have been submitted.

final evaluation
A detailed investigation of the proposals offered by the vendors remaining after the preliminary evaluation.

Evaluation Techniques

The exact procedure used to make the final evaluation and selection varies from one organization to the next. Some were first introduced in Chapter 2, including return on investment (ROI), earnings growth, market share, customer satisfaction, and total cost of ownership (TCO). Cabelas and Staples, for example, are using Web-style testimonials to get customer satisfaction information on its products and information systems.[33] Other companies, such as Backcountry, use live online chat to get customer satisfaction information.[34] General Motors and the Pentagon are adopting the Capability Maturity Model Integration (CMMI), first developed by the Carnegie Mellon Software Engineering Institute, to help evaluate new

Figure 13.6

The Stages in Preliminary and Final Evaluations

The number of possible alternatives decreases as the firm gets closer to making a final decision.

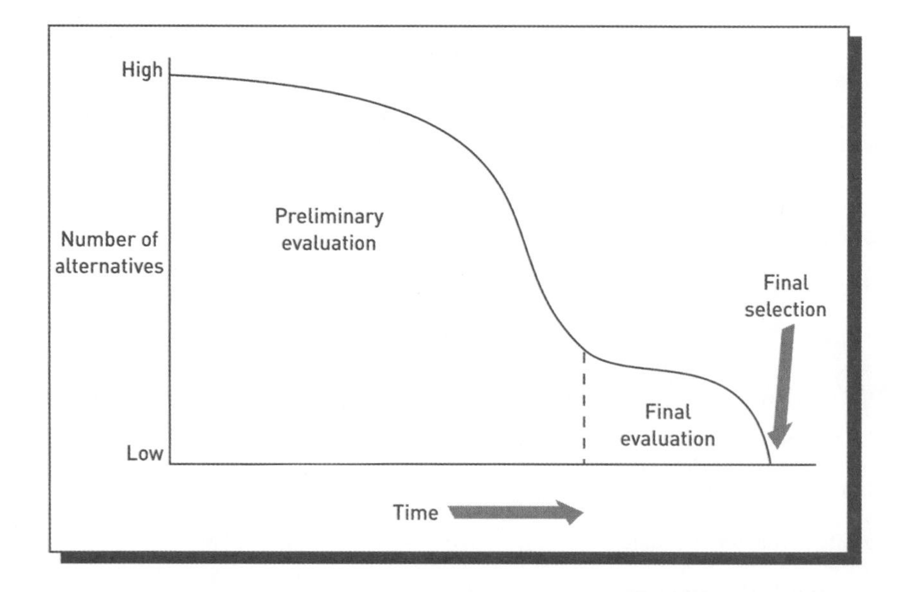

equipment and systems purchases.[35] In addition, four other approaches are commonly used: group consensus, cost/benefit analysis, benchmark tests, and point evaluation.

Group Consensus

group consensus
Decision making by a group that is appointed and given the responsibility of making the final evaluation and selection.

In **group consensus**, a decision-making group is appointed and given the responsibility of making the final evaluation and selection. Usually, this group includes the members of the development team who participated in either systems analysis or systems design. This approach might be used to evaluate which of several screen layouts or report formats is best.

Cost/Benefit Analysis

cost/benefit analysis
An approach that lists the costs and benefits of each proposed system. After they are expressed in monetary terms, all the costs are compared with all the benefits.

Cost/benefit analysis is an approach that lists the costs and benefits of each proposed system. After they are expressed in monetary terms, all the costs are compared with all the benefits. Table 13.3 lists some of the typical costs and benefits associated with the evaluation and selection procedure. This approach is used to evaluate options whose costs can be quantified, such as which hardware or software vendor to select.

Benchmark Tests

benchmark test
An examination that compares computer systems operating under the same conditions.

A **benchmark test** is an examination that compares computer systems operating under the same conditions. Most computer companies publish their own benchmark tests, but some forbid disclosure of benchmark tests without prior written approval. Thus, one of the best approaches is for an organization to develop its own tests and then use them to compare the equipment it is considering. This approach might be used to compare the end-user system response time on two similar systems. Several independent companies and journals also rate computer systems.

Point Evaluation

point evaluation system
An evaluation process in which each evaluation factor is assigned a weight, in percentage points, based on importance. Then each proposed system is evaluated in terms of this factor and given a score ranging from 0 to 100. The scores are totaled, and the system with the greatest total score is selected.

One of the disadvantages of cost/benefit analysis is the difficulty of determining the monetary values for all the benefits. An approach that does not employ monetary values is a **point evaluation system**. Each evaluation factor is assigned a weight, in percentage points, based on importance. Then each proposed information system is evaluated in terms of this factor and given a score, such as one ranging from 0 to 100, where 0 means that the alternative does not address the feature at all and 100 means that the alternative addresses that feature perfectly. The scores are totaled, and the system with the greatest total score is selected. When using point evaluation, an organization can list and evaluate literally hundreds of factors. Figure 13.7 shows a simplified version of this process. This approach is used when there are many options to be evaluated, such as which software best matches the needs of a particular business.

Table 13.3

Cost/Benefit Analysis Table

Costs	Benefits
Development costs	Reduced costs
Personnel	Fewer personnel
Computer resources	Reduced manufacturing costs
	Reduced inventory costs
	More efficient use of equipment
	Faster response time
	Reduced downtime or crash time
	Less spoilage
Fixed costs	**Increased Revenues**
Computer equipment	New products and services
Software	New customers
One-time license fees for software and maintenance	More business from existing customers
	Higher price as a result of better products and services
Operating costs	**Intangible benefits**
Equipment lease and/or rental fees	Better public image for the organization
Computer personnel (including salaries, benefits, etc.)	Higher employee morale
	Better service for new and existing customers
Electric and other utilities	The ability to recruit better employees
Computer paper, tape, and disks	Position as a leader in the industry
Other computer supplies	System easier for programmers and users
Maintenance costs	
Insurance	

		System A			System B		
Factor's importance		*Evaluation*		*Weighted evaluation*	*Evaluation*		*Weighted evaluation*
Hardware	35%	95	35%	33.25	75	35%	26.25
Software	40%	70	40%	28.00	95	40%	38.00
Vendor support	25%	85	25%	21.25	90	25%	22.50
Totals	100%			82.5			86.75

Figure 13.7

An Illustration of the Point Evaluation System

In this example, software has been given the most weight (40 percent), compared with hardware (35 percent) and vendor support (25 percent). When system A is evaluated, the total of the three factors amounts to 82.5 percent. System B's rating, on the other hand, totals 86.75 percent, which is closer to 100 percent. Therefore, the firm chooses system B.

Ryder's GPS System Development Nearly Out of Control

Ryder is a worldwide provider of transportation, logistics, and supply chain management solutions. Ryder employs 28,800 people worldwide and maintains a fleet of 159,400 vehicles. Kevin Bott is Ryder's senior vice president and CIO based in the company's headquarters in Miami, Florida.

When global positioning systems and wireless communications became mainstream technologies, Kevin Bott saw big potential for gaining more control over Ryder operations. Kevin proposed that Ryder install GPS receivers in all of its trucks to wirelessly send location information to headquarters. The vehicle location information could be used to better coordinate the movement of freight around the world.

Kevin formed a small team of experts from across the company to help design what was hoped would be a "lean" information system project with substantial benefits. In addition to beaming location information to headquarters, the team designed a system that would connect the system to the vehicle to track mileage, fuel use, and vehicle maintenance information. Kevin contracted with Teletrac for the GPS technology and Cingular Wireless for the transport of the data to headquarters.

GPS equipment was shipped for installation in 5,000 vehicles for a test pilot. Meanwhile, word about the project was spreading around corporate headquarters. Managers from every division were beginning to appreciate the possible benefits the system would provide. Managers and executives began inviting themselves to the weekly planning meetings for the new system. Kevin found that his "lean" system was beginning to be weighed down by too many ideas and demands.

Problems also surfaced as the systems were installed. Ryder supports a diverse fleet of many vehicle makes, models, and years. The GPS computer system had to be reworked to communicate with these different types of vehicles. At headquarters, new members of Kevin Bott's team were pulling the project in a variety of directions. For the sake of the project, Kevin began meeting only with those in the original core team. With their help, he made important decisions to steer the project back on track with reasonable goals. Kevin froze the design specifications so that further inessential alterations were no longer considered. The core team worked with primary vendor partners to resolve installation issues.

Ultimately, the project was completed with success, resulting in a new GPS system called RydeSmart. The upfront hardware investment was offset immediately by savings in the following efficiencies:

- A 10 to 15 percent reduction in fuel consumption through improved routing and reduction of unauthorized use and idle time
- Fifteen-minute savings in work hour per driver per day
- Increased efficiency in tracking disabled vehicles and repairing them
- Simplified compliance with government regulations
- Automated trip reports for fuel tax compliance
- Automated and wireless odometer reading and reporting
- Electronic log books that save the driver from having to record and report trip statistics

The new system allows drivers to complete more trips per day with less clerical work. Computerized rout optimization provides drivers with the shortest and least congested route to their destination.

RydeSmart has also boosted Ryder's market share. Customers like the fact that the system helps get products delivered efficiently and safely. Kevin Bott reports that "We have achieved savings in trip records processing, breakdown repair costs, idle-time fuel reductions, out-of-route mileage reductions, improved asset utilization, and more-efficient driver hours." Bott is credited with a brilliant system development effort, refusing to allow the project to be derailed by competing corporate goals and instead concentrating on tangible areas of savings.

Discussion Questions

1. How did Kevin Bott's lean information systems development project become bloated and nearly derailed?
2. What solution did Bott employ to regain control of the project?

Critical Thinking Questions

1. What internal political pressures do you suppose Bott had to face to gain control of the project? How could diplomacy make all interested parties feel appreciated?
2. How do you think drivers felt about having GPS installed in their vehicles that allowed Ryder headquarters to track their movements? As a CIO, how might you handle driver complaints?

SOURCES: McAdams, Jennifer, "Lean Projects," *Computerworld*, December 10, 2007, *www.computerworld.com/action/article.do?command=viewArticleBasic&articleId=305886&pageNumber=1*; RydeSmart Web site, *www.rydesmart.ryder.com*, accessed July 19, 2008; Ryder Web site, *www.ryder.com*, accessed July 19, 2008.

Freezing Design Specifications

Near the end of the design stage, some organizations prohibit further changes in the design of the system. Freezing systems design specifications means that the user agrees in writing that the design is acceptable. (See Figure 13.8.) Other organizations, however, allow or even encourage design changes. These organizations often use agile or rapid systems development approaches, introduced in Chapter 12.

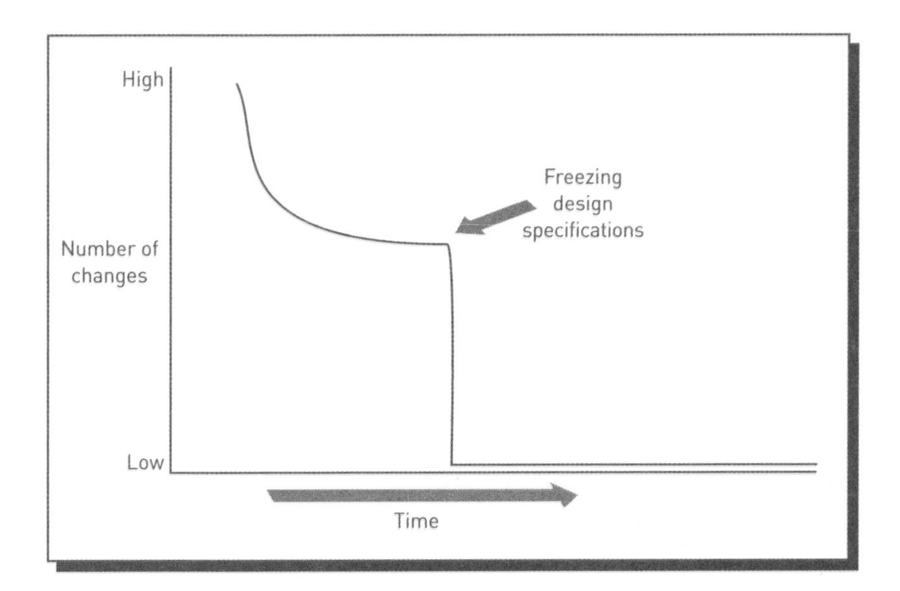

Figure 13.8

Freezing Design Specifications

The Contract

One of the most important steps in systems design is to develop a good contract if new computer facilities are being acquired. Finding the best terms where everyone makes a profit can be difficult. Most computer vendors provide standard contracts; however, such contracts are designed to protect the vendor, not necessarily the organization buying the computer equipment. Some organizations include penalty clauses in the contract, in case the vendor does not meet its obligation by the specified date. Typically, the request for proposal becomes part of the contract. This saves a considerable amount of time in developing the contract, because the RFP specifies in detail what is expected from the vendors.

The Design Report

System specifications are the final results of systems design. They include a technical description that details system outputs, inputs, and user interfaces as well as all hardware, software, databases, telecommunications, personnel, and procedure components and the way these components are related. The specifications are contained in a **design report**, which is the primary result of systems design. The design report reflects the decisions made for systems design and prepares the way for systems implementation. The contents of the design report are summarized in Figure 13.9.

design report
The primary result of systems design, reflecting the decisions made and preparing the way for systems implementation.

Figure 13.9

A Typical Table of Contents for a Systems Design Report

Johnson & Florin, Inc.
Systems Design Report

Contents

PREFACE
EXECUTIVE SUMMARY of SYSTEMS DESIGN
REVIEW of SYSTEMS ANALYSIS
MAJOR DESIGN RECOMMENDATIONS
- Hardware design
- Software design
- Personnel design
- Communications design
- Database design
- Procedures design
- Training design
- Maintenance design

SUMMARY of DESIGN DECISIONS
APPENDICES
GLOSSARY of TERMS
INDEX

SYSTEMS IMPLEMENTATION

systems implementation
A stage of systems development that includes hardware acquisition, software acquisition or development, user preparation, hiring and training of personnel, site and data preparation, installation, testing, start-up, and user acceptance.

After the information system has been designed, a number of tasks must be completed before the system is installed and ready to operate. This process, called **systems implementation**, includes hardware acquisition, programming and software acquisition or development, user preparation, hiring and training of personnel, site and data preparation, installation, testing, start-up, and user acceptance. The typical sequence of systems implementation activities is shown in Figure 13.10.

Virtualization, first introduced in Chapter 3, has had a profound impact on many aspects of systems implementation. As mentioned in Chapter 4, virtualization software can make computers act like or simulate other computers. The result is often called a *virtual machine.* Using virtualization software, servers and mainframe computers can run software applications written for different operating systems. Virtualization is also being used to implement hardware, software, databases, and other components of an information system. As discussed earlier, virtualization can be environmentally friendly, reducing power consumption and requiring less space for equipment. Virtualization, however, introduces important implementation considerations, including security and backup procedures.[36] We start our discussion of systems implementation with hardware acquisition.

Acquiring Hardware from an IS Vendor

To obtain the components for an information system, organizations can purchase, lease, or rent computer hardware and other resources from an IS vendor. An *IS vendor* is a company that offers hardware, software, telecommunications systems, databases, IS personnel, or other computer-related resources. Types of IS vendors include general computer manufacturers (such as IBM and Hewlett-Packard), small computer manufacturers (such as Dell and Toshiba), peripheral equipment manufacturers (such as Hewlett-Packard and Canon), computer dealers and distributors (such as Radio Shack and Best Buy), and chip makers such as Intel and AMD.[37] Hardware vendors can provide very small or very large systems. The U.S. Census Bureau, for example, will acquire over 1,000 small handheld devices to help it collect census data.[38] On the other hand, the Defense Advanced Research Projects Agency (DARPA) purchased super computers and systems from IBM and Cray, valued at $500 million.[39] DARPA hopes its efforts to build a supercomputer will have commercial applications.

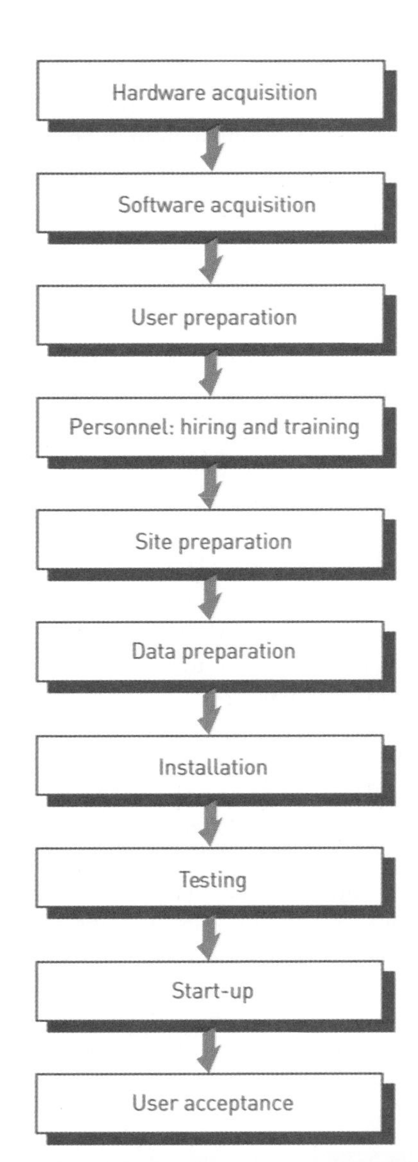

Figure 13.10

Typical Steps in Systems Implementation

Computer dealers, such as Best Buy, manufacture build-to-order computer systems and sell computers and supplies from other vendors.

(Source: © Justin Sullivan/Getty Images.)

As mentioned earlier, organizations often consider virtual machines, such as servers, in acquiring hardware.[40] Tellabs, for example, acquired virtualized Dell PowerEdge Servers for its operations.[41] According to a Tellabs representative, "In the past, we had been loading many of our servers with just one application, and their utilization rates were commonly 25 percent or less. We knew virtualization technology had evolved to where it could really help us consolidate."

In addition to buying, leasing, or renting computer hardware, companies can pay only for the computing services that they use. Called "pay-as-you-go," "on-demand," or "utility" computing, this approach requires an organization to pay only for the computer power it uses, as it would pay for a utility such as electricity. A bank, for example, can buy only the computer resources it needs from IBM and other companies. Hewlett-Packard offers its clients a "capacity-on-demand" approach, in which organizations pay according to the computer resources actually used, including processors, storage devices, and network facilities.

Companies can also purchase used computer equipment. This option is especially attractive to firms that are experiencing an economic slowdown. Companies often use traditional Internet auctions to locate used or refurbished equipment. Popular Internet auction sites sometimes sell more than $1 billion of computer-related equipment annually, and companies can purchase equipment for about 20 or 30 cents on the dollar. However, buyers need to beware: Prices are not always low, and equipment selection can be limited on Internet auction sites.

Acquiring Software: Make or Buy?

As we have discussed throughout this book, software can make the difference between profit and loss, life and death. The SmartBeam IMRT software program from Varian Medical Systems, for example, focuses radiation beams to kill cancer cells, spare good cells, and save lives.[42] According to the CEO of the company, "The old radiation treatment compared to the new is like comparing a flashlight to a laser beam."

As with hardware, application software can be acquired in several ways. As previously mentioned, it can be purchased from external developers or developed in-house. This decision is often called the **make-or-buy decision**. A comparison of the two approaches is shown in Table 13.4. Today, most software is purchased. SAP, the large international software company headquartered in Germany, produces modular software and sells to a variety of companies. The approach gives its customers using the software more flexibility in what they use and what they pay for SAP's modules. The key is how the purchased systems are integrated into an effective system. Software can also be developed or made. Allstate Insurance, for example decided to develop a new software program, called Next Gen, to speed claims processing and reinforce its "You're in good hands" slogan.[43] The company is expected to spend over $100 million on the new software.

make-or-buy decision
The decision regarding whether to obtain the necessary software from internal or external sources.

Table 13.4

Comparison of Off-the-Shelf and Developed Software

Factor	Off-the-Shelf (Buy)	Developed (Make)
Cost	Low Cost	High Cost
Needs	Might not get what you need	Custom software to satisfy your needs
Quality	Usually high quality	Quality can vary depending on the programming team
Speed	Can acquire it now	Can take years to develop
Competitive advantage	Other organizations can have the same software and same advantage	Can develop a competitive advantage with good software

Externally Acquired Software and Software as a Service (SaaS)

A company planning to purchase or lease software from an outside company has many options. Commercial off-the-shelf development is often used. The *commercial off-the-shelf (COTS)* development process involves the use of commonly available products from software vendors. It combines software from various vendors into a finished system. In many cases, it is necessary to write some original software from scratch and combine it with purchased or leased software. For example, a company can purchase or lease software from several software vendors and combine it into a finished software program.

Organizations are also acquiring more virtualization software from software vendors, including operating systems and application software.[44] Windows Server 2008, for example, provides virtualization tools that allow multiple operating systems to run on a single server. Virtualization software such as VMware is being used by businesses to safeguard private data. Kindred Healthcare, for example, uses VMware on its server to run hundreds of virtual Windows PC desktops that are accessed by mobile computers throughout the organization.[45]

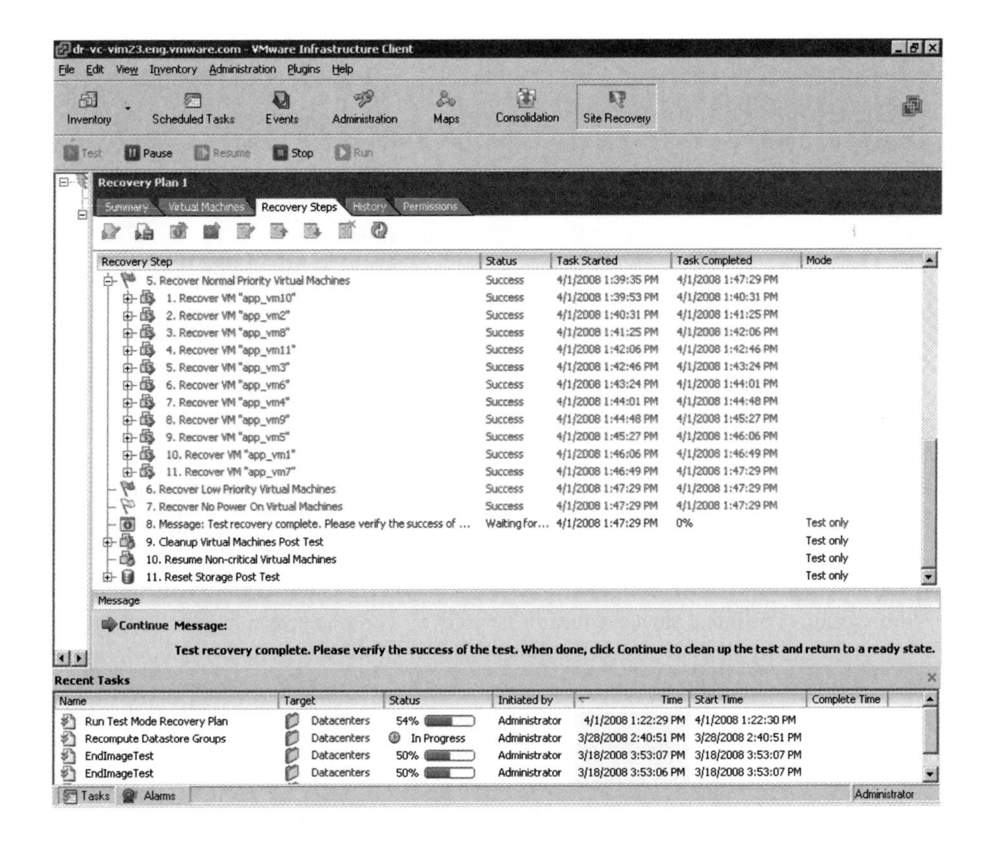

Businesses are using virtualization software such as VMware to safeguard private data.

(Source: Courtesy of VMware, Inc.)

As mentioned in Chapter 4, *Software as a Service (SaaS)* allows businesses to subscribe to Web-delivered application software by paying a monthly service charge or a per-use fee. Instead of acquiring software externally from a traditional software vendor, SaaS allows individuals and organizations to access needed software applications over the Internet. The Humane Society of the United States, for example, used SaaS to obtain and process credit-card contributions from donors.[46] The Humane Society also had to make sure that any software they used was compliant with the Payment Card Industry Data Security Standard. The Humane Society used a SaaS product called QualysGuard by Qualys (*www.qualys.com*) to meet its needs. According to the CIO, "SaaS opened our eyes to a new way of doing things. With QualysGuard, we didn't need to install any software or infrastructure." Companies such as Google are using the cloud computing approach to deliver word processing, spreadsheet programs, and other software over the Internet.[47] According to the research director for AMR Research, "It's the beginning of the first major challenge to Microsoft as the default enterprise interface in the last ten years."

In-House Developed Software

Another option is to make or develop software internally. This approach requires the company's IS personnel to be responsible for all aspects of software development. Some advantages inherent with in-house developed software include meeting user and organizational requirements and having more features and increased flexibility in terms of customization and changes. Software programs developed within a company also have greater potential for providing a competitive advantage because competitors cannot easily duplicate them in the short term. If software is to be developed internally, a number of tools and techniques can be used. In addition, in-house developed software is often constantly changing.[48] According the Chief Scientist at IBM's Rational Software Corporation, "Software has been and will remain fundamentally hard... Today, a typical system tends to be continuously evolving. You never turn it off." A few of the tools and techniques used to develop in-house software are briefly discussed below.

- **CASE and object-oriented approaches.** As mentioned in Chapter 12, CASE tools and the object-oriented approach are often used during software development. AXA Financial Services, for example, saved millions of dollars in developing a system by reusing software. JetBlue Airways worked with Visual Studio .NET to implement an inventory tracking system that used the object-oriented approach.
- **Cross-platform development.** One software development technique, called **cross-platform development**, allows programmers to develop programs that can run on computer systems that have different hardware and operating systems, or platforms. Web service tools, such as .NET by Microsoft, introduced in Chapter 7, are examples. With cross-platform development, for example, the same program can run on both a personal computer and a mainframe or on two different types of PCs.
- **Integrated development environment. Integrated development environments (IDEs)** combine the tools needed for programming with a programming language in one integrated package. An IDE allows programmers to use simple screens, customized pull-down menus, and graphical user interfaces. Visual Studio from Microsoft is an example of an IDE. Oracle Designer, which is used with Oracle's database system, is another example of an IDE. The popular Eclipse Workbench supports IDEs that can be used with the C and C++ programming languages. Eclipse Workbench includes a debugger and a compiler, along with other tools.
- **Documentation.** With internally developed software, documentation *is* always important. **Technical documentation** is used by computer operators to execute the program and by analysts and programmers to solve problems or modify the program. In technical documentation, the purpose of every major piece of computer code is written out and explained. Key variables are also described. **User documentation** is developed for the people who use the program. This type of documentation shows users, in easy-to-understand terms, how the program can and should be used. Incorporating a description of the benefits of the new application into user documentation can help stakeholders understand the reasons for the program and speed user acceptance.

cross-platform development
A development technique that allows programmers to develop programs that can run on computer systems having different hardware and operating systems, or platforms.

integrated development environments (IDEs)
A development approach that combines the tools needed for programming with a programming language into one integrated package.

technical documentation
Written details used by computer operators to execute the program and by analysts and programmers to solve problems or modify the program.

user documentation
Written descriptions developed for people who use a program, showing users, in easy-to-understand terms, how the program can and should be used.

Acquiring Database and Telecommunications Systems

Because databases are a blend of hardware and software, many of the approaches discussed earlier for acquiring hardware and software also apply to database systems including open-source databases.[49] MasterCard International needed to acquire additional storage capacity. Existing storage capacity was about to run out as the company expanded its business. Instead of adding incremental storage capacity, the company decided to use a large-scale storage area network (SAN). The results were immediate and apparent. *Virtual databases* and *database as a service (DaaS)* are also popular ways to acquire database capabilities.[50] Sirius XM Radio, Bank of America, and Southwest Airlines, for example, use the DaaS approach to manage many of their database operations from the Internet.[51] In another case, a brokerage company was able to reduce storage capacity by 50 percent by using database virtualization.[52]

With the increased use of e-commerce, the Internet, intranets, and extranets, telecommunications is one of the fastest-growing applications for today's organizations. Medical Missions for Children (MMC), for example, uses medical diagnosis through teleconferencing to treat children and improve their lives.[53] Like database systems, telecommunications systems require a blend of hardware and software. For personal computer systems, the primary piece of hardware is a modem. For client/server and mainframe systems, the hardware can include multiplexers, concentrators, communications processors, and a variety of network equipment. Communications software will also have to be acquired from a software company or developed in-house. Again, the earlier discussion on acquiring hardware and software also applies to the acquisition of telecommunications hardware and software.

User Preparation

User preparation is the process of readying managers, decision makers, employees, other users, and stakeholders for the new systems. This activity is an important but often ignored area of systems implementation. When a new operating system or application software package is implemented, user training is essential.[54] "It will take a lot more planning and definitely some thought toward training," said a TRW Automotive Holdings executive about installing Office 2007. In some cases, companies decide not to install the latest software because the amount of time and money needed to train employees is too much.[55] In one survey, over 70 percent of the respondents indicated that they were in no hurry to install a new operating system.[56] Additional user training was an important factor in delaying the installation of the new operating system for many companies. Because user training is so important, some companies provide training for their clients, including in-house, software, video, Internet, and other training approaches.[57]

user preparation
The process of readying managers, decision makers, employees, other users, and stakeholders for new systems.

Providing users with proper training can help ensure that the information system is used correctly, efficiently, and effectively.

(Source: © Stockbyte/Getty Images.)

IS Personnel: Hiring and Training

Depending on the size of the new system, an organization might have to hire and, in some cases, train new IS personnel.[58] An IS manager, systems analysts, computer programmers, data-entry operators, and similar personnel might be needed for the new or modified system.

As with users, the eventual success of any system depends on how it is used by the IS personnel within the organization. Training programs should be conducted for the IS personnel who will be using the computer system. These programs are similar to those for the users, although they can be more detailed in the technical aspects of the systems. Effective training will help IS personnel use the new system to perform their jobs and support other users in the organization.[59]

Site Preparation

The location of the new system needs to be prepared, a process called **site preparation**. For a small system, site preparation can be as simple as rearranging the furniture in an office to make room for a computer. With a larger system, this process is not so easy because it can require special wiring and air conditioning. One or two rooms might have to be completely renovated, and additional furniture might have to be purchased. A special floor might have

site preparation
Preparation of the location of a new system.

to be built, under which the cables connecting the various computer components are placed, and a new security system might be needed to protect the equipment. For larger systems, additional power circuits might also be required. Today, developing IS sites that are energy efficient is important for most systems development implementations.[60]

Data Preparation

data preparation, or data conversion
Ensuring all files and databases are ready to be used with new computer software and systems.

Data preparation, or **data conversion**, involves making sure that all files and databases are ready to be used with new computer software and systems. If an organization is installing a new payroll program, the old employee-payroll data might have to be converted into a format that can be used by the new computer software or system. After the data has been prepared or converted, the computerized database system or other software will then be used to maintain and update the computer files.

Installation

installation
The process of physically placing the computer equipment on the site and making it operational.

Installation is the process of physically placing the computer equipment on the site and making it operational. Although normally the manufacturer is responsible for installing computer equipment, someone from the organization (usually the IS manager) should oversee the process, making sure that all equipment specified in the contract is installed at the proper location. After the system is installed, the manufacturer performs several tests to ensure that the equipment is operating as it should.

Testing

unit testing
Testing of individual programs.

system testing
Testing the entire system of programs.

volume testing
Testing the application with a large amount of data.

integration testing
Testing all related systems together.

acceptance testing
Conducting any tests required by the user.

Good testing procedures are essential to make sure that the new or modified information system operates as intended. Inadequate testing can result in mistakes and problems. A $13 million systems development effort to build a vehicle title and registration system had to be shut down because inaccurate data led to vehicles being pulled over or stopped by mistake.[61] According to one state official, "We couldn't have people out there having their cars impounded because of inaccurate information in the … database." In some cases, one problem can cascade into additional problems or cause multiple systems to fail.[62] Problems with a project to consolidate data center servers, for example, resulted in more than 160,000 Internet sites being shut down. The company that was trying to consolidate its database servers was hosting the Internet sites. Some Internet sites were down for more than six days. Better testing may have prevented these types of problems.

Several forms of testing should be used, including testing each program (**unit testing**), testing the entire system of programs (**system testing**), testing the application with a large amount of data (**volume testing**), and testing all related systems together (**integration testing**), as well as conducting any tests required by the user (**acceptance testing**). Figure 13.11 lists the types of testing. In addition to these forms of testing, there are different types of testing.

Figure 13.11
Types of Testing

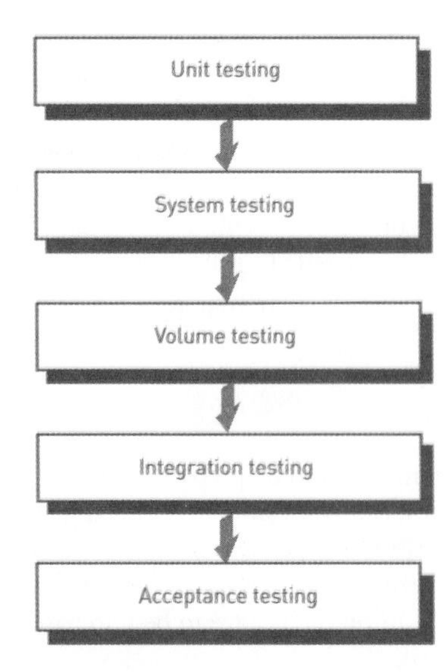

Alpha testing involves testing an incomplete or early version of the system; **beta testing** involves testing a complete and stable system by end users. Alpha-unit testing, for example, is testing an individual program before it is completely finished. Beta-unit testing, on the other hand, is performed after alpha testing, when the individual program is complete and ready for use by end users.

alpha testing
Testing an incomplete or early version of the system.

beta testing
Testing a complete and stable system by end users.

Unit testing is accomplished by developing test data that will force the computer to execute every statement in the program. In addition, each program is tested with abnormal data to determine how it will handle problems.

System testing requires the testing of all the programs together. It is not uncommon for the output from one program to become the input for another. So, system testing ensures that program output can be used as input for another program within the system. Volume testing ensures that the entire system can handle a large amount of data under normal operating conditions. Integration testing ensures that the new programs can interact with other major applications. It also ensures that data flows efficiently and without error to other applications. For example, a new inventory control application might require data input from an older order processing application. Integration testing would be done to ensure smooth data flow between the new and existing applications. Integration testing is typically done after unit and system testing. Metaserver, a software company for the insurance industry, has developed a tool called iConnect to perform integration testing for different insurance applications and databases.

Finally, acceptance testing makes sure that the new or modified system is operating as intended. Run times, the amount of memory required, disk access methods, and more can be tested during this phase. Acceptance testing ensures that all performance objectives defined for the system or application are satisfied. Involving users in acceptance testing can help them understand and effectively interact with the new system. Acceptance testing is the final check of the system before start-up.

Start-Up

Start-up, also called *cutover*, begins with the final tested information system. When start-up is finished, the system is fully operational. Start-up can be critical to the success of the organization. If not done properly, the results can be disastrous. In one case, a small manufacturing company that decided to stop an accounting service used to send out bills on the same day they were going to start their own program to send out bills to customers. The manufacturing company wanted to save money by using their own billing program developed by an employee. The new program didn't work, the accounting service wouldn't help because they were upset about being terminated, and the manufacturing company wasn't able to send out any bills to customers for more than three months. The manufacturing company almost went bankrupt.

start-up (also called *cutover*)
The process of making the final tested information system fully operational.

Various start-up approaches are available (see Figure 13.12). **Direct conversion** (also called *plunge* or *direct cutover*) involves stopping the old system and starting the new system on a given date. Direct conversion is usually the least desirable approach because of the potential for problems and errors when the old system is shut off and the new system is turned on at the same instant.

direct conversion (also called *plunge* or *direct cutover*)
Stopping the old system and starting the new system on a given date.

The **phase-in approach** is a popular technique preferred by many organizations. In this approach, sometimes called a *piecemeal approach*, components of the new system are slowly phased in while components of the old one are slowly phased out. When everyone is confident that the new system is performing as expected, the old system is completely phased out. This gradual replacement is repeated for each application until the new system is running every application. In some cases, the phase-in approach can take months or years.

phase-in approach (also called *piecemeal approach*)
Slowly replacing components of the old system with those of the new one. This process is repeated for each application until the new system is running every application and performing as expected; also called a *piecemeal approach*.

Pilot start-up involves running the new system for one group of users rather than all users. For example, a manufacturing company with many retail outlets throughout the country could use the pilot start-up approach and install a new inventory control system at one of the retail outlets. When this pilot retail outlet runs without problems, the new inventory control system can be implemented at other retail outlets. Google's personal health records application was tested as a pilot project in a Cleveland health clinic before it was made available to more healthcare facilities.[63] This pilot start-up approach let Google install the

pilot start-up
Running the new system for one group of users rather than all users.

Figure 13.12

Start-Up Approaches

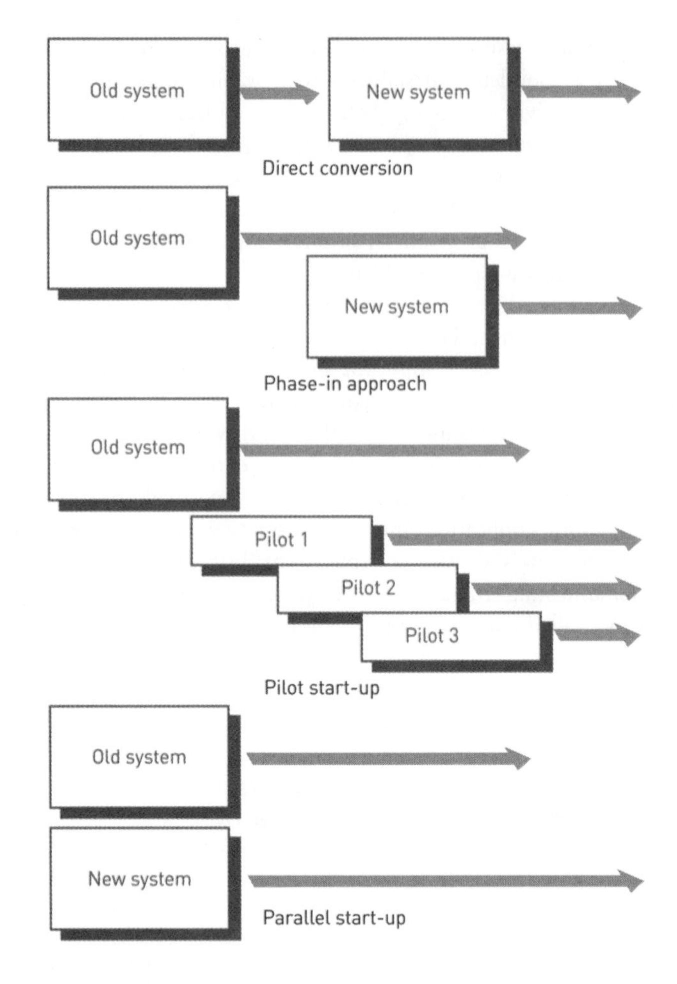

parallel start-up
Running both the old and new systems for a period of time and comparing the output of the new system closely with the output of the old system; any differences are reconciled. When users are comfortable that the new system is working correctly, the old system is eliminated.

user acceptance document
A formal agreement signed by the user that states that a phase of the installation or the complete system is approved.

application on a smaller scale before making the medical records application available to people around the world.

Parallel start-up involves running both the old and new systems for a period of time. The output of the new system is compared closely with the output of the old system, and any differences are reconciled. When users are comfortable that the new system is working correctly, the old system is eliminated.

User Acceptance

Most mainframe computer manufacturers use a formal **user acceptance document**—a formal agreement the user signs stating that a phase of the installation or the complete system is approved. This is a legal document that usually removes or reduces the IS vendor's liability for problems that occur after the user acceptance document has been signed. Because this document is so important, many companies get legal assistance before they sign the acceptance document. Stakeholders can also be involved in acceptance testing to make sure that the benefits to them are indeed realized.

SYSTEMS OPERATION AND MAINTENANCE

systems operation
Use of a new or modified system.

Systems operation involves all aspects of using the new or modified system in all kinds of operating conditions. Getting the most out of a new or modified system during its operation is the most important aspect of systems operations for many organizations. Throughout this book, we have seen many examples of information systems operating in a variety of settings and industries. Thus, we will not cover the operation of an information system in detail in this section. To provide adequate support, many companies use a formal help desk. A *help desk* consists of people with technical expertise, computer systems, manuals, and other resources needed to solve problems and give accurate answers to questions. With today's advances in telecommunications, help desks can be located around the world. If you are having trouble with your PC and call a toll-free number for assistance, you might reach a help desk in India, China, or another country. For most organizations, operations costs over the life of a system are much greater than the development costs.

Systems maintenance involves checking, changing, and enhancing the system to make it more useful in achieving user and organizational goals. Systems maintenance is important for individuals, groups, and organizations. Individuals, for example, can use the Internet, computer vendors, and independent maintenance companies, including YourTechOnline.com (*www.yourtechonline.com*), Geek Squad (*www.geeksquad.com*), PC Pinpoint (*www.pcpinpoint.com*), and others. Organizations often have personnel dedicated to maintenance.

systems maintenance
A stage of systems development that involves checking, changing, and enhancing the system to make it more useful in achieving user and organizational goals.

This maintenance process can be especially difficult for older software. A *legacy system* is an old system that might have been patched or modified repeatedly over time. An old payroll program in COBOL developed decades ago and frequently changed is an example of a legacy system. Legacy systems can be very expensive to maintain. At some point, it becomes less expensive to switch to new programs and applications than to repair and maintain the legacy system. Maintenance costs for older legacy systems can be 50 percent of total operating costs in some cases. Aspen Skiing Company, for example, decided to replace one of its legacy systems for a newer one.[64] According to the CIO, "We've been using a legacy application developed in-house. It's a very effective application from the standpoint of customer service. But it's extremely difficult to sustain it. So we had to go out and actually buy a new system to do what we do."

Software maintenance is a major concern for organizations. In some cases, organizations encounter major problems that require recycling the entire systems development process. In other situations, minor modifications are sufficient to remedy problems. Hardware maintenance is also important. Companies such as IBM are investigating *autonomic computing*, in which computers will be able to manage and maintain themselves.[65] The goal is for computers to be self-configuring, self-protecting, self-healing, and self-optimizing. Being self-configuring allows a computer to handle new hardware, software, or other changes to its operating environment. Being self-protecting means a computer can identify potential attacks, prevent them when possible, and recover from attacks if they occur. Attacks can include viruses, worms, identity theft, and industrial espionage. Being self-healing means a computer can fix problems when they occur, and being self-optimizing allows a computer to run faster and get more done in less time.

Getting rid of old equipment is an important part of maintenance. The options include selling it on Web auction sites such as eBay, recycling the equipment at a computer-recycling center, and donating it to a charitable organization, such as a school, library, or religious organization. When discarding old computer systems, it is always a good idea to permanently remove sensitive files and programs. Companies such as McAfee and Blancco have software to help people remove data and programs from old computers and transfer them to new ones. As mentioned in the section on environmental design, companies are finding ways to dispose of old equipment in a way that minimizes environmental damage.

Reasons for Maintenance

After a program is written, it will need ongoing maintenance. A Texas restaurant, for example, decided to make maintenance changes to its security system after its customers' credit card numbers were stolen.[66] Experience shows that frequent, minor maintenance to a program, if properly done, can prevent major system failures later. Some of the reasons for program maintenance are the following:

- Changes in business processes
- New requests from stakeholders, users, and managers
- Bugs or errors in the program

- Technical and hardware problems
- Corporate mergers and acquisitions
- Government regulations
- Change in the operating system or hardware on which the application runs
- Unexpected events, such as severe weather or terrorist attacks

Most companies modify their existing programs instead of developing new ones because existing software performs many important functions, and companies can have millions of dollars invested in their old legacy systems. So, as new systems needs are identified, the burden of fulfilling the needs most often falls on the existing system. Old programs are repeatedly modified to meet ever-changing needs. Yet, over time, repeated modifications tend to interfere with the system's overall structure, reducing its efficiency and making further modifications more burdensome.

Types of Maintenance

Software companies and many other organizations use four generally accepted categories to signify the amount of change involved in maintenance. A **slipstream upgrade** usually requires recompiling all the code, allowing the program to run faster and more efficiently. Many companies don't announce to users that a slipstream upgrade has been made. A slipstream upgrade usually requires recompiling all the code, so it can create entirely new bugs. This maintenance practice can explain why the same computers sometimes work differently with what is supposedly the same software. A **patch** is a minor change to correct a problem or make a small enhancement. It is usually an addition to an existing program. That is, the programming code representing the system enhancement is usually "patched into," or added to, the existing code. Although slipstream upgrades and patches are minor changes, they can cause users and support personnel big problems if the programs do not run as before.[67] Many patches come from off-the-shelf software vendors, such as Microsoft. According to the chief technology officer of a California law firm, "Overall, we were astounded with the quantity and size of the latest patches. This month's patches will cost us over 100 hours of IT time to test and apply." A new **release** is a significant program change that often requires changes in the documentation of the software. Finally, a new **version** is a major program change, typically encompassing many new features.

slipstream upgrade
An upgrade that usually requires recompiling all the code, allowing the program to run faster and more efficiently.

patch
A minor change to correct a problem or make a small enhancement. It is usually an addition to an existing program.

release
A significant program change that often requires changes in the documentation of the software.

version
A major program change, typically encompassing many new features.

The Request for Maintenance Form

Because of the amount of effort that can be spent on maintenance, many organizations require a **request for maintenance form** to authorize modification of programs. This form is usually signed by a business manager, who documents the need for the change and identifies the priority of the change relative to other work that has been requested. The IS group reviews the form and identifies the programs to be changed, determines the programmer who will be assigned to the project, estimates the expected completion date, and develops a technical description of the change. A cost/benefit analysis might be required if the change requires substantial resources.

request for maintenance form
A form authorizing modification of programs.

Performing Maintenance

Depending on organizational policies, the people who perform systems maintenance vary. In some cases, the team who designs and builds the system also performs maintenance. This ongoing responsibility gives the designers and programmers an incentive to build systems well from the outset: If problems occur, they will have to fix them. In other cases, organizations have a separate **maintenance team**. This team is responsible for modifying, fixing, and updating existing software.

maintenance team
A special IS team responsible for modifying, fixing, and updating existing software.

In the past, companies had to maintain each computer system or server separately. With hundreds or thousands of computers scattered throughout an organization, this task could be very costly and time consuming. Today, the maintenance function is becoming more automated. Some companies, for example, use maintenance tools and software that will allow them to maintain and upgrade software centrally.

A number of vendors have developed tools to ease the software maintenance burden. RescueWare from Relativity Technologies is a product that converts third-generation code such as COBOL to highly maintainable C++, Java, or Visual Basic object-oriented code.[68] RescueWare lets a programmer see the original system as a set of object views, which visually illustrate module functioning and program structures.

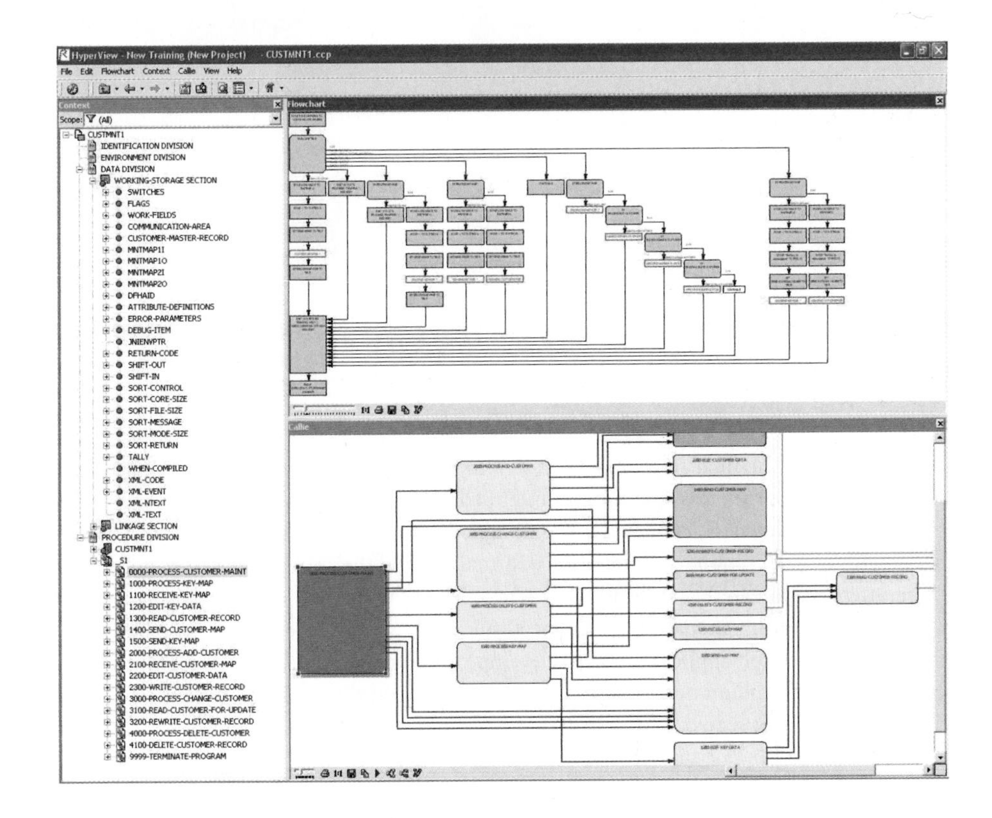

Relativity Technologies' Modernization Workbench is a PC-based software solution that enables companies to consolidate legacy or redundant systems into a single, more maintainable, and modern application.

(Source: Courtesy of Relativity Technologies.)

The Relationship Between Maintenance and Design

Programs are expensive to develop, but they are even more expensive to maintain. For older programs, the total cost of maintenance can be up to five times greater than the total cost of development. A determining factor in the decision to replace a system is the point at which it is costing more to fix than to replace. Programs that are well designed and documented to be efficient, structured, and flexible are less expensive to maintain in later years. Thus, there is a direct relationship between design and maintenance. More time spent on design up front can mean less time spent on maintenance later.

In most cases, it is worth the extra time and expense to design a good system. Consider a system that costs $250,000 to develop. Spending 10 percent more on design would cost an additional $25,000, bringing the total design cost to $275,000. Maintenance costs over the life of the program could be $1,000,000. If this additional design expense can reduce maintenance costs by 10 percent, the savings in maintenance costs would be $100,000. Over the life of the program, the net savings would be $75,000 ($100,000 – $25,000). This relationship between investment in design and long-term maintenance savings is shown in Figure 13.13.

The need for good design goes beyond mere costs. Companies risk ignoring small system problems when they arise, but these small problems can become large in the future. As mentioned earlier, because maintenance programmers spend an estimated 50 percent or more of their time deciphering poorly written, undocumented program code, they have little time to spend on developing new, more effective systems. If put to good use, the tools and techniques discussed in this chapter will allow organizations to build longer-lasting, more reliable systems.

Figure 13.13

The Value of Investment in Design

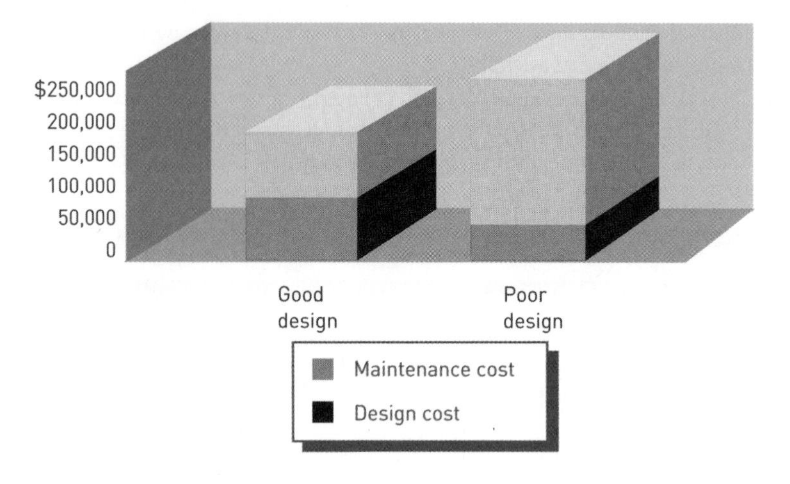

SYSTEMS REVIEW

systems review
The final step of systems development, involving the analysis of systems to make sure that they are operating as intended.

Systems review, the final step of systems development, is the process of analyzing systems to make sure that they are operating as intended. This process often compares the performance and benefits of the system as it was designed with the actual performance and benefits of the system in operation. In some cases, a formal audit of the application can be performed, using internal and external auditors.[69] Systems review can be performed during systems development, resulting in halting the new systems while they are being built because of problems.[70] A payroll application being developed for a national health service, for example, was almost $170 million over budget. As a result, work on the application that serves about 37,000 workers was halted so the entire project could be reviewed in detail.

Problems and opportunities uncovered during systems review trigger systems development and begin the process anew. For example, as the number of users of an interactive system increases, it is not unusual for system response time to increase. If the increase in response time is too great, it might be necessary to redesign some of the system, modify databases, or increase the power of the computer hardware.

Internal employees, external consultants, or both can perform systems review. When the problems or opportunities are industry-wide, people from several firms can get together. In some cases, they collaborate at an IS conference or in a private meeting involving several firms.

Types of Review Procedures

event-driven review
A review triggered by a problem or opportunity such as an error, a corporate merger, or a new market for products.

The two types of review procedures are event-driven and time-driven (see Table 13.5). An **event-driven review** is triggered by a problem or opportunity such as an error, a corporate merger, or a new market for products. In one case, a large insurance company operating in Louisiana was ordered by a Louisiana court to pay a client over $500,000 in wind damages and over $2 million in fines for not paying the claim in a timely fashion.[71] This helped trigger an event-driven review that resulted in new software claims programs.

Table 13.5

Examples of Review Types

Event Driven	Time Driven
Problem with an existing system	Monthly review
Merger	Yearly review
New accounting system	Review every few years
Executive decision that an upgraded Internet site is needed to stay competitive	Five-year review

In some cases, companies wait until a large problem or opportunity occurs before a change is made, ignoring minor problems. In contrast, some companies use a continuous improvement approach to systems development. With this approach, an organization makes changes to a system even when small problems or opportunities occur. Although continuous improvement can keep the system current and responsive, repeatedly designing and implementing changes can be both time consuming and expensive.

A **time-driven review** is performed after a specified amount of time. Many application programs are reviewed every six months to one year. With this approach, an existing system is monitored on a schedule. If problems or opportunities are uncovered, a new systems development cycle can be initiated. A payroll application, for example, can be reviewed once a year to make sure that it is still operating as expected. If it is not, changes are made.

time-driven review
Review performed after a specified amount of time.

Many companies use both approaches. A billing application, for example, might be reviewed once a year for errors, inefficiencies, and opportunities to reduce operating costs. This is a time-driven approach. In addition, the billing application might be redone after a corporate merger, if one or more new managers require different information or reports, or if federal laws on bill collecting and privacy change. This is an event-driven approach.

System Performance Measurement

Systems review often involves monitoring the system, called **system performance measurement**. The number of errors encountered, the amount of memory required, the amount of processing or CPU time needed, and other problems should be closely observed. If a particular system is not performing as expected, it should be modified, or a new system should be developed or acquired.

system performance measurement
Monitoring the system—the number of errors encountered, the amount of memory required, the amount of processing or CPU time needed, and other problems.

System performance products have been developed to measure all components of the information system, including hardware, software, database, telecommunications, and network systems. IBM Tivoli OMEGAMON can monitor system performance in real time.[72] Precise Software Solutions has system performance products that provide around-the-clock performance monitoring for ERP systems, Oracle database applications, and other programs.[73] HP also offers a software tool called Business Technology Optimization (BTO) software to help companies analyze the performance of their computer systems, diagnose potential problems, and take corrective action if needed.[74] When properly used, system performance products can quickly and efficiently locate actual or potential problems.

system performance products
Software that measures all components of the computer-based information system, including hardware, software, database, telecommunications, and network systems.

Measuring a system is, in effect, the final task of systems development. The results of this process can bring the development team back to the beginning of the development life cycle, where the process begins again.

SUMMARY

Principle

Designing new systems or modifying existing ones should always help an organization achieve its goals.

The purpose of systems design is to prepare the detailed design needs for a new system or modifications to the existing system. Logical systems design refers to the way that the various components of an information system will work together. The logical design includes data requirements for output and input, processing, files and databases, telecommunications, procedures, personnel and job design, and controls and security design. Physical systems design refers to the specification of the actual physical components. The physical design must specify characteristics for hardware and software design, database and telecommunications, and personnel and procedures design.

Logical and physical design can be accomplished using the traditional systems development life cycle or the object-oriented approach. Using the OO approach, analysts design key objects and classes of objects in the new or updated system. The sequence of events that a new or modified system requires is often called a scenario, which can be diagrammed in a sequence diagram.

A number of special design considerations should be taken into account during both logical and physical system design. Interface design and control relates to how users access and interact with the system. A sign-on procedure consists of identification numbers, passwords, and other safeguards needed for individuals to gain access to computer resources. If the system under development is interactive, the design must consider menus, help facilities, table lookup facilities, and restart procedures. A good interactive dialogue will ask for information in a clear manner, respond rapidly, be consistent among applications, and use an attractive format. Also, it will avoid use of computer jargon and treat the user with respect.

System security and control involves many aspects. Error prevention, detection, and correction should be part of the system design process. Causes of errors include human activities, natural phenomena, and technical problems. Designers should be alert to prevention of fraud and invasion of privacy.

Disaster recovery is an important aspect of systems design. Disaster planning is the process of anticipating and providing for disasters. A disaster can be an act of nature (a flood, fire, or earthquake) or a human act (terrorism, error, labor unrest, or erasure of an important file). The primary tools used in disaster planning and recovery are hardware, software, database, telecommunications, and personnel backup.

Security, fraud, and the invasion of privacy are also important design considerations. Most IS departments establish tight systems controls to maintain data security. Systems controls can help prevent computer misuse, crime, and fraud by employees and others. Systems controls include input, output, processing, database, telecommunications, and personnel controls.

Environmental design, also called green design, involves systems development efforts that slash power consumption, require less physical space, and result in systems that can be disposed in a way that doesn't negatively affect the environment. A number of companies are developing products and services to help save energy. Environmental design also deals with how companies are developing systems to dispose of old equipment. The U.S. government is also involved in environmental design. It has a plan to require federal agencies to purchase energy-efficient computer systems and equipment. The plan would require federal agencies to use the Electronic Product Environmental Assessment Tool (EPEAT) to analyze the energy usage of new systems. The U.S. Department of Energy rates products with the Energy Star designation to help people select products that save energy and are friendly to the environment.

Whether an individual is purchasing a personal computer or an experienced company is acquiring an expensive mainframe computer, the system could be obtained from one or more vendors. Some of the factors to consider in selecting a vendor are the vendor's reliability and financial stability, the type of service offered after the sale, the goods and services the vendor offers and keeps in stock, the vendor's willingness to demonstrate its products, the vendor's ability to repair hardware, the vendor's ability to modify its software, the availability of vendor-offered training of IS personnel and system users, and evaluations of the vendor by independent organizations.

If new hardware or software will be purchased from a vendor, a formal request for proposal (RFP) is needed. The RFP outlines the company's needs; in response, the vendor provides a written reply. In addition to responding to the company's stated needs, the vendor provides data on its operations. This data might include the vendor's reliability and stability, the type of postsale service offered, the vendor's ability to perform repairs and fix problems, the available vendor training, and the vendor's reputation. Financial options to consider include purchase, lease, and rent.

RFPs from various vendors are reviewed and narrowed down to the few most likely candidates. In the final evaluation, a variety of techniques—including group consensus, cost/benefit analysis, point evaluation, and benchmark tests—can be used. In group consensus, a decision-making group is appointed and given responsibility for making the final evaluation and selection. With cost/benefit analysis, all costs and benefits of the alternatives are expressed in monetary terms. Benchmarking involves comparing computer systems operating under the same condition. Point evaluation assigns weights to evaluation factors, and each alternative is evaluated in terms of each factor and given a score from 0 to 100. After the vendor is chosen, contract negotiations can begin.

At the end of the systems design step, the final specifications are frozen and no changes are allowed so that implementation can proceed. One of the most important steps in systems design is to develop a good contract if new computer facilities are being acquired. A final design report is developed at the end of the systems design phase.

Principle

The primary emphasis of systems implementation is to make sure that the right information is delivered to the right person in the right format at the right time.

The purpose of systems implementation is to install the system and make everything, including users, ready for its operation. Systems implementation includes hardware acquisition, software acquisition or development, user preparation, hiring and training of personnel, site and data preparation, installation, testing, start-up, and user acceptance. Hardware acquisition requires purchasing, leasing, or renting computer resources from an IS vendor. Hardware is typically obtained from a computer hardware vendor.

Software can be purchased from vendors or developed in-house—a decision termed the *make-or-buy decision*. Virtualization, first introduced in Chapter 3, has had a profound impact on many aspects of systems implementation. A purchased software package usually has a lower cost, less risk regarding the features and performance, and easy installation. The amount of development effort is also less when software is purchased. Software as a service (SaaS) is becoming a popular way to purchase software capabilities. Developing software can result in a system that more closely meets the business needs and has increased flexibility in terms of customization and changes. Developing software also has greater potential for providing a competitive advantage. Increasingly, companies are using service providers to acquire software, Internet access, and other IS resources.

Cross-platform development and integrated development environments (IDEs) make software development easier and more thorough. CASE tools are often used to automate some of these techniques. Technical and user documentation is always important in developing in-house software.

Database and telecommunications software development involves acquiring the necessary databases, networks, telecommunications, and Internet facilities. Companies have a wide array of choices, including newer object-oriented database systems. Virtual databases and database as a service (DaaS) are popular ways to acquire database capabilities.

Implementation must address personnel requirements. User preparation involves readying managers, employees, and other users for the new system. New IS personnel might need to be hired, and users must be well trained in the system's functions. Preparation of the physical site of the system must be done, and any existing data to be used in the new system will require conversion to the new format. Hardware installation is done during the implementation step, as is testing. Testing includes program (unit) testing, systems testing, volume testing, integration testing, and acceptance testing.

Start-up begins with the final tested information system. When start-up is finished, the system is fully operational. There are a number of different start-up approaches. Direct conversion (also called plunge or direct cutover) involves stopping the old system and starting the new system on a given date. With the phase-in approach, sometimes called a piecemeal approach, components of the new system are slowly phased in while components of the old one are slowly phased out. When everyone is confident that the new system is performing as expected, the old system is completely phased out. Pilot start-up involves running the new system for one group of users rather than all users. Parallel start-up involves running both the old and new systems for a period of time. The output of the new system is compared closely with the output of the old system, and any differences are reconciled. When users are comfortable that the new system is working correctly, the old system is eliminated. Many IS vendors ask the user to sign a formal user acceptance document that releases the IS vendor from liability for problems that occur after the document is signed.

Principle

Maintenance and review add to the useful life of a system but can consume large amounts of resources. These activities can benefit from the same rigorous methods and project management techniques applied to systems development.

Systems operation is the use of a new or modified system. Systems maintenance involves checking, changing, and enhancing the system to make it more useful in obtaining user and organizational goals. Maintenance is critical for the continued smooth operation of the system. The costs of performing maintenance can well exceed the original cost of acquiring the system. Some major causes of maintenance are new requests from stakeholders and managers, enhancement requests from users, bugs or errors, technical or hardware problems, newly added equipment, changes in organizational structure, and government regulations.

Maintenance can be as simple as a program patch to correct a small problem to the more complex upgrading of software with a new release from a vendor. For older programs, the total cost of maintenance can be greater than the total cost of development. Increased emphasis on design can often reduce maintenance costs. Requests for maintenance should be documented with a request for maintenance form, a document that formally authorizes modification of programs. The development team or a specialized maintenance team can then make approved changes. Maintenance can be greatly simplified with the object-oriented approach.

Systems review is the process of analyzing and monitoring systems to make sure that they are operating as intended. The two types of review procedures are the event-driven review and the time-driven review. An event-driven review is triggered by a problem or opportunity. A time-driven review is started after a specified amount of time.

Systems review involves measuring how well the system is supporting the mission and goals of the organization. System performance measurement monitors the system for number of errors, amount of memory and processing time required, and so on.

CHAPTER 13: SELF-ASSESSMENT TEST

Designing new systems or modifying existing ones should always help an organization achieve its goals.

1. ______________ details system outputs, inputs, and user interfaces; specifies hardware, software, databases, telecommunications, personnel, and procedures; and shows how these components are related.
2. Determining the hardware and software required for a new system is an example of ___________.
   a. logical design
   b. physical design
   c. interactive design
   d. object-oriented design
3. Disaster planning is an important part of designing security and control systems. True or False?
4. ______________ involves systems development efforts that slash power consumption and require less physical space.
5. Scenarios and sequence diagrams are used with ______________.
   a. object-oriented design
   b. point evaluation
   c. incremental design
   d. nominal evaluation
6. Near the end of the design stage, an organization prohibits further changes in the design of the system. This is called ______________.
7. In object-oriented systems design, a sequence of events is called a *scenario* and can be diagrammed in a sequence diagram. True or False?

The primary emphasis of systems implementation is to make sure that the right information is delivered to the right person in the right format at the right time.

8. ASP is an example of an IS vendor that offers hardware and software solutions. True or False?
9. ___________ software can make computers act like or simulate other computers, reducing costs and space requirements.
10. What type of documentation is used by computer operators to execute a program and by analysts and programmers?
    a. unit documentation
    b. integrated documentation
    c. technical documentation
    d. user documentation
11. ______________ testing involves testing the entire system of programs.
12. The phase-in approach to conversion involves running both the old system and the new system for three months or longer. True or False?

Maintenance and review add to the useful life of a system but can consume large amounts of resources. These activities can benefit from the same rigorous methods and project management techniques applied to systems development.

13. A(n) ______________ is a minor change to correct a problem or make a small enhancement to a program or system.
14. Many organizations require a request for maintenance form to authorize modification of programs. True or False?
15. A systems review that is caused by a problem with an existing system is called ___________.
    a. object review
    b. structured review
    c. event-driven review
    d. critical factors review
16. Corporate mergers and acquisitions can be a reason for systems maintenance. True or False?
17. Monitoring a system after it has been implemented is called ______________.

CHAPTER 13: SELF-ASSESSMENT TEST ANSWERS

(1) Systems design (2) b (3) True (4) environmental design (5) a (6) freezing design specifications (7) True (8) False (9) virtualization (10) c (11) System (12) False (13) patch (14) True (15) c (16) True (17) system performance measurement

REVIEW QUESTIONS

1. What is the purpose of systems design?
2. What is interactive processing? What design factors should be taken into account for this type of processing?
3. How can the object-oriented approach be used during systems design?
4. What is the difference between logical and physical design?
5. What is environmental design?
6. What are the different types of software and database backup? Describe the procedure you use to back up your homework files.
7. Identify specific controls that are used to maintain input integrity and security.
8. What is an RFP? What is typically included in one? How is it used?
9. What activities go on during the user preparation phase of systems implementation?
10. What is systems operation?
11. What are the major steps of systems implementation?
12. What are some tools and techniques for software development?
13. Give three examples of an IS vendor.
14. How can SaaS be used in software acquisition?
15. What are the steps involved in testing the information system?
16. What are some of the reasons for program maintenance? Explain the types of maintenance.
17. How is system performance measurement related to the systems review?

DISCUSSION QUESTIONS

1. Describe the participants in the systems design stage. How do these participants compare with the participants of systems investigation?
2. Assume that you are the owner of a company that is about to start marketing and selling bicycles over the Internet. Describe what environmental design steps you could use to reduce power consumption with your information system.
3. Assume that you want to start a new video rental business for students at your college or university. Go through logical design for a new information system to help you keep track of the videos in your inventory.
4. Assume that you are the owner of an online stock-trading company. Describe how you could design the trading system to recover from a disaster.
5. Identify some of the advantages and disadvantages of purchasing a database package instead of taking the DaaS approach.
6. Discuss the relationship between maintenance and systems design.
7. Is it equally important for all systems to have a disaster recovery plan? Why or why not?
8. Several approaches were discussed to evaluate a number of systems acquisition alternatives. No one approach is always the best. How would you decide which approach to use for evaluation when selecting a new personal computer and printer?
9. What are the advantages and disadvantages of the object-oriented approach to systems implementation?
10. You have been hired to oversee a major systems development effort to purchase a new accounting software package. Describe what is important to include in the contract with the software vendor.
11. Assume that you are starting an Internet site to sell clothing. Describe how you would design the interactive processing system for this site. Draw a diagram showing the home page for the Web site. Describe the important features of this home page.
12. Identify the various forms of testing. Why are there so many different types of tests?
13. What is the goal of conducting a systems review? What factors need to be considered during systems review?
14. Describe how you would select the best admissions software for your college or university. What features would be most important for school administrators? What features would be most important for students?
15. What issues might you expect to arise if you initiate the use of a request for maintenance form when none had been required previously? How would you deal with these issues?
16. Assume that you have a personal computer that is several years old. Describe the steps you would use to perform systems review to determine whether you should acquire a new PC.

PROBLEM-SOLVING EXERCISES

1. You have been hired to develop a new information system to provide online backup of term papers, student projects, and other important student files for colleges and universities across the country. Determine how you can use the principles of environmental design to develop the new information system. Use a graphics program, such as PowerPoint, to develop a set of slides that shows how much money you can save and how your design is friendly to the environment.
2. A project team has estimated the costs associated with the development and maintenance of a new system. One approach requires a more complete design and will result in a slightly higher design and implementation cost but a lower maintenance cost over the life of the system. The second approach cuts the design effort, saving some dollars but with a likely increase in maintenance cost.

    a. Enter the following data in the spreadsheet. Print the result.

The Benefits of Good Design

	Good Design	Poor Design
Design costs	$14,000	$10,000
Implementation cost	$42,000	$35,000
Annual maintenance cost	$32,000	$40,000

    b. Create a stacked bar graph that shows the total cost, including the design, implementation, and maintenance costs. Be sure that the chart has a title and that the costs are labeled on the chart.
    c. Use your word processing software to write a paragraph that recommends an approach to take and why.
3. You have been hired to design a new sales ordering program. The program needs a database that contains a customer table containing important customer information, an inventory table that contains current inventory levels, and an order table that contains customer number, inventory number, and order quantity. Develop a database that shows the fields in each table. Include ten sample records in each table.

TEAM ACTIVITIES

1. Assume that your project team has been working for three months to complete the systems design of a new Web-based customer ordering system. Two possible options seem to meet all users' needs. The project team must make a final decision on which option to implement. The following table summarizes some of the key facts about each option.

    a. What process would you follow to make this important decision?
    b. Who needs to be involved?
    c. What additional questions need to be answered to make a good decision?
    d. Based on the data, which option would you recommend and why?
    e. How would you account for project risk in your decision making?

Factor	Option 1	Option 2
Annual gross savings	$1.5 million	$3.0 million
Total development cost	$1.5 million	$2.2 million
Annual operating cost	$0.5 million	$1.0 million
Time required to implement	9 months	15 months
Risk associated with project (expressed in probabilities)		
Benefits will be 50% less than expected	20%	35%
Cost will be 50% greater than expected	25%	30%
Organization will not/cannot make changes necessary for system to operate as expected	20%	25%
Does system meet all mandatory requirements?	Yes	Yes

2. Your team has been hired to explore word processing, graphics, database, and spreadsheet capabilities by the owner of a new restaurant. The new owner has heard about cloud computing, SaaS, and DaaS. Your team should prepare a report on the advantages and disadvantages of using a traditional office suite from a company such as Microsoft compared to other approaches.
3. Your team has been asked to purchase and install a network system that includes five PCs, two printers, and a wireless network for a small business. Develop an RFP that is to be sent to four PC vendors that specifies all the equipment and software that is needed.

WEB EXERCISES

1. Use the Internet to find two different systems development projects that failed to meet cost or performance objectives. Summarize the problems and what should have been done. You might be asked to develop a report or send an e-mail message to your instructor about what you found.
2. Using the Web, search for information on hardware and software virtualization. Write a report on what you found. Under what conditions would you use virtualization?

CAREER EXERCISES

1. Describe what type of information system you would need in your chosen job. Your description should include logical and physical design. What specific steps would you include to be able to recover from a natural or man-made disaster, such as a hurricane or terrorist attack?
2. Explore two Internet career sites, such as *www.monster.com*. Using the principles of systems review, evaluate both sites and then describe what you would change and what you would retain for each career Internet site, assuming they were both about to undergo the systems development process for major improvements.

CASE STUDIES

Case One

Rogers Pulls an All-Nighter

Rogers Communications is one of Canada's largest telecom companies. It offers home phone services, wireless phone services, Internet service, and cable TV service. Rogers provides small shops in 93 malls across Canada to service its customers. The stores sell a wide variety of phones and other telecom devices and services.

Until recently, Rogers outsourced the information technology services for its mall stores to a third-party company. Francois Chevallier, vice president of retail systems for Rogers Retail, thought that Rogers could improve its systems and business practices if it took control of its own systems. "When you have different stores with different systems and management structures, the experience cannot be consistent. Our goal was to achieve that consistency and raise the bar," says Chevallier.

Chevallier proposed a massive upgrade of Rogers' retail systems that would provide consistency in business practices and customer experience and connect all data in a unified system accessible from headquarters. The project faced two big challenges. First, to avoid any interruption to service, systems in all 93 stores should be upgraded simultaneously when stores were closed, which is tricky when dealing with stores across four time zones. Second, the upgrade would take place in the middle of winter when weather was unpredictable, with periodic snow and wind storms and temperatures diving as low as -27 degrees Celsius (-17 F).

This challenging project would require more human resources than Chevallier had. He pulled in an outside company, Connections Canada Inc. (CCI), to assist with the project. The team decided that the upgrade would need to take place at all locations simultaneously over a six-hour period while the stores were closed, which meant working through the night. In order for the upgrade to go flawlessly, the team would need to invest in practice, training, and preparation. Unfortunately they had only four months to prepare.

They decided to create a virtual store, or staging facility, at CCI that mimicked the real Rogers' mall stores. The components of the new system were set up in the virtual store including all software and hardware: routers, computers, cash-registers, and PIN readers. The intention was to create and configure a system for each store using the staging facility and then ship the preconfigured components of the system to each store—a kind of "store-in-a-box."

Experts from different fields including information systems, human resources, operations, finance, supply chain, real estate, marketing, inventory management, and internal communications assisted in configuring the system to meet all organizational needs. The finance expert set up the banking environment for each store. The supply chain expert modified the supply-chain elements of the new system. In other words, each expert worked on his or her area of specialty. The team used a SharePoint intranet to allow all experts involved to communicate online. They spent weeks developing, testing, and adjusting the new system in the virtual store. The development continued until all experts were satisfied.

Once the system was designed and running smoothly in the store prototype, four actual stores were selected for testing. One at a time, the system was installed in each store. With each installation, lessons were learned and problems became fewer. The installation at the fourth store was carried out flawlessly.

The new system was ready for installation in the remaining stores. One technician was hired and trained for each of the store installations. Many backup technicians were trained as well in case the primary technician failed to show up. The store-in-a-box was shipped to each of the stores in shipping containers that could withstand the coldest Canadian winter temperatures. Communications were set up that would allow each technician to give a step-by-step report to the control center at headquarters. At headquarters, ten project managers would be tracking the progress of their districts and reporting to the primary project managers. If trouble arose, it could be addressed within minutes.

The installation went off without a hitch. The project was completed on schedule in four months at a cost of one million dollars. In the end, 500 Rogers employees were trained on the new system, which included an "intranet for resources and policies, a new supply chain model, integrated point-of-sale and merchandise management systems, and a foundation for good customer service." The team credits the success of this ambitious system development project to its detailed preparation, especially the staging facility, and tight communications and cooperation throughout the process.

Discussion Questions

1. What challenges did Rogers face in the installation of its new retail system?
2. What implementation techniques did Rogers employ to assure a smooth transition to the new system?

Critical Thinking Questions

1. Why did Rogers feel it necessary to upgrade all stores at the same time? What are the benefits and risks of that decision?
2. What role did communications technologies play in the success of this system upgrade?

SOURCES: Lau, Kathleen, "Rogers' IT overhaul: All in a day's work," itWorldCanada, May 25, 2007, *www.itworldcanada.com/Pages/Docbase/ViewArticle.aspx?id=idgml-7bce5860-7681-4e37-86f7-7b6fa8244639&sub=472646*; Rogers Communications Web site, *www.rogers.com*, accessed July 20, 2008.

Case Two

Northrop Grumman Builds Super Systems

Northrop Grumman is a global defense and technology company specializing in information services, electronics, aerospace, and shipbuilding for government and commercial purposes. Northrop Grumman scientists and engineers place heavy computational demands on their computer systems. Simulations and computer-aided design applications require a lot of processing power, which Northrop Grumman met for years with multiprocessor workstations and small clusters of servers. Each project in Northrop Grumman received enough computing power to support its needs. Unfortunately, budgets did not always accommodate the needs of new projects, and if they did, setup was time consuming, and sometimes deals were lost in the delay. Maintaining many disjoint systems was also a major challenge for Northrop Grumman's information system support staff.

Bradley Furukawa, VP and CIO, had an idea for a better system. Rather than custom-designing, building, and supporting many computer systems for many projects, Furukawa wanted to build one large supercomputer cluster that could be shared by all projects. Furukawa was given permission to try his theory out on Northrop Grumman's Space Technology unit.

Furukawa assembled a team of information systems specialists and scientists to create a massively parallel supercomputer from Linux-based blade servers. The cluster quickly grew to 979 processors supporting over 100 applications specially adapted for parallel processing. Furukawa saw his role in the development as more political than technical. "It was my job to make sure the funds were there, make sure [the project] stayed visible in front of the vice president and president, remove any administrative barriers ... and let the engineers and scientists do their thing."

Furukawa also had important diplomatic responsibilities. He worked to make sure that the designers of the system kept an "enterprise perspective" rather than focusing on their own project's needs. A governance board was created to decide how supercomputer resources were to be shared across projects in the enterprise. "No matter how much computing power you have, you can always max it out. They set the priorities," Furukawa says. The board was responsible for creating the controls to allocate system resources and monitor their use.

Furukawa's plan worked. The shared supercomputer cluster allows Northrop Grumman employees to complete work faster. Equally important, it allows new project proposals to get off the ground quickly. The salesforce now closes more new deals and keeps existing contracts funded. Customers appreciate the additional simulations and analysis provided by the increase in processing power.

Word of the success of the new supercomputer system reached top executives, who were so impressed that they funded additional growth for the system—which now has 1,800 processors running 400 custom designed scientific and engineering applications. Working to keep the money flowing, Furukawa has opened up the supercomputer to other areas of Northrop Grumman over the company's internal network. He wants to keep the system fully occupied to make a case for further expansion. The ultimate goal for the system is 3,000 CPUs.

Furukawa cites visibility as critical to the project's success. The team was quick to get scientists working on the supercomputer as soon as it was ready. As scientists and their projects benefited from the system, word of the success spread quickly. "Keeping the project on schedule and within budget really added to the success and credibility," says Clayton Kau, vice president and general manager of Northrop Grumman's space and defense products division.

Discussion Questions

1. What technique was used at Northrop Grumman to assist scientists and engineers in being more productive?
2. What role did the CIO play in the development of Northrop Grumman's new system?

Critical Thinking Questions

1. Why was it smart to establish a governance board to manage the new system?
2. What do you think the CIO's biggest challenges are in growing the new system? How is he meeting those challenges?

SOURCES: Mitchell, Robert, "Hot Projects," *Computerworld*, December 10, 2007, *www.computerworld.com/action/article.do?command=viewArticleBasic&articleId=30589 7&pageNumber=1*; Northrop Grumman's Web site, *www.northropgrumman.com*, accessed July 20, 2008.

Questions for Web Case

See the Web site for this book to read about the Whitmann Price Consulting case for this chapter. Following are questions concerning this Web case.

Whitmann Price Consulting: Systems Design, Implementation, Maintenance, and Review Considerations

Discussion Questions

1. What function(s) did the systems analysis report and the design report play in the creation of the AMCI system?
2. What precautions did Josh and Sandra take to make sure the AMCI system was stable and met customer needs prior to mass production?

Critical Thinking Questions

1. What did Josh and Sandra do that was contrary to "textbook execution" of the systems development life cycle? Why?
2. Why do you think Whitmann Price executives shelved the suggestions made by the systems analysts regarding extensions to the system?

NOTES

Sources for the opening vignette: SAP staff, "CARLSBERG POLSKA," SAP Customer Success Story, 2008, *http://download.sap.com/solutions/business-suite/erp/customersuccess/download.epd?context=325F35A1E0FF4678D2A54EC579718E6D1649CA7E39CB3F21A8220A1C60D05035163EBC9888E2DCB75C3690C84707498354B2855ABF89C603*; Carlsberg Polska Web site, *http://carlsbergpolska.pl*, accessed July 17, 2008; Carlsberg Group Web site, *www.carlsberggroup.com*, accessed July 17, 2008.

1 Thibodeau, Patrick, "NYSE Places Buy on Linux," *Computerworld*, December 17, 2007, p. 9.
2 Bulk, Frank, "Copper Costs Lots of Pretty Pennies," *Information Week*, May 26, 2008, p. 29.
3 Preston, Rob, "Will Cloud Computing Rain on IT's Parade?" *Information Week*, February 18, 2008, p. 52.
4 Cheng, Roger, "Those on the Go Get to Go Online," *The Wall Street Journal*, September 4, 2007, p. B3.
5 Hoover, Nicholas, "Microsoft Opens Live Mesh to More Users," *Information Week*, April 28, 2008, p. 22.
6 Gomes, Lee, "Good Site, Bad Site: Evolving Web Design," *The Wall Street Journal*, June 12, 2007. p. B3.
7 Shipley, Greg, "Risk=$," *Information Week*, March 31, 2008, p. 31.
8 Fonseca, Brian, "E-Discovery Rules Still Causing IT Headaches," *Computerworld*, January 7, 2008, p. 14.
9 Srivastava, Samar, "Search Software Gets Boost," *The Wall Street Journal*, May 16, 2007, p. B6.
10 Lei, Tao Ai, "Using Software to Save Time and Money," *The Straits Times*, May 6, 2008.
11 Vijayan, Jaikumar, "Offshore Worker Nabbed for Caterpillar Data Theft," *Computerworld*, September 17, 2007, p. 16.
12 Wiens, Jordan, "With Security, More Is Better," *Information Week*, March 10, 2008, p. 42.
13 Ambrosio, Johnanna, "Multiple Short Outages Can Add Up to Major Problems," *Computerworld*, May 12, 2008, p. 16.
14 Fonseca, Brian, "IT Facing Up to a Task of Working with Executives on Disaster Recovery," *Computerworld*, April 30, 2007, p. 10.
15 Pratt, Mary, "Surviving the Big One," *Computerworld*, March 31, 2008, p. 23.
16 Fonseca, Brian, "IT Facing Up to a Task of Working with Executives on Disaster Recovery," *Computerworld*, April 30, 2007, p. 10.
17 Mitchell, Robert, "On the Edge of Disaster," *Computerworld*, April 30, 2007, p. 25.
18 Rogers, James, "EMC's Disaster Plan," *Information Week*, March 3, 2008, p. 28.
19 Rotenberg, Marc, "The Grill," *Computerworld*, February 18, 2008, p. 19.
20 Foust, Dean, "Diagnosis: Identity Theft," *Business Week*, January 8, 2007, p. 30.
21 Hagel, John, "The Grill," *Computerworld*, January 7, 2008, p. 18.
22 Brandel, Mary, "Green IT Companies," *Computerworld*, February 18, 2008, p. 29.
23 Collett, Stacy, "Slimmed-Down Servers," *Computerworld*, March 3, 2008, p. 28.
24 Mitchell, Robert, "Power Pinch," *Computerworld*, April 30, 2007, p. 19.
25 Thibodeau, Patrick, "EPA Helps Put Data Centers on an Energy Diet," *Computerworld*, March 24, 2008, p. 12.
26 Weier, Mary Hayes, "So Your Company's Going Green," *Information Week*, May 5, 2008, p. 24.
27 Wittmann, Art, "Think Thin, Green, and Virtual," *Information Week*, May 26, 2008, p. 19.

28 Carlton, Jim, "To Cut Fuel Bills, Try High-Tech Help," *The Wall Street Journal*, March 11, 2008, p. B3.
29 Chea, Terence, "Ramping Up Against E-Waste," *Rocky Mountain News*, March 4, 2007, p. 8K.
30 Boehret, Katherine, "Where Computers Go When They Die," *The Wall Street Journal*, April 11, 2007, p. D1.
31 Thibodeau, Patrick, "Green PC Push By Feds May Seed Wider Adoption," *Computerworld*, January 14, 2008, p. 14.
32 *www.energystar.gov/index.cfm?fuseaction=find_a_product*, accessed June 15, 2008.
33 Mangalindan, Mylene, "New Marketing Style: Clicks and Mortar," *The Wall Street Journal*, December 21, 2007, p. B5.
34 Flandez, Raymund, "Online Retailers Get Chatty," *The Wall Street Journal*, December 18, 2007, p. B9.
35 Thibodeau, Patrick, "Heft of New Tech-Buying Guidelines," *Computerworld*, November 19, 2007, p. 12.
36 Babcock, Charles, "Virtualization's Tipping Point," *Information Week*, May 18, 2008, p. 14.
37 Nam, Susan, "The Journey Machine," *Forbes*, November 26, 2007, p. 152.
38 Hamblen, Matt, "Census to Start Small on Handheld Rollout," *Computerworld*, January 8, 2007, p. 7.
39 Thibodeau, Patrick, "DARPA Pushes to Bring Supercomputer to the Masses," *Computerworld*, January 8, 2007, p. 35.
40 Thibodeau, Patrick, "Marines Look for a Few Less Servers, Via Virtualization," *Network World*, November 6, 2007.
41 Dell Staff, "Pushing the Virtualization Envelope," *Information Week*, May 28, 2007, p. NWC 8.
42 Pratt, Mary, "Targeting Cancer," *Computerworld*, December 3, 2007, p. 42.
43 Weier, Mary Hayes, "Good Hands Aren't Enough," *Information Week*, April 23, 2007, p. 49.
44 Lawton, Christopher and Clark, Don, "Virtualization Is Pumping Up Servers," *The Wall Street Journal*, March 6, 2007, p. B4.
45 Staff, "Kindred Healthcare leverages VI3 and VDI to provide physicians with secure bedside access to patient records using wirelessly connected thin clients," Computerworld VMware Success Stories (video), *www.vmware.com/customers/stories/success_video.html?id=khealthcarevdi*, accessed February 2, 2008.
46 Magda, Beverly, "SaaS to the Rescue," *Information Week*, May 19, 2008, p. 31.
47 Delaney, Kevin, "Google Further Tests Microsoft's Domain," *The Wall Street Journal*, February 22, 2007, p. A4.
48 Boock, Grady, "The Grill," *Computerworld*, October 29, 2007, p. 28.
49 Singer, Michael, "Sun Sticks to Its Open Agenda," *Information Week*, May 5, 2008, p. 20.
50 IT Redux Web site, *http://itredux.com/office-20/database/?family=Database*, accessed March 23, 2008.
51 Lai, Eric, "Cloud database vendors: What, us worry about Microsoft?" *Computerworld*, March 12, 2008, *www.computerworld.com/action/article.do?command=viewArticleBasic&articleId=9067979&pageNumber=1.*
52 Fonseca, Brian, "Virtualization Cutting Storage Costs for Some Large Firms," *Computerworld*, April 23, 2007, *www.computerworld.com/action/article.do?command=viewArticleBasic&taxonomyId=154&articleId=290230&int src=hm_topic.*
53 Pratt, Mary, "Goodwill Mission," *Computerworld*, April 28, 2008, p. 24.
54 Guth, Robert, "Office 2007 Poses Hurdle," *The Wall Street Journal*, January 23, 2007, p. B2.
55 Mossberg, Walter, "Bold Redesign Improves Office 2007," *The Wall Street Journal*, January 4, 2007, p. B1.
56 Dornan, Andy, "No Hurry with Vista," *Information Week*, December 3, 2007, p. AB10.
57 Stackpole, Beth, "IT's Top Five Training Mistakes," *Computerworld*, April 28, 2008, p. 20.
58 Hoover, Nicholas, "A Year Later, IT Still Grapples with Vista Deployment Timing," *Information Week*, December 17, 2007, p. 32.
59 Kozberg, Donna Walters, "Career Watch," *Computerworld*, April 28, 2008, p. 35.
60 Carlton, Jim, "Computer Rooms Feel the Heat," *The Wall Street Journal*, April 10, 2007, p. B4.
61 Weiss, Todd, "Colorado DMV Puts Brakes on $13 M Registration System," *Computerworld*, April 9, 2007, p. 8.
62 Rosencrance, Linda, "Botched Data Center Move Takes Out 165K Web Sites," *Computerworld*, November 12, 2007, p. 12.
63 McGee, Marianne Kolbasuk, "Cleveland, Then the World," *Information Week*, May 26, 2008, p. 18.
64 Major, Paul, "The Grill," *Computerworld*, May 12, 2008, p. 18.
65 *www.research.ibm.com/autonomic*, accessed June 20, 2008.
66 Vijayan, Jaikumar, "Restaurant Chain Served Card Data to Hackers," *Computerworld*, May 19, 2008, p. 8.
67 Vijayan, Jaikumar, "Constant Patch Releases Forcing New IT Processes," *Computerworld*, February 25, 2008, p. 14.
68 *www.relativity.com/pages/home.asp*, accessed June 20, 2008.
69 Thurman, Mathias, "An Audit Can Be an Opportunity," *Computerworld*, January 14, 2008, p. 36.
70 Perhins, Bart, "Pulling the Plug on a Project," *Computerworld*, March 10, 2008, p. 42.
71 Weier, Mary Hayes, "Good Hands Aren't Enough," *Information Week*, April 23, 2007, p. 49.
72 *www-306.ibm.com/software/tivoli/products/omegamon-xe-cics*, accessed June 20, 2008.
73 *www.precise.com*, accessed June 20, 2008.
74 *https://h10078.www1.hp.com/cda/hpms/display/main/hpms_home.jsp?zn=bto&cp=1_4011_100__*, accessed June 20, 2008.

PART 5

Information Systems in Business and Society

CHAPTER

• 14 •

The Personal and Social Impact of Computers

PRINCIPLES	LEARNING OBJECTIVES
▪ Policies and procedures must be established to avoid waste and mistakes associated with computer usage.	▪ Describe some examples of waste and mistakes in an IS environment, their causes, and possible solutions. ▪ Identify policies and procedures useful in eliminating waste and mistakes. ▪ Discuss the principles and limits of an individual's right to privacy.
▪ Computer crime is a serious and rapidly growing area of concern requiring management attention.	▪ Explain the types of computer crime and impacts. ▪ Identify specific measures to prevent computer crime.
▪ Jobs, equipment, and working conditions must be designed to avoid negative health effects from computers.	▪ List the important negative effects of computers on the work environment. ▪ Identify specific actions that must be taken to ensure the health and safety of employees.
▪ Practitioners in many professions subscribe to a code of ethics that states the principles and core values that are essential to their work.	▪ Outline criteria for the ethical use of information systems.

Information Systems in the Global Economy eBay, United States

Battling Hackers and Fraudsters Day In and Day Out

Online giant eBay provides the world's largest online marketplace. Roughly 147 million people buy and sell all kinds of merchandise and services on eBay. It is estimated that more than $1,900 worth of goods is sold on eBay every second. Hundreds of thousands of people in the United States depend on eBay for their living. Some of those people earn their living illegally and unethically.

With so many transactions taking place on eBay, the company manages as much money as a global bank. In fact, eBay owns the largest online banking service, PayPal, which is used to facilitate transactions between online buyers and sellers. Unfortunately, managing the largest online bank and marketplace makes eBay a huge target for hackers and fraudsters. The level of information security implemented at eBay far exceeds the level of security used in a brick-and-mortar bank.

Fraud is a major challenge for eBay. Criminals gain illegal access to customer accounts and use the accounts and their good reputation to sell knock-offs—primarily imitations of high-quality items. Of all Internet-related crimes, auction fraud appears to be the biggest problem. The 2007 Internet Crime Report published by the FBI explains: "Internet auction fraud was by far the most reported offense, comprising 44.9 percent of referred complaints. Non-delivered merchandise and/or payment accounted for 19.0 percent of complaints. Check fraud made up 4.9 percent of complaints. Credit/debit card fraud, computer fraud, confidence fraud, and financial institutions fraud round out the top seven categories of complaints referred to law enforcement during the year."

In 2008, at least two eBay crimes made headlines. The first involves a Romanian hacker called Vladuz who hacked into eBay systems and masqueraded as an official eBay representative. The damage Vladuz caused is estimated at 1 million dollars. He was eventually apprehended and awaits trial. The second big story involves Jeremiah Mondello, a 23-year-old from Oregon. Mondello stole more than 40 eBay and PayPal accounts and used them to sell over a million dollars worth of counterfeit software, making more than $400,000 for himself. Mondello is now spending two years in prison, and must pay $225,000 in fines and devote 450 hours to community service once released from prison.

Even though many crimes are taking place on eBay as you read this sentence, eBay says that fraud remains a tiny fraction of the million or so transactions that take place each day. eBay keeps fraud under control by investing heavily in information security tools and practices.

eBay uses many types of security tools to address many kinds of threats. On its Web site, eBay states that PayPal uses "the world's most advanced proprietary fraud prevention systems to create a safe payment solution." The company also invests in an automated security system to keep hackers out of the network. The system uses more than a dozen scanning applications to monitor vulnerabilities on eBay's global network and on all partner networks that connect to eBay's extranet.

The security software that patrols eBay's systems provides continuous reports to security engineers. The software also creates reports for system administrators and executives that provide an overview of network conditions and illustrate the impact of information security investment. Additionally, the security software measures eBay's compliance with government regulations involving information security.

The battle to protect valuable and private information online is one in which all levels of management in businesses and governments are or should be fully engaged. Attacks are increasing from people around the world looking to benefit financially or politically. Businesses such as eBay enforce rigid policies and procedures to make sure that their networks, employees, and partners are operating in the most secure manner possible.

As you read this chapter, consider the following:

- What are the primary concerns of corporations regarding security, privacy, and ethics?
- What strategies can assist a company with issues of security and privacy, and at what cost?

Why Learn About the Personal and Social Impact of the Internet?

A wide range of nontechnical issues associated with the use of information systems and the Internet provide both opportunities and threats to modern organizations. The issues span the full spectrum—from preventing computer waste and mistakes, to avoiding violations of privacy, to complying with laws on collecting data about customers, to monitoring employees. If you become a member of a human resources, information systems, or legal department within an organization, you will likely be charged with leading the organization in dealing with these and other issues covered in this chapter. Also, as a user of information systems and the Internet, it is in your own self-interest to become well versed on these issues. You need to know about the topics in this chapter to help avoid or recover from crime, fraud, privacy invasion, and other potential problems. This chapter begins with a discussion of preventing computer waste and mistakes.

Earlier chapters detailed the significant benefits of computer-based information systems in business, including increased profits, superior goods and services, and higher quality of work life. Computers have become such valuable tools that today's businesspeople have difficulty imagining work without them. Yet the information age has also brought the following potential problems for workers, companies, and society in general:

- Computer waste and mistakes
- Computer crime
- Privacy issues
- Work environment problems
- Ethical issues

This chapter discusses some of the social and ethical issues as a reminder of these important considerations underlying the design, building, and use of computer-based information systems. No business organization, and, hence, no information system, operates in a vacuum. All IS professionals, business managers, and users have a responsibility to see that the potential consequences of IS use are fully considered.

Managers and users at all levels play a major role in helping organizations achieve the positive benefits of IS. These people must also take the lead in helping to minimize or eliminate the negative consequences of poorly designed and improperly utilized information systems. For managers and users to have such an influence, they must be properly educated. Many of the issues presented in this chapter, for example, should cause you to think back to some of the systems design and systems control issues discussed previously. They should also help you look forward to how these issues and your choices might affect your future use of information systems.

COMPUTER WASTE AND MISTAKES

Computer-related waste and mistakes are major causes of computer problems, contributing as they do to unnecessarily high costs and lost profits. Computer waste involves the inappropriate use of computer technology and resources. Computer-related mistakes refer to errors, failures, and other computer problems that make computer output incorrect or not useful, caused mostly by human error. This section explores the damage that can be done as a result of computer waste and mistakes.

Computer Waste

The U.S. government is the largest single user of information systems in the world. It should come as no surprise, then, that it is also perhaps the largest abuser. The government is not unique in this regard—the same type of waste and misuse found in the public sector also exists in the private sector. Some companies discard old software and computer systems when they still have value. Others waste corporate resources to build and maintain complex systems that are never used to their fullest extent.

A less dramatic, yet still relevant, example of waste is the amount of company time and money employees can waste playing computer games, sending unimportant e-mail, or accessing the Internet. Junk e-mail, also called *spam*, and junk faxes also cause waste. People receive hundreds of e-mail messages and faxes advertising products and services not wanted or requested. Not only does this waste time, but it also wastes paper and computer resources. Worse yet, spam messages often carry attached files with embedded viruses that can cause networks and computers to crash or allow hackers to gain unauthorized access to systems and data.

A spam filter is software that attempts to block unwanted e-mail. One approach to filtering spam involves building lists of acceptable and unacceptable e-mail addresses. The lists can be created manually or automatically based on how the users keep or discard their e-mail. Another approach is automatic rejection of e-mail based on the content of the message or the appearance of keywords in the message. Rejected e-mail automatically goes to the spam or junk e-mail folder of your e-mail service. CA Anti Spam, SpamEater Pro, ChoiceMail One, and Spam Buster are among the most highly rated anti-spam software and cost from around $20 to $50.[1] Many e-mail programs have built-in spam filters.

A word of caution: some spam filters might require first-time e-mailers to be verified before their e-mails are accepted. This can be disastrous for people in sales or customer service who are frequently receiving e-mails from people they do not know. In one case, a spam filter blocked e-mail to close a real estate deal valued at around $175,000. The deal never closed and the real estate was sold to someone else because of the blocked e-mail.

Image-based spam is a new tactic spammers use to circumvent spam-filtering software that rejects e-mail based on the content of messages and the use of keywords. The message is presented in a graphic form that can be read by people but not computers. The images in this form of spam can be quite offensive.

When waste is identified, it typically points to one common cause: the improper management of information systems and resources.

Computer-Related Mistakes

Despite many people's distrust of them, computers rarely make mistakes. Yet even the most sophisticated hardware cannot produce meaningful output if users do not follow proper procedures. Mistakes can be caused by unclear expectations and a lack of feedback. A programmer might also develop a program that contains errors. In other cases, a data-entry clerk might enter the wrong data. Unless errors are caught early and prevented, the speed of computers can intensify mistakes. As information technology becomes faster, more complex, and more powerful, organizations and computer users face increased risks of experiencing the results of computer-related mistakes. Consider these examples from recent news.

- Shares of Moody's Corporation rating agency fell more than 20 percent following a *Financial Times* report alleging a computer coding error boosted the investment ratings of a particular class of debt instrument four levels to Aaa (the highest possible rating) and wasn't immediately corrected after it was uncovered.[2]
- An internal investigation by NASA concluded that multiple programming errors caused the eventual loss of the Mars Global Surveyor orbiter spacecraft in January 2008. Fortunately, by the time of the loss, the orbiter had lasted four times longer than expected and was successful in mapping the surface of Mars and studying its atmosphere.[3]

- Computer problems are a frequent cause for airline flight cancellations and delays. For example, a computer glitch at All Nippon Airways Company grounded or delayed hundreds of domestic flights in Japan.[4] United Air Lines had to cancel 24 domestic flights and delay another 250 flights by more than 90 minutes when its systems for dispatching flights failed.[5]
- A computer error knocked out ATM and wire and ACH transfer services provided by Wells Fargo for two days.[6]

PREVENTING COMPUTER-RELATED WASTE AND MISTAKES

To remain profitable in a competitive environment, organizations must use all resources wisely. Preventing computer-related waste and mistakes should therefore be a goal. Today, nearly all organizations use some type of Computer Based Information System (CBIS). To employ IS resources efficiently and effectively, employees and managers alike should strive to minimize waste and mistakes. Preventing waste and mistakes involves (1) establishing, (2) implementing, (3) monitoring, and (4) reviewing effective policies and procedures.

Establishing Policies and Procedures

The first step to prevent computer-related waste is to establish policies and procedures regarding efficient acquisition, use, and disposal of systems and devices. Computers permeate organizations today, and it is critical for organizations to ensure that systems are used to their full potential. As a result, most companies have implemented stringent policies on the acquisition of computer systems and equipment, including requiring a formal justification statement before computer equipment is purchased, definition of standard computing platforms (operating system, type of computer chip, minimum amount of RAM, etc.), and the use of preferred vendors for all acquisitions.

Prevention of computer-related mistakes begins by identifying the most common types of errors, of which there are surprisingly few. Types of computer-related mistakes include the following:

- Data-entry or data-capture errors
- Errors in computer programs
- Errors in handling files, including formatting a disk by mistake, copying an old file over a newer one, and deleting a file by mistake
- Mishandling of computer output
- Inadequate planning for and control of equipment malfunctions
- Inadequate planning for and control of environmental difficulties (such as electrical and humidity problems)
- Installing computing capacity inadequate for the level of activity on corporate Web sites
- Failure to provide access to the most current information by not adding new Web links and not deleting old links

To control and prevent potential problems caused by computer-related mistakes, companies have developed policies and procedures that cover the acquisition and use of computers, with a goal of avoiding waste and mistakes. Training programs for individuals and workgroups as well as manuals and documents covering the use and maintenance computer systems also help prevent problems. Other preventive measures include approval of certain systems and applications before they are implemented and used to ensure compatibility and cost-effectiveness, and a requirement that documentation and descriptions of certain applications be filed or submitted to a central office, including all cell formulas for spreadsheets and a description of all data elements and relationships in a database system. Such standardization can ease access and use for all personnel.

Many companies have established strong policies to prevent employees from wasting time using computers inappropriately at work. A survey of 304 U.S. companies determined that over one-fourth of bosses have fired employees for inappropriate use of e-mail and one-third have fired workers for wasting valuable time on the Internet.[7] Three workers were terminated and another 26 workers for the Collier County, Florida government were given unpaid suspensions for inappropriate use of county computers.[8]

After companies have planned and developed policies and procedures, they must consider how best to implement them.

Implementing Policies and Procedures

Implementing policies and procedures to minimize waste and mistakes varies according to the business conducted. Most companies develop such policies and procedures with advice from the firm's internal auditing group or its external auditing firm. The policies often focus on the implementation of source data automation and the use of data editing to ensure data accuracy and completeness, and the assignment of clear responsibility for data accuracy within each information system. Some useful policies to minimize waste and mistakes include the following:

- Changes to critical tables, HTML, and URLs should be tightly controlled, with all changes authorized by responsible owners and documented.
- A user manual should be available covering operating procedures and documenting the management and control of the application.
- Each system report should indicate its general content in its title and specify the time period covered.
- The system should have controls to prevent invalid and unreasonable data entry.
- Controls should exist to ensure that data input, HTML, and URLs are valid, applicable, and posted in the right time frame.
- Users should implement proper procedures to ensure correct input data.

Training is another key aspect of implementation. Many users are not properly trained in using applications, and their mistakes can be very costly. When business intelligence tools were first installed at the Maryland Department of Transportation, inexperienced users began executing queries unrelated to their jobs. The large number of queries and report requests led to system performance problems, with excessive run times and slow response time to queries. The department implemented new policies for running requests and provided training to show users why it was important to access only the data required to do their work. System performance was improved through the implementation of these policies.[9]

Because more and more people use computers in their daily work, it is important that they understand how to use them. Training is often the key to acceptance and implementation of policies and procedures. Because of the importance of maintaining accurate data and of people understanding their responsibilities, companies converting to ERP and e-commerce systems invest weeks of training for key users of the system's various modules.

Monitoring Policies and Procedures

To ensure that users throughout an organization are following established procedures, the next step is to monitor routine practices and take corrective action if necessary. By understanding what is happening in day-to-day activities, organizations can make adjustments or develop new procedures. Many organizations implement internal audits to measure actual results against established goals, such as percentage of end-user reports produced on time, percentage of data-input errors detected, number of input transactions entered per eight-hour shift, and so on.

The Société Générale scandal in France is a classic example of an individual employee circumventing internal policies and procedures. A low-level trader on the arbitrage desk at the French bank created a series of fraudulent and unauthorized investment transactions that built a $72 billion position in European stock index futures.[10] Eventually the house of cards collapsed, causing the bank to lose over $7 billion—even though a compliance officer at the bank had been alerted months in advance not once, but twice that something unusual was going on.[11]

Reviewing Policies and Procedures

The final step is to review existing policies and procedures and determine whether they are adequate. During review, people should ask the following questions:

- Do current policies cover existing practices adequately? Were any problems or opportunities uncovered during monitoring?
- Does the organization plan any new activities in the future? If so, does it need new policies or procedures addressing who will handle them and what must be done?
- Are contingencies and disasters covered?

This review and planning allows companies to take a proactive approach to problem solving, which can enhance a company's performance, such as by increasing productivity and improving customer service. During such a review, companies are alerted to upcoming changes in information systems that could have a profound effect on many business activities.

Tokyo Electron, a global supplier of semiconductor production equipment, provides an excellent example of a firm thoroughly reviewing its policies and procedures. As a U.S. subsidiary of Tokyo Electron of Japan, Tokyo Electron U.S. Holdings was required to comply with the Sarbanes-Oxley Act. When Japan's Financial Instruments and Exchange Law, that country's equivalent of the Sarbanes-Oxley Act, went into effect, the firm used it as a motivation to re-examine its entire set of policies regarding user access to data and applications, financial control, and protection of intellectual property.[12]

Information systems professionals and users still need to be aware of the misuse of resources throughout an organization. Preventing errors and mistakes is one way to do so. Another is implementing in-house security measures and legal protections to detect and prevent a dangerous type of misuse: computer crime.

COMPUTER CRIME

Even good IS policies might not be able to predict or prevent computer crime. A computer's ability to process millions of pieces of data in less than one second can help a thief steal data worth millions of dollars. Compared with the physical dangers of robbing a bank or retail store with a gun, a computer criminal with the right equipment and know-how can steal large amounts of money from the privacy of a home. The following is a sample of recent computer crimes:

- Criminals illegally obtained information about the bank accounts of an undetermined number of Citibank customers. They created counterfeit ATM cards encoded with the stolen information to make some 9,000 fraudulent ATM withdrawals totaling millions of dollars. Avivah Litan, Gartner vice president, stated: "Criminals have found ways to basically bypass many of the controls banks have in place. So ATM and debit card fraud is expected to rise. In our surveys, banks themselves expect the rate of fraud to double over the next two years."[13]
- A Chilean hacker gathered personal data about 6 million people from various Chilean government sites including names, addresses, phone numbers, ID numbers, and e-mail addresses and posted them to a blog site for all to see. The hacker's motivation was to protest his country's weak data security.[14]
- A hacker is alleged to have broken into the computers holding the financial results of IMS Health to learn of the firm's disappointing results for the quarter prior to their public announcement. Taking advantage of this knowledge, the hacker purchased over $41,000 in sell options, figuring the stock would go down when results were announced. The investment resulted in profits of nearly $300,000.[15]
- A 15-year-old Pennsylvania student broke into an educational network and saved on a flash drive the names, addresses, and Social Security numbers of some 55,000 people. The student was arrested and charged with four offenses of unlawful duplication and theft.[16]

- When customers initially link their brokerage accounts to their bank account to allow the transfer of funds, firms such as E*Trade and Schwab.com use a test procedure to make micro-deposits of a few cents to a few dollars to the bank account to ensure that the account numbers and routing information are correct. A hacker took advantage of a backdoor to this procedure by opening tens of thousands of banking accounts with the brokerages and linked them to fraudulent brokerage accounts to collect the micro-deposits. The hacker stole more than $50,000 over six months.[17]

Although no one really knows how pervasive cybercrime is, according to the 2007 FBI Internet Crime Report, 206,844 complaints of crime were perpetrated over the Internet during 2007 with a dollar value of $240 million in losses.[18] Unfortunately, this represents a small fraction of total computer-related crimes as many crimes go unreported because companies don't want the bad press or don't believe that law enforcement could help. Such lack of publicity makes the job even tougher for law enforcement. Most companies that have been electronically attacked won't talk to the press. A big concern is loss of public trust and image—not to mention the fear of encouraging copycat hackers.

The Computer Security Institute, with the participation of the San Francisco Federal Bureau of Investigation (FBI) Computer Intrusion Squad, conducts an annual survey of computer crime and security. The aim of the survey is to raise awareness of security as well as to determine the scope of computer crime in the United States. The following are a few of the highlights of the 2007 Computer Crime and Security Survey based on responses from 494 companies and government agencies that are members of the Computer Security Institute[19]:

- Financial fraud, followed by virus attacks, is the leading cause of financial loss from computer incidents.
- For the respondents, the average annual loss from computer incidents was $350,424.
- A full 46 percent of the respondents said they had suffered a security incident, though only 29 percent of the respondents reported computer intrusions to law enforcement. (Surprisingly, 10 percent responded that they did not know if they had been subjected to an incident.)

The tenth annual *InformationWeek* Global Information Security survey reveals that the number one tactical security problem for U.S. companies in 2007 was creating and enhancing user awareness of security policies.[20]

Today, computer criminals are a new breed—bolder and more creative than ever. With the increased use of the Internet, computer crime is now global. It's not just on U.S. shores that law enforcement has to battle cybercriminals. Regardless of its nonviolent image, computer crime is different only because a computer is used. It is still a crime. Part of what makes computer crime so unique and difficult to combat is its dual nature—the computer can be both the tool used to commit a crime and the object of that crime.

THE COMPUTER AS A TOOL TO COMMIT CRIME

A computer can be used as a tool to gain access to valuable information and as the means to steal thousands or millions of dollars. It is, perhaps, a question of motivation—many people who commit computer-related crime claim they do it for the challenge, not for the money. Credit card fraud—whereby a criminal illegally gains access to another's line of credit with stolen credit card numbers—is a major concern for today's banks and financial institutions. In general, criminals need two capabilities to commit most computer crimes. First, the criminal needs to know how to gain access to the computer system. Sometimes, obtaining access requires knowledge of an identification number and a password. Second, the criminal must know how to manipulate the system to produce the desired result. Frequently, a critical computer password has been talked out of a person, a practice called **social engineering**. Or, the attackers simply go through the garbage—**dumpster diving**—for important pieces of information that can help crack the computers or convince someone at the company to give

social engineering
Using social skills to get computer users to provide information to access an information system or its data.

dumpster diving
Going through the trash cans of an organization to find secret or confidential information, including information needed to access an information system or its data.

them more access. In addition, over 2,000 Web sites offer the digital tools—for free—that will let people snoop, crash computers, hijack control of a machine, or retrieve a copy of every keystroke.

Although all the details have not been revealed, it appears that an outsider used social engineering skills to convince an MTV employee to download malicious software onto a corporate computer. Through this ruse, attackers gained access to over 5,000 employees' personal data including names, Social Security numbers, birth dates, and salary data.[21]

Also, with today's sophisticated desktop publishing programs and high-quality printers, crimes involving counterfeit money, bank checks, traveler's checks, and stock and bond certificates are on the rise. As a result, the U.S. Treasury redesigned and printed new currency that is much more difficult to counterfeit.

Cyberterrorism

Cyberterrorism has been a concern for countries and companies around the globe. The U.S. government considered the potential threat of cyberterrorism serious enough that it established the National Infrastructure Protection Center in February 1998. This function was transferred to the Homeland Security Department's Information Analysis and Infrastructure Protection Directorate to serve as a focal point for threat assessment, warning, investigation, and response for threats or attacks against the country's critical infrastructure, which provides telecommunications, energy, banking and finance, water systems, government operations, and emergency services. Successful cyberattacks against the facilities that provide these services could cause widespread and massive disruptions to the normal function of American society. International Multilateral Partnership Against Cyber Terrorism (IMPACT) is a global public and privately supported initiative against cyberterrorism.[22]

cyberterrorist
Someone who intimidates or coerces a government or organization to advance his political or social objectives by launching computer-based attacks against computers, networks, and the information stored on them.

A **cyberterrorist** is someone who intimidates or coerces a government or organization to advance his political or social objectives by launching computer-based attacks against computers, networks, and the information stored on them. Fortunately, only relatively few cases of cyberterrorism have been documented, including the following:

- The small Baltic nation of Estonia was subjected to a cyberterrorism attack for three weeks in 2007 that disabled government and corporate networks. The attack followed deadly riots by the nation's ethnic Russian minority in response to the relocation of a Soviet war memorial. Moscow has denied any involvement.[23]
- Pro-China cyberterrorists launched a brief denial-of-service attack on the CNN Web site, which they believe has been overly critical of China, to protest the news network's coverage of Tibet. The attack was cancelled after less than 30 minutes, but the group threatened to launch another attack in the near future.[24]

Identity Theft

identify theft
A crime in which an imposter obtains key pieces of personal identification information, such as Social Security or driver's license numbers, to impersonate someone else.

Identity theft is a crime in which an imposter obtains key pieces of personal identification information, such as Social Security or driver's license numbers, to impersonate someone else. The information is then used to obtain credit, merchandise, and/or services in the name of the victim or to provide the thief with false credentials. In 2007, 8.4 million adults in the United States were victims of identity fraud, according to Javelin Strategy & Research, which compiles a widely accepted survey.[25] The perpetrators of these crimes employ such an extensive range of methods that investigating them is difficult.

In some cases, the identity thief uses personal information to open new credit accounts, establish cellular phone service, or open a new checking account to obtain blank checks. In other cases, the identity thief uses personal information to gain access to the person's existing accounts. Typically, the thief changes the mailing address on an account and runs up a huge bill before the person whose identity has been stolen realizes there is a problem. The Internet has made it easier for an identity thief to use the stolen information because transactions can be made without any personal interaction.

Another popular method to get information is "shoulder surfing"—the identity thief simply stands next to someone at a public office, such as the Bureau of Motor Vehicles, and watches as the person fills out personal information on a form.

Consumers can help protect themselves by regularly checking their credit reports with major credit bureaus, following up with creditors if their bills do not arrive on time, not revealing any personal information in response to unsolicited e-mail or phone calls (especially Social Security numbers and credit card account numbers), and shredding bills and other documents that contain sensitive information.

The U.S. Congress passed the Identity Theft and Assumption Deterrence Act of 1998 to fight identity theft. Under this act, the Federal Trade Commission (FTC) is assigned responsibility to help victims restore their credit and erase the impact of the imposter. It also makes identity theft a federal felony punishable by a prison term ranging from 3 to 25 years.

Internet Gambling

Many people enjoy Internet gambling as a recreational and leisure activity. Baccarat, bingo, blackjack, pachinko, poker, roulette, and sports betting are all readily available online. The size of the online gambling market is not known, but one estimate is that $10–20 billion is wagered on online poker alone each year.[26] Although Internet gambling is legal in more than 70 countries, the legality of these online activities is far from clear in the United States.

- The Interstate Wire Act of 1961 has been interpreted by the Department of Justice as banning all Internet gambling. However, various courts have interpreted the Act as covering only sporting events and exempting casino games such as blackjack and poker.[27]
- The Unlawful Internet Gambling Enforcement Act of 2006 (UIGEA) made it illegal to transfer funds from banks or financial institutions to online gambling sites. However, it failed to clarify the issue of the legality of gambling online. The Act simply states that some gambling is unlawful under state or federal law without specifying details other than excluding several specific gambling activities from its purview.
- Various individual states have passed laws regulating Internet gambling. These laws regulate making bets online, taking bets online, and transferring money between the bettor and an online casino anywhere in the world.

CBSSports.com and Facebook were investigated briefly by the FBI for collaborating to make it easier for Facebook users to fill out brackets for the NCAA 2008 Basketball Tournament. Leslie Anne Wade, senior vice president at CBS stated: "These are new issues that are going to require new thought processes and new answers. [CBS will] look at it."[28]

The revenues generated by Internet gambling represent a major untapped source of income for the state and federal governments. A study prepared by PriceWaterhouseCoopers estimates that the taxation of Internet gambling would yield somewhere between $8.7 billion and $42.8 billion in additional federal revenues during its first ten years.[29]

THE COMPUTER AS THE OBJECT OF CRIME

A computer can also be the object of the crime, rather than the tool for committing it. Tens of millions of dollars worth of computer time and resources are stolen every year. Each time system access is illegally obtained, data or computer equipment is stolen or destroyed, or software is illegally copied, the computer becomes the object of crime. These crimes fall into several categories: illegal access and use, data alteration and destruction, information and equipment theft, software and Internet piracy, computer-related scams, and international computer crime.

Illegal Access and Use

Crimes involving illegal system access and use of computer services are a concern to both government and business. Since the outset of information technology, computers have been plagued by criminal hackers. Originally, a **hacker** was a person who enjoys computer

hacker
A person who enjoys computer technology and spends time learning and using computer systems.

technology and spends time learning and using computer systems. A **criminal hacker**, also called a **cracker**, is a computer-savvy person who attempts to gain unauthorized or illegal access to computer systems to steal passwords, corrupt files and programs, or even transfer money. In many cases, criminal hackers are people who are looking for excitement—the challenge of beating the system. Today, many people use the term hacker and cracker interchangeably. **Script bunnies** admire crackers but have little technical savvy. They are crackers who download programs called *scripts* that automate the job of breaking into computers. **Insiders** are employees, disgruntled or otherwise, working solo or in concert with outsiders to compromise corporate systems. The biggest threat for many companies is their own employees who hack into their computers, not external hackers. Insiders have extra knowledge that makes them especially dangerous—they know logon IDs, passwords, and company procedures that help them evade detection.

criminal hacker (cracker)
A computer-savvy person who attempts to gain unauthorized or illegal access to computer systems to steal passwords, corrupt files and programs, or even transfer money.

script bunny
A cracker with little technical savvy who downloads programs called scripts, which automate the job of breaking into computers.

insider
An employee, disgruntled or otherwise, working solo or in concert with outsiders to compromise corporate systems.

Some criminals have started phony VoIP phone companies and sold subscriptions for services to unsuspecting customers. Instead of establishing their own network, the criminals hack into the computers that route calls over the networks of legitimate VoIP providers and use this network to carry its customers' calls. One criminal obtained more than $1 million for more than 10 million minutes of VoIP service stolen from a legitimate VoIP service provider.[30]

Catching and convicting criminal hackers remains a difficult task. The method behind these crimes is often hard to determine. Even if the method behind the crime is known, tracking down the criminals can take a lot of time. It took years for the FBI to arrest one criminal hacker for the alleged theft of almost 20,000 credit card numbers that had been sent over the Internet.

Data and information are valuable corporate assets. The intentional use of illegal and destructive programs to alter or destroy data is as much a crime as destroying tangible goods. The most common of these programs are viruses and worms, which are software programs that, when loaded into a computer system, will destroy, interrupt, or cause errors in processing. Such programs are also called *malware*, and the growth rate for such programs is epidemic. Internet security firm McAfee estimates that 150 to 200 malware programs emerge each day.[31]

A **virus** is a computer program file capable of attaching to disks or other files and replicating itself repeatedly, typically without the user's knowledge or permission. Some viruses attach to files, so when the infected file executes, the virus also executes. Other viruses sit in a computer's memory and infect files as the computer opens, modifies, or creates the files. They are often disguised as games or images with clever or attention-grabbing titles such as "Boss, nude." Some viruses display symptoms, and some viruses damage files and computer systems. Computer viruses are written for several operating systems, including Windows, Macintosh, UNIX, and others.

virus
A computer program file capable of attaching to disks or other files and replicating itself repeatedly, typically without the user's knowledge or permission.

Virus writers can become very aggressive in their attacks. For example, a variant of the GPcode virus encrypts various file types including .doc, .txt, .pdf, .xls, and images, and then demands a ransom payment for the key required to decrypt the files.[32] An increasing problem is the purchase of computer equipment already infected with malware. Best Buy unknowingly sold digital picture frames that were infected with a computer virus during the manufacturing process.[33] Seagate Technology confirmed that many of its 500 GB hard drives left an Asian manufacturing plant infected with malware designed to steal online gaming passwords.

Worms are parasitic computer programs that replicate but, unlike viruses, do not infect other computer program files. Worms can create copies on the same computer or can send the copies to other computers via a network. Worms often spread via Internet Relay Chat (IRC).

worm
A parasitic computer program that can create copies of itself on the infected computer or send copies to other computers via a network.

A **Trojan horse** program is a malicious program that disguises itself as a useful application or game and purposefully does something the user does not expect. Trojans are not viruses because they do not replicate, but they can be just as destructive. Many people use the term to refer only to nonreplicating malicious programs, thus making a distinction between Trojans and viruses.

Trojan horse
A malicious program that disguises itself as a useful application or game and purposefully does something the user does not expect.

Although security is often cited as a strong point of the Mac computer, an increasing number of malware threats against the Mac OS X operating system have been uncovered. For example, an OS X Trojan horse is disguised either as an AppleScript known as ASthtv05 or bundled as an application named AStht_v06. When executed, this Trojan horse enables the attacker to remotely access the user's iSight camera, log keystrokes, retrieve screen shots, and manipulate file sharing settings.[34]

A *logic bomb* is a type of Trojan horse that executes when specific conditions occur. Triggers for logic bombs can include a change in a file by a particular series of keystrokes or at a specific time or date.

A *rootkit* is a set of programs that enable its user to gain administrator level access to a computer or network. Once installed, the attacker can gain full control of the system and even obscure the presence of the rootkit from legitimate system administrators. The Mebroot rootkit infects the master boot record, the first sector of the hard drive that the personal computer views before loading the operating system, making it all but invisible to security software and administrators. In an especially nefarious attack, hackers have created Web pages that when visited by users with certain browsers, release the Mebroot malware to infect the machine, a process known as a *drive-by download.*[35]

A *variant* is a modified version of a virus that is produced by the virus's author or another person who amends the original virus code. If changes are small, most antivirus products will also detect variants. However, if the changes are significant, the variant might go undetected by antivirus software.

The Storm worm is a Trojan horse that infects personal computers running the Microsoft operating systems. It began infecting computers via e-mail messages with a subject line about weather disasters in Europe, hence the name. Over time, and as users became wiser, the subject line of the malicious e-mail has changed several times. The e-mail contains an attachment that if opened loads a "cocktail" of various malware programs onto a personal computer. The result is that the computer is compromised and acts as a "zombie" computer under control of other computers. Such "zombies" are often used to send spam. It is estimated that as many as 40 million personal computers could have been infected by the Storm worm between January 2007 and February 2008.[36]

In some cases, a virus or a worm can completely halt the operation of a computer system or network for days or longer until the problem is found and repaired. In other cases, a virus or a worm can destroy important data and programs. If backups are inadequate, the data and programs might never be fully functional again. The costs include the effort required to identify and neutralize the virus or worm and to restore computer files and data, as well as the value of business lost because of unscheduled computer downtime.

The F-Secure Corporation provides centrally managed security solutions, and its products include antivirus, file encryption, and network security solutions for all major platforms—from desktops to servers and from laptops to handhelds. F-Secure is headquartered in Helsinki, Finland and provides real-time virus statistics on the most active viruses in the world at its Web site, *www.f-secure.com/virus-info/statistics.*

McAfee Security for Consumers is a division of Network Associates Inc. that delivers retail and online solutions designed to secure, protect, and optimize the computers of consumers and home office users. McAfee's retail desktop products include premier antivirus, security, encryption, and desktop optimization software. McAfee delivers software through an Internet browser to provide these services to users online through its Web site *www.mcafee.com,* one of the largest paid subscription sites on the Internet with over 2 million active paid subscribers. McAfee provides a real-time map of where the latest viruses are infecting computers worldwide at *http://us.mcafee.com/virusInfo/default.asp.* See Figure 14.1. The site also provides software for scanning your computer for viruses and tips on how to remove a virus.[37]

Using Antivirus Programs

As a result of the increasing threat of viruses and worms, most computer users and organizations have installed **antivirus programs** on their computers. Such software runs in the background to protect your computer from dangers lurking on the Internet and other possible sources of infected files. Some antivirus software is even capable of repairing common virus infections automatically, without interrupting your work. The latest virus definitions are downloaded automatically when you connect to the Internet, ensuring that your PC's

antivirus program
Software that runs in the background to protect your computer from dangers lurking on the Internet and other possible sources of infected files.

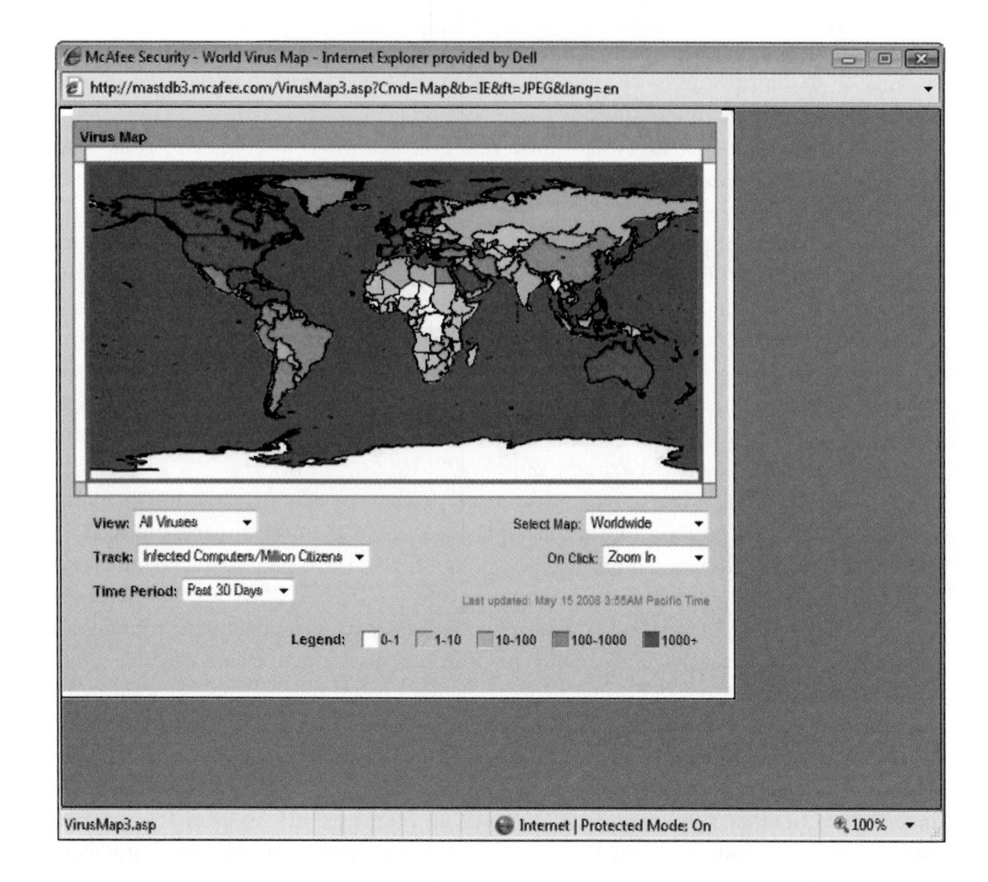

Figure 14.1

Global Virus Infections—Number of Infected Computers per Million Citizens

(Source: Courtesy of McAfee, Inc.)

protection is current. To safeguard your PC and prevent it from spreading viruses to your friends and coworkers, some antivirus software scans and cleans both incoming and outgoing e-mail messages. Some of the most highly rated antivirus software for 2007 includes BitDefender, Kaspersky, ESET Nod32, AVG Anti-Virus, F Secure Anti-Virus, Trend Micro, McAfee VirusScan, Norton AntiVirus, and CA Antivirus. This software can be purchased for $20 to $66.

Many e-mail services and ISP providers offer free antivirus protection. For example, AOL and MWEB (one of South Africa's leading ISPs) offer free antivirus software from McAfee.

Disk defragmentation reorganizes the contents of the disk to store the pieces of each file close together and contiguously. It also creates larger regions of free space. Tests have shown that antivirus scans run significantly faster on computers with regularly defragmented files and free space, reducing the time to do a complete scan by 18 to 58 minutes.[38] Consider running disk defragmentation software on a regular basis.

Proper use of antivirus software requires the following steps:

1. **Install antivirus software and run it often.** Many of these programs automatically check for viruses each time you boot up your computer or insert a disk or CD, and some even monitor all e-mail and file transmissions and copying operations.
2. **Update antivirus software often.** New viruses are created all the time, and antivirus software suppliers are constantly updating their software to detect and take action against these new viruses.
3. **Scan all removable media, including CDs, before copying or running programs from them.** Hiding on disks or CDs, viruses often move between systems. If you carry document or program files on removable media between computers at school or work and your home system, always scan them.
4. **Install software only from a sealed package or secure Web site of a known software company.** Even software publishers can unknowingly distribute viruses on their program disks or software downloads. Most scan their own systems, but viruses might still remain.

5. **Follow careful downloading practices.** If you download software from the Internet or a bulletin board, check your computer for viruses immediately after completing the transmission.
6. **If you detect a virus, take immediate action.** Early detection often allows you to remove a virus before it does any serious damage.

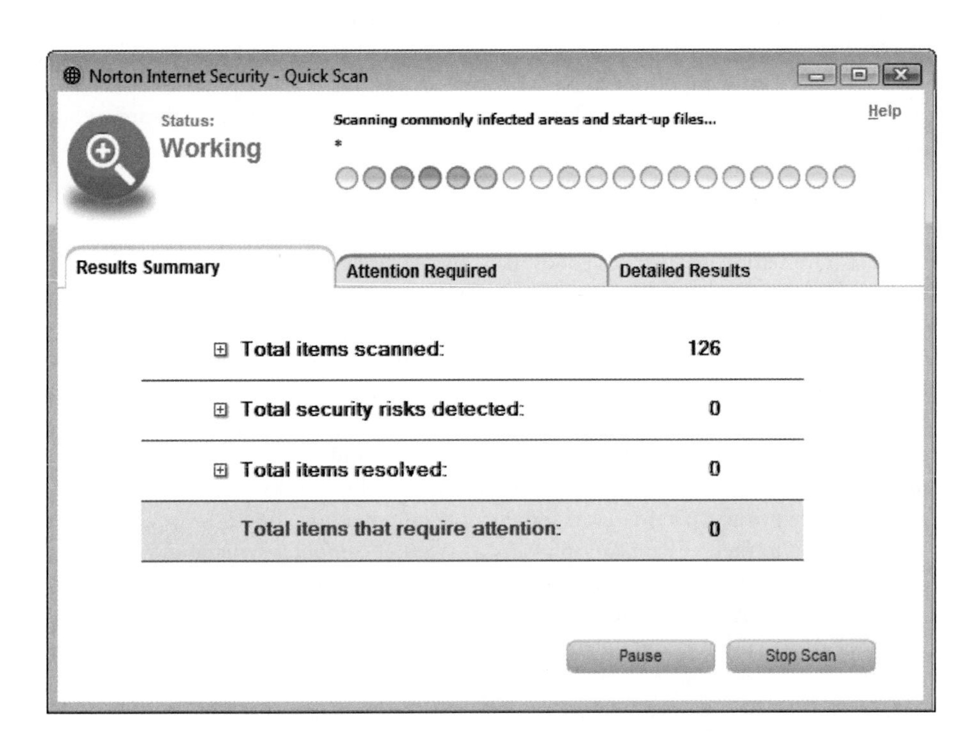

Antivirus software should be used and updated often.

Despite careful precautions, viruses can still cause problems. They can elude virus-scanning software by lurking almost anywhere in a system. Future antivirus programs might incorporate "nature-based models" that check for unusual or unfamiliar computer code. The advantage of this type of antivirus program is the ability to detect new viruses that are not part of an antivirus database.

Hoax, or false, viruses are another problem. Criminal hackers sometimes warn the public of a new and devastating virus that doesn't exist to create fear. Companies sometimes spend hundreds of hours warning employees and taking preventive action against a nonexistent virus. Security specialists recommend that IS personnel establish a formal paranoia policy to thwart virus panic among gullible end users. Such policies should emphasize that before users forward an e-mail alert to colleagues and higher-ups, they should send it to the help desk or the security team. The corporate intranet can be used to explain the difference between real viruses and fakes, and it can provide links to Web sites to set the record straight.

Be aware that virus writers also use known hoaxes to their advantage. For example, AOL4FREE began as a hoax virus warning. Then, a hacker distributed a destructive Trojan attached to the original hoax virus warning. Always remain vigilant and never open a suspicious attachment.

Spyware

Spyware is software installed on a personal computer to intercept or take partial control over the user's interaction with the computer without knowledge or permission of the user. Some forms of spyware secretly log keystrokes so that user name and passwords may be captured. Other forms of spyware record information about the user's Internet surfing habits and sites that have been visited. Still other forms of spyware change personal computer settings so that the user experiences slow connection speeds or is redirected to different Web pages than those expected. The number of personal computers infected with spyware has become epidemic, and users need to install anti-spyware software. The top four rated anti-spyware software for

spyware
Software that is installed on a personal computer to intercept or take partial control over the user's interaction with the computer without knowledge or permission of the user.

2008 includes Spy Sweeper, CounterSpy, Spyware Doctor, and SuperAntiSpyware, which cost in the range of $17 to $30.[39]

DirectRevenue was a major distributor of adware, a form of spyware that monitors the viewing habits of Internet users and displays targeted pop-up ads. The company received more than $80 million in ad revenue from its clients before it eventually ceased operations.[40] The company offered consumers free screensavers, games, and utility software but failed to disclose that downloading this software would load adware as well. Once it was installed, it was nearly impossible to identify, locate, and remove the adware. DirectRevenue agreed to a settlement with the FTC that barred future downloads of their adware without informed consent on the part of consumers. The firm was also fined $1.5 million.[41]

Information and Equipment Theft

Data and information are assets or goods that can also be stolen. People who illegally access systems often do so to steal data and information. To obtain illegal access, criminal hackers require identification numbers and passwords. Some criminals try different identification numbers and passwords until they find ones that work. Using password sniffers is another approach. A **password sniffer** is a small program hidden in a network or a computer system that records identification numbers and passwords. In a few days, a password sniffer can record hundreds or thousands of identification numbers and passwords. Using a password sniffer, a criminal hacker can gain access to computers and networks to steal data and information, invade privacy, plant viruses, and disrupt computer operations.

password sniffer
A small program hidden in a network or a computer system that records identification numbers and passwords.

In addition to theft of data and software, all types of computer systems and equipment have been stolen from offices. Portable computers such as laptops and portable storage devices (and the data and information stored in them) are especially easy for thieves to take. In many cases, the data and information stored in these systems are more valuable than the equipment and there is a risk that the data can be used in identity theft. In addition, the organization responsible receives a tremendous amount of negative publicity that can cause it to lose existing and potential future customers. Often, the responsible organization offers to pay for credit monitoring services for those people affected in an attempt to restore customer goodwill and avoid law suits.

Perhaps the worst single example in terms of number of people affected by theft of equipment was in May 2006, when the Department of Veterans Affairs announced that a laptop and hard drive containing some 26.5 million personal records of current and former members of the military were stolen.[42] Here are a few more examples of laptops stolen that contained personal information. In most cases, the laptops were left in plain view where others could see them and the data was not encrypted or protected in any manner.

- *August 2007*: The Connecticut Department of Revenue Services revealed that a laptop containing personally identifiable data about more than 106,000 taxpayers was missing.[43]
- *September 2007*: The Gap Inc. revealed that a laptop storing personal information on 800,000 job applicants was stolen from a contractor that managed job applicant data for the firm.[44]
- *December 2007*: Laptop computers were stolen from the Davidson County, Tennessee election office containing personal information for more than 337,000 registered voters.[45]
- *January 2008*: Horizon Blue Cross Blue Shield of New Jersey revealed that an employee laptop containing personal information of 300,000 clients was stolen.[46]

Many companies are putting into place tough measures to protect the data on laptops amid the epidemic of thefts. These policies include the following elements:

- Clear guidelines on what kind of data (and how much of it) can be stored on vulnerable laptops. In many cases, private data or company confidential data may not be downloaded to laptops that leave the office.
- Requiring that data stored on laptops be encrypted and doing spot checks to ensure that this policy is followed.

- Requiring that all laptops be secured using a lock and chain device so that they can not be easily removed from an office area.
- Providing training to employees and contractors on the need for safe handling of laptops and their data. For example, laptops should never be left in a position where they can be viewed by the public, such as on the front seat of an automobile.

In addition to the theft or loss of laptop computers, the U.S. Customs and Border Protection has increased the level of performing random inspection of electronic media. For years, U.S. agents have been taking and searching traveler's laptops, digital cameras, cell phones, PDAs, and other electronic devices. In some reported cases, the electronic devices of travelers entering the United States have been taken for two or more weeks to be inspected for evidence of child pornography or criminal or terrorist activity.[47]

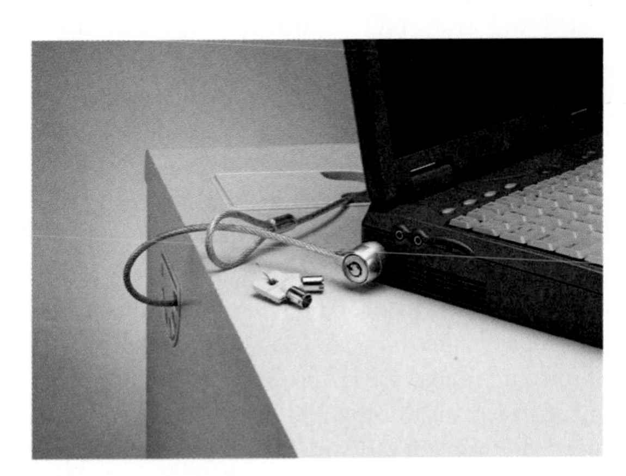

To fight computer crime, many companies use devices that disable the disk drive or lock the computer to the desk.

(Source: Courtesy of Kensington Technology Group.)

Safe Disposal of Personal Computers

Many companies donate personal computers they no longer need to schools, churches, or other organizations. Some sell them at a deep discount to their employees or put them up for sale on Internet auction sites such as eBay. However, care must be taken to ensure that all traces of any personal or company confidential data is completely removed. Simply deleting files and emptying the Recycle Bin does not make it impossible for determined individuals to view the data. Be sure to use disk-wiping software utilities that overwrite all sectors of your disk drive making all data unrecoverable. For example, Darik's Boot and Nuke (DBAN) is free and can be downloaded from the SourceForge Web site.

Patent and Copyright Violations

Works of the mind, such as art, books, films, formulas, inventions, music, and processes that are distinct and "owned" or created by a single person or group are called intellectual property. Copyright law protects authored works such as art, books, film, and music. Patent laws protect processes, machines, objects made by humans or machines, compositions of matter, and new uses of these items.

Each time you use a word processing program or access software on a network, you are taking advantage of someone else's intellectual property. Like books and movies—other intellectual properties—software is protected by copyright laws. Often, people who would never think of plagiarizing another author's written work have no qualms about using and copying software programs they have not paid for. Such illegal duplicators are called *pirates*; the act of unauthorized copying or distribution of copyrighted software is called **software piracy**.

software piracy
The act of unauthorized copying or distribution of copyrighted software

Software piracy often involves the copying, downloading, sharing, selling, or installing of multiple copies onto personal or work computers. When you purchase software, you are purchasing a license to use it; you do not own the actual software. The license states how many times you can install the software. If you make more copies of the software than the license permits, you are pirating.[48]

The Software and Information Industry Alliance (SIIA) was the original software antipiracy organization, formed and financed by many of the large software publishers. Microsoft financed the formation of a second antipiracy organization, the Business Software Alliance (BSA). The BSA, through intense publicity, has become the more prominent organization. Other software companies, including Apple, Adobe, Hewlett-Packard, and IBM, now contribute to the BSA. The BSA estimates that the software industry lost over $48 billion in 2007 in revenue to worldwide software piracy. "Worldwide, for every two dollars of software purchased legitimately, one dollar was obtained illegally."[49]

Penalties for software piracy can be severe. If the copyright owner brings a civil action against someone, the owner can seek to stop the person from using its software immediately and can also request monetary damages. The copyright owner can then choose between compensation for actual damages—which includes the amount it has lost because of the person's infringement, as well as any profits attributable to the infringement—and statutory damages, which can be as much as $150,000 for each program copied. In addition, the government can prosecute software pirates in criminal court for copyright infringement. If convicted, they could be fined up to $250,000 or sentenced to jail for up to five years, or both.[50]

The Web site *www.MDofPC.com* was caught selling unlicensed copies of software from Adobe, McAfee, Microsoft, and Symantec. Visitors to the site could pay online and then be given access to the software for downloading when their payment cleared. The BSA initiated an investigation that led eventually to a $36,000 fine against the operators of the site.[51] The Acorn Engineering Company agreed to pay $250,000 to BSA to settle claims that it had unlicensed copies of software from Adobe, Autodesk, and Microsoft on its computers.[52]

Another major issue in regards to copyright infringement is the downloading of music that is copyright protected. Estimates vary widely as to how much music piracy is costing the recording industry. An estimate from the Institute for Policy Innovation (an economic public policy organization) is that the recording industry loses about $5.3 billion and retailers lose about $1.0 billion, for a total direct loss of $6.3 billion. In addition, the U.S. government loses about $422 million in tax revenue.[53]

Operation Copycat is an ongoing undercover investigation into Warez groups, which are online organizations engaged in the illegal uploading, copying, and distribution of copyrighted works such as music, movies, games, and software, often even before they are released to the public. The investigation is led by the Computer Hacking and Intellectual Property (CHIP) Unit of the United States Attorney's Office and the FBI. Operation Copycat has resulted in 40 convictions over a period of three years from July 2005 to July 2008. Those convicted have typically been sentenced to over a year in prison, required to pay fines in excess of $200,000, and had to forfeit all computer and other equipment used in committing the offenses.[54]

The Motion Picture Association of America estimates that it loses over $18 billion per year from movie theft. It won a $100 million judgment against TorrentSpy for offering thousands of copyright-protected movies and TV shows.[55]

Patent infringement is also a major problem for computer software and hardware manufacturers. It occurs when someone makes unauthorized use of another's patent. If a court determines that a patent infringement is intentional, it can award up to three times the amount of damages claimed by the patent holder. It is not unusual to see patent infringement awards in excess of $10 million.

To obtain a patent or to determine if a patent exists in an area a company seeks to exploit requires a lengthy (typically longer than 25 months) search by the U.S. Patent Office. Indeed, the patent process is so controversial that there is a broad consensus among manufacturing firms, the financial community, consumer and public interest groups and government leaders demanding patent reform. Here are just a few examples of numerous recent lawsuits involving patent infringement.

- Personal computer manufacturers Acer, Apple, Dell, and Hewlett-Packard were sued for allegedly violating four patents related to a wireless communications privacy method and system held by Saxon Innovations.[56]

- Telecommunications network equipment maker Tellabs filed a patent infringement lawsuit against Fujitsu in regards to technology associated with optical and multiplexing systems and equipment.[57]
- Red Hat settled patent infringement claims over business process software with Firestar Software and DataTern.[58]

Computer-Related Scams

People have lost hundreds of thousands of dollars on real estate, travel, stock, and other business scams. Today, many of these scams are being perpetrated with computers. Using the Internet, scam artists offer get-rich-quick schemes involving bogus real estate deals, tout "free" vacations with huge hidden costs, commit bank fraud, offer fake telephone lotteries, sell worthless penny stocks, and promote illegal tax-avoidance schemes.

Over the past few years, credit card customers of various banks have been targeted by scam artists trying to get personal information needed to use their credit cards. The scam works by sending customers an e-mail including a link that seems to direct users to their bank's Web site. At the site, they are greeted with a pop-up box asking them for their full debit card numbers, their personal identification numbers, and their credit card expiration dates. The problem is that the Web site customers are directed to is a fake site operated by someone trying to gain access to their private information. As discussed previously, this form of scam is called *phishing*. According to the IT research firm Gartner, more than an estimated 124 million people in the U.S. received phishing e-mails and some 3.6 million of them lost a total of $3.2 billion during 2007.[59] One common phishing scam involves e-mails claiming to be from eBay's security team and warning recipients that they have a security issue to resolve. The e-mail includes a link urging the recipient to take action. Clicking the link takes the user to a page requesting personal information that, if provided, compromises the victim's identity.[60]

The following is a list of tips to help you avoid becoming a scam victim:

- Don't agree to anything in a high-pressure meeting or seminar. Insist on having time to think it over and to discuss things with your spouse, partner, or attorney. If a company won't give you the time you need to check out an offer and think things over, you don't want to do business with them. A good deal now will be a good deal tomorrow; the only reason for rushing you is if the company has something to hide.
- Don't judge a company based on appearances. Flashy Web sites can be created and published in a matter of days. After a few weeks of taking money, a site can vanish without a trace in just a few minutes. You might find that the perfect money-making opportunity offered on a Web site was a money maker for the crook and a money loser for you.
- Avoid any plan that pays commissions simply for recruiting additional distributors. Your primary source of income should be your own product sales. If the earnings are not made primarily by sales of goods or services to consumers or sales by distributors under you, you might be dealing with an illegal pyramid.
- Beware of shills, people paid by a company to lie about how much they've earned and how easy the plan was to operate. Check with an independent source to make sure that you aren't having the wool pulled over your eyes.
- Beware of a company's claim that it can set you up in a profitable home-based business but that you must first pay up front to attend a seminar and buy expensive materials. Frequently, seminars are high-pressure sales pitches, and the material is so general that it is worthless.
- If you are interested in starting a home-based business, get a complete description of the work involved before you send any money. You might find that what you are asked to do after you pay is far different from what was stated in the ad. You should never have to pay for a job description or for needed materials.
- Get in writing the refund, buy-back, and cancellation policies of any company you deal with. Do not depend on oral promises.
- Do your homework. Check with your state attorney general and the National Fraud Information Center before getting involved, especially when the claims about a product or potential earnings seem too good to be true.

If you need advice about an Internet or online solicitation, or if you want to report a possible scam, use the Online Reporting Form or Online Question & Suggestion Form features on the Web site for the National Fraud Information Center at *http://fraud.org*, or call the NFIC hotline at 1-800-876-7060.

International Computer Crime

Computer crime is also an international issue, and it becomes more complex when it crosses borders. As already mentioned, the software industry loses about $11–12 billion in revenue to software piracy annually, with about $9 billion of that occurring outside the United States.

With the increase in electronic cash and funds transfer, some are concerned that terrorists, international drug dealers, and other criminals are using information systems to launder illegally obtained funds. Computer Associates International developed software called CleverPath for Global Compliance for customers in the finance, banking, and insurance industries to eliminate money laundering and fraud. Companies that are required to comply with legislation such as the USA Patriot Act and Sarbanes-Oxley Act might lack the resources and processes to do so. The software automates manual tracking and auditing processes that are required by regulatory agencies and helps companies handle frequently changing reporting regulations. The application can drill into a company's transactions and detect transaction patterns that suggest fraud or other illegal activities based on built-in business rules and predictive analysis. Suspected fraud cases are identified and passed on to the appropriate personnel for action to thwart criminals and help companies avoid paying fines.

PREVENTING COMPUTER-RELATED CRIME

Because of increased computer use today, greater emphasis is placed on the prevention and detection of computer crime. Although all states have passed computer crime legislation, some believe that these laws are not effective because companies do not always actively detect and pursue computer crime, security is inadequate, and convicted criminals are not severely punished. However, all over the United States, private users, companies, employees, and public officials are making individual and group efforts to curb computer crime, and recent efforts have met with some success.

Crime Prevention by State and Federal Agencies

State and federal agencies have begun aggressive attacks on computer criminals, including criminal hackers of all ages. In 1986, Congress enacted the Computer Fraud and Abuse Act, which mandates punishment based on the victim's dollar loss.

For at least five years after the September 11, 2001 terrorist attacks, the U.S. Treasury Department and CIA executed a program called the Terrorist Finance Tracking Program that relied on data in international money transfers from the Society for Worldwide Interbank Financial Telecommunications. The goal of the program was to track and combat terrorist financing. The program was credited with helping to capture at least two terrorists; however, revelation of the secret program's existence stirred up controversy and rendered the program ineffective.

The Department of Defense also supports the Computer Emergency Response Team (CERT), which responds to network security breaches and monitors systems for emerging threats. Law enforcement agencies are also increasing their efforts to stop criminal hackers, and many states are now passing new, comprehensive bills to help eliminate computer crimes. A complete listing of computer-related legislation by state can be found at *www.onlinesecurity.com/forum/article46.php*. Recent court cases and police reports involving computer crime show that lawmakers are ready to introduce newer and tougher computer crime legislation.

International Cyber Espionage

In 2007 and 2008, businesses and government agencies in many countries experienced a spike in targeted attacks originating outside their borders, many from China. Analysis of the attacks lead security experts to believe that many governments are involved in cyber espionage—that is, the use of the Internet to spy on other governments. Not only is the Internet being leveraged for international espionage, but it is also being used for economic espionage. Economic espionage refers to the use of the Internet by nation-states to steal corporate information in an effort to gain economic advantages in multinational deals.

The SANS Institute, a leading information security research group, ranked cyber espionage number three on its list of the Top Ten Cyber Menaces for 2008. Number one was Web site attacks that exploit Web browser vulnerabilities to install malware on PCs, and number two was botnets. Consider the following examples of recent cyber espionage and economic espionage.

The U.K. government has accused the Chinese of hacking into the computer systems of "some of its leading companies." The Director-General of intelligence agency MI5 sent letters to 300 chief executives and security chiefs of financial institutions warning of a sharp rise in instances of electronic espionage. The organization believes that at least 20 foreign intelligence services are engaged in cyber espionage against "U.K. interests." The organization is most concerned about Russia and China.

One report describes how Chinese hackers infected the Rolls-Royce corporate network with a Trojan horse that sent secret corporate information from the network to a remote server. Shell Oil Company discovered a Chinese cyber spy ring in Houston, Texas, working to steal confidential pricing information from servers at its operations in Africa.

Attacks and hacks against the Pentagon's computer system, the Oak Ridge National Lab, and Los Alamos National Lab, where U.S. nuclear weapons technology is developed, have all been traced to China. Information about top scientists was stolen from the Oak Ridge Lab. Germany, France, and New Zealand have reported similar attacks originating in China.

One of the favorite tools for cyber and economic espionage is a rootkit that works at a low level in the computer system, intercepting messages between the operating system and security software. The rootkit avoids detection, while sending secure data out a backdoor of the network to the hacker's server over the Internet—often located in China, and sometimes to servers registered to the Chinese government. These rootkits can make their way into private networks by tricking employees into visiting Web sites or opening attachments containing the rootkit. The trick typically involves detailed and custom-designed phishing e-mail messages that use social-engineered knowledge to persuade the recipient that the e-mail is legitimate.

Although it would be easy to jump to the conclusion that the Chinese government is behind all of these attacks, experts are quick to point out that it is difficult to pinpoint the origin of an attack. The Internet makes it possible for hackers to launch attacks from any server in the world. If an attack originates in China and is engineered by a Chinese citizen, it still cannot be determined if that person is working for the government. The Chinese government vehemently denies any part in cyber espionage. Still, most governments hold the Chinese government accountable for not cracking down on hackers if not actually sponsoring them. It is estimated that 30 percent of malicious software is created in China. The next largest distributor of malware is Russia and Eastern Europe.

Even if the Chinese government is actively involved in cyber-espionage, as many governments are accusing, it is far from alone. A report developed by security firm McAfee states that "120 countries are developing ways to use the Internet as a weapon to target financial markets, government computer systems, and utilities." A number of experts are calling this the "cyber cold war."

Government agencies and businesses that may be targeted by cyber espionage and economic espionage are advised to use data-leak prevention products and database-monitoring tools. These tools lock down data and prevent copies from leaving the network. Some companies have gone as far as maintaining two separate networks, one for secure data and the other for Internet communications. This prevents malware from secretly funneling data out network backdoors to hackers.

Discussion Questions

1. How do cyber espionage and economic espionage differ?
2. What tricks are used by hackers to infiltrate systems and gain access to private information?

Critical Thinking Questions

1. What are the dangers if the cyber cold war turns into an actual cyber war?
2. Why are countries and businesses concerned about cyber-espionage that originates in China and Russia?

Sources: Messmer, Ellen, "Cyber espionage seen as growing threat to business, government," *Network World*, January 17, 2008, *www.networkworld.com/news/2008/011708-cyberespionage.html*; Kirk, Jeremy, "Shell, Rolls-Royce reportedly hacked by Chinese spies," *Computerworld*, December 3, 2007, *www.computerworld.com/action/article.do?command=viewArticleBasic&articleId=9050 538&intsrc=news_list*; Dahdah, Howard, "UK government accuses Chinese of IT espionage," *Computerworld*, December 3, 2007, *www.computerworld.com/action/article.do?command=viewArticleBasic&articleId=9050499&intsrc=news_list*; Reimer, Jeremy, "Chinese government at the center of five cyber attack claims," *Ars Technica*, September 14, 2007, *http://arstechnica.com/news.ars/post/20070914-chinese-government-at-the-center-of-five-cyber-attack-claims.html*; Griffiths, Peter, "World faces 'cyber cold war' threat," Reuters, November 29, 2007, *www.reuters.com/article/technologyNews/idUSL2932083320071129?feedType=RSS&fee dName=technologyNews*.

Crime Prevention by Corporations

Companies are also taking crime-fighting efforts seriously. Many businesses have designed procedures and specialized hardware and software to protect their corporate data and systems. Specialized hardware and software, such as encryption devices, can be used to encode data and information to help prevent unauthorized use. As discussed in Chapter 7, encryption is the process of converting an original electronic message into a form that can be understood only by the intended recipients. A key is a variable value that is applied using an algorithm to a string or block of unencrypted text to produce encrypted text or to decrypt encrypted text. Encryption methods rely on the limitations of computing power for their effectiveness—if breaking a code requires too much computing power, even the most determined code crackers will not be successful. The length of the key used to encode and decode messages determines the strength of the encryption algorithm.

As employees move from one position to another at a company, they can build up access to multiple systems if inadequate security procedures fail to revoke access privileges. It is clearly not appropriate for people who have changed positions and responsibilities to still have access to systems they no longer use. To avoid this problem, many organizations create role-based system access lists so that only people filling a particular role (e.g., invoice approver) can access a specific system.

Fingerprint authentication devices provide security in the PC environment by using fingerprint recognition instead of passwords. Laptop computers from Lenovo, Toshiba, and others have built-in fingerprint readers used to log on and gain access to the computer system and its data. The JetFlash 210 Fingerprint USB Flash Drive requires users to swipe their fingerprints and match them to one of up to 10 trusted users to access the data. The data on the flash drive can also be encrypted for further protection.[61]

Fingerprint authentication devices provide security in the PC environment by using fingerprint recognition instead of passwords.

(Source: Permission granted by Pay By Touch.)

Crime-fighting procedures usually require additional controls on the information system. Before designing and implementing controls, organizations must consider the types of computer-related crime that might occur, the consequences of these crimes, and the cost and complexity of needed controls. In most cases, organizations conclude that the trade-off between crime and the additional cost and complexity weighs in favor of better system controls. Having knowledge of some of the methods used to commit crime is also helpful in preventing, detecting, and developing systems resistant to computer crime (see Table 14.1). Some companies actually hire former criminals to thwart other criminals.

Although the number of potential computer crimes appears to be limitless, the actual methods used to commit crime are limited. The following list provides a set of useful guidelines to protect your computer from criminal hackers.

- Install strong user authentication and encryption capabilities on your firewall.
- Install the latest security patches, which are often available at the vendor's Internet site.
- Disable guest accounts and null user accounts that let intruders access the network without a password.

Methods	Examples
Add, delete, or change inputs to the computer system.	Delete records of absences from class in a student's school records.
Modify or develop computer programs that commit the crime.	Change a bank's program for calculating interest to make it deposit rounded amounts in the criminal's account.
Alter or modify the data files used by the computer system.	Change a student's grade from C to A.
Operate the computer system in such a way as to commit computer crime.	Access a restricted government computer system.
Divert or misuse valid output from the computer system.	Steal discarded printouts of customer records from a company trash bin.
Steal computer resources, including hardware, software, and time on computer equipment.	Make illegal copies of a software program without paying for its use.
Offer worthless products for sale over the Internet.	Send e-mail requesting money for worthless hair growth product.
Blackmail executives to prevent release of harmful information.	Eavesdrop on organization's wireless network to capture competitive data or scandalous information.
Blackmail company to prevent loss of computer-based information.	Plant logic bomb and send letter threatening to set it off unless paid considerable sum.

Table 14.1

Common Methods Used to Commit Computer Crimes

- Do not provide overfriendly logon procedures for remote users (e.g., an organization that used the word *welcome* on their initial logon screen found they had difficulty prosecuting a criminal hacker).
- Restrict physical access to the server and configure it so that breaking into one server won't compromise the whole network.
- Give each application (e-mail, File Transfer Protocol, and domain name server) its own dedicated server.
- Turn audit trails on.
- Consider installing caller ID.
- Install a corporate firewall between your corporate network and the Internet.
- Install antivirus software on all computers and regularly download vendor updates.
- Conduct regular IS security audits.
- Verify and exercise frequent data backups for critical data.

Using Intrusion Detection Software

An **intrusion detection system (IDS)** monitors system and network resources and notifies network security personnel when it senses a possible intrusion. Examples of suspicious activities include repeated failed logon attempts, attempts to download a program to a server, and access to a system at unusual hours. Such activities generate alarms that are captured on log files. Intrusion detection systems send an alarm, often by e-mail or pager, to network security personnel when they detect an apparent attack. Unfortunately, many IDSs frequently provide false alarms that result in wasted effort. If the attack is real, network security personnel must make a decision about what to do to resist the attack. Any delay in response increases the probability of damage from a criminal hacker attack. Use of an IDS provides another layer of protection in the event that an intruder gets past the outer security layers—passwords, security procedures, and corporate firewall.

intrusion detection system (IDS)
Software that monitors system and network resources and notifies network security personnel when it senses a possible intrusion.

A firm called Internet Security Systems (ISS) manages security for other organizations through its Managed Protection Services. The company's IDSs are designed to recognize 30 of the most-critical threats, including worms that go after Microsoft software and those that exploit Apache Web servers and other programs. When an attack is detected, the service automatically blocks it without requiring human intervention. Taking the manual intervention step out of the process enables a faster response and minimizes damage from a criminal hacker. To encourage customers to adopt its service, ISS guaranteed up to $50,000 in cash if the prevention service failed.

Security Dashboard

security dashboard
Software that provides a comprehensive display on a single computer screen of all the vital data related to an organization's security defenses including threats, exposures, policy compliance and incident alerts.

Many organizations employ **security dashboard** software to provide a comprehensive display on a single computer screen of all the vital data related to an organization's security defenses including threats, exposures, policy compliance, and incident alerts. The goal is to reduce the effort required for monitoring and to identify threats earlier. Data comes from a variety of sources including firewalls, applications, servers, and other software and hardware devices.

Figure 14.2

The Computer Network Defence Internet Operational Picture

The Computer Network Defence Internet Operational Picture, a security dashboard designed for the United Kingdom government and military networks, displays near real-time information on new and emerging cyber threats.

Associated Newspapers publishes six of the United Kingdom's largest newspapers that deliver timely information to some 6 million daily subscribers. Its journalists work in many countries and time zones. The organization implemented a security dashboard to cut the potential of interruption to its news cycle and raise the security protection of its news stories. As Mark Callaby, IT Security Officer, states: "We have a diverse IT infrastructure, which makes it difficult to track the current status of system patches and identify potential vulnerabilities. We needed to improve our ability to detect spyware, as well as establish a centralized view of our infrastructure and its security status."[62]

Using Managed Security Service Providers (MSSPs)

Keeping up with computer criminals—and with new regulations—can be daunting for organizations. Criminal hackers are constantly poking and prodding, trying to breach the security defenses of companies. Also, such recent legislation as HIPAA, Sarbanes-Oxley, and the USA Patriot Act requires businesses to prove that they are securing their data. For most small and mid-sized organizations, the level of in-house network security expertise needed to protect their business operations can be quite costly to acquire and maintain. As a result, many are outsourcing their network security operations to managed security service providers (MSSPs) such as Counterpane, Guardent, Internet Security Services, Riptech, and Symantec. MSSPs monitor, manage, and maintain network security for both hardware and software. These companies provide a valuable service for IS departments drowning in reams of alerts and false alarms coming from virtual private networks (VPNs); antivirus, firewall, and intrusion detection systems; and other security monitoring systems. In addition, some provide vulnerability scanning and Web blocking/filtering capabilities.

Filtering and Classifying Internet Content

To help parents control what their children see on the Internet, some companies provide *filtering software* to help screen Internet content. Many of these screening programs also prevent children from sending personal information over e-mail or through chat groups. This stops children from broadcasting their name, address, phone number, or other personal information over the Internet. The two approaches used are filtering, which blocks certain Web sites, and rating, which places a rating on Web sites. According to the 2004 Internet Filter Review, the five top-rated filtering software packages are, in order: ContentProtect, Cybersitter, Net Nanny, CyberPatrol, and FilterPack.

Business organizations also implement filtering software to prevent employees from visiting nonwork-related Web sites, particularly those related to gambling or those containing pornographic or other offensive material. Before implementing Web site blocking, the users must be informed about the company's policies and why they exist. It is best if the organization's Internet users, management, and IS organization work together to define the policy to be implemented. The policy should be clear about the repercussions to employees who attempt to circumvent the blocking measures.

The Internet Content Rating Association (ICRA) is a nonprofit organization whose members include Internet industry leaders such as America Online, Bell South, British Telecom, IBM, Microsoft, UUNet, and Verizon. Its specific goals are to protect children from potentially harmful material, while also safeguarding free speech on the Internet. Using the ICRA rating system, Web authors fill out an online questionnaire describing the content of their site—what is and isn't present. The broad topics covered include chat capabilities, the language used, nudity and sexual content, violence depicted, and other areas such as alcohol, drugs, gambling, and suicide. Based on the authors' responses, ICRA then generates a content label (a short piece of computer code) that the authors add to their site. Internet users (and parents) can then set their browser to allow or disallow access to Web sites based on the objective rating information declared in the content label and their own subjective preferences. Reliance on Web site authors to do their own rating has its weaknesses, though. Web site authors can lie when completing the ICRA questionnaire so that their site receives a content label that doesn't accurately reflect the site's content. In addition, many hate group and sexually explicit sites don't have an ICRA rating, so they will not be blocked unless a browser is set to block all unrated sites. Also, this option would block out so many acceptable sites that it could make Web surfing useless. For these reasons, at this time, site labeling is at best a complement to other filtering techniques.

The U.S. Congress has made several attempts to limit children's exposure to online pornography including the Communications Decency Act (enacted 1996) and the Child Online Protection Act (enacted 1998). Within two years of their being enacted, the U.S. Supreme Court found that both these acts violated the First Amendment (freedom of speech) and ruled them to be unconstitutional. The Children's Internet Protection Act (CIPA) was signed into law in 2000 and later upheld by the Supreme Court in 2003. Under CIPA, schools and libraries subject to CIPA do not receive the discounts offered by the "E-Rate" program unless they certify that they have certain Internet safety measures in place to block or filter "visual depictions that are obscene, child pornography, or are harmful to minors."[63] (The E-Rate program provides many schools and libraries support to purchase Internet access and computers).

The Yorba Linda Library Commission applied for and began receiving discounts on Internet access through the federally funded E-Rate program in July 2007. The estimated savings exceeds $10,000. As a result, the library plans to install filter software on its 28 computers for the safety of computer users under the age of 18.[64]

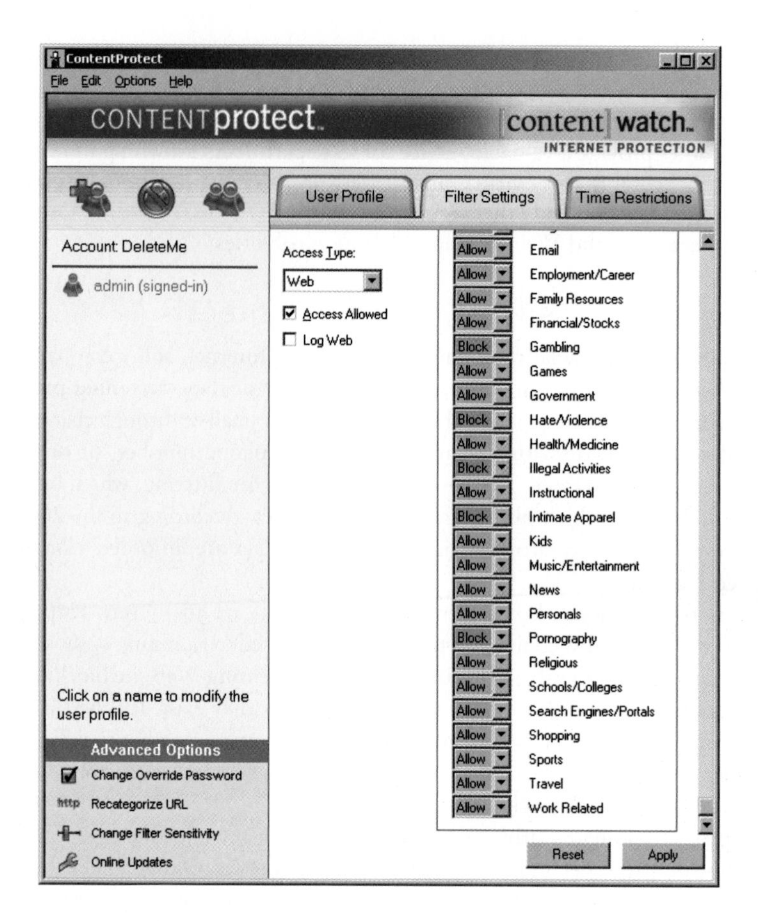

ContentProtect is a filtering software program that helps block unwanted Internet content from children and young adults.

(Source: Courtesy of ContentWatch Inc.)

Internet Libel Concerns

With the increased popularity of networks and the Internet, libel becomes an important legal issue. A publisher, such as a newspaper, can be sued for libel, which involves publishing an intentionally false written statement that is damaging to a person's reputation. Generally, a bookstore cannot be held liable for statements made in newspapers or other publications it sells. Online services, such as CompuServe and America Online, might exercise some control over who puts information on their service but might not have direct control over the content of what is published by others on their service. So, can online services be sued for libel for content that someone else publishes on their service? Do online services more closely resemble a newspaper or a bookstore? This legal issue has not been completely resolved, but some court cases have been decided. The *Cubby, Inc. v. CompuServe* case ruled that CompuServe was similar to a bookstore and not liable for content put on its service by others. In this case, the judge stated, "While CompuServe can decline to carry a given publication altogether, in reality, after it does decide to carry a given publication, it will have little or no editorial control over that publication's content." This case set a legal precedent that has been applied in similar, subsequent cases. Companies should be aware that publishing Internet content to the world can subject them to different countries' laws in the same way that exporting physical products does.

Geolocation tools match the user's IP address with outside information to determine the actual geographic location of the online user where the customer's computer signal enters the Internet. This enables someone to identify the user's actual location within approximately 50 miles. Internet publishers can now limit the reach of their published speech to avoid potential legal risks. Use of such technology is also dividing the global Internet into separate content regions, with readers in Brazil, Japan, and the United States all receiving variations of the same information from the same publisher.

Individuals, too, must be careful what they post on the Internet to avoid libel charges. In many cases, disgruntled former employees are being sued by their former employers for material posted on the Internet.

Preventing Crime on the Internet

As mentioned in Chapter 7, Internet security can include firewalls and many methods to secure financial transactions. A firewall can include both hardware and software that act as a barrier between an organization's information system and the outside world. Some systems have been developed to safeguard financial transactions on the Internet.

To help prevent crime on the Internet, the following steps can be taken:

1. Develop effective Internet usage and security policies for all employees.
2. Use a stand-alone firewall (hardware and software) with network monitoring capabilities.
3. Deploy intrusion detection systems, monitor them, and follow up on their alarms.
4. Monitor managers and employees to make sure that they are using the Internet for business purposes.
5. Use Internet security specialists to perform audits of all Internet and network activities.

Even with these precautions, computers and networks can never be completely protected against crime. One of the biggest threats is from employees. Although firewalls provide good perimeter control to prevent crime from the outside, procedures and protection measures are needed to protect against computer crime by employees. Passwords, identification numbers, and tighter control of employees and managers also help prevent Internet-related crime.

PRIVACY ISSUES

Another important social issue in information systems involves privacy. In 1890, U.S. Supreme Court Justice Louis Brandeis stated that the "right to be left alone" is one of the most "comprehensive of rights and the most valued by civilized man." Basically, the issue of privacy deals with this right to be left alone or to be withdrawn from public view. With information systems, privacy deals with the collection and use or misuse of data. Data is constantly being collected and stored on each of us. This data is often distributed over easily accessed networks and without our knowledge or consent. Concerns of privacy regarding this data must be addressed. For example, the U.S. Department of Health and Human Services has received over 26,000 complaints of medical privacy breaches since new privacy rules went into effect in 2003.[65]

With today's computers, the right to privacy is an especially challenging problem. More data and information are produced and used today than ever before. When someone is born, takes certain high school exams, starts a job, enrolls in a college course, applies for a driver's license, purchases a car, serves in the military, gets married, buys insurance, gets a library card, applies for a charge card or loan, buys a house, or merely purchases certain products, data is collected and stored somewhere in computer databases. A difficult question to answer is, "Who owns this information and knowledge?" If a public or private organization spends time and resources to obtain data on you, does the organization own the data, and can it use the data in any way it desires? Government legislation answers these questions to some extent for federal agencies, but the questions remain unanswered for private organizations.

Privacy and the Federal Government

The federal government is the largest collector of data in the United States. Over 4 billion records exist on citizens, collected by about 100 federal agencies, ranging from the Bureau of Alcohol, Tobacco, and Firearms to the Veterans Administration. Other data collectors include state and local governments and commercial and nonprofit organizations of all types and sizes. The government must be on guard at all times to safeguard this data. For example, two workers were fired at the State Department when electronic monitoring detected unauthorized accessing of the personal passport information of three 2008 presidential candidates.[66]

The European Union has a data-protection directive that requires firms transporting data across national boundaries to have certain privacy procedures in place. This directive affects virtually any company doing business in Europe, and it is driving much of the attention being given to privacy in the United States.

Privacy at Work

The right to privacy at work is also an important issue. Currently, the rights of workers who want their privacy and the interests of companies that demand to know more about their employees are in conflict. A recent poll uncovered that 78 percent of companies monitor their employees while at work in one form or another.[67] According to another recent survey, nearly one-third of companies have fired an employee for violating corporate e-mail policies.[68] Statistics such as these have raised employee concerns. For example, workers might find that they are being closely monitored via computer technology. These computer-monitoring systems tie directly into workstations; specialized computer programs can track every keystroke made by a user. This type of system can determine what workers are doing while at the keyboard. The system also knows when the worker is not using the keyboard or computer system. These systems can estimate what people are doing and how many breaks they are taking. Needless to say, many workers consider this close supervision very dehumanizing.

E-Mail Privacy

E-mail also raises some interesting issues about work privacy. Federal law permits employers to monitor e-mail sent and received by employees. Furthermore, e-mail messages that have been erased from hard disks can be retrieved and used in lawsuits because the laws of discovery demand that companies produce all relevant business documents. On the other hand, the use of e-mail among public officials might violate "open meeting" laws. These laws, which apply to many local, state, and federal agencies, prevent public officials from meeting in private about matters that affect the state or local area.

E-mail has changed how workers and managers communicate in the same building or around the world. E-mail, however, can be monitored and intercepted. As with other services—such as cellular phones—the convenience of e-mail must be balanced with the potential of privacy invasion.

(Source: © Gary Conner/Photo Edit.)

Instant Messaging Privacy

Using instant messaging (IM) to send and receive messages, files, and images introduces the same privacy issues associated with e-mail. As with e-mail, federal law permits employers to monitor instant messages sent and received by employees. Do not send personal or private IMs at work. Other significant privacy issues depend on the instant messaging client that you use. For example, at one time AOL and ICQ stated in their privacy policy that "You waive any right to privacy" and that they may use your instant messages in any way they see fit. Here are a few other tips:

- Choose a nonrevealing, nongender-specific, unprovocative IM screen name (Sweet Sixteen, 2hot4u, UCLAMBA, all fail this test).
- Don't send messages you would be embarrassed to have your colleagues or significant other read.
- Do not open files or click links in messages from people you do not know.
- Never send sensitive personal data such as credit card numbers, bank account numbers, or passwords via IM.

Privacy and Personal Sensing Devices

RFID tags, essentially microchips with antenna, are embedded in many of the products we buy such as medicine containers, clothing, computer printers, car keys, library books, and tires. RFID tags generate radio transmissions that if appropriate measures are not taken, can lead to potential privacy concerns. Once these tags are associated with the individual who purchased the item, someone can potentially track individuals by the unique identifier associated with the RFID chip.

Several states have reacted to the potential for abuse of RFID tags by going so far as passing legislation prohibiting the implantation of RFID chips under people's skin without their approval.[69]

Privacy and the Internet

Some people assume that there is no privacy on the Internet and that you use it at your own risk. Others believe that companies with Web sites should have strict privacy procedures and be accountable for privacy invasion. Regardless of your view, the potential for privacy invasion on the Internet is huge. People wanting to invade your privacy could be anyone from criminal hackers to marketing companies to corporate bosses. Your personal and professional information can be seized on the Internet without your knowledge or consent. E-mail is a prime target, as discussed previously. Sending an e-mail message is like having an open conversation in a large room—people can listen to your messages. When you visit a Web site on the Internet, information about you and your computer can be captured. When this information is combined with other information, companies can know what you read, what products you buy, and what your interests are.

Most people who buy products on the Web say it's very important for a site to have a policy explaining how personal information is used, and the policy statement must make people feel comfortable and be extremely clear about what information is collected and what will and will not be done with it. However, many Web sites still do not prominently display their privacy policy or implement practices completely consistent with that policy. The real issue that Internet users need to be concerned with is—what do content providers want with their personal information? If a site requests that you provide your name and address, you have every right to know why and what will be done with it. If you buy something and provide a shipping address, will it be sold to other retailers? Will your e-mail address be sold on a list of active Internet shoppers? And if so, you should realize that it's no different than the lists compiled from the orders you place with catalog retailers. You have the right to be taken off any mailing list.

A potential solution to some consumer privacy concerns is the screening technology called the **Platform for Privacy Preferences (P3P)** being proposed to shield users from sites that don't provide the level of privacy protection they desire. Instead of forcing users to find and read through the privacy policy for each site they visit, P3P software in a computer's browser will download the privacy policy from each site, scan it, and notify the user if the policy does not match his preferences. (Of course, unethical marketers can post a privacy policy that does not accurately reflect the manner in which the data is treated.) The World Wide Web Consortium, an international industry group whose members include Apple, Commerce One, Ericsson, and Microsoft, is supporting the development of P3P. Version 1.1 of the P3P was released in February 2006 and can be found at *www.w3.org/TR/2006/WD-P3P11-20060210/Overview.html.*

Platform for Privacy Preferences (P3P)
A screening technology that shields users from Web sites that don't provide the level of privacy protection they desire.

The Children's Online Privacy Protection Act (COPPA) was passed by Congress in October 1998. This act was directed at Web sites catering to children, requiring them to post comprehensive privacy policies on their sites and to obtain parental consent before they collect any personal information from children under 13 years of age. Web site operators who violate the rule could be liable for civil penalties of up to $11,000 per violation.[70] The Act has made an impact in the design and operations of Web sites that cater to children. For example, Lions Gate Entertainment, the operator of the *www.thebratzfilm.com* Web site, had to modify its site after the Council of Better Business Bureaus determined the site failed to meet the COPPA requirements. The Web site requested personally identifiable information to register for the Bratz Newsletter and register for a chance to win a trip to the premiere of *The Bratz Movie* without first obtaining verifiable parental consent.[71]

A social network service employs the Web and software to connect people for whatever purpose. There are thousands of such networks, which have become popular among teenagers. Some of the more popular social networking Web sites include Bebo, Classmates.com, Facebook, Hi5, Imbee, MySpace, Namesdatabase.com, Tagged, and XuQa. Most of these Web sites allow one to easily create a user profile that provides personal details, photos, even videos that can be viewed by other visitors to the Web site. Some of the Web sites have age restrictions or require that a parent register their preteen by providing a credit card to validate the parent's identity. Teens can provide information about where they live, go to school, their favorite music, and interests in hopes of meeting new friends. Unfortunately, they can also meet ill-intentioned strangers at these sites. Many documented encounters involve adults masquerading as teens attempting to meet young people for illicit purposes. Parents are advised to discuss potential dangers, check their children's profiles, and monitor their activities at such Web sites.

Fairness in Information Use

Selling information to other companies can be so lucrative that many companies will continue to store and sell the data they collect on customers, employees, and others. When is this information storage and use fair and reasonable to the people whose data is stored and sold? Do people have a right to know about data stored about them and to decide what data is stored and used? As shown in Table 14.2, these questions can be broken down into four issues that should be addressed: knowledge, control, notice, and consent.

In the past few decades, significant laws have been passed regarding a person's right to privacy. Others relate to business privacy rights and the fair use of data and information.

Fairness Issues	Database Storage	Database Usage
The right to know	Knowledge	Notice
The ability to decide	Control	Consent

Knowledge. Should people know what data is stored about them? In some cases, people are informed that information about them is stored in a corporate database. In others, they do not know that their personal information is stored in corporate databases.

Control. Should people be able to correct errors in corporate database systems? This is possible with most organizations, although it can be difficult in some cases.

Notice. Should an organization that uses personal data for a purpose other than the original purpose notify individuals in advance? Most companies don't do this.

Consent. If information on people is to be used for other purposes, should these people be asked to give their consent before data on them is used? Many companies do not give people the ability to decide if information on them will be sold or used for other purposes.

Table 14.2

The Right to Know and the Ability to Decide Federal Privacy Laws and Regulations

The Privacy Act of 1974

The major piece of legislation on privacy is the Privacy Act of 1974 (PA74). PA74 applies only to certain federal agencies. The act, which is about 15 pages long, is straightforward and easy to understand. The purpose of this act is to provide certain safeguards for people against an invasion of personal privacy by requiring federal agencies (except as otherwise provided by law) to do the following:

- Permit people to determine what records pertaining to them are collected, maintained, used, or disseminated by such agencies
- Permit people to prevent records pertaining to them from being used or made available for another purpose without their consent
- Permit people to gain access to information pertaining to them in federal agency records, to have a copy of all or any portion thereof, and to correct or amend such records
- Ensure that they collect, maintain, use, or disseminate any record of identifiable personal information in a manner that ensures that such action is for a necessary and lawful purpose, that the information is current and accurate for its intended use, and that adequate safeguards are provided to prevent misuse of such information
- Permit exemptions from this act only in cases of an important public need for such exemption, as determined by specific law-making authority
- Be subject to civil suit for any damages that occur as a result of willful or intentional action that violates anyone's rights under this act

PA74, which applies to all federal agencies except the CIA and law enforcement agencies, also established a Privacy Study Commission to study existing databases and to recommend rules and legislation for consideration by Congress. PA74 also requires training for all federal employees who interact with a "system of records" under the act. Most of the training is conducted by the Civil Service Commission and the Department of Defense. Another interesting aspect of PA74 concerns the use of Social Security numbers—federal, state, and local governments and agencies cannot discriminate against people for not disclosing or reporting their Social Security number.

Gramm-Leach-Bliley Act

This act was passed in 1999 and required all financial institutions to protect and secure customers' nonpublic data from unauthorized access or use. Under terms of this act, it was assumed that all customers approve of the financial institutions' collecting and storing their personal information. The institutions were required to contact their customers and inform them of this fact. Customers were required to write separate letters to each of their individual financial institutions and state in writing that they wanted to opt out of the data collection and storage process. Most people were overwhelmed with the mass mailings they received from their financial institutions and simply discarded them without ever understanding their importance.

USA Patriot Act

As discussed previously, the 2001 Uniting and Strengthening America by Providing Appropriate Tools Required to Intercept and Obstruct Terrorism Act (USA Patriot Act) was passed in response to the September 11 terrorism acts. Proponents argue that it gives necessary new powers to both domestic law enforcement and international intelligence agencies. Critics argue that the law removes many of the checks and balances that previously allowed the courts to ensure that law enforcement agencies did not abuse their powers. For example, under this act, Internet service providers and telephone companies must turn over customer information, including numbers called, without a court order if the FBI claims that the records are relevant to a terrorism investigation. Also, the company is forbidden to disclose that the FBI is conducting an investigation.

Other Federal Privacy Laws

In addition to PA74, other pieces of federal legislation relate to privacy. A federal law that was passed in 1992 bans unsolicited fax advertisements. This law was upheld in a 1995 ruling by the Ninth U.S. Circuit Court of Appeals, which concluded that the law is a reasonable way to prevent the shifting of advertising costs to customers. Table 14.3 lists additional laws related to privacy.

Law	Provisions
Fair Credit Reporting Act of 1970 (FCRA)	Regulates operations of credit-reporting bureaus, including how they collect, store, and use credit information
Tax Reform Act of 1976	Restricts collection and use of certain information by the Internal Revenue Service
Electronic Funds Transfer Act of 1979	Outlines the responsibilities of companies that use electronic funds transfer systems, including consumer rights and liability for bank debit cards
Right to Financial Privacy Act of 1978	Restricts government access to certain records held by financial institutions
Freedom of Information Act of 1970	Guarantees access for individuals to personal data collected about them and about government activities in federal agency files
Education Privacy Act	Restricts collection and use of data by federally funded educational institutions, including specifications for the type of data collected, access by parents and students to the data, and limitations on disclosure
Computer Matching and Privacy Act of 1988	Regulates cross-references between federal agencies' computer files (e.g.,to verify eligibility for federal programs)
Video Privacy Act of 1988	Prevents retail stores from disclosing video rental records without a court order
Telephone Consumer Protection Act of 1991	Limits telemarketers' practices
Cable Act of 1992	Regulates companies and organizations that provide wireless communications services, including cellular phones
Computer Abuse Amendments Act of 1994	Prohibits transmissions of harmful computer programs and code, including viruses
Gramm-Leach-Bliley Act of 1999	Requires all financial institutions to protect and secure customers' nonpublic data from unauthorized access or use
USA Patriot Act of 2001	Requires Internet service providers and telephone companies to turn over customer information, including numbers called, without a court order, if the FBI claims that the records are relevant to a terrorism investigation

Table 14.3

Federal Privacy Laws and Their Provisions

INFORMATION SYSTEMS @ WORK

Controlling Privacy in Finland's Largest Information System

Arek Oy, Ltd develops information systems and provides system services to pension insurance providers in Finland. The government of Finland has created laws to ensure that anyone earning a paycheck in Finland receives a pension upon retirement.

Finnish employers are required to maintain records on every employee, including the employee's name, national ID number, date of birth, work history, and other private information, along with an account of every paycheck issued to the employee. Employers share that information with one of many pension insurance companies. Arek Oy was created by the Finnish Centre for Pensions (ETK) and the country's authorized pension insurance providers to develop and manage the information systems that collect, store, and deliver employee information to the pension insurance industry.

Arek Oy was established in 2004 to perform an important task. The mission of the new company was to develop the largest information system used in Finland. The company had 30 months to complete the task, which may seem generous until you consider the size of the system. The goal of the pension insurance information system was to manage employment records of every person that works in Finland. If Arek Oy could not provide a flawless system by the deadline, they would put workers' pensions at risk, acquire hefty fines from the government, and ruin their own reputation, which would most likely mean the end of Arek Oy.

What made the systems development especially challenging was that Arek Oy had to apply many privacy rules and regulations as defined by the Finnish government. Today's privacy-sensitive culture makes database development and maintenance a time and resource-consuming affair for businesses and governments around the globe.

In general, sensitive employee data must be hidden from the eyes of all but approved parties. The systems engineers for Arek Oy were not allowed to see the data stored in the pension database. Special data privacy solutions were employed to mask personal identification information in database records—a practice called "de-identification." Arek Oy set up a safe sandbox for development that provided realistic, fictionalized data for developers to use when testing the systems. These types of systems are referred to as test-data management systems; they promote information privacy by allowing database developers to create reliable systems without accessing the actual private data that the system will manage.

Government privacy regulations, although important to customers and citizens, are particularly burdensome to businesses and information system developers. To assist developers in complying with privacy laws, database management systems provided by major information systems companies such as IBM have compliance embedded in their systems. Arek Oy reduced its stress and responsibility by adopting such a system to use for its pension insurance information system.

As you might guess, Arek Oy was successful in meeting its deadline for Finland's largest information system. It has deployed a database management system that includes a safe sandbox for test-data management that meets the high privacy standards of the Finnish government. The many pension insurance companies that work with the system can develop database applications using the secure and private environment that Arek Oy has provided.

Considering the time and effort that Arek Oy invested in complying with government privacy regulations, it's clear why many companies not governed by regulations are hesitant to commit resources to privacy practices. In most cases it isn't a matter of not caring, but of providing the best quality system for the least amount of money. The Arek Oy case provides a good example of the benefits and costs of government regulations.

Discussion Questions

1. What challenges did Arek Oy face in the Finnish pension systems development project?
2. What techniques did the company use to meet project requirements and government regulations?

Critical Thinking Questions

1. Besides government regulations, what other pressure might persuade a business to employ strict privacy practices?
2. What are the risks involved for a company that takes shortcuts and allows systems developers to see private data?

Sources: IBM Staff, "Arek Oy deploys IBM Optim to deliver the largest information management system in Finland," IBM Case Studies, May 30, 2008, *www-01.ibm.com/software/success/cssdb.nsf/CS/LWIS-7F5QWZ?OpenDocument&Site=default&cty=en_us*; Arek Oy Web site, *www.arek.fi*, accessed August 2, 2008.

Corporate Privacy Policies

Even though privacy laws for private organizations are not very restrictive, most organizations are very sensitive to privacy issues and fairness. They realize that invasions of privacy can hurt their business, turn away customers, and dramatically reduce revenues and profits. Consider a major international credit card company. If the company sold confidential financial information on millions of customers to other companies, the results could be disastrous. In a matter of days, the firm's business and revenues could be reduced dramatically. Therefore, most organizations maintain privacy policies, even though they are not required by law. Some companies even have a privacy bill of rights that specifies how the privacy of employees, clients, and customers will be protected. Corporate privacy policies should address a customer's knowledge, control, notice, and consent over the storage and use of information. They can also cover who has access to private data and when it can be used.

Multinational companies face an extremely difficult challenge in implementing data-collection and dissemination processes and policies because of the multitude of differing country or regional statutes. For example, Australia requires companies to destroy customer data (including backup files) or make it anonymous after it's no longer needed. Firms that transfer customer and personnel data out of Europe must comply with European privacy laws that allow customers and employees to access data about them and let them determine how that information can be used.

A few examples of corporate privacy policies are shown in Table 14.4.

Table 14.4

Corporate Privacy Policies

Company	URL
Starwood Hotels & Resorts	*www.starwoodhotels.com/corporate/privacy_policy.html*
United Parcel Service	*www.ups.com/content/corp/privacy_policy.html*
Visa	*www.corporate.visa.com/ut/privacy.jsp*
Walt Disney Internet Group	*http://disney.go.com/corporate/privacy/pp_wdig.html*

A good database design practice is to assign a single unique identifier to each customer—so that each has a single record describing all relationships with the company across all its business units. That way, the organization can apply customer privacy preferences consistently throughout all databases. Failure to do so can expose the organization to legal risks—aside from upsetting customers who opted out of some collection practices. Again, the 1999 Gramm-Leach-Bliley Financial Services Modernization Act required all financial service institutions to communicate their data privacy rules and honor customer preferences.

Individual Efforts to Protect Privacy

Although numerous state and federal laws deal with privacy, the laws do not completely protect individual privacy. In addition, not all companies have privacy policies. As a result, many people are taking steps to increase their own privacy protection. Some of the steps that you can take to protect personal privacy include the following:

- **Find out what is stored about you in existing databases.** Call the major credit bureaus to get a copy of your credit report. You are entitled to a free credit report every 12 months (see *freecreditreport.com*). You can also obtain a free report if you have been denied credit in the last 60 days. The major companies are Equifax (800-685-1111, *www.equifax.com*), TransUnion (800-916-8800, *www.transunion.com*), and Experian (888-397-3742, *www.experian.com*). You can also submit a Freedom of Information Act request to a federal agency that you suspect might have information stored on you.
- **Be careful when you share information about yourself.** Don't share information unless it is absolutely necessary. Every time you give information about yourself through an 800, 888, or 900 call, your privacy is at risk. Be vigilant in insisting that your doctor, bank, or financial institution not share information about you with others without your written consent.

- **Be proactive to protect your privacy.** You can get an unlisted phone number and ask the phone company to block caller ID systems from reading your phone number. If you change your address, don't fill out a change-of-address form with the U.S. Postal Service; you can notify the people and companies that you want to have your new address. Destroy copies of your charge card bills and shred monthly statements before disposing of them in the garbage. Be careful about sending personal e-mail messages over a corporate e-mail system. You can also get help in avoiding junk mail and telemarketing calls by visiting the Direct Marketing Association Web site at *www.the-dma.org*. Go to the Web site and look under Consumer Help-Remove Name from Lists.
- **When purchasing anything from a Web site, make sure that you safeguard your credit card numbers, passwords, and personal information.** Do not do business with a site unless you know that it handles credit card information securely. (Look for a seal of approval from organizations such as the Better Business Bureau Online or TRUSTe. When you open the Web page where you enter credit card information or other personal data, make sure that the Web address begins with *https* and check to see if a locked padlock icon appears in the Address bar or status bar). Do not provide personal information without reviewing the site's data privacy policy. Many credit card companies issue single-use credit card numbers on request. Charges appear on your usual bill, but the number is destroyed after a single use, eliminating the risk of stolen credit card numbers.

THE WORK ENVIRONMENT

The use of computer-based information systems has changed the makeup of the workforce. Jobs that require IS literacy have increased, and many less-skilled positions have been eliminated. Corporate programs, such as reengineering and continuous improvement, bring with them the concern that, as business processes are restructured and information systems are integrated within them, the people involved in these processes will be removed.

However, the growing field of computer technology and information systems has opened up numerous avenues to professionals and nonprofessionals of all backgrounds. Enhanced telecommunications has been the impetus for new types of business and has created global markets in industries once limited to domestic markets. Even the simplest tasks have been aided by computers, making cash registers faster, smoothing order processing, and allowing people with disabilities to participate more actively in the workforce. As computers and other IS components drop in cost and become easier to use, more workers will benefit from the increased productivity and efficiency provided by computers. Yet, despite these increases in productivity and efficiency, information systems can raise other concerns.

Health Concerns

Organizations can increase employee effectiveness by paying attention to the health concerns in today's work environment. For some people, working with computers can cause occupational stress. Anxieties about job insecurity, loss of control, incompetence, and demotion are just a few of the fears workers might experience. In some cases, the stress can become so severe that workers might sabotage computer systems and equipment. Monitoring employee stress can alert companies to potential problems. Training and counseling can often help the employee and deter problems.

Heavy computer use can affect one's physical health as well. A job that requires sitting at a desk and using a computer for many hours a day qualifies as a sedentary job. Such work can double the risk of seated immobility thromboembolism (SIT), the formation of blood clots in the legs or lungs. People leading a sedentary lifestyle are also likely to experience an undesirable weight gain which can lead to increased fatigue and greater risk of type 2 diabetes, heart problems, and other serious ailments.

Other work-related health hazards involve emissions from improperly maintained and used equipment. Some studies show that poorly maintained laser printers can release ozone into the air; others dispute the claim. Numerous studies on the impact of emissions from display screens have also resulted in conflicting theories. Although some medical authorities believe that long-term exposure can cause cancer, studies are not conclusive at this time. In any case, many organizations are developing conservative and cautious policies.

Most computer manufacturers publish technical information on radiation emissions from their CRT monitors, and many companies pay close attention to this information. San Francisco was one of the first cities to propose a video display terminal (VDT) bill. The bill requires companies with 15 or more employees who spend at least four hours a day working with computer screens to give 15-minute breaks every two hours. In addition, adjustable chairs and workstations are required if employees request them.

In addition to the possible health risks from radio-frequency exposure, cell phone use has raised a safety issue—an increased risk of traffic accidents as vehicle operators become distracted by talking on their cell phones (or operating their laptop computers, car navigation systems, or other computer devices) while driving. As a result, some states have made it illegal to operate a cell phone while driving.

Carpal tunnel syndrome (CTS) is an aggravation of the pathway for the nerves that travel through the wrist (carpal tunnel). CTS involves wrist pain, a feeling of tingling and numbness, and difficulty grasping and holding objects. In the late 1990s, many worker compensation claims were filed by people whose job required them to work at a keyboard many hours a day. However, a 2001 study by the Mayo Clinic found that heavy computer users (up to seven hours per day) had the same rate of carpal tunnel as the general population. It appears that CTS is caused by factors other than the repetitive motion of typing on a keyboard.[72]

Avoiding Health and Environmental Problems

Many computer-related health problems are caused by a poorly designed work environment. The computer screen can be hard to read, with glare and poor contrast. Desks and chairs can also be uncomfortable. Keyboards and computer screens might be fixed in place or difficult to move. The hazardous activities associated with these unfavorable conditions are collectively referred to as *work stressors.* Although these problems might not be of major concern to casual users of computer systems, continued stressors such as repetitive motion, awkward posture, and eye strain can cause more serious and long-term injuries. If nothing else, these problems can severely limit productivity and performance.

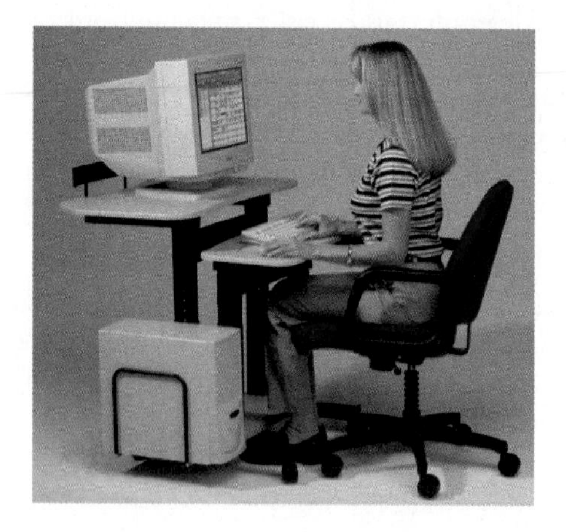

Research has shown that developing certain ergonomically correct habits can reduce the risk of adverse health effects when using a computer.

(Source: Courtesy of Balt, Inc.)

ergonomics
The science of designing machines, products, and systems to maximize the safety, comfort, and efficiency of the people who use them.

The science of designing machines, products, and systems to maximize the safety, comfort, and efficiency of the people who use them, called **ergonomics**, has suggested some approaches to reducing these health problems. The slope of the keyboard, the positioning and design of display screens, and the placement and design of computer tables and chairs have been carefully studied. Flexibility is a major component of ergonomics and an important feature of computer devices. People come in many sizes, have differing preferences, and require different positioning of equipment for best results. Some people, for example, want to place the keyboard in their laps; others prefer it on a solid table. Because of these individual differences, computer designers are attempting to develop systems that provide a great deal of flexibility. In fact, the revolutionary design of Apple's iMac computer came about through concerns for users' comfort. After using basically the same keyboard design for over a decade, Microsoft introduced a new split keyboard called the Natural Ergonomic Keyboard 4000. The keyboard provides improved ergonomic features such as improved angles that reduce motion and how much you must stretch your fingers when you type. The design of the keyboard also provides more convenient wrist and arm postures which make typing more convenient for users.

Computer users who work at their machines for more than an hour per day should consider using LCD screens, which are much easier on your eyes than CRT screens. If you stare at a CRT screen all day long, your eye muscles can become fatigued from the screen flicker and bright backlighting of the monitor. LCD screens provide a much better viewing experience for your eyes by virtually eliminating flicker while still being bright without harsh incandescence. Also, remember to blink! We tend to focus hard on the screen and blink much less than normal. The result is red, dry, itchy eyes. A few drops of artificial tears and changing focus away from the screen periodically to rest the eyes has been found to help.

In addition to steps taken by hardware manufacturing companies, computer users must also take action to reduce repetitive stress injury (RSI) caused by overuse of the computer through repeated movements that affects muscles, tendons, or nerves in the arms, hands, or upper back. For example, when working at a workstation, the top of the monitor should be at or just below eye level. Your wrists and hands should be in line with your forearms, with your elbows close to your body and supported. Your lower back needs to be well supported. Your feet should be flat on the floor. Take an occasional break to get away from the keyboard and screen. Stand up and stretch while at your workplace. Do not ignore pain or discomfort. Many workers ignore early signs of RSI, and as a result, the problem becomes much worse and more difficult to treat.

It is estimated that nearly 2 billion personal computers have been sold worldwide. This creates a tremendous disposal problem because personal computers and monitors contain lead, mercury, cadmium, and other metals defined as hazardous according to federal laws that govern their disposal. Congress is considering placing an "e-fee" that would be paid like a sales tax on personal computers, computer monitors, TVs, and some other electronic devices to cover the cost of their safe disposal. The annual cost could be in the neighborhood of $300 million. In the meantime, most personal computer manufacturers have implemented recycling programs and many are trying to redesign their products to reduce material that cannot be easily recycled. Many firms also specialize in the recycling of old personal computers. Unfortunately, some recycling programs ultimately send electronics waste to developing nations in Africa and Asia where it is disposed in environmentally unfriendly ways.[73]

ETHICAL ISSUES IN INFORMATION SYSTEMS

As you've seen throughout this book in the "Ethical and Societal Issues" boxes, ethical issues deal with what is generally considered right or wrong. As we have seen, laws do not provide a complete guide to ethical behavior. Just because an activity is defined as legal does not mean that it is ethical. As a result, practitioners in many professions subscribe to a **code of ethics** that states the principles and core values that are essential to their work and, therefore, govern their behavior. The code can become a reference point for weighing what is legal and what is ethical. For example, doctors adhere to varying versions of the 2000-year-old Hippocratic Oath, which medical schools offer as an affirmation to their graduating classes.

code of ethics
A code that states the principles and core values that are essential to a set of people and, therefore, govern their behavior.

Some IS professionals believe that their field offers many opportunities for unethical behavior. They also believe that unethical behavior can be reduced by top-level managers developing, discussing, and enforcing codes of ethics. Various IS-related organizations and associations promote ethically responsible use of information systems and have developed useful codes of ethics. The Association for Computing Machinery (ACM) is the oldest computing society, founded in 1947, and boasts more than 80,000 members in more than

100 countries. The ACM has a code of ethics and professional conduct that includes eight general moral imperatives that can be used to help guide the actions of IS professionals. These guidelines can also be used for those who employ or hire IS professionals to monitor and guide their work. These imperatives are outlined in the following list:

As an ACM member I will ...

1. Contribute to society and human well-being.
2. Avoid harm to others.
3. Be honest and trustworthy.
4. Be fair and take action not to discriminate.
5. Honor property rights including copyrights and patents.
6. Give proper credit for intellectual property.
7. Respect the privacy of others.
8. Honor confidentiality.

(Source: ACM Code of Ethics and Professional Conduct, *www.acm.org/constitution/code.html* accessed August 10, 2008.)

The mishandling of the social issues discussed in this chapter—including waste and mistakes, crime, privacy, health, and ethics—can devastate an organization. The prevention of these problems and recovery from them are important aspects of managing information and information systems as critical corporate assets. Increasingly, organizations are recognizing that people are the most important component of a computer-based information system and that long-term competitive advantage can be found in a well-trained, motivated, and knowledgeable workforce.

SUMMARY

Principle

Policies and procedures must be established to avoid waste and mistakes associated with computer usage.

Computer waste is the inappropriate use of computer technology and resources in both the public and private sectors. Computer mistakes relate to errors, failures, and other problems that result in output that is incorrect and without value. Waste and mistakes occur in government agencies as well as corporations. At the corporate level, computer waste and mistakes impose unnecessarily high costs for an information system and drag down profits. Waste often results from poor integration of IS components, leading to duplication of efforts and overcapacity. Inefficient procedures also waste IS resources, as do thoughtless disposal of useful resources and misuse of computer time for games and personal processing jobs. Inappropriate processing instructions, inaccurate data entry, mishandling of IS output, and poor systems design all cause computer mistakes.

A less dramatic, yet still relevant, example of waste is the amount of company time and money employees can waste playing computer games, sending unimportant e-mail, or accessing the Internet. Junk e-mail, also called *spam*, and junk faxes also cause waste.

Preventing waste and mistakes involves establishing, implementing, monitoring, and reviewing effective policies and procedures. Careful programming practices, thorough testing, flexible network interconnections, and rigorous backup procedures can help an information system prevent and recover from many kinds of mistakes. Companies should develop manuals and training programs to avoid waste and mistakes. Company policies should specify criteria for new resource purchases and user-developed processing tools to help guard against waste and mistakes. Spam filters that block unwanted mail should be installed.

Principle

Computer crime is a serious and rapidly growing area of concern requiring management attention.

Some crimes use computers as tools (e.g., to manipulate records, counterfeit money and documents, commit fraud via telecommunications links, and make unauthorized electronic transfers of money). Identity theft is a crime in which an imposter obtains key pieces of personal identification information to impersonate someone else. The information is then used to obtain credit, merchandise, and services in the name of the victim, or to provide the thief with false credentials.

A cyberterrorist is someone who intimidates or coerces a government or organization to advance his political or social objectives by launching computer-based attacks against computers, networks, and the information stored on them. A criminal hacker, also called a *cracker*, is a computer-savvy person who attempts to gain unauthorized or illegal access to computer systems to steal passwords, corrupt files and programs, and even transfer money. Script bunnies are crackers with little technical savvy. Insiders are employees, disgruntled or otherwise, working solo or in concert with outsiders to compromise corporate systems. The greatest fear of many organizations is the potential harm that can be done by insiders who know system logon IDs, passwords, and company procedures.

Computer crimes target computer systems and include illegal access to computer systems by criminal hackers, alteration and destruction of data and programs by viruses (system, application, and document), and simple theft of computer resources. A virus is a program that attaches itself to other programs. A worm functions as an independent program, replicating its own program files until it destroys other systems and programs or interrupts the operation of computer systems and networks. Malware is a general term for software that is harmful or destructive. A Trojan horse program is a malicious program that disguises itself as a useful application and purposefully does something the user does not expect. A logic bomb is designed to "explode" or execute at a specified time and date. A variant is a modified version of a virus that is produced by the virus's author or another person by amending the original virus code. A password sniffer is a small program hidden in a network or computer system that records identification numbers and passwords. Spyware is software installed on a personal computer to intercept or take partial control over the user's interactions with the computer without knowledge or permission of the user.

Identity theft is a crime in which an imposter steals personal identification information to obtain credit, merchandise, or services in the name of the victim. Although Internet gambling is popular, its legality is questionable within the United States.

Because of increased computer use, greater emphasis is placed on the prevention and detection of computer crime. Antivirus software is used to detect the presence of viruses, worms, and logic bombs. Use of an intrusion detection system (IDS) provides another layer of protection in the event that an intruder gets past the outer security layers—passwords, security procedures, and corporate firewall. It monitors system and network resources and notifies network security personnel when it senses a possible intrusion. Many small and mid-sized organizations are outsourcing their network security operations to managed security service providers (MSSPs), which monitor, manage, and maintain network security hardware and software.

Software piracy might represent the most common computer crime. It is estimated that the software industry loses nearly $48 billion in revenue each year to software piracy. Computer scams have cost people and companies thousands of dollars. Computer crime is also an international issue.

Security measures, such as using passwords, identification numbers, and data encryption, help to guard against illegal computer access, especially when supported by effective control procedures. Encryption enables users of an unsecured public network such as the Internet to securely and privately exchange data through the use of a public and a private cryptographic key pair that is obtained and shared through a trusted authority. The use of biometrics, involving the measurement of a person's unique characteristics, such as fingerprints, irises, and retinal images, is another way to protect important data and information systems. Virus-scanning software identifies and removes damaging computer programs. Law enforcement agencies armed with new legal tools enacted by Congress now actively pursue computer criminals.

Although most companies use data files for legitimate, justifiable purposes, opportunities for invasion of privacy abound. Privacy issues are a concern with government agencies, e-mail use, corporations, and the Internet. The Children's Internet Protection Act was enacted to protect minors using the Internet. The Privacy Act of 1974, with the support of other federal laws, establishes straightforward and easily understandable requirements for data collection, use, and distribution by federal agencies; federal law also serves as a nationwide moral guideline for privacy rights and activities by private organizations. The USA Patriot Act, passed only five weeks after the September 11 terrorist attacks, requires Internet service providers and telephone companies to turn over customer information, including numbers called, without a court order, if the FBI claims that the records are relevant to a terrorism investigation. Also, the company is forbidden to disclose that the FBI is conducting an investigation. Only time will tell how this act will be applied in the future. The Gramm-Leach-Bliley Act requires all financial institutions to protect and secure customers' nonpublic data from unauthorized access or use. Under terms of this act, it is assumed that all customers approve of the financial institutions collecting and storing their personal information.

A business should develop a clear and thorough policy about privacy rights for customers, including database access. That policy should also address the rights of employees, including electronic monitoring systems and e-mail. Fairness in information use for privacy rights emphasizes knowledge, control, notice, and consent for people profiled in databases. People should know about the data that is stored about them and be able to correct errors in corporate database systems. If information on people is to be used for other purposes, they should be asked to give their consent beforehand. Each person has the right to know and the ability to decide. Platform for Privacy Preferences (P3P) is a screening technology that shields users from Web sites that don't provide the level of privacy protection they desire.

Principle

Jobs, equipment, and working conditions must be designed to avoid negative health effects from computers.

Computers have changed the makeup of the workforce and even eliminated some jobs, but they have also expanded and enriched employment opportunities in many ways. Jobs that involve heavy use of computers contribute to a sedentary lifestyle, which increases the risk of health problems. Some critics blame computer systems for emissions of ozone and electromagnetic radiation. Use of cell phones while driving has been linked to increased car accidents.

The study of designing and positioning computer equipment, called *ergonomics*, has suggested some approaches to reducing these health problems. Ergonomic design principles help to reduce harmful effects and increase the efficiency of an information system. The slope of the keyboard, the positioning and design of display screens, and the placement and design of computer tables and chairs are essential for good health. RSI prevention includes keeping good posture, not ignoring pain or problems, performing stretching and strengthening exercises, and seeking proper treatment. Although they can cause negative health consequences, information systems can also be used to provide a wealth of information on health topics through the Internet and other sources.

Principle

Practitioners in many professions subscribe to a code of ethics that states the principles and core values that are essential to their work.

Ethics determine generally accepted and discouraged activities within a company and society at large. Ethical computer users define acceptable practices more strictly than just refraining from committing crimes; they also consider the effects of their IS activities, including Internet usage, on other people and organizations. The Association for Computing Machinery developed guidelines and a code of ethics. Many IS professionals join computer-related associations and agree to abide by detailed ethical codes.

CHAPTER 14: SELF-ASSESSMENT TEST

Policies and procedures must be established to avoid waste and mistakes associated with computer usage.

1. Business managers and end users must work with IS professionals to implement and follow proper IS usage policies to ensure effective use of company resources. True or False?
2. Computer-related waste and mistakes are major causes of computer problems, contributing to unnecessarily high ____ and lost ____.
3. Unwanted e-mail is often referred to as __________.

Computer crime is a serious and rapidly growing area of concern requiring management attention.

4. According to the 2007 FBI Internet Crime Report, the dollar amount of Internet crime reported exceeded $250 million. True or False?
5. ______ is using one's skills to get computer users to provide you with information to access an information system or its data.
6. The vast majority of organizations conduct some form of computer security audit. True or False?
7. ____________________ is a crime in which an imposter obtains key pieces of personal identification information, such as Social Security or driver's license numbers, to impersonate someone else.
8. Internet gambling in the United States is completely legal. True or False?
9. A logic bomb is a type of Trojan horse that executes when specific conditions occur. True or False?
10. Malware capable of spreading itself from one computer to another is called a __________.
    a. logic bomb
    b. Trojan horse
    c. virus
    d. worm
11. A(n) ___________ is a modified version of a virus that is produced by the virus's author or another person amending the original virus code.
12. The Business Software Alliance estimates that the software industry lost over $48 billion in 2007 due to worldwide software piracy. True or False?
13. Phishing is a computer scam that seems to direct users to a bank's Web site but actually captures key personal information about its victims. True or False?

Jobs, equipment, and working conditions must be designed to avoid negative health effects from computers.

14. CTS, or ____________________, is the aggravation of the pathway of nerves that travel through the wrist.

Practitioners in many professions subscribe to a code of ethics that states the principles and core values that are essential to their work.

15. Just because an activity is defined as legal, it does not mean that it is ethical. True or False?

CHAPTER 14: SELF-ASSESSMENT TEST ANSWERS

(1) True (2) costs, profits (3) spam (4) False (5) Social engineering (6) True (7) Identity theft (8) False (9) True (10) d (11) variant (12) True (13) True (14) carpal tunnel syndrome (15) True

REVIEW QUESTIONS

1. What is a spam filter? How does such a program work? What is the issue with image-based spam?
2. How can antivirus software reduce computer waste?
3. Outline a four-step process to prevent computer waste and mistakes.
4. Why are all computer crimes not reported to law enforcement agencies?
5. What is a virus? What is a worm? How are they different?
6. What is a variant? What dangers are associated with such malware?
7. What is phishing? What actions can you take to reduce the likelihood that you will be a victim of this crime?
8. Outline measures you should take to protect yourself against viruses and worms.
9. What does intrusion detection software do? What are some of the issues with the use of this software?
10. Identify at least five tips to follow to avoid becoming a victim of a computer scam.

11. What is the difference between a patent and a copyright? What copyright issues come into play when downloading software or music from a Web site?
12. What is the difference between the Children's Online Privacy Protection Act and the Children's Internet Protection Act?
13. What is ergonomics? How can it be applied to office workers?
14. What specific actions can you take to avoid spyware?
15. What is a code of ethics? Give an example.

DISCUSSION QUESTIONS

1. Imagine that your friend regularly downloads copies of newly released, full-length motion pictures for free from the Internet and makes copies for others for a small fee. Do you think that this is ethical? Is it legal? Would you express any concerns with him?
2. Outline an approach, including specific techniques (e.g., dumpster diving, phishing, social engineering) that you could employ to gain personal data about the members of your class.
3. Your 12-year-old niece shows you a dozen or so photos of herself and a brief biography including address and cell phone number she plans to post on MySpace. What advice might you offer her about posting personal information and photos?
4. Imagine that you are a hacker and have developed a Trojan horse program. What tactics might you use to get unsuspecting victims to load the program onto their computer?
5. Discuss the importance of educating employees in preventing computer waste and computer crime. Imagine that you are given the assignment of developing a computer education program for your employer. What topics would you cover in the course?
6. Briefly discuss the potential for cyberterrorism to cause a major disruption in our daily life. What are some likely targets of a cyberterrorist? What sort of action could a cyberterrorist take against these targets?
7. You are the new head of corporate security for a large Fortune 1000 company and are alarmed at the number of laptop computers your firm's employees lose each month. What actions would you take to cut down on the potential for loss of personal and/or company confidential data?
8. Do you believe that the National Security Agency should be able to collect the telephone call records of U.S. citizens without the use of search warrants? Why or why not?
9. Using information presented in this chapter on federal privacy legislation, identify which federal law regulates the following areas and situations: cross-checking IRS and Social Security files to verify the accuracy of information, customer liability for debit cards, your right to access data contained in federal agency files, the IRS obtaining personal information, the government obtaining financial records, and employers' access to university transcripts.
10. Briefly discuss the difference between acting morally and acting legally. Give an example of acting legally and yet immorally.

PROBLEM-SOLVING EXERCISES

1. Access the Web sites for the Recording Industry Association of America (RIAA), Motion Picture Association of America (MPAA), and Business Software Alliance (BSA) to get estimates of the amount of piracy worldwide for at least three years. Use a graphics package to develop a bar chart to show the amount of music, motion picture, and software piracy over a three-year time period. Compare the amount of piracy to the total music, motion picture, and software revenue for the same time period.
2. Using spreadsheet software and appropriate forecasting routines, develop a forecast for the amount of piracy for next year. Document any assumptions you make in developing your forecast.
3. Using your word processing software, write a few brief paragraphs summarizing the trends you see from reviewing the data for the past few years. Then cut and paste the information from Exercise 1 and your forecast from Exercise 2 into your report.

TEAM ACTIVITIES

1. Visit your school's library and interview the librarians about the use of Internet software filters. What level and kinds of complaints are made about the use of filtering software? Who is responsible for updating the list of sites that are deemed "off limits" for minors? What is their opinion about the need for and effectiveness of the software filter?
2. Have each member of your team access ten different Web sites and summarize their findings in terms of the existence of data privacy policy statements: Did the site have such a policy? Was it easy to find? Was it complete and easy to understand? Did you find any sites using the P3P standard or ICRA rating method?

WEB EXERCISES

1. The Computer Emergency Response Team Coordination Center (CERT/CC) is located at the Software Engineering Institute (SEI), a federally funded research and development center at Carnegie Mellon University in Pittsburgh, Pennsylvania. Do research on the center and write a brief report summarizing its activities.
2. Search the Web for a site that provides software to detect and remove spyware. Write a short report for your instructor summarizing your findings.
3. Do research on the Web to discover what role the Business Software Alliance plays in the protection of software. Document some of the tactics it uses to identify and punish organizations that it determines to practice software piracy.

CAREER EXERCISES

1. You are a senior member of a marketing organization for a manufacturer of children's toys. A recommendation has been made to develop a Web site to promote and sell your firm's products as well as learn more about what parents and their children are looking for in new toys. Develop a list of laws and regulations that will affect the design of the Web site. Describe how these will limit the operation of your new Web site.
2. You have just begun a new position in customer relations for a mid-sized bank. Within your first week on the job, several customers have expressed concern about potential theft of customer data from the bank's computer databases and identity theft. Who would you talk with to develop a satisfactory response to address your customers' concerns? What key points would you need to verify with bank employees?

CASE STUDIES

Case One

IT Consumerization and Web 2.0 Security Challenges

In recent years, the direction of investment in information technologies has shifted. The shift is in reaction to the fact that in 2004, independent consumers passed business and government in their consumption of digital electronics devices. More digital devices, such as notebooks, cell phones, and media players, are being designed for consumers rather than businesses. New and popular technologies are now being introduced into the workplace by employees rather than systems analysts. This is a trend that some refer to as IT consumerization. Unfortunately, consumer devices and systems are introducing a host of new systems vulnerabilities.

A big concern regarding IT consumerization is the free flow of communications and data sharing. Today's Web 2.0 technologies make it all too easy for employees to share information that they shouldn't. A study in the United Kingdom revealed that three-quarters of U.K. businesses have banned the use of instant messaging services such as AIM, Windows Live Messenger, and Yahoo Messenger. The primary concern is the loss of sensitive business information. Even though the

IM services could prove useful for business communications, most businesses are concerned about security rather than interested in innovative communication.

Consider the Apple iPhone. Some businesses that have supported RIM's Blackberry smartphone are feeling pressure from their employees to support the iPhone as well. Systems security experts are hesitant to comply due to concerns over information privacy. For example, the iPhone 3G does not include data encryption native to the device. If the phone is lost or stolen, private corporate information is vulnerable. Systems analysts are stuck trying to serve both a demanding workforce and corporate security needs.

CTO Gary Hodge at U.S. Bank is concerned about Web 2.0 applications. "We always said outside the corporation was untrusted and inside the corporation was trusted territory. Web 2.0 has changed all that. We've had to expose the internal workings of the corporation. There's a whole rash of new devices coming out to enable people to compute when they want to, with the iPhones and smartphones." Hodge worries that smartphone manufacturers haven't paid enough attention to security. CTOs and CIOs are feeling as though they are losing control of their systems and data.

Dmitri Alperovitch, principal research scientist for Secure Computing, is also concerned about security and Web 2.0. The concern stems from the browser becoming a computing platform itself. Although businesses have learned to protect traditional operating systems, they have little power when the browser is acting like an operating system. Web 2.0 sites and social networking sites allow anyone to create applications and post files and content. This increases the risks of transmitting malware and revealing corporate secrets. Gary Dobbins, director of information security at the University of Notre Dame, has simple and effective advice for information security: "Never trust the browser."

In banking, minor lapses in security can have devastating results. Bank CIOs see Web 2.0 as expanding their security perimeter. Web 2.0 gives them a much larger area to watch. Because of this, many banks are taking a hard line. For example, U.S. Bank only allows employees to access business-related content on their PCs. The bank restricts the use of any type of portable storage including USB drives and CDs. Every electronic transmission that leaves the bank is monitored.

For Gary Hodge, investing in information security at U.S. Bank isn't a matter of ROI, but rather a survival necessity. "We protect money. It's new for us to have to protect vast amounts of information," Hodge said. "We spend millions of dollars on security but it doesn't generate any new revenue. I haven't been able to show anybody a return on investment. It comes down to can we secure the organization at the right risk and the right cost. You can't spend all the money. You have to figure out what level of risk you're willing to tolerate."

Discussion Questions

1. What are the differences in information security needs for a bank versus a retail store?
2. Why are IT consumerization and Web 2.0 challenging business information security?

Critical Thinking Questions

1. Do you think that over time consumer devices may become as secure as banking systems? Why or why not?
2. Do you think the "hard line" taken by U.S. Bank in regards to information security policies is justified? Why or why not? Would you be willing to work in that environment?

Sources: Stokes, Jon, "Analysis: IT consumerization and the future of work," *Ars Technica*, July 6, 2008, *http://arstechnica.com/news.ars/post/20080706-analysis-it-consumerization-and-the-future-of-work.html*; Skinner, Carrie-Ann, "U.K businesses ban IM over security concerns," *Computerworld*, July 15, 2008, *www.computerworld.com/action/article.do?command=viewArticleBasic&articleId=9110159*; Brodkin, Jon, "U.S. Bank suffers Web 2.0 security headaches," *Network World*, April 30, 2008, *www.networkworld.com/news/2008/043008-interop-bank-web-2-security.html*; Hamblen, Matt, "iPhone 3G, business must wait," *Macworld UK*, June 16, 2008, *www.macworld.co.uk/ipod-itunes/news/index.cfm?newsid=21659*.

Case Two

San Francisco WAN Held Captive

Sometimes in protecting a network, the ones to watch are within the organization. That's the lesson learned by the City of San Francisco. The city's network administrator for its multimillion dollar wide area network (WAN) seized control of the network and denied other system administrators access for ten days while jailed.

The network administrator, who had been experiencing conflicts with his supervisor, created a super password that effectively locked out all administrators but himself to the network's switches and routers. When he refused to reveal the password, he was arrested and held on a $5 million bond. The network that he held captive connects various city offices around San Francisco and supports 60 percent of the municipal government's information traffic. During the system administrator's incarceration, the city network continued functioning without incident.

The system administrator's lawyer argued the defendant felt that none of the people who requested the password were qualified to have it. The defendant claimed his supervisor was undermining his work. The defendant wanted to uncover the problems in the city's Department of Telecommunication Information Services (DTIS). His intent was to "expose the utter mismanagement, negligence, and corruption at DTIS, which if left unchecked, will in fact place the City of San Francisco in danger," his motion read. It is assumed that drastic budget cuts that resulted in losing 200 of 350 employees at DTIS were behind the stress that ultimately drove the administrator to extreme measures.

The network administrator finally revealed the super password to the network when after ten days in prison, San Francisco mayor Gavin Newsom visited him. The two had a lengthy private discussion that concluded with the mayor receiving the password, saving the city the hundreds of thousands of dollars it would have cost to sequentially reset hundreds of switches and routers around the city.

This case points to several important lessons for businesses to observe regarding system administration. Rick

Cook of *Computerworld* suggests that perhaps policies used by nuclear power plants, NASA, and the military might have prevented San Francisco from losing control of its network. Nuclear power plants deny access to systems at the slightest sign of suspicious activity. In San Francisco's case, by the time the suspicious activity was noticed, it was too late. The system administrator obviously did not have proper oversight and supervision. If the city used a system that logged administrator activities and assigned security officers to review them regularly, the damage could have been prevented.

In the military, two people are required to take simultaneous actions to launch nuclear missiles. Similar requirements could be implemented with important system actions such as managing switches and routers.

A most important preventive step is called identity management and access control (IM/AC). Identity management requires usernames and passwords, which most networks do effectively. Access control, however, is often undermanaged in important networks. Access controls prevent users from accessing systems and commands for which they do not have authority.

Through a combination of close supervision, duplication of responsibilities, and identity management and access control, the San Francisco WAN kidnapping might have been avoided. Unfortunately, security measures come at some cost. Obviously, with budget cutbacks, the city could not afford the level of security needed for such an important network. As global economies become strained and economies increasingly depend on the stability of secure information systems, San Francisco's dilemma could be played out at a much grander scale unless security for information systems becomes as important as for nuclear power plants and missiles.

Discussion Questions

1. What was the cause of the problems for San Francisco's WAN?
2. How might these problems have been prevented?

Critical Thinking Questions

1. Should information system security be considered as important as security at a nuclear facility, as suggested in this article? Why or why not?
2. Did this system administrator's actions create the effect that he obviously intended? Were his actions justified and ethical?

Sources: Cook, Rick, "Opinion: How to protect your network from rogue IT employees," *Computerworld*, July 21, 2008, *www.computerworld.com/action/article.do?command=viewArticleBasic&taxonomyName=security&articleId=9110385&taxonomyId=17&intsrc=kc_feat*; McMillan, Robert, "San Francisco IT admin locks up city network," *Computerworld*, July 21, 2008, *www.computerworld.com/action/article.do?command=viewArticleBasic&articleId=322438*; McMillan, Robert, and Venezia, Paul, "San Francisco's mayor gets back keys to the network," *Computerworld*, July 23, 2008, *www.computerworld.com/action/article.do?command=viewArticleBasic&taxonomyName=security&articleId=9110520&taxonomyId=17&intsrc=kc_top*.

QUESTIONS FOR WEB CASE

See the Web site for this book to read about the Whitmann Price Consulting case for this chapter. Following are questions concerning this Web case.

Whitmann Price Consulting: The Personal and Social Impact of Computers

Discussion Questions

1. Why do you think extending access to the Whitmann Price network beyond the business's walls dramatically elevated the risk to information security?
2. What was the primary tool used to minimize that risk, and how does it work?

Critical Thinking Questions

1. Why does information security usually come at the cost of user convenience?
2. List the security policies put in place for the AMCI system and the rationale that you think is behind them.

NOTES

Sources for the opening vignette: Qualys staff, "eBay, Inc. - Securing the World's Online Marketplace with QualysGuard," Qualys Case Study, May 10, 2008, *www.bitpipe.com/detail/RES/1210427689_288.html*; Gross, Grant, "Oregon man sentenced to four years for piracy, ID theft," *Computerworld*, July 24, 2008, *www.computerworld.com/action/article.do?command=viewArticleBasic&articleId=9110621&source=rss_news10*; Goodin, Dan, "Notorious eBay hacker arrested in Romania," *The Register*, April 18, 2008, *www.theregister.co.uk/2008/04/18/vladuz_arrested*; Sullivan, Bob, "How far has Vladuz hacked into eBay?," MSNBC Red Tape Chronicles, March 2, 2008, *http://redtape.msnbc.com/2007/03/how_far_has_vla.html*; FBI staff, "2007 IC3 Annual Report," FBI, 2008, *www.ic3.gov/media/annualreport/2007-IC3Report.pdf*; About eBay Web Page, *http://news.ebay.com/about.cfm*, accessed July 26, 2008.

1 "Spam Filter Review 2008," Top Ten Reviews, *http://spam-filter-review.toptenreviews.co*, accessed June 23, 2008.
2 Westbrook, Jesse, "SEC Asks About Ratings Errors on Structured Products," *Bloomberg.com*, June 26, 2008.
3 Songini, Marc L., "Computer Glitch Led to Mars Global Surveyor's Demise," *Computerworld*, April 27, 2007.
4 Williams, Martyn, "Computer Glitch Hits Hundreds of Japan Flights," *Computerworld*, May 28, 2007.

5 McMillan, Robert and Mullins, Robert, "United Flights Grounded by Computer Glitch," *Computerworld*, June 21, 2007.
6 Colliver, Victoria and Muscat, Sabine, "Wells Fargo ATM, Other Glitches Last Longer Than First Reported," *San Francisco Chronicle*, August 22, 2007.
7 Barak, Sylvie, "Getting Fired for Using the Internet Becomes Commonplace," *The Inquirer*, March 3, 2008.
8 Stackel, I.M., "Another Collier Employee Fired for Inappropriate E-mails," *NaplesNews.com*, July 2, 2008.
9 Havenstein, Heather, "IT Officials Are Clearing BI Hurdles to Expand Systems," *Computerworld*, May 28, 2007.
10 Gumbel, Peter, "4 Things I Learned from Societe Generale," *CNN Money*, February 1, 2008.
11 Schwartz, Nelson D. and Bennhold, Katrin, "Societe Generale Scandal: 'A Suspicion That This Was Inevitable,'" *International Herald Tribune*, February 5, 2008.
12 Vance, Jeff, "Using Policy and Compliance Tools to Reduce Insider Threats," at *www.cioupdate.com*, accessed June 27, 2008.
13 Miller, Chuck, "ATM Hackers Net Millions Using Stolen Information," *Secure Computing*, June 20, 2008.
14 Barak, Sylvie, "Hacker Exposes Six Million Chilean's Data to Make a Point," *Secure Computing*, May 13, 2008.
15 Thomson, Iain, "Ukrainian Hacker May Get to Keep Profits," *Secure Computing*, February 19, 2008.
16 Thurston, Richard, "U.S. School Network Falls Victim to Child Hacker," *Secure Computing*, May 27, 2008.
17 Carr, Jim, "Californian Indicted in US $50,000 Scam of E*Trade, Schwab.com," *Secure Computing*, May 30, 2008.
18 Regan, Keith, "Web Crime Spikes in 2007, Losses Near $240 M," *Electronic Commerce Times*, April 4, 2008.
19 "CSI Survey 2007," *GoCSI.com*, accessed June 27, 2008.
20 Greenemeier, Larry, "The Threat Within: Employees Pose the Biggest Security Risk," *InformationWeek*, July16, 2007.
21 Kaplan, Dan, "MTV Breach Impacts 5,000 Employees, Successful Social-Engineering Blamed," *Secure Computing*, March 11, 2008.
22 Ko, Carol, "Malaysia to Build Centre to Study Cyberterrorism," *IT World Canada*, June 12, 2008.
23 Associated Press, "Stung by Cyber Warfare, Estonia, NATO Allies to Sign Deal on Cyber Defense Center," *International Herald Tribune*, May 14, 2008.
24 McMillan, Robert, "CNN Cyberattack Called Off," *CIO*, April 19, 2008.
25 "2007 Identity Fraud Survey Report: Identity Fraud is Dropping, Continued Vigilance Necessary," Javelin Strategy & Research, February 2007.
26 Varrone, Carl, "What Nobody Else is Saying About Online Poker," Dog Ear Publishing, 2007, p. 126.
27 Chiang, Jennifer W., "Don't Bet On It: How Complying with Federal Internet Gambling Law is Not Enough," *Shidler Journal for Law, Commerce + Technology*, June 6, 2007.
28 Havenstein, Heather, "Report: FBI Looks into Facebook March Madness Betting Pools," *Computerworld*, March 17, 2008.
29 James, Clement, "US House Committee Votes on Web Gambling Ban," *vnunet.com*, June 23, 2008.
30 Sowa, Tom, "A Hacker's Wrong Turn," *SpokesmanReview.com*, August 20, 2007.
31 "Newly Discovered Malware," McAfee Web site *http://vil.nai.com/vil/newly_discovered_viruses.aspx*, accessed July 4, 2008.
32 Hulme, George, "New Ransom-Ware Virus Resurfaces," *Information Week*, June 9, 2008.
33 Keizer, Gregg, "Best Buy Sold Infected Digital Picture Frames, *Computerworld*, January 2, 2008.
34 Nichols, Shaun, "Twin Trojans Attack Macs," *Secure Computing*, June 23, 2008.
35 Kirk, Jeremy, "'Mebroot' Proves to be a Tough Rootkit to Crack," *Computerworld*, March 4, 2008.
36 Leyden, John, "Dodgy Drug Sales Underpin Storm Worm," *The Register*, June 12, 2008.
37 "About McAfee," McAfee Web site, *http://us.mcafee.com/root/aboutUs.asp*, accessed on June 28, 2008.
38 "Antivirus Software and Disk Defragmentation," *Tech Republic*, June 23, 2008.
39 "Anti-Spyware Software Reviews for 2008," Top Ten Reviews, *http://anti-spyware-review.toptenreviews.com*, accessed July 1, 2008.
40 Lemos, Robert, "Spyware Purveyor DirectRevenue Closes Down," *SecurityFocus*, October 25, 2007.
41 "DirectRevenue LLC Settles FTC Charges," Federal Trade Commission Web site, February 16, 2007, *www.tfc.gov/opa/2007/02*, accessed July 1, 2008.
42 Keizer, Gregg, "VA Loses Another Hard Drive, Vet Data At Risk," *Information Week*, February 5, 2007.
43 Vijayan, Jaikumar, "Another Day, Another Laptop Theft: Now, It's Connecticut's Revenue Agency," *Computerworld*, August 31, 2007.
44 Fonseca, Brian, "Personal Data on 800,000 Gap Job Applicants Exposed in Laptop Theft," *Computerworld*, September 26, 2007.
45 Claburn, Thomas, "Record Number of Data Breaches Reported in 2007," *Information Week*, December 31, 2007.
46 McGee, Marianne Kolbasuk, "Laptop Stolen with Personal Data on 300,000 Health Insurance Clients," *Information Week*, January 30, 2008.
47 Puzzanghera, Jim, "Laptop Seizures at Customs Raise Outcry," *Los Angeles Times*, June 26, 2008.
48 "What is Software Piracy?" Business Software Alliance Web site, *www.bsa.org/country.aspx*, accessed July 3, 2008.
49 "Worldwide Software Piracy Rate Holds Steady at 35%; Global Losses up 15%," Business Software Alliance, *www.bsa.org/country*, accessed July 3, 2008.
50 US Code: Title 17 Copyrights, Cornell University Law School, *www.law.cornell.edu/uscode/17*, accessed July 3, 2008.
51 "Judgment Entered Against PA Company for Widespread Unauthorized Distribution of Software," Business Software Alliance Web site, April 24, 2008, *www.bsa.org/country*, accessed July 3, 2008.
52 "Los Angeles-Area Engineering Company Pays $250,000 to The Business Software Alliance," April 14, 2008, *www.bsa.org/country*, accessed July 3, 2008.
53 Jones, K.C., "Music Piracy Costs U.S. Economy $12.5 Billion, Report Reveals," *InformationWeek*, August 22, 2007.
54 "Two Site Operators Receive Prison Terms for Criminal Copyright Infringement," United States Department of Justice Web site, May 14, 2008 accessed at *www.usdoj.gov/usaso/can/press* on July 3, 2008.
55 "Jones, K.C., "TorrentSpy Ordered to Pay $100 Million in Piracy Case," *InformationWeek*, May 8, 2008.
56 Claburn, Thomas, "Acer, Apple, Dell and HP Sued for Patent Infringement," *Information Week*, June 30, 2008.
57 "Tellabs Sues Fujitsu Claiming Patent Infringement," *PC World*, June 12, 2008.
58 Kanaracus, Chris, "Red Hat Settles Patent Suits with Firestar, DataTern," *PC World*, June 11, 2008.
59 Rogers, John, "Gartner: US $3.2 Billion Lost to Phishing Attacks in One Year," *SC Magazine*, December 19, 2007.
60 Carr, Jim, "Phishing Scam Uses AOL Address to Target eBay Users," *SC Magazine*, November 13, 2007.
61 "Jet Flash 210 Fingerprint USB Flash Drive," Transcend Web site at *www.transcendusa.com/Products/ModDetail.asp?ModNo=108&SpNo=2&LangNo=0*, accessed July 4, 2008.
62 "Customer Success Stories: Associated Newspapers," accessed at Computer Associates Web site, http://ca.com/us/success/Collateral.aspx?CID=147880 on June 28,2008.
63 Children's Internet Protections Act, Pub. L. 106-552, accessed at http://ifea.net/cipa.html on June 22, 2008.
64 Welch, Erin, "Child-Proofing Internet Access," *The Orange County Register*, January 29, 2008.
65 Houser, Mark, "UPMC Admits Privacy Violation," *Pittsburgh Tribune-Review*, April 13, 2007.
66 Jones, K.C., "Obama, Clinton, McCain Passport Breaches Expose Human, Not Tech Weakness," *InformationWeek*, March 21, 2008.
67 "3/4 of Companies Monitor Employee Web Browsing," *Yahoo! Tech*, May 9, 2007.
68 Singel, Ryan, "Nearly Ten Percent of Companies Have Fired Bloggers, Survey Claims," *Wired*, July 19, 2007.
69 Lewan, Todd, "Microchips Everywhere: A Future Vision," *The Seattle Times*, January 29, 2008.
70 "Children's Online Privacy Protection Act – 15 USC 6501 – 6506," *www.softforyou.com/add/COPPA.pdf*, accessed June 23, 2008.
71 Bean, Linda, "CARU Reviews Site Operated by Lions Gate Entertainment," *CARU News*, October 24, 2007.
72 Simon, Ellen, "How Are Your Wrists Feeling?" *Cincinnati Enquirer*, May 28, 2008, p. E1.
73 Gross, Grant, "E-Waste Recycling Faces Challenges, Critics Say," *Computerworld*, April 20, 2008.

acceptance testing Conducting any tests required by the user.

accounting MIS An information system that provides aggregate information on accounts payable, accounts receivable, payroll, and many other applications.

ad hoc DSS A DSS concerned with situations or decisions that come up only a few times during the life of the organization.

Advanced Encryption Standard (AES) An extremely strong data encryption standard sponsored by the National Institute of Standards and Technology based on a key size of 128 bits, 192 bits, or 256 bits.

alpha testing Testing an incomplete or early version of the system.

analog signal A variable signal continuous in both time and amplitude so that any small fluctuations in the signal are meaningful.

antivirus program Software that runs in the background to protect your computer from dangers lurking on the Internet and other possible sources of infected files.

application flowcharts Diagrams that show relationships among applications or systems.

application program interface (API) An interface that allows applications to make use of the operating system.

application service provider (ASP) A company that provides software, support, and the computer hardware on which to run the software from the user's facilities over a network.

arithmetic/logic unit (ALU) The part of the CPU that performs mathematical calculations and makes logical comparisons.

ARPANET A project started by the U.S. Department of Defense (DoD) in 1969 as both an experiment in reliable networking and a means to link DoD and military research contractors, including many universities doing military-funded research.

artificial intelligence (AI) The ability of computers to mimic or duplicate the functions of the human brain.

artificial intelligence systems People, procedures, hardware, software, data, and knowledge needed to develop computer systems and machines that demonstrate the characteristics of intelligence.

asking directly An approach to gather data that asks users, stakeholders, and other managers about what they want and expect from the new or modified system.

asynchronous communications A form of communications where the receiver gets the message after some delay—sometimes hours or days after the message is sent.

attribute A characteristic of an entity.

auditing Analyzing the financial condition of an organization and determining whether financial statements and reports produced by the financial MIS are accurate.

backward chaining The process of starting with conclusions and working backward to the supporting facts.

backbone One of the Internet's high-speed, long-distance communications links.

batch processing system A form of data processing where business transactions are accumulated over a period of time and prepared for processing as a single unit or batch.

benchmark test An examination that compares computer systems operating under the same conditions.

best practices The most efficient and effective ways to complete a business process.

beta testing Testing a complete and stable system by end users.

blade server A server that houses many individual computer motherboards that include one or more processors, computer memory, computer storage, and computer network connections.

Bluetooth A wireless communications specification that describes how cell phones, computers, faxes, personal digital assistants, printers, and other electronic devices can be interconnected over distances of 10–30 feet at a rate of about 2 Mbps.

bot A software tool that searches the Web for information such as products and prices.

brainstorming A decision-making approach that often consists of members offering ideas "off the top of their heads."

bridge A telecommunications device that connects one LAN to another LAN using the same telecommunications protocol.

broadband communications A telecommunications system in which a very high rate of data exchange is possible.

business intelligence The process of gathering enough of the right information in a timely manner and usable form and analyzing it to have a positive impact on business strategy, tactics, or operations.

business-to-business (B2B) e-commerce A subset of e-commerce where all the participants are organizations.

business-to-consumer (B2C) e-commerce A form of e-commerce in which customers deal directly with an organization and avoid intermediaries.

byte (B) Eight bits that together represent a single character of data.

cache memory A type of high-speed memory that a processor can access more rapidly than main memory.

Cascading Style Sheet (CSS) A file or portion of an HTML file that defines the visual appearance of content in a Web page.

CASE repository A database of system descriptions, parameters, and objectives.

central processing unit (CPU) The part of the computer that consists of three associated elements: the arithmetic/logic unit, the control unit, and the register areas.

centralized processing Processing alternative in which all processing occurs at a single location or facility.

certificate authority (CA) A trusted third-party organization or company that issues digital certificates.

certification A process for testing skills and knowledge, which results in a statement by the certifying authority that confirms an individual is capable of performing a particular kind of job.

change model A representation of change theories that identifies the phases of change and the best way to implement them.

channel bandwidth The rate at which data is exchanged over a telecommunications channel, usually measured in bits per second (bps).

character A basic building block of information, consisting of uppercase letters, lowercase letters, numeric digits, or special symbols.

chat room A facility that enables two or more people to engage in interactive "conversations" over the Internet.

chief knowledge officer (CKO) A top-level executive who helps the organization use a KMS to create, store, and use knowledge to achieve organizational goals.

choice stage The third stage of decision making, which requires selecting a course of action.

click fraud A problem arising in a pay-per-click online advertising environment where additional clicks are generated beyond those that come from actual, legitimate users.

clickstream data The data gathered based on the Web sites you visit and the items you click.

client/server An architecture in which multiple computer platforms are dedicated to special functions such as database management, printing, communications, and program execution.

clock speed A series of electronic pulses produced at a predetermined rate that affects machine cycle time.

closed shops IS departments in which only authorized operators can run the computers.

cloud computing Using a giant cluster of computers to serve as a host to run applications that require high-performance computing.

code of ethics A code that states the principles and core values that are essential to a set of people and, therefore, govern their behavior.

cold site A computer environment that includes rooms, electrical service, telecommunications links, data storage devices, and the like; also called a *shell*.

command-based user interface A user interface that requires you to give text commands to the computer to perform basic activities.

compact disc read-only memory (CD-ROM) A common form of optical disc on which data, once it has been recorded, cannot be modified.

competitive advantage A significant and (ideally) long-term benefit to a company over its competition.

competitive intelligence One aspect of business intelligence limited to information about competitors and the ways that knowledge affects strategy, tactics, and operations.

compiler A special software program that converts the programmer's source code into the machine-language instructions consisting of binary digits.

computer literacy Knowledge of computer systems and equipment and the ways they function; it stresses equipment and devices (hardware), programs and instructions (software), databases, and telecommunications.

computer network The communications media, devices, and software needed to connect two or more computer systems or devices.

computer programs Sequences of instructions for the computer.

computer-aided software engineering (CASE) Tools that automate many of the tasks required in a systems development effort and encourage adherence to the SDLC.

computer-assisted manufacturing (CAM) A system that directly controls manufacturing equipment.

computer-based information system (CBIS) A single set of hardware, software, databases, telecommunications, people, and procedures that are configured to collect, manipulate, store, and process data into information.

computer-integrated manufacturing (CIM) Using computers to link the components of the production process into an effective system.

concurrency control A method of dealing with a situation in which two or more people need to access the same record in a database at the same time.

consumer-to-consumer (C2C) e-commerce A subset of e-commerce that involves consumers selling directly to other consumers.

content streaming A method for transferring multimedia files over the Internet so that the data stream of voice and pictures plays more or less continuously without a break, or very few of them; enables users to browse large files in real time.

continuous improvement Constantly seeking ways to improve business processes to add value to products and services.

control unit The part of the CPU that sequentially accesses program instructions, decodes them, and coordinates the flow of data in and out of the ALU, registers, primary storage, and even secondary storage and various output devices.

coprocessor The part of the computer that speeds processing by executing specific types of instructions while the CPU works on another processing activity.

cost center A division within a company that does not directly generate revenue.

cost/benefit analysis An approach that lists the costs and benefits of each proposed system. After they are expressed in monetary terms, all the costs are compared with all the benefits.

counterintelligence The steps an organization takes to protect information sought by "hostile" intelligence gatherers.

creative analysis The investigation of new approaches to existing problems.

criminal hacker (cracker) A computer-savvy person who attempts to gain unauthorized or illegal access to computer systems to steal passwords, corrupt files and programs, or even transfer money.

critical analysis The unbiased and careful questioning of whether system elements are related in the most effective ways.

critical path Activities that, if delayed, would delay the entire project.

critical success factors (CSFs) Factors that are essential to the success of a functional area of an organization.

cross-platform development A development technique that allows programmers to develop programs that can run on computer systems having different hardware and operating systems, or platforms.

culture A set of major understandings and assumptions shared by a group.

customer relationship management (CRM) system A system that helps a company manage all aspects of customer encounters, including marketing and advertising, sales, customer service after the sale, and programs to retain loyal customers.

cybermall A single Web site that offers many products and services at one Internet location.

cyberterrorist Someone who intimidates or coerces a government or organization to advance his political or social objectives by launching computer-based attacks against computers, networks, and the information stored on them.

data Raw facts, such as an employee number, total hours worked in a week, inventory part numbers, or sales orders.

data administrator A nontechnical position responsible for defining and implementing consistent principles for a variety of data issues.

data analysis The manipulation of collected data so that the development team members who are participating in systems analysis can use the data.

data cleanup The process of looking for and fixing inconsistencies to ensure that data is accurate and complete.

data collection Capturing and gathering all data necessary to complete the processing of transactions.

data correction The process of reentering data that was not typed or scanned properly.

data definition language (DDL) A collection of instructions and commands used to define and describe data and relationships in a specific database.

data dictionary A detailed description of all the data used in the database.

data editing The process of checking data for validity and completeness.

Data Encryption Standard (DES) An early data encryption standard developed in the 1970s that uses a 56-bit private key algorithm.

data entry Converting human-readable data into a machine-readable form.

data input Transferring machine-readable data into the system.

data item The specific value of an attribute.

data manipulation The process of performing calculations and other data transformations related to business transactions.

data manipulation language (DML) The commands that are used to manipulate the data in a database.

data mart A subset of a data warehouse.

data mining An information-analysis tool that involves the automated discovery of patterns and relationships in a data warehouse.

data model A diagram of data entities and their relationships.

data preparation, or data conversion Ensuring all files and databases are ready to be used with new computer software and systems.

data storage The process of updating one or more databases with new transactions.

data store Representation of a storage location for data.

data warehouse A database that collects business information from many sources in the enterprise, covering all aspects of the company's processes, products, and customers.

database An organized collection of facts and information.

database administrator (DBA) A skilled IS professional who directs all activities related to an organization's database.

database approach to data management An approach whereby a pool of related data is shared by multiple application programs.

database management system (DBMS) A group of programs that manipulate the database and provide an interface between the database and the user of the database and other application programs.

data-flow diagram (DFD) A model of objects, associations, and activities that describes how data can flow between and around various objects.

data-flow line Arrows that show the direction of data element movement.

decentralized processing Processing alternative in which processing devices are placed at various remote locations.

decision room A room that supports decision making, with the decision makers in the same building, combining face-to-face verbal interaction with technology to make the meeting more effective and efficient.

decision support system (DSS) An organized collection of people, procedures, software, databases, and devices used to support problem-specific decision making.

decision-making phase The first part of problem solving, including three stages: intelligence, design, and choice.

delphi approach A decision-making approach in which group decision makers are geographically dispersed; this approach encourages diversity among group members and fosters creativity and original thinking in decision making.

demand report A report developed to give certain information at someone's request.

design report The primary result of systems design, reflecting the decisions made and preparing the way for systems implementation.

design stage The second stage of decision making, in which alternative solutions to the problem are developed.

desktop computer A relatively small, inexpensive, single-user computer that is highly versatile.

deterrence controls Rules and procedures to prevent problems before they occur.

dialogue manager A user interface that allows decision makers to easily access and manipulate the DSS and to use common business terms and phrases.

digital audio player A device that can store, organize, and play digital music files.

digital camera An input device used with a PC to record and store images and video in digital form.

digital certificate An attachment to an e-mail message or data embedded in a Web site that verifies the identity of a sender or Web site.

digital rights management (DRM) Refers to the use of any of several technologies to enforce policies for controlling access to digital media such as movies, music, and software.

digital signal A signal that represents bits.

digital subscriber line (DSL) A telecommunications service that delivers high-speed Internet access to homes and small businesses over the existing phone lines of the local telephone network.

digital video disc (DVD) A storage medium used to store digital video or computer data.

direct access A retrieval method in which data can be retrieved without the need to read and discard other data.

direct access storage device (DASD) A device used for direct access of secondary storage data.

direct conversion (also called *plunge* or *direct cutover*) Stopping the old system and starting the new system on a given date.

direct observation Watching the existing system in action by one or more members of the analysis team.

disaster recovery The implementation of the disaster plan.

disaster recovery plan (DRP) A formal plan describing the actions that must be taken to restore computer operations and services in the event of a disaster.

disk mirroring A process of storing data that provides an exact copy that protects users fully in the event of data loss.

distance learning The use of telecommunications to extend the classroom.

distributed database A database in which the data can be spread across several smaller databases connected via telecommunications devices.

distributed processing Processing alternative in which computers are placed at remote locations but are connected to each other via a network.

document production The process of generating output records and reports.

documentation The text that describes the program functions to help the user operate the computer system.

domain The allowable values for data attributes. Also, the area of knowledge addressed by an expert system.

domain expert The person or group who has the expertise or knowledge the expert system is trying to capture.

downsizing Reducing the number of employees to cut costs.

drill-down report A report providing increasingly detailed data about a situation.

dumpster diving Going through the trash cans of an organization to find secret or confidential information, including information needed to access an information system or its data.

dynamic Web pages Web pages containing variable information that are built to respond to a specific Web visitor's request.

economic feasibility The determination of whether the project makes financial sense and whether predicted benefits offset the cost and time needed to obtain them.

economic order quantity (EOQ) The quantity that should be reordered to minimize total inventory costs.

effectiveness A measure of the extent to which a system achieves its goals; it can be computed by dividing the goals actually achieved by the total of the stated goals.

efficiency A measure of what is produced divided by what is consumed.

e-Government The use of information and communications technology to simplify the sharing of information, speed formerly paper-based processes, and improve the relationship between citizens and government.

electronic bill presentment A method of billing whereby a vendor posts an image of your statement on the Internet and alerts you by e-mail that your bill has arrived.

electronic business (e-business) Using information systems and the Internet to perform all business-related tasks and functions.

electronic cash An amount of money that is computerized, stored, and used as cash for e-commerce transactions.

electronic commerce (e-commerce) Conducting business activities (e.g., distribution, buying, selling, marketing, and servicing of products or services) electronically over computer networks such as the Internet, extranets, and corporate networks.

electronic data interchange (EDI) An intercompany, application-to-application communication of data in a standard format, permitting the recipient to perform a standard business transaction, such as processing purchase orders.

electronic document distribution A process that enables the sending and receiving of documents in a digital form without being printed (although printing is possible).

electronic exchange An electronic forum where manufacturers, suppliers, and competitors buy and sell goods, trade market information, and run back-office operations.

electronic funds transfer (EFT) A system of transferring money from one bank account directly to another without any paper money changing hands.

electronic retailing (e-tailing) The direct sale from business to consumer through electronic storefronts, typically designed around an electronic catalog and shopping cart model.

empowerment Giving employees and their managers more responsibility and authority to make decisions, take certain actions, and have more control over their jobs.

encryption The process of converting an original message into a form that can only be understood by the intended receiver.

end-user systems development Any systems development project in which business managers and users assume the primary effort.

enterprise data modeling Data modeling done at the level of the entire enterprise.

enterprise resource planning (ERP) system A set of integrated programs capable of managing a company's vital business operations for an entire multisite, global organization.

enterprise sphere of influence The sphere of influence that serves the needs of the firm in its interaction with its environment.

enterprise system A system central to the organization that ensures information can be shared across all business functions and all levels of management to support the running and managing of a business.

entity A generalized class of people, places, or things for which data is collected, stored, and maintained.

entity symbol Representation of either a source or destination of a data element.

entity-relationship (ER) diagrams Data models that use basic graphical symbols to show the organization of and relationships between data.

environmental design Also called *green design*, it involves systems development efforts that slash power consumption, require less physical space, and result in systems that can be disposed in a way that doesn't negatively affect the environment.

ergonomics The science of designing machines, products, and systems to maximize the safety, comfort, and efficiency of the people who use them.

event-driven review A review triggered by a problem or opportunity such as an error, a corporate merger, or a new market for products.

exception report A report automatically produced when a situation is unusual or requires management action.

execution time (E-time) The time it takes to execute an instruction and store the results.

executive support system (ESS) Specialized DSS that includes all hardware, software, data, procedures, and people used to assist senior-level executives within the organization.

expert system Hardware and software that stores knowledge and makes inferences, similar to a human expert.

explanation facility Component of an expert system that allows a user or decision maker to understand how the expert system arrived at certain conclusions or results.

Extensible Markup Language (XML) The markup language for Web documents containing structured information, including words, pictures, and other elements.

external auditing Auditing performed by an outside group.

extranet A network based on Web technologies that links selected resources of a company's intranet with its customers, suppliers, or other business partners.

feasibility analysis Assessment of the technical, economic, legal, operational, and schedule feasibility of a project.

feedback Output that is used to make changes to input or processing activities.

field Typically a name, number, or combination of characters that describes an aspect of a business object or activity.

file A collection of related records.

File Transfer Protocol (FTP) A protocol that describes a file transfer process between a host and a remote computer and allows users to copy files from one computer to another.

final evaluation A detailed investigation of the proposals offered by the vendors remaining after the preliminary evaluation.

financial MIS An information system that provides financial information not only for executives but also for a broader set of people who need to make better decisions on a daily basis.

five-forces model A widely accepted model that identifies five key factors that can lead to attainment of competitive advantage, including (1) the rivalry among existing competitors, (2) the threat of new entrants, (3) the threat of substitute products and services, (4) the bargaining power of buyers, and (5) the bargaining power of suppliers.

flat organizational structure An organizational structure with a reduced number of management layers.

flexible manufacturing system (FMS) An approach that allows manufacturing facilities to rapidly and efficiently change from making one product to making another.

forecasting Predicting future events to avoid problems.

forward chaining The process of starting with the facts and working forward to the conclusions.

front-end processor A special-purpose computer that manages communications to and from a computer system serving hundreds or even thousands of users.

full-duplex channel A communications channel that permits data transmission in both directions at the same time, so a full-duplex channel is like two simplex channels.

game theory The use of information systems to develop competitive strategies for people, organizations, or even countries.

Gantt chart A graphical tool used for planning, monitoring, and coordinating projects.

gateway A telecommunications device that serves as an entrance to another network.

genetic algorithm An approach to solving large, complex problems in which a number of related operations or models change and evolve until the best one emerges.

geographic information system (GIS) A computer system capable of assembling, storing, manipulating, and displaying geographic information, that is, data identified according to its location.

gigahertz (GHz) Billions of cycles per second.

graphical user interface (GUI) An interface that uses icons and menus displayed on screen to send commands to the computer system.

grid chart A table that shows relationships among the various aspects of a systems development effort.

grid computing The use of a collection of computers, often owned by multiple individuals or organizations, to work in a coordinated manner to solve a common problem.

group consensus Decision making by a group that is appointed and given the responsibility of making the final evaluation and selection.

group consensus approach A decision-making approach that forces members in the group to reach a unanimous decision.

group support system (GSS) Software application that consists of most elements in a DSS, plus software to provide effective support in group decision making; also called *group decision support system* or *computerized collaborative work system.*

hacker A person who enjoys computer technology and spends time learning and using computer systems.

handheld computer A single-user computer that provides ease of portability because of its small size.

hardware Any machinery (most of which uses digital circuits) that assists in the input, processing, storage, and output activities of an information system.

half-duplex channel A communications channel that can transmit data in either direction, but not simultaneously.

help facility A program that provides assistance when users want to know more about a program or feature or what type of response is expected.

heuristics Commonly accepted guidelines or procedures that usually find a good solution.

hierarchy of data Bits, characters, fields, records, files, and databases.

highly structured problems Problems that are straightforward and require known facts and relationships.

hot site A duplicate, operational hardware system or immediate access to one through a specialized vendor.

HTML tags Codes that let the Web browser know how to format text—as a heading, as a list, or as body text—and whether images, sound, and other elements should be inserted.

human resource MIS An information system that is concerned with activities related to employees and potential employees of an organization, also called a personnel MIS.

hyperlink Highlighted text or graphics in a Web document that, when clicked, opens a new Web page or section of the same page containing related content.

Hypertext Markup Language (HTML) The standard page description language for Web pages.

identify theft A crime in which an imposter obtains key pieces of personal identification information, such as Social Security or driver's license numbers, to impersonate someone else.

IF-THEN statements Rules that suggest certain conclusions.

image log A separate file that contains only changes to applications.

implementation stage A stage of problem solving in which a solution is put into effect.

incremental backup Making a backup copy of all files changed during the last few days or the last week.

inference engine Part of the expert system that seeks information and relationships from the knowledge base and provides answers, predictions, and suggestions similar to the way a human expert would.

informatics A specialized system that combines traditional disciplines, such as science and medicine, with computer systems and technology.

information A collection of facts organized in such a way that they have additional value beyond the value of the individual facts.

information center A support function that provides users with assistance, training, application development, documentation, equipment selection and setup, standards, technical assistance, and troubleshooting.

information service unit A miniature IS department.

information system (IS) A set of interrelated components that collect, manipulate, store, and disseminate data and information and provide a feedback mechanism to meet an objective.

information systems literacy Knowledge of how data and information are used by individuals, groups, and organizations.

information systems planning Translating strategic and organizational goals into systems development initiatives.

infrared transmission A wireless communications technology that operates at a frequency of 300 GHz and above that requires line-of-sight transmission and operates over short distances—such as a few yards.

input The activity of gathering and capturing raw data.

insider An employee, disgruntled or otherwise, working solo or in concert with outsiders to compromise corporate systems.

installation The process of physically placing the computer equipment on the site and making it operational.

instant messaging A method that allows two or more people to communicate online using the Internet.

institutional DSS A DSS that handles situations or decisions that occur more than once, usually several times per year or more. An institutional DSS is used repeatedly and refined over the years.

instruction time (I-time) The time it takes to perform the fetch-instruction and decode-instruction steps of the instruction phase.

integrated development environment (IDE) A development approach that combines the tools needed for programming with a programming language into one integrated package.

integration testing Testing all related systems together.

intellectual property Includes works of the mind such as books, films, music, processes, and software, which are distinct somehow and are owned and/or created by a single entity.

intelligence stage The first stage of decision making, in which potential problems or opportunities are identified and defined.

intelligent agent Programs and a knowledge base used to perform a specific task for a person, a process, or another program; also called *intelligent robot* or *bot*.

intelligent behavior The ability to learn from experiences and apply knowledge acquired from experience, handle complex situations, solve problems when important information is missing, determine what is important, react quickly and correctly to a new situation, understand visual images, process and manipulate symbols, be creative and imaginative, and use heuristics.

internal auditing Auditing performed by individuals within the organization.

Internet The world's largest computer network, consisting of thousands of interconnected networks, all freely exchanging information.

Internet Protocol (IP) A communication standard that enables traffic to be routed from one network to another as needed.

Internet service provider (ISP) Any company that provides Internet access to people or organizations.

intranet An internal network based on Web technologies that allows people within an organization to exchange information and work on projects.

intrusion detection system (IDS) Software that monitors system and network resources and notifies network security personnel when it senses a possible intrusion.

Java An object-oriented programming language from Sun Microsystems based on C++ that allows small programs (applets) to be embedded within an HTML document.

joining Manipulating data to combine two or more tables.

joint application development (JAD) A process for data collection and requirements analysis in which users, stakeholders, and IS professionals work together to analyze existing systems, propose possible solutions, and define the requirements of a new or modified system.

just-in-time (JIT) inventory A philosophy of inventory management in which inventory and materials are delivered just before they are used in manufacturing a product.

kernel The heart of the operating system, which controls the most critical processes.

key A field or set of fields in a record that is used to identify the record.

key-indicator report A summary of the previous day's critical activities; typically available at the beginning of each workday.

knowledge The awareness and understanding of a set of information and ways that information can be made useful to support a specific task or reach a decision.

knowledge acquisition facility Part of the expert system that provides convenient and efficient means of capturing and storing all the components of the knowledge base.

knowledge base A component of an expert system that stores all relevant information, data, rules, cases, and relationships used by the expert system.

knowledge engineer A person who has training or experience in the design, development, implementation, and maintenance of an expert system.

knowledge user The person or group who uses and benefits from the expert system.

LCD display Flat display that uses liquid crystals—organic, oil-like material placed between two polarizers—to form characters and graphic images on a backlit screen.

learning systems A combination of software and hardware that allows the computer to change how it functions or reacts to situations based on feedback it receives.

legal feasibility The determination of whether laws or regulations may prevent or limit a systems development project.

linking Data manipulation that combines two or more tables using common data attributes to form a new table with only the unique data attributes.

local area network (LAN) A network that connects computer systems and devices within a small area, such as an office, home, or several floors in a building.

logical design A description of the functional requirements of a system.

lookup tables Tables containing data that computer programs can develop and use to simplify and shorten data entry.

machine cycle The instruction phase followed by the execution phase.

magnetic disk A common secondary storage medium, with bits represented by magnetized areas.

magnetic stripe card A type of card that stores limited amounts of data by modifying the magnetism of tiny iron-based particles contained in a band on the card.

magnetic tape A secondary storage medium; Mylar film coated with iron oxide with portions of the tape magnetized to represent bits.

mainframe computer A large, powerful computer often shared by hundreds of concurrent users connected to the machine via terminals.

maintenance team A special IS team responsible for modifying, fixing, and updating existing software.

make-or-buy decision The decision regarding whether to obtain the necessary software from internal or external sources.

management information system (MIS) An organized collection of people, procedures, software, databases, and devices that provides routine information to managers and decision makers.

market segmentation The identification of specific markets to target them with advertising messages.

marketing MIS An information system that supports managerial activities in product development, distribution, pricing decisions, and promotional effectiveness.

massively parallel processing systems A form of multiprocessing that speeds processing by linking hundreds or thousands of processors to operate at the same time, or in parallel, with each processor having its own bus, memory, disks, copy of the operating system, and applications.

material requirements planning (MRP) A set of inventory-control techniques that help coordinate thousands of inventory items when the demand of one item is dependent on the demand for another.

megahertz (MHz) Millions of cycles per second.

menu-driven system A system in which users simply pick what they want to do from a list of alternatives.

meta tag A special HTML tag, not visible on the displayed Web page, that contains keywords representing your site's content, which search engines to use to build indexes pointing to your Web site.

metropolitan area network (MAN) A telecommunications network that connects users and their devices in a geographical area that spans a campus or city.

microcode Predefined, elementary circuits and logical operations that the processor performs when it executes an instruction.

middleware Software that allows different systems to communicate and exchange data.

MIPS Millions of instructions per second, a measure of machine cycle time.

mission-critical systems Systems that play a pivotal role in an organization's continued operations and goal attainment.

mobile commerce (m-commerce) Transactions conducted anywhere, anytime.

model base Part of a DSS that provides decision makers access to a variety of models and assists them in decision making.

model management software Software that coordinates the use of models in a DSS.

modem A telecommunications hardware device that converts (modulates and demodulates) communications signals so they can be transmitted over the communication media.

monitoring stage The final stage of the problem-solving process, in which decision makers evaluate the implementation.

Moore's Law A hypothesis stating that transistor densities on a single chip double every two years.

MP3 A standard format for compressing a sound sequence into a small file.

multicore microprocessor A microprocessor that combines two or more independent processors into a single computer so they can share the workload and improve processing capacity.

multiplexer A device that combines data from multiple data sources into a single output signal that carries multiple channels, thus reducing the number of communications links needed and therefore, lowering telecommunications costs.

multiprocessing The simultaneous execution of two or more instructions at the same time.

natural language processing Processing that allows the computer to understand and react to statements and commands made in a "natural" language, such as English.

Near Field Communication (NFC) A very short-range wireless connectivity technology designed for cell phones and credit cards.

net present value The net amount by which project savings exceed project expenses after allowing for the cost of capital and the passage of time.

network operating system (NOS) Systems software that controls the computer systems and devices on a network and allows them to communicate with each other.

network-attached storage (NAS) Storage devices that attach to a network instead of to a single computer.

network-management software Software that enables a manager on a networked desktop to monitor the use of individual computers and shared hardware (such as printers), scan for viruses, and ensure compliance with software licenses.

networks Computers and equipment that are connected in a building, around the country, or around the world to enable electronic communications.

neural network A computer system that can simulate the functioning of a human brain.

nominal group technique A decision-making approach that encourages feedback from individual group members, and the final decision is made by voting, similar to the way public officials are elected.

nonprogrammed decision A decision that deals with unusual or exceptional situations.

object-oriented database A database that stores both data and its processing instructions.

object-oriented database management system (OODBMS) A group of programs that manipulate an object-oriented database and provide a user interface and connections to other application programs.

object-oriented systems development (OOSD) An approach to systems development that combines the logic of the systems development life cycle with the power of object-oriented modeling and programming.

object-relational database management system (ORDBMS) A DBMS capable of manipulating audio, video, and graphical data.

on-demand computing Contracting for computer resources to rapidly respond to an organization's varying workflow. Also called on-demand business and utility computing.

online analytical processing (OLAP) Software that allows users to explore data from a number of perspectives.

online transaction processing (OLTP) A form of data processing where each transaction is processed immediately, without the delay of accumulating transactions into a batch.

open shops IS departments in which people, such as programmers and systems analysts, are allowed to run the computers, in addition to authorized operators.

open-source software Software that is freely available to anyone in a form that can be easily modified.

operating system (OS) A set of computer programs that controls the computer hardware and acts as an interface with application programs.

operational feasibility The measure of whether the project can be put into action or operation.

optical disc A rigid disc of plastic onto which data is recorded by special lasers that physically burn pits in the disc.

optimization model A process to find the best solution, usually the one that will best help the organization meet its goals.

organization A formal collection of people and other resources established to accomplish a set of goals.

organizational change How for-profit and nonprofit organizations plan for, implement, and handle change.

organizational culture The major understandings and assumptions for a business, corporation, or other organization.

organizational learning The adaptations to new conditions or alterations of organizational practices over time.

organizational structure Organizational subunits and the way they relate to the overall organization.

output Production of useful information, usually in the form of documents and reports.

outsourcing Contracting with outside professional services to meet specific business needs.

parallel computing The simultaneous execution of the same task on multiple processors to obtain results faster.

parallel start-up Running both the old and new systems for a period of time and comparing the output of the new system closely with the output of the old system; any differences are reconciled. When users are comfortable that the new system is working correctly, the old system is eliminated.

password sniffer A small program hidden in a network or a computer system that records identification numbers and passwords.

patch A minor change to correct a problem or make a small enhancement. It is usually an addition to an existing program.

perceptive system A system that approximates the way a person sees, hears, and feels objects.

personal area network (PAN) A network that supports the interconnection of information technology within a range of 33 feet or so.

personal productivity software The software that enables users to improve their personal effectiveness, increasing the amount of work they can perform and enhancing its quality.

personal sphere of influence The sphere of influence that serves the needs of an individual user.

personalization The process of tailoring Web pages to specifically target individual consumers.

phase-in approach (also called *piecemeal approach*) Slowly replacing components of the old system with those of the new one. This process is repeated for each application until the new system is running every application and performing as expected; also called a *piecemeal approach*.

phishing A practice that entails sending bogus messages purportedly from a legitimate institution to pry personal information from customers by convincing them to go to a "spoofed" Web site.

physical design The specification of the characteristics of the system components necessary to put the logical design into action.

pilot start-up Running the new system for one group of users rather than all users.

pipelining A form of CPU operation in which multiple execution phases are performed in a single machine cycle.

pixel A dot of color on a photo image or a point of light on a display screen.

planned data redundancy A way of organizing data in which the logical database design is altered so that certain data entities are combined, summary totals are carried in the data records rather than calculated from elemental data, and some data attributes are repeated in more than one data entity to improve database performance.

plasma display A plasma display uses thousands of smart cells (pixels) consisting of electrodes and neon and xeon gases which are electrically turned into plasma (electrically charged atoms and negatively charged particles) to emit light.

Platform for Privacy Preferences (P3P) A screening technology that shields users from Web sites that don't provide the level of privacy protection they desire.

point evaluation system An evaluation process in which each evaluation factor is assigned a weight, in percentage points, based on importance. Then each proposed system is evaluated in terms of this factor and given a score ranging from 0 to 100. The scores are totaled, and the system with the greatest total score is selected.

point-of-sale (POS) device A terminal used in retail operations to enter sales information into the computer system.

policy-based storage management Automation of storage using previously defined policies.

portable computer A computer small enough to be carried easily.

predictive analysis A form of data mining that combines historical data with assumptions about future conditions to predict outcomes of events, such as future product sales or the probability that a customer will default on a loan.

preliminary evaluation An initial assessment whose purpose is to dismiss the unwanted proposals; begins after all proposals have been submitted.

primary key A field or set of fields that uniquely identifies the record.

primary storage (main memory; memory) The part of the computer that holds program instructions and data.

private branch exchange (PBX) A telephone switching exchange that serves a single organization.

problem solving A process that goes beyond decision making to include the implementation stage.

procedures The strategies, policies, methods, and rules for using a CBIS.

process A set of logically related tasks performed to achieve a defined outcome.

process symbol Representation of a function that is performed.

processing Converting or transforming data into useful outputs.

productivity A measure of the output achieved divided by the input required.

profit center A department within an organization that focuses on generating profits.

Program Evaluation and Review Technique (PERT) A formalized approach for developing a project schedule that creates three time estimates for an activity.

programmed decision A decision made using a rule, procedure, or quantitative method.

programmer A specialist responsible for modifying or developing programs to satisfy user requirements.

programming languages Sets of keywords, symbols, and a system of rules for constructing statements by which humans can communicate instructions to be executed by a computer.

project deadline The date the entire project is to be completed and operational.

project milestone A critical date for the completion of a major part of the project.

project organizational structure A structure centered on major products or services.

project schedule A detailed description of what is to be done.

projecting Manipulating data to eliminate columns in a table.

public network services Systems that give personal computer users access to vast databases and other services, usually for an initial fee plus usage fees.

quality The ability of a product (including services) to meet or exceed customer expectations.

quality control A process that ensures that the finished product meets the customers' needs.

questionnaires A method of gathering data when the data sources are spread over a wide geographic area.

Radio Frequency Identification (RFID) A technology that employs a microchip with an antenna that broadcasts its unique identifier and location to receivers.

random access memory (RAM) A form of memory in which instructions or data can be temporarily stored.

rapid application development (RAD) A systems development approach that employs tools, techniques, and methodologies designed to speed application development.

read-only memory (ROM) A nonvolatile form of memory.

record A collection of related data fields.

redundant array of independent/ inexpensive disks (RAID) A method of storing data that generates extra bits of data from existing data, allowing the system to create a "reconstruction map" so that if a hard drive fails, the system can rebuild lost data.

reengineering (process redesign) The radical redesign of business processes, organizational structures, information systems, and values of the organization to achieve a breakthrough in business results.

register A high-speed storage area in the CPU used to temporarily hold small units of program instructions and data immediately before, during, and after execution by the CPU.

relational model A database model that describes data in which all data elements are placed in two-dimensional tables, called *relations*, which are the logical equivalent of files.

release A significant program change that often requires changes in the documentation of the software.

reorder point (ROP) A critical inventory quantity that determines when to order more inventory.

replicated database A database that holds a duplicate set of frequently used data.

report layout A technique that allows designers to diagram and format printed reports.

request for maintenance form A form authorizing modification of programs.

request for proposal (RFP) A document that specifies in detail required resources such as hardware and software.

requirements analysis The determination of user, stakeholder, and organizational needs.

restart procedures Simplified processes to access an application from where it stopped.

return on investment (ROI) One measure of IS value that investigates the additional profits or benefits that are generated as a percentage of the investment in IS technology.

revenue center A division within a company that generates sales or revenues.

reverse 911 service A communications solution that delivers emergency notifications to users in a selected geographical area.

rich Internet application Software that has the functionality and complexity of traditional application software, but does not require local installation and runs in a Web browser.

robotics Mechanical or computer devices that perform tasks requiring a high degree of precision or that are tedious or hazardous for humans.

router A telecommunications device that forwards data packets across two or more distinct networks toward their destinations, through a process known as routing.

rule A conditional statement that links conditions to actions or outcomes.

satisficing model A model that will find a good—but not necessarily the best—problem solution.

scalability The ability to increase the capability of a computer system to process more transactions in a given period by adding more, or more powerful, processors.

schedule feasibility The determination of whether the project can be completed in a reasonable amount of time.

scheduled report A report produced periodically, or on a schedule, such as daily, weekly, or monthly.

schema A description of the entire database.

screen layout A technique that allows a designer to quickly and efficiently design the features, layout, and format of a display screen.

script bunny A cracker with little technical savvy who downloads programs called scripts, which automate the job of breaking into computers.

search engine A valuable tool that enables you to find information on the Web by specifying words that are key to a topic of interest, known as keywords.

secondary storage (permanent storage) Devices that store larger amounts of data, instructions, and information more permanently than allowed with main memory.

Secure Sockets Layer (SSL) A communications protocol is used to secure sensitive data during e-commerce.

security dashboard Software that provides a comprehensive display on a single computer screen of all the vital data related to an organization's security defenses including threats, exposures, policy compliance and incident alerts.

selecting Manipulating data to eliminate rows according to certain criteria.

semistructured or unstructured problems More complex problems in which the relationships among the pieces of data are not always clear, the data might be in a variety of formats, and the data is often difficult to manipulate or obtain.

sequential access A retrieval method in which data must be accessed in the order in which it is stored.

sequential access storage device (SASD) A device used to sequentially access secondary storage data.

server A computer designed for a specific task, such as network or Internet applications.

service-oriented architecture (SOA) A modular method of developing software and systems that allows users to interact with systems, and systems to interact with each other.

shared workspace A common work area where authorized project members and colleagues can share documents, issues, models, schedules, spreadsheets, photos, and all forms of information to keep each other current on the status of projects or topics of common interest.

shareware and freeware Software that is very inexpensive or free, but whose source code cannot be modified.

sign-on procedure Identification numbers, passwords, and other safeguards needed for someone to gain access to computer resources.

simplex channel A communications channel that can transmit data in only one direction.

single-user license A software license that permits only one person to use the software, typically on only one computer.

site preparation Preparation of the location of a new system.

slipstream upgrade An upgrade that usually requires recompiling all the code, allowing the program to run faster and more efficiently.

smart card A credit card–sized device with an embedded microchip to provide electronic memory and processing capability.

smartphone A phone that combines the functionality of a mobile phone, personal digital assistant, camera, Web browser, e-mail tool, and other devices into a single handheld device.

social engineering Using social skills to get computer users to provide information to access an information system or its data.

software The computer programs that govern the operation of the computer.

software as a service (SaaS) A service that allows businesses to subscribe to Web-delivered business application software by paying a monthly service charge or a per-use fee.

software piracy The act of unauthorized copying or distribution of copyrighted software

software suite A collection of single application programs packaged in a bundle.

source data automation Capturing and editing data where it is initially created and in a form that can be directly input to a computer, thus ensuring accuracy and timeliness.

speech-recognition technology Input devices that recognize human speech.

spyware Software that is installed on a personal computer to intercept or take partial control over the user's interaction with the computer without knowledge or permission of the user.

stakeholders People who, either themselves or through the organization they represent, ultimately benefit from the systems development project.

start-up (also called *cutover*) The process of making the final tested information system fully operational.

static Web pages Web pages that always contain the same information.

statistical sampling Selecting a random sample of data and applying the characteristics of the sample to the whole group.

steering committee An advisory group consisting of senior management and users from the IS department and other functional areas.

storage area network (SAN) The technology that provides high-speed connections between data-storage devices and computers over a network.

storefront broker A company that acts as an intermediary between your Web site and online merchants who have the products and retail expertise.

strategic alliance (strategic partnership) An agreement between two or more companies that involves the joint production and distribution of goods and services.

strategic planning Determining long-term objectives by analyzing the strengths and weaknesses of the organization, predicting future trends, and projecting the development of new product lines.

structured interview An interview where the questions are written in advance.

supercomputers The most powerful computer systems with the fastest processing speeds.

switch A telecommunications device that uses the physical device address in each incoming message on the network to determine to which output port it should forward the message to reach another device on the same network.

synchronous communications A form of communications where the receiver gets the message instantaneously, when it is sent.

syntax A set of rules associated with a programming language.

system A set of elements or components that interact to accomplish goals.

system performance measurement Monitoring the system—the number of errors encountered, the amount of memory required, the amount of processing or CPU time needed, and other problems.

system performance products Software that measures all components of the computer-based information system, including hardware, software, database, telecommunications, and network systems.

system performance standard A specific objective of the system.

system testing Testing the entire system of programs.

systems analysis The systems development phase that determines what the information system must do to solve the problem by studying existing systems and work processes to identify strengths, weaknesses, and opportunities for improvement.

systems analyst A professional who specializes in analyzing and designing business systems.

systems controls Rules and procedures to maintain data security.

systems design The systems development phase that defines how the information system will do what it must do to obtain the problem solution.

systems development The activity of creating or modifying business systems.

systems implementation The systems development phase involving the creation or acquisition of various system components detailed in the systems design, assembling them, and placing the new or modified system into operation.

systems investigation The systems development phase during which problems and opportunities are identified and considered in light of the goals of the business.

systems investigation report A summary of the results of the systems investigation and the process of feasibility analysis and recommendation of a course of action.

systems maintenance A stage of systems development that involves checking, changing, and enhancing the system to make it more useful in achieving user and organizational goals.

systems operation Use of a new or modified system.

systems request form A document filled out by someone who wants the IS department to initiate systems investigation.

systems review The final step of systems development, involving the analysis of systems to make sure that they are operating as intended.

team organizational structure A structure centered on work teams or groups.

technical documentation Written details used by computer operators to execute the program and by analysts and programmers to solve problems or modify the program.

technical feasibility Assessment of whether the hardware, software, and other system components can be acquired or developed to solve the problem.

technology acceptance model (TAM) A model that describes the factors leading to higher levels of acceptance and usage of technology.

technology diffusion A measure of how widely technology is spread throughout the organization.

technology infrastructure All the hardware, software, databases, telecommunications, people, and procedures that are configured to collect, manipulate, store, and process data into information.

technology infusion The extent to which technology is deeply integrated into an area or department.

technology-enabled relationship management Occurs when a firm obtains detailed information about a customer's behavior, preferences, needs, and buying patterns and uses that information to set prices, negotiate terms, tailor promotions, add product features, and otherwise customize its entire relationship with that customer.

telecommunications The electronic transmission of signals for communications, which enables organizations to carry out their processes and tasks through effective computer networks.

telecommunications medium Any material substance that carries an electronic signal and serves as an interface between a sending device and a receiving device.

telecommunications protocol A set of rules that governs the exchange of information over a communications medium.

telecommuting A work arrangement whereby employees work away from the office using personal computers and networks to communicate via e-mail with other workers and to pick up and deliver results.

thin client A low-cost, centrally managed computer with essential but limited capabilities and no extra drives, such as a CD or DVD drive, or expansion slots.

time-driven review Review performed after a specified amount of time.

total cost of ownership (TCO) The measurement of the total cost of owning computer equipment, including desktop computers, networks, and large computers.

traditional approach to data management An approach whereby separate data files are created and stored for each application program.

traditional organizational structure An organizational structure in which major department heads report to a president or top-level manager.

transaction Any business-related exchange, such as payments to employees, sales to customers, and payments to suppliers.

transaction processing cycle The process of data collection, data editing, data correction, data manipulation, data storage, and document production.

transaction processing system (TPS) An organized collection of people, procedures, software, databases, and devices used to record completed business transactions.

transaction processing system audit A check of a firm's TPS systems to prevent accounting irregularities and/or loss of data privacy.

Transmission Control Protocol (TCP) The widely used Transport-layer protocol that most Internet applications use with IP.

Trojan horse A malicious program that disguises itself as a useful application or game and purposefully does something the user does not expect.

tunneling The process by which VPNs transfer information by encapsulating traffic in IP packets over the Internet.

ultra wideband (UWB) A wireless communications technology that transmits large amounts of digital data over short distances of up to 30 feet using a wide spectrum of frequency bands and very low power.

unified communications A technology solution that provides a simple and consistent user experience across all types of communications such as instant messaging, fixed and mobile phone, e-mail, voice mail, and Web conferencing.

Uniform Resource Locator (URL) An assigned address on the Internet for each computer.

unit testing Testing of individual programs.

unstructured interview An interview where the questions are not written in advance.

user acceptance document A formal agreement signed by the user that states that a phase of the installation or the complete system is approved.

user documentation Written descriptions developed for people who use a program, showing users, in easy-to-understand terms, how the program can and should be used.

user interface The element of the operating system that allows you to access and command the computer system.

user preparation The process of readying managers, decision makers, employees, other users, and stakeholders for new systems.

users People who will interact with the system regularly.

utility programs Programs that help to perform maintenance or correct problems with a computer system.

value chain A series (chain) of activities that includes inbound logistics, warehouse and storage, production, finished product storage, outbound logistics, marketing and sales, and customer service.

version A major program change, typically encompassing many new features.

videoconferencing A telecommunications system that combines video and phone call capabilities with data or document conferencing.

virtual organizational structure A structure that employs individuals, groups, or complete business units in geographically dispersed areas that can last for a few weeks or years, often requiring telecommunications or the Internet.

virtual private network (VPN) A private network that uses a public network (usually the Internet) to connect multiple remote locations.

virtual reality The simulation of a real or imagined environment that can be experienced visually in three dimensions.

virtual reality system A system that enables one or more users to move and react in a computer-simulated environment.

virtual tape A storage device that manages less frequently needed data so that it appears to be stored entirely on tape cartridges, although some parts of it might actually be located on faster hard disks.

virtual workgroups Teams of people located around the world working on common problems.

virus A computer program file capable of attaching to disks or other files and replicating itself repeatedly, typically without the user's knowledge or permission.

vision systems The hardware and software that permit computers to capture, store, and manipulate visual images.

voice mail Technology that enables users to send, receive, and store verbal messages for and from other people around the world.

Voice over Internet Protocol (VoIP) A collection of technologies and communications protocols that enables your voice to be converted into packets of data that can be sent over a data network such as the Internet, a WAN or LAN.

voice-to-text service A service that captures voice mail messages, converts them to text, and sends them to an e-mail account.

volume testing Testing the application with a large amount of data.

Web 2.0 The Web as a computing platform that supports software applications and the sharing of information between users.

Web auction An Internet site that matches buyers and sellers.

Web browser Web client software such as Internet Explorer, Firefox, and Safari used to view Web pages.

Web log (blog) A Web site that people can create and use to write about their observations, experiences, and feelings on a wide range of topics.

Web page construction software Software that uses Web editors and extensions to produce both static and dynamic Web pages.

Web services Software modules supporting specific business processes that users can interact with over a network (such as the Internet) on an as-needed basis.

Web site development tools Tools used to develop a Web site, including HTML or visual Web page editor, software development kits, and Web page upload support.

wide area network (WAN) A telecommunications network that ties together large geographic regions.

Wi-Fi Protected Access (WPA) A security protocol that offers significantly improved protection over WEP.

Wired equivalent privacy (WEP) An early attempt at securing wireless communications based on encryption using a 64- or 128-bit key that is not difficult for hackers to crack.

wireless mesh A way to route communications between network nodes (computers or other devices) by allowing for continuous connections and reconfiguration around blocked paths by "hopping" from node to node until a connection can be established.

workgroup Two or more people who work together to achieve a common goal.

workgroup application software Software that supports teamwork, whether in one location or around the world.

workgroup sphere of influence The sphere of influence that serves the needs of a workgroup.

workstation A more powerful personal computer that is used for technical computing, such as engineering, but still fits on a desktop.

World Wide Web A collection of tens of millions of server computers that work together as one in an Internet service using hyperlink technology to provide information to billions of users.

Worldwide Interoperability for Microwave Access (WiMAX) The common name for a set of IEEE 802.16 wireless metropolitan area network standards that support different types of communications access.

worm A parasitic computer program that can create copies of itself on the infected computer or send copies to other computers via a network.

Subject

A boldface page number indicates a key term and the location of its definition in the text.

Numbers

A

B

D

I

J

K

L

N

O

P

Q

R

S

T

X

Y

Z

Company Names

Numbers

A

B

C

D

E

N

O

P

Q

R

S

T